BAT ECOLOGY

BAT ECOLOGY

Edited by

Thomas H. Kunz and M. Brock Fenton

The University of Chicago Press

Chicago and London

The University of Chicago Press, Chicago 60637
The University of Chicago Press, Ltd., London
© 2003 by The University of Chicago
All rights reserved. Published 2003
Paperback edition 2005
Printed in the United States of America

12 11 10   3 4 5

ISBN: 0-226-46206-4 (cloth)
ISBN: 0-226-46207-2 (paperback)

Library of Congress Cataloging-in-Publication Data

Bat ecology / edited by Thomas H. Kunz and M. Brock Fenton.
    p. cm.
    Includes bibliographical references (p.   ) and index.
    ISBN 0-226-46206-4 (cloth : alk. paper)
    1. Bats—Ecology.   I. Kunz, Thomas H.   II. Fenton, M. Brock
(Melville Brockett), 1943–

QL737 .C5 B3594 2003
599.4'17—dc21

                                                2002011945

To our wives
Margaret and Eleanor

CONTENTS

PART II. FUNCTIONAL ECOLOGY

PART III. MACROECOLOGY

This book represents the first comprehensive treatment of bat ecology in more than 20 years. These past 2 decades have witnessed an enormous growth of interest in the ecological study of bats. In part, this reflects the increased recognition and appreciation of bats for their role in ecosystem services (as insect predators, seed dispersers, and pollinators), for their astonishing ecological and evolutionary diversity, and for their role as models for studies in ecophysiology, ecomorphology, trophic interactions, biogeography, emerging diseases, and conservation. This extraordinary interest and attention comes at a time when bats are being threatened from the adverse affects of increased human population growth and the associated deforestation, conversion of natural habitats to forest and agricultural monocultures, water, soil, and air pollution, and introductions of xenobiotics and exotic species.

Since the publication of *Ecology of Bats* (Plenum Press, 1982), many technological, analytical, and empirical advances have fostered an increase in our knowledge of these flying mammals. Advances in technology and analytical techniques have been largely influenced by new developments in electronic instrumentation and information technology. Empirical advances in bat ecology have benefited from these technological developments but also from research based on models and testable hypotheses. Generous funding from both government and non-government sources will continue to play an important role in making advances in this important field of study.

We chose a multi-author format for this book because of the diversity and complexity of bat ecology, and the belief that a multi-author approach provides a more comprehensive and authoritative coverage than a single-author volume. This book is primarily targeted at professional ecologists, students, conservation biologists, natural resource managers, and scientists with a broad interest in ecology, behavior, evolution, and conservation.

Authors were invited to summarize and synthesize published works in their respective disciplines, focusing largely on discoveries made during the past 20 years. The topics treated in this book represent major areas of current research on bat ecology. In large measure the variation and depth of treatment in each chapter reflects the breadth of topics and the amount of information currently available for review.

This book is organized into three sections: "Life History and Social Biology" (chaps. 1-5), "Functional Ecology" (chaps. 6-10), and "Macroecology" (chaps. 11-15). This organization provides a general framework for the book and accommodates expected overlap in material treated in various chapters. For example, body mass is an important independent variable in the analyses presented in several chapters, and conservation biology is a common theme in others. The section on life history and social biology includes chapters on the roosting ecology of cavity and foliage roosting bats, sensory ecology and communication, sexual selection and sperm competition, ecology of migration, and life-history strategies. The section on functional ecology includes chapters on ecomorphology, interactions between echolocating bats and their insect prey, bats and flowers, bats and fruit, and physiological ecology and energetics. The section on macroecology includes chapters dealing with evolution and ecological diversity, trophic structures and ecological organization, geographic distribution patterns, emerging viral infections, and conservation ecology.

Bats spend over half their lives subjected to the conditions in their roost environment, so it is not surprising that roosts play an important role in the ecology and evolution of bats. In chapter 1, Kunz and Lumsden summarize current knowledge on the diversity of cavity and foliage roosting bats and discuss the relationships between roosting habits, morphological specializations for roosting, colony size, roost fidelity, social organization, and foraging behavior. Variation in roost use is also discussed in the context of diet, reproductive status, and migratory habits. Because sustainable bat populations are strongly influenced by the availability of roosts, the importance of conservation is emphasized.

How bats use acoustic, visual, olfactory, tactile, and thermal cues while foraging, during mother-infant interactions, in courtship and mating, and in other social contexts has been the subject of considerable research. In chapter 2, Altringham and Fenton use case studies from seven species to illustrate the range of conditions under which bats live. They examine the role of sound production in bats, how sound is used by echolocating species to detect, track, and assess airborne prey, and how acoustic cues are used by other animal-eating bats, some of which depend on prey-generated sounds to detect and evaluate potential prey. They also consider how bats use olfactory cues for making decisions about what to eat. The role of vision in the foraging behavior of bats is reviewed in the context of feeding behavior in frugivorous, nectarivorous, and sanguinivorous species. Thermal detection appears to play a crucial role in prey detection in vampire bats, and tactile cues may play a variety of social roles in courtship and mating. Finally, these authors discuss multimodal perception, in which bats rely on different sensory cues while foraging and as they interact in their roost environments.

There are many examples in bats with male mating territories in which females roost and mate with more than one male. In many of these species, males produce audible vocalizations, scent mark females and territories, and engage in aerial displays that may provide information important for female mate choice. In chapter 3, Wilkinson and McCracken exlore the ecological determinants of social group stabilty and the potential roles of female choice and sperm competition in the evolution of bat mating systems. Female bats clearly have the opportunity to mate with multiple males and apparently do so in many species. Multiple mating by females and characteristics of the reproductive physiology of bats suggest that sperm competition is an important and underappreciated factor in the evolution of mating systems. Using data from more than 100 species, the authors test whether mating systems in bats are influenced by testes size.

Annual migrations from a few to more than 1,000 kilometers characterize the life cycles of many species of bats. Perhaps best known are the insectivorous species in temperate regions that migrate from their summer ranges to winter hibernacula in response to harsh climates and reduced food availability. In chapter 4, Fleming and Eby review current knowledge about ecological and behavioral aspects of bat migration. They address important questions relating to who migrates and why, ecological conditions that favor the evolution of migratory behavior, physiological consequences of migration, the amount of time and energy that bats devote to annual migrations, life-history and population consequences of migration, and how reproduction, social behavior, and the genetic structure of bats may affect migration, as well as the conservation implications of migration. They also compare the migratory ecology of bats with that of birds to search for evolutionary commonalities and differences.

Bats are unique among mammals not only because they fly but also because they are small and have life-history characteristics that are generally attributable to larger species. In chapter 5, Barclay and Harder review current data on life-history characteristics of bats (e.g., life span, gestation length, litter size, size at birth, age at sexual maturity). They assess how variation within traits is correlated with various ecological factors (e.g., seasonality and use of torpor/hibernation or migration) and how proximate constraints (such as nutrient availability) may limit this variation. They consider how the life histories of bats match patterns described for mammals in general, and with shrews in particular. They review information on bat mortality rates and use these data to test three alternate hypotheses. They postulate that a combination of low extrinsic mortality, a constraint on number of ova produced, and limited availability of resources help explain the unusual life histories of bats.

Although bats are often rare and sometimes difficult to study in the field, they can provide important models for studying diversity of form and function. In chapter 6, Swartz, Freeman, and Stockwell examine recent advances in the study of the relationships among morphology, behavior, and ecology of bats. They highlight case studies of particular relevance to understanding how the morphology of bats relates to their ecology, with special attention to the structure of the feeding apparatus, particularly the teeth and skull in relation to dietary preferences, and the morphology of the wing in relation to flight performance. In their analysis, they consider the role of body size as an

important determinant of mechanical and ecological function. They emphasize the importance of recognizing ontogenetic, intersexual, and other intraspecific variation, the need for integrative analyses that span field and laboratory, and the value of computer modeling in studies of ecomorphology.

Not surprisingly, insects constitute a large portion of the diet of insectivorous species. In chapter 7, Jones and Rydell review the types of insects eaten
by bats and when and whether bats specialize on certain taxa. They discuss the
importance of echolocation in prey detection and capture and how echolocation may constrain the types of insects eaten. They discuss the ways in which
bats capture insects and why and when they rely on echolocation. In the context of echolocation, they attempt to answer in what manner diet and prey size
of insectivorous bats are influenced by the call structure of echolocation, and
whether bats select insect prey based on predictions from optimal foraging
theory. Because insects are important prey for bats, many insects have evolved
defensive mechanisms thought to reduce the risks of predation. The authors
discuss how insects defend themselves against bats and review which insect
species can hear bats and what kinds of evasive behaviors they take to avoid
being eaten. They also review research on the evolution of moth ears in relation to the echolocation calls of bats.

In theory, a flower can influence a pollinator's nectar-feeding behavior by
rationing the amount of its nectar reward. In fact, a flower cannot increase its
reward (nectar volume) indefinately because such an increase would reduce
the number of visits that pollinators would make to flowers and, thus, reduce
the reproductive success of the plant. In chapter 8, von Helverson and Winter
evaluate the characteristics of flowers pollinated by New World glossophagine bats by asking whether bat flowers make themselves easier to find
and thus save the pollinator energetic costs and time, the mode by which
flower bats reduce their costs for flower visiting, how various species avoid or
reduce competition, in what ways species assemblages of nectarivorous bats
are structured, and the adaptations of bat flowers and how their evolution has
been influenced by pollinators.

Fruit-eating bats play a major role in tropical ecosystems by dispersing
seeds and, thus, promoting the regeneration and maintenance of forests. In
chapter 9, Dumont summarizes the types of fruits in the diet of fruit-eating
bats, including their physical and nutritional qualities. She then investigates
three interrelated mechanisms of resource partitioning within fruit bat assemblages—the physical properties of fruit, morphological variation in masticatory systems and wings, and variation among bats in fruit processing strategies. She also evaluates the different foraging strategies used by fruit-eating
bats and concludes that vertical stratification appears to be a mechanism for
resource partitioning in all bat assemblages. However, she notes that fruit-
eating bats in the Old World seem to rely more on "steady state" and "big
bang" crops than do fruit-eating bats in the New World tropics. She argues

that while the complexity of the resource-partitioning system prohibits explanation from a single subset of variables, there are associations among groups of variables that provide a glimpse of underlying themes in patterns of resource use.

In chapter 10, Speakman and Thomas begin by explaining the fundamental principles that govern heat flow and temperature regulation. They examine the energetic implications of different thermoregulatory strategies and metabolic rates and of different life-history traits of bats. They evaluate resting metabolism and thermoregulatory strategies, patterns of basal metabolic rate (BMR), daily torpor, and hibernation. They examine the energetic cost of living, including the cost of flight, reproduction, daily energy and water budgets. They evaluate how adjustments are made in BMR and discuss the functional relationships between field metabolic rate and production.

In recent years, it has become increasingly clear that phylogenetic hypotheses provide a critical framework for interpreting patterns of evolution in biological systems. In chapter 11, Simmons and Conway review the ecological diversity in bats from an evolutionary perspective (e.g., patterns of variation in diet, foraging and flight behavior, echolocation call structure, roosting habits, hibernation/migration patterns, social behavior, and reproductive patterns). Recent phylogenies of bats are used as a framework for examining the possible evolutionary origins of different ecologies, with an emphasis on exploring the mosaic pattern of distribution of different ecological traits in different clades. The role that phylogenetic history may have played in the evolution of ecological diversity in bats is explored in terms of constraints and opportunities.

As a group, bats exploit a tremendous variety of food resources and achieve remarkable abundance and diversity in certain habitats. Thus, it is not surprising that bats influence and are influenced by coexisting species. In chapter 12, Patterson, Willig, and Stevens examine elements underlying the ecological organization of bats. They review literature that describes the principal trophic strategies of bats and the salient features of each major resource. They identify prevailing patterns of niche partitioning and special opportunities or constraints that appear to be associated with each feeding strategy. Although bats respond to these opportunities and constraints at multiple levels of organization—individual variation, abundance and range limits of species, and the amalgamation of local assemblages into faunas—the mechanisms that underlie these ecological patterns remain elusive. As the authors appropriately note, bat ecologists must be prepared to compare, contrast, and ultimately integrate ecological patterns with those shown by other taxonomic groups.

The Chiroptera is a species-rich mammalian order whose members exhibit diverse ecological adaptations and occupy a broad range of biomes and habitats on each continent except Antarctica. In chapter 13, Willig, Patterson, and Stevens summarize what is known about continental distributions and species richness of bats on a spatial scale—with a focus on North and South America, largely because detailed information on distributions for other parts of the world are limited. They evaluate empirical data and hypotheses that may affect variation in range size in bats, including body mass, taxonomic affiliation, geometry of land masses (including elevation), and climatic variables, and conclude that range sizes of bats are larger compared to other mammals, vary independently of body size, are positively associated with species abundance, are related to elevation and latitude, and are more constrained by ecological than phylogenetic considerations. They argue that quantitative analyses that combine information on all aspects of diversity at the local and regional, scales, combined with phylogenetic analyses, are needed to help understand factors that determine the distribution and diversity of bats at hemisperical and global scales.

What important microbiological agents are associated with bats? Do bats have life-history strategies that protect individuals from or predispose them to infection? Do these infectious agents have an impact on the mortality of bats (or other species)? In chapter 14, Messenger, Rupprecht, and Smith review the biological attributes of bats in relation to disease transmission, the possible associations of bats with the emergence of new viral diseases (including Marburg, Eboloa, Hendra, Menangle, and Nipah viruses), and the discovery and transmission of rabies-related viruses. They consider the ecology of bats regarding rabies transmission and explore the rise of terrestrial epizootics and the possible relationship between bat rabies and rabies in other mammals. Finally, they discuss the myths, mysteries, and misconceptions associated with rabies in bats.

The unusual combination of life-history traits manifested by bats (small body size, long life span, nocturnality, and volancy) means that it is difficult to draw parallels with conservation efforts directed at other taxa. In chapter 15, Racey and Entwistle address a series of questions concerning the conservation status of the Microchiroptera and Megachiroptera and their impact on mammalian biodiversity. These questions include: Can population size be determined? Are there edge-of-range effects? Can changes in population size be determined? Can habitat preferences be determined? What features of habitats are important? Are habitat factors important for trophic or navigational reasons? Can the conclusions and management implications of local autecological studies be elevated to national or range scales? Does roost enhancement have an effect at the population level? Does habitat enhancement have an effect at the population level? What should be the direct conservation criteria for bats—rarity? Species richness? The authors also consider the effects of habitat fragmentation and roost availability and the direct and biogeographical evidence for the importance in ecosystem conservation.

The support of many individuals and organizations has made this book possible. In particular, we thank our respective institutions (Boston University and York University) for their support and the government and nongovernment organizations who have, over the years, supported our research on the ecology of bats (National Science Foundation, National Geographic Society, and the Lubee Foundation [Kunz] and the Natural Sciences and Engineering Research Council of Canada [Fenton]). We are especially grateful to each author who provided chapters for this book. Their ideas and contributions have made it possible to present an up-to-date, comprehensive, and authoritative view of bat ecology. We are also grateful to the many individuals who reviewed one or more chapter manuscripts. These individuals include Steve Austed, Hector Arita, Johanna Bloss, Danny Brass, Robert Barclay, Paul Faure, Peg Eby, Jim Fullard, Trish Freeman, Ted Fleming, Tom Griffiths, Larry Heaney, David Hoskins, John Kirsch, Al Kurta, Tom Lacher, Gary McCracken, Brian McNab, Ulla Norberg, Colin O'Donnell, Dixie Pierson, Don Thomas, Jens Rydell, Bill Schutt, Charles Trimarchi, Merlin Tuttle, Ruth Utzurrum, Christian Voigt, John Winkelman, and Don Wilson. We also wish to thank Rafael Avila-Flores, Bethany Bernasconi, Stefania Biscardi, Jen Blasko, Jazmine Oprecio, Erin Ruppert, Genevieve Spanser, and Hanna ter Hofstede for assistance in preparing the indexes. Finally, we wish to thank Susan Abrams whose "woof-woof" gave us the encouragement to pursue this project, Yvonne Zipter for copy editing, Jennifer Howard for management of illustrations, Dennis Anderson for book design, and Christie Henry who enthusiastically supported this project from its onset.

Thomas H. Kunz and M. Brock Fenton

PART I

LIFE HISTORY AND SOCIAL BIOLOGY

CHAPTER 1

Ecology of Cavity and Foliage Roosting Bats

Thomas H. Kunz and Linda F. Lumsden

Introduction

Bats occupy a wide variety of roosts in both natural and manmade structures. More than half of the approximately 1,100 species of living bats use plants exclusively or opportunistically as roosts. Others seek shelter in caves, rock crevices, mines, tombs, buildings, bridges, and other manmade structures. Some bats roost in cavities constructed by other animals, including bird nests and nests of ants and termites, whereas others roost in exposed places on branches and the trunks of trees. Thus, it is not surprising that the conditions and events associated with roosting have played a major role in the ecology and evolution of bats (Kunz 1982). Roosts are important sites for mating, hibernation, and rearing young. They often facilitate complex social interactions, offer protection from inclement weather, promote energy conservation, and minimize risks of predation.

Roosting habits of bats are influenced by the diversity and abundance of roosts, the distribution and abundance of food, and an energy economy influenced by body size and the physical environment. Roosting ecology is ultimately tempered by constraints of phylogenetic inertia and a compromise of opposing selective pressures derived from both roost and nonroost sources (Kunz 1982). Morphological, physiological, and behavioral characteristics of bats commonly regarded as adaptations for roosting include flattened skulls, suction pads and disks on feet and wrists, cryptic markings and postures, clustering, torpor, and synchronous nightly departures. These traits reflect compromises imposed by manner of flight, body size, predator pressure, energy economy, and variations in the physical environment (Kunz 1982).

Bats seek shelter in a wide variety of roost types, ranging along a continuum from ephemeral to permanent. At one extreme, roosting sites in caves, mines, and some rock crevices offer the advantages of relative permanency, thermal stability, and protection from climatic extremes but may be patchy in distribution. At the other extreme, spaces beneath exfoliating bark and foliage generally are ephemeral and more subject to environmental extremes but are more abundant and ubiquitous (Kunz 1982). Associations between bats and roosts range from being obligatory to opportunistic, with selection of a particular type of roost dependent on its availability (Kunz 1996).

3

We focus this review primarily on bats that seek shelter in tree cavities, in foliage, on exposed branches and boles of trees, natural cavities, or structures modified by bats. This emphasis is timely because forests have become increasingly threatened by anthropogenic factors (e.g., timber management, deforestation, and associated habitat alteration and loss). Knowledge of roost requirements is a prerequisite to understanding the impact of disturbance on bat populations and to providing focus to conservation efforts (Fenton and Rautenbach 1998; Pierson 1998). Bats that occupy habitats that are highly susceptible to disturbance and loss are of special concern to conservation biologists. Only in recent history have manmade structures, such as mines, bridges, and buildings, provided alternative habitats comparable to caves and tree cavities.

Types of Roosts
Tree Cavities

Tree cavities are important roost resources in both temperate and tropical regions (Barclay and Brigham 1996; Boonman 2000; Kunz 1982, 1996; Pierson 1998; Rosevear 1965; Sedgeley and O'Donnell 1999b; Simmons and Voss 1998; Tuttle 1976; Verschuren 1957, 1966; figs. 1.1 and 1.2). In North America, Europe, Asia, and Australia, tree cavities are used mostly by members of the Vespertilionidae. In Africa, they are used mostly by members of the Vespertilionidae and Hipposideridae. In the Neotropics, tree cavities are used primarily by members of the Phyllostomidae and Emballonuridae. Although cavity-roosting habits are most common among microchiropterans, they may be more common in some small megachiropterans than previously recognized (Bonaccorso 1998; Flannery 1995; Rainey 1998).

Within temperate regions, the proportions of species assemblages that use tree cavities vary geographically (Humphrey 1975). In western North America and Australia, where extensive areas of native forests remain, a relative high proportion of bat species roost in tree cavities (Barclay and Brigham 1996; Churchill 1998; Pierson 1998; fig. 1.1A). By contrast, in western Europe and eastern North America, where natural tree cavities have been depleted by extensive clearing of forests and misguided forest management practices, many cavity-roosting species now rely considerably on manmade structures such as buildings (Kunz and Reynolds, in press), bridges (Kunz 1982), and bat houses (Mayle 1990; Tuttle and Hensley 1993). In some areas, tree cavities provide suitable roosting habitats for bats on a year-round basis. At higher latitudes, tree cavities may be too cold during winter months, and thus bats must seek alternative roosts, usually in caves or other subterranean structures (Mayle 1990). In lowland tropical regions, where caves are absent, tree cavities provide one of the primary roosting habitats for bats. On Barro Colorado Island, Panama, tree cavities are commonly occupied by small harem groups of *Artibeus jamaicensis* (Morrison 1979; fig. 1.1B).

Figure 1.1. *A*, Small cavity in dead branch of a river red gum (*Eucalyptus camaldulensis*) occupied by a colony of *Chalinolobus gouldii* in temperate southeastern Australia (photo by L. Lumsden); *B*, tree cavity used by a small harem group of *Artibeus jamaicensis* on Barro Colorado, Panama Canal Zone (photo by D. W. Morrison); *C*, a partially rotted basal cavity in a large, buttressed tree (*Pradosia cochlearia*) in a lowland tropical rainforest offers shelter to *Glossophaga soricina, Micronycteris megalotis,* and *Carollia perspicillata* in French Guiana (from Simmons and Voss 1998; photo by R. Voss); *D*, basal and bole cavities in a baobob tree (*Adansonia digitata*) used as day-roosting sites by *Cardioderma cor* in west Africa (photo by T. A. Vaughan). *Arrows* denote roost openings.

Figure 1.2. *A,* Partially rotted, fallen tree with a cavity used as a day roost by *Furipterus horrens* (from Simmons and Voss 1998; photo by R. Voss); *B,* underside of a broken tree trunk (*arrow*) used as a day roost by *Peropteryx macrotis, P. kappleri,* and *Cormura brevirostris* (from Simmons and Voss 1998; photo by R. Voss). *Arrows* denote roost areas.

Buttress cavities form semidarkened spaces on the exterior of lowland tropical trees (Kaufman 1988; Richards 1996; Whitmore 1998), and provide ideal roosting habitats for tropical bats (Simmons and Voss 1998; Tuttle 1976). In strangler figs, adjacent buttresses may fuse or anastomose to form deep, vertical cavities adjacent to the bole (fig. 1.1C). Such cavities and the spaces between adjacent buttresses are often used as day roosts by *Saccopteryx bilineata* (Bradbury and Emmons 1974; Bradbury and Vehrencamp 1976). Cavities that form in baobab trees (*Adansonia digitata*) are commonly used by *Cardioderma cor* (Vaughan 1976; fig. 1.1D) and several other micrchiropteran species in Africa (Verschuren 1957).

Cavities may form in the boles, trunks, or branches of live and dead trees. In general, large dead trees (snags) remain standing for longer periods than do small trees, with snags in old-growth Douglas fir (*Pseudotsuga menziesii*) thought to take 250 yr after death to completely decompose (Cline et al. 1980). In many areas, these snags provide important roosting sites for bats (e.g., Brigham et al. 1997; Crampton and Barclay 1998; Lumsden et al. 2002b; Mattson et al. 1996; Ormsbee and McComb 1998; Zielinski and Gellman 1999).

In old-growth temperate and tropical forests, basal cavities sometimes form in the interior of living trees, when the heartwood is exposed to fire (Finney 1991). These basal cavities share some characteristics with caves, including stable temperatures and humidity, pronounced light gradients, protection from rain, relatively spacious internal flight space (Gellman and Zielinski 1996), and extended longevity, with some trees living up to 2,000 yr (Becking 1982). Because basal cavities may persist for a major portion of a tree's life, they are considered to be important resources for cavity-roosting bats (Gellman and Zielinski 1996). That hollow trees are prevalent in nutrient-poor soils, especially in tropical regions, led Janzen (1976) to suggest that rotted cavities may be selected as a mechanism to trap minerals and nitrogen from the accumulation of animal feces. If this hypothesis is correct, deposits of nitrogen-rich guano from bats may play an important role in forest dynamics (see Gellman and Zielinski 1996; Pierson 1998).

Cavity formation in trees results from a range of processes, including fungal infection, insect attack, excavation by termites and woodpeckers, lightning strikes, fire, and natural damage to branches (Bennett et al. 1994; Betts 1996; Gibbons 1994; Kalcounis and Brigham 1998; Mackowski 1984; Pierson 1998). The number and size of cavities vary with the diameter, age, and height of the tree, with larger and older trees having experienced longer periods of exposure to processes of cavity formation and development than smaller trees (Bennett et al. 1994; Lindenmayer et al. 1993; Sedgeley and O'Donnell 1999b). These relationships vary among species, with some trees forming cavities at earlier stages of growth than others (Bennett et al. 1994; Cline et al. 1980; Gibbons 1994; Raphael and Morrison 1987). In Australia and New Zealand, where no vertebrate species are known to excavate tree cavities, roosts used by bats are more often found in older trees. Tree cavities that form due to physical forces or actions of invertebrates generally do not develop until trees are large and old, usually after 100 or more years (Mackowski 1984; Mawson and Long 1994).

When trees with cavities and buttresses die and fall, they may continue to provide roosting spaces for bats (fig. 1.2). In the tropics, cavities in fallen tree trunks are commonly used as roosts by members of the Emballonuridae (e.g., *Emballonura monticola, Cormura brevirostris, Peropteryx leucoptera, Saccopteryx bilineata* [Bernard 1999; Lekagul and McNeely 1977; Reid 1997; Tuttle 1970]) and Phyllostomidae (e.g., *Carollia perspicillata, Lonchophylla thomasi, Micronycteris hirsuta, M. megalotis, Mimon crenulatum, M. bennettii, Trachops cirrhosus* [LaVal 1977; Reid 1997; Simmons and Voss 1998; Tuttle 1970]). In Central America, fallen trees appear to be the primary roosting habitat for *Furipterus horrens* (Reid 1997), with up to 59 individuals recorded from a single roost (LaVal 1977).

Selection of Tree-Cavity Roosts

Knowledge of how bats use tree cavities has increased in recent years, largely due to the use of radiotelemetry. Early research on cavity-roosting bats concentrated mostly on relatively large species (e.g., Barclay et al. 1988; Fenton 1983), but with radio transmitters currently weighing less than 0.5 g, knowledge of small cavity-roosting species (<6 g), such as *Myotis californicus* (Brigham et al. 1997) and *Vespadelus pumilus* (Law and Anderson 2000) have markedly increased.

Several patterns have begun to emerge in the types of roosts used by cavity-roosting bats. Recent studies have focused on whether bats select particular roost attributes relative to their abundance (see references in table 1.1). To this end, roost and available habitat features have been measured and compared statistically. Because cavity-roosting bats generally do not modify their roost environment, they select roost sites from those that form from physical means, invertebrate activity, or facilitated by cavity excavators. In north temperate regions, several species of bats occupy abandoned woodpecker cavities (e.g., Betts 1996; Gaisler et al. 1979; Kalcounis and Brigham 1998; Pierson 1998; van Heerdt and Sluiter 1965), and in fact some species select these cavities in preference to those that form by physical or invertebrate action. For example, 85% of the 81 *Nyctalus noctula* roosts reported by Boonman (2000) in the Netherlands were observed in cavities excavated by woodpeckers. In these and similar situations, roost selection was strongly influenced by the preference of the original excavators for nesting sites and the decay characteristics of the tree (Kalcounis and Hecker 1996).

Selection pressures (in particular due to microclimate variables and predators) that govern the choice of tree and site characteristics should be similar regardless of the species of bat (Vonhof and Barclay 1996). Vonhof and Barclay

(1996) and Brigham et al. (1997) predicted that cavity-roosting bats should require a number of large dead trees, in specific stages of decay and that project above the canopy in relatively open areas (fig. 1.3).

To assess the generality of these predictions, we summarize results from 26 studies (table 1.1) that have statistically assessed selection of various roost tree and site attributes. Some species in these studies roosted both in tree cavities and beneath exfoliating bark, and these were not separated in the assessment of roost selection. Thus, we include information on tree cavities and spaces beneath exfoliating bark. Roosts located beneath exfoliating bark are treated in more detail below.

Selection of roosts can occur at a number of different levels (cavity, tree, stand, and landscape), and various studies have addressed one or more of these criteria. Most studies have assessed selection at the level of the tree and/or the stand, with fewer studies providing information on cavity selection and landscape characteristics.

Cavity size and shape may directly influence the number of bats present, their social structure, and roost microclimate (Kunz 1982). Several investigators have demonstrated differences in the types of openings used by bats to gain access to roosts (Lumsden et al. 2002b; Vonhof and Barclay 1996), but few studies have measured "available" cavities to assess whether bats actually select particular types of cavities. A notable exception is the study by Sedgeley and O'Donnell (1999a) in New Zealand, who compared characteristics of 84 tree cavities used by *Chalinolobus tuberculatus* with 57 other available but unoccupied cavities. Roosts were predominantly formed in knotholes with medium-sized openings and had thick cavity walls with dry, medium-sized internal spaces. Roosts typically were high above the ground, and the areas that surrounded the openings were uncluttered by adjacent vegetation. Sedgeley and O'Donnell (1999a) suggested that these characteristics facilitated easy access to the roost and provided good insulation. The microclimate (temperature and humidity) in available cavities was also compared, with roost cavities being the most stable, as well as having higher humidities and temperatures that continued well into the night (Sedgeley 2001).

Cavity roosts may be selected by bats to increase their protection from predators and competitors. Predation on bats may occur within the roost or as they depart at dusk (Fenton et al. 1994; Speakman 1991). Bats that enter torpor on a daily (or extended) basis may not be sufficiently alert to escape from predators and, hence, need to select roosts that deny entry to predators. In temperate regions of Australia, predation by birds and arboreal animals, including goannas, pythons, and marsupial carnivores, may exert a strong selection pressure on bats to choose tree cavities with openings not much larger than their own body size (Tidemann and Flavel 1987). However, small openings to tree cavities may not be as important in areas that lack predators (e.g., in New Zealand [Sedgeley and O'Donnell 1999a]) or in tropical regions, where some

Table 1.1. Selection of commonly measured habitat variables of roost sites used by temperate tree-roosting microchiropteran species (expanded from Sedgeley and O'Donnell 1999b)

Species	Country	Sex	No Roosts	Type of Cavity	Tree sp.	Dead/ Decayed	Large Diam.	Tall Roost tree	% Bark	Low Canopy Cover/ Clutter around Roost	Aspect of Roost Entr.	Roost Stand	Other	References
Chalinolobus gouldii	Australia	M & F	89	H	...	0	+	...	...	...	0	+	...	Lumsden et al. 2002a, 2002b
Chalinolobus tuberculatus	New Zealand	M & F	291	H	+	+	+	+	...	+	0	+	1, 5, 6, 8	Sedgeley and O'Donnell 1999a 1999b
Eptesicus fuscus	Canada	F (Mat)	15	H	+	+	+	+	+	...	...	...	3, 10	Vonhof 1996
Eptesicus fuscus	Canada	F (Mat)	27	H	+	...	...	...	...	...	+	...	7	Kalcounis and Brigham 1998
Falsistrellus tasmaniensis	Australia	M	8	H	0	...	+	+	...	...	...	+	...	Herr 1998
Lasiomycteris noctivagans	USA	M & F	35	H UB	...	+	+	...	...	...	+	+	14	Mattson et al. 1996
Lasiomycteris noctivagans	Canada	F (Mat)	12	H UB	+	0	+	+	0	...	...	...	10, 12	Vonhof 1996
Lasiomycteris noctivagans	Canada	F	36	CR	+	+	+	0	...	...	...	...	7	Barclay et al. 1988
Lasiomycteris noctivagans	USA	F (Mat)	17	H	0	+	0	+	0	0	0	0	2, 8, 9, 10, 11	Betts 1998
Lasiomycteris noctivagans	USA	M & F	15	?	+	+	+	+	...	...	...	+	...	Campbell et al. 1996
Myotis californicus	Canada	F (Mat)	19	UB H	+	+	+	0	0	+	0	+	11, 12	Brigham et al. 1997
Myotis daubentonii	The Netherlands	...	27	H	+	...	...	...	...	...	0	+	15	Boonman 2000
Myotis evotis	Canada	M & F	19	UB	+	...	+	...	0	+	+	+	6	Vonhof and Barclay 1997
Myotis evotis	USA	F	73	UB	+	+	0	+	0	+	0	+	16	Waldien et al. 2000
Myotis septentrionalis	USA	F (Mat)	32	UB H	+	0	0	...	0	...	...	...	...	Foster and Kurta 1999
Myotis septentrionalis	USA	F (Mat)	47	?	...	+	+	+	+	0	...	+	...	Sasse and Pekins 1996
Myotis sodalis	USA	F (Mat)	23	UB	+	...	+	+	0	...	...	...	...	Kurta et al. 1996
Myotis sodalis	USA	F (Mat)	54	?	...	+	0	...	0	+	...	...	...	Callahan et al. 1997
Myotis volans	USA	F (Mat)	41	?	0	+	0	+	0	...	...	0	9, 14	Ormsbee and McComb 1998

Variables Assessed for Selection

Species	Location	Sex	n	Roost type								Note	Reference
Nyctalus noctula	The Netherlands	...	81	H	+	...	...	...	...	...	0	6, 15	Boonman 2000
Nyctophilus bifax	Australia	F (Mat)	52	H UB*	...	+	+	...	...	...	...	...	Lunney et al. 1995
Nyctophilus geoffroyi	Australia	M & F	139	H UB	+	+	...	...	...	+	+	...	Lumsden et al. 2002a, 2002b
Nyctophilus gouldi	Australia	M & F	38	H UB	+	+	...	...	...	...	...	...	Lunney et al. 1988
Vespadelus darlingtoni	Australia	M & F	10	H	+	+	...	...	...	...	+	...	Herr and Klomp 1999
Vespadelus pumilus	Australia	M & F	91	H	+	+	...	...	...	+	...	...	Law and Anderson 2000
Species that were combined in analysis:													
Chalinolobus morio *Nyctophilus geoffroyi* *Vespadelus regulus* *Vespadelus darlingtoni*	Australia	M & F	23	H UB	0	+	...	...	...	...	...	...	Taylor and Savva 1988
Lasionycteris noctivagans *Myotis lucifugus*	Canada	F	27	H	+	0	+	0	+	...	...	4, 6	Crampton and Barclay 1998
Lasionycteris noctivagans *Eptesicus fuscus* *Myotis evotis* *Myotis volans*	Canada	M & F	21	UB H	+	+	+	0	+	...	...	13	Vonhof and Barclay 1996
Myotis lucifugus *Myotis septentrionalis*	Canada	F	8	H UB	0	0	−	0	...	...	...	5	Grindal 1999
Idionycteris phyllotis *Myotis evotis* *Myotis occultus* *Myotis thysanodes*	USA	F (Mat)	121	UB H	...	+	0	+	...	+	+	16	Rabe et al. 1998

(continued on next page)

Table 1.1. (continued)

Species	Country	Sex	No Roosts	Type of Cavity	Variables Assessed for Selection								Other	References
					Tree sp.	Dead/ Decayed	Large Diam.	Tall Roost tree	% Bark	Low Canopy Cover/ Clutter around Roost	Aspect of Roost Entr.	Roost Stand		
Myotis volans														
Eptesicus fuscus														
Antrozous pallidus														
Myotis auriculus														
% of studies that showed positive selection					78	84	73	76	20	78	36	86		

Note. Sex = M & F if both sexes investigated or F (Mat) if roosts largely from females and young during the maternity season. Types of cavity were coded thus: CR = crevice in bark; H = hollow; UB = under bark; ? = reference does not specify if roosts are in cavities or under bark. Variables assessed for selection are as follows: + = positive selection; 0 = used at random; − = negative selection; ellipses dots = the characteristic was not measured or not compared statistically with availability. "Other" refers to other measured variables, included here only if selection was shown: 1 = trunk height and surface; 2 = % stem remaining; 3 = number of limbs; 4 = amount of rot; 5 = number of cavities per tree; 6 = cavity type and dimension; 7 = cavity microclimate; 8 = height of cavities aboveground; 9 = height of roost tree in relation to canopy; 10 = distance to nearest tree of same or greater height; 11 = distance to nearest tree as tall as roost; 12 = distance to nearest neighboring tree; 13 = distance to nearest available/cavity tree; 14 = distance from water; 15 = distance from forest edge; 16 = slope.

* Also includes a small number of foliage roosts. Note that only the November sample is included in this analysis as the May sample was largely from foliage roosts.

Figure 1.3. *A*, Maternity roost of *Nyctophilus geoffroyi* in a snag of a river red gum (*Eucalyptus camaldulensis*) in Australia; *arrow* denotes roost opening (photo by L. Lumsden); *B*, Douglas fir snags (*Pseudotsuga menziesii*) provide roosting cavities for a maternity colony of *Lasionycteris noctivagans* in British Columbia, Canada (photo by M. Vonhof).

species roost in cavities with large basal openings (e.g., *Desmodus rotundus* [Wilkinson 1985], *Rhinolophus hildebrandti* [Fenton and Rautenbach 1986], *Noctilio albiventris* [Fenton et al. 1993], and *Nycteris thebaica* [Aldridge et al. 1990]).

In the Neotropics, where *Saccopteryx bilineata* typically roosts in relatively accessible buttress cavities (Bradbury and Emmons 1974; Bradbury and Vehrencamp 1976) they may remain active to avoid predators (Genoud and Bonaccorso 1986). Alertness, however, does not guarantee protection from predators, as observed by Arendt (1986) when a St. Lucia boa (*Boa constrictor*) captured a *Brachyphylla cavernarum* that unsuccessfully retreated upward into a large tree cavity. That some tropical species select cavities with large openings may reflect the fact that thermal constraints generally are less than in temperate regions where well-insulated cavities should offer important thermoregulatory advantages.

A number of variables have been measured to assess selection at the level of the roost tree and in the immediate vicinity of the tree (roost stand; table 1.1). As predicted by Vonhof and Barclay (1996) and Brigham et al. (1997), most, but not all, bat species that have been studied select trees that are large in diameter, taller than surrounding trees, and relatively uncluttered by adjacent vegetation. Vonhof and Barclay (1996) and Betts (1998) suggest three benefits of such roosts: (1) increased conspicuousness and hence ease of bats finding the roost tree; (2) reduced predation risk; and (3) maintenance of an optimal microclimate. Tall trees with an open canopy generally experience elevated exposures to solar radiation that may increase the energetic benefits to bats.

In general, bats are not likely to discriminate between tree species per se but, rather, select trees based on the specific characteristics of the cavities associated with a particular species (Sedgeley and O'Donnell 1999b). Notwithstanding, the species of tree was an important variable in several studies (e.g., Boonman 2000; Sedgeley and O'Donnell 1999b; Vonhof 1996). This may reflect the fact that some species provide better insulation than others. Using infrared thermal imaging, Rieger (1996) showed that beech trees (*Fagus sylvatica*) used as roosts by *Myotis daubentonii* remained warmer during the day and night than other tree species. The size of the tree appears to influence the microclimate in the cavity roost, and thus large trees are often selected as roosts (table 1.1). Slender trunks offer less insulation against extreme temperatures than do large ones (Alder 1994; Gellman and Zielinski 1996; Sluiter et al. 1973).

The amount of bark present on a tree also appears to influence the internal microclimate of the roost, with thicker bark providing the greatest insulation (Nicolai 1986). Dead trees are generally less well insulated than live ones owing to a lack of bark and a lower water content (Maeda 1974). Dead trees often contain more cavities than do live ones; and many of the studies summarized in table 1.1 noted that most roost trees were dead, although this varied among tree species and areas. Some trees, such as conifers, generally do not form cavities until they begin to decay, whereas species such as eucalypts form cavities when they are alive and healthy. Thus, in the case of eucalypts, although dead trees are not categorically avoided, certain bat species roost primarily in live trees (Lumsden et al. 2002b).

Not only do bats select particular trees as roost sites, they may also select particular parts of the forest in which to roost. Several studies have compared variables within the roost stand to other areas of the forest, with the majority showing selection for one or more of these variables (table 1.1). For example, in North America, areas around roosts of *Lasionycteris noctivagans* had more roost-type trees, a lower canopy cover, shorter understory, and less vegetative cover than did random plots (Campbell et al. 1996). In southeastern Australia, *Nyctophilus geoffroyi* and *Chalinolobus gouldii* selected areas of forest that contained high densities of their respective preferred roost trees (Lumsden et al. 2002a).

Selection can also occur at the landscape scale with roosts of some species being closer to water (Boonman 2000; Ormsbee and McComb 1998), closer to the forest edge (Boonman 2000; Sedgeley and O'Donnell 1999b), or associated with other landscape elements (Lumsden et al. 2002a). In the Netherlands, Boonman (2000) found that *Nyctalus noctula* and *Myotis daubentonii* roosted closer to the edge of forested areas than was expected from randomly chosen cavities and suggested that these trees may experience greater exposure to solar radiation, resulting in warmer cavities. Moreover, bats that foraged outside the forested area were able to reduce the time and energy spent flying through the forest.

How dependent bats are on certain characteristics of roosts can be explored by determining ways that a single species reacts to the availability of roost resources. Dependence on certain characteristics would be indicated if some variables were consistently selected in different environments. If variables were used selectively it might indicate that the bats were more flexible in their use of these characteristics. Selection of roosts by *Vespadelus pumilus* was investigated at two sites in eastern Australia with different disturbance histories: old-growth and regenerating forest (Law and Anderson 2000). At sites where numerous large, old trees were available, bats selected those in preference to smaller trees. In the regenerating forest, the remaining dead trees and large trees in an adjacent area were preferentially used as roosts. In addition, understory trees, such as blackwood (*Acacia melanoxylon*), which forms cavities at a smaller tree diameter were used as roost sites. Although these understory trees were present in the old-growth forest, they were not used as roosts when more suitable cavities were available.

Trees selected by bats may not only vary regionally and by area but also intraspecifically by sex and season. Roosts selected by maternity colonies may be different from those used during the nonbreeding period. Adult males and nonreproductive females tend to select cooler roost sites at temperatures that allow them to enter torpor, thus minimizing energy expenditure (Hamilton and Barclay 1994; Kerth et al. 2000). During the nonbreeding season in south-eastern Australia, both males and females of *Nyctophilus geoffroyi* occupy a wide range of structures, including buildings, under bark, and within cavities in relatively small trees. Within the same area, females selected large dead trees during the maternity period that were more than twice the diameter of those selected at other times (Lumsden et al. 2002b). The location of a roost within a given landscape may also vary between maternity and nonbreeding periods. Maternity roosts of *Chalinolobus tuberculatus* (Sedgeley and O'Donnell 1999b) and *N. geoffroyi* (Lumsden et al. 2002a) were located closer to the forest edge than were nonbreeding roosts.

Several studies have investigated roost use during the breeding season, but they have seldom separated data by reproductive status or age (e.g., Brigham et al. 1997; Callahan et al. 1997; Kalcounis and Brigham 1998; Ormsbee and McComb 1998). While it can be expected that pregnant and lactating females both would require a warm microclimate to enhance the rapid growth of the fetus and young, distinguishing use of roosts between these groups can reveal differences in their requirements at these times. For example, Kerth et al. (2000) found that pregnant females of *Myotis bechsteinii* preferred significantly cooler roosts than did lactating females.

Are Tree Cavities Limited Resources for Bats?

Long-term studies on roost selection and detailed information on the avail-ability of roosts are needed to determine whether roosts are limiting to bats. Because many species of bats show strong selection for particular types of roosts, tree cavities may be limiting depending on their relative abundance (Crampton and Barclay 1998). From observations on cavity and tree selec-tion by *Chalinolobus tuberculatus,* Sedgeley and O'Donnell (1999a) determined that only 1.3% of trees contained cavities that were suitable as roosts. Based on the density of suitable trees, they calculated that more than 3,000 potential roost trees were present in their study area. Although colonies of *C. tubercula-tus* shifted roost sites almost every day and rarely reused roosts (O'Donnell 2000), they concluded that roosts were relatively abundant (Sedgeley and O'Donnell 1999a).

Little information is available on competition between different bat species and whether they partition available roost resources. Several studies have compared different species in areas where they may have access to the same tree cavities. In some situations, no differences in roost charactersitics were found between sympatric species (Crampton and Barclay 1998; Vonhof 1996), whereas significant differences were found in other situations, which had larger sample sizes (Boonman 2000; Lumsden et al. 2002b). Perkins (1996) suggested that local distribution, species composition, and population size of bats in managed forests in North America were related to interspecific competition for limited roost sites. There is also evidence that other vertebrate and invertebrate species compete for roosts and, at times, may evict bats from tree cavities (e.g., Maeda 1974; Mason et al. 1972; Sedgeley and O'Donnell 1999a; Start 1998; Tidemann and Flavel 1987).

Spaces beneath Exfoliating Bark

Spaces that form beneath exfoliating bark (fig. 1.4) also provide alternate roosting sites to cavities in branches and tree trunks for some species. In North America, crevices beneath exfoliating bark are used predominantly by *Myotis* spp. (e.g., *M. californicus* [Brigham et al. 1997], *M. evotis* [Vonhof and Barclay 1996, 1997], *M. septentrionalis* [Foster and Kurta 1999], *M. sodalis* [Humphrey et al. 1977; Kurta et al. 1993, 1996], and *M. volans* [Vonhof and Barclay 1996]; fig. 1.4A) and by *Lasionycteris noctivagans* (Mattson et al. 1996; Vonhof and Barclay 1996). In Australia, spaces beneath exfliating bark are predominantly used as roosts by *Nyctophilus* spp. (e.g., *N. arnhemensis* [Churchill 1998], *N. bifax* [Lunney et al. 1995], *N. geoffroyi* [Hosken 1996; Lumsden et al. 2002b; Taylor and Savva 1988], *N. gouldi* [Lunney et al. 1988; Tidemann and Flavel 1987], and *N. timoriensis* [Churchill 1998]). In the Neotropics, spaces beneath

Figure 1.4. *A*, Exfoliating bark (*arrow*) on a Douglas fir snag (*Pseudotsuga menziesii*) used as a day roost by *Myotis volans* in British Colombia (photo by M. Vonhof); *B*, exfoliating bark (*arrow*) on a living grignon tree (*Ocotea rubra*) used as day roosts by *Artibeus obscurus* in French Guiana (from Simmons and Voss 1998; photo by R. Voss).

exfoliating bark are used as day roosts by *Artibeus obscurus* (Simmons and Voss 1998; fig. 1.4B). To our knowledge, no species are known to roost exclusively beneath exfoliating bark.

Exfoliating bark is present on some live trees but is more common on dead trees. Roosts beneath exfoliating bark are generally more ephemeral than tree cavities, with some lasting only a few months, although others may persist for several years (Humphrey et al. 1977). For some North American tree species (e.g., shagbark hickory [*Carya ovata*] and grignon [*Ocoetea rubra*]), exfoliating bark typically persists for the life of the tree and, thus, potentially provides more roost permanency than does bark on dead trees. One of the risks of roosting beneath exfoliating bark, however, is illustrated by an observation made by Kurta (1995), who witnessed a strip of bark falling into water from a dead tree with a torpid male *Eptesicus fuscus* still attached.

Roost sites beneath exfoliating bark may have different thermal properties than tree cavities and usually are climatically less stable (Crampton and Barclay 1998). The roost microclimate may be influenced by the insulative properties of the bark (Nicolai 1986), the size of the tree, and the amount of solar exposure it receives (Kurta et al. 1996; Vonhof and Barclay 1997). Where roost sites have been compared to available spaces beneath exfoliating bark, bats generally select sites that are thermally stable (Kurta et al. 1996; Vonhof and Barclay 1997).

Little empirical data are available on predation rates on bats that occupy exfoliating bark roosts. Wunder and Carey (1996) suggested that bats that roost beneath exfoliating bark may be more susceptible to predation than those that roost in tree cavities. However, bats are more likely to escape from beneath exfoliating bark if there are alternate routes of exit. Further investigations are needed to determine if predation is an important variable in the selection of roosts beneath exfoliating bark. Despite the apparent disadvantages of roosting in ephemeral sites, including suboptimal microclimates and, perhaps, increased risks of predation, a number of species actively select roosts beneath exfoliating bark, even in situations where tree cavities are abundant (e.g., Foster and Kurta 1999; Lumsden et al. 2002b). This suggests that, for some species, there are benefits to roosting in these situations or that at least the occupants incur no additional costs.

Interspecific competition for roosts by bats may be less for those that roost beneath exfoliating bark than those that roost in tree cavities. Only a small percentage of bat species use roosts beneath exfoliating bark, despite the fact that these roosts may be more abundant in some situations. In a study conducted in a wetland area of North America, with numerous dead trees, two species of *Myotis* that roosted beneath exfoliating bark were examined, with characteristics of roost trees compared to evaluate possible competition between the two species (Foster and Kurta 1999; Kurta et al. 1993, 1996). *Myotis sodalis* roosted exclusively beneath exfoliating bark on dead green ash (*Fraxinus* sp.) trees with low canopy cover. Although *M. septentrionalis* roosts had similar characteristics, this species exhibited a broader roosting niche. A greater variety of cavity types, tree species, decay stages, and amount of canopy cover was used by *M. septentrionalis*. Where roosts were abundant, there was little overlap in the actual trees that were used by these two species, although Foster and Kurta (1999) suggested that interspecific interactions may exist in areas with lower availability. In addition to competition between bat species, there may be competition with other vertebrates and invertebrates (e.g., Kurta and Foster 1995).

Rather than concealing themselves beneath exfoliating bark, some species roost in narrow crevices between heavily furrowed folds of bark or in the narrow spaces formed between two touching tree trunks. In Canada, *L. noctivagans* has been observed roosting in such crevices during their northward migration (Barclay et al. 1988). The bats were well concealed, as they were wedged into the crevices with both their dorsal and ventral surfaces in contact with the substrate. These roosts provided shelter from rain, wind, and possibly predators and were cool enough to allow bats to enter torpor. Despite appearing to be opportunistic in choice of roost sites, individual *L. noctivagans* selected older willow trees with large circumferences as they provided more crevices than did younger trees.

Exposed Boles

A few species, most notably members of the Emballonuridae, roost in relatively exposed areas on the sides of tree boles, beneath fallen trees, and on branches and similar structures (bridges) suspended over water (Bradbury 1977b; Bradbury and Emmons 1974; Bradbury and Vehrencamp 1976; Goodwin and Greenhall 1961; Simmons and Voss 1998). Emballonurids often assume stereotypical postures, but are largely inconspicuous to observers because of their cryptic coloration and mottled pelage. For example, *Rhynconycteris naso* and *Saccopteryx leptura* commonly roost on the open boles of trees in small harem groups, often spaced in a vertical row, one above the other (fig. 1.5). Colonies of *R. naso* range from three to 45 individuals, whereas those of *S. leptura* are generally smaller, ranging from two to nine individuals. Colonies of both species use several alternative roosts, between which the entire group may move as a unit (Bradbury and Vehrencamp 1976). Other Neotropical emballonurids, such as *Centronycteris maximiliani*, *Cormura brevirostris*, *Peropteryx kappleri*, *P. macrotis*, and *P. leucoptera*, are known to roost in small groups beneath trunks and buttresses of fallen trees (LaVal 1977; Simmons and Voss 1998).

Cavities in Bird Nests

A limited number of bat species (primarily members of the Vespertilionidae, Molossidae, and Emballunuridae) use abandoned bird nests as roost sites (Dalquest and Walton 1970; Schulz 1997). For some bats, bird nests constructed

Figure 1.5. Small harem group of *Saccopteryx bilineata* roosting on the exposed bole of a lowland rainforest tree in Panama (photo by M. D. Tuttle). The bats are cryptic as viewed against the lichen-covered bark.

from plant material serve as primary roost resources, but, for others, they are used only opportunistically. Schulz (1997) reviewed use of abandoned bird nests by bats and found that three types of nests were primarily used as roosts: hanging nests; open cup-shaped nests; and enclosed mud nests (some of which include plant material mixed with soil).

In tropical areas of Africa, Asia, and Australia, members of the genera *Kerivoula* and *Murina* commonly roost in suspended bird nests (Schulz 1997). These nests are dome-shaped structures constructed from long pieces of vegetation such as bark, rootlets, leaves, ferns, and palm fibers (fig. 1.6). Schulz (2000) suggested that roosts located in suspended bird nests may be less sus-

Figure 1.6. Suspended bird nest used as a day roost by *Kerivoula papuensis* in Australia, with a modified opening at the base (photo by C. Taylor).

ceptible to predation from terrestrial, climbing, and aerial predators than other types of roosts.

In Australia, the majority of *Kerivoula papuensis* roosts have been found in the abandoned hanging nests of brown gerygones (*Gerygone mouki*) and yellow-throated scrub wrens (*Sericornis citreogularis* [Schulz 1995, 2000]). These roosts were mostly occupied by single males or females, with a maximum group size of eight individuals. No breeding activity of bats was recorded in these nests, suggesting that they may not provide the required roost microclimate for raising young (Schulz 2000). When birds originally construct these nests they have a side entrance. However, each nest used as a roost by *K. papuensis* contained a hole in the base (fig. 1.6). This led Schulz (2000) to suggest that bats modify these nests to make them more suitable as roosting sites.

This modification allows an unobstructed view of the ground below and a quick exit route in response to disturbance. He suggested that such modifications were similar to those made by tent-making bats observed in some Neotropical and Paleotropical species (Kunz et al. 1994). In particular, they resembled the stem tents made by *Cynopterus sphinx* in vines (Balasingh et al. 1995) and flower/fruit clusters (Bhat and Kunz 1995; Storz et al. 2000a, 2000b).

Murina florium has also been recorded roosting in suspended bird nests in Australia, with seven of the 11 roosts observed in nests of the yellow-throated scrub wren and the fern wren (*Oreoscopus gutturalis* [Schulz and Hannah 1998]). The scrub wren nests, also used by *K. papuensis*, all included a basal hole modification; however, the nests of the fern wren, which were not used by *K. papuensis*, were unmodified.

In Africa, *Kerivoula argentata* and *K. lanosa* roost alone, or in small groups of up to six individuals, in the disused hanging nests of masked and spectacled weaverbirds (*Ploceus velatus* and *P. ocularis*) and scarlet-chested sunbirds (*Nectarinia senegalensis* [Skinner and Smithers 1990]). In India, *Kerivoula picta* roost in nests of the baya weaverbird (*Ploceus philippinus* [Sharma 1986]). All roosts of *K. picta* were in incomplete nests, with no bats found in the large number of completed nests that were examined.

In contrast to hanging nests, mud nests are predominantly made from soil but may include plant parts. Generally, these nests are used opportunistically by bat species that typically roost in other structures. Two basic forms of mud nests include open-cup structures and enclosed bottle-shaped structures, which are made by swallows and swiftlets. These are constructed in both natural and artificial situations, such as along stream banks, under rock overhangs, under bridges, and in road culverts. In the southwestern United States, *Myotis velifer* has been observed roosting throughout the year in open-cup nests of barn swallows (*Hirundo rustica*) and cave swallows (*Petrochelidon fulva* [Jackson et al. 1982; Pitts and Scharninghausen 1986; Ritzi et al. 1998]) and enclosed, bottle-shaped nests constructed by cliff swallows (*P. pyrrhonota* [Buchanan 1958; Manning et al. 1987; Pitts and Scharninghausen 1986]). *Tadarida brasiliensis* also roosts in the bottle-shaped nests constructed by cliff swallows, although it appears to use these roosts more opportunistically (Buchanan 1958; Pitts and Scharninghausen 1986).

In Australia, Schulz (1998) surveyed 5,175 bottle-shaped mud nests constructed by fairy martins (*Hirundo ariel*) and located 204 bats representing eight species (all vespertilionids) at a frequency of 3.9 bats per 100 nests. *Myotis macropus/M. moluccarum* and *Vespadelus troughtoni* were the most frequently recorded species, making up 83% of the individuals present. Group size ranged from one to 21 individuals.

Cavities in Bamboo Culm

Roosting within the culm of bamboo has been reported primarily for four species of vespertilionids from Southeast Asia: *Eudiscopus denticulus, Glischropus tylopus, Tylonycteris pachypus,* and *T. robustula* (Kock and Kovac 2000; Kofron 1994; Medway and Marshall 1970). Each is a small species (3.5–10 g) with thickened thumb pads (see fig. 1.15) that assist in gripping the smooth inner surface of the internode cavities of bamboo culm. *Tylonycteris pachypus, T. robustula,* and *E. denticulus* have extremely flattened skulls, an adaptation that facilitates access to roosts with small openings. Anecdotal records of other Asian bats roosting in bamboo culm include *Glischropus javanus* (Chasen 1939), *Kerivoula papillosa* (Bates and Harrison 1997), *Myotis hasseltii* (Lekagul and McNeely 1977), *Pipistrellus mimus* (Lekagul and McNeely 1977), and *P. paterculus* (Bates and Harrison 1997).

In their study on roosting habits of *T. pachypus* and *T. robustula* in Malaysia, Medway and Marshall (1970, 1972) located 448 roosting groups in the internodes of standing, green culms of the bamboo *Gigantochloa scortechinii.* Openings to the roosting chambers were all narrow vertical slits formed by the pupation chambers and emergence holes of the leaf beetle *Lasiochila goryi.* Although there was broad overlap in physical dimensions of roosts used by both bat species, there were significant differences in diameter of the culm and the length and width of the opening. The smaller *T. pachypus* was able to use roosts with narrower openings, from which *T. robustula* was excluded. Both species showed selection in the types of roosts used compared to their availability in the area. Bats selected roosts that had openings located in the lower half of the cavity. Although 28% of roosts were used by both species on different occasions, only once were the two species found roosting together. Group sizes ranged from one to 20 individuals for *T. pachypus* (average 4.9) and from one to 32 for *T. robustula* (average 3.5). Maternity groups sometimes consisted entirely of females, but many contained a single adult male (Medway and Marshall 1970). The number of potential roosts in the study area greatly exceeded the number used by bats, and individuals frequently changed roost sites on a daily basis.

In Thailand, *Eudiscopus denticulus* also roosts in small groups in internode cavities of the bamboo *Gigantochloa* sp. (Kock and Kovac 2000). Bats roost in the upper part of the cavity, apparently clinging to the inner surface with the adhesive disks on their feet and wrists. Kofron (1994) observed four groups of *Glischropus tylopus* roosting in dead bamboo culms in Brunei, northwest Borneo. Each group formed a social unit, consisting primarily of adult females and young. All roosts were approximately 2 m above the ground in bamboo culms that were 4.6–4.8 cm in diameter, and with openings less than 2.5 cm wide. Openings were either made or modified by beetles or by a gnawing rodent.

Foliage

Trees and understory plants provide roosting opportunities for a wide range of species, many of which are frugivorous or nectarivorous. Old World fruit bats, particularly members of the genus *Pteropus*, often roost on exposed branches within the tree canopy (fig. 1.7A; see review in Pierson and Rainey 1992). Most species are strongly colonial, although some are solitary or roost in small family groups. In large colonies, roost sites (camps) may include a number of adjacent trees (Pierson and Rainey 1992; Ratcliffe 1932). *Pteropus* spp. typically show long-term fidelity to traditional roost sites, particularly those that remain undisturbed (Pierson and Rainey 1992; Ratcliffe 1932; Wiles 1987). Several camps in Australia have been used repeatedly for more than 80 yr (Lunney and Moon 1997).

Old World megachiropterans roost in a wide range of habitats. *Pteropus vampyrus*, one of the largest megachiropterans, exemplifies this variation. In Malaysia, *P. vampyrus* has been recorded in lowland coastal areas, roosting in rainforest, mangroves, and coconut groves (Davis 1962; Goodwin 1979; Lim 1966; Payne et al. 1985). In Indonesia, on the island of Pulau Rambut, this bat roosts in kedoya (*Amoora aphanamixis*) and kepuh trees (*Sterculia foetida* [Wiriosoepartha et al. 1986]) and, on Sumatra, in cultivated kapok trees (*Ceiba pentandra* [Davison 1992]).

The specific criteria that *Pteropus* spp. use to select roost sites are poorly understood. However, canopy structure, physical features of the surrounding area, and location with respect to feeding habitat appear to be important variables. Ratcliffe (1932) specified seclusion and protection from the sun. Richards (1990a) evaluated 17 variables to characterize roost sites and roost environments of *P. conspicillatus* and found that distance to nearest rainforest accounted for 75.1% of the sample variance. Pierson and Rainey (1992) suggested that protection from strong winds and access to updrafts were important criteria for roost selection. Among six variables evaluated in a multivariate model, canopy height and height of emergent trees were the most important variables in explaining roosting sites used by *P. alecto* and *P. scapulatus* (Tidemann et al. 1999).

In the relatively intact forests of tropical Australia, the presence of water appears to be a major factor influencing roost selection by *P. alecto*, with roosts typically located in riparian habitats, especially during the dry season (Loughland 1998; Tidemann et al. 1999). When ambient temperatures increase during the rainy season, roosts are mostly located in rainforest habitat, with smaller numbers in mangrove and bamboo forests. In rainforest habitat, *P. alecto* roosts in the shade beneath the dense forest canopy, but during the dry season individuals that roost in the deciduous bamboo forest are exposed directly to sunlight (Palmer and Woinarski 1999). Tidemann et al. (1999) suggested that initial colonization of roosts by *P. alecto* was random when roost vegetation is

Figure 1.7. *A*, Colony of *Pteropus poliocephalus* roosting in a partially defoliated emergent tree in Australia; *B*, small group of *P. poliocephalus* roosting on defoliated branches (photos by P. Eby); *C*, mother and young *Lasiurus cinereus* roosting on a branch of a spruce tree in North America (photo by M. D. Tuttle); *D*, three young *Lasiurus borealis* roosting on a branch of a maple tree in North America (photo by T. H. Kunz).

not limiting, but factors such as human predation, cyclones, and fires may influence their persistence.

For many island species of the Pteropodidae, and to a lesser extent populations in mainland areas, the influence of external forces such as habitat loss and hunting pressure may mask the influence of habitat preferences on roost location in pteropodids. For example, existing roosts of both *P. livingstonii* in the Comores Islands and *P. samoensis* in American Samoa are concentrated in montane forests on steep slopes (Brooke et al. 2000; Craig et al. 1994; Reason and Trewhella 1994). These current roosting habits may reflect avoidance of increased hunting pressure in highly fragmented lowland forests more than a preference for steep montane vegetation.

The emergent trees used by highly gregarious species, such as *Pteropus* and *Eidolon*, are often defoliated by their occupants (Banack 1996; Bonaccorso 1998; Brooke et al. 2000; Flannery 1995; Goodwin 1979; Jones 1972; Kitchener et al. 1990; Nelson 1965; Okon 1974; fig. 1.7*B*). Bonaccorso (1998) suggested that defoliated branches in roost trees might facilitate visual observations of social partners, rivals, or approaching aerial predators. Reduced foliage cover also exposes roosting animals to sunlight, wind, and rain. When exposed to the hot sun during the day, individuals often cool themselves by flapping their wings, licking their chest and wings, and panting (Nelson 1965; Neuweiler 1969; Ochoa and Kunz 1999). In cool weather or during heavy rain, large pteropodids almost completely wrap themselves with their wings (Lekagul and McNeely 1977; Ochoa and Kunz 1999).

Some species of small, foliage-roosting megachiropterans are associated with specific types of plants. For example, in peninsular Malaysia, the day roosts of *Cynopterus horsfieldi* appear associated with the availability of the epiphytic bird's nest fern (*Asplenium nidus*). Leaf shelters are formed when older, dry leaves of this fern droop downward creating a skirt of dead leaves, beneath which *C. horsfieldi* roosts (Tan et al. 1997).

Little is known about factors that influence the distribution and abundance of foliage-roosting microchiropterans, although availability and dispersion of plants and characteristics of foliage appear important (Kunz 1982, 1996; Morrison 1980). The leaved canopies of both temperate and tropical trees provide roost sites for a number of species. The roosting habits of foliage-roosting microchiropterans seem to reflect the structure and composition of local forest communities (Constantine 1958, 1959, 1966; Menzel et al. 1998; Morrison 1980; Schulz and Hannah 1998; Simmons and Voss 1998).

Foliage-roosting microchiropterans roost in a range of situations, including epiphytes, beneath unmodified live and dead leaves and in the foliage of tree ferns. In northern Australia, *Murina florium* roosts in vertically suspended clusters of dead leaves (Schulz and Hannah 1998). These roosts are typically located in the rainforest understory close to breaks in the forest canopy. Most roosts are occupied by singletons, however, a group of 12 bats was observed in the curled base of a suspended dead palm leaf (*Archontophoenix* sp. [Schulz and Hannah 1998]).

Members of the vespertilionid genus *Lasiurus* typically roost among leaves in densely foliated tree canopies (Constantine 1958, 1966; Hutchinson and Lacki 2000; Kurta and Lehr 1995; Mager and Nelson 2001; McClure 1942; Menzel et al. 1998; Saugey et al. 1998; Shump and Shump 1982; Webster et al. 1980). Roosts have been recorded from a range of both conifers (fig. 1.7C) and hardwoods (fig. 1.7D) and usually consist of individuals or small family groups.

Within their roosts, lasiurines typically cling to leaf petioles or the tips of small branches (Menzel et al. 1998). Constantine (1966) identified three types of shelters used by *Lasiurus borealis* and *L. cinereus* in agricultural areas in the midwestern United States (Iowa): an inverted bowl-shaped canopy made of tree branches infiltrated with grapevines that provided structural rigidity; a dense, leafy overhang of new succulent growth; and a dense tuft of older, nonsucculent leaves.

The roosting habits of *L. borealis* and *L. seminolus* in the southeastern United States (Georgia) were quantitatively investigated by Menzel et al. (1998) to determine which variables influenced roost site selection and how this varied among the two species. Roost trees of both species were significantly taller and had larger diameters than did surrounding trees. *Lasiurus borealis* roosted predominantly in hardwoods, which tended to have smaller diameters than the conifers used by *L. seminolus*. Moreover, the understory and canopy vegetation in the vicinity of *L. borealis* roosts showed a greater diversity than those occupied by *L. seminolus*. Mager and Nelson (2001) quantified roost variables for *L. borealis* and found that this species showed a preference for large trees, nearly 80% of which exceeded 30 cm in diameter. They also suggested, as did Constantine (1966) and Menzel et al. (1998), that canopies of more mature trees provided a high degree of protective cover for foliage-roosting bats.

The roosting height of *L. borealis* within a tree may vary geographically and between tree species (Constantine 1958, 1966; McClure 1942; Mager and Nelson 2001; Menzel et al. 1998), although roosting height may reflect methodological differences among studies. Constantine (1966), who based his research on direct observations, found that solitary *L. borealis* generally roosted relatively low in the tree, whereas small family groups were more often observed on branches higher in the canopy. He suggested that selection of high roosts by adult females with young may provide greater concealment from terrestrial predators, avoid disturbances from activity on the ground that might dislodge young, and permit young a greater opportunity to conduct initial flights (Constantine 1966). Using radiotelemetry, Mager and Nelson (2001) reported that most roosts of *L. borealis* were more than 5 m above the ground and

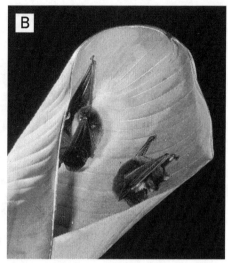

Figure 1.8. *A,* Furled banana leaf (*Musa* sp.) used as a day roost by *Thyroptera tricolor; B,* two *T. tricolor* clinging to the inner, moist surface near the opening of a furled banana leaf used as a day roost (photos by M. D. Tuttle).

located within 1.5 m of the edge of the crown, with few branches nearby that would obstruct flight.

Furled Leaves

Some foliage-roosting microchiropterans in the Old and New World occupy cylindrical spaces formed by furled leaves (fig. 1.8). Species that utilize these roosts have highly specialized foot and thumb pads that enable them to cling to the smooth upper surfaces of leaves (*Myotis bocagei* [Baagøe 1978; Brosset 1976], *M. mystacinus* [Medway 1983], *Myzopoda aurita* [Findley and Wilson 1974], *Pipistrellus nanus* [Happold and Happold 1990, 1996; LaVal and LaVal 1977], *Thyroptera discifera* and *T. tricolor* [Findley and Wilson 1974]). These morphological specializations may make it impossible for them to cling to rough surfaces, but the typical head-up posture inside a furled leaf may be ideal for detecting potential predators (Findley and Wilson 1974).

The growth form of wild banana leaves (*Musa* spp. and *Strelitzia* spp.) and similar-shaped leaves (*Heliconia* spp.) and the geographic distribution of these species may have an important effect on the local distribution of bats that roost in furled leaves. The Neotropical disk-winged bats, *T. discifera* and *T. tricolor* routinely occupy furled leaves of *Heliconia* spp. and *Calathea* spp., both within the forest and in forest clearings (Findley and Wilson 1974; Goodwin and Greenhall 1961; Wimsatt and Villa-R 1970). Occupied leaves are predominantly found in areas that are shaded for part of the day. Findley and Wilson (1974) found that most *T. tricolor* roosts were in furled leaves with the diameter of the opening between 50 and 100 mm. In Natal, South Africa, *Pipistrellus nanus* typically roosts in furled leaves of domesticated banana plants (*Musa* spp.), but may have used leaves of wild banana (*Strelitzia nicolai*) before domesticated varieties were introduced (LaVal and LaVal 1977).

Furled leaves are highly ephemeral roosts, with the leaves typically opening (unfurling) in 1–3 d. This necessitates the frequent movement to nearby furled leaves (Brosset 1976; Findley and Wilson 1974; Happold and Happold 1990, 1996; LaVal and LaVal 1977). Despite being from different biogeographic regions and phylogenetic groups, the behavior of different species that roost in furled leaves is remarkably similar, suggesting behavioral convergence in these distantly related taxa.

Furled leaves are typically occupied by singletons or small groups of bats. In Malaysia, colony sizes of *Myotis mystacinus* range from one to eight individuals (Medway 1983), whereas in Costa Rica, groups of *T. tricolor* range from one to nine, with each roosting group occupying an exclusive area (Findley and Wilson 1974). In Malawi (east-central Africa) males of *P. nanus* often roost alone or in small multifemale groups, with exclusive maternity colonies formed during the breeding season (Happold and Happold 1990).

Tents
Diversity of Bats That Roost in "Tents"

Among the 19 species of bats known to roost in or construct tents (table 1.2), 15 are members of the New World family Phyllostomidae, three are members of the Old World family Pteropodidae, and one is a member of the Vespertilionidae (Kunz et al. 1994; Hodgkison et al., in press). One direct observation (Balasingh et al. 1995) and several indirect observations support the hypothesis that some tents are indeed made by bats (Brooke 1990; Chapman 1932; Choe and Timm 1985; Kunz et al. 1994; Kunz and McCracken 1996; Tan et al. 1997; Timm 1987). Notwithstanding, it remains unclear whether all bat species that have been observed roosting in tents are responsible for their construction.

Tent-making and tent-roosting behaviors are restricted to relatively small-bodied species because larger bats (especially the larger megachiropterans) could not be supported or adequately concealed by altered leaves, or they would not be sufficiently agile to fly in the physically cluttered environments often associated with tents in the forest understory (Kunz et al. 1994). Among the 15 species of Neotropical bats known to roost in or thought to construct tents, body masses range from 8 to 50 g, and each feeds largely on small fruits, although some also may feed on nectar, pollen, leaves, and insects (Gardner 1977). In the Paleotropics, the three megachiropterans species (*Balionycteris*

Table 1.2. Biogeographic and phylogenetic patterns in tent architecture and tent-roosting/tent-making behavior in Paleotropical and Neotropical bats

Biogeographic Region and Suborder/Family/Subfamily/Species	Architectural Style of Tent							
	1	2	3	4	5	6	7	8
Paleotropics	X	X		X				X
Megachiroptera:								
Pteropodidae:								
Pteropodinae:								
Balionycteris maculata								X
Cynopterus:								
C. brachyotis	X	X		X				X
C. sphinx		X						X
Microchiroptera:								
Vespertilionidae:								
Scotophilus kuhlii	X							
Neotropics	X	X	X	X	X	X	X	
Microchiroptera:								
Phyllostomidae:								
Carollinae:								
Rhinophylla pumilio			X	X				
Stenodermatinae:								
Artibeus:								
A. anderseni								X
A. cinereus		X		X	X	X	X	
A. glaucus					X			
A. gnomus				X				
A. jamaicensis		X	X	X				
A. phaeotis				X	X		X	
A. toltecus				X				
A. watsoni		X	X	X	X		X	
Ectophylla alba				X			X	
Mesophylla macconnelli				X	X	X		
Urodema:								
U. bilobatum		X	X	X	X	X	X	X
U. magnirostrum		X						
Vampyressa:								
V. nymphaea		X						
V. pusilla				X				

Note. Architectural styles of tents are as follows: 1 = conical; 2 = palmate umbrella; 3 = pinnate; 4 = apical; 5 = bifid; 6 = pyramid; 7 = boat; 8 = stem (after Kunz et al., 1994).

maculata, Cynopterus brachyotis and *C. sphinx*) that occupy tents range in body mass from approximately 18 to 65 g (Hodgkison et al., in press; T. H. Kunz, personal observation; Storz et al. 2001) and feed almost exclusively on understory fruits, flowers, nectar, and leaves (Bhat 1994; Boon and Corlett 1989; Elangovan et al. 1999, 2000; Hodgkison 2002; Marshall 1985; Ruby et al. 2000; Tan et al. 1998).

In the only published report of a tent-roosting vespertilionid, Rickart et al. (1989) found singletons and small groups of *Scotophilus kuhlii* beneath modified and collapsed fronds of *Livistona rotundifolia* in the Philippines. Although roosting groups of this bat were common, it is unclear whether this species constructed these tents or parasitized existing ones that were made by other species (Kunz et al. 1994). *Cynopterus brachyotis* and *C. sphinx* are sympatric with *S. kuhlii* and both species are known to roost in or construct umbrella tents similar to those occupied by *S. kuhlii.*

Tent-Roost Architecture

The eight architectural styles of tents (table 1.2; fig. 1.9) generally reflect the size and shape of the leaf and the number of veins and/or plications chewed or the plant part that is modified (e.g., leaves, stems, root masses). Kunz et al. (1994) suggested that the resulting architectural style of bat tents in leaves is more a consequence of leaf shape than of the behavioral repertoire of the bat constructing the tent. An exception to this may be the tents of *Ectophylla alba*, all of which are constructed from one basic leaf shape (oblong) modified into a characteristic boat-shaped structure (Brooke 1990).

Existing classifications of tent architecture (table 1.2) offer a convenient way to organize the diversity of plant taxa used by bats in tent construction (table 1.3). Although future observations will no doubt require modification of this scheme, it provides a useful framework for comparing the architectural styles of tents constructed by species that are from different biogeographic regions and that have different evolutionary histories. The relatively limited number of leaf forms and plant structures in the subcanopy of Neotropical and Paleotropical forests should limit the kinds of architectural styles of tents that can be constructed by bats. For example, circular, semicircular, and ovoid cuts appear to be the only ways that bats can effectively modify palmate-shaped leaves into effective tents. Similarly, oblong leaves of *Heliconia* spp. and *Musa* spp. are most commonly modified into tents when cuts are made parallel to the midrib to form a boat-shaped tent, although the paradox tent appears to be an exception (Kunz et al. 1994; Timm 1987). Chewed basal veins in ovoid leaves of *Philodendron* spp. and similar leaf forms invariably form apical tents, whereas the chewed petioles and/or midribs in a rosette of six to 14 leaves on understory saplings form conical tents. When bats sever the innermost stems of pendulous flower/fruit clusters, vines, and root masses, this invariably creates cavities and bell-shaped enclosures known as stem tents.

Some variation in tent architecture may occur if a leaf is modified by more than one bat species. For example, a large epiphyte, *Anthurium jenmanii*, was observed alternately occupied by small harem groups of *Artibeus cinereus* and *Mesophylla macconnelli* (T. H. Kunz and G. F. McCracken, unpublished data). This large, oblong leaf was modified with a long cut parallel to the midrib, characteristic of boat-style tents typically used by *A. cinereus*. Superimposed on this tent leaf were two J-shaped cuts that were similar to those observed in bifid and paradox tents. The only types of leaf architecture reported for

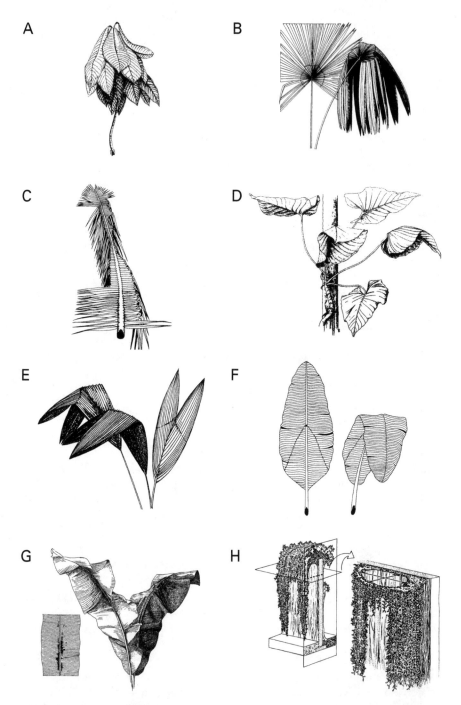

Figure 1.9. Eight architectural styles of tents constructed by tent-making bats. *A*, Conical tent (unidentiified sapling); *B*, palmate umbrella tent (*Sabal mauritiiformis*); *C*, pinnate tent (*Scheelea rostrata*); *D*, apical tent (*Philodendron ornatum*); *E*, bifid tent (*Cocos nucifera*); *F*, paradox tent (*Musa* sp.); *G*, boat tent (*Heliconia* sp.); and *H*, stem tent (*Polyalthia longifolia*) (illustrations by E. Rooks, from Kunz et al. 1994).

Table 1.3. Species of bats that have been observed in tents and the plant species from which the tents were constructed

Bat Suborder/ Family/Species	Plant Family	Plant Species	Tent Style	References
Megachiroptera:				
Pteropodidae:				
Balionycteris maculatus	Aspleniaceae	*Asplenium nidus*	8	Hodgkison et al., in press
	Zingiberaceae	*Hedychium longicornutum*	8	Hodgkison et al., in press
Cynopterus:				
C. brachyotis	Anacardiaceae	Unidentified sapling	1	Kunz et al. 1994
	Araceae	*Scindapsus aureus*	4	Tan et al. 1997
		Philodendron lacerum	8	Tan et al. 1997
	Liliaceae	*Dracaena fragrans*	1	Tan et al. 1997
	Orchidaceae	*Cymbidium finlaysonianum*	8	Tan et al. 1997
	Palmae	*Borassus flabellifer*	2	Tan et al. 1997
		Corypha utan	2	Boon and Corlett 1989; Tan et al. 1997
		Licuala grandis	2	Tan et al. 1997
		Livistona chinensis	2	Tan et al. 1997
		Livistona rotundifolia	2	Tan et al. 1997
		Livistona saribus	2	Tan et al. 1997
		Prichardia pacifica	2	Tan et al. 1997
		Sabal palmetto	2	Tan et al. 1997
		Unidentified palm	2	Lekagul and McNeely 1977
C. sphinx	Annonaceae	*Polyalthia longifolia*	8	Balasingh et al. 1995; Bhat 1994; Storz et al. 2000a, 2000b
	Compositae	*Vernonia scandens*	8	Baslasingh et al. 1993, 1995
	Palmae	*Borassus flabellifer*	2	Bhat 1994
		Caryota urens	8	Bhat 1994; Bhat and Kunz 1995; Phillips 1924; Storz et al. 2000a, 2000b
		Corypha sp.	2	Goodwin 1979
		Corypha umbraculifera	2	Bhat 1994
		Livistona chinensis	2	Bhat 1994
		Livistona rotundifolia	2	Kunz et al. 1994
		Roystonea regia	2	Bhat 1994

(continued on next page)

Table 1.3. (continued)

Bat Suborder/ Family/Species	Plant Family	Plant Species	Tent Style	References
Microchiroptera:				
Phyllostomidae:				
Artibeus:				
A. anderseni	Heliconiaceae	Heliconia sp.	7	Timm 1987
A. cinereus	Araceae	Anthurium jenmanii	6	Kunz et al. 1994
		Philodendron fragrantissiumum	4	Kunz et al. 1994
		Philodendron ornatum	4	Kunz et al. 1994
		Philodendron simsii	4	Kunz et al. 1994
		Xanthosoma undipes	4	Kunz et al. 1994
		Xanthosoma sp.	4	Timm 1987
	Heliconiaceae	Heliconia sp.	7	Kunz et al. 1994
	Musaceae	Musa sp.	?	Goodwin and Greenhall 1961
	Palmae	Astrocaryum sciophilum	5	Simmons and Voss 1998
		Coccoloba latifolia	4	Kunz et al. 1994
		Cocos nucifera	5	Kunz et al. 1994
		Manicaria saccifera	5	Kunz et al. 1994; Goodwin and Greenhall 1961
		Unspecified	?	Goodwin and Greenhall 1961
A. glaucus	Araceae	Xanthosoma sp.	4	Timm 1987
A. gnomus	Araceae	Monstera lechleriana	4	Timm 1987; Timm and Clauson 1990
		Philodendron sp. A	4	Charles-Dominique 1993
		Philodendron sp.	4	Simmons and Voss 1998
	Sterculiaceae	Sterculia sp.	4	Charles-Dominique 1993
	Strelitziaceae	Phenakospermum guyannense	4	Simmons and Voss 1998
A. jamaicensis	Araceae	Philodendron fragrantissimum	4	Kunz et al. 1994
	Palmae	Coccothrinax barbadensis	2	Kunz et al. 1994
		Sabal mauritiiformis	2	Kunz et al. 1994
		Scheelea rostrata	3	Foster and Timm 1976
A. phaeotis	Rubiaceae	Pentagonia donnel-smithii	4	Brooke 1990; Kunz et al. 1994
	Araceae	Philodendron mediacostatum	4	Kunz et al. 1994
	Heliconiaceae	Heliconia imbricata	7	Barbour 1932
	Musaceae	Musa × paradisiaca	7	Timm 1987
	Palmae	Unidentified	5	Kunz et al. 1994

Bat species	Family	Plant species	No.	References
A. toltecus	Araceae	*Anthurium caperatum*	4	Timm 1984
A. watsoni	Araceae	*Anthurium ravenii*	4	Boinski and Timm 1985; Choe and Timm 1985
	Cyclanthaceae	*Asplundia* sp.	2	Boinski and Timm 1985; Kunz et al. 1994
		Carludovica palmata	2	Boinski and Timm 1995; Choe and Timm 1985; Kunz et al., 1994
		Carludovica drudei	2	Kunz et al. 1994
		Cyclanthus bipartitus	2	Boinski and Timm 1985; Kunz et al. 1984
	Heliconiaceae	*Heliconia imbricata/pogonantha*	7	Brooke 1990
		Heliconia latispatha	7	Brooke 1990
		Heliconia imbricata	7	Boinski and Timm 1985; Choe and Timm 1985; Timm 1987
		Heliconia sp.	7	Boinski and Timm 1985; Choe and Timm 1985
	Marantaceae	*Calathea insignis*	4	Choe and Timm 1985
	Musaceae	*Musa × paradisiaca*	7	Boinski and Timm 1985; Choe and Timm 1985; Timm 1987
	Palmae	*Astrocaryum strandleyanum*	5	A. Gardner, personal communication
		Asterogyne martiana	5	Boinski and Timm 1995, Timm 1987; Kunz et al. 1994; Stoner 2000
		Bactris wendlandiana	3	Timm 1987
		Cocos nucifera	5	Boinski and Timm 1985, Choe and Timm 1985
		Geonoma congesta	5	Timm 1987
		Geonoma cuneata	5	Chapman 1932; Timm 1987
		Geonoma oxycarpa	5	Ingles 1953
		Geonoma sp.	5	Choe and Timm 1985; Kunz et al. 1994; Timm 1987
		Welfia georgii	5	Boinski and Timm 1985; Choe and Timm 1985
Ectophylla alba	Araceae	*Anthurium cuneatissimum*	4	Brooke 1987
		Philodendron macrophylla	4	Brooke 1987
		Philodendron mediacostatum	4	Brooke 1987
		Rhodospatha wendlandii	4	Brooke 1987
		Singonium potophyllum	4	Brooke 1987
	Heliconiaceae	*Heliconia imbricata*	7	Brooke 1987; Timm and Mortimer 1976
		Heliconia latispatha	7	Brooke 1987; Timm and Mortimer 1976
		Heliconia pogonantha	7	Brooke 1987; Timm and Mortimer 1976
		Heliconia sarapensis	7	Kunz et al. 1994
		Heliconia tortuosa	7	Timm and Mortimer 1976
		Heliconia sp.	7	Timm and Mortimer 1976

(continued on next page)

Table 1.3. *(continued)*

Bat Suborder/ Family/Species	Plant Family	Plant Species	Tent Style	References
Mesophylla macconnelli	Araceae	*Anthurium jenmanii*	6	Kunz et al. 1994
		Philodendron fragrantissimum	4	Kunz et al. 1994
		Xanthosoma sp.	4	Koepcke 1984
	Palmae	*Astrocaryum macrocalyx*	5	Foster 1992
		Astrocaryum sciophilum	5	Charles-Dominique 1993; Simmons and Voss 1998
		Geonoma sp.	5	Koepcke 1984
		Unidentified palm	?	Emmons 1990
Rhinophylla pumilio	Araceae	*Philodendron melinonii*	4	Charles-Dominique 1993
		Philodendron ornatum	4	Charles-Dominique 1993
		Rhodospatha latifolia	4	Charles-Dominique 1993
	Heliconiaceae	*Heliconia* sp.	4	Zortéa 1995
	Musaceae	*Musa* sp.	4	Zortéa 1995
	Palmae	*Astrocaryum sciophilum*	5	Charles-Dominique 1993; Simmons and Voss 1998
		Atalea ataleoides	5	Charles-Dominique 1993
	Sterculiaceae	*Sterculia* sp.	4	Charles-Dominique 1993
	Strelitziaceae	*Phenakospermum guyannense*	4	Simmons and Voss 1998
Uroderma:				
U. bilobatum	Araceae	*Philodendron fragrantissimum*	2	Kunz et al. 1994
	Cyclantaceae	*Carludovica palmata*	2	Kunz et al. 1994; Timm 1987
	Flacourtiaceae	*Carpotroche platyptera*	1	Kunz et al. 1994
	Heliconiaceae	*Heliconia imbricata*	7	Kunz et al. 1994
		Heliconia latispatha	7	Kunz et al. 1994
	Moraceae	*Cecropia* sp.	?	Buchanan 1969
	Musaceae	*Musa* sp.	6	Kunz et al. 1994; Timm 1987
		Musa sp.	?	Buchanan 1969
	Palmae	*Coccothrynax barbadensis*	2	Kunz and McCracken 1996; Kunz et al. 1994
		Cocos nucifera	3	Lewis 1992; Timm and Clauson 1990; Timm and Lewis 1991
		Cocos nucifera	5	Barbour 1932
		Cocos nucifera	?	Goodwin and Greenhall 1961
		Livistona chinensis	2	Barbour 1932
		Manicaria saccifera	5	Kunz et al. 1994

Bat species	Plant family	Plant species	Tent style	Reference
		Pritchardia pacifica	2	Barbour 1932
		Scheelea rostrata	3	Timm 1987
		Sabal mauritiiformis	2	Buchanan 1969; Goodwin and Greenhall 1961; Kunz et al. 1994; Timm 1987
	Polygonaceae	Unidentified	?	Bloedel 1955
		Coccoloba manzanillensis	1	Chapman 1932
	Strelitziaceae	*Phenakospermum guyannense*	4	Simmons and Voss 1998
U. magnirostrum	Palmae	*Astrocaryum murumuru*	3	Timm 1987
Vampyressa:				
V. nympheae	Rubiaceae	*Pentagonia donnel-smithii*	4	Brooke 1987
V. pusilla	Araceae	*Philodendron macrophylla*	4	Kunz et al. 1994
		Philodendron sp.	4	Timm 1984
	Heliconiaceae	*Heliconia spathocircinatha*	4 (7)	Zortéa and De Brito 2000
		Heliconia richardiana	4 (7)	Zortéa and De Brito 2000
		Rhodospatha wendlandii	4	Kunz et al. 1994
Vespertilionidae:				
Scotophilus kuhlii	Palmae	*Livistona rotundifolia*	2	Rickart et al. 1989

Note. Some bats may roost in tents that they do not construct. Tent style: 1 = conical; 2 = palmate umbrella; 3 = pinnate; 4 = apical; 5 = bifid; 6 = paradox; 7 = boat; 8 = stem (after Kunz et al. 1994).

M. macconnelli (Foster 1992; Koepcke 1984) include the J-shaped cuts (characteristic of bifid tents) and those in which the basal veins have been chewed (as in apical tents).

Considering the relatively temporary nature of tents and the number of veins, stems, or roots that must be modified to construct such structures, the cost of construction should vary depending on the architectural style and number of tents that are made (Kunz et al. 1994). The number of leaf veins that some tent-making bats modify varies from three to four in apical tents of *Philodendron* spp. (Kunz et al. 1994) to more than 60 in palmate umbrella tents constructed in *Sabal* sp. and *Prichardia* sp. or bifid and pinnate tents in *Cocus nucifera* (Kunz et al. 1994; Timm 1987). Tents constructed in the leaves of succulent aroids such as *Philodendron* spp. are probably completed in a single night. By contrast, tents constructed in palmate leaves, where large numbers of veins and plications are chewed, may take several days to complete (Barbour 1932; Kunz and McCracken 1996). Construction of stem tents by *C. sphinx* may take several months to complete (Balasingh et al. 1995).

Another small (12–18 g) megachiropteran (*Balionycteris maculata*) in peninsular Malaysia is known to modify root masses of the epiphytic ginger (*Aslpenium nidus*) and bird's nest fern (*Hedychium longicornutum*) by severing roots and excavating soil to create roosting cavities (Hodgkison et al., in press; fig. 1.10).

Convergence in Tent-Roosting Behavior

The striking similarities in tent architecture among the Old World pteropodids and the New World phyllostomids (fig. 1.11) supports an interpretation of convergence in tent-making and tent-roosting behavior. For example, oval-shaped or round tent crowns in palmate leaves of *Livistona rotundifolia* (attributed to the Old World megachiropteran *C. sphinx* and *C. brachyotis*) and similar-shaped tents in palmate leaves of *Coccothrinax barbadensis* (attributed to the New World microchiropteran *Artibeus jamaicensis*) are nearly identical (Kunz et al. 1994).

Convergence in bat-tent architecture may be a consequence of similarity in leaf morphology, but observations that members of taxa with divergent evolutionary histories (Megachiroptera and Microchiroptera) are capable of modifying similar leaf forms in nearly identical ways supports an interpretation of behavioral convergence (Kunz et al. 1994). Because tent-making species are relatively small (<65 g) and largely frugivorous, they are probably subjected to similar selection pressures relating to foraging and roosting behavior in the structurally similar forests in which they have evolved.

Among the Megachiroptera, the close ancestral affiliations and similar roosting habits of *C. sphinx* and *C. brachyotis* suggest that tent-making behavior may have evolved only once in this genus and, thus, could be expected in other members of this taxon (Kunz et al. 1994). The roosting habits of other

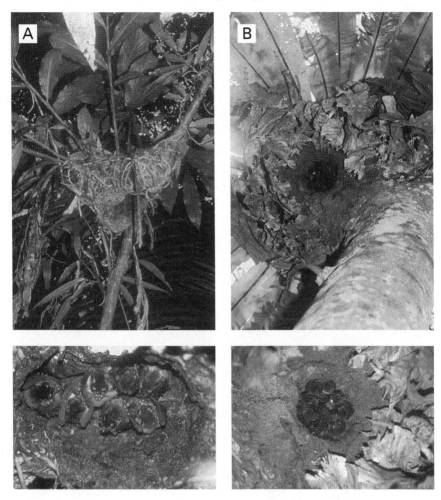

Figure 1.10. *A*, Excavated root mass of the epiphytic ginger, *Aslpenium nidus* (*top*), occupied by a harem group of *Balionycteris maculata* (*bottom*); *B*, excavated root mass of a bird's nest fern, *Hedychium longicornutum* (*top*), occupied by a small harem group of *B. maculata* (*bottom*) (photos by R. Hodgkison).

species of *Cynopterus* have not been fully described, but future investigations are needed to test the hypothesis that tent-making behavior is a trait shared by other members of the genus (Kunz et al. 1994).

With the exception of *Rhinophylla pumilio* (subfamily Caroliinae), all New World tent-roosting species belong to the subfamily Stenodermatinae (tribe Stenodermatini). However, existing phylogenies for the Phyllostomidae (Baker et al. 1989, 2000; Lim 1993; Owen 1987, 1988; reviewed in Wetterer et al. 2000)

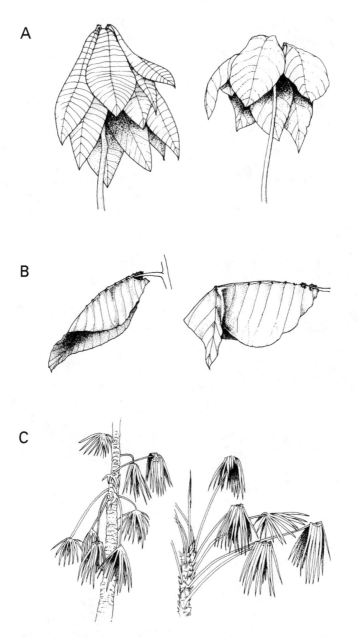

Figure 1.11. Convergence in bat-tent architecture in the Paleotropics (Indonesia) (*left*) and Neotropics (Costa Rica) (*right*). *A,* Conical tents, unidentified Anacardiacea (*left*), *Carpotroche platyptera* (*right*); *B,* apical tents, *Dillenia beccariana* (*left*) and *Pentogonia donnel-smithii* (*right*); and *C,* palmate umbrella tents, *Livistonia rotundifolia* (*left*) and *Carludovica palmata* (*right*) (illustrations by E. Rooks, after Kunz et al. 1994).

are not sufficiently developed to establish whether tent-making/tent-roosting bats share a common ancestor. Of the 18 genera recognized in the tribe Stenodermatini, Owen's (1987) proposed assignment of *Demanura* is herein included with *Artibeus*, and the genus *Mesophylla* is judged distinct from *Ectophylla* (Lim 1993).

Tent-making/roosting behavior in the Phyllostomidae has been reported for six genera—*Artibeus* (*Dermanura*), *Ectophylla*, *Mesophylla*, *Rhinophylla*, *Uroderma*, and *Vampyressa*. Using Lim's (1993) phylogeny for the Phyllostomidae, tent-making/roosting habits may have evolved three or four times (see Kunz et al. 1994). A similar conclusion is made if we accept Wetterer et al.'s (2000) phylogeny of the Phyllostomidae. If we use Baker et al.'s (2000, fig. 2) phylogeny on which to map tent-roosting/tent-making bats, five clades are recognized with tent-roosting/tent-making traits. One clade includes *Artibeus* (*Dermanura*), a second includes *Ectophylla*, a third includes *Mesophylla* and *Vampyressa pusilla*, a fourth includes *Uroderma* and *Vampyressa bidens*, and a fifth clade includes *Rhinophylla*.

Given the provisional nature of each phylogeny, especially at the species level, one cannot rule out the possibility that tent-making/roosting behavior in the Phyllostomidae may have evolved only once, especially if the trait was present in a taxon ancestral to these groups (Kunz et al. 1994). Alternatively, its apparent absence in other members of the Stenodermatinae (e.g., *Chiroderma*, *Enchisthenes*, *Platyrrhinus*, *Surnira*, and *Vampyrodes*) may reflect the subsequent loss of these traits or, perhaps, that they have not yet developed in these taxa.

A complicating factor in attempting to assess phylogenetic relationships among tent-roosting/tent-making species is that some bats may parasitize tents initially made by others (Brooke 1987; Charles-Dominique 1993; Foster 1992; Kunz et al. 1994; Simmons and Voss 1998; Timm 1987). Charles-Dominique (1993) suggested that *Rhinophylla pumilio* occupied tents that were constructed by other species—perhaps *Mesophylla macconnelli* or *Artibeus gnomus*—and thus may not construct their own tents (also see Simmons and Voss 1998). Further studies should be designed to distinguish among tent-making and tent-roosting species.

Why Do Bats Construct and/or Roost in Tents

Relative to other types of foliage roosts, tents provide several potential benefits to their occupants. These include greater protection from inclement weather (rain, wind, and direct solar radiation), increased protection from visually oriented predators, avoidance of parasites, proximity to food resources, and increased energy economy (Bhat and Kunz 1995; Brooke 1987, 1990; Kunz et al. 1994; Kunz and McCracken 1996; Tan et al. 1999; Timm 1987; Timm and Lewis 1991; Timm and Mortimer 1976). Tent-making behavior presumably evolved because individuals benefit from the investment made in their construction and because the benefits outweigh the costs that bats may have incurred in their construction (Kunz et al. 1994).

Several investigators have suggested that the mating system of tent-roosting bats is based on roost-defense polygyny (Balasingh et al. 1995; Brooke 1990; Kunz and McCracken 1996; reviewed in McCracken and Wilkinson 2000). When males modify leaves and other plant parts into tents, they create spaces that can be potentially defended against intruding males more effectively than they could if they occupied unmodified foliage (Balasingh et al. 1995; Kunz and McCracken 1996). If we assume that tents are constructed by males, and males recruit females for mating (Balasingh et al. 1995), important benefits of tent construction would be that the tents are attractive to females and that either the tent or females can be defended by the harem males (Balasingh et al. 1995; Kunz and McCracken 1996; Storz et al. 2000a). If tent architecture is perceived by females as an extension of a male's phenotype, then a male would benefit if females selected those traits when choosing a mate.

What characteristics of plants do bats select when they modify leaves and other plant parts into tents? If tent quality is an important factor in female choice, tents constructed in the most physically vigorous leaves may be an important criterion for attracting females (Stoner 2000). Height above the ground, age, and angle of the leaf are important factors. *Artibeus watsoni* and *A. cinereus* typically select leaves for tent construction that are approximately 1.5–5 m above the ground, even though other leaves may be available (Choe and Timm 1985; Stoner 2000). Brooke (1987) noted that *Vampyressa nymphaea* showed a preference for young leaves, presumably because they were softer (more succulent) and thus easier to modify. *Ectophylla alba* appears to select *Heliconia* leaves for tent construction based on height above the ground and angle of the petiole (Brooke 1990; Timm and Mortimer 1976). Thus, males may judge the suitability of leaves and other plant parts for tent construction based on their degree of protection from inclement weather and predators, but they may also increase the potential for polygyny and their ability to defend tents (and female occupants) against intrusions by other males.

The hypothesis that tents provide protection from rain was supported experimentally by Choe (1994), who placed cotton balls beneath unmodified leaves. He found that the cotton balls placed beneath tents were less saturated with rainwater compared to those beneath unmodified leaves. Some plant species in which tents are constructed have characteristics that may increase protection from potential predators. Stoner (2000) found that *A. watsoni* chose understory palms (*Astergyne martiana*) that were taller and had thicker petioles and longer leaves than a random selection of plants and suggested that larger leaves provided safer roosting sites from terrestrial predators. Large leaves and the presence of spines on the rachis may increase protection from visually oriented predators (Charles-Dominique 1993; Simmons and Voss 1998; Stoner 2000). In addition, bats that roost in tents with long slender petioles are more likely to be alerted to approaching predators (e.g., snakes and scansorial primates), whose body masses and activity may disturb the surrounding foliage (Boinski and Timm 1985; Timm and Mortimer 1976).

There have been no systematic studies comparing rates of predation on bats that occupy unmodified foliage versus tents. Because tents offer their occupants unobstructed views of their environment below, bats should be alert to a predator approaching from the ground and may be able to escape before being attacked (Brooke 1990; Kunz and McCracken 1996). Tent-roosting bats generally are alert in their tents—which may further reduce risks of predation. In addition, the construction and occupancy of a number of tents within a bat's home range, as observed in *M. macconnelli,* may confer advantages to bats if it introduces a degree of uncertainty into the search tactics of a potential predator (Foster 1992). That tent-making behavior originated solely as an antipredator strategy (see Boinski and Timm 1985), however, is doubtful, since other benefits are more compelling.

Some visually oriented predators may learn to recognize tents and, thus, use an acquired search image to increase their chances of locating tents in which bats roost. Anecdotal observations have revealed that tent-roosting bats are occasionally taken by squirrel monkeys (*Saimiri oerstedii* and *S. sciureus* [Boinski and Timm 1985; Souza et al. 1997]) and by white-faced capuchin monkeys (*Cebus capucinus* [Perry 1994]). Moreover, some bats that were disturbed by squirrel monkeys were subsequently preyed on by double-toothed kites (*Harpagus bidentatus* [Boinski and Timm 1985]). In India, tents of *Cynopterus sphinx* are sometimes harassed by known or suspected predators, such as jungle crows (*Corvus macrorhynchos*), house crows (*C. splendens*), rat snakes (*Ptyas mucosus*), or spectacled cobras (*Naja naja* [Storz et al. 2000b]). However, based on the evolutionary success of tent-roosting/tent-making bats, the benefits of tent-roosting appear to outweigh the disadvantages of occasional predation.

Theoretically, enclosed spaces should trap metabolic heat and thus increase the temperature inside an occupied roost, thereby reducing a bat's daily energy expenditure. Although thermoregulatory benefits may accrue to cavity-roosting bats in temperate regions (Kurta 1985), the hypothesis that tents in tropical areas provide thermoregulatory benefits to their occupants remains to be tested. Relative to the enclosed spaces of a tree or building cavity, leaves generally have low insulative properties. However, even the partially enclosed space of a tent cavity could reduce both heat and water losses due to convective forces. If wind velocity in foliage roosts is 18%–30% less than the velocity outside of roosts (see Walsberg 1986), then tent-roosting bats could benefit energetically.

Timm (1987) suggested that tents might play a role in controlling ectoparasites, but this would only benefit bats if they regularly moved to alternate sites when parasite infestations increased. This may be the situation for some species but not for others. Several species, including groups of *Ectophylla alba, Cynopterus sphinx,* and *C. brachyotis* may occupy the same tent continuously for several weeks during the maternity period (Balasingh et al. 1995; Brooke 1990; Tan et al. 1997). Abandonment of tents may be more associated with the breakup of the maternity roost than with a buildup of parasites. Timm's (1987) hypothesis may apply only to those species that regularly move among several alternate tents (e.g., small stenodermines that roost in apical tents).

Several authors have qualitatively examined the amount of foliage clutter around occupied and unoccupied tents, noting that occupied tents had the least amount of adjacent clutter (Brooke 1990; Kunz and McCracken 1996). The open space beneath tents not only provides a clear view of approaching predators (Brooke 1987, 1990) but also offers bats unimpeded access to and from the tent as they depart to and return from foraging bouts (Balasingh et al. 1995; Kunz 1982; Kunz et al. 1994; Kunz and McCracken 1996). Easy access to tents on the wing may also be one tent characteristic that females use to judge tent quality. For example, Balasingh et al. (1995) found that the largest harems of *C. sphinx* occupied stem tents that had the least amount of clutter around the opening.

Cavities in Arboreal Termite and Ant Nests

At least four species of Neotropical microchiropterans (Phyllostomidae: *Tonatia silvicola, T. carrikeri, T. brasiliense,* and *Phyllostomus hastatus*), one Australian vespertilionid (*Murina florium*), and one megachiropteran (*Balionycteris maculata*) have been observed roosting in excavated arboreal ant nests (Clague et al. 1999; Goodwin and Greenhall 1961; Handley 1966, 1976; Hodgkison et al., in press; Kalko et al. 1999; McCarthy et al. 1992; fig. 1.12). Whether Neotropical bats modify these nests or exploit cavities created by trogans (*Trogan* sp.), orange-chinned parakeets (*Brotogeris* sp.), or other species (Kalko et al. 1999) remains to be determined.

In Panama, termite nests occupied by *T. silvicola* were active and located approximately 5–7 m above the ground (Kalko et al. 1999). Each nest was excavated from below with a space approximately 30–40 cm deep (Kalko et al. 1999). The size of day-roosting groups of *T. silvicola* was typically small, ranging from four to five individuals, often including a single male. Some day roosts were used at night (as feeding roosts), and males tended to forage near these sites. The proximity of a male *T. silvicola* to its day roost and the presence of several females suggest the possibility of harem formation (Kalko et al. 1999).

In peninsular Malaysia, *Balionycteris maculata* also has been observed roosting in nests of the arboreal ant *Crematogaster abvenini* (fig. 1.12) and are excavated similarly to the way some tent-making bats modify fruit/flower clusters, leaves, vines, and root masses. These excavated structures are often occupied by single adult males and small harem groups of *B. maculata* ranging from four to eight bats, although groups as large as 20 individuals have been observed (Hodgkison et al., in press; Lim 1966). It is not known how or when these excavations occur, although Hodgkison et al. (in press) postulated that bats use their teeth and/or claws to scrape away soil and nest material, perhaps similarly to the way they excavate soil around root masses (see fig. 1.10).

Figure 1.12. Arboreal ant nest (*Crematogaster ebinina*) excavated by and used as a day roost by *Balionycteris maculata* (photo by R. Hodgkison). Excavated basal opening is shown with an arrow.

Morphological and Behavioral Adaptations for Roosting

The opportunistic and sometimes obligate use of cavities and foliage as roosts frequently involves specialized behavioral and morphological adaptations of bats (Fenton 1992a; Kunz 1982, 1996). The cranium and postcranial skeleton may be under strong selection pressures from a bat's roosting environment (Bennett 1993; Fenton 1992a; Kunz 1982). Some microchiropterans that seek shelter in tree crevices and bamboo culm have evolved modified cranial, pelvic, and pectoral structures that coincide with an extraordinary ability to crawl through small spaces (Fenton 1992a). For example, *Tylonycteris pachypus* and *T. robustula* have strongly flattened crania that facilitate access to the interior spaces of bamboo culm (Medway and Marshall 1972).

Many bats have evolved specialized thumbs and feet for roosting. Megachiropterans often assume pendant postures by hanging from one or both feet, facilitated by a specialized locking mechanism (Bennett 1993). Species in the genera *Pteropus, Eidolon,* and *Epomophorus* are especially adept at using their thumbs and claws for climbing among branches (Nelson 1965; Neuweiler 1969; Wickler and Seibt 1976). By contrast, some foliage-roosting microchiropterans, such as *Lavia frons,* are less agile in roosting situations and select relatively open sites for roosting that are relatively free of surrounding vegetation (Wickler and Uhrig 1969).

Some microchiropterans have specialized thumb and footpads or suction pads that predispose them for certain roosting situations (fig. 1.13). The highly specialized thumb and footpads in *Myzopoda aurita, Thyroptera discifera,* and *T. tricolor* make it possible for them to cling to the smooth upper surfaces of furled leaves (Schliemann and Mags 1978; Wilson 1978; Wilson and Findley 1977). Similarly, the modifications of the foot and thumb pads of *Thyroptera pachypus* and *T. robustula* facilitate entry and exit through small openings on smooth bamboo culm and allow them to cling to the interior of the culm cavity (Kunz 1982; Medway and Marshall 1972; Schliemann and Hoeber 1978).

Cryptic markings are pronounced in several species of both Old and New World foliage-roosting bats (Bonaccorso 1998; Kunz 1982), and these traits presumably confer a selective advantage in certain roosting situations. Some foliage roosting bats are well concealed in their roosting places by either real or apparent color. The reddish and yellowish coloration of some lasiurine bats (e.g., *Lasiurus borealis* and *L. intermedius*) that roost in the foliage of deciduous trees may confer protection from predators. The greenish wings of *Paranyctimene raptor* provides an effective camouflage in foliage (Bonaccorso 1998). The white pelage of *Ectophylla alba* assumes a greenish appearance as the light that transmits through leaves casts a green light on these tent-making bats (Brooke 1990)—perhaps making them less readily detected by visually oriented predators.

The so-called painted bats of the genus *Kerivoula* typically have long, thick, woolly pelage that ranges in color from yellow to bright orange and scarlet (Allen 1939; Dobson 1877; Fenton 1992a; Nowak 1994; Schulz 2000). Similarly, the pelage of other temperate and tropical plant-roosting bats are colored with hues of yellow, orange, and red, resembling fruits and leaves. The contrasting lighter colors around the head and neck (mantle) of some megachiropterans suggests a type of countershading that may confer a certain degree of crypsis (Bonaccorso 1998; Dobson 1877; Flannery 1995; Novick 1977). Similarly, the mottled and woolly pelage of *Rhynconycteris naso,* which roosts in small groups on the exposed boles of tropical trees (Bradbury and

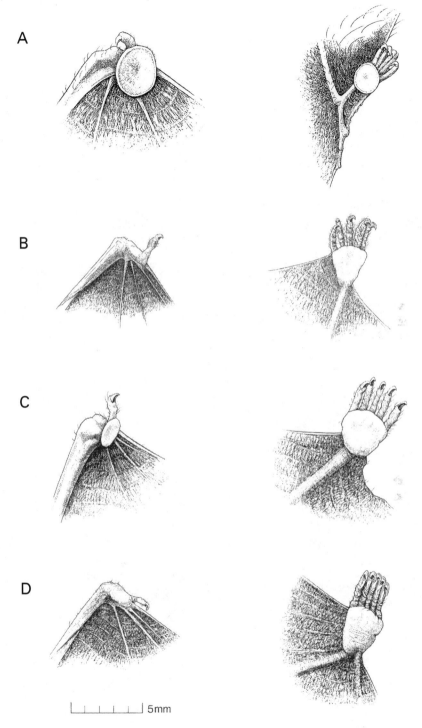

Figure 1.13. Morphological specializations of wrists and feet in (*A*) *Thyroptera tricolor*, (*B*) *Pipistrellus nanus*, (*C*) *Tylonycteris pachypus*, and (*D*) *Glischropus tylopus*, used for roosting on the smooth inner surfaces of bamboo culms or furled leaves. (Illustrations by P. Esty, from Kunz 1982.)

Figure 1.14. *A,* Cryptic pelage marking of *Epomophorus* sp. roosting in foliage. White ear and shoulder patches disguise the appearance of the bats as observed from below in the sun-flecked canopy (from Fenton 1992b; photo by M. B. Fenton); *B,* contrasting metacarpels and phalanges of *Cynopterus brachyotis* disrupt the uniform pattern of wings against the veins and plications of palm fronds (photo by T. H. Kunz).

Vehrencamp 1976), may be of benefit by camouflaging these bats from potential predators (see fig. 1.5).

Other foliage-roosting bats may be concealed with disruptive marking on their pelage or wings, such as the spotted pattern on the wings and ears of *Nyctimene albiventer* (Bonaccorso 1998). Similarly, the reticulate markings on the wings of *Chalinolobus varigatus*, the white dorsal stripe and patches on the head and shoulders of the vespertilionid *Scotomanes ornatus*, and the white ear spots, mottled wings, and contrasting metacarpels and phalanges of *Epomophorus wahlbergi, Cynopterus sphinx,* and *C. brachyotis* may function as disruptive patterns in roosting situations (fig. 1.14) and, thus, potentially protect them from visually oriented predators (Fenton 1992b; Lekagul and McNeely 1977; Nowak 1994; Rosevear 1965).

Concealment of bats in foliage also may be enhanced by certain roosting habits. The near motionless postures sometimes observed in *Syconycteris australis* (Bonaccorso 1998; Law 1993), *Pteropus poliocephalus* (Nelson 1965), *Epomops franqueti* and *Micropteropus pusillus* (Jones 1972), *Lavia frons* (Kingdon 1974), *Epomophorus gambianus* (Marshall and McWilliam 1982), *Nyctimene major* (Nowak 1994), *N. robinsoni* (Spencer and Fleming 1989), and *Lasiurus borealis* and *L. seminolus* (Constantine 1958, 1959, 1966), where they are shrouded by their wings, giving them the appearance of dead leaves, may reduce detection by potential predators.

Schulz (1999) described an unusual behavior in the foliage-roosting species *Murina florium*, which he termed "leaf wrapping." An individual would drag a leaf over its body and hold it in place using the hindfoot and thumb, thus obscuring the body from view. This behavior may provide shelter from sunlight and rain and reduce vulnerability to predation.

Colony Size

Colony sizes of bats roosting in cavities or foliage vary considerably. Although intraspecific differences exist, a number of species that roost in cavities typically form small colonies of fewer than 10 individuals (table 1.4). The size of colonies may reflect the social structure of a particular species, such as harem formation in *Artibeus jamaicensis* (Morrison 1979) or monogamous pairs in *Vampyrum spectrum* (Vehrencamp et al. 1977). Other species form larger groups, for example, in Kenya, up to 80 individuals of the megadermatid *Cardioderma cor* roost in hollow baobab trees (Vaughan 1976; see fig. 1.1D). Some cavity-roosting species form very large groups, such as *Hipposideros caffer*, with colonies of more than 1,000 individuals (Kingdon 1974), and *Mystacina tuberculata*, where up to 4,500 individuals have been found in a single tree cavity in New Zealand (Lloyd and McQueen 1997).

Bats that roost beneath exfoliating bark are often solitary or form small groups (e.g., Crampton and Barclay 1998; Mattson et al. 1996; Menzel et al.

Table 1.4. Mean colony size, roost fidelity, and roost area of selected species of bats roosting in tree cavities or beneath exfoliating bark

Species	Country	Mean Colony Size	Roost Fidelity	Roost Area	References
Mystacinidae:					
Mystacinia sp.	New Zealand	A, C	B	. . .	O'Donnell et al. 1999
Mystacinia tuberculata	New Zealand	C	. . .	B	Lloyd and McQueen 1997; Daniel and Williams 1984
Noctilionidae:					
Noctilio albiventris	Costa Rica	. . .	C	B	Fenton et al. 1993
Phyllostomidae:					
Artibeus jamaicensis	Panama	A	C	A	Morrison 1979
Desmodus rotundus	Costa Rica	A	C	B	Wilkinson 1985
Vampyrum spectrum	Costa Rica	A	C	. . .	Vehrencamp et al. 1977
Rhinolophidae:					
Rhinolophus hildebrandti	South Africa	B	C	. . .	Fenton and Rautenbach 1986
Vespertilionidae:					
Chalinolobus tuberculatus	New Zealand	B	A	A	O'Donnell and Sedgeley 1999
Eptesicus fuscus	Canada, USA	B	B	A	Betts 1996; Brigham 1991; Kalcounis and Brigham 1998
Lasionycteris noctivagans	Canada, USA	A, B	B	A	Betts 1996; Crampton and Barclay 1998; Mattson et al. 1996; Vonhof and Barclay 1996
Myotis californicus	Canada	B	A	A	Brigham et al. 1997
Myotis daubentonii	Switzerland	B	A	B	Rieger 1996
Myotis evotis	Canada	A	A	A	Vonhof and Barclay 1996; Waldien et al. 2000
Myotis lucifugus	Canada	B	B	B	Barclay and Cash 1985; Crampton and Barclay 1998
Myotis septentrionalis	USA	A, B	A	A	Foster and Kurta 1999; Sasse and Pekins 1996
Myotis sodalis	USA	A, B	B, C	A	Humphrey 1977; Kurta et al. 1993, 1996
Myotis volans	Canada, USA	A, C	B	A	Chung-MacCoubrey 1996; Ormsbee 1996; Vonhof and Barclay 1996
Nyctalus lasiopterus	Japan	B	B	. . .	Maeda 1974
Nyctalus noctula	Germany	A, C	A	B	Kronwitter 1988; van Heerdt and Sluiter 1965; Sluiter et al. 1973
Nycticeius humeralis	USA	A	A	. . .	Menzel et al. 2001
Nyctophilus bifax	Australia	A	A	A	Lunney et al. 1995
Nyctophilus geoffroyi	Australia	A	A	A	Hosken 1996
Nyctophilus gouldi	Australia	A	A	A	Lunney et al. 1988; Tidemann and Flavel 1987
Scotophilus borbonicus	Zimbabwe	A	A	A	Fenton 1983; Fenton and Rautenbach 1986
Vespadelus darlingtoni	Australia	B		A	Herr and Klomp 1999
Vespadelus pumilus	Australia	A, B	A	A	Law and Anderson 2000

Note. Mean colony sizes: A < 10; B = 11–100; C > 100 individuals. Roost fidelity: A = shifting roost on average every 1–2 d; B = shifting every 3–10 d; C = shifting > 10 d. Roost area is the mean distance between consecutive roosts: A < 400 m; B > 400 m. Ellipses dots indicate that the characteristic was not measured.

2001; Vonhof and Barclay 1997). In some situations, the physical area beneath exfoliating bark may limit the number of individuals, although colony sizes of up to 50 individuals have been reported (e.g., *Myotis sodalis* [Humphrey et al. 1977; Kurta et al. 1996] and *M. volans* [Baker and Phillips 1965]).

Maternity roosts generally contain more individuals than do nonbreeding roosts, with increased energy conservation expected owing to clustering behavior of females and young (Roverud and Chappell 1991). At times, lactating females may resort to solitary roosting to facilitate entry into torpor after unsuccessful foraging bouts or if their energetic balance is close to a critical threshold (O'Donnell and Sedgeley 1999). Males more often than females roost alone in cavities (Law and Anderson 2000).

Clustering behavior is important for temperate species that hibernate in tree cavities. *Nyctalus noctula,* one of the few European species that hibernates in tree cavities, can tolerate lower temperatures when clustered. Sluiter et al. (1973) found that clusters of approximately 100 individuals could survive for at least 53 d when the roost temperature was below 0°C. In contrast, single individuals were usually forced to arouse from hibernation at these temperatures.

Degrees of coloniality vary among species of *Pteropus* (Pierson and Rainey 1992). In Western and American Samoa, *P. tonganus* forms year-round colonies comprising several thousand individuals (Brooke et al. 2000; Cox 1983; Wilson and Engbring 1992). The closely related *P. samoensis* occupies the same islands, but it is solitary or roosts in loose aggregations of up to a dozen individuals (Brooke et al. 2000; Cox 1983; Rainey 1998; Wilson and Engbring 1992). In the Philippines and peninsular Malaysia, roosting groups of *P. vampyrus* typically consist of fewer than 100 individuals (Lim 1966). However, a colony of approximately 15,000 was observed in a mangrove forest in southwestern Borneo (Lyon 1911), and in Indonesia one colony was estimated to include as many as 21,000 individuals (Wiriosoepartha et al. 1986). In eastern Australia, colonies of *P. poliocephalus* range from a few hundred to 200,000 individuals (Eby 1991; Eby et al. 1999; fig. 1.15A). Estimates of the number of individuals in large camps are relatively crude and may be subject to observer error (see Garnett et al. 1999).

Some foliage-roosting megachiropterans exhibit a gregarious phase during one season but may roost alone or form small groups at other times. In some regions, the reduction or breakup of large aggregations of pteropodids has been attributed to the decrease in abundance of flower blossoms and fruits, suggesting that the stimulus for dispersal and nongregarious behavior reflects depletion of local food resources (Fleming and Eby, this volume; Palmer and Woinarski 1999; Richards 1995; Thomas 1983).

Mixed species roosts are common in some species of *Pteropus*. In Papua New Guinea, *P. conspicillatus* has been observed roosting with *P. hypomelanus* (Bonaccorso 1998). In the Philppines, *P. vampyrus* has been observed sharing

Figure 1.15. Two types of stem tents (*top*) constructed by *Cynopterus sphinx*, occupied by small harem groups (*bottom*) in (*A*) kital palm fruit cluster (*Caryota urens*) and (*B*) mast tree (*Polyalthea longifolia*). *Arrow* in bottom right photo show positions of harem male (photos by K. H. Tan and J. Balasingh, respectively).

roosts with *Acerodon jubatus* in groups ranging from several hundred to about 150,000 individuals (Heideman and Heaney 1992; Mudar and Allen 1986). In Australia, *P. poliocephalus*, *P. alecto*, and *P. scapulatus* sometimes form mixed groups that exceed 50,000 individuals (Eby et al. 1999; Ratcliffe 1932; Tidemann et al. 1999; fig. 1.16*B*). In contrast to the large members of the Pteropodidae,

Figure 1.16. *A,* Colony of *Pteropus poliocephalus* roosting on branches of a partially defoliated trees in eastern Australia (photo by P. Birt); *B,* mixed colony *P. poliocephalus, P. alecto,* and *P. scapulatus* occupying defoliated trees in southeast Queensland, Australia (photo by T. Allofs).

smaller species of fruit and nectar feeding bats of both the New and Old Worlds are often solitary or form small groups (Bonaccorso 1998; Fenton et al. 1985; Morrison 1980; Tan et al. 1999).

Roost Fidelity

Bats exhibit a wide range of fidelity to their roosts. This variation is often reflected in the type of roost, life history stage, and form of social organization (Bradbury 1977b; Lewis 1995; McCracken and Wilkinson 2000). Bats that roost in relatively permanent roosts generally exhibit higher levels of fidelity compared to those that roost in more ephemeral situations (Lewis 1995). Although many species of cavity- and foliage-roosting bats show low fidelity to specific roost sites, they often exhibit high levels of fidelity to roost areas, with individuals moving among several alternate roosts. A number of reasons for roost lability have been proposed, including reduced probability of predation, lower ectoparasite loads, familiarity with different roost microclimates, disturbance, and decreased commuting costs to foraging areas (Lewis 1995).

Available data on roost fidelity for species that roost in tree cavities or beneath exfoliating bark generally can be grouped into three categories (table 1.4): species that, on average, shift roost sites every 1–2 d; those that shift every 3–10 d; and species that remain in the same roost for more than 10 d. These data indicate that 12 species (50%) shift roost sites almost every day. Six species, while still moving regularly, remain in the same roost for up to 10 d. Five species remain faithful to one roost for long periods: *Artibeus jamaciensis, Desmodus rotundus, Noctilio albiventris, Rhinolophus hildebrandti, Vampyrum spectrum.* Interestingly, each of the latter species often roost in large basal cavities in live trees. A number of species in which individuals shift roost sites regularly alternate between a core number of preferred roost trees (e.g., *Nyctalus noctula* [Kronwitter 1988], *Myotis daubentonii* [Rieger 1996], and *M. sodalis* [Callahan et al. 1997; Kurta et al. 1996]). In contrast, a new roost tree was occupied almost every day by colonies of *Chalinolobus tuberculatus,* with all individuals usually abandoning the roost together (O'Donnell and Sedgeley 1999).

Bats that roost beneath exfoliating bark often show low levels of fidelity, which may be associated with the relatively ephemeral nature of these sites (Kurta et al. 1996; Lewis 1995). Movements to alternate roosts may occur every 1–2 d (e.g., Grindal 1999; Kurta et al. 1996; Mattson et al. 1996; Menzel et al. 2001), although longer occupations have been reported, especially during the maternity period (Humphrey et al. 1977). Kurta et al. (1996) found that pregnant female *M. sodalis* shifted roosts more often than did lactating or nonreproductive females. They suggested that pregnant females moved to reacquaint themselves with suitable roosts after returning from overwintering sites, whereas lactating females reduced shifting due to the added cost of moving young.

Several species roost beneath exfoliating bark during the nonbreeding season but abandon these sites in favor of tree cavities during the maternity period, as observed in *Lasionycteris noctivagans* (Mattson et al. 1996; Vonhof and Barclay 1996), *Nyctophilus geoffroyi* (Lumsden et al. 2002b), and *Nycticeus humeralis* (Menzel et al. 2001). Other species, such as *M. sodalis* (Callahan et al. 1977; Humphrey et al. 1977; Kurta et al. 1993, 1996) and *M. californicus* (Brigham et al. 1997), do, however, occupy roosts beneath exfoliating bark throughout the maternity period.

Lactating females typically shift roosts less often than nonbreeding individuals do, and this may reflect the energetic costs of moving nonvolant young between roosts or the limited availability of roosts suitable for rearing young (Kurta et al. 1996; Mattson et al. 1996; Menzel et al. 2001; Vonhof and Barclay 1996). In contrast, females in maternity roosts of some species, such as *C. tuberculatus* (O'Donnell and Sedgeley 1999), shift as often as nonbreeding individuals, suggesting that, for these species, the benefits of shifting roosts outweigh the energetic costs. Roost sites were abundant in O'Donnell and Sedgeley's (1999) study, which may indicate that the availability of suitable maternity roosts in other areas influences the rate of roost switching.

When individuals shift roosts, it is usually to one nearby. A number of studies have found that the mean distance between consecutive tree cavity or bark roosts is fewer than 400 m (table 1.4), although the distance between others may reach several kilometers (Kronwitter 1988; Rieger 1996). Distances between consecutive roosts are usually less than the distances between all roost sites and foraging areas, suggesting that being closer to foraging areas is not the reason for cavity-dwelling bats shifting among different roosts (Kurta et al. 1996; Lumsden et al. 2002a; O'Donnell 2001; Wilkinson 1985).

Roost fidelity in foliage-roosting vespertilionids, such as *Lasiurus borealis* and *L. seminolis*, is generally low, with individuals rarely using the same roost on consecutive days (Mager and Nelson 2001; Menzel et al. 1998). Menzel et al. (1998) reported that *L. borealis* and *L. seminolus* spent an average of 1.2 and 1.7d, respectively, in any one roost. Although both species moved frequently, roosts of *L. borealis* were located within a significantly larger area (2.6 ha) than those of *L. seminolis* (0.2 ha). Mager and Nelson (2001) found that 82% of the *L. borealis* roosts that they observed on consecutive days were less that 100 m apart.

Seasonal shifts in the roosting habitats of some foliage-roosting bats are associated with annual cycles of temperature or rainfall. In Australia, male and female *Syconycteris australis* shift from winter roosts in the warmer and moister rainforest edge and littoral rainforests to summer roosts in the cooler interior forests (Law 1993). Law postulated that this behavior allowed *S. australis* to avoid the relatively cool temperatures inside the forests in winter, as well as the hot temperatures of the forest exterior from spring through autumn. The moderate temperatures, high moisture–laden sea breezes, and buffering capacity of the forest appear to protect these bats from seasonal extremes.

Similarly, the foliage-roosting microchiropteran *Nyctophilus bifax* in eastern Australia shifts its roost location from the forest interior in late spring to the forest edge in late autumn, suggesting that local movements of this species make it possible for individuals to avoid high temperatures in spring–autumn and low temperatures in winter (Lunney et al. 1995).

Traditional roosts of most pteropodids are used seasonally, often in response to climatic variation and the availability of food (Fleming and Eby, this volume; Law 1993; Loughland 1998; Nelson 1965; Okon 1974; Parry-Jones and Augee 1991; Ratcliffe 1932; Thomas 1983; Vardon and Tidemann 1999). In West Africa, three species (*Eidolon helvum, Myonycteris torquata,* and *Nanonycteris veldkampi*) migrate in response to the seasonal availability of food (Thomas 1983). *Pteropus poliocephalus* migrates distances of up to 750 km along the east coast of Australia in response to seasonal fruiting and flowering phenologies (Eby 1991; Spencer et al. 1991). Although some colonies of *P. conspicillatus* move seasonally between sites, most remain in the same general areas showing a strong affinity for rainforest trees year round (Richards 1990a). In contrast, *P. alecto* moves between roosts seasonally, from bamboo and mangrove habitats in the dry season to rainforest in the wet season (Palmer and Woinarski 1999).

Day roosts of the African species *Epomophorus wahlbergi* are commonly located in riverine or gallery forests (Fenton 1992b; Fenton et al. 1985; Wickler and Seibt 1976). Typically, three to six individuals roost in groups spaced a few centimeters apart. Roost switching appears to be common, with entire groups moving several meters or more every 5–6 d. By contrast, *Epomophorus gambianus* forms colonies ranging from a few to 50–100 well-spaced individuals in the crowns of trees (Jones 1972; Marshall and McWilliam 1982). In Australia and Papau New Guinea, *Nyctimene robinsoni* and *Syconycteris australis* roost alone in dense rainforest foliage (Spencer and Fleming 1989; Law 1993; Winkelmann et al. 2000). *Syconycteris australis* shows little fidelity to specific roosts, and roost sites are seldom occupied for more than a single day (Law 1993).

Influence of Roost Selection on Social Organization and Mating Systems

Our understanding of how the roosting environment influences social biology and mating systems of bats is in its infancy. Research on several species has revealed a wide range of mating systems in foliage-roosting bats, including leks in *Hypsignathus monstrosus* (Bradbury 1977a); multimale/ multifemale groups in *Epomophorous wahlbergi* (Wickler and Seibt 1976); seasonally variable, single male/multifemale aggregations in *Pteropus seychellensis* (Cheke and Dahl 1981); year-round harems with labile female groups in *P. mariannus* and *P. tonganus* (Grant and Banack 1999; Wiles 1987), *Cynop-*

terus sphinx (Balasingh et al. 1995; Storz et al. 2000b), and *C. brachyotis* (Tan et al. 1997); seasonally variable, multimale/multifemale groups in *P. poliocephalus* (Nelson 1965); year-round, multimale/multifemale groups in *P. giganteus* (Neuweiler 1969); and monogamy in *P. samoensis* (Banack 1996; Craig et al. 1994).

One hypothesis for the evolution of polygyny is that resources are limiting (Clutton-Brock 1989; Emlen and Oring 1977). When breeding females are distributed among defensible roosts (mating territories), the potential for polygyny will depend on the site fidelity (or group cohesion) of the females. In American Samoa, observations on *P. tonganus* indicate that this species roosts in large colonies, ranging upward to several thousand individuals. Within these colonies year-round harem groups are formed, averaging 5.3 females per male, with males, either singly or in groups, occupying roosts on trees surrounding reproductive females (Grant and Banack 1999). Harems typically roost in trees that are devoid of leaves. When bats roost in live trees, males typically remove leaves from roost areas, and scent mark specific roost sites, which they defend from conspecifics. Harem boundaries are delineated by the branching patterns of the roost trees. Grant and Banack (1999) suggested that the mating system of *P. tonganus* had elements of both resource defense (tree roosts) and female defense polygyny. In contrast, the sympatric *P. samoensis* typically roosts alone or in monogamous pairs (Banack 1996; Craig et al. 1994; Pierson and Rainey 1992).

Cynopterus sphinx is one of the best-studied megachiropterans with respect to the effect of roost resources on its mating system. Observations suggest that tents are constructed by single males and that these males are joined by one or more females once a tent is completed (Balasingh et al. 1995; Bhat and Kunz 1995; Storz et al. 2000a, 2000b). Solitary bats that have been observed constructing tents are invariably males (Storz et al. 2000b), and tent construction appears to occur several weeks or months before the arrival of females (Storz and Kunz 2000). The unique characteristics of stem tents (fig. 1.15, also figs. 1.9H and 1.10) make it possible for males to defend these semienclosed structures and their female occupants from incursions by other males (Balasingh et al. 1995; Bhat and Kunz 1995; Hodgkison et al., in press; Kunz and McCracken 1996; Storz et al. 2000a, 2000b).

The question of whether female tent-making bats select roostmates on the basis of tent quality or some characteristic of the male deserves further study. Based on an analysis of several tent variables, neither Balasingh et al. (1995) nor Storz et al. (2000a, 2000b) found a significant relationship between these variables and group size in *C. sphinx*. Notwithstanding, group size varies seasonally, with larger harem groups forming in the dry season than in the wet season (Storz et al. 2000b). Other observations suggest that variation in harem size may be largely influenced by transient movements of females among adjacent roosts (Storz et al. 2000a).

Tents that *C. sphinx* construct in flower/fruit clusters of the kital palm may persist for several years, and harem males may retain breeding tenure for up to four breeding seasons at these sites (Storz et al. 2000b). Males and females generally remain faithful to one colony (a spatial array of several tents), but individuals often switch roosts from one breeding period to another (Storz et al. 2000b). Limited availability of roosts is not sufficient to explain harem group formation in female *C. sphinx* (Storz et al. 2000b). Because males construct and occupy tents in the absence of females, Storz et al. (2000b) suggested that the male mating strategy was based on territorial defense of roosts rather than on direct defense of labile female groups. Thus, the mating system of *C. sphinx*, as in other polygynous bat species (see McCracken and Wilkinson 2000), can be attributed largely to variation in behavioral cohesiveness of reproductively active females (Balasingh et al. 1995; Storz et al. 2000b).

The relationship between roosting ecology, roost fidelity, and mating systems has also been studied extensively in *Artibeus jamaicensis*, as this species is known to form harems in tree cavities (Morrison 1979; Morrison and Handley 1991; Morrison and Morrison 1981), caves (Kunz et al. 1983; Ortega and Arita 2000), and tents (Kunz and McCracken 1996; Foster and Timm 1976; Timm 1987). Morrison (1979) postulated a mating system based on resource defense polygyny, largely based on his observations that tree cavities (see fig. 1.1*B*) were limited resources on Barro Colorado Island, Panama. He suggested that a polygynous mating system would not be expected in caves and foliage roosts because these resources were not limiting. Kunz et al. (1983) tested this hypothesis by examining the group size and social organization of *A. jamaicensis* in a cave environment in Puerto Rico. Their observations indicated that caves with solution cavities offered resources to *A. jamaicensis* that were potentially limiting as well as defensible, supporting the hypothesis that such caves provided potential for the evolution of polygyny in this species. Subsequently, Kunz and McCracken (1996) observed social groups of tent-roosting *A. jamaicensis* in palmate umbrella tents in Trinidad and, similarly, described this as a polygynous mating system based on the ability of males to defend these roosts. Adult males roosted singly or in larger groups composed of one male (rarely two) and from one to 10 females. During parturition and lactation, females and young roosted separately from males.

Available data on other foliage-roosting and cavity-roosting species strongly suggest a predominance of female-biased social groups. Multi-female/single male groups of *Ectophylla alba* form following parturition (Brooke 1990). The presence of single males in the tent immediately following parturition, but not at other times, suggests a postpartum estrus in this species. Whether males defend the leaf as a resource to gain access to females has not been determined (Brooke 1990). Observations that some males are found in bachelor groups suggests that harem males may prevent other males from gaining access to the tents or to females during the receptive period of females.

Patterns of year-round harem maintenance have been reported for other polygynous species (McCracken and Wilkinson 2000), including *Saccopteryx bilineata*, which typically roosts in the semidarkened spaces of buttress cavities that form in large, tropical trees (Bradbury and Emmons 1974; Bradbury and Vehrencamp 1976; Voigt et al. 2001). Colonies range upward to 42 individuals, and within these colonies, males defend territories of 1–3 m^2 of vertical surface in the buttress cavity during the day. Females are distributed among these territories, forming harems of one to eight individuals per male (Bradbury and Vehrencamp 1976). Movements by males between colonies are less frequent than by females. Odor appears to play an important role in this and other harem-forming emballonurids (Voigt and Helversen 1999). Harem males of *S. bilineata* disseminate odors by salting and fanning them from a propatagial sac while in the roost and during energetically costly hovering displays before females. Females subsequently choose among the territorial males with whom they may mate (Voigt and Helversen 1999).

The mating system of *Pipistrellus nanus* also appears to be strongly influenced by its roosting habits. When roosting in furled leaves, males are seldom found together with females and their young during parturition and lactation (Happold and Happold 1996). At other times of the year, adults roost both singly and in small groups of up to 12 individuals. Where adult males roost with females, there is usually only a single male present. Males may exclude other males from their roosts, but it is not clear how and why this exclusion occurs. Group composition is relatively labile, but some males attract more females than do others.

Pipistrellus nanus exhibits little fidelity to specific roosts but, instead, shows fidelity to clumps of banana plants (Happold and Happold 1996). Males show higher fidelity to the clumps of banana plants than do females, but no relationship was found between the reliability of clumps and number of females recruited. Females seem to be attracted to males independent of roost quality, and males and females roost with several potential mates in succession, with no evidence for sustained bonding (Happold and Happold 1996). Furled banana leaves were not limiting in this area, with less than 50% of the available roosts occupied at any time. Because competition was minimal for these roosts, males did not have a mechanism for demonstrating their relative fitness to females, and thus Happold and Happold (1996) suggested a promiscuous mating system for this species.

Myotis bocagei also roosts in furled banana leaves (*Musa* spp.) and forms harems with stable female membership, with male tenure extending up to 1 yr or more (Brosset 1976). Differences in mating system between *M. bocagei* and *P. nanus* may reflect the fact that females of *M. bocagei* are not simultaneously receptive, as they are in *P. nanus* (Happold and Happold 1996). Intraspecific differences also occur, as when *P. nanus* roosts in thatched roofs, males defend territories and form harems (O'Shea 1980). Thus, the mating system in *P. nanus* varies depending on the availability of semipermanent thatch roosts, on the one hand (O'Shea 1980), and the relative ephemeral roosts in banana leaves, on the other (Happold and Happold 1996).

Relationships between Roosting and Foraging Behavior
Night Roosts

Night roosts play an important functional role in the foraging ecology of bats. They are used as resting places between foraging bouts, promote digestion and energy conservation, provide retreats from predators and inclement weather, provide places to ingest food transported from nearby feeding areas, function as feeding perches for sit-and-wait predators, and serve as places that promote social interactions and information transfer (Kunz 1982).

A wide range of structures are used by bats as night roosts, including buildings, bridges, caves, mines, or rock surfaces (reviewed in Kunz 1982). Comparatively little is known about night roosts in tree cavities or in vegetation. Some bats return to their day roost at night (e.g., *Nyctalus noctula* [Kronwitter 1988], *Pipistrellus nanus* [O'Shea 1980], and *Tonatia silvicola* [Kalko et al. 1999]). Harem males of *Cynopterus sphinx* and *C. brachyotis* typically roost in tents during the day and night, although females use separate feeding roosts at night (Balasingh et al. 1995; Bhat and Kunz 1995; Elangovan et al. 1999; Tan et al. 1997).

Cavity-roosting bats sometimes select night roosts separate from their day roost (e.g., *Eptesicus fuscus* [Brigham 1991] and *Scotophilus (borbonicus) leucogaster* [Fenton 1983]). Night roosts may be located in the vicinity of feeding areas to reduce the energetic cost of returning to the day roost and to minimize the associated risk of predation (Kunz 1982). The distance to foraging areas and the prevailing weather conditions may influence whether individuals return to the day roost following feeding bouts (Shiel et al. 1999). Brigham (1991) compared the proportion of times that separate night roosts were used by *E. fuscus* for individuals foraging at different distances from the diurnal roost. When bats foraged less than 1 km from the day roost, they used alternate night roosts 25% of the time. In contrast, when bats foraged up to 4 km away, they used night roosts 60% of the time.

The proportion of the night spent in night roosts varies both daily and seasonally in relation to reproductive condition, prey density, and ambient temperature. Long night-roosting periods in *Myotis lucifugus* that roost in buildings are generally associated with cool nights and low prey densities (Anthony et al. 1981). Females return to the day roost most frequently when they have dependent young that require feeding during the night (Racey 1982). Once young bats are independent, females increasingly use separate night roosts (Anthony et al. 1981).

Feeding roosts are used by bats that "capture" food elsewhere and retreat to a roost to consume the food. In this way, plant-visiting species that use feeding roosts away from source trees make an important contribution to seed dispersal (Boon and Corlett 1989; Morrison 1978a; Richards 1990b). Feeding roosts are used extensively by both mega- and microchiropteran frugivores. Small, canopy-feeding, foliage-roosting bats may harvest fruits and carry them to nearby feeding roosts (Bhat 1994; Boon and Corlett 1989; Charles-Dominique 1993; Elangovan et al. 1999; Law 1993; Spencer and Fleming 1989). This may be a form of antipredator behavior (Fenton et al. 1985; Kunz 1982; Spencer and Fleming 1989) or perhaps avoidance behavior in response to conspecifics at food sources (Richards 1990b, 1995). Heithaus and Fleming (1978) estimated that individual *Carollia perspicillata* made 40–50 trips a night between fruiting trees and feeding roosts.

Feeding roosts of small species are often located in dense vegetation and are thus relatively inaccessible to terrestrial predators (Morrison 1978a). In contrast, larger pteropodids, which are less vulnerable to predation, generally remain in the food tree where they consume the fruit and nectar located at that site (Richards 1990b; Thomas and Fenton 1978). *Pteropus conspicillatus* in northern Australia exhibits what Richards (1990b) termed a "raider vs. resident strategy" at feeding sites. Dominant individuals establish feeding territories in fruit trees early in the evening until the tree is full of feeding bats. Later in the night, when other bats attempt to join the feeding group, these individuals are evicted by the residents. Fruit taken by the raiders before escaping from the aggression of residents is taken elsewhere to be consumed (Richards 1990b).

Some insectivorous and carnivorous microchiropterans use feeding roosts, often when their prey is too large to consume in flight. These "sit-and-wait" predators use feeding perches, where individuals hang from vantage points while scanning their surroundings for both vertebrate and invertebrate prey (Audet et al. 1991; Csada 1996; Schulz 1986). Forays to pursue prey are typically brief, with a bat returning to its original perch to consume the item (Vaughan and Vaughan 1986). Species that use this foraging strategy are predominantly gleaning bats with low wing loading and low aspect ratios (e.g., *Hipposideros diadema* [Pavey 1998], *Lavia frons* [Vaughan and Vaughan 1986], *Macroderma gigas* [Tidemann et al. 1985], *Megaderma lyra* [Audet et al. 1991], *Nycteris grandis* [Fenton et al. 1990], *Rhinolophus hildebrandti* [Fenton and Rautenbach 1986], *R. rouxi* [Neuweiler et al. 1987], *Tonatia silvicola* [Kalko et al. 1999], and *Trachops cirrhosus* [Kalko et al. 1999]). Hunting from feeding perches may be a way to minimize foraging costs and allow capture of heavy prey relative to the bat's body mass without incurring high costs of flight (Audet et al. 1991).

Nocturnal calling roosts are common among male epomophorine bats in Africa (e.g., *Epomophorus wahlbergi* [Wickler and Seibt 1976] and *Hypsignathus monstrosus* [Bradbury 1977a]). Males typically space themselves apart hanging from small branches and undertake extensive periods of calling and displaying to attract females. Several species of microchiropterans use calling or "singing" roosts. Males of *Nyctalus noctula* in Europe spend considerable time during the mating season, calling from their roost to attract females (Alcalde 1999; Kronwitter 1988). Singing roosts used by *Mystacina tuberculata* in New Zealand are usually located in small trees within 100 m of day roosts (Daniel 1990; O'Donnell et al. 1999). Single bats call from the openings of small cavities for periods of 10–40 min. The repetitive, high-intensity song is audible to the human ear for up to 50 m. Other bats visit these calling sites, some of which are allowed to enter the cavity while others are repelled aggressively.

Relationships among Roosts and Foraging Areas

The relationship among roosts and foraging areas is influenced by several factors, including roost type, the availability of roosts, food and water, flight morphology, colony size, and reproductive cycles. Small microchiropterans often commute less than several kilometers between roost sites and foraging areas (e.g., Brigham et al. 1997; Campbell et al. 1996; Sasse and Pekins 1996; Schulz and Hannah 1998). However, radiotelemetry studies indicate that many species of bats fly greater distances and have larger home ranges than previously recognized and sometimes forage at distances ranging from 10 to 30 km from roost sites (Arlettaz 1999; Barclay 1989; O'Donnell 2001; Pierson 1998). In these situations, especially where roost sites are limiting, bats select optimal roosting sites and optimal foraging areas and appear not to minimize commuting distances to reduce energetic costs (Brigham 1991; Fenton et al. 1985; Lumsden et al. 2002a). Although home ranges may be large, individuals often concentrate their nightly activity in localized areas rather than foraging over the entire range (O'Donnell 2001; O'Donnell et al. 1999; Robinson and Stebbings 1997).

Flight morphology and body size have been used to interpret foraging ranges of bats, with larger species and those with high aspect ratios (narrow, pointed wings) likely to commute greater distances to forage (Fenton 1997; Jones et al. 1995). However, some species do not conform to these predictions. For example, in New Zealand, *Chalinolobus tuberculatus* (10 g) and *Mysticina tuberculata* (15 g) both fly considerably further from their roost than predicted, up to 19 and 24 km, respectively (O'Donnell 2001; O'Donnell et al. 1999). Different flight patterns may be used for commuting from roosts sites compared to those for foraging. Some species (e.g., *Myotis emarginatus* and *M. myotis* in Europe [Arlettaz 1999; Audet 1990; Krull et al. 1991] and *Nyctophilus geoffroyi* and *Hipposideros* spp. in Australia [Lumsden et al. 2002a; Pavey and Burwell 2000]) use a slow maneuverable flight pattern while foraging but employ a faster and more direct flight when commuting, a strategy that may enable them to commute greater distances.

Proximity to food resources may also be an important determinant in roost site selection for some species. Bats that use roost types that are abundant (e.g., foliage) are more likely to move their day roosts in response to food availability. The roosting and foraging behavior of small, solitary, foliage-roosting, and fruit-eating pteropodids differ from what is often observed in the larger, gregarious species. *Micropteropus pusilla*, *Epomops buettikorteri*, and *Epomophorus wahlbergi* from Africa roost alone or in small groups and feed near their roosts in the forest canopy (Fenton et al. 1985; Thomas 1982; Wickler and Seibt 1976). *Nyctimene robinsoni* and *Syconycteris australis* from Australasia forage mostly in the subcanopy of rainforests, and generally roost near food trees used on the previous night (Law 1993; Spencer and Fleming 1989; Winkelmann et al. 2000).

In contrast, large gregarious megachiropterans such as *Eidolon*, *Pteropus*, and *Acerodon* often form large colonies that are located many kilometers from feeding areas (Eby 1996; Nelson 1965; Palmer and Woinarski 1999; Richards 1995; Tidemann, et al. 1999). Large pteropodids, in particular, are highly mobile, and their roosting and foraging behavior appear to be adaptations to fruit and nectar sources that are patchy in time and space (Fleming and Eby, this volume; Palmer and Woinarski 1999).

To examine the influence of food availability on the foraging distances of a New World plant-visiting bat, Morrison (1978b) compared the commuting distances of *Artibeus jamaicensis* in two areas with different densities of figs (*Ficus* spp). On Barro Colorado Island, Panama, where figs were abundant, females commuted 0.6 km, whereas they commuted 8 km in Chamela, Mexico, where figs were much less common. Cavity-roosting females returned to the same roost even after shifting foraging areas, and Morrison (1978b) suggested that, if suitable roosts were scarce, commuting a longer distance to a food resource may cost less than searching for a suitable roost near the new foraging site. After Hurricane Hugo devastated Puerto Rico in 1989, population levels of *Stenoderma rufum* declined by 70%, and, in response to the lower food availability, individuals home ranges increased fivefold. This increased the cost of commuting and foraging, in terms of time and energy, and may have affected their reproductive success (Gannon and Willig 1994).

Proximity to water appears to be an important consideration in roost site selection by bats (Campbell et al. 1996; Mattson et al. 1996; Ormsbee and McComb 1998). Some species of insectivorous bats concentrate foraging directly over or in the vicinity of water (e.g., Bogdanowicz 1994; Brigham 1991; Jaberg et al. 1998), and hence the energetic costs of commuting may be reduced when roosts are located nearby. Water is also required for drinking by many species (Kurta et al. 1989, 1990), which is reflected in the concentration of bats observed at watering holes in arid and semiarid regions (e.g.,

Lumsden and Bennett 1995; Szewczak et al. 1998). Some bats also visit watering holes to obtain nutrients such as calcium and sodium, which may otherwise be limiting resources, especially during pregnancy and lactation (Barclay 1995).

The number of individuals in a roost may influence the distance that bats need to travel to find food. Refuging theory (Hamilton and Watt 1970) predicts that some individuals in a refuge should commute farther than others to minimize competition. Solitary roosting species should have little competition for food close to their roost, whereas food resources may be limiting around large communal roosts. In Western and American Somoa, *Pteropus samoensis*, which roost singly or in small groups, forage largely in the same area in which they roost (Brooke et al. 2000; Cox 1983; Rainey 1998; Wilson and Engbring 1992). In contrast, the sympatric *P. tonganus*, which roosts in colonies of several thousand individuals, commutes long distances from roosting sites to foraging areas (Richmond et al. 1998; Wilson and Engbring 1992).

Reproductive condition may influence foraging behavior, with higher energetic demands on females during the breeding season. Lactating females typically forage closer to their roost sites than do pregnant or nonbreeding females, presumably to reduce the time and energetic cost of returning to suckle young during the night (e.g., *Chalinolobus tuberculatus* [O'Donnell 2001], *Nyctalus leisleri* [Shiel et al. 1999], *Nyctophilus bifax* [Lunney et al. 1995], and *Plecotus auritus* [Fuhrmann and Seitz 1992]). However, some species fly similar or greater distances (e.g., *Chalinolobus gouldii* [Lumsden et al. 2002a], *Lasiurus cinereus* [Barclay 1989], and *Nyctophilus geoffroyi* [Lumsden et al. 2002a]), suggesting that for these species the energetic benefits of foraging in optimal feeding areas outweigh the commuting costs.

The time spent by females in the day roost during the night varies throughout the lactation period in response to the thermoregulatory abilities of the young and the type of roost selected. For example, the exposed foliage roosts used by *L. cinereus* may require females to spend more time with their young early in lactation to keep them warm (Barclay 1989). As young develop, the time that females spend roosting declines and the length of foraging bouts increases. Once young *Chalinolobus tuberculatus* commence flying, females may commute greater distances to foraging areas to reduce competition for food resources within the vicinity of the roost while the flight capabilities of the young develop (O'Donnell 2001).

Distances between roosts and foraging areas of the large pteropodids also can differ between sexes. In northern Australia, females of *P. alecto* are restricted to one roost when they have dependent young (Palmer and Woinarski 1999; Palmer et al. 2000). Females in large colonies may not have the flexibility to change to alternate roosts, especially if the cost of transport would place the young and mother at risk of predation. By contrast, males are less constrained in their roost selection. If food patches are widely spaced due to forest fragmentation, this may be an important factor in limiting areas that can be used as roosting sites (Eby 1991; Palmer and Woinarski 1999).

Conservation of Cavity and Foliage Roosting Bats

Bat populations continue to decline in many parts of the world (Hutson et al. 2001; Kunz and Pierson 1994; Kunz and Racey 1998; Mickleburgh et al. 1992; Racey and Entwistle, this volume; Wilson and Graham 1992). Factors that contribute to these declines vary regionally, but deforestation and conversion of native habitats to intensive agriculture or other human developments pose the greatest threats. Deforestation has reduced the availability of many important roost resources, and loss of such roosts is having an enormous impact on the density and distribution of local bat faunas.

In the Indo-Pacific region, Asia and Australia, where more than 60% of megachiropterans live on islands and in coastal habitats, deforestation is the most important factor contributing to their decline (Law 1996; Racey and Entwistle, this volume; Rainey 1998; Utzurrum 1998). Pressures resulting from unchecked human population growth in some countries, and the custom of land tenure in others (Whewell 1992), have contributed considerably to the loss of tropical forests (Arita and Ortega 1998; Marinho-Filho and Sazima 1998; Utzurrum 1998). Some roosting and foraging habitats of bats are being severely altered or are being made uninhabitable (Rainey 1998; Robertson 1992). For example, mangrove forests, which serve as important habitats for several island and coastal species of pteropodids, are largely being destroyed for the woodchip industry (Robertson 1992; Start and Marshall 1976), aquaculture, or commercial development (A. Zubaid, personal communication).

Reductions in some populations of pteropodids on tropical islands have been attributed to the direct and indirect impacts of periodic typhoons (Craig et al. 1994; Lemke 1992; Pierson et al. 1996; Rainey 1998). In addition to bats being killed during these storms, food and roosting resources may be severely depleted, resulting in the fragmentation of camps into smaller groups and a reduction in the recruitment of young animals (Banack 1996; Rainey 1998; Richards 1990a; Robertson 1992). On some islands in the western Indian Ocean, the combined effect of deforestation and typhoons has severely threatened the existence of already endangered faunas (Cheke and Dahl 1981; Craig et al. 1994; Reason and Trewhella 1994). Overhunting by humans and predation by the introduced brown tree snake (*Boiga irregularis*) have also had an impact on local populations of some species of *Pteropus* (Craig et al. 1994; Mickleburgh et al. 1992; Rainey 1998).

The capacity of colonial *Pteropus* to roost in modified habitats varies widely. Species such as *P. niger* on Mauritius (Cheke and Dahl 1981) and *P. samoensis* in American Samoa (Pierson et al. 1996) roost mostly in primary forests. In the highly fragmented landscapes of eastern Australia, *P. poliocephalus, P. alecto,*

and *P. scapulatus* commonly roost in small patches of remnant lowland vegetation surrounded by agricultural land or urban development (Lunney and
Moon 1997; Parry-Jones and Augee 2001). While these species show high
fidelity to traditional camps, they may use alternate roosts when existing vegetation is cleared or made uninhabitable (Lunney and Moon 1997; Tidemann
et al. 1999).

Conservation strategies for many temperate species must address seasonal
and geographical variation in their roosting habits (Pierson 1998; Racey 1998;
Racey and Entwistle, this volume). Some temperate species hibernate in caves
and rock crevices in winter (Barbour and Davis 1969) but roost in tree cavities
or beneath exfoliating bark in warm months (Barclay and Brigham 1996).
Thus, protecting only their summer roosts is inadequate for the conservation
of these species. For example, population declines reported for the endangered *Myotis sodalis* were intially attributed to disturbances at caves, where
this species hibernates. But while most of the critical hibernacula have now
been protected, populations have continued to decline. Only recently have
efforts been made to focus attention on protecting summer roosts and foraging habitats. Species dependence on cavity roosts may vary geographically.
For example, *Eptesicus fuscus* is commonly associated with buildings in eastern North America (Kunz and Reynolds, in press) but appears to depend
more on tree cavities in western North America (Betts 1996; Brigham 1991;
Vonhof 1996).

Protection of forests that provide important roost and food resources for
bats should be an important conservation goal (Pierson 1998). Densities of
tree-cavity-roosting species are greater in old-growth stands, where structural
diversity provides a range of roosting options (Barclay and Brigham 1996).
Current forest management practices that favor even-age monospecific
stands, short rotation times, and selective removal of dead and dying trees reduce the availability of roosting habitat of cavity-roosting species (Barclay and
Brigham 1996). Unless efforts are made to reverse these forest management
practices, expected declines in bat populations and species richness are likely
to have serious consequences for ecosystem function (Pierson 1998).

In western Europe, deforestation and habitat fragmentation have long
threatened the conservation of bats and other wildlife (Bright 1993). In Great
Britian, few contiguous forest habitats exist, which limits the availability of
suitable roosting sites for tree-cavity bats. In other parts of Europe, however,
some forests are sufficiently mature or protected to provide suitable habitat for
species that roost in tree cavities (e.g., *Nyctalus noctula* [Boonman 2000] and
Myotis bechsteinii and *M. daubentonii* [Kerth et al. 2000]). Notwithstanding,
these and other European bat species also roost in manmade structures (Entwistle et al. 1997; Kerth et al. 2000; Schober and Grimmberger 1989; Swift
1998) or in boxes designed for birds (Benzal 1991) and bats (Stebbings and

Walsh 1985; Swift 1998). The successful use of bat houses in Europe and North America (Tuttle and Hensley 1993), in providing roosting habitats for bats that traditionally roost in tree cavities, is a conservation management practice that is expected to increase as forest management practices reduce the availability of natural tree cavities.

Many cavity-roosting species move among roost sites on a regular basis, using a number of trees within a defined roost area (Barclay and Brigham 1996, Lewis 1995; O'Donnell and Sedgeley 1999). As a result, high densities of suitable roost trees are required (Kerth et al. 2000; Lumsden and Bennett 2000). Such roosts are most likely to occur in large, contiguous stands of mature forest. Suitable roosts are liable to be present in lower abundance in areas managed for timber production and in agricultural areas. Little is known about the impact of reduced densities of roost trees or the affect of roosting in suboptimal tree roosts. However, Brigham and Fenton (1986) demonstrated that reproductive success in E. fuscus was lower in suboptimal building roosts, and this also may be the case for bats that roost in tree cavities. The use of artificial bat houses (which may mimic natural tree cavities) provide a model for testing the effects of different roost variables on reproductive success (Kerth et al. 2000).

Forest management practices that focus on retaining high densities of roost trees should benefit cavity-roosting species. Management should focus on maintaining potential roost trees that are easily accessible to bats and have moderate to high levels of exposure to solar radiation (e.g., snags that protrude above the canopy, in canopy gaps, and adjacent to forest clearings [Barclay and Brigham 1996; Waldien et al. 2000]). The concentration of snags in clusters may be beneficial to some species (Waldien et al. 2000). Methods that promote the creation of snags and accelerate cavity formation need to be considered as management options (Bull and Partridge 1986; Carey and Sanderson 1981; Lewis 1998). Because small cavities form as precursors to larger ones, bats may be among the first cavity-using species to colonize rejuvenating forests (Barclay and Brigham 1996; Tidemann and Flavel 1987). In part, this reflects the strong commuting abilities of bats, which gives them an advantage over nonvolant mammals in exploiting newly available habitat (Rawlinson et al. 1992; Whitaker and Jones 1994). Species that roost beneath exfoliating bark may be able to recolonize forests earlier than species dependent on large cavities (see Kunz 1996; Mackowski 1984). In addition, as many species have large home ranges and travel considerable distances between roost sites and foraging areas, the environment in which they live needs to be evaluated and managed at the landscape scale (Barclay and Brigham 1996; Lumsden et al., in press a).

In some regions, such as Africa and India, the collection of firewood by people has increasingly become an important factor in forest destruction and modification. When human populations exceed the carrying capacity of local ecosystems, natural habitats are readily converted into deserts and wasteland (Racey and Entwistle, this volume). Overgrazing by elephants and other wildlife that are restricted to reserves in parts of Africa has led to irreversible changes in woodlands on which bats depend for both roosts and food (Fenton and Rautenbach 1998). Although some fruit-eating and insectivorous bats in tropical regions may be relatively common in urban and agricultural areas, far removed from forested areas, the numbers of endemic species in these regions have become severely reduced. In the Philippines, where there is a relatively high degree of endemism, endemic fruit bats have not been reported from urban orchards and agricultural areas (Utzurrum 1995, 1998). Notwithstanding, some species such as *Cynopterus brachyotis* and *C. sphinx* have become relatively common and exploit a wide variety of human-altered habitats for roosting, including buildings in urban areas, as well as trees in orchards and gardens (Balasingh et al. 1995; Bhat 1994; Bhat and Kunz 1995; Boon and Corlett 1989; Tan et al. 1997; Utzurrum 1998).

Plantings of ornamental trees in some urban areas have provided new roosting (and feeding) opportunities for some foliage-roosting species (e.g., *Cynopterus sphinx* [Balasingh et al. 1995; Bhat and Kunz 1995], *C. brachyotis* [Tan et al. 1997], *C. horsfield* [Tan et al. 1999], *Lasiurus xanthinus* [Constantine 1998], and *Scotophilus kuhlii* [Rickart et al. 1989]). The use of palm "skirts" as a roost resource appears to be important for *L. xanthinus* in the southeastern United States, as recent range extensions have been reported where ornamental palms have been planted (Constantine 1998). Similarly, plantings of ornamental palms along boulevards and in residential areas in the Old World tropics may not only provide opportunities for range expansion, but may also support increased densities of bats that use these roost resources (Tan et al. 1997).

Future conservation efforts for bats in both temperate and tropical ecosystems must include protection of roost resources. In large measure, this can best be accomplished by focusing on protecting large forest reserves and suitable corridors for seasonal migrants (Fleming and Eby, this volume). Many of the megachiropterans that form large aggregations at traditional roost sites disperse seasonally over great distances, and their protection requires management strategies that focus on the protection of movement corridors as well as on local habitats that provide both roosts and food resources. The protection of forest habitats that provide such resources should be a high priority for conservation biologists (Racey and Entwistle, this volume). Because island and coastal ecosystems have experienced the most severe pressures from deforestation and commercial development, increased efforts are needed to protect these fragile ecosystems (Rainey 1998).

Island and peninsular species that show high levels of endemism (Corbet and Hill 1992; Heaney 1986; Pierson and Rainey 1992; Rodriguez-Duran and

Kunz 2001) and have evolved highly dependent, and sometimes obligate, relationships with the local flora are most vulnerable (Kunz 1996). The loss of "keystone" or "pivotal" species can lead to an ever-increasing cascade of extinctions (Cox et al. 1991; Howe 1984; Mills et al. 1993; Myers 1986; Rainey et al. 1995). The enumeration (Crome and Richards 1988; Francis 1990; Zubaid 1993) and protection of biological diversity in temperate and tropical forest ecosystems, especially on islands and in coastal ecosystems, remains one of the greatest challenges to conservation biologists.

Acknowledgments

We are grateful to S. Churchill, P. Eby, A. Kurta, C. O'Donnell, E. D. Pierson, and M. Schulz for suggestions on the manuscript and to T. Allofs, J. Balasingh, P. Birt, P. Eby, P. Espy, M. B. Fenton, R. Hodgkison, L. Lumsden, D. H. Morrison, E. Rooks, C. Taylor, K. H. Tan, M. D. Tuttle, T. A. Vaughan, M. Vonhof, and R. Voss for sharing their photographs and illustrations that were used in this chapter. We also thank Cari Watkins, who assisted in the preparation of the final manuscript and illustrations. Kunz's research on roosting ecology has been supported by the U.S. National Science Foundation, the National Geographic Society, and the Lubee Foundation, and Lumsden's research has been supported by the Arthur Rylah Institute.

Literature Cited

Alcalde, J. T. 1999. New ecological data on the noctule bat (*Nyctalus noctula* Schreber, 1774) (Chiroptera, Vespertiliondae) in two towns of Spain. Mammalia, 63:273–280.

Alder, H. 1994. Erste erfahrungen mit dem data logger: Ereigniszahlung vor baumhohlenquartieren von wasserfledermausen, *Myotis daubentoni*, bei gleichzeitiger messung mikroklimatischer werte. Mitteilungen der Naturforschenden Gesellschaft Schaffhausen, 39:119–133.

Aldridge, H. D. J. N., M. Orbist, H. G. Merriam, and M. B. Fenton. 1990. Roosting, vocalizations, and foraging by the African bat *Nycteris thebaica*. Journal of Mammalogy, 71:242–246.

Allen, G. M. 1939. Bats. Harvard University Press, Cambridge, Mass.

Anthony, E. L. P., M. H. Stack, and T. H. Kunz. 1981. Night roosting and the nocturnal time budget of the little brown bat, *Myotis lucifugus*: effects of reproductive status, prey density, and environmental conditions. Oecologia, 51:151–156.

Arendt, W. J. 1986. Bat predation by the St. Lucia boa (*Boa constrictor orophias*). Caribbean Journal of Science, 22:219–220.

Arita, H. T., and J. Ortega. 1998. The middle American bat fauna: conservation in the Neotropical-Nearctic border. Pp. 295–308 *in:* Bat Biology and Conservation (T. H. Kunz and P. A. Racey, eds.). Smithsonian Institution Press, Washington, D.C.

Arlettaz, R. 1999. Habitat selection as a major resource partitioning mechanism between the two sympatric sibling bat species *Myotis myotis* and *Myotis blythii*. Journal of Animal Ecology, 68:460–471.

Audet, D. 1990. Foraging behavior and habitat use by a gleaning bat, *Myotis myotis* (Chiroptera: Vespertilionidae). Journal of Mammalogy, 71:420–427.

Audet, D., D. Krull, G. Marimuthu, S. Sumithran, and J. B. Singh. 1991. Foraging behavior of the Indian false vampire bat, *Megaderma lyra* (Chiroptera: Megadermatidae). Biotropica, 23:63–67.

Baagøe, J. S. 1978. Observations on the biology of the banana bat, *Pipistrellus nanus*. Pp. 275–282 *in:* Proceedings of the Fourth International Bat Research Conference (R. J. Olembo, J. B. Castelino, and F. A. Mutere, eds.). Kenya National Academy for Advancement of Arts and Sciences, Kenya Literature Bureau, Nairobi.

Baker, R. H., and C. J. Phillips. 1965. Mammals from El Navado de Colima, Mexico. Journal of Mammalogy, 46:691–693.

Baker, R. J., C. S. Hood, and R. Honeycutt. 1989. Phylogenetic relationships and classification of the higher categories of the New World bat family Phyllostomidae. Systematic Zoology, 38:228–238.

Baker, R. J., C. A. Porter, J. C. Patton, and T. A.Van Den Bussche. 2000. Systematics of bats of the family Phyllostomidae based on Rag2 DNA sequences. Occasional Papers, Museum of Texas Tech University, 202:1–16.

Balasingh, J., S. Isaac, and R. Subbaraj. 1993. Tent-roosting by the frugivorous bat *Cynopterus sphinx* (Vahl 1797) in southern India. Current Science, 65:418.

Balasingh, J., J. Koilraj, and T. H. Kunz. 1995. Tent construction by the short-nosed fruit bat, *Cynopterus sphinx* (Chiroptera: Pteropodidae) in southern India. Ethology, 100:210–229.

Banack, S. A. 1996. Flying foxes, genus *Pteropus,* in the Samoan Islands: interactions with forest communities. Ph.D. Dissertation. University of California, Berkeley.

Barbour, R. W., and W. H. Davis. 1969. Bats of America. University Press of Kentucky, Lexington.

Barbour, T. 1932. A peculiar roosting habit of bats. Quarterly Review of Biology, 7:307–312.

Barclay, R. M. R. 1989. The effect of reproductive condition on the foraging behavior of female hoary bats, *Lasiurus cinereus.* Behavioral and Ecological Sociobiology, 24:31–37.

Barclay, R. M. R. 1995. Does energy or calcium availability constrain reproduction in bats? Symposia of the Zoological Society of London, no. 67:245–258.

Barclay, R. M. R., and R. M. Brigham, eds. 1996. Bats and Forests Symposium, October 19–21, 1995, Victoria, British Columbia, Canada. British Colombia Ministry of Forests, Victoria.

Barclay, R. M. R., and K. J. Cash. 1985. A noncommensal maternity roost of the little brown bat (*Myotis lucifugus*). Journal of Mammalogy, 66:782–783.

Barclay, R. M. R., P. A. Faure, and D. R. Farr. 1988. Roosting behavior and roost selection by migrating silver-haired bats (*Lasionycteris noctivagans*). Journal of Mammalogy, 69:821–825.

Bates, P. J. J., and D. L. Harrison. 1997. Bats of the Indian Subcontinent. Harrison Zoological Museum, Sevenoaks, Kent.

Becking, R. W. 1982. Pocket Flora of the Redwood Forest. Island Press, Covelo, Calif.

Bennett, A. F., L. F. Lumsden, and A. O. Nicholls. 1994. Tree hollows as a resource for wildlife in remnant woodlands: spatial and temporal patterns across the northern plains of Victoria, Australia. Pacific Conservation Biology, 1:222–235.

Bennett, M. 1993. Structural modifications involved in the fore and hind limb grip of some flying foxes (Chiroptera: Pteropodidae). Journal of Zoology (London), 229:237–248.

Benzal, J. 1991. Population dynamics of the brown long-eared bat (*Plecotus auritus*) occupying bird boxes in a pine forest plantation in central Spain. Netherlands Journal of Zoology, 41:241–249.

Bernard, E. 1999. Notes on a colony of *Peropteryx leucoptera* (Emballonuridae) in Brazil. Bat Research News, 40:37–38.

Betts, B. J. 1996. Roosting behaviour of silver-haired bats (*Lasionycteris noctivagans*) and big brown bats (*Eptesicus fuscus*) in northeast Oregon. Pp. 55–61 *in:* Bats and Forests Symposium, October 19–21, 1995, Victoria, British Columbia, Canada (R. M. R. Barclay and R. M. Brigham, eds.). British Colombia Ministry of Forests, Victoria.

Betts, B. J. 1998. Roosts used by maternity colonies of silver-haired bats in northeastern Oregon. Journal of Mammalogy, 79:643–650.

Bhat, H. R. 1994. Observations of the food and feeding behavior of *Cynopterus sphinx* Vahl (Chiroptera, Pteropodidae) at Pune, India. Mammalia, 58:363–370.

Bhat, H. R., and T. H. Kunz. 1995. Altered flower/fruit clusters of the kitul palm used as roosts by the short-nosed fruit bat, *Cynopterus sphinx* (Chiroptera: Pteropodidae). Journal of Zoology (London), 235:597–604.

Bloedel, P. 1955. Observations on life histories of Panama bats. Journal of Mammalogy, 36:232–235.

Bogdanowicz, W. 1994. Myotis daubentonii. Mammalian Species, 475:1–9.

Boinski, S., and R. M. Timm. 1985. Predation by squirrel monkeys and double-toothed kites on tent-making bats. American Journal of Primatology, 9:121–127.

Bonaccorso, F. J. 1998. Bats of Papau New Guinea. Conservation International, Washington, D.C.

Boon, P. P., and R. T. Corlett. 1989. Seed dispersal by the lesser short-nosed fruit bat (*Cynopterus brachyotis,* Pteropodidae, Megachiroptera). Malayan Nature Journal, 42:251–256.

Boonman, M. 2000. Roost selection by noctules (*Nyctalus noctula*) and Daubenton's bats (*Myotis daubentonii*). Journal of Zoology (London), 251:385–389.

Bradbury, J. W. 1977a. Lek mating behavior in the hammer-headed bat. Zeitschrift für Tierpsychologie, 45:225–255.

Bradbury, J. W. 1977b. Social organization and communication. Pp. 1–72 *in:* Biology of Bats (W. A. Wimsatt, ed.). Academic Press, New York.

Bradbury, J. W., and L. H. Emmons. 1974. Social organization of some Trinidad bats. I. Emballonuridae. Zeitschrift für Tierpsychologie, 36:137–183.

Bradbury, J. W., and S. L. Vehrencamp. 1976. Social organization and foraging in emballonurid bats. I. Field studies. Behavioral Ecology and Sociobiology, 1:337–381.

Brigham, R. M. 1991. Flexibility in foraging and roosting behaviour by the big brown bat (*Eptesicus fuscus*). Canadian Journal of Zoology, 69:117–121.

Brigham, R. M., and M. B. Fenton. 1986. The influence of roost closure on the roosting and foraging behaviour of *Eptesicus fuscus* (Chiroptera: Vespertilionidae). Canadian Journal of Zoology, 64:1128–1133.

Brigham, R. M., M. J. Vonhof, R. M. R. Barclay, and J. C. Gwilliam. 1997. Roosting behavior and roost-site preferences of forest-dwelling California bats (*Myotis californicus*). Journal of Mammalogy, 78:1231–1239.

Bright, P. 1993. Habitat fragmentation and predictions for British mammals. Mammal Review, 230:101–111.

Brooke, A. P. 1987. Tent selection, roosting ecology and social organization of the tent-making bat, *Ectophylla alba*, in Costa Rica. Journal of Zoology (London), 221:11–19.

Brooke, A. P. 1990. Tent construction and social organization in *Vampyressa nymphaea* (Chiroptera: Phyllostomidae) in Costa Rica. Journal of Tropical Ecology, 3:171–175.

Brooke, A. P., C. Solek, and A. Tualaulelei. 2000. Roosting behavior of colonial and solitary flying foxes in American Samoa (Chiroptera: Pteropodidae). Biotropica, 32:338–350.

Brosset, A. 1976. Social organization in the African bat, *Myotis bocagei*. Zeitschrift für Tierpsychologie, 42:50–56.

Buchanan, F. M. 1969. Bats of the Arima Valley, Trinidad, W.I. Privately published. Asa Wright Nature Centre, Trinidad and Tobago, 53 pp.

Buchanan, M. 1958. *Tadarida* and *Myotis* occupying cliff swallow nests. Journal of Mammalogy, 39:434–435.

Bull, E .L., and A. D. Partridge. 1986. Methods of killing trees for use by cavity nesters. Wildlife Society Bulletin, 14:142–146.

Callahan, E. V., R. D. Drobney, and R. L. Clawson. 1997. Selection of summer roosting sites by Indiana bats (*Myotis sodalis*) in Missouri. Journal of Mammalogy, 78:818–825.

Campbell, L. A., J. G. Hallett, and M. A. O'Connell. 1996. Conservation of bats in managed forests: use of roosts by *Lasionycteris noctivagans*. Journal of Mammalogy, 77:976–984.

Carey, A. B., and H. R. Sanderson. 1981. Routing to accelerate tree-cavity formation. Wildlife Society Bulletin, 9:14–21.

Chapman, F. M. 1932. A home-making bat. Natural History, 32:555–556.

Charles-Dominique, P. 1993. Tent use by the bat *Rhinophylla pumilio* (Phyllostomidae: Carolliinae) in French Guiana. Biotropica, 25:111–116.

Chasen, F. N. 1939. Four new mammals from Java. Treubia, 17:185–188.

Cheke, A. S., and J. F. Dahl. 1981. The status of bats on western Indian Ocean islands, with special reference to *Pteropus*. Mammalia, 45:205–238.

Choe, J. C. 1994. Ingenious design of tent roosts by Peter's tent-making bat, *Uroderma bilobatum* (Chiroptera: Phyllostomidae). Journal of Natural History, 28:731–737.

Choe, J. C., and R. M. Timm. 1985. Roosting site selection by *Artibeus watsoni* (Chiroptera: Phyllostomidae) on *Anthurium ravenii* (Araceae) in Costa Rica. Journal of Tropical Ecology, 1:241–247.

Chung-MacCoubrey, A. L. 1996. Bat species composition and roost use in pinyon-juniper woodlands of New Mexico. Pp. 118–123 *in:* Bats and Forests Symposium, October 19–21, 1995, Victoria, British Columbia, Canada (R. M. R. Barclay and R. M. Brigham, eds.). British Colombia Ministry of Forests, Victoria.

Churchill, S. 1998. Australian Bats. Reed New Holland, Sydney.

Clague, C. I., R. B. Coles, O. J. Whybird, H. J. Spencer, and P. Flemons. 1999. The occurrence and distribution of the tube-nosed insectivorous bat (*Murina florium*) in Australia. Proceedings of the Linnean Society of New South Wales, 121:175–191.

Cline, S. P., A. B. Berg, and H. M. Wight. 1980. Snag characteristics and dynamics in Douglas-fir forests, western Oregon. Journal of Wildlife Management, 44:773–786.

Clutton-Brock, T. H. 1989. Mammalian mating systems. Proceedings of the Royal Society of London B, 236:339–372.

Constantine, D. G. 1958. Ecological observations on lasiurine bats in Georgia. Journal of Mammalogy, 39:64–70.

Constantine, D. G. 1959. Ecological observations on lasiurine bats in the North Bay area of California. Journal of Mammalogy, 40:13–15.

Constantine, D. G. 1966. Ecological observations of lasiurine bats in Iowa. Journal of Mammalogy, 47:34–41.

Constantine, D. G. 1998. Range extensions of ten species of bats in California. Bulletin of the Southern California Academy of Sciences, 97:49–75.

Corbet, G. B., and J. E. Hill. 1992. Mammals of the Indo-Malayan Region: A Systematic Review. British Museum Publications, Oxford University Press, London.

Cox, P. A. 1983. Observations on the natural history of Samoan bats. Mammalia, 47:519–523.

Cox, P. A., T. Elmqvist, E. D. Pierson, and W. E. Rainey. 1991. Flying foxes as strong interactors in South Pacific island ecosystems: a conservation hypothesis. Conservation Biology, 5:448–454.

Craig, P., P. W. Trail, and T. E. Morrell. 1994. The decline of fruit bats in American Samoa due to hurricanes and overhunting. Biological Conservation, 69:261–266.

Crampton, L. H., and R. M. R. Barclay. 1998. Selection of roosting and foraging habitat by bats in different-aged aspen mixedwood stands. Conservation Biology, 12:1347–1358.

Crome, F. H. J., and G. C. Richards. 1988. Bats and gaps: microchiropteran community structure in a Queensland rainforest. Ecology, 69:1960–1969.

Csada, R. 1996. Cardioderma cor. Mammalian Species, 519:1–4.

Dalquest, W. W., and D. W. Walton. 1970. Diurnal retreats of bats. Pp. 162–187 in: About Bats (B. H. Slaughter and D. W. Walton, eds.). Southern Methodist University Press, Dallas.

Daniel, M. J. 1990. Order Chiroptera. Pp. 114–137 in: The Handbook of New Zealand Mammals (C. M. King, ed.). Oxford University Press, Auckland.

Daniel, M. J., and G. R. Williams. 1984. A survey of the distribution, seasonal activity and roost sites of New Zealand bats. New Zealand Journal of Ecology, 7:9–25.

Davis, D. D. 1962. Mammals of the lowland rainforest of North Borneo. Bulletin of the National Museum, Singapore, 31:1–129.

Davison, G. W. H. 1992. Pteropus vampyrus malaccensis. Pp. 142–143 in: Old World Fruit Bats: An Action Plan for the Family Pteropodidae (S. P. Mickleburgh, A. M. Hutson, and P. A. Racey, eds.). International Union for Conservation of Nature and Natural Resources Survival Commission, Gland, Switzerland.

Dobson, G. E. 1877. Protective mimicry among bats. Nature, 15:354.

Eby, P. 1991. Seasonal movements of grey-headed flying foxes, Pteropus poliocephalus (Chiroptera: Pteropodidae), from two maternity camps in northern New South Wales. Wildlife Research, 18:547–559.

Eby, P. 1996. Interactions between the grey-headed flying fox, Pteropus poliocephalus (Chiroptera: Pteropodidae) and its diet plants—seasonal movements and seed dispersal. Ph.D. Thesis, University of New England, Armidale, New South Wales.

Eby, P., G. Richards, L. Collins, and K. Parry-Jones. 1999. The distribution, abundance and vulnerability to population reduction of the grey-headed flying fox Pteropus poliocephalus in New South Wales. Australian Zoologist, 31:240–253.

Elangovan, V., G. Marimuthu, and T. H. Kunz. 1999. Temporal patterns of individual and group foraging behaviour in the short-nosed fruit bat, Cynopterus sphinx, in south India. Journal of Tropical Ecology, 15:681–687.

Elangovan, V., G. Marimuthu, and T. H. Kunz. 2000. Nectar feeding behavior of the short-nosed fruit bat, *Cynopterus sphinx* (Pteropodidae). Acta Chiropterologica, 2:1–5.

Emlen, S. T., and L. W. Oring. 1977. Ecology, sexual selection, and the evolution of mating systems. Science (Washington, D.C.), 197:215–223.

Emmons, L. H. 1990. Neotropical Rainforest Mammals: A Field Guide. University of Chicago Press, Chicago.

Entwistle, A. C., P. A. Racey, and J. R. Speakman. 1997. Roost selection by the brown long-eared bat (*Plecotus auritus*). Journal of Applied Ecology, 34:399–408.

Fenton, M. B. 1983. Roosts used by the African insectivorous bat, *Scotophilus leucogaster* (Chiroptera: Vespertilionidae). Biotropica, 15:129–132.

Fenton, M. B. 1992a. Bats. Facts on File, New York.

Fenton, M. B. 1992b. Pelage patterns and crypsis in roosting bats: *Taphozous mauritianus* and *Epomophorus* species. Koedoe, 35:49–55.

Fenton, M. B. 1997. Science and the conservation of bats. Journal of Mammalogy, 78:1–14.

Fenton, M. B, D. Audet, D. C. Dunning, J. Long, C. B. Merriam, D. Pearl, D. M. Syne, B. Adkins, S. Pedersen, and T. Wohlgenant. 1993. Activity patterns and roost selection by *Noctilio albiventris* (Chiroptera: Noctilionidae) in Costa Rica. Journal of Mammalogy, 74:607–613.

Fenton, M. B., R. M. Brigham, A. M. Mills, and I. L. Rautenbach. 1985. The roosting and foraging areas of *Epomophorus wahlbergi* (Pteropodidae) and *Scotophilus viridis* (Vespertilionidae) in Kruger National Park, South Africa. Journal of Mammalogy, 66:461–468.

Fenton, M. B., and I. L. Rautenbach. 1986. A comparison of the roosting and foraging behaviour of three species of African insectivorous bats (Rhinolophidae, Vespertilionidae and Molossidae). Canadian Journal of Zoology, 64:2860–2867.

Fenton, M. B., and I. L. Rautenbach. 1998. Impacts of ignorance and human and elephant populations on the conservation of bats in African woodlands. Pp. 261–270 *in:* Bat Biology and Conservation (T. H. Kunz and P. A. Racey, eds.). Smithsonian Institution Press, Washington, D.C.

Fenton, M. B., I. L. Rautenbach, S. E. Smith, C. M. Swanepoel, J. Grosell, and J. van Jaarsveld. 1994. Raptors and bats: threats and opportunities. Animal Behaviour, 48:9–18.

Fenton, M. B., C. M. Swanepoel, R. M. Brigham, J. Cebek, and M. B. C. Hickey. 1990. Foraging behavior and prey selection by large slit-faced bats (*Nycteris grandis;* Chiroptera: Nycteridae). Biotropica, 22:2–8.

Findley, J. S., and D. E. Wilson. 1974. Observations on Neotropical disk-winged bats, *Thyroptera tricolor* Spix. Journal of Mammalogy, 55:562–571.

Finney, M. A. 1991. Ecological effects of prescribed and simulated fire on the coast redwood (*Sequoia sempervirens*). Ph.D. Dissertation. University of California, Berkeley.

Flannery, T. 1995. Mammals of the South-West Pacific and Moluccan Islands. Reed Books, Chatswood, New South Wales.

Foster, M. S. 1992. Tent roosts of Macconnelli's bat (*Vampyressa macconnelli*). Biotropica, 24:447–454.

Foster, M. S., and R. M. Timm. 1976. Tent-making by *Artibeus jamaicensis* (Chiroptera: Phyllostomatidae) with comments on plants used by bats for tents. Biotropica, 8:65–269.

Foster, R. W., and A. Kurta. 1999. Roosting ecology of the northern bat (*Myotis septentrionalis*) and comparisons with the endangered Indiana bat (*Myotis sodalis*). Journal of Mammalogy, 80:659–672.

Francis, C. M. 1990. Trophic structure of bat communities in the understorey of lowland dipterocarp rain forest in Malaysia. Journal of Tropical Ecology, 6:421–431.

Fuhrmann, M., and A. Seitz. 1992. Nocturnal activity of the brown long-eared bat (*Plecotus auritus* L., 1758): data from radio-tracking in the Lenneberg forest near Mainz (Germany). Pp. 538–548 *in*: Wildlife Telemetry: Remote Monitoring and Tracking of Animals (I. G. Priede and S. M. Swift, eds.). Ellis Horwood, New York.

Gaisler, J., V. Hanak, and J. Dungel. 1979. A contribution to the population ecology of *Nyctalus noctula* (Mammalia: Chiroptera). Prirodovedne Prace Ustavu Ceskoslovenske Akademie Ved Brne, 13:1–38.

Gannon, M. R., and M. R. Willig. 1994. The effects of Hurricane Hugo on bats of the Luquillo Experimental Forest of Puerto Rico. Biotropica, 26:320–331.

Gardner, A. L. 1977. Feeding habits. Pp. 393–350 *in*: Biology of Bats of the New World Family Phyllostomatidae (R. J. Baker, J. K. Jones, Jr., and D. C. Carter, eds.). Pt. 2. Special Publications, the Museum, Texas Tech University, 13. Texas Tech Press, Lubbock.

Garnett, S., O. Whybird, and H. Spencer. 1999. The conservation status of the spectacled flying fox *Pteropus conspicillatus* in Australia. Australian Zoologist, 31:38–50.

Gellman, S. T., and W. J. Zielinski. 1996. Use by bats of old-growth redwood hollows on the north coast of California. Journal of Mammalogy, 77:255–265.

Genoud, M., and F. J. Bonaccorso. 1986. Temperature regulation, rate of metabolism, and roost temperature in the greater white-lined bat *Saccopteryx bilineata* (Emballonuridae). Physiological Zoology, 59:49–54.

Gibbons, P. 1994. Sustaining key old-growth characteristics in native forests used for wood production: retention of trees with hollows. Pp. 59–84 *in*: Ecology and Sustainability of Southern Temperate Ecosystems (T. W. Norton and S. R. Dovers, eds.). Commonwealth Scientific and Industrial Research Organization, East Melbourne.

Goodwin, G. G., and A. M. Greenhall. 1961. A review of bats of Trinidad and Tobago. Bulletin of the American Museum of Natural History, 122:187–302.

Goodwin, R. E. 1979. The bats of Timor: systematics and ecology. Bulletin of the American Museum of Natural History, 163:73–122.

Grant, G. S., and S. A. Banack. 1999. Harem structure and reproductive behavior of *Pteropus tonganus* in American Samoa. Australian Mammalogy, 21:111–120.

Grindal, S. D. 1999. Habitat use by bats, *Myotis* spp., in western Newfoundland. Canadian Field-Naturalist, 113:258–263.

Hamilton, I. M., and R. M. R. Barclay. 1994. Patterns of daily torpor and day-roost selection by male and female big brown bats (*Eptesicus fuscus*). Canadian Journal of Zoology, 72:744–749.

Hamilton, W. J., III, and K. E. F. Watt. 1970. Refuging. Annual Review of Ecology and Systematics, 1:263–286.

Handley, C. O., Jr. 1966. Checklist of the mammals of Panama. Pp. 753–795 *in*: Ectoparasites of Panama (R. L. Wenzel and V. J. Tipton, eds.). Field Museum of Natural History, Chicago.

Handley, C. O., Jr. 1976. Mammals of the Smithsonian Venzuelan Project. Brigham Young University Science Bulletin, Biological Series, 20:1–91.

Happold, D. C. D., and M. Happold. 1990. The domiciles, reproduction, social organisation and sex ratios of the banana bat *Pipistrellus nanus* (Chiroptera, Vespertilionidae) in Malawi. Zeitschrift für Säugetierkunde, 55:145–160.

Happold, D. C. D., and M. Happold. 1996. The social organization and population dynamics of leaf-roosting banana bats, *Pipistrellus nanus* (Chiroptera, Vespertilionidae), in Malawi, east-central Africa. Mammalia, 60:517–544.

Heaney, L. R. 1986. Biogeography of the mammals of Southeast Asia: estimates of colonization, extinction, and speciation. Biological Journal of the Linnean Society,
28:127–165.

Heideman, P. D., and L. R. Heaney. 1992. *Pteropus vampyrus lanensis*. Pp. 141–142 *in:* Old
World Fruit Bats: An Action Plan for the Family Pteropodidae (S. P. Mickleburgh,
A. M. Hutson, and P. A. Racey, eds.). International Union for Conservation of Nature
and Natural Resources Survival Commission, Gland, Switzerland.

Heithaus, E. R., and T. H. Fleming. 1978. Foraging movements of a frugivorous bat, *Carollia perspicillata* (Phyllostomatidae). Ecological Monographs, 48:127–143.

Herr, A. 1998. Aspects of the ecology of insectivorous forest-dwelling bats (Microchiroptera) in the western slopes of the Australian alps. Ph.D. Thesis. Charles Sturt University, Albury, Australia.

Herr, A., and N. I. Klomp. 1999. Preliminary investigation of roosting habitat preferences of the large forest bat *Vespadelus darlingtoni* (Chiroptera, Vespertilionidae).
Pacific Conservation Biology, 5:208–213.

Hodgkison, R. 2001. The ecology of fruit bats (Chiroptera: Pteropodidae) in a Malaysian lowland dipterocarp forest, with particular reference to the spotted-winged
fruit bat (*Balionycteris maculata,* Thomas). Ph.D. Dissertation. University of Aberdeen,
Aberdeen.

Hodgkison, R., S. T. Balding, A. Zubaid, and T. H. Kunz. In press. Roosting ecology and
social organization of the spotted-winged fruit bat, *Balionycteris maculata,* Thomas
(Chiroptera: Pteropodidae) in a Malaysian lowland dipterocarp forest. Journal of
Tropical Ecology.

Hosken, D. J. 1996. Roost selection by the lesser long-eared bat, *Nyctophilus geoffroyi,* and
the greater long-eared bat, *N. major* (Chiroptera: Vespertilionidae) in *Banksia* woodlands. Journal of the Royal Society of Western Australia, 79:211–216.

Howe, H. F. 1984. Implications of seed dispersal by animals for tropical reserve management. Biological Conservation, 30:261–281.

Humphrey, S. R. 1975. Nursery roosts and community diversity of Nearctic bats. Journal of Mammalogy, 56:321–346.

Humphrey, S. R., A. R. Richter, and J. B. Cope. 1977. Summer habitat and ecology of the
endangered Indiana bat, *Myotis sodalis*. Journal of Mammalogy, 58:334–346.

Hutchinson, J. T., and M. J. Lacki. 2000. Selection of day roosts by red bats in mixed mesophytic forests. Journal of Wildlife Management, 64:87–94.

Hutson, A. M., S. P. Mickleburgh, and P. A. Racey. 2001. Microchiropteran Bats: Global
Status Survey and Conservation Action Plan. IUCN/SSC Chiroptera Specialist
Group. International Union for Conservation of Nature and Natural Resources,
Gland, Switzerland.

Ingles, L. G. 1953. Observations on Barro Colorado Island mammals. Journal of Mammalogy, 34:266–268.

Jaberg, C., C. Leuthold, and J.-D. Blant. 1998. Foraging habitats and feeding strategy of the parti-coloured bat *Vespertilio murinus* L., 1758 in western Switzerland. Myotis, 36:51–61.

Jackson, J. A., B. J. Schardien, C. D. Cooley, and B. E. Rowe. 1982. Cave *Myotis* roosting in barn swallow nests. Southwestern Naturalist, 27:463–464.

Janzen, D. H. 1976. Why tropical trees have rotten cores. Biotropica, 8:110.

Jones, C. 1972. Comparative ecology of three pteropid bats in Rio Muni, West Africa. Journal of Zoology (London), 167:353–370.

Jones, G., P. L. Duverge, and R. D. Ransome. 1995. Conservation biology of an endangered species: field studies of greater horseshoe bats. Symposia of the Zoological Society London, no. 67:309–324.

Kalcounis, M. C., and R. M. Brigham. 1998. Secondary use of aspen cavities by tree-roosting big brown bats. Journal of Wildlife Management, 62:603–611.

Kalcounis, M. C., and K. R. Hecker. 1996. Intraspecific variation in roost-site selection by little brown bats (*Myotis lucifugus*). Pp. 81–90 *in*: Bats and Forests Symposium, October 19–21, 1995 Victoria, British Columbia, Canada (R. M. R. Barclay and R. M. Brigham, eds.). British Colombia Ministry of Forests, Victoria.

Kalko, E. K. V., D. Friemel, C. O. Handley, Jr., and H.-U. Schnitzler. 1999. Roosting and foraging behavior of two Neotropical gleaning bats, *Tonatia silvicola* and *Trachops cirrhosus* (Phyllostomidae). Biotropica, 31:344–353.

Kaufman, L. 1988. The role of developmental crises in the formation of buttresses: a unified hypothesis. Evolutionary Trends in Plants, 2:39–51.

Kerth, G., K. Weissmann, and B. Konig. 2000. Day-roost selection in female Bechstein's bats (*Myotis bechsteinii*): a field experiment to determine the influence of roost temperature. Oecologia (Berlin), 126:1–9.

Kingdon, J. 1974. East African Mammals. Vol. 2. Pt. A. Insectivores and Bats. Academic Press, London.

Kitchener, D. J., Boeadi, I. Charlton, and Maharadatunkamsi. 1990. Wild mammals of Lombok Island: Nusa Tenggara, Indonesia: systematics and natural history. Records of the Western Australian Museum, Supplement no. 33:1–129.

Kock, D., and D. Kovac. 2000. *Eudiscopus denticulus* (Osgood 1932) in Thailand with notes on its roost (Chiroptera: Vespertilionidae). Zeitschrift für Säugetierkunde, 65:1121–1123.

Koepcke, J. 1984. "Blattzelte" als Schlafplätze der Fledermaus *Ectophylla macconnelli* (Thomas, 1901) (Phyllostomidae) im tropischen Regenwald von Peru. Säugetierkundliche Mitteilungen, 31:123–126.

Kofron, C. P. 1994. Bamboo-roosting of the thick-thumbed pipistrelle bat (*Glischropus tylopus*) in Borneo. Mammalia, 58:306–309.

Kronwitter, F. 1988. Population structure, habitat use, and activity patterns of the noctule bat, *Nyctalus noctula* Schreib. 1774 (Chiroptera: Vespertilionidae) revealed by radio-tracking. Myotis, 26:23–85.

Krull, D., A. Schumm, W. Metzner, and G. Neuweiler. 1991. Foraging areas and foraging behavior in the notch-eared bat, *Myotis emarginatus* (Vespertilionidae). Behavioral Ecology and Sociobiology, 28:247–253.

Kunz, T. H. 1982. Roosting ecology of bats. Pp. 1–55 *in*: Ecology of Bats (T. H. Kunz, ed.). Plenum Press, New York.

Kunz, T. H. 1996. Obligate and opportunistic interactions of Old-World tropical bats and plants. Pp. 37–65 *in*: Conservation and Faunal Biodiversity in Malaysia

(Z. A. A. Hasan and Z. Akbar, eds.). Penerbit Universiti Kebangsaan Malaysia, Bangi.

Kunz, T. H., P. V. August, and C. D. Burnett. 1983. Harem social organization in cave-roosting *Artibeus jamaicensis* (Chiroptera: Phyllostomidae). Biotropica, 15:133–138.

Kunz, T. H., M. S. Fujita, A. P. Brooke, and G. F. McCracken. 1994. Convergence in tent architecture and tent-making behavior among Neotropical and Paleotropical bats. Journal of Mammalian Evolution, 2:57–78.

Kunz, T. H., and G. F. McCracken. 1996. Tents and harems: apparent defence of foliage roosts by tent-making bats. Journal of Tropical Ecology, 12:121–137.

Kunz, T. H., and E. D. Pierson. 1994. Bats of the world—an introduction. *In:* Bats of the World (R. W. Nowak, ed.). Johns Hopkins University Press, Baltimore.

Kunz, T. H., and P. A. Racey, eds. 1998. Bat Biology and Conservation. Smithsonian Institution Press, Washington, D.C.

Kunz, T. H., and D. S. Reynolds. In press. Bat colonies in buildings. *In:* Monitoring Trends in Bat Populations of the U.S. and Territories: Problems and Prospects (T. J. O'Shea and M. A. Bogan, eds.). U.S. Geological Survey, Biological Resources Division, Information and Technology Report, Washington, D.C.

Kurta, A. 1985. External insulation available to a non-nesting mammal, the little brown bat (*Myotis lucifugus*). Comparative Biochemistry and Physiology A, Comparative Physiology, 8:413–420.

Kurta, A. 1995. Bark roost of a male big brown bat (*Eptesicus fuscus*). Bat Research News, 35:63.

Kurta, A., G. P. Bell, K. A. Nagy, and T. H. Kunz. 1989. Water balance of free-ranging little brown bats (*Myotis lucifugus*) during pregnancy and lactation. Canadian Journal of Zoology, 67:2468–2472.

Kurta, A., and R. Foster. 1995. The brown creeper (Aves: Certhiidae): a competitor of bark-roosting bats? Bat Research News, 36:6–7.

Kurta, A., D. King, J. A. Teramino, J. M. Stribley, and K. J. Williams. 1993. Summer roosts of the endangered Indiana bat (*Myotis sodalis*) on the northern edge of its range. American Midland Naturalist, 129:132–138.

Kurta, A., T. H. Kunz, and K. A. Nagy. 1990. Energetics and water flux of free-ranging big brown bats (*Eptesicus fuscus*) during pregnancy and lactation. Journal of Mammalogy, 71:59–65.

Kurta, A., and G. C. Lehr. 1995. Lasiurus ega. Mammalian Species, 515:1–7.

Kurta, A., K. J. Williams, and R. Mies. 1996. Ecological, behavioral, and thermal observations of a peripheral population of Indiana bats (*Myotis sodalis*). Pp. 102–117 *in:* Bats and Forests Symposium, October 19–21, 1995, Victoria, British Columbia, Canada (R. M. R. Barclay and R. M. Brigham, eds.). British Columbia Ministry of Forests, Victoria.

LaVal, R. K. 1977. Notes on some Costa Rican bats. Brenesia, 10/11:77–83.

LaVal, R. K., and M. L. LaVal. 1977. Reproduction and behavior of the African banana bat, *Pipistrellus nanus*. Journal of Mammalogy, 58:403–410.

Law, B. S. 1993. Roosting and foraging ecology of the Queensland blossom bat, *Syconycteris australis*, in north-eastern New South Wales: flexibility in response to seasonal variation. Wildlife Research, 20:419–431.

Law, B. S. 1996. The ecology of bats in south-east Australian forests and potential impacts of forestry practices: a review. Pacific Conservation Biology, 2:363–374.

Law, B. S., and J. Anderson. 2000. Roost preferences and foraging ranges of the eastern forest bat, *Vespadelus pumilus* under two disturbance histories in northern New South Wales. Austral Ecology, 24:352–367.

Lekagul, B., and J. A. McNeely. 1977. Mammals of Thailand. Association for the Conservation of Wildlife, Bangkok.

Lemke, T. O. 1992. State of the Marianas fruit bat (*Pteropus mariannus*) in the northern Mariana islands north of Saipan. Pp. 68–93 *in:* Pacific Island Flying Foxes: Proceedings of an International Conservation Conference (D. E. Wilson and G. L. Graham, eds.). U.S. Fish and Wildlife Service, Biological Report 90 (23), Washington, D.C.

Lewis, J. C. 1998. Creating snags and wildlife trees in commercial forest landscapes. Western Journal of Applied Forestry, 13:97–101.

Lewis, S. E. 1992. Behavior of Peter's tent making bat, *Uroderma bilobatum*, at maternity roosts in Costa Rica. Journal of Mammalogy, 73:541–546.

Lewis, S. E. 1995. Roost fidelity of bats: a review. Journal of Mammalogy, 76:481–496.

Lim, B. K. 1993. Cladistic reappraisal of Neotropical stenodermatine bat phylogeny. Cladistics, 9:147–165.

Lim, L. B. 1966. Abundance and distribution of Malaysian bats in different ecological habitats. Federation Museum Journal, 9:60–73.

Lindenmayer, D. B., R. B. Cunningham, C. F. Donnelly, M. T. Tanton, and H. A. Nix. 1993. The abundance and development of cavities in *Eucalyptus* trees: a case study in the montane forests of Victoria, southeastern Australia. Forest Ecology and Management, 60:77–104.

Lloyd, B., and S. McQueen. 1997. Roosting behaviour of *Mystacina tuberculata* in central North Island, New Zealand. Australasian Bat Society Newsletter, 7:57.

Loughland, R. A. 1998. Mangal roost selection by the flying-fox *Pteropus alecto* (Megachiroptera: Pteropodidae). Marine and Freshwater Research, 49:351–352.

Lumsden, L. F., and A. F. Bennett. 1995. Bats of a semiarid environment in south-eastern Australia: biogeography, ecology and conservation. Wildlife Research, 22:217–240.

Lumsden, L. F., and A. F. Bennett. 2000. Bats in rural landscapes: a significant but largely unknown faunal component. Pp. 42–50 *in:* Balancing Conservation and Production in Grassy Landscapes: Proceedings of the Bushcare Grassy Landscapes Conference, Clare, South Australia, 19–21 August 1999 (T. Barlow and R. Thorburn, eds.). Environment Australia, Canberra.

Lumsden, L. F., A. F. Bennett, and J. E. Silins. 2002a. Location of roosts of the lesser long-eared bat *Nyctophilus geoffroyi* and Gould's wattled bat *Chalinolobus gouldii* in a fragmented landscape in south-eastern Australia. Biological Conservation, 106:237–249.

Lumsden, L. F., A. F. Bennett, and J. E. Silins. 2002b. Selection of roost sites by the lesser long-eared bat (*Nyctophilus geoffroyi*) and Gould's wattled bat (*Chalinolobus gouldii*) in south-eastern Australia. Journal of Zoology (London), 257:207–218.

Lunney, D., J. Barker, T. Leary, D. Priddel, R. Wheeler, P. O'Connor, and B. Law. 1995. Roost selection by the north Queensland long-eared bat *Nyctophilus bifax* in littoral rainforest in the Iluka World Heritage Area, New South Wales. Australian Journal of Ecology, 20:532–537.

Lunney, D., J. Barker, D. Priddel, and M. O'Connell. 1988. Roost selection by Gould's long-eared bat, *Nyctophilus gouldi* Tomes (Chiroptera: Vespertilionidae), in logged forests on the south coast of New South Wales. Australian Wildlife Research, 15:375–384.

Lunney, D., and C. Moon. 1997. Flying foxes and their camps in the remnant rainforests of northeast New South Wales. Pp. 247–277 *in:* Australia's Ever-Changing Forests. III. Proceedings of the Third National Conference on Australian Forest History (J. Darvel, ed.). Centre for Resource and Environmental Studies, Canberra.

Lyon, M. W. 1911. Mammals collected by Dr. W. L. Abbott on Borneo. Proceedings of the United States National Museum, 40:53–146.

Mackowski, C. M. 1984. The ontogeny of hollows in Blackbutt (*Eucalyptus pilularis*) and its relevance to the management of forests for possums, gliders and timber. Pp. 553–567 *in:* Possums and Gliders (A. P. Smith and I. D. Hume, eds.). Surrey Beatty & Sons, Chipping Norton, Sydney.

Maeda, K. 1974. Eco-ethologie de la grande noctule, *Nyctalus lasiopterus,* a Sappora, Japon. Mammalia, 38:461–487.

Mager, K. J., and T. A. Nelson. 2001. Roost-site selection by eastern red bats (*Lasiurus borealis*). American Midland Naturalist, 145:120–126.

Manning, R. W., J. K. J. Jones, Jr., R. R. Hollander, and C. Jones. 1987. Notes on distribution and natural history of some bats on the Edward Plateau and in adjacent areas of Texas. Texas Journal of Science, 39:279–285.

Marinho-Filho, J., and I. Sazima. 1998. Brazilian bats and conservation biology. Pp. 282–294 *in:* Bat Biology and Conservation (T. H. Kunz and P. A. Racey, eds.). Smithsonian Institution Press, Washington, D.C.

Marshall, A. G. 1985. Old World phytophagous bats (Megachiroptera) and their food supply: a survey. Zoological Journal of the Linnean Society, 29:115–135.

Marshall, A. G., and A. N. McWilliam. 1982. Ecological observations on epomophorine fruit-bats (Megachiroptera) in West Africa savanna woodland. Journal of Zoology (London), 198:53–67.

Mason, C. F., R. E. Stebbings, and G. P. Winn. 1972. Noctules (*Nyctalus noctula*) and starlings (*Sturnus vulgaris*) competing for roosting holes. Journal of Zoology (London), 166:467.

Mattson, T. A., S. W. Buskirk, and N. L. Stanton. 1996. Roost sites of the silver-haired bat (*Lasionycteris noctivagans*) in the Black Hills, South Dakota. Great Basin Naturalist, 56:247–253.

Mawson, P. R., and J. L. Long. 1994. Size and age parameters of nest trees used by four species of parrot and one species of cockatoo in south-west Australia. Emu, 94:149–155.

Mayle, B. A. 1990. A biological basis for bat conservation in British woodlands—a review. Mammal Review, 20:159–195.

McCarthy, T. G. A., A. L. Gardner, and C. O. Handley, Jr., 1992. Tonatia carrikeri. Mammalian Species, 407:1–4.

McClure, H. E. 1942. Summer activities of bats (genus *Lasiurus*) in Iowa. Journal of Mammalogy, 23:430–434.

McCracken, G. F., and G. S. Wilkinson. 2000. Bat mating systems. Pp. 321–362 *in:* The Reproductive Biology of Bats (E. G. Crichton and P. H. Krutzsch, eds.). Academic Press, New York.

Medway, Lord. 1983. The Wild Mammals of Malaya. 2d ed. Oxford University Press, Kuala Lumpur.

Medway, Lord, and A. G. Marshall. 1970. Roost-site selection among flat-headed bats (*Tylonycteris* spp.). Journal of Zoology (London), 161:237–245.

Medway, Lord, and A. G. Marshall. 1972. Roosting associations of flat-headed bats, *Ty-*

lonycteris species (Chiroptera: Vespertilionidae) in Malaysia. Journal of Zoology (London), 168:463–482.

Menzel, M. A., T. C. Carter, B. R. Chapman, and J. Laerm. 1998. Quantitative comparison of tree roosts used by red bats (*Lasiurus borealis*) and Seminole bats (*L. seminolus*). Canadian Journal of Zoology, 76:630–634.

Menzel, M. A., T. C. Carter, W. M. Ford, and B. R. Chapman. 2001. Tree-roost characteristics of subadult and female adult evening bats (*Nycticeius humeralis*) in upper coastal plain of South Carolina. American Midland Naturalist, 145:112–119.

Mickleburgh, S., P. A. Racey, and A. M. Hutson, eds. 1992. Old World Fruit Bat Action Plan. International Union for Conservation of Nature and Natural Resources, Gland, Switzerland.

Mills, L. S., M. E. Soulé, and D. F. Doak. 1993. The keystone-species concept in ecology and conservation. BioScience, 43:219–224.

Morrison, D. W. 1978a. Foraging ecology and energetics of the frugivorous bat *Artibeus jamaicensis.* Ecology, 59:716–723.

Morrison, D. W. 1978b. Influence of habitat on the foraging distances of the fruit bat, *Artibeus jamaicensis.* Journal of Mammalogy, 59:622–624.

Morrison, D. W. 1979. Apparent male defense of tree hollows in the fruit bat, *Artibeus jamaicensis.* Journal of Mammalogy, 60:11–15.

Morrison, D. W. 1980. Foraging and day-roosting dynamics of canopy fruit bats in Panama. Journal of Mammalogy, 61:20–29.

Morrison, D. W., and C. O. Handley, Jr. 1991. Roosting behavior. Pp. 131–136 *in:* Demography and Natural History of the Common Fruit Bat, *Artibeus jamaicensis,* on Barro Colorado Island, Panama (C. O. Handley, Jr., D. E. Wilson, and A. L. Gardner, eds.). Smithsonian Contributions in Zoology, no. 511, Smithsonian Institution Press, Washington, D.C.

Morrison, D. W., and S. H. Morrison. 1981. Economics of harem maintenance by a Neotropical bat. Ecology, 62:864–866.

Mudar, K. M., and M. S. Allen. 1986. A list of bats from northeastern Luzon, Philippines. Mammalia, 50:219–225.

Myers, N. 1986. Tropical deforestation and a megaextinction spasm. Pp. 394–409 *in:* Conservation Biology: The Science of Scarcity and Diversity (M. E. Soulé, ed.). Sinauer Associates, Sunderland, Mass.

Nelson, J. E. 1965. Behaviour of Australian Pteropodidae (Megachiroptera). Animal Behaviour, 8:544–557.

Neuweiler, G. 1969. Vehaltensbeobachtungen an einer indischen Flughundkolonie (*Pteropus g. giganteus* Brünn). Zeitschrift für Tierpsychologie, 26:166–199.

Neuweiler, G., W. Metzner, U. Heilmann, R. Rubsamen, M. Eckrich, and H. H. Costa. 1987. Foraging behavior and echolocation in the rufous horseshoe bat (*Rhinolophus rouxi*) of Sri Lanka. Behavioral Ecology and Sociobiology, 20:53–67.

Nicolai, V. 1986. The bark of trees: thermal properties, microclimate and fauna. Oecologia (Berlin), 69:148–160.

Novick, A. 1977. Acoustic orientation. Pp. 73–287 *in:* Biology of Bats. Vol. 3 (W. A. Wimsatt, ed.). Academic Press, New York.

Nowak, R. M. 1994. Walker's Bats of the World. Johns Hopkins University Press, Baltimore.

Ochoa, H., and T. H. Kunz. 1999. Behavioral thermoregulation in the island flying fox *Pteropus hypomelanus.* Journal of Thermal Biology, 24:15–20.

O'Donnell, C. F. J. 2000. Cyptic local populations in a temperate rainforest bat *Chalinolobus tuberculatus* in New Zealand. Animal Conservation, 3:287–297.

O'Donnell, C. F. J. 2001. Home range and use of space by *Chalinolobus tuberculatus*, a temperate rainforest bat from New Zealand. Journal of Zoology (London), 253: 253–264.

O'Donnell, C. F. J., J. Christie, C. Corben, J. A. Sedgeley, and W. Simpson. 1999. Rediscovery of short-tailed bats (*Mystacina* sp.) in Fiordland, New Zealand: preliminary observations of taxonomy, echolocation calls, population size, home range, and habitat use. New Zealand Journal of Ecology, 23:21–30.

O'Donnell, C. F. J., and J. A. Sedgeley. 1999. Use of roosts by the long-tailed bat, *Chalinolobus tuberculatus*, in temperate rainforest in New Zealand. Journal of Mammalogy, 80:913–923.

Okon, E. E. 1974. Fruit bats at Ife: their roosting and food preferences. Nigerian Field, 39:33–40.

Ormsbee, P. C. 1996. Characteristics, use, and distribution of day roosts selected by female *Myotis volans* (long-legged myotis) in forested habitat of the Central Oregon Cascades. Pp. 124–131 *in:* Bats and Forests Symposium, October 19–21, 1995 Victoria, British Columbia, Canada (R. M. R. Barclay and R. M. Brigham, eds.). British Colombia Ministry of Forests, Victoria.

Ormsbee, P. C., and W. C. McComb. 1998. Selection of day roosts by female long-legged myotis in the central Oregon Cascade Range. Journal of Wildlife Management, 62:596–603.

Ortega, J., and H. T. Arita. 2000. Defense of females by dominant males of *Artibeus jamaicensis* (Chiroptera: Phyllostomidae). Ethology, 106:395–407.

O'Shea, T. J. 1980. Roosting, social organization and annual cycle in a Kenya population of the bat *Pipistrellus nanus*. Zeitschrift für Tierphychologie, 53:171–195.

Owen, R. D. 1987. Phylogenetic analysis of the bat subfamily Stenodermatinae (Mammalia: Chiroptera). Special Publications, the Museum, Texas Tech University, 26:1–65.

Owen, R. D. 1988. Phenetic analysis of the bat subfamily Stenodermatinae (Chiroptera: Phyllostomidae). Journal of Mammalogy, 69:795–810.

Palmer, C., O. Price, and C. Bach. 2000. Foraging ecology of the black flying fox (*Pteropus alecto*) in the seasonal tropics of the Northern Territory, Australia. Wildlife Research, 27:169–178.

Palmer, C., and J. C. Z. Woinarski. 1999. Seasonal roosts and foraging movements of the black flying fox (*Pteropus alecto*) in the Northern Territory: resource tracking in a landscape mosaic. Wildlife Research, 26:823–838.

Parry-Jones, K. A., and M. L. Augee. 1991. Food selection by grey-headed flying foxes (*Pteropus poliocephalus*) occupying a summer colony site near Gosford, New South Wales. Wildlife Research, 18:111–124.

Parry-Jones, K. A., and M. L. Augee. 2001. Factors affecting the occupation of a colony site in Sydney NSW by the grey-headed flying-fox, *Pteropus poliocephalus* (Pteropodidae). Austral Ecology, 26:47–55.

Pavey, C. R. 1998. Colony sizes, roost use and foraging ecology of *Hipposideros diadema reginae*, a rare bat from tropical Australia. Pacific Conservation Biology, 4:232–239.

Pavey, C. R., and C. J. Burwell. 2000. Foraging ecology of three species of hipposiderid bats in tropical rainforest in north-east Australia. Wildlife Research, 27:283–287.

Payne, J., C. M. Francis, and K. Phillips. 1985. A Field Guide to the Mammals of Borneo. Sabah Society, Kota Kinabalu, Sabah, Malaysia.

Perkins, J. M. 1996. Does competition for roosts influence bat distribution in a managed forest? Pp. 164–172 *in:* Bats and Forests Symposium, October 19–21, 1995 Victoria, British Columbia, Canada (R. M. R. Barclay and R. M. Brigham, eds.). British Colombia Ministry of Forests, Victoria.

Perry, S. 1994. Begging and transfer of coati meat by white-faced capuchin monkeys, *Cebus capucinus*. Primates, 35:409–415.

Phillips, W. W. 1924. A Guide to the Mammals of Ceylon. Pt. 1. Ceylon Journal of Science, 13:1–63.

Pierson, E. D. 1998. Tall trees, deep holes, and scarred landscapes: conservation biology of North American bats. Pp. 309–325 *in:* Bat Biology and Conservation (T. H. Kunz and P. A. Racey, eds.). Smithsonian Institution Press, Washington, D.C.

Pierson, E. D., T. Elmqvist, W. E. Rainey, and P. A. Cox. 1996. Effects of tropical cyclonic storms on flying fox populations on the South Pacific islands of Samoa. Conservation Biology, 10:438–451.

Pierson, E. D., and W. E. Rainey. 1992. The biology of flying foxes of the genus *Pteropus:* a review. Pp. 1–17 *in:* Pacific Island Flying Foxes: Proceedings of an International Conservation Conference (D. E. Wilson and G. Graham, eds.). U.S. Fish and Wildlife Service, Biological Report, 90 (23), Washington, D.C.

Pitts, R. M., and J. J. Scharninghausen. 1986. Use of cliff swallow and barn swallow nests by the cave bat, *Myotis velifer,* and the free-tailed bat, *Tadarida brasiliensis.* Texas Journal of Science, 38:265–266.

Rabe, M. J., T. E. Morrell, H. Green, J. deVos, and C. R. Miller. 1998. Characteristics of ponderosa pine snag roosts used by reproductive bats in northern Arizona. Journal of Wildlife Management, 62:612–621.

Racey, P. A. 1982. Ecology of bat reproduction. Pp. 57–104 *in:* Ecology of Bats (T. H. Kunz, ed.). Plenum Press, New York.

Racey, P. A. 1998. Ecology of European bats in relation to their conservation. Pp. 249–260 *in:* Bat Biology and Conservation (T. H. Kunz and P. A. Racey, eds.). Smithsonian Institution Press, Washington, D.C.

Rainey, W. E. 1998. Conservation of bats on remote Indo-Pacific Islands. Pp. 326–341 *in:* Bat Biology and Conservation (T. H. Kunz and P. A. Racey, eds.). Smithsonian Institution Press, Washington, D.C.

Rainey, W. E., E. D. Pierson, T. Elmlqvist, and P. A. Cox. 1995. The role of pteropodids in oceanic island ecosystems of the Pacific. Symposia of the Zoological Society of London, no. 67:47–62.

Raphael, M. G., and M. L. Morrison. 1987. Decay and dynamics of snags in the Sierra Nevada, California. Forest Science, 33:774–783.

Ratcliffe, F. 1932. Notes on the fruit bats (*Pteropus* spp.) of Australia. Journal of Animal Ecology, 1:32–57.

Rawlinson, P. A., R. A. Zann, S. van Balen, and I. W. B. Thornton. 1992. Colonization of the Krakatau islands by vertebrates. GeoJournal, 28:225–231.

Reason, P. F., and W. J. Trewhella. 1994. The status of *Pteropus livingstonii* in the Comores. Oryx, 28:107–114.

Reid, F. A. 1997. A Field Guide to the Mammals of Central America and Southeast Mexico. Oxford University Press, New York.

Richards, G. C. 1990a. The spectacled flying-fox, *Pteropus conspicillatus* (Chiroptera: Pteropodidae) in north Queensland. 1. Roost sites and distribution patterns. Australian Mammalogy, 13:17–24.

Richards, G. C. 1990b. The spectacled flying-fox, *Pteropus conspicillatus*, (Chiroptera: Pteropodidae) in north Queensland. 2. Diet, seed dispersal and feeding ecology. Australian Mammalogy, 13:25–31.

Richards, G. C. 1995. A review of ecological interactions of fruit bats in Australian ecosystems. Symposia of the Zoological Society of London, no. 67:76–96.

Richards, P. W. 1996. The Tropical Rain Forest. 2d ed. Cambridge University Press, Cambridge.

Richmond, J. Q., S. A. Banack, and G. S. Grant. 1998. Comparative analysis of wing morphology, flight behaviour, and habitat use in flying foxes (Genus: *Pteropus*). Australian Journal of Zoology, 46:283–289.

Rickart, E. A., P. D. Heideman, and R. C. B. Utzurrum. 1989. Tent-roosting by *Scotophilus kuhlii* (Chiroptera: Vespertilionidae) in the Philippines. Journal of Tropical Ecology, 5:433–436.

Rieger, V. I. 1996. Wie nutzen wasserfledermause, *Myotis daubentonii* (Kuhl, 1817), ihre tagesquartiere? (How do Daubenton's bats, *Myotis daubentonii* (Kuhl, 1817), use their day roosts?). Zeitschrift für Säugetierkunde, 61:202–214.

Ritzi, C. M., C. W. Walker, and R. L. Honeycutt. 1998. Utilization of cave swallow nests by the cave myotis, *Myotis velifer*, in central Texas. Texas Journal of Science, 50:175–176.

Robertson, P. B., 1992. Small islands, natural catastrophes, and rapidly disappearing forests: a high vulnerability recipe for island populations of flying foxes. Pp. 41–45 *in:* Pacific Island Flying Foxes: Proceedings of an International Conservation Conference (D. E. Wilson and G. L. Graham, eds.). U.S. Fish and Wildlife Service, Biological Report 90 (23), Washington, D.C., 176 pp.

Robinson, M. F., and R. E. Stebbings. 1997. Home range and habitat use by the serotine bat, *Eptesicus serotinus*, in England. Journal of Zoology (London), 243:117–136.

Rodriguez-Duran, A., and T. H. Kunz. 2001. Biogeography of West Indian bats: an ecological perspective. Pp. 353–366 *in:* Biogeography of the West Indies (C. A. Woods, ed.). CRC Press, Boca Rotan, Fla.

Rosevear, D. R. 1965. The Bats of West Africa. British Museum (Natural History), London.

Roverud, R. C., and M. A. Chappell. 1991. Energetic and thermoregulatory aspects of clustering behavior in the Neotropical bat *Noctilio albiventris*. Physiological Zoology, 64:1527–1541.

Ruby, J., P. T. Nathan, J. Balasingh, and T. H. Kunz. 2000. Chemical composition of leaves and fruits eaten by short-nosed fruit bat, *Cynopterus sphinx* (Megachiroptera). Journal of Chemical Ecology, 26:2825–2841.

Sasse, D. B., and P. J. Pekins. 1996. Summer roosting ecology of northern long-eared bats (*Myotis septentrionalis*) in the White Mountain National Park. Pp. 91–101 *in:* Bats and Forests Symposium, October 19–21, 1995 Victoria, British Columbia, Canada (R. M. R. Barclay and R. M. Brigham, eds.). British Colombia Ministry of Forests, Victoria.

Saugey, D. A., B. G. Crump, R. L. Vaughn, and G. A. Heidt. 1998. Notes on the natural history of *Lasiurus borealis* in Arkansas. Journal of the Arkansas Academy of Science, 52:92–98.

Schliemann, H., and M. Hoeber. 1978. The structure and function of the pads on the thumb and foot of *Tylonycteris*. Pp. 39–50 *in:* Proceedings of the Fourth International Bat Research Conference (R. J. Olembo, J. B. Castelino, and F. A. Mutere, eds.). Kenya National Academy for Advancement of Arts and Sciences, Kenya Literature Bureau, Nairobi.

Schliemann, H., and B. Mags. 1978. Myzopoda aurita. Mammalian Species, 116:1–2.

Schober, W., and E. Grimmberger. 1989. A Guide to the Bats of Britain and Europe. Hamlyn, London.

Schulz, M. 1986. Vertebrate prey of the ghost bat, *Macroderma gigas,* at Pine Creek, Northern Territory. Macroderma, 2:59–62.

Schulz, M. 1995. Utilisation of suspended bird nests by the golden-tipped bat (*Kerivoula papuensis*) in Australia. Mammalia, 59:280–283.

Schulz, M. 1997. Bats in bird nests in Australia: a review. Mammal Review, 27:69–76.

Schulz, M. 1998. Bats and other fauna in disused fairy martin *Hirundo ariel* nests. Emu, 98:184–191.

Schulz, M. 1999. Leaf wrapping behavior in the flute-nosed bat *Murina florium.* Bat Research News, 40:6–8.

Schulz, M. 2000. Roosts used by the golden-tipped bat, *Kerivoula papuensis* (Chiroptera: Vespertilionidae). Journal of Zoology (London), 250:467–478.

Schulz, M., and D. Hannah. 1998. Relative abundance, diet and roost selection of the tube-nosed insect bat, *Murina florium,* on the Atherton Tablelands, Australia. Wildlife Research, 25:261–271.

Sedgeley, J. A. 2001. Quality of cavity microclimate as a factor influencing selection of maternity roosts by a tree-dwelling bat, *Chalinolobus tuberculatus,* in New Zealand. Journal of Applied Ecology, 38:425–438.

Sedgeley, J. A., and C. F. J. O'Donnell. 1999a. Factors influencing roost cavity selection by a temperate rainforest bat (*Chalinolobus tuberculatus,* Vespertilionidae) in New Zealand. Journal of Zoology (London), 249:437–446.

Sedgeley, J. A., and C. F. J. O'Donnell. 1999b. Roost selection by the long-tailed bat, *Chalinolobus tuberculatus,* in temperate New Zealand rainforest and its implications for the conservation of bats in managed forests. Biological Conservation, 88:261–276.

Sharma, S. K. 1986. Painted bats and nests of Baya weaver bird. Journal of the Bombay Natural History Society, 81:196.

Shiel, C. B., R. E. Shiel, and J. S. Fairley. 1999. Seasonal changes in the foraging behaviour of Leisler's bats (*Nyctalus leisleri*) in Ireland as revealed by radio-telemetry. Journal of Zoology (London), 249:347–358.

Shump, K. A., and A. U. Shump. 1982. Lasiurus cinereus. Mammalian Species, 185:1–5.

Simmons, N. B., and R. S. Voss. 1998. The mammals of Paracou, French Guiana: a Neotropical lowland rain forest fauna. Pt. 1. Bats. Bulletin of the American Museum of Natural History, no. 237, 219 pp.

Skinner, J. D., and R. H. N. Smithers. 1990. The Mammals of the Southern African Subregion. University of Pretoria, Pretoria.

Sluiter, J. W., A. M. Voute, and P. F. van Heerdt. 1973. Hibernation of *Nyctalus noctula.* Periodicum Biologorum, 75:181–188.

Souza, L. L., S. Ferrari, and A. L. Pina. 1997. Feeding behavior and predation of a bat by *Saimiri sciureus* in a semi-natural Amazonian environment. Folia Primatologica, 68:194–198.

Speakman, J. R. 1991. The impact of predation by birds on bat populations in the British Isles. Mammal Review, 21:123–142.

Spencer, H. J., and T. H. Fleming. 1989. Roosting and foraging behaviour of the Queensland tube-nosed bat, *Nyctimene robinsoni* (Pteropodidae). Preliminary radio-tracking observations. Australian Wildlife Research, 16:413–420.

Spencer, H. J., C. Palmer, and K. Parry-Jones. 1991. Movements of fruit bats in eastern Australia, determined by using radiotracking. Wildlife Research, 18:463–468.

Start, A. N. 1998. Do rainbow lorikeets evict bats? Western Australian Naturalist, 22:123–124.

Start, A. N., and A. G. Marshall. 1976. Nectarivorous bats as pollinators of trees in West Malaysia. Pp. 141–150 *in:* Tropical Trees: Variation, Breeding, and Conservation (J. Burley and B. T. Styles, ed.). Academic Press, London.

Stebbings, R. E., and J. T. Walsh. 1985. Bat Boxes. Fauna and Flora Preservation Society, London, 15 pp.

Stoner, K. 2000. Leaf selection by the tent-making bat *Artibeus watsoni* in *Asterogyne martiana* palms in southwestern Costa Rica. Journal of Tropical Ecology, 16:151–157.

Storz, J. F., J. Balasingh, H. R. Bhat, P. T. Nathan, D. P. S. Doss, A. A. Prakash, and T. H. Kunz. 2001. Clinal variation in body size and sexual dimorphism in an Indian fruit bat, *Cynopoterus sphinx* (Chiroptera: Pteropodidae). Biological Journal of the Linnean Society, 72:17–31.

Storz, J. F., J. Balasingh, P. T. Nathan, K. Emmanuel, and T. H. Kunz. 2000a. Dispersion and site fidelity in a tent-roosting population of the short-nosed fruit bat (*Cynopterus sphinx*). Journal of Tropical Ecology, 16:1–18.

Storz, J. F., H. R. Bhat, and T. H. Kunz. 2000b. Social structure of a polygynous tent-making bat, *Cynopterus sphinx* (Megachiroptera). Journal of Zoology (London), 251:151–165.

Storz, J. F., and T. H. Kunz. 2000. Cynopterus sphinx. Mammalian Species, 613:1–8.

Swift, S. M. 1998. Long-eared Bats. T & A. D. Poyser, Ltd., London.

Szewczak, J. M., S. M. Szewczak, M. L. Morrison, and L. S . Hall. 1998. Bats of the White and Inyo Mountains of California-Nevada. Great Basin Naturalist, 58:66–75.

Tan, K. H., A. Zubaid, and T. H. Kunz. 1997. Tent construction and social organization in *Cynopterus brachyotis* (Muller) (Chiroptera: Pteropodidae) in peninsular Malaysia. Journal of Natural History, 31:1605–1621.

Tan, K. H., A. Zubaid, and T. H. Kunz. 1998. Food habits of *Cynopterus brachyotis* (Muller) (Chiroptera: Pteropodidae) in peninsular Malaysia. Journal of Tropical Ecology, 14:299–307.

Tan, K. H., A. K. Zubaid, and T. H. Kunz. 1999. Roost selection and social organisation in *Cynopterus horsfieldi* (Chiroptera: Pteropodidae). Malayan Nature Journal, 53:295–298.

Taylor, R. J., and N. M. Savva. 1988. Use of roost sites by four species of bats in state forest in south-eastern Tasmania. Australian Wildlife Research, 15:637–645.

Thomas, D. W. 1982. The ecology of an African savanna fruit bat community: resource partitioning and role in seed dispersal. Ph.D. Thesis. University of Aberdeen, Aberdeen.

Thomas, D. W. 1983. The annual migrations of three species of West African fruit bats (Chiroptera: Pteropodidae). Canadian Journal of Zoology, 61:2266–2272.

Thomas, D. W., and M. B. Fenton. 1978. Notes on the dry season roosting and foraging behaviour of *Epomophorus gambianus* and *Rousettus aegyptiacus* (Chiroptera: Pteropodidae). Journal of Zoology (London), 186:403–406.

Tidemann, C. R., and S. C. Flavel. 1987. Factors affecting choice of diurnal roost sites by tree-hole bats (Microchiroptera) in south-eastern Australia. Australian Wildlife Research, 14:459–473.

Tidemann, C. R., D. M. Priddel, J. E. Nelson, and J. D. Pettigrew. 1985. Foraging behaviour of the Australian ghost bat, *Macroderma gigas* (Microchiroptera: Megadermatidae). Australian Journal of Zoology, 33:705–713.

Tidemann, C. R., M. J. Vardon, R. A. Loughland, and P. J. Brocklehurst. 1999. Dry season camps of flying-foxes (*Pteropus* spp.) in Kakadu World Heritage Area, North Australia. Journal of Zoology (London), 247:155–163.

Timm, R. M. 1994. Tent construction by *vampyressa* in Costa Rica. Journal of Mammalogy, 65:166–167.

Timm, R. M. 1987. Tent construction by the bats of the genera *Artibeus* and *Uroderma*. Fieldiana, Zoology (Special issue: Studies in Neotropical mammalogy: essays in honor of Philip Hershkovitz, ed. B. D. Patterson and R. M. Timm), n.s., no. 39: 187–212.

Timm, R. M., and B. L. Clauson. 1990. A roof over their feet. Natural History, 3:55–58.

Timm, R. M., and S. E. Lewis. 1991. Tent construction and use by *Uroderma bilobatum* in coconut palms (*Cocos nucifera*) in Costa Rica. Bulletin of the American Museum of Natural History (Special issue: Contributions to mammalogy in honor of Karl F. Koopman, ed. T. A. Griffiths and D. Klingener), 206:251–260.

Timm, R. M, and J. Mortimer. 1976. Selection of roost sites by Honduran white bats, *Ectophylla alba* (Chiroptera: Phyllostomatidae). Ecology, 57:385–389.

Tuttle, M. D. 1970. Distribution and zoogeography of Peruvian bats, with comments on natural history. University of Kansas Science Bulletin, 49:45–86.

Tuttle, M. D. 1976. Collecting techniques. Pp. 71–88 *in:* Biology of Bats of the New World Family Phyllostomatidae (R. J. Baker, J. K. Jones, Jr., and D. C. Carter, eds.). Pt. 1. Special Publications, the Museum, Texas Tech University, 13. Texas Tech Press, Lubbock.

Tuttle, M. D., and D. Hensley. 1993. Bat houses: the secrets of success. Bats, 11:3–14.

Utzurrum, R. C. B. 1995. Feeding ecology of Philippine fruit bats: patterns of resource use and seed dispersal. Symposia of the Zoological Society of London, no. 67:63–78.

Utzurrum, R. C. B. 1998. Geographic patterns, ecological gradients, and the maintenance of tropical fruit bat diversity. The Philippine model. Pp. 342–353 *in:* Bat Biology and Conservation (T. H. Kunz and P. A. Racey, eds.). Smithsonian Institution Press, Washington, D.C.

van Heerdt, P. F., and J. W. Sluiter. 1965. Notes on the distribution and behavior of the noctule bat (*Nyctalus noctula*) in the Netherlands. Mammalia, 29:463–477.

Vardon, M. J., and C. R. Tidemann. 1999. Flying-foxes (*Pteropus alecto* and *P. scapulatus*) in the Darwin region, north Australia: patterns in camp size and structure. Australian Journal of Zoology, 47:411–423.

Vaughan, T. A. 1976. Nocturnal behavior of the African false vampire bat (*Cardioderma cor*). Journal of Mammalogy, 57:227–248.

Vaughan, T. A., and R. P. Vaughan. 1986. Seasonality and the behavior of the African yellow-winged bat. Journal of Mammalogy, 67:91–102.

Vehrencamp, S. L., F. G. Stiles, and J. W. Bradbury. 1977. Observations on the foraging behavior and avian prey of the Neotropical carnivorous bat, *Vampyrum spectrum*. Journal of Mammalogy, 58:469–478.

Verschuren, J. 1957. Ecologie, biologie, et systématique des Chiroptères. Exploration du Parc National de la Garamba. no. 7 (Mission H. de Saeger). Institut des Parcs Nationaux du Congo Belge, Brussels.

Verschuren, J. 1966. Introduction à la écologie et a la biologie des Chiroptères. Pp. 25–65 *in:* Exploration du Parc National Albert, no. 2 (Mission F. Bouliere et J. Verschuren, 1957–1961). Fasc. 2. Institut des Parcs Nationaux du Congo Belge, Brussels.

Voigt, C. C., and O. von Helversen. 1999. Storage and display of odour by male *Saccopteryx bilineata* (Chiroptera, Emballonuridae). Behavioral Ecology and Sociobiology, 47:29–40.

Voigt, C. C., O. von Helversen, R. Michener, and T. H. Kunz. 2001. The economics of harem maintenance in the sac-winged bat, *Saccopteryx bilineata* (Emballonuridae). Behavioral Ecology and Sociobiology, 50:31–36.

Vonhof, M. J. 1996. Roost-site preferences of big brown bats (*Eptesicus fuscus*) and silver-haired bats (*Lasionycteris noctivagans*) in the Pend d'Oreille Valley in southern British Columbia. Pp. 62–80 *in:* Bats and Forests Symposium, October 19–21, 1995 Victoria, British Columbia, Canada (R. M. R. Barclay and R. M. Brigham, eds.). British Colombia Ministry of Forests, Victoria.

Vonhof, M. J., and R. M. R. Barclay. 1996. Roost-site selection and roosting ecology of forest-dwelling bats in southern British Columbia. Canadian Journal of Zoology, 74:1797–1805.

Vonhof, M. J., and R. M. R. Barclay. 1997. Use of tree stumps as roosts by the western long-eared bat. Journal of Wildlife Management, 61:674–684.

Waldien, D. L., J. P. Hayes, and E. B. Arnett. 2000. Day-roosts of female long-eared *Myotis* in western Oregon. Journal of Wildlife Management, 64:785–796.

Walsberg, G. E. 1986. Thermal consequences of roost-site selection: the relative importance of three modes of heat conservation. Auk, 103:1–7.

Webster, W. D., J. K. Jones, and R. J. Baker. 1980. Lasiurus intermedius. Mammalian Species, 132:1–3.

Wetterer, A. L., M. V. Rockman, and N. B. Simmons. 2000. Phylogeny of phyllostomid bats (Mammalia: Chiroptera): data from diverse morphological systems, sex chromosomes, and restriction sites. Bulletin of the American Museum of Natural History, 248:1–200.

Whewell, G. D. 1992. Flying foxes in the Solomon Islands. Pp. 102–104 *in:* Pacific Island Flying Foxes: Proceedings of an International Conservation Conference (D. E. Wilson and G. L. Graham, eds.). U.S. Fish and Wildlife Service, Biological Report 90 (23), Washington, D.C., 176 pp.

Whitaker, R. J., and S. J. Jones. 1994. The role of frugivorous bats and birds in the rebuilding of a tropical forest ecosystem, Krakatau, Indonesia. Journal of Biogeography, 21:245–258.

Whitmore, T. C. 1998. An Introduction to Tropical Rain Forests. 2d ed. Oxford University Press, Oxford.

Wickler, W., and U. Seibt. 1976. Field studies of the African fruit bat, *Epomophorus wahlbergi*, with special reference to male calling. Zeitschrift für Tierpsychologie, 40:345–376.

Wickler, W., and D. Uhrig. 1969. Verhalten und Ökologische Nische der Gelflügelfled-ermaus, *Lavia frons* (Geoffroy) (Chiroptera, Megadermatidae). Zeitschrift für Tier-psychologie, 26:726–736.

Wiles, G. J. 1987. Current research and future management of Mariannus fruit bats (Chi-roptera: Pteropodidae) on Guam. Australian Mammalogy, 10:93–95.

Wilkinson, G. S. 1985. The social organization of the common vampire bat. I. Pattern and cause of association. Behavioral Ecology and Sociobiology, 17:111–121.

Wilson, D. E., 1978. Thyroptera discifera. Mammalian Species, 104:1–3.

Wilson, D. E., and J. Engbring. 1992. The flying foxes *Pteropus samoensis* and *Pteropus tonganus:* status in Fiji and Samoa. Pp. 74–101 *in:* Pacific Island Flying Foxes: Pro-ceedings of an International Conservation Conference (D. E. Wilson and G. L. Gra-ham, eds.). U.S. Fish and Wildlife Service, Biological Report 90 (23), Washington, D.C.

Wilson, D. E., and J. S. Findley. 1977. Thyroptera tricolor. Mammalian Species, 71:1–3.

Wilson, D. E., and G. L. Graham, eds. 1992. Pacific Island Flying Foxes: Proceedings of an International Conservation Conference. U.S. Fish and Wildlife Service, Biological Report 90 (23), Washington, D.C.

Wimsatt, W. A., and B. Villa-R. 1970. Locomotor adaptations in the disc-winged bat, *Thyroptera tricolor.* American Journal of Anatomy, 129:89–119.

Winkelmann, J. R., F. J. Bonaccorso, and T. L. Strickler. 2000. Home range of the south-ern blossom bat, *Syconycteris australis,* in Papau New Guinea. Journal of Mammal-ogy, 81:408–414.

Wiriosoepartha, A. S., A. S. Mukhtar, and M. Bismark. 1986. Habitat and population study of flying foxes *Pteropus vampyrus* in relation with coastal birds conservation in Pulau Rambut Nature Reserve. Buletin Penelitan Hutan, 479:17–27 (in Malay, En-glish summary).

Wunder, L., and A. B. Carey. 1996. Use of the forest canopy by bats. Northwest Science, 70:79–85.

Zielinski, W. J., and S. T. Gellman. 1999. Bat use of remnant old-growth redwood stands. Conservation Biology, 13:160–167.

Zortéa, M. 1995. Observations on tent-using in the carolline bat *Rhinophylla pumilio* in southeastern Brazil. Chiroptera Neotropical, 1:2–4.

Zortéa, M., and F. A. De Brito. 2000. Tents used by *Vampyressa pusilla* (Phyllostomidae) in southeastern Brazil. Journal of Tropical Ecology, 16:475–480.

Zubaid, A. 1993. A comparison of the bat fauna between primary and fragmented sec-ondary forest in peninsular Malaysia. Mammalia, 57:202–206.

CHAPTER 2

Sensory Ecology and Communication in the Chiroptera

John D. Altringham and M. Brock Fenton

Introduction

Bats, like other animals, collect, communicate, and interpret information using a wide range of sensory cues, including hearing, vision, olfaction, and touch (Smith 1977). This diversity of channels for communication offers the potential for a mixture of signals that vary in range of effective operation as well as in endurance. Thus, for example, while some olfactory signals are effective over short range and persist for short periods of time, others are more enduring and detectable over greater distances. The same blend of features characterizes some of the acoustic and visual signals used by bats. As we shall see, multi-modal communication is a striking feature of bats, combining at least visual, acoustic, and olfactory signals.

The propensity of bats to roost collectively with many conspecifics, as well as their often enormous ears, together give the impression that bats are highly gregarious, if not social animals, depending heavily on sound for orientation and communication. Impressions about the role of sound are supported by the literature: witness the number of citations about acoustic orientation (Novick 1977) compared to those for vision, olfaction, and taste (Bloss 1999; Suthers 1970). The differences in publication dates between these reviews does not account for the discrepancy. A computer-based literature search covering the period from 1981 to June 1998 revealed an even larger discrepancy, with approximately 10 references on sound and echolocation for every one on other senses. Echolocation has preoccupied many biologists who work with bats, and yet, as Grinnell (1980) demonstrated, the research opportunities presented by this behavior were not vigorously pursued for over 30 yr after Griffin's (1958) initial work. The work on echolocation ranges from anatomy and neurobiology to behavior.

Are bats, in fact, typically gregarious? The answer is an emphatic "no." Although some species roost with hundreds, thousands, or even more conspecifics, many others live in small groups or are solitary (Altringham 1996; Fenton 2001; Kunz 1982). Furthermore, the day-roosting habits of most species are unknown or poorly understood (e.g., Kunz and Lumsden, this volume;

Nowak 1991). A clear demonstration of this reality is provided by Simmons and Voss (1998), who located the day roosts of only 30 of the 78 Neotropical species they studied. To further complicate the issue, switching between day roosts (e.g., Lewis 1995) means that the number of bats constituting a "colony" may not be reflected by the number of bats in any one roost at any one time. This is particularly obvious in highly social species like *Desmodus rotundus*, where the members of a colony may occupy several roosts on any one day (Wilkinson 1985).Thus, while some bats are gregarious, living in large groups that may regularly use a particular roost, many are solitary and a few live in distinct social groups (Kunz and Lumsden, this volume).

Is hearing the dominant sense in the orientation and communication of bats? Not necessarily. Sound plays a vital role in the lives of microchiropteran bats, all of which appear to be able to echolocate (Altringham 1996; Fenton et al. 1995). But echolocation does not always play a central role in foraging because some echolocating bats use prey-generated sounds to locate, track, and evaluate prey (Fenton et al. 1995). We remain largely ignorant about the role that echolocation plays in the lives of microchiropterans feeding on fruit, nectar, and pollen or blood (Fenton et al. 1995), although recent studies have begun to address this problem (Joermann et al. 1988; Kalko and Condon 1998; Thies et al. 1998). The recent discovery of an acoustic guide in a bat-pollinated flower (Helversen and Helversen 1999) illustrates the potential for further remarkable discoveries. Morphological evidence suggests that the earliest known fossil bats (Eocene) were echolocators (Simmons and Geisler 1998), perhaps indicating that echolocation is a primitive feature of the Microchiroptera (e.g., Fenton et al. 1995). The role of other senses in the lives of bats has received little attention, and, with some notable exceptions, our knowledge about vision and olfaction has not advanced greatly since Suthers's (1970) review, compared to the enormous strides made in studies on echolocation.

The purpose of this chapter is to review the sensory ecology and communication behavior of bats. To begin, we illustrate the range of conditions in which bats operate and the variety of cues affecting their behavior by presenting a brief compilation of observations about the behavior of seven well-studied, gregarious species. Then we review what is known about the auditory, visual, olfactory, tactile, and thermoperceptual abilities of bats before proceeding to consider the role of sound (from echolocation to communication) in behavior and multimodal communication. We finish by identifying some challenges for future work.

Seven Species

Epomophorus wahlbergi

Epomophorus wahlbergi (Family Pteropodidae) is a 60–80-g megachiropteran (fig. 2.1a) that roosts by day in small groups in well-lighted places, usually in

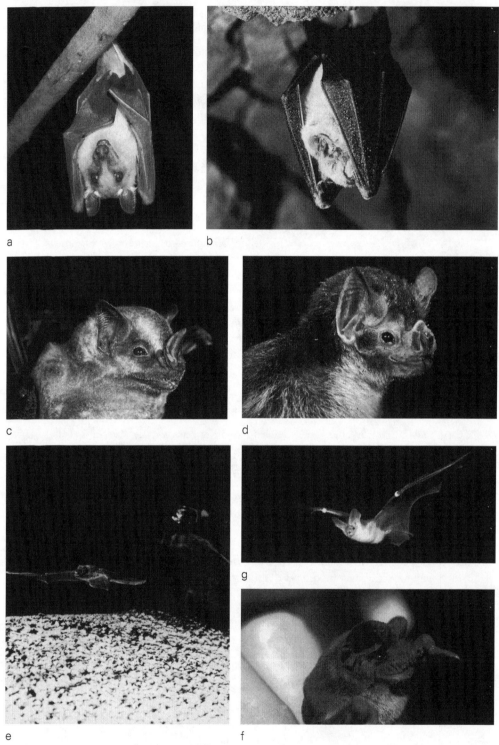

Figure 2.1. Photographs of the seven species of bats reviewed in the text: *Epomophorus wahlbergi* (*a*), *Rhinolophus ferrumequinum* (*b*), *Artibeus jamaicensis* (*c*), *Desmodus rotundus* (*d*), *Pipistrellus pipistrellus* (*e*), *Antrozous pallidus* (*f*), and *Tadarida brasiliensis* (*g*). (Photographs *a, c, d, f,* and *g* by M. B. Fenton; *b* and *e* by Jens Rydell.)

foliage but occasionally in shelter caves (Fenton 1992) or under thatched roofs (Wickler and Siebt 1976). Apart from mothers and their dependent young, *E. wahlbergi* does not roost in physical contact with one another. Individuals often roost in the same place day after day (Fenton et al. 1985), and roosts may be used repeatedly by these groups (Wickler and Siebt 1976). Figs are the mainstay of the diet, and many individuals may visit fruiting trees. The bats orient by vision. They do not echolocate, and other species in the genus depend on olfaction to identify ripe fruit (Thomas 1984). During the mating season at night, males call repeatedly and monotonously from roosts and are visited by females. Although this mate-attraction call is conspicuous and carries well, the sound is not as striking as that of the related lekking species *Hypsignathus monstrosus* (Bradbury 1977a; Wickler and Siebt 1976). Vision appears to be the dominant sense in orientation, while olfaction plays a central role in food selection. Acoustic signals appear to be vital in advertising for mates. For more details about its biology, see Acharya (1992).

Rhinolophus ferrumequinum

Rhinolophus ferrumequinum (Family Rhinolophidae) is a medium-sized (20–30 g) bat (fig. 2.1b) that roosts in caves, mines, cellars and similar structures, and buildings. Roosts are usually close to foraging sites and free from human disturbance, but ambient light levels vary from complete darkness to light shade. In summer, maternity colonies of up to several hundred females, juveniles, and immature *R. ferrumequinum* may form, but they frequently fragment and disperse to different roost sites, often in response to changes in weather and roost temperatures (Ransome 1991). There is no evidence of a hierarchy among bats in summer roosts, and aggression is minimal. Communication between individuals in the tight clusters appears to involve both vocalizations and tactile stimuli. Single males occupy small territories in late summer and then mate with up to eight visiting females. There may be large aggregations of adult and immature bats in hibernacula consisting of a variety of complex age and sex compositions (Ransome 1991). In spring, adult females return to the territories of the males visited in the autumn before forming maternity colonies. *Rhinolophus ferrumequinum* feeds on a wide range of insect prey, detected and tracked using echolocation. Laboratory experiments show that this species can identify many insect species from their echo spectra (von der Emde and Schnitzler 1990), and field studies suggest that it forages selectively (Jones 1990). Echolocation plays a central role in the orientation and foraging behavior of this species, and acoustic signals are involved in mother-young interactions. The roles of vision and olfaction in the lives of this species are not well known.

Artibeus jamaicensis

Artibeus jamaicensis (Family Phyllostomidae) is a medium-sized (30–50 g) microchiropteran (fig. 2.1c) that roosts alone or in small groups. This bat uses a variety of roosts, most commonly in dark caves and hollow trees (Kunz et al. 1983; Morrison 1978; Ortega and Arita 2000), but also in well-lighted tents (Kunz et al. 1994). Individual bats may be faithful to a particular roost or use several. Within a roost, individuals may be in physical contact, particularly groups including a single male with females (Kunz et al. 1983). *Artibeus jamaicensis* feeds heavily on figs and may travel several kilometers from its roost to a tree with ripe figs (Morrison 1978). Although capable of echolocaton, *A. jamaicensis* appears to use olfaction to assess the ripeness of fruit (Kalko et al. 1996). When captured and handled, this bat produces piercing "distress calls," which often attract conspecifics (August 1979). Vision appears to play a dominant role in orientation behavior, and olfaction is used in the selection of food. The role of echolocation remains enigmatic, but other vocalizations are used in different aspects of the lives of this species. For more details about its biology, see Ortega and Castro-Arellano (2001).

Desmodus rotundus

Desmodus rotundus (Family Phyllostomidae) (fig. 2.1*d*) usually roosts in dark areas such as hollow trees and caves. This 30–40-g bat sometimes forms colonies of more than 100 individuals that roost together occasionally, but they often distribute themselves around several alternative roosts (Wilkinson 1985). Individuals in a roost are frequently in physical contact, and those returning to a cluster typically initiate naso-naso contact with members already in the group, thus facilitating the exchange of olfactory information. This species has a well-developed social structure, which includes reciprocal sharing of food (Wilkinson 1984). *Desmodus rotundus* has relatively large eyes for microchiropterans, the ability to echolocate, infrared sensors (Kürten and Schmidt 1982), and well-developed hearing (Schmidt et al. 1991). Individual *D. rotundus* may travel several kilometers one way to find prey, apparently depending on a combination of passive hearing and infrared cues to locate a sleeping host and suitable bite site. Vision probably plays a dominant role in orientation, but this species uses a range of other cues (olfactory, acoustic, and infrared) in foraging. The role of echolocation in this species remains unclear, but olfaction is clearly vital in a range of behavioral situations. For more information about its biology, see Greenhall et al. (1983).

Pipistrellus pipistrellus

Pipistrellus pipistrellus (Family Vespertilionidae) is one of the most common European species; it is small (4–10 g; fig. 2.1*e*) and typically roosts in buildings. Pipistrelles use echolocation to detect, track, and assess airborne prey. *Pipistrellus pipistrellus* has recently been described as two species (*P. pipistrellus* and *P. pygmaeus*) that exist as two phonic types with different echolocation calls corresponding to distinct gene pools (Barratt et al. 1997) that do not appear to interbreed (Park et al. 1996). One phonic type typically forms large, stable maternity roosting groups, the other smaller and more nomadic ones. In both types, females roost in close body contact, often in small cavities, typically in buildings. Males of *P. pipistrellus* use songflights probably to attract females (Lundberg and Gerell 1986). Social calls are used at foraging sites, apparently to advertise feeding territories at low prey densities (Barlow and Jones 1997b), and at maternity colonies (J. Altringham, unpubublished observation) for reasons not yet known. Different colonies appear to have distinct odors, and their members can discriminate between individuals from the groups (de Fanis and Jones 1995a). We found no published data on their visual acuity. Echolocation dominates foraging behavior, and vocalizations are used in a range of other behavioral situations. Olfactory cues are vital for many aspects of its behavior, but the role of vision in orientation in this species remains poorly understood.

Antrozous pallidus

Antrozous pallidus (Family Antrozoidae) is a 15–20-g bat typically occurring in rocky desert regions. *Antrozous pallidus* (fig. 2.1*f*) roosts in caves, crevices, mines, hollow trees, and buildings in groups of 10–200 individuals (Hermanson and O'Shea 1983; Vaughan and O'Shea 1976). It is a social and vocal species (Brown 1976), making a number of different calls in the day roost and in night roosting clusters (O'Shea and Vaughan 1977). There is evidence for social learning in foraging behavior (Gaudet and Fenton 1984). Individuals that roost together alternate occupying the exposed outer positions in clusters, suggesting reciprocal altruism (Trune and Slobodchikoff 1978). Individuals and groups may change roost sites frequently, with lactating females maintaining group cohesion while switching (Lewis 1996). *Antrozous pallidus* gleans much of its food from the ground or from plant surfaces (Vaughan and O'Shea 1976). Prey are often located using prey-generated sounds (Bell 1982), as well as by echolocation. The acoustic medium plays a prominent role in this species, ranging from foraging to communication. Olfactory cues are also important, but the role of vision remains unclear. More details of their biology are presented in Hermanson and O'Shea (1983).

Tadarida brasiliensis

Tadarida brasiliensis (Family Molossidae) is a 10–15-g bat (fig. 2.1*g*) that often lives in colonies numbering from thousands to millions (Wilkins 1989). By day, roosts range from the total darkness of caves to crevices that may admit some light. Within roosts, individuals are typically in physical contact with one another, packing the available roosting space. This bat uses echolocation for detecting, tracking, and evaluating airborne prey, feeding almost exclusively on flying insects. Other vocalizations mediate interactions between females and their nursing young (Balcombe and McCracken 1992). The escape response of this species is visually guided (Mistry 1990), but there are few details about the acuity of its vision. Like many species, *T. brasiliensis* undertakes substantial seasonal migrations in part of its range. We know little about which senses this bat uses for navigation over long distances, reflecting the situation for other migrating species. Acoustic cues play a dominant role in the lives of bats from foraging (echolocation) to mother-young interactions. Olfaction also plays a vital role, but we know little about the use of vision in this bat. For more details of its biology, see Wilkins (1989).

The summaries above illustrate the variety of conditions under which bats roost, the range of social structures and behaviors, and the importance of acoustic, visual, and olfactory channels in communication. It is clear that the sensory world of bats extends well beyond the acoustic. Lack of detailed information on most of the approximately 1,100 species of bats suggests that we have only scratched the surface.

The Senses

Hearing

Sounds that move through air lose energy by spreading loss and by attenuation. While the former applies to any sounds, the extent of attenuation is frequency dependent, particularly affecting sounds >20 kHz (Lawrence and Simmons 1982). Attenuation is of particular concern for echolocating microchiropterans (Lawrence and Simmons 1982), as it can affect the range over which vocalizations are transmitted.

The structure of a bat's auditory system is typically mammalian, and most species hear well over a frequency range of at least 10–50 kHz. In this respect, bats differ little from most other small mammals that have similar audiograms in this range (Neuweiler 1990). However, echolocating bats often have more acute hearing over the frequencies that are most useful to them, whereas non-echolocating species such as *Cynopterus sphinx* are less sensitive and perhaps more closely resemble the typically mammalian auditory condition (see fig. 2.2). For example, the auditory systems of gleaners such as *Plecotus auritus*, *Megaderma lyra* and *Macroderma gigas* are most sensitive between 10 and 20 kHz (Coles et al. 1989; Guppy and Coles 1988; Kulzer et al. 1984; Marimuthu and Neuweiler 1987; Neuweiler et al. 1984). Hearing in these three species is particularly sensitive, more so than in any other studied mammals, and all of them use prey-generated sounds to detect, locate, and assess prey. On the other hand, *M. lyra* also uses echolocation to detect frogs on ponds (Marimuthu et al. 1995). Few studies have measured auditory sensitivity (behaviorally or neuronally) below 10 kHz or, for that matter, the frequencies of sounds emitted by the known prey of bats, but there is evidence that low-frequency hearing and prey-generated sounds are closely matched.

While gleaning, *Antrozous pallidus* may hunt by listening for sounds produced by their prey, responding to the sounds of wing fluttering by moths

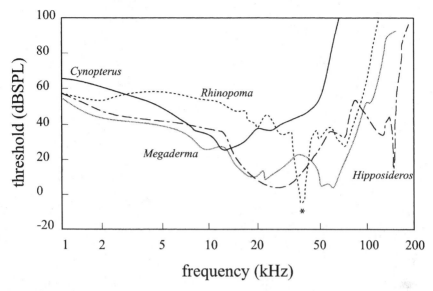

Figure 2.2. Audiograms from a nonecholocating megachiropteran, *Cynopterus sphinx;* the gleaner *Megaderma lyra,* which uses short, broadband, frequency-modulated echolocation calls; and two species with long, narrowband echolocation calls. One is a low duty-cycle echolocator, *Rhinopoma hardwickei,* the other a high duty-cycle echolocator, *Hipposideros speoris.* The asterisk (*) identifies the area of maximum sensitivity, the acoustic fovea of *H. speoris,* which coincides with the 138 kHz that dominates the echolocation calls of this species. (Adapted from Neuweiler et al. 1984.)

(2–14 kHz, with most energy at 5–9 kHz [Bell 1982]) or to the sounds of crickets (Fuzessery et al. 1993). The lower-frequency limit for accurate cricket localization was 3–8 kHz, matching the 5 kHz determined from a physiological survey of the inferior colliculus (Fuzessery et al. 1993). The bats located these sounds with an accuracy of ±1° in the horizontal and vertical planes because of an acute sensitivity to interaural time differences (Fuzessery 1997).

In behavioral experiments, *Trachops cirrhosus* is highly sensitive to sounds <5 kHz, matching the frequencies that dominate the calls of the frogs they typically eat (Ryan et al 1983). In the laboratory, *Cardioderma cor* seems to depend more on prey-generated sounds than on echolocation or vision when searching for frogs (Ryan and Tuttle 1987). Specializations in the inferior colliculus (Schmidt et al. 1991) and of the external ears (Obrist et al. 1993) of *Desmodus rotundus* make this species sensitive to breathing noises at 20–35 kHz, strongly suggesting it also makes use of prey-generated sounds to find a victim.

The evolution of echolocation has lead to numerous adaptive changes in the hearing of bats. The most striking changes occur in high duty cycle bats whose echolocation calls are dominated by one frequency (the constant frequency, or CF, bats), a specialization for detecting movement or flutter by Doppler shift compensation. Bats that employ such a mechanism (members of

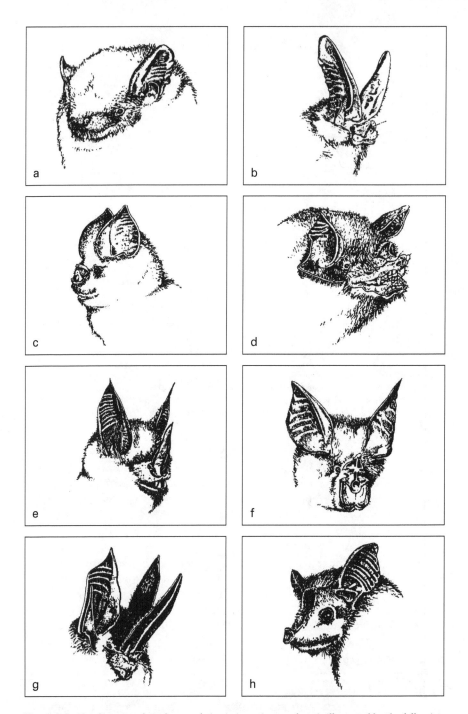

Figure 2.3. The diversity of ear form and size in insectivorous bats is illustrated by the following species: *Chalinolobus variegatus* (*a*), *Corynorhinus townsendii* (*b*) (Vespertilionidae), *Hipposideros caffer* (*c*) (Hipposideridae), *Pteronotus parnellii* (*d*) (Mormoopidae), *Mimon bennettii* (*e*) (Phyllostomidae), *Rhinolophus hildebrandti* (*f*) (Rhinolophidae), *Lonchorhina aurita* (*g*) (Phyllostomidae), and *Rhinopoma hardwickei* (*h*) (Rhinopomatidae). (Drawings by Max Licht from Fenton 1994b.)

the Rhinolophidae and Hipposideridae and the mormoopid *Pteronotus parnellii*) have an "acoustic fovea" tuned to the CF echo frequency (Neuweiler 1990; fig. 2.2). Hearing acuity and directionality are enhanced by the pinnae and matched to the requirements of the bat (Obrist et al. 1993). In general, the pinnae of echolocating species, including echolocating megachiropterans, make hearing more directional than it is in nonecholocating megachiropterans (fig. 2.3). One element of the acoustic fovea is the mechanical tuning of the pinnae to the frequencies that dominate the echolocation calls (Obrist et al. 1993). Similar tuning of the pinnae also occurs in low duty cycle bats that use echolocation calls dominated by narrow-band components, as in the molossids (Obrist et al. 1993). Bats that hunt using prey-generated sounds typically have large pinnae and maximum gain at lower frequencies (Obrist et al. 1993).

Research on *Phyllostomus discolor* demonstrated how hearing sensitivity reflected echolocation and communication behavior (Esser and Daucher 1996). Using pure tones in a forced-choice experiment, Esser and Daucher (1996) characterized the behavioral audiogram of this species. The audiogram had peaks in sensitivity around 85 kHz, but covered the bandwidth of the echolocation call, as well as 10–55 kHz, which is the bandwidth of the communication calls used by this species. Compared to other mammals, *P. discolor* has a marked ability for frequency discrimination, apparently related to its use of distinctive frequency modulated communication calls (Esser and Kiefer 1996).

Vision

There is considerable variation in the size of the eyes of bats (fig. 2.4), with megachiropterans having large, prominent eyes (fig. 2.1*a*), and some microchiropterans rather small and inconspicuous eyes (fig. 2.4). Larger eyes appear to be associated with larger visual centers of the brains. Visual acuity is variable with minimal separable angles of 0.4° in large-eyed species, rising to 6° in those with small eyes (e.g., Manske and Schmidt 1976; Suthers 1966). In *Macrotus californicus,* a gleaner, optomotor responses indicated single-point visual acuity of 3 min 38 s of arc, compared to 15 min of arc for *Eptesicus fuscus* (an aerial feeder) and *Antrozous pallidus* (a gleaner). Both gleaners retained their visual acuity at light levels as low as 2×10^{-4} mL (approximately 2×10^{-3} lux), comparable to ground luminance on a clear, moonless night (Bell and Fenton 1986).

The eyes of bats appear to be adapted for use over distances greater than those over which echolocation is most effective, but vision may not provide the same level of acuity as echolocation (Suthers 1970; Suthers and Wallis 1970). A clear indication of this comes from a comparison of the distance at which *Eptesicus fuscus* first detects a 19-mm diameter sphere: 5 m using echolocation (Kick 1982) and 1 m using vision (15 min of arc acuity [Fenton et al. 1998]). The same bat would first see a tree-sized object at about 300 m, but, using echolocation, would first detect it at <40 m. By comparison, *Macrotus californicus*

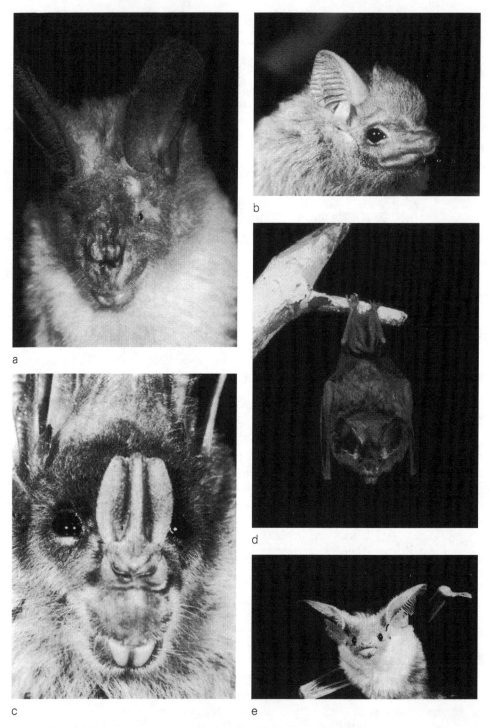

Figure 2.4. Bats show considerable variation in eye size, as illustrated by *Nycteris grandis* (*a*) (Nycteridae), *Balantiopteryx plicata* (*b*) (Emballonuridae), *Megaderma lyra* (*c*) (Megadermatidae), *Micronycteris schmidtorum* (*d*) (Phyllostomidae), and *Otonycteris hemprichii* (*e*) (Vespertilionidae). The *M. schmidtorum* shows the degree of binocularity possible even with a relatively long-snouted species. (Photographs by M. B. Fenton.)

would first detect a 19-mm diameter object at about 18 m (3 min 38 s of arc). There are no comparable data for echolocation but the short, low-intensity, high-frequency calls should translate into a very short (<2 m) detection range.

Vision must be of some importance in foraging to microchiropterans with large and forward-pointing eyes, such as megadermatids (fig. 2.4c), some phyllostomids (fig. 2.4d), and even vespertilionids such as *Otonycteris hemprichii* (fig. 2.4e). Of particular note may be the bat's posture and the position of its eyes (fig. 2.4). As noted by Bell and Fenton (1986), *M. californicus* shows little evidence of binocular overlap when looking down its nose, but when this bat looks straight up, there is potential for binocular cues. The available data are sparse and conflicting. Experiments on the megadermatid *Carioderma cor* (Ryan and Tuttle 1987) and the antrozoid *Antrozous pallidus* (Fuzessery et al. 1993) suggest that vision is not important in prey location. In contrast, *M. californicus* uses vision to locate its prey (Bell 1985; Bell and Fenton 1986). In the open desert habitats of southern California and adjacent Mexico, vision may be of more practical value than it is in densely vegetated habitats with less predictable lighting. Since both *M. californicus* and *A. pallidus* are bats of the desert, habitat does not explain the differences in their use of vision. As we have seen, the two species differ markedly in their visual acuity.

Like most nocturnal mammals, the retinas of bats are dominated by rods, but some species of megachiropterans appear to have cones (Suthers 1970). Megachiropterans have a tapetum lucidum, the layer at the back of the retina that reflects light back through the retinal cells, enhancing visual performance under conditions of low light. The eyes of bats are adapted to operate best at low light levels, with thresholds approaching those of owls (Neuweiler 1967) and exceeding the performance of the human eye at low light levels. In the Megachiroptera and frugivorous phyllostomids, visual and olfactory centers of the brain are large relative to those of other bats, suggesting that vision and olfaction play a significant role in their lives (Barton et al. 1995). Tolerance to light, as determined by electroretinography (Hope and Bhatnagar 1979), is highly variable and probably related to behavior (Fenton 1985). For example, species that roost in foliage are probably more tolerant of bright light than those that roost in dark places such as caves, tree cavities, and crevices.

Chase (1981) showed that *Anoura geoffroyi* preferentially used visual cues to escape from a maze but depended on echolocation in the absence of visual cues. When visual and echolocation cues gave conflicting information, this bat preferentially responded to visual ones. The same bats relied on olfactory and echolocation cues when selecting food (Suthers et al. 1969). In another study, Chase (1983) performed similar experiments on eight microchiropteran species and obtained similar results, namely, visual cues were preferentially used in an escape response. Two species were tested at different times over the 24-h cycle, and Chase found reversals in their use of a light or dark passages for escape. The bats preferred the dark passage when they would normally be settling into their cave roosts, and a lighted passage at times when they would usually be foraging. Mistry (1990) obtained similar results with *Tadarida brasiliensis.*

Vision also plays a vital role in other behaviors, perhaps the most noticeable being the timing of emergence from day roosts (Laidlaw and Fenton 1971; Mueller 1968). Jones and Rydell (1994) suggested that gleaning species emerge later than aerial insectivores because of the greater risks of predation intrinsic to their foraging mode.

Olfaction

The phyllostomid *Carollia perspicillata* can recognize minute concentrations of specific odor components, which can attract it to food when no other cues are available (Laska 1990b). This bat can identify specific fruit odors even after those odors are diluted or chemically masked (Laska 1990a). Hessel and Schmidt (1994) demonstrated the importance of odor to *C. perspicillata* but showed that individual experience and training can greatly influence whether visual, olfactory, or acoustic cues are used under laboratory conditions. These findings emphasize the importance of multimodal cues.

Like many other mammals (e.g., Ewer 1968), many species of bats have distinctive odors (Dapson et al. 1977). For example, *Noctilio leporinus* has a distinctive odor derived from oily secretions produced in the subaxillary region (Brooke and Decker 1996). The chemical composition of the lipids in the secretions varies between sexes and roosts. Males from the same roost are more similar to each other than to other males or to females. Females scent mark themselves with the secretions of other females in their group (Brooke 1997). Adult males of *Sturnira lilium* have a noticeable "spicy" odor associated with their shoulder glands (Scully et al. 2000), and the "sweet" smell of adult males of *Saccopteryx bilineata* emanates from their wing sacs (Voigt and Helversen 1999).

For some species, odors play a vital role in the recognition of pups by their mothers (e.g., Gustin and McCracken 1987) and also can allow colony recognition (Bloss et al. 2002; de Fanis and Jones 1995b) and discrimination between males and females (Bouchard 2001). Strong, species-specific, penetrating odors (Quay 1970; Schmidt 1985), for example, of *Noctilio leporinus* (Brooke and Decker 1996), *Taphozous nudiventris* (Brosset 1962), or *Antrozous pallidus,* may result from some combination of glandular secretion, diet, and bacterial fermentation (Gorman and Towbridge 1989).

Most mammalian scent glands include both sebaceous (flask-shaped) and sudoriferous (tube-shaped) elements that synthesize different odoriferous molecules (Müller-Schwarze 1983; fig. 2.5). Variations in their chemical products can affect both the odor (i.e., sweet, musky, spicy) and color (i.e., white, pink) of scent gland secretions (Albone 1984; Quay 1970; Strauss and Ebling 1970). Bacterial fermentation also can contribute to an individual's odor, and

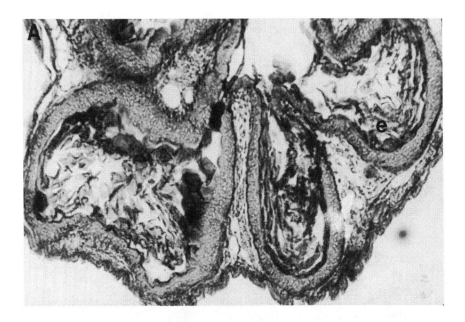

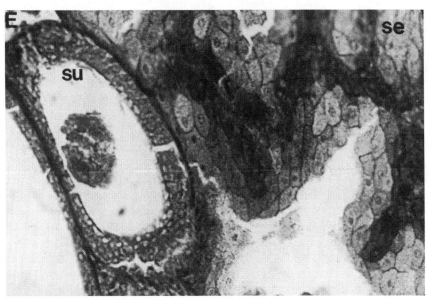

Figure 2.5. Compared here are histological sections of the wing sac of a male *Saccopteryx bilineata* (*a*) (Emballonuridae) and the gular gland of a male *Molossus ater* (*b*) (Molossidae). Here *e* = cornified epidermis, *su* = sudoriferous gland, and *se* = sebaceous gland. Note the absence of sudoriferous and sebaceous glands in the wing sac of *S. bilineata*.

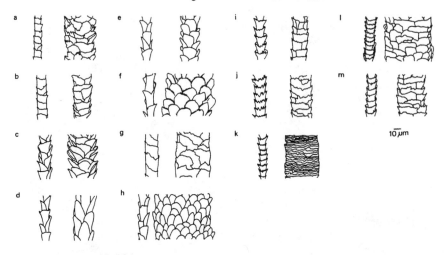

Figure 2.6. This comparison of body hairs (from the mid dorsum, *left*) and scent-dispersing hairs (*right*) illustrates the range of osmetrichia reported from bats. Included are *Aethalops alecto,* male body and ruff (*a*) and female body and ruff (*b*); *Eonycteris spelaea,* male body and ruff (*c*); *Epomphorus anurus,* male body and epaulette (*d*); *Epomops franqueti,* male body and epaulette (*e*); *Myonycterus torquatus,* male body and ruff (*f*); *Ptenochirus jagori,* male body and ruff (*g*); *Rousettus angolensis,* male body and ruff (*h*); *Chaerephon chapini,* male body and crest (*i*); *Mops condylura,* male body and crest (*j*); *Chaerephon major,* male body and crest (*k*); *Chaerephon nigeriae,* male body and crest (*l*); and *Chaerophon pumilus,* male body and crest (*m*). (Illustrations from Hickey and Fenton 1987.)

it is affected by the environment in which the bat roosts and by the gland itself. The combination of roosting in warm, humid locations and having specialized scent glands provides ideal conditions for the proliferation of bacteria, in turn affecting the odor of the secretions.

Many species of bats are sexually dimorphic in odor, with males having a much stronger scent than females (e.g., *N. leporinus* [Brooke and Decker 1996]). Sometimes this coincides with sexual dimorphism in glandular development, for example, in *Tadarida brasiliensis* (Gustin and McCracken 1987; Gutierrez and Agustin 1973), *Molossus ater* (Scully et al. 2000), hipposiderids and Old World emballonurids (Brosset 1962). The females of these species and many others have rudimentary scent glands or lack them (Quay 1970).

The scent glands of mammals may be associated with specialized hairs, "osmetrichia" (fig. 2.6), which tend to be larger in diameter than body hairs and have "pinecone-like" cuticular scale patterns (Hickey and Fenton 1987; Stoddart 1980). Osmetrichia appear to act like paint brushes, assisting in holding glandular secretions and applying them to objects or conspecifics (Flood 1985; Stoddart 1980). The osmetrichia of pteropodid and molossid bats may enhance the communicative effect by enriching both olfactory and visual cues (Bouchard 2001; Hickey and Fenton 1987). Bats such as *Sturnira lilium* (Scully et al. 2000), *Taphozous melanopogon* (Brosset 1962), *Epomops* spp. (Bradbury

1977a), and many molossids (Hickey and Fenton 1987) have conspicuous hairs associated with glandular regions (fig. 2.7).

Another indication of the differential importance of olfactory cues is the variation in the size of the vomeronasal organ (Jacobson's organ) in bats (Bhatnagar 1980). While these structures have not been reported in the Megachiroptera, they are highly variable in the Microchiroptera and most pronounced in the three genera of vampire bats (*Desmodus, Diaemus,* and *Diphylla* [Bhatnagar 1980]).

Touch

Touch has not been studied directly in bats. The tactile systems of bats, however, appear to be typically mammalian, suggesting that touch must be important to them in a variety of roosting situations.

Thermoperception

Desmodus rotundus appears to be the only mammal with heat-sensing organs: small, directional pits around the nose leaf (Kürten and Schmidt 1982). These organs are insulated from the surrounding tissues and maintained 9°C lower than adjacent areas of the face, an essential feature if they are to detect animals with skin temperatures like their own. These heat sensors are sensitive enough to assist *D. rotundus* in choosing an appropriate bite site. Although not yet described, we expect that heat-sensing organs are present in the other species of vampire bats (*Diaemus youngii* and *Diphylla ecaudata*).

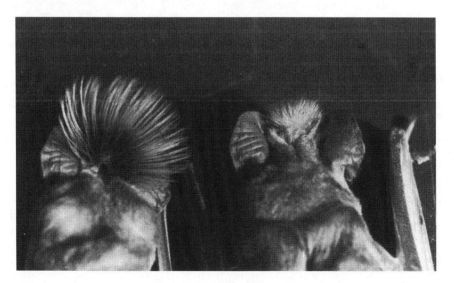

Figure 2.7. Males of *Chaerephon chapini* (*left*) have a conspicuous inter-aural crest that is lacking in females (*right*). The crest hairs of males are osmetrichia (see fig. 2.6, *i*).

Multimodal Perceptions

In an increasing number of situations, it is obvious that bats simultaneously depend on different sensory cues or that their choice of cues will depend on the situation and their prior experience. For example, Thies et al. (1998) demonstrated that two fruit-eating phyllostomids (*Carollia* spp.) initially located their food using olfactory cues from ripe *Piper* fruit, before switching to echolocation for the final approach and removal of the fruit. Odor, followed by shape, were the most important cues in tempting the bats to take various forms of real or artificial fruit. Shape appeared to be determined by echolocation, although the role of vision was not overtly addressed. Changes in call structure as the bats make their final approach are consistent with the need for improved spatial localization. These findings support previous experiments on *Carollia* species (Hessel and Schmidt 1994; Laska 1990a, 1990b; Rieger and Jakob 1988). Another phyllostomid, *Artibeus jamaicensis*, also detects its food primarily by scent (Kalko et al. 1996). In contrast, *Phyllostomus hastatus* uses echolocation to detect the fruits of *Gurania spinulosa*, which hang in the open on leafless branches. *Phyllostomus hastatus*, however, probably uses olfaction to choose between ripe and unripe fruit at close range (Kalko and Condon 1998). Clearly the relative importance and specific uses of odor depend on the task.

Myotis lucifugus uses both echolocation and vision in obstacle avoidance (Bradbury and Nottebohm 1969). *Phyllostomus discolor* relied to a large extent on spatial memory as individuals approached a laboratory "roost," but the emphasis switched to vision and echolocation as they closed in on their target, with vision being dominant (Höller and Schmidt 1996). By manipulating the roost, these studies demonstrated that bats have considerable flexibility over which sensory modality is best for a given task, but learning performance is influenced by visual similarity between real and dummy landing lattices. It is perhaps surprising, therefore, that *Megaderma lyra*, a bat with very large, forward-pointing eyes, could fly blindfolded through an obstacle course, but refused to take flight when its ears were temporarily plugged (Lee et al. 1992).

Vision must be important in navigation although the evidence is equivocal. Davis (1966) reviewed the earlier work and concluded that vision may play a significant role in short-range homing. Layne (1967) provided further evidence supporting this view. The best evidence for the use of vision comes from the work by Williams and colleagues (Williams et al. 1966; Williams and Williams 1967, 1970) who radio tracked *Phyllostomus hastatus* fitted with masks. Bats with no masks, or masks through which they could see, made rapid, direct flights back to their roosts over tens of kilometers. Bats with masks that obscured their vision could still fly rapidly but were more erratic, returning to the roost later than unimpaired bats. When released over 20 km from the roost, masked bats usually failed to return. Little work has been done since, and short- and long-range homing abilities and homing mechanisms remain unresolved. Some bats make long migrations, often exceeding 1,000 km (e.g., Fleming and Eby, this volume; Griffin 1970; Tuttle 1976), but how they navigate over such distances is unknown. *Eptesicus fuscus* will orient itself with respect to an artificial post-sunset glow when released in a planetarium (Buchler and Childs 1982).

Sound: Echolocation to Communication

Just as bats may simultaneously use information collected by olfaction and echolocation or echolocation and vision, signals may serve dual or multiple functions. For example, the signal a bat uses in echolocation also is available for communication. Here we use a general definition of communication (communication occurs when the signal[s] from one animal change the behavior of another), so that interspecific interactions mediated by echolocation calls (bat-bat or bat-insect [see Jones and Rydell, this volume]) qualify as "communication." Key issues in the use of sound in communication can be summarized in three questions: What form do communication sounds take? What messages can be communicated and what functions do they serve? How sophisticated is this communication?

The echolocation calls of any bat have the potential to communicate (intentionally or unintentionally) its identity, location, and activity. Perhaps the most direct demonstration of communication has come from playback experiments. Barclay (1982) used field playbacks of echolocation calls from *Myotis lucifugus* and attracted conspecifics from up to 50 m away. *Myotis lucifugus* was also attracted to playback presentations of the echolocation calls of *Eptesicus fuscus* and vice versa. Subadults of *M. lucifugus* were more responsive to presentations of echolocation calls than were adults. These results (Barclay 1982) suggest that conspecifics at least, and perhaps other species, may eavesdrop on other bats to locate potential foraging sites. Other playback studies have demonstrated that *Lasiurus borealis* responds to presentations of conspecifics, particularly to feeding buzzes (Balcombe and Fenton 1988). Playback experiments have also shown that echolocation calls can be used for spacing individuals while they forage. *Euderma maculatum,* typically a solitary aerial forager, responded to playback calls either by flying away from or by flying toward the loudspeaker and emitting irritation buzzes (Leonard and Fenton 1984).

Changes in echolocation call structure with the presence or absence of conspecifics also suggests a communicative function. *Rhinopoma hardwickei,* when foraging in open habitats, produces narrow-band echolocation calls adjusting the structure and frequencies of the calls according to the presence of conspecifics (Habersetzer 1981). When leaving or entering a roost, however, this bat produces a pure-tone, multiharmonic call lasting up to 100 ms (Haberset-zer 1981), making it of little use in echolocation. Some changes in echolocation calls and the production of other vocalizations are probably related to com-munication rather than to echolocation tasks. Obrist (1995) demonstrated how changes in the features of echolocation calls reflected the setting in which the bats were hunting, including the presence of obstacles and conspecifics. Bar-clay (1983) noted marked pulse-to-pulse variation in the echolocation calls of a number of Neotropical emballonurids, perhaps reflecting their role in com-munication. Similar pulse-to-pulse variation has also been reported from molossids (e.g., Fenton et al. 1998; Kössl et al. 1999).

Intraspecific, individual differences in echolocation calls have been re-ported for a number of species (e.g., Habersetzer 1981; Jones et al. 1992; Mas-ters et al. 1995; Obrist 1995; Park et al. 1996). These variations have the poten-tial to transfer additional information between bats and/or play a role in anti-interference mechanisms in echolocation. Masters et al. (1995) found evi-dence for differences between the echolocation calls of individual *E. fuscus*, for family resemblance in call structure, and for differences between adults and juveniles, all of which could be exploited in communication. Pearl and Fenton (1996) found significant differences in call structure between colonies of *M. lu-cifugus*, but effects due to the setting in which the calls were recorded could not be eliminated. Call structure appears to have both inherited and learned com-ponents, although their relative importance appears to vary from species to species.

Variation in structure characterizes the vocalizations of bats and thus can affect the information available for eavesdroppers (Fenton 1994a). Echoloca-tion calls vary in structure (bandwidth, duration, interpulse interval) reflect-ing the changing situation that may confront a bat through a sequence from detection through pursuit of airborne prey (e.g., Obrist 1995). Notwithstand-ing, echolocation calls of bats are significantly less variable than social calls (Fenton 1994a), which is directly relevant to the communication potential. There is, however, a continuum from echolocation to social calls.

Modifications of echolocation calls may enhance their communication function. A good example was provided by Suthers (1965), who found that, when on a collision course with another conspecific, *Noctilio leporinus* dropped the terminal portion of its echolocation call by an octave, effectively "honking" to alert the other bat that they were on an impending collision course. In re-sponse to the honks (fig. 2.8), one or both bats changed their course. Honking has now been noted in a number of other aerial-feeding species (e.g., Fenton and Bell 1979; 1981).

Many bats emit calls during flight that differ considerably in structure from echolocation calls. They may occur between echolocation calls and often are broadband calls with low-frequency components and multiple syllables (e.g., *Lasiurus cinereus* [Belwood and Fullard 1984]). Such vocalizations are fre-

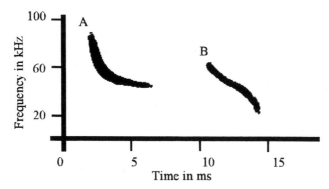

Figure 2.8. Compared here is an echolocation call (A) and a "honk," or a modified echolocation call (B), of *Myotis volans* (after Fenton and Bell 1979).

quently referred to as "social calls," but their functions are not well understood. Almost invariably, these vocalizations are much longer in duration than echolocation calls, precluding their use in echolocation by low duty cycle bats (which cannot tolerate overlap between pulse and echo).

Several studies suggest that social calls advertise a bat's presence and facilitate mutual avoidance or even territoriality. The "song" of *Cardioderma cor* is audible to humans and is produced as individuals fly about their feeding areas (Vaughan 1976). The use of this song is highly seasonal, and feeding success appears to be related to singing at times of low food abundance (Vaughan 1976), suggesting that it serves a role in mutual avoidance. Like other gleaning bats, *C. cor* produces short, low-intensity echolocation calls dominated by higher-frequency sounds (>20 kHz), a combination of factors that certainly limits the range over which the calls will carry. The song calls of *C. cor* are much lower in frequency (<20 kHz) and therefore are audible over a greater distance. The same situation applies to the songs of the megadermatid *Macroderma gigas* (Tidemann et al. 1985).

The most convincing demonstration of a territorial function for social calls was provided by Barlow and Jones (1997b). Two sympatric species of *Pipistrellus* emit distinctly different social calls (fig. 2.9). At low insect densities, the rate of social-call production increased with decreasing insect density. Playback of social calls of one species significantly reduced the activity of that species (determined from the number of recorded bat passes) but did not affect the activity of the other species. These social calls are often associated with chases (e.g., Lundberg and Gerell 1986; Miller and Degn 1981), which are more common at low than at high insect densities (Lundberg and Gerell 1986; Racey and Swift 1985). These experiments suggest that the social calls of *P. pipistrellus* advertise the presence of a foraging individual. The chases suggest that the bats actively defend patches of prey from conspecifics, which demands only

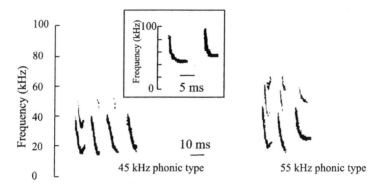

Figure 2.9. Typical songflight calls and search phase echolocation calls (inset) used by two sympatric species of pipistrelle bats (45 and 55 kHz phonic types) in the United Kingdom (from Barlow and Jones 1997a).

that individuals recognize the call of a conspecific. Defense of a fixed territory may involve the recognition by the trespasser of the territory holder.

In *Lasiurus borealis*, conspecific chases are not territorial but, rather, occur when more than one bat pursues the same insect, a situation that was revealed from observations of marked individuals (Hickey and Fenton 1990) and playback experiments (Balcombe and Fenton 1988). When hunting moths that responded to their echolocation calls, pairs of *L. borealis* were significantly more successful than were solitary bats (Acharya and Fenton 1999). Since the same marked individuals repeatedly foraged at the same sites (Hickey and Fenton 1990), this suggests cooperative hunting (Fenton 2003).

Social calls may also attract conspecifics to a foraging site. One of the best-documented examples is provided by *Phyllostomus hastatus*, which uses social calls to maintain group cohesion. *Phyllostomus hastatus* forms stable groups of unrelated females, often for life. These groups roost together (McCracken and Bradbury 1981), and, on emergence from their roost, individuals emit "screech" calls to recruit others and maintain contact when commuting and foraging (Wilkinson and Boughman 1998; fig. 2.10). Screech calls are loud, broadband (4–18 kHz), and noisy (Boughman 1997). Calls of individuals within a group are statistically indistinguishable, but the calls of different groups are quite distinctive. In captivity, individuals can recognize their group mates by their unique calls (Boughman and Wilkinson 1998). Boughman (1997) argued that these calls are well suited to their role, since rapid-onset, noisy calls are more easily located than pure tones are. Transfer experiments demonstrated that over time females modfied their screech calls to match those of their group mates (Boughman 1998).

Several other species produce low-frequency social calls (e.g., *Nycteris thebaica* [Aldridge et al. 1990] and *Macroderma gigas* [Tidemann et al. 1985]), whose functions remain unstudied. In *M. gigas,* chirping often coincides with the resumption of feeding after inactive periods during foraging (Tidemann et al. 1985). *Nycteris thebaica* produces two kinds of birdlike "chirps," both of which are low-frequency vocalizations readily audible to humans (Aldridge et al. 1990). The narrow-band calls are produced by *N. thebaica* at hunting perches and attract conspecifics. The broadband calls were produced as the bats left their day roosts. The role of chirps in the behavior of this bat remains unknown (Aldridge et al. 1990). *Megaderma lyra* emits "grumbling" sequences in flight, presumably to reduce aggression (Leippert 1994).

Acoustic communication need not involve vocalizations. The sounds of chewing proved important in observational learning in *M. lucifugus, E. fuscus,* and *Antrozous pallidus* (Gaudet and Fenton 1984). Captive *Nycteris grandis* also responded to the sounds of chewing, adjusting their feeding behavior to minimize conflict with other bats in the same roost (Fenton et al. 1983).

Bats, like many other animals, use alarm calls and distress calls in what appear to be stressful situations. Nelson (1964) described two different alarm calls in *Pteropus poliocephalus,* one short, loud, and harsh call was specifically associated with the appearance of an eagle and caused other individuals to emit the same call. The use of different alarm calls for different threats has been reported in other species, the best studied being vervet monkeys (Cheney and Seyfarth 1990).

Broadcasting their broadband distress calls caused *M. lucifugus* to dive at the loudspeaker (Fenton et al. 1976). A few vocalizing *P. pipistrellus* confined to a cage near its maternity roost caused a 20–80-fold increase in the number of bat passes around the cage (Russ et al. 1998). Recorded distress calls also attracted bats, but the response was less marked and waned rapidly (Fenton et al. 1976). Distress calls of pipistrelles are multisyllable, downward sweeps from 40 to 18 kHz, with maximum energy at around 27 kHz (fig. 2.9). This

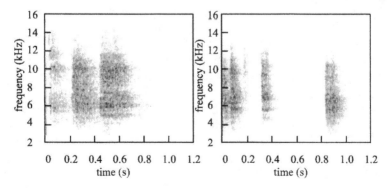

Figure 2.10. Sample screech calls from two groups of *Phyllostomus hastatus* used to maintain contact by foraging and commuting groups of individuals (from Boughman and Wilkinson 1998).

lower-frequency call is somewhat less directional and attenuates less than echolocation calls, making it audible over greater distances. August (1979) found that playbacks of recorded and simulated distress calls of *Artibeus jamaicensis* attracted conspecifics and other stenodermine phyllostomids. His innovative synthetic calls made it clear that the calls attracted the attention of other bats.

We can say little about the information conveyed by their messages, but the calls of bats can be quite complex. The vocal repertoires of several species of bats were reviewed by Fenton (1985). A spectacular example is provided by *Pteronotus parnellii*, which use 33 discrete syllables either singly or in various combinations, apparently governed by several syntactical rules (Kanwal et al. 1994). The longevity of bats, the propensity of some species to form stable, social units, and their capacity for extracting detailed information from acoustic signals suggests considerable capacity for elaborate vocal communication.

Multimodal Communication

In mammals, scent-marking behavior works synergistically with odoriferous secretions to optimize communication (e.g., Ewer 1968; Gorman and Trowbridge 1989). An example of specialized marking behavior is "salting," which is used by male *Saccopteryx bilineata* to mark roosts and females (Voigt and Helversen 1999). Bats such as *Molossus ater, M. sinaloe, Hipposideros ater*, and *H. fulvus* can evert their glands and deposit a thick, strong-smelling secretion, and there is a record of a male *Molossus molossus* everting its gland and completely soaking a female conspecific (Schmidt 1985). In captivity, male *M. molossus* scent mark females, subordinate males, and specific areas within cages (Haussler et al. 1981; Schmidt 1985). The same is true of the African molossids *Chaerephon pumilus* and *Mops condylura* (Bouchard 2001).

Following the trend of signals with multimedia components, the communication behaviors associated with mating in bats involve vocal, visual, olfactory, and tactile cues (Fenton 1985) and are as diverse as mating systems in bats (Altringham 1996; Bradbury 1977b; McCracken and Wilkinson 2000), as are the associated communication behaviors that involve vocal, visual, olfactory, and tactile cues in mate attraction and mating (Fenton 1985). A specific call to attract potential mates appears to be used by several species, but direct confirmation of function, for example, by playback experiments, is lacking. The most obvious examples of mate attraction have been reported from the epomophorine bats of Africa, and the best known among these is *Hypsignathus monstrosus*. Males of this species, which have enlarged larynges, call from tree roosts using a repetitive "honk" accompanied by wing flapping (Bradbury 1977a). Call rate increases as females pass by, but we do not know what features of the calls or, for that matter, other aspects of the display, give one male greater success than another. Among other epomophorine bats, males of *Epomophorus wahlbergi* also call repeatedly, increasing their call rate in the presence of females (Wickler and Seibt 1976). Males of *Epomops franqueti* display conspicuous epaulettes while calling and increase their call rate in the presence of females (Bradbury 1981). Epomophorine bats use low-frequency calls, presumably because they can attract females from greater ranges than if they used higher-frequency calls.

Several microchiropterans also use low-frequency calls to attract mates. Males of *Nyctalus noctula* vocalize from tree cavities presumably to attract mates (Sluiter and van Heerdt 1966), using a call well within the range of human hearing. Two pipistrelles, *Pipistrellus pipistrellus* and *P. nathusii* use multisyllable calls in the range of 45–14 kHz (Barlow and Jones 1997a; Gerell-Lundberg and Lundberg 1994; fig. 2.9), the former in a songflight, the latter produced either while in flight or from a stationary position outside the roost. In both species, calls were repeated at around 1.6 times a second. Experiments that confirm their function and determine which factors indicate fitness have not yet been conducted. *Megaderma lyra* also has a songflight (Leippert 1984).

Once a female has been attracted by the calls and is close to a male, other signals may come into play. Visually conspicuous sexual dimorphism is unusual in bats, but some striking examples suggest that visual signals are important in mating. For example, nothing is known of the mating behavior of *Chaerophon chapini,* but the impressive erectile crest of the male (see fig. 2.7) must surely provide a visual cue to conspecifics. As noted above, the males of many species of epomophorine bats are larger than females and have prominent epaulettes, usually eversible tufts of long, white hairs (Bradbury 1977a; Wickler and Seibt 1976). The wing flapping associated with calling in these bats may be a visual signal but may also serve to disperse glandular secretions. In sexually mature males of *Sturnira lilium,* shoulder patches are conspicuous both visually and by the scent produced by underlying glands (Scully et al. 2000), strongly suggesting a role in mate selection. There are many other examples of sexually dimorphic apocrine and exocrine glands in bats (Quay 1970; Scully et al. 2000), but only a few are associated with visually conspicuous structures.

As in some other mammals (Hrdy 1977), mating activity itself in bats can attract conspecifics. The best example reported to date are the calls given by *M. lucifugus* during copulation (Barclay and Thomas 1979). These long, low-frequency vocalizations are readily audible to humans and to bats in the underground sites where this species congregates to mate (Barclay et al. 1979; Thomas et al. 1979). Copulation calls attract conspecifics, which can disrupt copulations (Barclay and Thomas 1979; Thomas et al. 1979). During the mating season, clusters of female *M. lucifugus* form around adult males. Individuals joining the cluster engage in naso-naso contact with bats already present, indicating an important role for olfactory cues.

Some Neotropical emballonurid bats (e.g., *Saccopteryx* spp, *Peropteryx* sp., *Balantiopteryx* sp. and *Cormura* sp. [Reid 1997]) have wing sacs that are prominent in males but less so in females and subadult males. The salting behavior of *S. bilineata* sometimes occurs when males hover directly in front of females with whom they share a roost (Voigt and Helversen 1999). These displays are usually accompanied by vocalizations (Voigt and Helversen 1999). Although referred to as "glands," the wing sacs are holding areas for products generated elsewhere, namely, urine, saliva, and secretions from the genital areas and from the gular glands (Quay 1970; Scully et al. 2000; Voigt and Helversen 1999). The males of many emballonurids and molossids have well-developed gular glands that may, as in the case of *Taphozous melanopogon*, be associated with visually conspicuous features (in this case, a black "beard" [Fenton 2001]).

As noted above, osmetrichia are frequently associated with sexually dimorphic glands in molossids and pteropodids (Bouchard 2001; Hickey and Fenton 1987). The gland and the specialized hairs may together present a combined visual and olfactory display. The hairs associated with other glands are not specialized as osmetrichia and may only provide a visual signal. S. Bouchard (personal communication) has observed that courting male *Chaerephon pumilus* display their crests to females, erecting them by lateral movements of the ears. Here again, naso-naso contact is a regular part of behavioral interactions among males, females and males, and females.

Many bats have markings on their fur and/or wings that may be visually conspicuous but not sexually dimorphic. Included here are basically light-dark contrasting features such as the ear tufts of some epomophorine bats, the dorsal spots or stripes on some vespertilionids (e.g., *Euderma maculatum*, *Chalinolobus egeria*, or *Chalinolobus beatus*) or the dorsal or facial stripes of some Neotropical emballonurids (*Saccopteryx* spp. or *Rhynchonycteris naso*) or stenodermine phyllostomids, respectively. In some cases—for example, ear spots and stripes—the function may be a type of disruptive coloration that enhances crypsis (Fenton 1992).

The context of a signal may be crucial. Barlow and Jones (1997a) compared social and songflight calls in the two phonic types of *P. pipistrellus*. Social calls are emitted by both sexes in a wide range of situations. Songflight calls are produced by males in the mating season and are emitted repetitively during back and forth flights past the roost. There were small but significant differences in the 45-kHz-type calls. These could be functionally important or simply reflect the fact that one sample comes from males only, the other from males and females with different echolocation calls (Park et al. 1996). In a related study, Barlow and Jones (1997b) demonstrated that social calls are used by individual *P. pipistrellus* to advertise their feeding sites. The very similar songflight calls could be used to attract mates and/or to drive away rival males.

Mother-Young Communication

A considerable body of evidence suggests that most, if not all, bats usually recognize and suckle their own young (de Fanis and Jones 1995a; Fenton 1985). In virtually all reported cases, recognition is based on vocalizations and/or scent. The vocal signals that mediate recognition can be the female's echolocation calls, isolation calls from the infant, or other communication calls. Olfactory recognition may come from chemicals on mothers and/or offspring, from the female's milk or saliva, or from glandular secretions with which the female anoints the young.

In most species studied, both vocal and olfactory cues are used and naso-naso contact is part of the interaction. For example, both cues, along with spatial memory, are used by female *Tadarida brasiliensis* to find offspring in the roost (Balcombe and McCracken 1992; Gustin and McCracken 1987; McCracken and Gustin 1991). Lactating females could identify their own muzzle scents and those of their offspring, but pups showed no preference for scent from their mother compared to that of a randomly chosen lactating female. Both females and young had large, active sebaceous glands around their chins and muzzles. Each young presumably carries its own odor, which may resemble that of its mother, as well as scent transferred to it from the mother's glands.

When searching for pups, female bats also produce "directive" calls that consist of intense, rapidly repeated, distinctive sounds (fig. 2.11). These calls are stereotyped within individuals but statistically different between them, and the calls attract young bats. It is not known to what extent young bats can identify the directive calls of their mothers.

Young *Plecotus auritus* do not show a preference for their mother's echolocation calls over those of a randomly chosen female (de Fanis and Jones 1995a). Young produce stereotyped isolation calls that are recognized by their mothers, and these appear to be a major factor in location and identification of their offspring (fig. 2.11). Females use either their young's isolation calls or scent as recognition cues. The young can distinguish their mothers by her echolocation calls or by her scent. Isolation calls have been described for a number of bat species.

In some species (e.g., some vespertilionids and phyllostomids [Bradbury 1977b; Gould 1977; Matsumara 1981]) early communication calls may be precursors to echolocation calls. However, this has not been supported by subsequent work (e.g., Esser and Lud 1997 [*Phyllostomus discolor*] and Scherrer and Wilkinson 1993 [*Nycticeius humeralis*]). Call duration decreases and frequency increases as young bats develop. In other vespertilionids (Jones et al. 1991; Moss 1988), echolocation and communication calls have separate origins. Isolation calls are typically low in frequency, long, and loud, with multiple harmonics and sinusoidal frequency modulation (fig. 2.11). For example, females

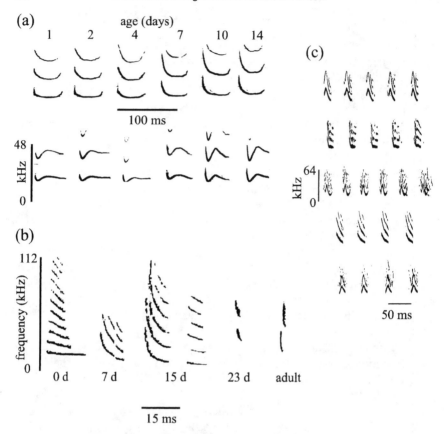

Figure 2.11. Isolation calls of (*a*) *Nycticeius humeralis* (from Scherrer and Wilkinson 1993) and (*b*) *Plecotus auritus* (from de Fanis and Jones 1995a) showing changes during development. *c*, Directive calls from *Tadarida brasiliensis* (from Balcombe and McCracken 1992).

of *P. discolor* use a distinctive sinusoidal FM pattern in their directive calls toward young (Esser and Lud 1997). Playback experiments using manipulated calls showed that this bat has sufficient auditory resolution to identify individuals from their directive calls.

The isolation calls of *Nycticeius humeralis* are not only distinctive enough to enable mothers to identify their own pups, but they also have a significant heritable component that may contain information about family identity (Scherrer and Wilkinson 1993), perhaps explaining the incidence of communal nursing in this species.

Some Challenges

This review leaves many questions unanswered, but we have identified five topics for special mention. Each has received little attention to date and offers promise for future studies of the sensory ecology of bats.

1. Some species use both sound and light for orientation even at night. How do vision and echolocation interact in these and other species? Of special note, here, are species such as the relatively large-eyed *Saccopteryx* spp. (Bradbury 1977b) or *Eptesicus nilssoni*, which forages in the midnight sun at high latitudes (Rydell 1992). Identifying other candidates for research on this topic will be easy, but designing experiments that reveal how bats resolve situations where vision and echolocation provide a conflicting picture about a situation will be much more challenging.

2. The variety of roosting situations used by bats means that they are exposed to different levels of ambient light by day. Some species consistently roost in dark locations such as caves (Kunz 1982), whereas others use more brightly lighted settings such as foliage (Kunz and Lumsden, this volume). Individuals of other species (e.g., *Artibeus jamaicensis*) sometimes alternate between darked and well-lighted roosts. It follows that the eyes of these bats should show differences in sensitivity to bright light. As noted above, electro-retinography is one way to approach this topic, and collecting more details about the visual performance of bats will also further our knowledge of this area.

3. How do bats find suitable roost sites, particularly those species that appear to have special roost requirements (see Kunz and Lumsden, this volume)? Included here would be species such as *Tylonycteris* spp. that roost in bamboo culms (Medway and Marshall 1970) or bats that roost beneath rocks (e.g., *Sauromys petrophilus*, *Platymops setiger*, and *Neoplatymops mattogrossensis* [Peterson 1965]). A variation on this question involves species that switch roosts frequently. *Thyroptera tricolor* has small home ranges where there are high densities of potential roosts in furled leaves (Vonhof 1998), which may only complicate the challenge of finding a suitable roost on any given night. Other bats known to roost in furled leaves include *Pipistrellus nanus* and *Myotis bocagei* (see Kunz and Lumsden, this volume). While the echolocation calls (Barclay 1982) or other vocalizations like the "rallying calls" of *Antrozous pallidus* can show how bats may be attracted by conspecifics (Vaughan and O'Shea 1976), it remains unclear how bats first locate roosts. There is some evidence that bats rely on spatial memory to locate known roosts (e.g., Höller 1995; Höller and Schmidt 1996; McCracken 1993), and spatial memory is probably involved in many other tasks (Neuweiler 1967).

4. Do the striking patterns of some bats play any role in intraspecific communication? Bats such as *Saccopteryx* spp. species and *Rhynchonycteris naso* have dorsal stripes that may help to camouflage them against their exposed

roosting sites. The same may be true of the ear tufts of epomophorines and facial stripes or back stripes of stenodermines, respectively (see Kunz and Lumsden, this volume). *Pteropus capistratus* may be the most spectacular example of a disruptive pattern. However, it is difficult to suggest anything other than a communication function for the spectacular black-and-white patterning of *Euderma maculatum* and *Chalinolobus superba* or the bright yellow wings of *Lavia frons*. Although color vision appears to be restricted to the megachiropterans, several species of microchiropterans have distinctly colorful markings. The yellow nose leaf and tragus of *Ectophylla alba* and the bright red markings on the wings of some species of *Kerivoula* and *Myotis welwitschii* are obvious examples. So, too, is the sexual dimorphism in *Lasiurus borealis*, where males are a brighter red than females. It remains to be determined if these and other markings on bats serve in intraspecific communication or confer some protection against predators (Kunz and Lumsden, this volume).

5. Finally, although vision has been shown to be important in homing, no research has been done to assess the importance of systems used by birds, such as magnetic cues. The availability of small (<1 g) radio transmitters may open up new opportunities for exploring the navigational behavior of bats.

Acknowledgments

We thank two anonymous reviewers and Sylvie Bouchard for their critical reviews of earlier versions of this manuscript. We are grateful to Jens Rydell for permission to use two of his photographs. Other permissions for illustrations are given in the captions. Altringham's work on bats has been supported by the Natural Environment Research Council (United Kingdom) and Fenton's by the Natural Sciences and Engineering Research Council (Canada).

Literature Cited

Acharya, L. 1992. Epomophorus wahlbergi. Mammalian Species, 394:1–4.

Acharya, L., and M. B. Fenton. 1999. Bat attacks and moth defensive behaviour around streetlights. Canadian Journal of Zoology, 77:27–33.

Albone, E. S. 1984. Mammalian semiochemistry: the investigation of chemical signals between mammals. John Wiley & Sons Ltd., Toronto.

Aldridge, H. D. J. N., M. Obrist, H. G. Merriam, and M. B. Fenton. 1990. Roosting, vocalizations and foraging by the African bat, *Nycteris thebaica*. Journal of Mammalogy, 71:242–246.

Altringham, J. D. 1996. Bats: Biology and Behaviour. Oxford University Press, Oxford.

August, P. V. 1979. Distress calls in *Artibeus jamaicensis*: ecology and evolutionary implications. Pp. 151–160 *in*: Vertebrate Ecology in the Northern Neotropics (J. F. Eisenberg, ed.). Smithsonian Institution Press, Washington, D.C.

Balcombe, J. P., and M. B. Fenton. 1988. Eavesdropping by bats: the influence of echolocation call design and foraging strategy. Ethology, 79:158–166.

Balcombe, J. P., and G. F. McCracken. 1992. Vocal recognition in Mexican free-tailed bats: do pups recognise mothers? Animal Behaviour, 43:79–88.

Barclay, R. M. R. 1982. Interindividual use of echolocation calls: eavesdropping by bats. Behavioral Ecology and Sociobiology, 10:271–275.

Barclay, R. M. R. 1983. Echolocation calls of emballonurid bats from Panama. Journal of Comparative Physiology A, 151:515–520.

Barclay, R. M. R., M. B. Fenton, and D. W. Thomas. 1979. Social behavior of *Myotis lucifugus*. II. Vocal communication. Behavioral Ecology and Sociobiology, 6:137–146.

Barclay, R. M. R., and D. W. Thomas. 1979. Copulation call of *Myotis lucifugus*: a discrete, situation-specific communication signal. Journal of Mammalogy, 60:632–634.

Barlow, K. E., and G. Jones. 1997a. Differences in songflight calls and social calls between two phonic types of the vespertilionid bat *Pipistrellus pipistrellus*. Journal of Zoology (London), 241:315–324.

Barlow, K. E., and G. Jones. 1997b. Function of pipistrelle social calls: field data and a playback experiment. Animal Behaviour, 53:991–999.

Barratt, E. M., R. Deaville, T. M. Burland, M. W. Bruford, G. Jones, P. A. Racey and R. K. Wayne. 1997. DNA answers the call of pipistrelle bat species. Nature, 387:138–139.

Barton, R. A., A. Purvis, and P. H. Harvey. 1995. Evolutionary radiation of visual and olfactory brain systems in primates, bats and insectivores. Philosophical Transactions of the Royal Society of London B, 348:381–392.

Bell, G. P. 1982. Behavioral and ecological aspects of gleaning by a desert insectivorous bat, *Antrozous pallidus* (Chiroptera: Vespertilionidae). Behavioral Ecology and Sociobiology, 10:217–223.

Bell, G. P. 1985. The sensory basis of prey location by the California leaf-nosed bat *Macrotus californicus* (Chiroptera: Phyllostomidae). Behavioral Ecology and Sociobiology, 16:343–347.

Bell, G. P., and M. B. Fenton. 1986. Visual acuity, sensitivity and binocularity in a gleaning insectivorous bat, *Macrotus californicus* (Chiroptera: Phyllostomidae). Animal Behaviour, 34:409–414.

Belwood, J. J., and J. H. Fullard. 1984. Echolocation and foraging behaviour in the Hawaiian hoary bat, *Lasiurus cinereus seminotus*. Canadian Journal of Zoology, 62:2113–2120.

Bhatnagar, K. P. 1980. The chiropteran vomeronasal organ: its relevance to the phylogeny of bats. Pp. 289–316 *in:* Proceedings of the Fifth International Bat Research Conference (D. E. Wilson and A. L. Gardner, eds.). Texas Tech University Press, Lubbock.

Bloss, J. 1999. Olfaction and the use of chemical signals by bats. Acta Chiropterologica, 1:35–45.

Bloss, J., T. E. Acree, J. M. Bloss, W. R. Hood, and T. H. Kunz. 2002. The potential use of chemical cues for colony-mate recognition in big brown bats, *Eptesicus fuscus*. Journal of Chemical Ecology, 28:799–814.

Bouchard, S. 2001. Sex discrimination and roostmate recognition by olfactory cues in the bats, *Mops condylurus* and *Chaerophon pumilus*. Journal of Zoology (London), 254:109–117.

Boughman, J. W. 1997. Greater spear-nosed bats give group-distinctive calls. Behavioral Ecology and Sociobiology, 40:61–70.

Boughman, J. W. 1998. Vocal learning by greater spear-nosed bats. Proceedings of the Royal Society of London B, 265:227–233.

Boughman, J. W., and G. S. Wilkinson. 1998. Greater spear-nosed bats discriminate group mates by vocalizations. Animal Behaviour, 55:1717–1732.

Bradbury, J. W. 1977a. Lek mating behaviour in the hammer-headed bat. Zeitschrift für Tierpsychologie, 45:225–255.

Bradbury, J. W. 1977b. Social organization and communication. Pp. 1–72 *in:* Biology of Bats (W. A. Wimsatt, ed.). Vol. III. Academic Press, New York.

Bradbury, J. W. 1981. The evolution of leks. Pp. 138–169 *in:* Natural Selection and Social Behaviour (R. D. Alexander and D. W. Tinkle, eds.). Chiron Press, New York.

Bradbury, J. W., and F. Nottebohm. 1969. The use of vision by the little brown bat, *Myotis lucifugus*, under controlled conditions. Animal Behaviour, 17:480–485.

Brooke, A. P. 1997. Organisation and foraging behavior of the fishing bat, *Noctilio leporinus* (Chiroptera: Noctilionidae). Ethology, 103:421–436.

Brooke, A. P., and D. M. Decker. 1996. Lipid compounds in secretions of the fishing bat, *Noctilio leporinus* (Chiroptera: Noctilionidae). Journal of Chemical Ecology, 22:1411–1428.

Brosset, A. 1962. The bats of central and western India. Journal of the Bombay Natural History Society, 59:583–624.

Brown, P. E. 1976. Vocal communication in the pallid bat, *Antrozous pallidus*. Zeitschrift für Tierpsychologie, 41:34–54.

Buchler, E. R., and S. B. Childs. 1982. Use of post-sunset glow as an orientation cue by the big brown bat (*Eptesicus fuscus*). Journal of Mammalogy, 63:243–247.

Chase, J. 1981. Visually guided escape response in microchiropteran bats. Animal Behaviour, 29:708–713.

Chase, J. 1983. Differential response to visual and acoustic cues during escape in the bat *Anoura geoffroyi*. Animal Behaviour, 31:526–531.

Cheney, D. L., and R. M. Seyfarth. 1990. How Monkeys See the World. University of Chicago Press, Chicago.

Coles, R. B., A. Guppy, M. E. Anderson, and P. Schlegel. 1989. Frequency sensitivity and directional hearing in the gleaning bat, *Plecotus auritus* (Linnaeus 1758). Journal of Comparative Physiology A, 165:269–280.

Dapson, R. W., E. H. Studier, M. J. Buckingham, and A. L. Studier. 1977. Histochemistry of odoriferous secretions from integumentary glands in three species of bats. Journal of Mammalogy, 58:531–535.

Davis, R. 1966. Homing performance and homing ability in bats. Ecological Monographs, 36:201–237.

de Fanis, E., and G. Jones. 1995a. Post-natal growth, mother infant interactions and development of vocalizations in the vespertilionid bat, *Plecotus auritus*. Journal of Zoology (London), 235:85–97.

de Fanis, E., and G. Jones. 1995b. The role of odour in the discrimination of conspecifics by pipistrelle bats. Animal Behaviour, 49:835–839.

Esser, K. H., and A. Daucher. 1996. Hearing in the FM bat *Phyllostomus discolor*—a behavioral audiogram. Journal of Comparative Physiology A, 178:779–785.

Esser, K. H., and R. Kiefer. 1996. Detection of frequency-modulation in the FM bat *Phyllostomus discolor*. Journal of Comparative Physiology A, 178, 787–796.

Esser, K. H., and B. Lud. 1997. Discrimination of sinusoidally frequency-modulated sound signals mimicking species-specific communication calls in the FM bat *Phyllostomus discolor.* Journal of Comparative Physiology A, 180:513–522.

Ewer, R. F. 1968. Ethology of Mammals. Plenum Press, New York.

Fenton, M. B. 1985. Communication in the Chiroptera. Indiana University Press, Bloomington.

Fenton, M.B. 1992. Pelage patterns and crypsis in roosting bats: *Taphozous mauritianus* and *Epomophorous* spp. Koedoe, 35:49–55.

Fenton, M. B. 1994a. Assessing signal variability and reliability: "to thine ownself be true." Animal Behaviour 47:757–764.

Fenton, M. B. 1994b. Echolocation: its impact on the behaviour and ecology of bats. Ecoscience, 1:21–30.

Fenton, M. B. 2001. Bats. Rev. ed. Facts on File, New York.

Fenton, M. B. 2003. Aerial-feeding bats: getting the most out of echolocation. *In:* Echolocation in Bats and Dolphins (J. Thomas, C. Moss and M. Vater, eds.). University of Chicago Press, Chicago (in press).

Fenton, M. B., D. Audet, M. K. Obrist, and J. Rydell. 1995. Signal strength, timing and self-deafening: the evolution of echolocation in bats. Palaeobiology, 21:229–242.

Fenton, M. B., and G. P. Bell. 1979. Echolocation and feeding behaviour in four species of *Myotis* (Chiroptera). Canadian Journal of Zoology, 57:1271–1277.

Fenton, M. B., and G. P. Bell. 1981. Recognition of species of insectivorous bats by their echolocation calls. Journal of Mammalogy, 62:233–243.

Fenton, M. B., J. J. Belwood, J. H. Fullard, and T. H. Kunz. 1976. Responses of *Myotis lucifugus* (Chiroptera: Vespertilionidae) to calls of conspecifics and to other sounds. Canadian Journal of Zoology, 54:1443–1448.

Fenton, M. B., R. M. Brigham, A. M. Mills, and I. L. Rautenbach. 1985. The roosting and foraging areas of *Epomophorous wahlbergi* (Pteropodidae) and *Scotophilus viridis* (Vespertilionidae) in Kruger National Park, South Africa. Journal of Mammalogy, 66:461–468.

Fenton, M. B., C. L. Gaudet, and M. L. Leonard. 1983. Feeding behaviour of the bats *Nycteris grandis* and *Nycteris thebaica* (Nycteridae) in captivity. Journal of Zoology (London), 200:347–354.

Fenton, M. B., C. V. Portfors, I. L. Rautenbach, and J. M. Waterman. 1998. Compromises: sound frequencies used in echolocation by aerial feeding bats. Canadian Journal of Zoology, 76:1174–1182.

Flood, P. 1985. The bats: order Chiroptera. Pp. 19–36 *in:* Social Odors in Mammals (R. Brown and D. Macdonald, eds). Vol. 2. Oxford University Press, New York.

Fuzessery, Z. M. 1997. Acute sensitivity of interaural time differences in the inferior colliculus of a bat that relies on passive sound localization. Hearing Research, 109:46–62.

Fuzessery, Z. M., B. Buttenhof, B. Andrews, and J. M. Kennedy. 1993. Passive sound localization of prey by the pallid bat (*Antrozous p. pallidus*). Journal of Comparative Physiology A, 171:767–777.

Gaudet, C. L., and M. B. Fenton. 1984. Observational learning in three species of insectivorous bats (Chiroptera). Animal Behaviour, 32:385–388.

Gerell-Lundberg, K, and R. Lundberg. 1994. The mating behaviour of the pipistrelle and the Nathusius' pipistrelle (Chiroptera)—a comparison. Folia Zoologica, 43: 315–324.

Gorman, M. L., and B. J. Trowbridge. 1989. The role of odor in the social lives of carnivores. Pp. 57–88 *in:* Carnivore Behavior, Ecology and Evolution (J. L. Gittleman, ed.). Cornell University Press, Ithaca, N.Y.

Gould, E. 1977. Echolocation and communication. Pp. 247–279 *in:* Biology of Bats of the New World Family Phyllostomatidae (R. J. Baker, J. K. Jones, Jr., and D. C. Carter, eds.). Pt.. 2. Special Publications, the Museum, Texas Tech University, 13. Texas Tech Press, Lubbock.

Greenhall, A. M., G. Joermann, U. Schmidt, and M. R. Seidel. 1983. Desmodus rotundus. Mammalian Species, 202:1–6.

Griffin, D. R. 1958. Listening in the Dark. Yale University Press, New Haven, Conn.

Griffin, D. R. 1970. Migration and homing in bats. Pp. 233–246 *in:* Biology of Bats (W. A. Wimsatt, ed.). Vol. I. Academic Press, New York.

Grinnell, A. D. 1980. Dedication. Pp. xix–xxiv *in:* Animal Sonar Systems (R.-G. Busnel and J. F. Fish, eds.). NATO Advanced Study Institutes Series, ser. A, vol. 28. Plenum Press, New York.

Guppy, A., and R. B. Coles. 1988. Acoustical and neural aspects of hearing in the Australian gleaning bats *Macroderma gigas* and *Nyctophilus gouldi.* Journal of Comparative Physiology A, 162:653–668.

Gustin, M. K., and G. F. McCracken. 1987. Scent recognition between females and pups in the bat *Tadarida brasiliensis mexicana.* Animal Behaviour, 35:13–19.

Gutierrez, M., and A. Agustin. 1973. Fine structure of the gular gland of the free-tailed bat, *Tadarida brasiliensis.* Journal of Morphology, 141:293–306.

Habersetzer, J. 1981. Adaptive echolocation sounds in the bat *Rhinopoma hardwickei,* a field study. Journal of Comparative Physiology A, 144:559–566.

Haussler, U. L., E. Moller, and U. Schmidt. 1981. Zur Haltung und Jugendentwicklung von *Molossus molossus* (Chiroptera). Zeitschhriftft für Säugentierkunde, 46:337–351.

Helversen, D. von, and O. von Helversen. 1999. Acoustic guide in bat-pollinated flower. Nature, 398:759–760.

Hermanson, J. W., and T. J. O'Shea. 1983. Antrozous pallidus. Mammalian Species, 213:1–8.

Hessel, K., and U. Schmidt. 1994. Multimodal orientation in *Carollia perspicillata* (Phyllostomidae). Folia Zoologica, 43:339–346.

Hickey, M. B. C., and M. B. Fenton. 1987. Scent-dispersing hairs (osmetrichia) in some Pteropodidae and Molossidae (Chiroptera). Journal of Mammalogy, 68:376–384.

Hickey, M. B. C., and M. B. Fenton. 1990. Foraging by red bats (*Lasiurus borealis*): do intraspecific chases mean territoriality? Canadian Journal of Zoology, 68:2477–2482.

Höller, P. 1995. Orientation of the bat *Phyllostomus discolor* (Phyllostomidae) on the return flight to its resting place. Ethology, 100:72–83.

Höller, P., and U. Schmidt. 1996. The orientation behavior of the lesser spear-nosed bat, *Phyllostomus discolor* (Chiroptera) in a model roost. Journal of Comparative Physiology A, 179:245–254.

Hope, G. M., and K. P. Bhatnagar. 1979. Effect of light adaptation on electrical responses of the retinas of four species of bats. Experientia, 35:1191–1193.

Hrdy, S. B. 1977. The Langurs of Abu. Harvard University Press, Cambridge, Mass.

Joermann, G., U. Schmidt, and C. Schmidt. 1988. The mode of orientation during flight and approach to landing in two phyllostomid bats. Ethology, 78:332–340.

Jones, G. 1990. Prey selection by the greater horseshoe bat (*Rhinolophus ferrumequinum*): optimal foraging by echolocation? Journal of Animal Ecology, 59:587–602.

Jones, G., T. Gordon, and J. Nightingale. 1992. Sex and age differences in the echolocation calls of the lesser horseshoe bat, *Rhinolophus hipposideros*. Mammalia, 56:189–193.

Jones, G., P. M. Hughes, and J. M. V. Rayner. 1991. The development of vocalizations in *Pipistrellus pipistrellus* (Chiroptera: Vespertilionidae) during post-natal growth and the maintenance of individual vocal signatures. Journal of Zoology (London), 225:71–84.

Jones, G., and J. Rydell. 1994. Foraging strategy and predation risk as factors influencing emergence time in echolocating bats. Philosophical Transactions of the Royal Society of London B, 346:445–455.

Kalko, E. K. V., and M. A. Condon. 1998. Fruit display in *Gurania spinulosa* (Cucurbitaceae) and the sensory modalities of *Phyllostomus hastatus* (Microchiroptera: Phyllostomidae). Functional Ecology, 12:364–372.

Kalko, E. K. V., E. A. Herre and C. O. Handley, Jr. 1996. The relation of fig fruit syndromes to fruit eating bats in the New and Old World tropics. Journal of Biogeography, 23:565–576.

Kanwal, J. S., S. Matsumura, K. Ohlemiller, and N. Suga. 1994. Analysis of acoustic elements and syntax in communication sounds emitted by moustached bats. Journal of the Acoustical Society of America, 96:1229–1254.

Kick, S. 1982. Target detection by the echolocating bat, *Eptesicus fuscus*. Journal of Comparative Physiology A, 145:432–435.

Kössl, M., E. Mora, F. Coro, and M. Vater. 1999. Two-toned echolocation calls from *Molossus molossus* in Cuba. Journal of Mammalogy, 80:924–928.

Kunz, T. H. 1982. Roosting ecology of bats. Pp. 1–55 *in:* Ecology of Bats (T. H. Kunz, ed.). Plenum Press, New York.

Kunz, T. H., P. V. August, and C. D. Burnett. 1983. Harem social organization in cave roosting *Artibeus jamaicensis* (Chiroptera: Phyllostomidae). Biotropica, 15:133–138.

Kunz, T. H., M. S. Fujita, A. P. Brooke, and G. F. McCracken. 1994. Convergence in tent architecture and ·tent-making behavior among Neotropical and Paleotropical bats. Journal of Mammalian Evolution, 2:57–78.

Kürten, L., and U. Schmidt. 1982. Thermoperception in the common vampire bat (*Desmodus rotundus*). Journal of Comparative Physiology A, 146:223–228.

Laidlaw, G. W. J., and M. B. Fenton. 1971. Control of nursery colony populations of bats by artificial light. Journal of Wildlife Management, 35:843–846.

Laska, M. 1990a. Olfactory discrimination ability in short-tailed fruit bat, *Carollia perspicillata* (Phyllostomatidae: Chiroptera). Journal of Chemical Ecology, 16:3291–3299.

Laska, M. 1990b. Olfactory sensitivity to odor components in the short-tailed fruit bat, *Carollia perspicillata* (Phyllostomidae, Chiroptera). Journal of Comparative Physiology A, 166:3291–3295.

Lawrence, B. D., and J. A. Simmons. 1982. Measurements of atmospheric attenuation at ultrasonic frequencies and the significance for echolocation by bats. Journal of the Acoustical Society of America, 71:585–590.

Layne, J. V. 1967. Evidence for the use of vision in diurnal orientation of the bat *Myotis austroriparius*. Animal Behaviour, 15:409–415.

Lee, D. N., F. R. van der Weel, T. Hitchcock, E. Matejowsky, and J. D. Pettigrew. 1992. Common principles of guidance by echolocation and vision. Journal of Comparative Physiology A, 171:563–571.

Leonard, M. L., and M. B. Fenton. 1984. Echolocation calls of *Euderma maculatum* (Chiroptera: Vespertilionidae): use in orientation and communication. Journal of Mammalogy, 65:122–126.

Leippert, D. 1994. Social behaviour on the wing in the false vampire, *Megaderma lyra*. Ethology, 98:111–127.

Lewis, S. E. 1995. Roost switching in bats: a review. Journal of Mammalogy, 76:481–496.

Lewis, S. E. 1996. Low roost-site fidelity in pallid bats: associated factors and effect on group stability. Behavioral Ecology and Sociobiology, 39:335–344.

Lundberg, K., and R. Gerell. 1986. Territorial advertisement and mate attraction in the bat *Pipistrellus pipistrellus*. Ethology, 71:115–124.

Manske, U., and U. Schmidt. 1976. Visual acuity of the vampire bat *Desmodus rotundus* and its dependence upon light intensity. Zeitschrift für Tierpsychologie, 42:215–221.

Marimuthu, G., J. Habersetzer, and D. Leippert. 1995. Active acoustic gleaning from the water surface by the Indian false vampire bat, *Megaderma lyra*. Ethology, 99:61–74.

Marimuthu, G., and G. Neuweiler. 1987. The use of acoustical cues for prey detection by the Indian false vampire bat, *Megaderma lyra*. Journal of Comparative Physiology A, 160:509–515.

Masters, W. M., K. A. S. Raver, and K. A. Kazial. 1995. Sonar signals of big brown bats, *Eptesicus fuscus*, contain information about individual identity, age and family affiliation. Animal Behaviour, 50:1243–1260.

Matsumura, G. 1981. Mother-infant communication in a horseshoe bat (*Rhinolophus ferrumequinum nippon*): vocal communication in three week old infants. Journal of Mammalogy, 62:20–28.

McCracken, G. F. 1993. Locational memory and female-pup reunions in Mexican free-tailed maternity colonies. Animal Behaviour, 45:811–813.

McCracken, G. F., and J. W. Bradbury. 1981. Social organization and kinship in the polygynous bat *Phyllostomus hastatus*. Behavioral Ecology and Sociobiology, 8:11–34.

McCracken, G. F., and M. K. Gustin. 1991. Nursing behaviour in Mexican free-tailed bat maternity colonies. Ethology, 89:305–321.

McCracken, G. F., and G. S. Wilkinson. 2000. Bat mating systems. Pp. 321–362 *in:* Reproductive Biology of Bats (E. G. Crichton and P. H. Krutzsch, eds.). Academic Press, San Diego, Calif.

Medway, Lord, and A. G. Marshall. 1970. Roost-site selection among flat-headed bats (*Tylonycteris* species). Journal of Zoology (London), 161:237–245.

Miller, L. A., and H. J. Degn. 1981. The acoustic behavior of four species of vespertilionid bats studied in the field. Journal of Comparative Physiology A, 142:67–74.

Mistry, S. 1990. Characteristics of the visually-guided escape response of the Mexican free-tailed bat, *Tadarida brasiliensis mexicana*. Animal Behaviour, 39:314–320.

Morrison, D. W. 1978. Foraging ecology and energetics of the frugivorous bat, *Artibeus jamaicensis*. Ecology, 59:716–723.

Moss, C. F. 1988. Ontogeny of vocal signals in the big brown bat, *Eptesicus fuscus*.

Pp. 115–120 *in:* Animal Sonar: Processes and Performance (P. E. Nachtigall and P. W. B. Moore, eds.). Plenum Press, New York.

Mueller, H. C. 1968. The role of vision in vespertilionid bats. American Midland Naturalist, 79:524–525.

Müller-Schwarze, D. 1983. Scent glands in mammals and their functions. Pp. 150–197 *in:* Advances in the Study of Mammalian Behavior (J. E. Eisenberg and D. G. Kleiman, eds.). American Society of Mammalogists, Lawrence, Kans.

Nelson, J. E. 1964. Vocal communication in Australian flying foxes (Pteropodidae: Megachiroptera). Zeitschrift für Tierpsychologie, 26:857–870.

Neuweiler, G. 1967. Interactions of other sensory systems with the sonar system. Pp. 509–534 *in:* Animal Sonar Systems (R.-G. Busnel, ed.). Vol. I. NATO Advanced Study Institute Series. Laboratoire de Physiologie Acoustique, Jouy-en-Josas, France.

Neuweiler, G. 1990. Auditory adaptations for prey capture in echolocating bats. Physiological Reviews, 70:615–641.

Neuweiler, G., S. Singh, and K. Sripathi. 1984. Audiograms of a south Indian bat community. Journal of Comparative Physiology A, 154:133–142.

Novick, A. 1977. Acoustic orientation. Pp. 74–287 *in:* Biology of Bats (W. A. Wimsatt, ed.). Vol. III. Academic Press, New York.

Nowak, R. M. 1991. Walker's Mammals of the World. 5th ed. John Hopkins University Press, Baltimore.

Obrist, M. K. 1995. Flexible bat echolocation—the influence of individual, habitat and conspecifics on sonar signal design. Behavioral Ecology and Sociobiology, 36:207–219.

Obrist, M. K., M. B. Fenton, J. L. Eger, and P. A. Schlegel. 1993. What ears do for bats: a comparative study of pinna sound pressure transformation in Chiroptera. Journal of Experimental Biology, 180:119–152.

Ortega, J., and H. T. Arita. 2000. Defense of females by dominant males of *Artibeus jamaicensis* (Chiroptera: Phyllostomidae). Ethology, 106:395–407.

Ortega, J., and I. Castro-Arellano. 2001. Artibeus jamaicensis. Mammalian Species, 662:1–9.

O'Shea, T. J., and T. A. Vaughan. 1977. Nocturnal and seasonal activities of the pallid bat, *Antrozous pallidus*. Journal of Mammalogy, 58:269–284.

Park, K. J., J. D. Altringham, and G. Jones. 1996. Assortative roosting in two phonic types of the bat *Pipistrellus pipistrellus:* further evidence for two cryptic species. Proceedings of the Royal Society B, 263:1495–1499.

Pearl, D. L., and M. B. Fenton. 1996. Echolocation calls provide information about group identity in the little brown bat, *Myotis lucifugus*. Canadian Journal of Zoology, 74:2184–2192.

Peterson, R. L. 1965. A review of the flat-headed bats of the family Molossidae from South American and Africa. Life Science Contributions, Royal Ontario Museum, 64:1–32.

Quay, W. B. 1970. Integument and derivatives. Pp. 2–56 *in:* Biology of Bats (W. A. Wimsatt, ed.). Vol. II. Academic Press, New York.

Racey, P. A., and S. M. Swift. 1985. Feeding ecology of *Pipistrellus pipistrellus* (Chiroptera: Vespertilionidae) during pregnancy and lactation. I. Foraging behaviour. Journal of Animal Ecology, 54:205–215.

Ransome, R. D. 1991. The greater horseshoe bat. Pp. 88–94 in: The Handbook of British Mammals (G. B. Corbet and S. Harris, eds.). Vol. II. Blackwell, Oxford.

Reid, F. A. 1997. A field guide to the mammals of Central America and southeast Mexico. Oxford University Press, New York.

Rieger, J. M., and E. M. Jakob. 1988. The use of olfaction in food location by frugivorous bats. Biotropica, 20:161–164.

Russ, J. M., P. A. Racey, and G. Jones. 1998. Intraspecific responses to distress calls of the pipistrelle bat Pipistrellus pipistrellus. Animal Behaviour, 55:705–713.

Ryan, M. J., and M. D. Tuttle. 1987. The role of prey-generated sounds, vision and echolocation in prey localization by the African bat Cardioderma cor (Megadermatidae). Journal of Comparative Physiology A, 161:59–66.

Ryan, M. J., M. D. Tuttle, and R. M. R. Barclay. 1983. Behavioral responses of the frog-eating bat, Trachops cirrhosus, to sonic frequencies. Journal of Comparative Physiology A, 150:413–418.

Rydell, J. 1992. Occurrence of bats in northernmost Sweden and their feeding ecology in summer. Journal of Zoology (London), 227:517–529.

Scherrer, J. A., and G. S. Wilkinson. 1993. Evening bat isolation calls provide evidence for heritable signatures. Animal Behaviour, 46:847–860.

Schmidt, U. 1985. The bats: order Chiroptera. Pp. 17–234 in: Social Odours in Mammals (R. Brown and D. Macdonald, eds.). Oxford University Press, New York.

Schmidt, U., P. Schlegel, H. Schweizer, and G. Neuweiler. 1991. Audition in vampire bats, Desmodus rotundus. Journal of Comparative Physiology A, 168:45–51.

Scully, W. M. R., M. B. Fenton, and A. S. M. Saleuddin. 2000. A histological examination of holding sacs and scent glandular organs of some bats (Emballonuridae, Hipposideridae, Phyllostomidae, Vespertilionidae and Molossidae. Canadian Journal of Zoology, 78:613–623.

Simmons, N. B., and J. H. Geisler. 1998. Phylogenetic relationships of Icaronycteris, Archaeonycteris, Hassianycteris and Palaeochiropteryx to extant bat lineages, with comments on the evolution of echolocation and foraging strategies in Microchiroptera. Bulletin of the American Museum of Natural History, 235:1–182.

Simmons, N. B., and R. S. Voss. 1998. The mammals of Paracou, French Guiana: a Neotropical lowland rainforest fauna. Pt. 1. Bats. Bulletin of the American Museum of Natural History, 237:1–219.

Sluiter, J. W., and P. F. van Heerdt. 1966. Seasonal habits of the noctule bat (Nyctalus noctula). Archives Neerlandaise de Zoologie, 16:423–439.

Smith, W. J. 1977. The behavior of communicating. Harvard University Press, Cambridge, Mass.

Stoddart, D. M. 1980. The ecology of vertebrate olfaction. Chapman & Hall, New York.

Strauss, J. S., and F. J. Ebling. 1970. Control and function of skin glands in mammals. Memoirs of the Society for Endocrinology, 18:341–371.

Suthers, R. A. 1965. Acoustic orientation by fish-catching bats. Journal of Experimental Zoology, 158:319–348.

Suthers, R. A. 1966. Optomotor responses by echolocating bats. Science, 152:1102–1104.

Suthers, R. A. 1970. Vision, olfaction and taste. Pp. 265–309 in: Biology of Bats (W. A. Wimsatt, ed.). Vol. 2. Academic Press, New York.

Suthers, R. A., and N.E. Wallis. 1970. Optics of the eyes of echolocating bats. Vision Research, 10:1165–1173.

Suthers, R. A., J. Chase, and B. Braford. 1969. Visual form discrimination by echolocating bats. Biological Bulletin, 137:535–546.

Thies, W., E. K. V. Kalko, and H.-U. Schnitzler. 1998. The roles of echolocation and olfaction in two Neotropical fruit-eating bats, Carollia perspicillata and C. castanea, feeding on Piper. Behavioral Ecology and Sociobiology, 42:397–409.

Thomas, D. W. 1984. Fruit intake and energy budgets of frugivorous bats. Physiological Zoology, 57:457–467.

Thomas, D. W., M. B. Fenton, and R. M. R. Barclay. 1979. Social behavior of Myotis lucifugus. I. Mating behavior. Behavioral Ecology and Sociobiology, 6:129–136.

Tidemann, C. R., D. M. Priddel, J. E. Nelson, and J. D. Pettigrew. 1985. Foraging behaviour of the Australian ghost bat, Macroderma gigas (Microchiroptera: Megadermatidae). Australian Journal of Zoology, 33:705–713.

Trune, D. R., and C. N. Slobodchikoff. 1978. Position of immatures in pallid bat clusters: a case of reciprocal altruism? Journal of Mammalogy, 59:193–195.

Tuttle, M. D. 1976. Population ecology of the grey bat (Myotis grisescens): philopatry, timing and patterns of movement, weight loss during migration and seasonal adaptive strategies. Occasional Papers, Museum of Natural History, University of Kansas, 54:1–38.

Vaughan, T. A. 1976. Nocturnal behavior of the African false vampire (Cardioderma cor). Journal of Mammalogy, 57:227–248.

Vaughan, T. A., and T. J. O'Shea. 1976. Roosting ecology of the pallid bat, Antrozous pallidus. Journal of Mammalogy, 57:227–248.

Voigt, C. C., and O. von Helversen. 1999. Storage and display of odour by male Saccopteryx bilineata (Chiroptera, Emballonuridae). Behavioral Ecology and Sociobiology, 47:29–40.

von der Emde, G., and H.-U. Schnitzler. 1990. Classification of insects by echolocating greater horseshoe bats. Journal of Comparative Physiology A, 167:423–430.

Vonhof, M. J. 1998. Social organization of the Neotropical disk-winged bat, Thyroptera tricolor. Bat Research News, 39:191.

Wickler, W., and U. Seibt. 1976. Field studies on the African fruit bat Epomophorous wahlbergi (Sundevall), with special reference to male calling. Zeitschrift für Tierpsychologie, 40:345–376.

Wilkins, K. T. 1989. Tadarida brasiliensis. Mammalian Species, 331:1–10.

Wilkinson, G. S. 1984. Reciprocal food sharing in the vampire bat. Nature, 308:181–184.

Wilkinson, G. S. 1985. The social organization of the common vampire bat. I. Pattern and cause of association. Behavioral Ecology and Sociobiology, 17:111–121.

Wilkinson, G. S., and J. W. Boughman. 1998. Social calls coordinate foraging in greater spear-nosed bats. Animal Behaviour, 55:337–350.

Williams, T. C., and J. M. Williams. 1967. Radio tracking of homing bats. Science, 155:1435–1436.

Williams, T. C., and J. M. Williams. 1970. Radio tracking of homing and flights of a Neotropical bat, Phyllostomus hastatus. Animal Behaviour, 18:302–309.

Williams, T. C., J. M. Williams, and D. R. Griffin. 1966. The homing ability of the Neotropical bat Phyllostomus hastatus. Science, 155:1435–1436.

CHAPTER 3

Bats and Balls: Sexual Selection and Sperm Competition in the Chiroptera

Gerald S. Wilkinson and Gary F. McCracken

Introduction

Darwin (1871) recognized that competition between individuals for access to mates can lead to the evolution of exaggerated ornaments or weapons when these traits influence mating success. He referred to this process as sexual selection and distinguished two mechanisms—male competition and female choice—by which it could occur. Recognition that sexual selection can also operate after mating was not widely appreciated until Parker's (1970) seminal work led to the realization that competition among mates can occur after copulation.

Some of the earliest and subsequently most consistent evidence for sexual selection by sperm competition has come from comparisons of testes size between species that differ in mating system. If sperm are costly to produce (e.g., Dewsbury 1982; Nakatsura and Kramer 1982; Olsson et al. 1997), and a male's probability of fertilization is proportional to the relative representation of his sperm in the female reproductive tract, as appears to be true for many mammals (Martin et al. 1974), then sexual selection should favor greater sperm production capability—that is, larger testes—when females are promiscuous. In internally fertilizing species, larger testes are predicted when females have opportunities to mate with multiple males, while smaller testes are expected in monogamous or highly polygynous species in which one male controls access to a female or group of females. These predictions have been upheld for many vertebrate groups, including primates (Harcourt et al. 1981, Short 1979), whales (Brownell and Ralls 1986), other mammals excluding bats (Kenagy and Trombulak 1986), birds (Møller 1988), and fishes (Stockley et al. 1997).

Sexual selection by sperm competition might be expected to be intense in bats for several reasons. In most other mammals, sperm are capable of fertilizing eggs for only a few days after copulation (Birkhead and Møller 1993; Gomendio et al. 1998). A short sperm lifespan sets a narrow window of time during which females must mate with more than one male for sperm competition to occur. In contrast, many female bats are capable of storing viable sperm for up to 6 mo (Racey 1979). Although sperm storage was first observed in a hibernating temperate species (Wimsatt 1942), sperm storage is now known to occur in a variety of tropical and temperate species (Racey and Entwistle 2000). Many bats also exhibit seasonal variation in testes size (Gustafson 1979, Heideman et al. 1992; Krutzsch 1979; O'Brien 1993) with changes up to 40-fold in mass (Racey and Tam 1974). Only males in good body condition initiate spermatogenesis (Entwistle et al. 1998, Speakman and Racey 1986). These observations are consistent with an energetic cost to sperm production and suggest that sperm is limiting for males, a necessary requirement for sexual selection to operate on sperm production.

Several lines of evidence suggest that female bats also remate often enough, at least in some species, for sperm to compete. Estimates of roost aggregation size positively covary with relative testes mass across 31 microchiropteran (Hosken 1997b) and 17 megachiropteran bat species (Hosken 1998b). These results have been interpreted as indicating that testes have become larger relative to body size in those species that form large colonies because sperm competition risk increases with group size in bats (Hosken 1997b, 1998b). Although few direct observations are available, females of several species have been observed to mate with multiple males during estrus, including *Desmodus rotundus* (Wilkinson 1985), *Tadarida brasiliensis* (A. Keeley and B. Keeley, unpublished manuscript), and *Phyllostomus discolor* and *Molossus ater* (Rasweiler 1987). Females of some sperm-storing species also may mate both before and during hibernation, for example, *Corynorhinus townsendii* (Pearson et al. 1952), *Myotis lucifugus* (Thomas et al. 1979), and *Nyctalus noctula* (Gebhard 1995). Sperm genotyping has demonstrated that females of *N. noctula* may have sperm from up to five males in their uteruses prior to conception (F. Mayer, personal communication), consistent with genetic evidence for multiple paternity (Mayer 1995). Paternity also differs from observed mating patterns in at least two species, *M. lucifugus* (Watt and Fenton 1995) and *Nyctophilus geoffroyi* (Hosken 1998a), as might be expected if ejaculate size, the competitive ability of sperm, or female storage/use of sperm differs across males. Thus, there is clearly opportunity for sperm competition to occur in bats.

While there has been substantial research on bat mating systems (reviewed in McCracken and Wilkinson 2000), sperm competition has received little study in bats (Fenton 1984; Hosken 1997b, 1998b) compared to the attention it has received in other mammals (Ginsberg and Huck 1989; Gomendio et al. 1998; Møller and Birkhead 1989). As yet, no one has tested if mating system influences testes size in bats. In this chapter we compile measurements of bat testes mass from published sources, capture records made by ourselves or colleagues, or measurements of museum specimens. Because testes mass covaries with body mass in other mammals (Kenagy and Trombulak 1986), we determine if the allometric relationship between testes mass and body mass in bats differs from other groups of mammals. Using information compiled in McCracken and Wilkinson (2000), we then categorize species according to male-female association and potential for female promiscuity to test whether variation in testes size among species is consistent with the sexual selection hypothesis. In addition, we consider several alternative explanations for testes size including colony size (Hosken 1997b, 1998b), breeding seasonality (Hosken 1998b; Kenagy and Trombulak 1986; Stockley and Purvis 1993), sperm storage (Hosken 1997b), and genetic potential for gametic dysfunction (Cohen 1967, 1973; Manning and Chamberlain 1994). Finally, we compare testes size between populations for two species of bats to determine if intraspecific differences exist and could be consistent with sexual selection.

Methods

We obtained testis mass from either direct measurements or estimation of testicular volume. Testicular volume was approximated as a prolate spheroid, that is: $(4/3)\pi r^2 \times \text{length}/2 = 0.5236 \times \text{length} \times \text{width}^2$ (Myers 1977). The product-moment correlation between mass and estimated volume for seven *Myotis* testes was $r = 0.996$ (Myers 1977) and for nine bat species $r = 0.964$ (Hosken 1998b). We obtained mass and volume estimates for testes from five species—*Lasiurus ega, Eptesicus furinalis, Pteropus tonganus, Eidolon helvum,* and *Mops condylura*—which range in body mass from 7.5 to 570 g. The correlation between $\log_{10}$ testis mass and $\log_{10}$ testis volume for these five species was also 0.996 ($P < 0.001$). Furthermore, the regression equation was $\log_{10}$ testis mass = 0.021 ($\pm$ 0.054) +1.024 ($\pm$ 0.051) $\log_{10}$ testis volume, indicating that testis volume can be used as an unbiased estimate of testis mass. For some species, only testis length was reported. From 10 species, we calculated the average testis width to be 0.66 $\times$ testis length, and then used this factor to estimate testis width from testis length when direct measurements of testis width were not available.

Pronounced seasonal variation in the size and visibility of the testes (Gustafson 1979; Krutzsch 1979) complicates comparative study of testes size in bats. Therefore, when seasonal variation in testes size was available (e.g., *Myotis albescens, M. nigricans, Eptesicus furinalis, Lasiurus ega* [Myers 1977]; *Nyctalus noctula* [Racey 1974], *Plecotus auritus* [Entwistle et al. 1998], *Pipistrellus pipistrellus* [Racey and Tam 1974], *Coleura afra* [McWilliam 1987], *Mops condylura* [Vivier and van der Merwe 1996], *Haplonycteris fischeri* [Heideman 1989], *Anoura geoffroyi* [Heideman et al. 1992], and *Pteropus poliocephalus* [McGuckin and Blackshaw 1991]), we report mean testis mass from the period of maximal testicular size. When such information was not available, we only measured scrotal testes, which often indicates spermatogenesis (Entwistle et al. 1998; Hosken et al. 1998). We include sample sizes when available for each species.

To allow comparison with previous allometric studies (Kenagy and Trombulak 1986), we doubled testis mass to estimate the combined mass of a pair of testes and used body mass of the same males scored for testis mass whenever possible. Because testes mass covaried allometrically with body mass, we use the residuals from a regression of $\log_{10}$ combined testes mass (CTM) on $\log_{10}$ body mass (BM) to test alternative hypotheses regarding testes size evolution. Sperm storage was scored from reviews of bat reproductive biology (Bernard 1989; Bernard and Cumming 1997; Gustafson 1979; Krutzsch 1979; Racey 1979; Racey and Entwistle 2000). We assume that females do not store sperm if spermatogenesis or copulations coincide with ovulation. Annual reproductive opportunities were categorized as one, two, or continuous (Bernard and Cumming 1997; Krutzsch 1979). Two sources (Baker 1979; Hsu and Benirschke 1977) were consulted for chromosome complement number. Estimates of colony size were taken from papers on sperm competition in microchiropteran (Hosken 1997b) and megachiropteran (Hosken 1998b) bats.

To determine if mating system influences testes size, we use two variables to categorize bat mating systems: (1) number of females associating with males during mating periods and (2) female opportunity for multiple mating. The number of females roosting with or visiting males was categorized into single-male/single-female, single-male/multiple-female, or multiple-male/multiple-female associations. Female mating opportunities were dichotomized as promiscuous or not. Single-male/single-female systems include those species in which male-female pairs typically roost together for extended periods during and beyond the mating period, as in *Vampyrum spectrum* (Vehrencamp et al. 1977) and *Lavia frons* (Vaughan and Vaughan 1986). Single-male/multiple-female systems include species typically classified as harem-holding systems, for example, *Phyllostomus hastatus* (McCracken and Bradbury 1981) and *Saccopteryx bilineata* (Bradbury and Emmons 1974), in which males associate with and defend a group of females for extended periods. This category also includes species in which female group associations are more labile, for example, *Carollia perspicillata* (Fleming 1988; Williams 1986), *Artibeus jamaicensis* (Kunz et al. 1983; Morrison 1979), and *Pipistrellus pipistrellus* (Gerell and Lundberg 1985; Lundberg and Gerell 1986), and species in which multiple females visit and mate, but do not roost, with displaying males, for example, epomophorine bats (Bradbury 1977, 1981). Multiple-male/multiple-female groups include species in which males and females roost together during the mating period, as in *Myotis lucifugus* (Thomas et al. 1979), *Rhynchonycteris naso* (Bradbury and Vehrencamp 1976), and *Pteropus poliocephalus* (Nelson 1965). We classified the mating system of 45 species into one of these three categories based on roosting association (table 3.1).

Table 3.1. Male body mass (BM), combined testes mass (CTM), roosting associations, opportunity for promiscuity (Prom.), sperm storage by sex, number of reproductive events per year (R/yr), and diploid chromosome complement (2n) for bats used in the study

Family/Species	BM (g)	CTM (mg)	% BM	N	Roost Assn.	Prom.	Sperm Store	R/yr	2n	References
Emballonuridae:										
Coleura afra	10.5	60	0.57	...	SM-MF	No	n	2	...	McWilliam 1987
Rhynchonycteris naso	3.8	17	0.45	4	MM-MF	No	n	1	...	Plumpton and Jones 1992
Saccopteryx bilineata	7.4	11	0.14	26	SM-MF	No	n	1	...	C. Voigt, personal communication
Saccopteryx leptura	4.1	25	0.61	6	SM-SF	No	n	1	...	N. Simmons, personal communication
Taphozous georgianus	30.0	76	0.25	...	...	...	m	1	...	Jolly and Blackshaw 1988; Kitchener 1973
Taphozous hildegardeae	33.0	138	0.42	4	...	...	f	1	...	McWilliam 1988b
Taphozous longimanus	36.0	77	0.21	...	...	...	...	C	42	Singh 1997
Saccolaimus peli	100.0	575	0.58	2	SM-SF	...	...	1	...	G. Wilkinson, AMNH
Megadermatidae:										
Cardioderma cor	26.0	80	0.31	12	SM-SF	...	...	2	...	G. Wilkinson, AMNH
Lavia frons	23.4	26	0.11	12	SM-SF	No	...	1	...	G. Wilkinson, AMNH
Macroderma gigas	108.0	168	0.16	1	...	...	...	1	...	G. Wilkinson, AMNH
Megaderma lyra	38.6	155	0.40	5	...	...	...	1	...	G. Wilkinson, AMNH
Megaderma spasma	22.6	113	0.50	10	...	...	...	1	...	G. Wilkinson, AMNH
Molossidae:										
Mormopterus planiceps	8.0	185	2.32	4	...	No	f	...	...	Krutzsch and Crighton 1987
Otomops martiensseni	45.0	143	0.32	76	SM-MF	No	...	1	...	Mutere 1973
Tadarida aegyptiaca	14.9	132	0.89	...	...	...	n	1	...	Hosken 1997b
Tadarida brasiliensis	11.1	91	0.81	146	MM-MF	Yes	n	1	48	A. Nicklaus, personal communication
Mops condylura	27.9	138	0.49	8	...	...	n	2	...	Vivier and van der Merwe 1996
Chaerophon pumila	10.8	38	0.36	1	SM-MF	No	n	C	...	G. Wilkinson
Mormoopidae:										
Pteronotus davyi	9.2	174	1.89	2	...	...	...	...	...	Adams 1989
Noctilionidae:										
Noctilio albiventris	31.2	64	0.21	7	...	...	n	...	34	Hood and Pitocchelli 1983
Noctilio leporinus	64.8	177	0.27	18	SM-MF	No	n	1	...	A. Brooke, personal communication

Phyllostomidae:

Species										Reference
Anoura cultrata	17.1	95	0.58	...	...	...	n	...	30	Tamsitt and Nagorsen 1982
Anoura geoffroyi	15.1	50	0.33	12	...	...	n	...	30	Heideman et al. 1992
Ardops nichollsi	16.9	95	0.58	...	...	...	n	...	30	Jones and Genoways 1973
Artibeus jamaicensis	47.6	314	0.66	15	SM-MF	...	n	2	30	Taft and Handley 1991
Artibeus jamaicensis	40.0	51	0.13	32	SM-MF	No	n	2	30	J. Ortega, personal communication
Artibeus literatus	79.0	229	0.29	...	...	...	n	2	30	G. Wilkinson
Artibeus phaeotis	12.0	50	0.42	...	...	...	...	...	30	Timm 1985
Brachyphyllum cavernum	40.1	64	0.16	...	SM-MF	No	...	...	32	Swanepoel and Genoways 1983
Carollia perspicillata	18.5	125	0.68	...	SM-MF	No	...	2	20	Fleming 1988
Centurio senex	22.9	80	0.35	...	...	...	...	2	28	Snow et al. 1980
Chrotopterus auritus	92.1	182	0.20	4	SM-SF	...	...	1	28	G. Wilkinson, AMNH
Desmodus rotundus	33.7	83	0.25	68	SM-MF	No	n	C	28	G. Wilkinson, personal communication
Ectophylla alba	5.5	70	1.45	25	SM-MF	...	n	2	30	A. Brooke, personal communication
Glossophaga leachii	8.4	29	0.35	...	...	...	n	...	32	Webster and Jones 1984
Leptonycteris curasoae	29.0	265	0.91	...	...	...	n	2	...	T. Fleming, personal communication
Macrotus waterhousi	16.0	37	0.23	...	...	...	n	...	46	Krutzsch et al. 1976
Monophyllus plethodon	15.5	42	0.27	...	...	...	n	...	32	Homan and Jones 1975
Phyllostomus discolor	44.9	649	1.45	56	SM-MF	No	n	2	32	G. Wilkinson
Phyllostomus hastatus	91.8	308	0.34	10	SM-MF	No	n	1	32	G. Wilkinson
Sturnira bidens	17.5	58	0.33	...	...	...	n	...	30	Molinari and Soriano 1987
Sturnira lilium	20.4	121	0.59	11	...	...	n	2	30	Gannon et al. 1989
Tonatia carrikeri	23.0	409	1.78	4	...	...	n	...	26	McCarthy et al. 1992
Uroderma bilobatum	14.9	98	0.66	...	SM-MF	...	n	2	44	Baker and Clark 1987
Vampyrodes caraccioli	26.9	332	1.24	2	...	...	n	...	30	Willis et al. 1990
Vampyrum spectrum	158.8	203	0.13	6	SM-SF	No	...	1	...	G. Wilkinson, AMNH
Pteropidae:										
Acerodon mackloti	518.0	6500	1.25	...	SM-MF	...	n	...	...	Hosken 1998b
Cynopterus sphinx	62.0	173	0.28	...	...	14	f	C	...	Storz et al. 2000
Dobsonia peroni	224.0	3280	1.46	...	...	...	...	...	...	Hosken 1998b
Eidolon helvum	325.0	5800	1.78	20	MM-MF	...	n	1	...	Mutere 1967

(continued on next page)

Table 3.1. (continued)

Family/Species	BM (g)	CTM (mg)	% BM	N	Roost Assn.	Prom.	Sperm Store	R/yr	2n	References
Eonycteris spelea	60.0	1596	2.66	...	MM-MF	...	n	C	...	Hosken 1998b
Epomophorus anurus	80.0	1800	2.25	...	...	...	...	2	...	Hosken 1998b
Epomops buttikoferi	178.0	358	0.20	6	SM-MF	...	n	1	...	Kofron and Chapman 1994
Haplonycteris fischeri	18.0	190	1.06	5	...	...	n	...	...	Heideman 1989
Hypsignathus monstrosus	310.0	471	0.15	2	SM-MF	...	f	2	35	G. Wilkinson, AMNH
Macroglossus minimus	15.9	574	3.61	...	...	...	...	...	...	Hosken 1998b
Macroglossus sobrinus	23.0	634	2.76	...	...	...	...	...	...	Hosken 1998b
Micropteropus pusillus	32.5	460	1.42	...	...	...	n	2	...	Hosken 1998b
Pteropus alecto	800.0	10200	1.28	...	MM-MF	Yes	...	...	...	Hosken 1998b
Pteropus giganteus	1021.0	16800	1.65	...	MM-MF	...	...	1	38	Hosken 1998b
Pteropus oetinus	470.0	11000	2.34	...	...	...	...	1	...	Hosken 1998b
Pteropus poliocephalus	860.0	17000	1.98	7	MM-MF	Yes	m	1	38	McGuckin and Blackshaw 1991
Pteropus scapulatus	400.0	14000	3.50	6	MM-MF	Yes	...	1	...	O'Brien et al. 1993
Pteropus tonganus	568.3	5400	0.95	2	SM-MF	...	m	...	...	A. Brooke, personal communication
Rousettus aegyptiacus	142.0	7000	4.93	...	...	...	n	2	...	Hosken 1998b
Rousettus amplexicaudatus	91.8	1480	1.61	...	...	...	...	...	...	Hosken 1998b
Rhinolophidae:										
Hipposideros galeritus	6.5	60	0.92	...	SM-SF	...	f	...	...	Hosken 1997b
Hipposideros speoris	10.0	80	0.80	...	...	...	f	...	...	Hosken 1997b
Rhinolophus clivosus	16.2	350	2.16	6	...	...	f	...	...	Bernard 1983
Rhinolophus hipposideros	7.0	38	0.54	...	...	...	f	...	...	Hosken 1997b
Rhinonycteris aurantius	9.8	12	0.12	...	...	...	...	...	...	Hosken 1997b
Rhinopomatidae:										
Rhinopoma hardwickei	11.0	251	2.28	1	...	...	f	1	...	Karim and Banerjee 1989
Rhinopoma microphyllum	31.0	1255	4.05	...	...	...	...	1	...	Anand Kumar 1965
Vespertilionidae:										
Antrozous pallidus	32.0	330	1.03	2	...	...	f	1	46	Beasley and Zucker 1984
Chalinolobis gouldi	14.0	29	0.21	...	...	...	f	1	...	Hosken 1997b
Eptesicus furinalis	7.5	154	2.05	12	...	...	f	2	...	Myers 1977
Euderma maculatum	15.0	66	0.44	...	...	...	f	1	...	Watkins 1977

Species										Reference
Lasiurus ega	11.9	45	0.39	8	...	...	f	1	...	Myers 1977
Lasiurus seminolus	9.3	13	0.14	...	...	...	f	...	28	Wilkins 1987
Miniopterus australis	7.5	40	0.53	...	SM-MF	No	f	1	...	Medway 1971
Miniopterus minor	7.0	50	0.71	3	MM-MF	Yes	...	1	...	McWilliam 1988a
Miniopterus schreibersi	13.0	76	0.58	9	...	...	n	...	...	Krutzsch and Crichton 1990
Myotis adversus	10.0	143	1.43	10	SM-MF	Yes	...	2	...	G. Wilkinson, NMNH
Myotis albescens	5.5	368	6.69	13	MM-MF	Yes	f	2	...	Myers 1977
Myotis bocagei	8.0	80	1.01	2	SM-MF	No	f	...	...	G. Wilkinson, NMNH
Myotis evotis	6.8	192	2.83	2	...	...	f	...	...	Manning and Jones 1989
Myotis lucifugus	6.8	108	1.59	6	MM-MF	Yes	f	1	...	Gustafson and Damassa 1985
Myotis nigricans	4.0	26	0.65	22	SM-MF	No	n	C	...	Myers 1977
Nyctalus noctula	28.0	540	1.93	2	SM-MF	Yes	f	1	...	Racy 1974
Nyctophilus geoffroyi	6.5	46	0.71	...	...	...	f	1	...	Hosken 1977b
Nyctophilus gouldi	14.0	217	1.55	1	...	...	f	1	...	Phillips and Inwards 1985
Nyctophilus major	13.5	60	0.44	...	...	...	f	1	...	Hosken 1997b
Pipistrellus mimus	3.6	53	1.47	...	...	...	f	1	38	Hosken 1997b
Pipistrellus nanus	3.5	31	0.90	7	SM-MF	Yes	f	1	...	Bernard et al. 1997; Happold and Happold 1990
Pipistrellus pipistrellus	5.0	230	4.60	...	SM-MF	Yes	f	1	...	Racey and Tam 1974
Pipistrellus rusticus	4.0	310	7.75	7	...	...	f	1	...	van der Merwe and Rautenbach 1990
Pipistrellus subflavus	5.0	59	1.18	3	...	...	f	1	30	Krutzsch and Crichton 1986
Plecotus auritus	7.2	44	0.62	55	MM-MF	...	f	1	...	Entwistle et al. 1998
Corynorhinus rafinesquii	9.0	754	8.38	...	MM-MF	Yes	f	1	...	Pearson et al. 1952
Rhogeesa tumida	5.1	77	1.51	1	...	...	...	1	...	G. Wilkinson
Scotophilus borbonicus	22.0	290	1.32	...	...	...	n	1	...	G. Wilkinson
Scotophilus heathii	45.0	387	0.86	6	...	...	f	1	...	Singh and Krishna 1996
Tylonycteris pachypus	4.1	140	3.41	...	MM-MF	Yes	f	1	...	Medway 1972
Tylonycteris robustula	8.4	200	2.38	...	MM-MF	Yes	f	1	...	Medway 1972

Note. MM-MF = multiple males roosting with multiple females; SM-MF = single male roosting with multiple females; SM-SF = single male roosting with single female; n = neither; m = male; and f = female; AMNH = American Museum of Natural History; NMNH = National Museum of Natural History. Ellipses dots indicate variables for which no data were available.

We would prefer to classify promiscuous systems as those in which females routinely mate with multiple males. However, such information is not available for most bats. Therefore, we operationally classified a species as promiscuous when available evidence indicates that females do not maintain consistent association with a single male at a roost throughout the mating period. Promiscuous mating occurs in many species of vespertilionid bats in which males will mate with females before hibernation in the fall and while females are in torpor during the winter, as in *Myotis lucifugus* (Thomas et al. 1979) and *Nyctalus noctula* (Gebhard 1995), but it has also been observed in nonhibernal mating species, such as *Tadarida brasiliensis* (A. Keeley and B. Keeley, unpublished manuscript). Promiscuous mating also occurs in single-male/multiple-female systems in which females move among different males during the mating season, for example, *Pipistrellus pipistrellus* (Gerell and Lundberg 1985; Lundberg and Gerell 1986) and *Pipistrellus nanus* (Happold and Happold 1996; O'Shea 1980), as well as in multiple-male/multiple-female systems. We were able to classify the mating systems of 30 species as promiscuous or not (table 3.1).

Close phylogenetic relationships can confound comparative analyses. If a variable, such as testes size, does not undergo rapid evolutionary change, and species are treated as independent observations in a comparative analysis, then the significance associated with a test result may be overestimated. One of the best methods for dealing with this problem is to compare phylogenetically independent contrasts (Harvey and Pagel 1991) to determine if correlated evolution has occurred. Unfortunately, this method requires a robust phylogeny, preferably with estimated branch lengths, for all species in the analysis. While a well-supported phylogeny for familial relationships has recently been proposed (Simmons and Geisler 1998), no species-level phylogeny is currently available for the order. Therefore, to determine if phylogenetic effects contribute to any association, we conduct three additional analyses. First, we use analysis of covariance (ANCOVA) to determine if bat families exhibit different allometric relationships between testes size and body size and to quantify the extent to which familial relationships contribute to observed associations between testes size. Second, we identified, a priori, and compared multiple species within genera or subfamilies that differ in mating system or degree of female promiscuity. We then use species averages within these groups in a paired comparison to determine if there is associated change in testes size. Third, for two species, *Artibeus jamaicensis* and *Tadarida brasiliensis*, we obtained data on testes size for two or more populations. Independent evidence indicates that these populations may differ in opportunities for multiple mating by females, either as a consequence of female movement patterns or mating assemblage size. Thus, this final comparison allows us to determine if intraspecific variation in testes size shows patterns of change that would be predicted by sexual selection due to sperm competition.

Results

Allometric Relationships between Testes Mass and Body Mass

We obtained or estimated testes mass from a total of 104 species of bats (table 3.1), including 84 species from nine families of microchiropteran bats and 20 species of megachiropteran bats (family Pteropodidae). Male body mass varied from 3.8 g for *Rhynchonycteris naso* to 1,021 g for *Pteropus giganteus*. Expressed as a percentage of body mass, combined testes mass ranged from 0.12% for *Lavia frons* to 8.4% for *Corynorhinus rafinesquii* (see also fig. 3.1).

Body mass explained 61% of CTM variation ($F = 159.9$, df = 1, 102, $P < 0.0001$) among bat species (fig. 3.2). In most mammals, larger species have progressively smaller testes, relative to body size, that is, the exponent in the allometric relationship of combined testes mass (CTM) on body mass (BM) is less than 1. In contrast, bat testes increase proportionally with body size—that is, $CTM = (0.009 \pm 0.002)BM^{(0.947 \pm 0.075)}$. The 95% confidence interval on the bat body mass exponent exceeds the exponents for all other orders of mammals (Kenagy and Trombulak 1986): rodents (0.77), primates (0.68), insectivores (0.74), and carnivores (0.59).

Analysis of variance revealed a highly significant effect of family on residual CTM ($F = 6.22$, df = 9, 94, $P < 0.0001$) but no significant interaction between family and body mass. Combined testes mass of rhinopomatid, pteropodid, and vespertilionid bats exceeded allometric expectations, while the testes of phyllostomid, noctilionid, megadermatid, and emballonurid bats were smaller than expected for their body sizes (fig. 3.3). Therefore, in the

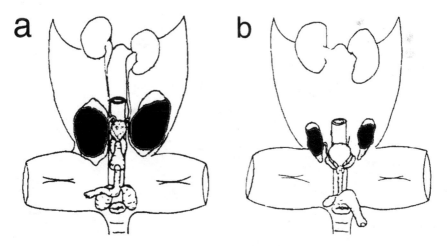

Figure 3.1. Anatomical perspective of the difference in testes size for breeding season males in (*a*) *Rhinopoma kinneari* (combined testes mass [CTM] approximately 4% body mass) and (*b*) *Taphozous longimanus* (CTM = 0.02%). (Redrawn from Krutzsch 1979.)

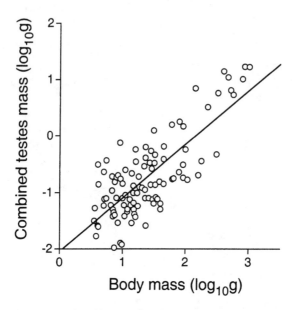

Figure 3.2. Plot of $\log_{10}$ combined testes mass on body mass for 104 bat species.

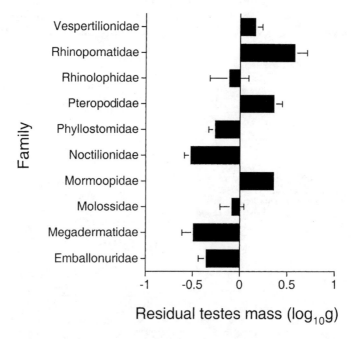

Figure 3.3. Mean ± SE residual combined testes mass, from the regression in figure 3.2, plotted for each family of bats in the data set.

analyses presented below we remove effects of body size by using the residuals from a regression of $\log_{10}$ CTM on $\log_{10}$ BM for all bat species. Then, if any variable exhibits significant differences in residual CTM, we conduct a two-way ANOVA to determine if the effect remains significant after removing familial differences.

Relative Testes Mass and Mating System

Comparison of residual CTM among the bat mating systems shows strong evidence for sexual selection by sperm competition (ANOVA: $F = 12.9$, df = 2, 43, $P < 0.0001$). Species in which multiple males roost with multiple females showed the largest relative testes, single-male/multiple-female species were intermediate in size, and the smallest relative testes occurred in single-male/single-female species (fig. 3.4). Fisher's partial least squares difference post hoc tests revealed that the two single-male categories did not differ from each other ($P = 0.12$), while the multiple-male category differed strongly from the other two ($P < 0.001$, for both comparisons). The effect of mating system appears to be independent of phylogeny, as family was not significant ($F = 1.59$, df = 7, 36, $P = 0.17$) while mating system remained significant ($F = 5.04$, df = 2, 36, $P = 0.012$) in a two-way ANOVA.

A comparison of residual CTM between promiscuous or non-promiscuous species revealed that promiscuous species have much larger testes ($t = 6.36$,

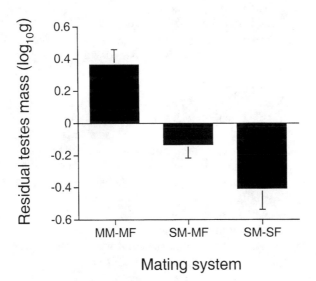

Figure 3.4. Mean ± SE residual combined testes mass for bats with mating period roosting associations that consist of multiple males and multiple females (*MM-MF*), single males and multiple females (*SM-MF*), and single males with single females (*SM-SF*).

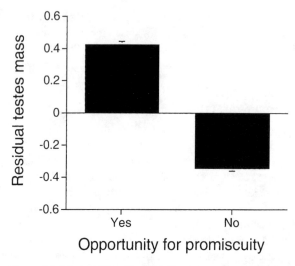

Figure 3.5. Mean ± SE residual combined testes mass for species categorized by their opportunity for promiscuity. Promiscuity indicates that females do not maintain consistent association with a single male at a roost throughout the mating period.

$P < 0.0001$; fig. 3.5). This result also appears to be independent of phylogeny, given that family was non-significant ($F = 1.03$, df = 6, 22, $P = 0.43$) while promiscuity remained significant ($F = 6.55$, df = 1, 22, $P = 0.018$) in a two-way ANOVA. Promiscuity appears to be a better predictor of residual testes mass than mating system, as promiscuity ($F = 15.6$, df = 1, 26, $P = 0.0005$), but not mating system ($F = 1.16$, df = 2, 26, $P = 0.33$), was significant in a two-way ANOVA. Mating system and promiscuity do not, however, occur independently across species. We classified 11 of 12 multiple-male species as promiscuous and 15 of 20 single-male species as non-promiscuous ($\chi^2 = 14.22$, $P = 0.0008$). Thus, any conclusion that residual testes size is influenced more by promiscuity than by mating system must be qualified given that this analysis was dependent on only a few species.

We used eight matched sets of closely related species—that is, either from the same genus or subfamily (table 3.2)—to determine if testes size exhibits correlated change with mating system. We predicted that bats in multiple-male roosting groups should have larger testes than bats in single-male roosting assemblages. In addition, we predicted that species with more stable roosting associations among females should exhibit smaller relative testes than species in which females move more frequently among males. For example, *P. hastatus* form more stable female groups than does *P. discolor* (Wilkinson 1987) and multiple males of *T. robustula* roost with multiple females less often than

Table 3.2. Phylogenetically independent paired comparisons used to test for effect of promiscuity and mating system on testes size

Multiple-Male Group/More Promiscuous	Single-Male Group/Less Promiscuous
Rhynchonycteris naso	*Saccopteryx leptura, Saccopteryx bilineata*
Tadarida brasiliensis	*Chaerophon pumila, Otomops martiensseni*
Phyllostomus discolor	*Phyllostomus hastatus*
P. discolor, P. hastatus	*Vampyrum spectrum*
Pteropus poliocephalus	*Pteropus tonganus*
Miniopteros minor	*Miniopteros australis*
Myotis albescens, Myotis lucifugus	*Myotis nigricans, Myotis bocagei, Myotis adversus*
Corynorhinus rafinesquii	*Plecotus auritus*
Tylonycteris pachypus	*Tylonycteris robustula*

Note. Multiple-male group/more promiscuous = either roost in multiple male, multiple female aggregations, or females exhibit less roosting group stability than closely related species. Single-male group/less promiscuous = either roost in single male, single female aggregations, or females show more roosting group stability than closely related species.

in *T. pachypus* (Medway 1972). Despite the relatively small sample size, the paired difference in average residual testes mass was highly significant (paired t-test: $t = 4.16$, $P = 0.0032$; fig. 3.6). Thus, testes size appears to have evolved rapidly in response to change in female opportunity for multiple mating.

Relative Testes Mass and Colony Size

We used estimates of colony size from Hosken (1997b, 1998b) to determine if residual testes mass covaries with roosting aggregation size for 48 species of bats. A least squares regression between $\log_{10}$ colony size and residual testes mass was significant ($F = 6.62$, df $= 1, 46$, $P = 0.013$), but colony size explained only 12.6% of the variation in residual testes size. When family and $\log_{10}$ colony size were included in an ANCOVA, both family ($F = 3.15$, df $= 6, 40$, $P = 0.013$) and colony size ($F = 4.72$, df $= 1, 40$, $P = 0.036$) were significant. Colony size did not, however, explain significant variation in residual testes mass when either promiscuity or mating system was included in a two-way ANOVA (respectively: $F = 0.06$, df $= 1, 10$, $P = 0.81$; $F = 0.80$, df $= 1, 18$, $P = 0.38$). Thus, colony size appears to be less important than the mating system for explaining variation in residual testes mass.

Relative Testes Mass, Sperm Storage, and Seasonality

Testes size might also be expected to increase if females store sperm because stored sperm should have more opportunity to interact. At the same time, if sperm suffer a constant rate of mortality during storage, then the duration of sperm storage should determine the amount of sperm that must be produced independently of any effects of sperm competition. To address these alternatives we used information on sperm storage patterns for 72 species. A comparison of residual testes size among these species reveals that sperm-storing

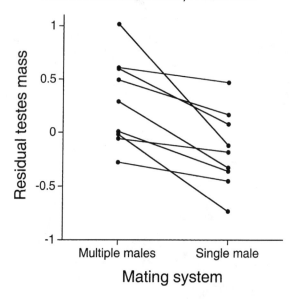

Figure 3.6. Mean ± SE residual combined testes mass for nine paired comparisons of species that differ in mating system (see table 3.2).

species do have relatively larger testes than species that do not store sperm ($t = 3.10$, $P = 0.003$). But, a two-way ANOVA using family and sperm storage shows that sperm storage is not significant when family is included in the model ($F = 0.81$, df $= 1, 63$, $P = 0.37$). Furthermore, if sperm competition was the only determinant of testes size, then species in which sperm is stored longer in females should have larger testes than species in which sperm is stored longer in males. In contrast, if sperm are produced to counter storage decay, then there should be no difference in testes size between male and female sperm-storing species. Although only three of 36 sperm-storing species exhibit longer male than female sperm storage—for example, *Taphozous georgianus*, in which sperm is produced by males in the fall and then stored in the cauda epididymides until copulation in the spring (Jolly and Blackshaw 1988)—we found no difference in residual testes mass between species with longer male than female sperm storage ($t = 0.46$, $P = 0.65$). Thus, while sperm storage may select for increased testes size, it exhibits strong phylogenetic patterns and does not depend on the sex where most storage occurs.

Males that breed only once per year might be expected to allocate more resources into sperm production and, therefore, testes size than species that breed two or more times per year (Kenagy and Trombulak 1986; Stockley and Purvis 1993). In contrast to this prediction, our information from 72 species reveals that the number of breeding episodes per year does not explain a significant fraction of the variation in residual combined testes mass (ANOVA: $F = 0.76$, df $= 2, 69$, $P = 0.47$). Species in which males breed once per year do not differ in residual testes size from those that breed continuously or twice per year.

Relative Testes Mass and Chromosome Complement

Alternatively, perhaps testes are larger in species prone to gamete dysfunction to insure some gametes survive until fertilization (Cohen 1967). One prediction from this hypothesis is that species with higher chromosome complements should produce more sperm to compensate for increased frequencies of recombination-related chromosome damage. Although support for this prediction has been obtained across a diverse assemblage of species (Cohen 1967, 1973; Manning and Chamberlain 1994), we found no evidence for it among bats. For the 31 species for which we have data, the product-moment correlation between diploid chromosome number and residual CTM was 0.134 ($P = 0.46$).

Intraspecific Variation in Testes Size

All of the analyses presented above ignore within-species variation. To test for geographic variation in testes size we compared CTM of *Artibeus jamaicensis* from Mexico (J. Ortega, personal communication) and Panama (Taft and Handley 1991) and of *Tadarida brasiliensis* from four colonies in Texas (A. Keeley, personal communication; fig. 3.7). Length and width of testes were measured on reproductively active males during the breeding season for both species. The *A. jamaicensis* measurements were taken during the mating season in caves on the Yucatan Peninsula, Mexico (Ortega and Arita 1999, 2000) or from a colony that had been captured in Panama and maintained at the National Zoological Park (Taft and Handley 1991). The *T. brasiliensis* measurements were taken between March 21 and March 30, 1998, in southwestern Texas during which time mating was observed.

Both species exhibited highly significant differences in testes size, independent of body size, between populations. *Artibeus jamaicensis* testes in Panama were about six times larger than those in Mexico ($t = 9.25$, $P < 0.0001$; fig. 3.6a). These two populations are approximately 1,600 km apart. The testes of *T. brasiliensis* also exhibited highly significant differences between sites ($F = 26.7$, df $= 3, 142$, $P < 0.0001$; fig. 3.6b). The results of Fisher's partial least squares difference post hoc tests are shown in figure 3.6b. Males at Eckert James River Cave had significantly larger testes than any other site, and Davis Cave males had larger testes than McNeil Bridge males. However, males at Davis Cave did not differ from males at South Grape Creek nor did males at South Grape Creek differ from McNeil Bridge. These caves are within 80 km

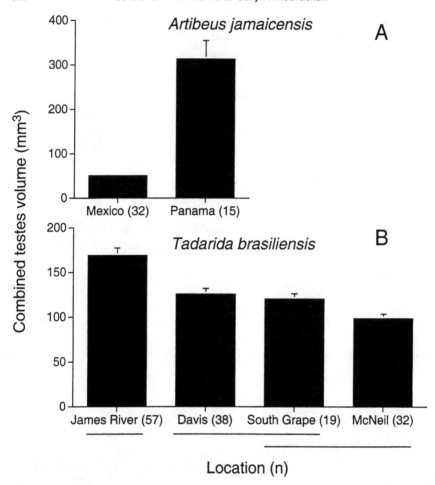

Figure 3.7. Mean ± SE combined testes volume for (*A*) two populations of *Artibeus jamaicensis* and (*B*) four colonies of *Tadarida brasiliensis*. Significant differences between sites according to Fisher's partial least squares differences post hoc comparisons are indicated by discontinuous lines beneath colony names.

of each other with the exception of McNeil Bridge, which is approximately 160 km from EckertJames River Cave, 100 km from Davis Cave, and 150 km from South Grape Creek.

Discussion

The range and maximal size of combined testes mass exhibited by different bat species is extraordinary, ranging from 0.12% to 8.4% of body mass. In a survey of 133 mammal species, which included only one bat, rodents exhibited the greatest range in testes size—from 0.15% for desert woodrats to 8.4% for gerbils (Kenagy and Trombulak 1986). Primates, which have been widely used as evidence that sexual selection influences testes size (Harcourt et al. 1981; Harvey and Harcourt 1984), exhibit combined testes mass ranging only from 0.02% to 0.75% of body mass. Thus, bats show variation in relative testes size that exceeds any other mammalian order. Allocation of body mass to testicular tissue in *Corynorhinus rafinesquii* not only equals or exceeds any other mammal but also exceeds any bird. The largest testes yet reported for birds occur in the alpine accentor with combined testes mass 7.7% of body mass (Nakamura 1990).

Our results strongly implicate sexual selection by sperm competition as the evolutionary explanation for both the range and maximal size of testes in bats. Species in which multiple males roost with multiple females during the breeding period have much larger testes than those in which single males attend single or multiple females. In addition, the fidelity of females to males at particular roosts has an independent effect on testes size suggesting that females are more likely to mate with multiple males when they move among roosts. The strong effect of promiscuity in our analysis suggests that female behavior facilitates sperm competition and causes sexual selection for testes size in bats.

Although we made considerable effort to obtain testes mass and volume during periods of maximal testes size when spermatogenesis is most likely to occur, little information on the reproductive cycle or mating behavior has been published for single-male/single-female species, such as *Vampyrum spectrum* and *Saccolaimus peli*. Since few specimens of these species were available for measurement, we could have underestimated mean testes size if the specimens we measured were collected during periods of testicular regression. However, we minimized this risk by restricting measurements to specimens with descended testes. Furthermore, residual testes size did not differ between single-male/single-female and single-male/multiple-female species, and the data for many species in the latter category could not have suffered this potential source of measurement error. Thus, while seasonal variation in testes size undoubtedly contributes unexplained variation in our data, it cannot account for the effects of mating system and female promiscuity that we obtained.

Several lines of evidence indicate that evolution of testes size is not constrained by phylogeny. Even though we did not conduct a formal independent contrasts analysis, the effects of the mating system on residual testes size did not disappear when differences among families were removed in the analysis. Furthermore, comparisons of residual testes size using species from the same subfamily or genus indicate that testis size can evolve rapidly and independently of body size (fig. 3.6). These results are consistent with the evolution of testes size occurring in response to the mating system.

Evolution of testes size may also be occurring within species, as indicated by significant variation in testes size between populations of both *Artibeus jamaicensis* and *Tadarida brasiliensis*. The differences between populations of either species cannot be ascribed to measurement bias because the differences are too large (*A. jamaiciensis*) or because the same person conducted all of the measurements (*T. brasiliensis*). Within species differences in testes size of *Miniopterus australis* have previously been attributed to seasonal variation caused by altitudinal differences between roost sites (Dwyer 1968). Seasonal variation is not likely, however, to explain the testes size differences we observed, at least for *T. brasiliensis* because measurements were taken over a 9-d period from nearby roosts and persist even when colonies measured on successive days are compared.

Nevertheless, other explanations for intraspecific variation in testes size are possible. For example, maximal testicular volume might be attained at different times at each *T. brasiliensis* colony. Alternatively, males at small colonies may be actively excluded from larger colonies, perhaps because they differ in age. Age has been found to contribute to population differences in relative testes size in the greenfinch, *Carduelis chloris* (Merilä and Sheldon 1999). At the same time, some of the intraspecific differences in testes size are consistent with sexual selection. *Artibeus jamaicensis* in Panama has been reported to form harems in hollow trees (Morrison 1979; Morrison and Morrison 1981). Females show some fidelity to one or more preferred tree cavities, but they also move between cavities. Thus, female groups are labile and a given female may visit more than one male roost during the breeding season. In contrast, in the Mexican cave-roosting population of *A. jamaicensis,* female group composition was stable from March to September through the breeding season (Ortega and Arita 1999). Thus, we expect that opportunities for multiple mating by females are greater in Panama than in Mexico.

The mating system of *T. brasiliensis* has not been well characterized, although males have been observed displaying to females using vocalizations (Lollar 1995), and females have been observed mating with multiple males in the field (A. Keeley, personal communication). The most obvious difference between sites is the size of the roosting population. Eckert James River Cave contains a large maternity colony with an estimated 4–6 million individuals present during the summer. The Davis Cave colony is substantially smaller, perhaps approaching 1 million bats, while South Grape Creek and McNeil Bridge each contain about 100,000 bats. Because testes size increases with colony size across bat species (Hosken 1997b, 1998b), the differences in testes size between these four colonies might also reflect differences in risk of sperm competition.

Bats also exhibit a steeper allometric relationship between testes mass and body mass than any other mammalian order. An explanation for this difference is not obvious. It seems intuitive that testes mass should increase as body mass increases, but what scaling relationship should be expected? Presumably, the answer depends on the volume in which sperm can interact within the female and how that volume scales with body mass. One possibility is that female bats store more sperm in their reproductive tracts than other mammals relative to body size. Bats store sperm in a variety of locations including the oviduct, uterus, and utero-tubal junction (Fenton 1984; Uchida and Mori 1987). In a study by Gomendio and Roldan (1993), residual sperm number in mammals has been found to correlate positively with uterine horn length, relative to body size. Unfortunately, this study did not include any bats. In contrast, after removing effects of body size, no significant relationship was found between uterine horn width and testes mass in 13 microchiropteran bats (Hosken 1997b) or between uterine horn length and testes mass in seven species of bats (Hosken 1998b). However, a highly significant relationship between body mass and uterine horn length was found with and without phylogenetic correction (Hosken 1998a). Whether the allometric relationship between female reproductive tract size and body size in bats differs from other mammals needs to be ascertained to determine if this factor can account for the difference in testes mass–body mass allometry observed between bats and other mammals.

Sperm storage has been proposed as a mechanism to promote sperm competition or permit selective use of sperm by females (Bernard and Cumming 1997; Birkhead and Møller 1993). Reports of phagocytosis of sperm in the female reproductive tract (Hosken 1997a; Racey et al. 1987; Rasweiler 1987) are, therefore, intriguing. Phagocytosis is often interpreted as a mechanism to remove dead sperm, but it might also function to select sperm. Nonrandom paternity by males (Hosken 1998a; Watt and Fenton 1995) might be due to selective sperm use by females, although other alternatives are also possible. At the same time, we did not find a difference in testes size between species that exhibit longer storage of sperm in females than males. While the relationship between testes size and sperm storage may reflect selection to insure adequate numbers of surviving sperm, the significant effect of family on the relationship between sperm storage and testes size indicates that variation in sperm storage cannot explain differences in testes size between closely related species.

Nevertheless, sperm storage may influence selection on other traits. A number of bat species that store sperm form sperm plugs (Fenton 1984). These are often interpreted as a mechanism for ensuring paternity and reducing sperm competition. For example, sperm plugs may ensure that sufficient sperm are corralled at the cervix to get past this barrier to storage, making subsequent removal impossible. This interpretation seems inadequate given that some species with large residual CTM—for example, *Nyctalus noctula* (table 3.1)—produce sperm plugs and in some species, females are capable of expelling plugs. An interesting proposition that deserves comparative study is that penile and/or bacula morphology represent counteradaptations by male bats to facilitate sperm or sperm plug removal (Bernard and Cumming 1997; Eberhard 1985).

This review has focused on testes size and ignored the possibility that sperm size might also influence sperm competition. However, bats exhibit variation in sperm morphology across species (Forman and Genoways 1979). In an early study of 232 mammals (Cummins and Woodall 1985), a negative relationship between sperm length and body mass was found for all mammals except bats, where the relationship was positive. The negative relationship was interpreted as indicating a trade-off between sperm size and number because more sperm need to be produced when the female reproductive tract is larger and dilution is greater. However, more recent work has shown that sperm size is independent of body mass in bats (Hosken 1997b), as well as in 300 species of mammals after correction for phylogenetic effects (Gage 1998). Thus, variation in sperm size is not explained by body size and warrants explanation. Hosken (1997b) failed to find a significant relationship between sperm length and estrous duration or group size in microchiropteran bats. He also failed to find a significant relationship between sperm length and uterine horn length, although only five species were used in this analysis (Hosken 1998b). In contrast, sperm length has been shown to correlate with oviduct length across mammals independently of phylogenetic effects (Gomendio and Roldan 1993). Future studies should consider the possibility that sperm length might influence sperm competitive ability or mobility (Birkhead et al. 1999, Gomendio and Roldan 1991; Stockley et al. 1997) and differ among mating systems in bats.

Acknowledgments

We thank Anne Brooke, Ted Fleming, Annika Keeley, Jorge Ortega, Tom O'Shea, Nancy Simmons, and Christian Voigt for sharing unpublished data or observations with us. We are also grateful to Nancy Simmons and Linda Gordon for giving us access to the American Museum of Natural History and National Museum of Natural History collections, respectively, and to David Hosken and Jay Storz for comments that helped to improve the manuscript.

Literature Cited

Adams, J. K. 1989. Pteronotus davyi. Mammalian Species, 346:1–5.
Anand Kumar, T. C. 1965. Reproduction in the rat-tailed bat *Rhinopoma kinneari*. Journal of Zoology (London), 147:147–155.
Baker, R. J. 1979. Karyology. Pp. 107–156 *in:* Biology of Bats of the New World Family Phyllostomatidae (R. J. Baker, J. K. Jones, Jr., and D. C. Carter, eds.). Pt. 3. Texas Tech Press, Lubbock.
Baker, R. J., and C. L. Clark. 1987. Uroderma bilobatum. Mammalian Species, 279:1–4.

Beasley, L. J., and I. Zucker. 1984. Photoperiod influences the annual reproductive cycle of the male pallid bat (*Antrozous pallidus*). Journal of Reproduction and Fertility, 70:567–573.

Bernard, R. T. F. 1983. Reproduction of *Rhinolophus clivosus* (Microchiroptera) in Natal, South Africa. Zeitschrift für Säugetierkunde, 48:321–329.

Bernard, R. T. F. 1989. The adaptive significance of reproductive delay phenomena in some South African Microchiroptera. Mammal Review, 19:27–34.

Bernard, R. T. F., and G. S. Cumming. 1997. African bats: evolution of reproductive patterns and delays. Quarterly Review of Biology, 72:253–274.

Bernard, R. T. F., D. C. D. Happold, and M. Happold. 1997. Sperm storage in a seasonally reproducing African vespertilionid, the banana bat (*Pipistrellus nanus*) from Malawi. Journal of Zoology (London), 241:161–174.

Birkhead, T. R., J. G. Martinez, T. Burke, and D. P. Froman. 1999. Sperm mobility determines the outcome of sperm competition in the domestic fowl. Proceedings of the Royal Society of London B, 266:1759–1764.

Birkhead, T. R., and A. P. Møller. 1993. Sexual selection and the temporal separation of reproductive events: sperm storage data from reptiles, birds and mammals. Biological Journal of the Linnean Society, 50:295–311.

Bradbury, J. W. 1977. Lek mating behavior in the hammer-headed bat. Zeitschrift für Tierpsychologie, 45:225–255.

Bradbury, J. W. 1981. The evolution of leks. Pp. 138–169 *in:* Natural Selection and Social Behavior (R. D. Alexander and D. W. Tinkle, eds.). Chiron Press, New York.

Bradbury, J. W., and L. H. Emmons. 1974. Social organization of some Trinidad bats. I. Emballonuridae. Zeitschrift für Tierpsychologie, 36:137–183.

Bradbury, J. W., and S. L. Vehrencamp. 1976. Social organization and foraging in Emballonurid bats. I. Field studies. Behavioral Ecology and Sociobiology, 1:337–381.

Brownell, R. L., Jr., and K. Ralls. 1986. Potential for sperm competition in baleen whales. Report to the International Whaling Commission, 8:97–112.

Cohen, J. 1967. Correlation between sperm "redundancy" and chiasma frequency. Nature, 215:862–863.

Cohen, J. 1973. Cross-overs, sperm redundancy and their close correlation. Heredity, 31:408–413.

Cummins, J. M., and P. F. Woodall. 1985. On mammalian sperm dimensions. Journal of Reproduction and Fertility, 75:153–175.

Darwin, C. 1871. The Descent of Man, and Selection in Relation to Sex. John Murray, London.

Dewsbury, D. A. 1982. Ejaculate cost and male choice. American Naturalist, 119:601–610.

Dwyer, P. D. 1968. The biology, origin and adaptation of *Miniopterus australis* (Chiroptera) in New South Wales. Australian Journal of Zoology, 16:49–68.

Eberhard, W. G. 1985. Sexual Selection and Animal Genitalia. Harvard University Press, Cambridge, Mass.

Entwistle, A. C., P. A. Racey, and J. R. Speakman. 1998. The reproductive cycle and determination of sexual maturity in male brown long-eared bats, *Plecotus auritus* (Chiroptera: Vespertilionidae). Journal of Zoology (London), 244:63–70.

Fenton, M. B. 1984. Sperm competition? the case of vespertilionid and rhinolophid bats.

Pp. 573–587 *in:* Sperm Competition and the Evolution of Animal Mating Systems (R. L. Smith, ed.). Academic Press, New York.

Fleming, T. H. 1988. The Short-Tailed Fruit Bat. A Study of Plant-Animal Interactions. University of Chicago Press, Chicago.

Forman, G. L., and H. H. Genoways. 1979. Sperm morphology. Pp. 177–204 *in:* Biology of Bats of the New World Family Phyllostomatidae (R. J. Baker, J. K. Jones, Jr., and D. C. Carter, eds.). Pt. 3. Texas Tech Press, Lubbock.

Gage, M. J. G. 1998. Mammalian sperm morphometry. Proceedings of the Royal Society of London B, 265:97–103.

Gannon, M. R., M. R. Willig, and J. K. Jones, Jr. 1989. Sturnira lilium. Mammalian Species, 333:1–5.

Gebhard, J. 1995. Observations on the mating behaviour of *Nyctalus noctula* (Schreber, 1774) in the hibernaculum. Myotis, 32:123–129.

Gerell, R., and K. Lundberg. 1985. Social organization in the bat *Pipistrellus pipistrellus.* Behavioral Ecology and Sociobiology, 16:177–184.

Ginsberg, J. R., and U. W. Huck. 1989. Sperm competition in mammals. Trends in Ecology and Evolution, 4:74–79.

Gomendio, M., A. H. Harcourt, and E. R. S. Roldan. 1998. Sperm competition in mammals. Pp. 667–756 *in:* Sperm Competition and Sexual Selection (T. R. Birkhead and A. P. Møller, eds.). Academic Press, New York.

Gomendio, M., and E. R. S. Roldan. 1991. Sperm competition influences sperm size in mammals. Proceedings of the Royal Society of London B, 243:181–185.

Gomendio, M., and E. R. S. Roldan. 1993. Coevolution between male ejaculates and female reproductive biology in eutherian mammals. Proceedings of the Royal Society of London B, 252:7–12.

Gustafson, A. W. 1979. Male reproductive patterns in hibernating bats. Journal of Reproduction and Fertility, 56:317–331.

Gustafson, A. W., and D. A. Damassa. 1985. Annual variations in plasma sex steroid-binding protein and testosterone concentrations in the adult male little brown bat: relations to the asynchronous recrudescence of the testis and accessory reproductive organs. Biology of Reproduction, 33:1126–1137.

Happold, D. C. D., and M. Happold. 1996. The social organization and population dynamics of leaf-roosting bats, *Pipistrellus nanus* (Chiroptera, Vespertilionidae), in Malawi, east-central Africa. Mammalia, 60:517–544.

Happold, D. C. M., and M. Happold. 1990. The domiciles, reproduction, social organisation and sex ratios of the banana bat *Pipistrellus nanus* (Chiroptera, Vespertilionidae) in Malawi, Central Africa. Zeitschrift für Säugetierkunde, 55:145–160.

Harcourt, A. H., P. H. Harvey, S. G. Garson, and R. V. Short. 1981. Testis weight, body weight and breeding system in primates. Nature, 293:55–57.

Harvey, P. H., and A. H. Harcourt. 1984. Sperm competition, testes size and breeding systems in primates. Pp. 589–600 *in:* Sperm Competition and the Evolution of Animal Mating Systems (R. L. Smith, ed.). Academic Press, New York.

Harvey, P. H., and M. D. Pagel. 1991. The Comparative Method in Evolutionary Biology. Oxford University Press, Oxford.

Heideman, P. D. 1989. Delayed development in Fischer's pygmy fruit bat, *Haplonycteris fischeri*, in the Philippines. Journal of Reproduction and Fertility, 85:363–382.

Heideman, P. D., P. Deoraj, and F. H. Bronson. 1992. Seasonal reproduction of a tropical bat, *Anoura geoffroyi*, in relation to photoperiod. Journal of Reproduction and Fertility, 96:765–773.

Homan, J. A., and J. K. Jones, Jr. 1975. Monophyllus plethodon. Mammalian Species, 58:1–2.

Hood, C. S., and J. Pitocchelli. 1983. Noctilio albiventris. Mammalian Species, 197:1–5.

Hosken, D. J. 1997a. Reproduction and the female reproductive cycle of *Nyctophilus geoffroyi* and *N. major* (Chiroptera: Vespertilionidae) from south-western Australia. Australian Journal of Zoology, 45:489–504.

Hosken, D. J. 1997b. Sperm competition in bats. Proceedings of the Royal Society of London B, 264:385–392.

Hosken, D. J. 1998a. Sperm fertility and skewed paternity during sperm competition in the Australian long-eared bat *Nyctophilus geoffroyi*. Journal of Zoology (London), 245:93–100.

Hosken, D. J. 1998b. Testes mass in megachiropteran bats varies in accordance with sperm competition theory. Behavioral Ecology and Sociobiology, 44:169–177.

Hosken, D. J., M. A. Blackberry, T. B. Stewart, and A. F. Stucki. 1998. The male reproductive cycle of three species of Australian vespertilionid bat. Journal of Zoology (London), 245:261–270.

Hsu, T. C., and K. Benirschke. 1977. An Atlas of Mammalian Chromosomes. Springer-Verlag, New York.

Jolly, S. E., and A. W. Blackshaw. 1988. Testicular migration, spermatogenesis, temperature regulation and environment of the sheath-tailed bat, *Taphozous georgianus*. Journal of Reproduction and Fertility, 84:447–455.

Jones, J. K., Jr., and H. H. Genoways. 1973. Ardops nichollsi. Mammalian Species, 24:1–2.

Karim, K. B., and S. Banerjee. 1989. Reproduction in the Indian mouse-tailed bat, *Rhinopoma hardwickei hardwickei* (Chiroptera, Rhinopomatidae). Reproduction, Fertility and Development, 1:255–264.

Kenagy, G. J., and S. C. Trombulak. 1986. Size and function of mammalian testes in relation to body size. Journal of Mammalogy, 67:1–22.

Kitchener, D. J. 1973. Reproduction in the common sheath-tailed bat, *Taphozous georgianus* (Thomas)(Microchiroptera: Emballonuridae), in western Australia. Australian Journal of Zoology, 21:375 389.

Kofron, C. P., and A. Chapman. 1994. Reproduction and sexual dimorphism of the West African fruit bat, *Epomops buettikoferi*, in Liberia. African Journal of Ecology, 32:308–316.

Krutzsch, P. H. 1979. Male reproductive patterns in nonhibernating bats. Journal of Reproduction and Fertility, 56:333–344.

Krutzsch, P. H., and E. G. Crichton. 1986. Reproduction of the male eastern pipistrelle, *Pipistrellus subflavus*, in the north-eastern United States. Journal of Reproduction and Fertility, 76:91–104.

Krutzsch, P. H., and E. G. Crichton. 1987. Reproductive biology of the male little mastiff bat, *Mormopterus planiceps* (Chiroptera: Molossidae), in southeast Australia. American Journal of Anatomy, 178:352–368.

Krutzsch, P. H., and E. G. Crichton. 1990. Reproductive biology of the male bent-winged bat, *Miniopterus schreibersii* (Vespertilionidae) in southeast South Australia. Acta Anatomica, 139:109–125.

Krutzsch, P. H., R. H. Watson, and C. D. Lox. 1976. Reproductive biology of the male leaf-nosed bat, *Macrotus waterhousii* in southwestern United States. Anatomical Record, 184:611–636.

Kunz, T. H., P. V. August, and C. D. Burnett. 1983. Harem social organization in cave roosting *Artibeus jamaicensis* (Chiroptera: Phyllostomidae). Biotropica, 15:133–138.

Lollar, A. 1995. Notes on the mating behavior of a captive colony of *Tadarida brasiliensis*. Bat Research News, 35:5.

Lundberg, K., and R. Gerell. 1986. Territorial advertisement and mate attraction in the bat *Pipistrellus pipistrellus*. Ethology, 71:115–124.

Manning, J. T., and A. T. Chamberlain. 1994. Sib competition and sperm competitiveness: an answer to "Why so many sperms?" and the recombination/sperm number correlation. Proceedings of the Royal Society of London B, 256:177–182.

Manning, R. W., and J. K. Jones, Jr. 1989. Myotis evotis. Mammalian Species, 329:1–5.

Martin, P. A., T. L. Reimers, J. R. Lodge, and P. J. Dzuik. 1974. The effect of ratios and numbers of spermatozoa mixed from two males on proportion of offspring. Journal of Reproduction and Fertility, 39:251–258.

Mayer, F. 1995. Genetic population structure of the noctule bat *Nyctalus noctula:* a molecular approach and first results. Symposia of the Zoological Society of London, no. 67:387–396.

McCarthy, T. J., A. J. Gardner, and C. O. Handley, Jr. 1992. Tonatia carrikeri. Mammalian Species, 407:1–4.

McCracken, G. F., and J. W. Bradbury. 1981. Social organization and kinship in the polygynous bat *Phyllostomus hastatus*. Behavioral Ecology and Sociobiology, 8:11–34.

McCracken, G. F., and G. S. Wilkinson. 2000. Bat mating systems. Pp. 321–362 *in:* Reproductive Biology of Bats (E. G. Crichton and P. H. Krutzsch, eds.). Academic Press, New York.

McGuckin, M. A., and A. W. Blackshaw. 1991. Seasonal changes in testicular size, plasma testosterone concentration and body weight in captive flying foxes (*Pteropus poliocephalus* and *P. scapulatus*). Journal of Reproduction and Fertility, 92:339–346.

McWilliam, A. N. 1987. The reproductive and social biology of *Coleura afra* in a seasonal environment. Pp. 324–350 *in:* Recent Advances in the Study of Bats (M. B. Fenton, P. Racey, and J. M. V. Rayner, eds.). Cambridge University Press, Cambridge.

McWilliam, A. N. 1988a. The reproductive cycle of male long-fingered bats, *Miniopterus minor* (Chiroptera: Vespertilionidae), in a seasonal environment of the African tropics. Journal of Zoology (London), 216:119–129.

McWilliam, A. N. 1988b. The reproductive cycle of male tomb bats *Taphozous hildegardeae* (Chiroptera: Emballonuridae), in a seasonal environment of the African tropics. Journal of Zoology (London), 215:433–442.

Medway, Lord. 1971. Observations of social and reproductive biology of the bent-winged bat *Miniopterus australis* in northern Borneo. Journal of Zoology (London), 165:261–273.

Medway, Lord. 1972. Roosting associations of flat-headed bats, *Tylonycteris* species (Chiroptera: Vespertilionidae) in Malaysia. Journal of Zoology (London), 168: 463–482.

Merilä, J., and B. C. Sheldon. 1999. Testis size variation in the greenfinch *Carduelis chlo-*

ris: relevance for some recent models of sexual selection. Behavioral Ecology and Sociobiology, 45:115–123.

Molinari, J., and P. J. Soriano. 1987. Sturnira bidens. Mammalian Species, 276:1–4.

Møller, A. P. 1988. Testes size, ejaculate quality and sperm competition in birds. Biological Journal of the Linnean Society, 33:273–283.

Møller, A. P., and T. R. Birkhead. 1989. Copulation behaviour in mammals: evidence that sperm competition is widespread. Biological Journal of the Linnean Society, 38:119–131.

Morrison, D. W. 1979. Apparent male defense of tree hollows in the bat, *Artibeus jamaicensis.* Journal of Mammalogy, 60:11–15.

Morrison, D. W., and S. H. Morrison. 1981. Economics of harem maintenance by a Neotropical bat. Ecology, 62:864–866.

Mutere, F. A. 1967. The breeding biology of equatorial vertebrates: reproduction in the fruit bat, *Eidolon helvum,* at latitude 0°20′ N. Journal of Zoology (London), 153:153–161.

Mutere, F. A. 1973. A comparative study of reproduction in two populations of the insectivorous bats, *Otomops martiensseni,* at latitudes 1°5′ S and 2°30′ S. Journal of Zoology (London), 171:79–92.

Myers, P. 1977. Patterns of reproduction of four species of vespertilionid bats in Paraguay. University of California Publications in Zoology, 107:1–41.

Nakamura, M. 1990. Cloacal protuberance and copulatory behavior of the alpine accentor (*Prunella collaris*). Auk, 107:284–295.

Nakatsura, K., and D. L. Kramer. 1982. Is sperm cheap? limited fertility and female choice in the lemon tetra (Pisces, Characidae). Science, 216:753–755.

Nelson, J. E. 1965. Behaviour of Australian Pteropodidae (Megachiroptera). Animal Behaviour, 13:544–557.

O'Brien, G. M. 1993. Seasonal reproduction in flying foxes, reviewed in the context of other tropical mammals. Reproductive Fertility and Development, 5:499–521.

O'Brien, G. M., J. D. Curlewis, and L. Martin. 1993. Effect of photoperiod on the annual cycle of testis growth in a tropical mammal, the little red flying fox, *Pteropus scapulatus.* Journal of Reproduction and Fertility, 98:121–127.

O'Shea, T. J. 1980. Roosting, social organization and the annual cycle in a Kenya population of the bat *Pipistrellus nanus.* Zeitschrift für Tierpyschologie, 53:171–195.

Olsson, M., T. Madsen, and R. Shine. 1997. Is sperm really so cheap? costs of reproduction in male adders, *Vipera berus.* Proceedings of the Royal Society of London B, 264:455–459.

Ortega, J., and H. T. Arita. 1999. Structure and social dynamics of harem groups in *Artibeus jamaicensis* (Chiroptera: Phyllostomidae). Journal of Mammalogy, 80:1173–1185.

Ortega, J., and H. T. Arita. 2000. Defense of females by dominant males of *Artibeus jamaicensis* (Chiroptera: Phyllostomidae). Journal of Mammalogy, 106:395–407.

Parker, G. A. 1970. Sperm competition and its evolutionary consequences in the insects. Biological Review of the Cambridge Philosophical Society, 45:525–567.

Pearson, O. P., M. R. Koford, and A. K. Pearson. 1952. Reproduction of the lump-nosed bat (*Corynorhinus rafinesquei*) in California. Journal of Mammalogy, 33:273–320.

Phillips, W. R., and S. J. Inwards. 1985. The annual activity and breeding cycles of

Gould's long-eared bat, *Nyctophilus gouldi* (Microchiroptera: Vespertilionidae). Australian Journal of Zoology, 33:111–126.

Plumpton, D. L., and J. K. Jones, Jr. 1992. Rhynchonycteris naso. Mammalian Species, 413:1–5.

Racey, P. A. 1974. The reproductive cycle in male noctule bats, *Nyctalus noctula.* Journal of Reproduction and Fertility, 41:169–182.

Racey, P. A. 1979. The prolonged storage and survival of spermatozoa in Chiroptera. Journal of Reproduction and Fertililty, 56:391–402.

Racey, P. A., and A. C. Entwistle. 2000. Life history and reproductive strategies of bats. Pp. 363–414 *in:* Reproductive Biology of Bats (E. G. Crichton and P. H. Krutzsch, eds.). Academic Press, San Diego, Calif.

Racey, P. A., and W. H. Tam. 1974. Reproduction in male *Pipistrellus pipistrellus* (Mammalia: Chiroptera). Journal of Zoology (London), 172:101–122.

Racey, P. A., T. A. Uchida, T. Mori, M. I. Avery, and M. B. Fenton. 1987. Sperm-epithelium relationships in relation to the time of insemination in little brown bats (*Myotis lucifugus*). Journal of Reproduction and Fertility, 80:445–454.

Rasweiler, J. J. 1987. Prolonged receptivity to the male and the fate of spermatozoa in the female black mastiff bat, *Molossus ater.* Journal of Reproduction and Fertility, 79:643–654.

Short, R. V. 1979. Sexual selection and its component parts, somatic and genital selection, as illustrated by man and the great apes. Advances in the Study of Behavior, 9:131–158.

Simmons, N. B., and J. H. Geisler. 1998. Phylogenetic relationships of *Icaronycteris, Archaeonycteris, Hassianycteris,* and *Paleonycteryx* to extant bat lineages, with comments on the evolution of echolocation and foraging strategies in Microchiroptera. Bulletin of the American Museum of Natural History, 235:1–182.

Singh, K., and A. Krishna. 1996. Seasonal changes in circulating serum concentration and in vitro testicular secretion of testosterone and androstenedione in the male vespertilionid bat (*Scotophilus heathi*). Journal of Experimental Zoology, 276:43–52.

Singh, U. P. 1997. Reproductive biology of the male sheath-tailed bat, *Taphozous longimanus* (Emballonuridae) from India. Biomedical Environmental Sciences, 10:14–26.

Snow, J. L., J. K. Jones, Jr., and W. D. Webster. 1980. Centurio senex. Mammalian Species, 138:1–3.

Speakman, J. R., and P. A. Racey. 1986. The influence of body condition on sexual development of male brown long-eared bats (*Plecotus auritus*) in the wild. Journal of Zoology (London), 210:515–525.

Stockley, P., M. J. G. Gage, G. A. Parker, and A. P. Møller. 1997. Sperm competition in fishes: the evolution of testis size and ejaculate characteristics in fish. American Naturalist, 149:933–954.

Stockley, P., and A. Purvis. 1993. Sperm competition in mammals: a comparative study of male roles and relative investment in sperm production. Functional Ecology, 7:560–570.

Storz, J. F., H. R. Bhat, and T. H. Kunz. 2000. Social structure of a polygynous tent-making bat, *Cynopterus sphinx* (Megachiroptera). Journal of Zoology (London), 251:151–165.

Swanepoel, P., and H. H. Genoways. 1983. Brachyphylla cavernarum. Mammalian Species, 205:1–6.

Taft, L., and C. O. Handley, Jr. 1991. Reproduction in a captive colony. Pp. 19–42 *in:* Demography and Natural History of the Common Fruit Bat, *Artibeus jamaicensis,* on Barro Colorado Island, Panama (C. O. Handley, Jr., D. E. Wilson, and A. L. Gardner, eds.). Smithsonian Institution Press, Washington, D.C.

Tamsitt, J. R., and D. Nagorsen. 1982. Anoura cultrata. Mammalian Species, 179:1–5.

Thomas, D. W., M. B. Fenton, and R. M. R. Barclay. 1979. Social behavior of the little brown bat *Myotis lucifugus.* I. Mating behavior. Behavioral Ecology and Sociobiology, 6:129–136.

Uchida, T. A., and T. Mori. 1987. Prolonged storage of spermatozoa in hibernating bats. Pp. 351–366 *in:* Recent Advances in the Study of Bats (M. B. Fenton, P. Racey, and J. M. V. Raynor, eds.). Cambridge University Press, Cambridge.

van der Merwe, M., and I. L. Rautenbach. 1990. Reproduction in the rusty bat, *Pipistrellus rusticus,* in the northern Transvaal bushveld, South Africa. Journal of Reproduction and Fertility, 89:537–542.

Vaughan, T. A., and R. P. Vaughan. 1986. Seasonality and the behavior of the African yellow-winged bat. Journal of Mammalogy, 67:91–102.

Vehrencamp, S. L., F. G. Stiles, and J. W. Bradbury. 1977. Observations on the foraging behavior and avian prey of the Neotropical carnivorous bat, *Vampyrum spectrum.* Journal of Mammalogy, 58:469–478.

Vivier, L., and M. van der Merwe. 1996. Reproductive pattern in the male Angolan free-tailed bat, *Tadarida (Mops) condylura* (Microchiroptera: Molossidae) in the eastern Transvaal, South Africa. Journal of Zoology (London), 239:465–476.

Watkins, L. C. 1977. Euderma maculatum. Mammalian Species, 77:1–4.

Watt, E. M., and M. B. Fenton. 1995. DNA fingerprinting provides evidence of discriminate suckling and non-random mating in little brown bats *Myotis lucifugus.* Molecular Ecology, 4:261–264.

Webster, W. D., and J. K. Jones, Jr. 1984. Glossophaga leachii. Mammalian Species, 226:1–3.

Wilkins, K. T. 1987. Lasiurus seminolus. Mammalian Species, 280:1–5.

Wilkinson, G. S. 1985. The social organization of the common vampire bat. II. Mating system, genetic structure, and relatedness. Behavioral Ecology and Sociobiology, 17:123–134.

Wilkinson, G. S. 1987. Altruism and cooperation in bats. Pp. 299–323 *in:* Recent Advances in the Study of Bats (M. B. Fenton, P. Racey, and J. M. V. Rayner, eds.). Cambridge University Press, Cambridge.

Williams, C. F. 1986. Social organization of the bat *Carollia perspicillata* (Chiroptera: Phyllostomidae). Ethology, 71:265–282.

Willis, K. B., M. R. Willig, and J. K. Jones, Jr. 1990. Vampyrodes caraccioli. Mammalian Species, 359:1–4.

Wimsatt, W. A. 1942. Survival of spermatozoa in the female reproductive tract of the bat. Anatomical Record, 83:299–307.

CHAPTER 4

Ecology of Bat Migration

Theodore H. Fleming and Peggy Eby

Introduction

Annual or seasonal migrations occur in all classes of vertebrates. Among higher vertebrates, the ecology of migration has been most thoroughly studied in birds, in which a substantial fraction of all species are migratory. According to Gill (1995), about 5 billion land birds of 187 species migrate from Europe and Asia to Africa each year. A similar number of about 200 species of land birds migrates from North America to the Caribbean Islands and Central and South America annually. These migrations include species of all body sizes and dietary classes—from 3-g hummingbirds to 13-kg trumpeter swans. Diurnal migratory passages of thousands of hawks or millions of passenger pigeons are (or were) spectacular sights. Distances that birds migrate are often prodigious, sometimes encompassing thousands to tens of thousands of kilometers (Weidensaul 1999). Migration is therefore an extremely important event in the lives of many species of birds.

In contrast, probably because most species are terrestrial, a much smaller fraction of mammals is migratory. Long-distance migrations of more than 1,000 km are restricted primarily to baleen whales. The eastern Pacific population of the gray whale (*Eschrichtius robustus*), for instance, annually migrates more than 18,000 km from the Bering Sea to the west coast of Baja California and back. Among terrestrial mammals, seasonal migrations of a few hundred kilometers are restricted to a few species of large ungulates, including wildebeests, gazelles, and zebras in the savannas of east Africa and caribou in the Arctic tundra. Even among bats, migrations of more than a few hundred kilometers are relatively uncommon. Nonetheless, migration appears to play an important role in the lives of many species of bats, particularly those living in highly seasonal environments. In this chapter we review current knowledge about ecological and behavioral aspects of migration in bats. To do this, we address the following questions: (1) Who migrates and why? What ecological conditions favor the evolution of migratory behavior in bats? (2) What are the physiological consequences of migration? How much time and energy do bats devote to annual migrations? (3) What are the life-history and population consequences of migration? To what extent are reproduction, social behavior, and the genetic structure of bats affected by migration? (4) What are the conservation implications of migration? What special needs exist for migratory bats to prevent their extinction? Whenever appropriate, in attempting to answer these questions we will compare the migratory ecology of bats with that of birds to look for evolutionary commonalities and differences.

Definition of Migration

We begin this chapter by defining migration as a seasonal, usually two-way, movement from one place or habitat to another to avoid unfavorable climatic conditions and/or to seek more favorable energetic conditions. Such movements often involve significant physiological adjustments (e.g., substantial fat deposition). We wish to distinguish between migration and dispersal, the latter of which can be defined as a one-way movement from one location to another. Dispersal usually, but not always, involves movement away from an animal's place of birth. In most mammals, young males are more likely to disperse than are young females; the opposite is true for most birds (Greenwood 1980). As we will see, migration in bats also tends to be sex biased, with females more likely to migrate (or at least to migrate significantly longer distances) than males. A strong sex bias in migratory tendency is less common in birds. We also discount roost shifts by individuals that occur within warm or cold seasons as migratory movements. Such shifts have been well documented in various species of temperate bats (e.g., in North America: *Myotis velifer* and *M. lucifugus*; in Europe: *Myotis myotis, Miniopterus schreibersii, Rhinolophus hipposideros* [Gaisler 1979; Griffin 1970; Humphrey and Cope 1976; Tinkle and Patterson 1965; Tuttle 1976]), but they do not represent migrations as we have defined this term. Finally, we do not consider short-distance (<50 km) habitat shifts to be migratory movements. Although such shifts are common in tropical birds (e.g., Karr and Freemark 1983) and in some tropical bats (e.g., *Saccopteryx bilineata* and *Phyllostomus discolor* in Costa Rica, *Artibeus jamaicensis* in Panama [Bonaccorso 1979; Bradbury and Vehrencamp 1976; Heithaus et al. 1975]), they usually encompass only short-distance movements and do not involve major physiological adjustments. Thus, tropical bats are not comparable to the longer-distance movements that many temperate (but not all) bats make during migration.

Why Migrate?

Migration, particularly when it involves intercontinental movements as it does in many species of birds, is costly in terms of time and energy and can involve substantial risks of mortality from many sources, including predators, food shortages, and inclement weather. Tuttle and Stevenson (1977), for example, reported that most public recoveries of banded gray bats (*Myotis grisescens*) occur at peak migration times in April and September and concluded that migration was a stressful event, particularly for young bats. For migratory behavior to evolve, its benefits to individuals should exceed its costs. In terms of game theory, migratory phenotypes (and genotypes) within populations should have higher fitness under certain conditions than sedentary phenotypes (Dingle 1980; Kaitala et al. 1993). Possible benefits of migration include increased access to food and physiologically favorable microclimatic conditions and reduced exposure to predators, parasites, and disease. Because climatic seasonality generally increases and annual plant (and animal) productivity generally decreases with increasing latitude, we expect conditions favoring the evolution of migratory behavior also to vary with latitude. Selection for migratory behavior should be more intense at higher latitudes than at lower latitudes, and a higher proportion of temperate bats should exhibit migratory behavior than do tropical species.

To make these generalizations more concrete, we illustrate latitudinal trends in average air temperatures in the coldest month (usually January north of the equator) and the ratio of highest:lowest average monthly air temperatures (i.e., seasonal range of temperatures) for 23 locations in the New World in figure 4.1. Maritime (= coastal) and continental (= inland) sites are differentiated in that figure. As expected, maritime sites north of the tropics tend to be milder than continental sites during winter, and winter air temperatures in the maritimes decrease with latitude more gradually than continental air temperatures above 25° N (fig. 4.1A). Average winter air temperatures are <0°C at most sites above 40° N. Ratios of seasonal extremes in average air temperature, however, are not markedly >1.0 until about 40° N; seasonal ratios at maritime sites at high latitudes are equal to or greater than those at continental sites (fig. 4.1B). In terms of air temperature, mid-latitudes in the northern hemisphere are more seasonal than are higher or lower latitudes. These data suggest that migration and/or hibernation should be especially common in bats that spend their summers at latitudes of 40° N, at least in North America. Relationships between migration and latitude in temperate bats are confounded by the use of torpor during winter. Some species of bats that occupy latitudes >60° during summer are sedentary and hibernate at these latitudes, while others migrate prior to hibernating (Rydell et al. 1994; Strelkov 1969).

Temperate bats thus have at least two reasons to migrate (or hibernate)—to avoid low winter air temperatures that are physiologically stressful and/or to avoid low resource (insect) levels; these two factors, of course, are not statistically independent but are positively correlated. Low levels of winter food resources probably are the ultimate reason why temperate bats migrate and/or hibernate. But what about tropical bats? What conditions in the tropics favor the evolution of migratory behavior? It cannot be due to low cold season temperatures, because, except at high elevations where few tropical bats live (Graham 1983; McNab 1982), "winter" air temperatures are unlikely to be physiologically stressful (fig. 4.1A). Instead, seasonal fluctuations in food availability,

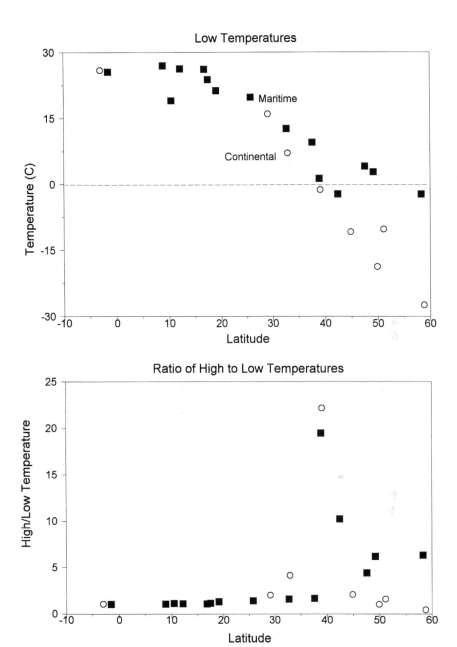

Figure 4.1. *A*, Latitudinal trends in average air temperatures in the coldest month (usually January); *B*, the ratio of highest:lowest average monthly air temperatures for 23 cities in the New World. Latitudes south of the equator have negative values. From south to north, cities include Manaus, Belem, Panama City, Caracas, Managua, Acapulco, Belize City, Veracruz, Miami, Hermosillo, San Diego, Dallas, San Francisco, Washington, Kansas City, Boston, Minneapolis, Seattle, Vancouver, Winnipeg, Calgary, Juneau, and Churchill. Source of data: www.worldclimate.com. *Dark squares* = maritime; *open circles* = continental.

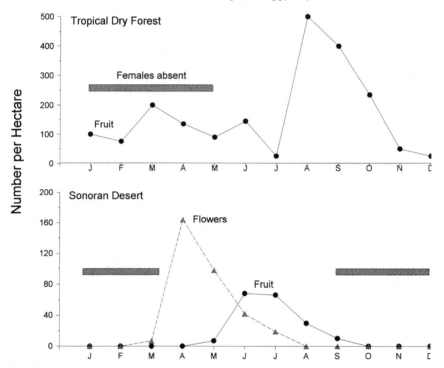

Figure 4.2. Seasonal variation, for phyllostomid bats, in the availability of fruit at Santa Rosa National Park, Costa Rica (tropical dry forest) and fruit and flowers at Bahia Kino, Sonora, Mexico (Sonoran Desert). (Sources of data: Fleming 1988; Fleming et al. 1996.)

probably correlated with seasonal variation in rainfall, are likely to "drive" migratory behavior in tropical bats.

Detailed data on seasonal variation in the availability of food resources for tropical bats are scarce, but two such data sets are shown in figure 4.2. In the lowland tropical dry forests of western Costa Rica, fruit levels are lowest during the prolonged dry season (November to mid-May), a time when many females of the frugivorous bat *Carollia perspicillata* migrate from this habitat (Fleming 1988). In the Sonoran Desert of northwestern Mexico (a northern extension of the tropical dry forest), bat flowers and fruits are available only from late March through late August, the period when females of the nectarivore/frugivore *Leptonycteris curasoae* migrate into this habitat from south-central Mexico (Cockrum 1991; Fleming et al. 1996). Smythe (1982) and Wolda (1992), among others, discuss general trends in the seasonality of Neotropical insects, but whether these fluxes lead to migration in tropical insectivorous bats is currently unknown.

Because temperate and tropical bats live in regions that differ profoundly in the occurrence of physiologically stressful temperatures and seasonal fluxes in resource availability, their life histories, including the propensity to migrate, differ significantly. For example, the lives of many temperate bats are centered on seasonal movements from warm summer roosts (which are energetically favorable for gestation and rearing young) to cool winter roosts (which are energetically favorable for deep torpor) and hibernation (see Speakman and Thomas, this volume). Migration is thus strongly associated with hibernation in many of these bats. In addition, although they regularly extract (and return) energy and materials from their summer habitats, temperate bats typically extract (and return) little energy and materials from their winter habitats during hibernation. As a result, the functional impact (as insect predators and nutrient dispersers) of many temperate bats on their ecosystems changes drastically on a seasonal basis. In contrast, no tropical bats are known to hibernate, although some undergo seasonal bouts of torpor (see Speakman and Thomas, this volume). Thus, periods of seasonal inactivity may be rare or absent in tropical species, and seasonal migration is driven much less by movements between physiologically favorable roost sites than by movements along gradients of resource availability. Unlike temperate species, tropical bats constantly extract (and return) energy and materials from all habitats in which they reside. Because of this basic dichotomy in lifestyles, we will discuss migratory behavior of temperate and tropical bats separately.

Who Migrates?

As summarized by Findley (1993), most of the approximately 1,100 extant species of bats live in the tropics. Only about 125 species (13%) reside in the Nearctic (40 species) or Palearctic (ca. 85 species) regions. Early banding studies of bats and previous reviews of their migratory behavior (e.g., Gaisler 1979; Griffin 1958, 1970; Schmidt-Koenig 1975; Strelkov 1969) reflected a temperate-zone bias, primarily because of the geographic distribution of bat researchers. This bias still exists in the literature so that migratory behavior appears to be common in bats. However, if most families of bats had a tropical origin (Hand and Kirsch 1998; Hill and Smith 1984), and if most contemporary species live under tropical or subtropical conditions in which selection pressures for migration are weak or nonexistent, then it is likely that the ancestral and still most common condition in bats is a sedentary lifestyle. We predict that only species living in seasonal tropical and temperate habitats are likely to migrate.

Temperate Bats

Seasonal movements of bats vary widely in scale. To facilitate our discussion of migration, we recognize three categories of spatial behavior in temperate bats (table 4.1; also see Roer 1995). As the tendency to migrate and the scale of migration vary within as well as between species (see discussions of partial

Table 4.1. Selected morphological and behavioral characteristics of European bats grouped into three migration categories

Species	Migration Category	Maximum Movement (km)	Mass (g)	Aspect Ratio	Wing Loading	Wing Tip Index	Foraging Strategy	Hibernation Roost	Northern Limit to Distribution (°latitude)
Nyctalus lasiopterus	Long distance	u/k	40–75	...	...	...	Aerial hawk	Trees/building	55
Nyctalus leiseri	Long distance	1,245	11–20	7.9	19.3	1.19	Aerial hawk	Trees/building	57
Nyctalus noctula	Long distance	1,600	17–40	7.4	16.1	0.99	Aerial hawk	Trees/building	61
Pipistrellus nathusii	Long distance	1,900	5–10	7.2	9.8	1.15	Aerial hawk	Trees/building	61
Pipistrellus pipistrellus	Long distance	1,160	4–9	7.5	8.1	1.74	Aerial hawk	Mixed	63
Vespertilio murinus	Long distance	1,440	12–21	7	10.2	...	Aerial hawk	Tall buildings	63
Barbastella barbastellus	Regional	290	6–13	6	9.1	2.33	Hawk	Subterranean	60
Miniopterus schreibersii	Regional	550	8–16	7	10.2	1.03	Aerial hawk	Subterranean	49
Myotis blythii	Regional	600	15–32	...	...	...	Hawk	Subterranean	49
Myotis brandtii	Regional	230	4–10	...	...	1.17	Hawk and glean	Subterranean	65
Myotis dasycneme	Regional	330	14–22	6.8	10.4	1.2	Trawl/hawk	Subterranean	61
Myotis daubentonii	Regional?	260	6–12	6.3	7	2.05	Trawl/hawk	Subterranean	63
Myotis myotis	Regional	390	27–45	6.3	11.2	1.89	Glean	Subterranean	54
Myotis mystacinus	Regional	240	4–8	6	7.1	2.26	Slow hawk/glean	Subterranean	64
Eptesicus nilssonii	Sedentary	115	8–16	6.6	8.1	...	Aerial hawk	Building	73
Eptesicus serotinus	Sedentary	330	15–35	6.5	12.2	1.57	Aerial hawk	Mixed	53
Myotis bechsteinii	Sedentary	35	7–13	6	9	4.37	Glean	Subterranean	60
Myotis emarginatus	Sedentary	60	6–14	5.9	7.1	2.05	Glean	Subterranean	51
Myotis nattereri	Sedentary	90	6–12	6.4	6.1	1.38	Glean	Subterranean	58
Pipistrellus kuhlii	Sedentary	u/k	5–9	6.3	8.5	...	Aerial hawk	Subterranean	50
Plecotus auritus	Sedentary	66	5–12	5.7	7.1	1.43	Glean	Mixed	64
Plecotus austriacus	Sedentary	62	7–14	6.1	7.9	1.82	Glean	Mixed	52
Rhinolophus blasii	Sedentary	40	11–16	...	...	...	Glean	Subterranean	47
Rhinolophus euryale	Sedentary	130	10–18	6.2	8.1	...	Glean	Subterranean	47
Rhinolophus ferrumequinum	Sedentary	180	14–34	6.1	12.2	2.13	Hawk/perch hunt	Subterranean	52
Rhinolophus hipposideros	Sedentary	150	4–9	5.7	7.1	2.59	Hawk/glean	Subterranean	54
Rhinolophus mehelyi	Sedentary	90	12–20	...	11.6	...	u/k	Subterranean	43
Tadarida teniotus	u/k	u/k	25–50	Very high	Very high	0.84	Aerial hawk	Subterranean	44

| *Myotis capaccinii* | u/k | u/k | 6–15 | … | 10.5 | … | Trawl/hawk | Subterranean | 46 |
| *Pipistrellus savii* | u/k | u/k | 5–10 | … | … | … | Aerial hawk | Subterranean | 43 |

Sources. The data for maximum movement and for northern limit to distribution are from Mitchell-Jones et al. (1999); for mass, Stebbings (1988); for aspect ratio, wing loading, and wing tip index, Norberg and Rayner (1987); for foraging strategy, Mitchell-Jones et al. (1999) and Norberg and Rayner (1987); and for hibernation roost, Mitchell-Jones (1999) and Stebbings (1988).

Note. Migration groupings denote the predominant migration strategies of species; see text for definitions. Movement data are from banding studies. Predominant types of hibernation roost are as follows: trees/building = tree hollows, buildings, foliage, and so on; subterranean = caves, mines, tunnels, vaults, and so on. u/k = unknown.

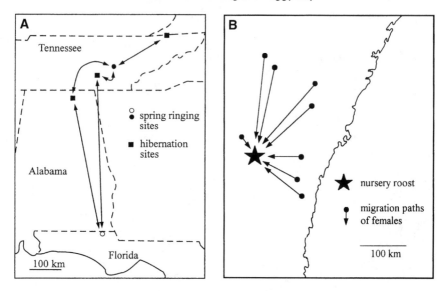

Figure 4.3. The migration paths of two regional migrants: (A) *Myotis grisescens* in the United States and (B) *Miniopterus schreibersii* in southeast Australia. (Sources of data: Dwyer 1966; Tuttle 1976.)

migration and sex-biased migration below), these categories describe the populations that undertake the longest migrations. Sedentary species are those that breed and hibernate in the same local areas, usually moving <50 km between their summer and winter roosts. In Europe, several species of *Myotis, Rhinolophus,* and *Plecotus* are relatively sedentary bats (Entwistle et al. 2000; Hooper and Hooper 1956; Kerth and Koenig 1999; Strelkov 1969). In North America, *Eptesicus fuscus, Corynorhinus rafinesquii, Antrozous pallidus,* and *Macrotus californicus* are examples of sedentary species (Barbour and Davis 1969). Regional migrants are those that migrate moderate distances (typically 100–500 km) between their summer and winter roosts (fig. 4.3). European examples include *Miniopterus schreibersii* and *Myotis dasycneme* (Strelkov 1969). North American examples include several species of *Myotis* (e.g., *M. lucifugus, M. grisescens*) as well as *Pipistrellus subflavus* and possibly *Lasiurus seminolus* (Barbour and Davis 1969). Long-distance migrants are those in which migratory behavior is highly developed so that individuals sometimes travel over 1,000 km between summer and winter roosts. European examples include *Nyctalus noctula* and *Pipistrellus nathusii* (fig. 4.4; Brosset 1990; Petit and Mayer 2000; Strelkov 1969). North American examples include species of *Lasiurus* (e.g., *L. cinereus, L. borealis*), *Lasionycteris noctivagans,* and the subtropical/tropical seasonal migrants *Leptonycteris curasoae* (fig. 4.5), *L. nivalis, Choeronycteris mexicana,* and *Tadarida brasiliensis* (fig. 4.6). None of these species are known to undergo one-way migrations of more than about 1,800 km. Compared with migrations undertaken by similar-sized (or smaller) birds, no bat species are truly long-distance migrants.

Membership in these migration categories depends on phylogeny, flight morphology, foraging ecology, and roost type (table 4.1). Among vespertilionids, sedentary species occur in tribes Eptesicini and Plecotini of the Vespertilioninae and short-distance migrants occur within the Myotinae, according to phylogenies presented in Simmons (1998) and Volleth and Heller (1994). The distribution of long-distance migrants, however, is different. European long-distance migrants (*N. noctula* and *P. nathusii*) occur within tribe Pipistrellini of

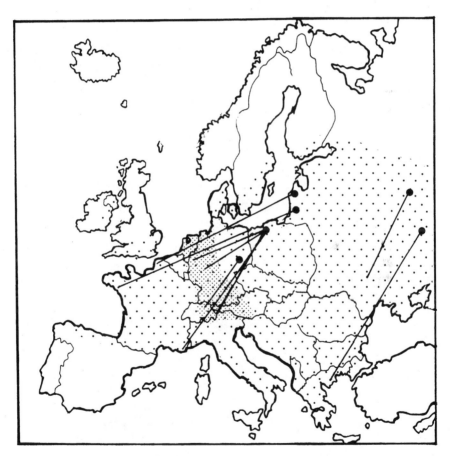

Figure 4.4. Paths of autumn migration in the European long-distance migrant *Pipistrellus nathusii*. Dark shading indicates the approximate area occupied by sedentary populations. (Sources of data: Arnold et al. 1996; Brosset 1990; Kapteyn and Lina 1994; Strelkov 1969, 1997.)

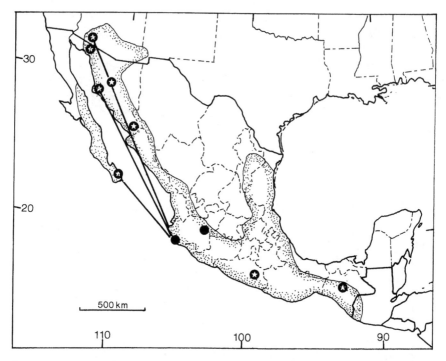

Figure 4.5. Migratory connections between a mating site (*solid dot*) and northern maternity roosts (*starred dots*) in *Leptonycteris curasoae*, as revealed by mitochondrial DNA analysis. (Redrawn with permission from Ceballos et al. [1997].)

the Vespertilioninae, whereas the New World long-distance migrants *Lasiurus* spp. and *Lasionycteris noctivagans* are not closely related to each other or to European long-distance migrants. *Lasiurus* is a member of the Myotinae, but *Lasionycteris* appears to be closely related to *Eptesicus* in the Vespertilioninae (S. Hoofer, personal communication) Thus, long-distance migration appears to have evolved independently at least three times in the Vespertilionidae. Likewise, although they belong to the subfamily Glossophaginae of the Phyllostomidae, species of *Leptonycteris* and *Choeronycteris mexicana* are not closely related (Simmons and Wetterer 2002) and have independently evolved long-distance migration.

Because sustained flight during migration has a high cost in energy and time, the wings of many migratory birds and bats are adapted for rapid, energetically efficient flight. Migratory bats have relatively narrow, aerodynamically efficient wings with pointed wing tips (i.e., high aspect ratios, high wing loading, and low wing tip indices, T_i [Norberg and Rayner 1987]). These characteristics are most distinct in species that migrate long distances. The wing

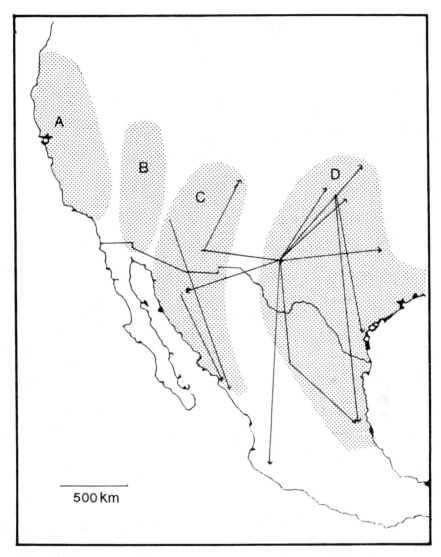

Figure 4.6. Distributions and migration paths of four populations of *Tadarida brasiliensis*, including two sedentary populations (*A* and *B*) and two populations that migrate between Mexico and the United States (*C* and *D*). Lines indicate recoveries of banded bats. (Source of data: Cockrum 1969.)

Table 4.2. Wing characteristics of European bats relative to their migration categories

	Long-Distance Migrants	Regional Migrants	Sedentary Species	P
Aspect ratio	7.4 ± 0.15 (7–7.9)	6.3 ± 0.16 (5.9–7)	6.2 ± 0.98 (5.7–6.6)	<0.0001
Wing tip index (T_i)	12.2 ± 0.16 (0.99–1.74)	1.75 ± 0.18 (1.17–2.33)	2.2 ± 0.39 (1.38–4.37)	0.12
Wing loading	12.7 ± 2.2 (8.1–19.3)	8.87 ± 0.68 (7–11.2)	8.9 ± 0.63 (6.1–12.2)	0.045

Note. Data are means ± standard error and (range) taken from Norberg and Rayner (1987). Significant differences between migration categories were tested by GLIM.

structure of European long-distance migrants differs significantly from that of regional migrants and sedentary species; aspect ratio and wing loading are significantly higher, and there is a trend toward lower T_i (table 4.2). North American bats that migrate long distances also have high aspect ratio wings and pointed wing tips (Norberg and Rayner 1987; Sahley et al. 1993).

While these traits enhance flight speed and efficiency, they concomitantly influence foraging patterns by inhibiting slow flight and the capacity to hover—flight characteristics that are essential for foliage gleaning or foraging in highly cluttered habitats (Fenton 1990; Norberg and Rayner 1987). Thus, long-distance migrants are typically aerial hawkers that forage in open habitats, and their diets are composed primarily of aerial insects (table 4.1; also see Schnitzler and Kalko 1998). It is unlikely, however, that the wings of long-distance migrants are adapted for migration per se (Winkler and Leisler 1992). Wing shape is more likely to be influenced by foraging strategy and diet, and long-distance migrants may be preadapted for energetically efficient, sustained flight because of their aerial hawking mode of foraging (Norberg and Rayner 1987).

In temperate regions, populations of sedentary bats, regional migrants, and long-distance migrants all breed at high latitudes. While the first two groups remain at high latitudes during winter, long-distance migrants overwinter in milder climatic zones. There is a clear relationship between the migration characteristics of temperate bats and their choice of hibernacula (table 4.1). In general, long-distance migrants hibernate aboveground in trees and buildings, either as individuals or in small groups, whereas regional migrants and sedentary species generally hibernate as aggregations in caves, mines, or tunnels (Roer 1995; Strelkov 1969). Exceptions include the occasional use of underground roosts by N. noctula (Ryberg 1947).

Temperate bats are highly selective of hibernation sites, and increased winter mortality occurs when individuals hibernate outside optimum conditions (Entwistle et al. 1997; Raesly and Gates 1986; Ransome 1971). Bats that hibernate in poorly insulated trees and buildings are generally tolerant of low and fluctuating air temperatures in their hibernacula; however, few species apparently tolerate the temperature extremes or the long hibernation periods experienced at high latitudes in continental regions (Strelkov 1969). In northern Europe, long-distance migrants such as *N. noctula* and *P. nathusii* leave breeding habitats where minimum winter temperatures fall below −10°C and where mean monthly temperatures remain below 0°C for 4–5 mo of the year to hibernate in areas with significantly milder winter temperatures. The majority of known hibernation sites of these species occur in areas where mean temperatures remain above 0°C.

In the southern hemisphere, South America is the only continent (excluding Antarctica) where the landmass extends south of 45° S and where seasonal shifts in temperature reach a magnitude similar to those in temperate North America and Europe. Little is known about the migratory habits of bats in southern Chile and Argentina (Redford and Eisenberg 1992). Of the three species of bats found in southern Chile, *Lasiurus borealis* and *Myotis chiloensis* are present only in summer (Johnson et al. 1992). On the basis of patterns found in the northern hemisphere, we predict that tree-roosting *L. borealis* migrates long distances to hibernate in a milder climatic zone whereas *M. chiloensis* migrates shorter distances to suitable hibernacula within the same climate zone.

Not all individuals of a migratory species undergo seasonal migrations. Partial migration is common in many bats (and birds), and several species that migrate substantial distances also have resident or sedentary populations in some parts of their range (table 4.3). Partial migration occurs in several European long-distance migrants. Sedentary and migratory populations of these species often coexist at hibernation sites but occupy separate summer habitats (Roer 1995; Strelkov 1969, 1997). For example, migratory *P. nathusii* hibernates in parts of Germany, Austria, Switzerland, and the Netherlands that additionally support relatively small sedentary populations (fig. 4.4; Arnold et al. 1996; Kapteyn and Lina 1994; Strelkov 1969, 1997). After hibernation, migrants depart for summer breeding sites in central and northern Europe. Similarly, breeding populations of *N. noctula* have been documented in areas used by hibernating migrants in southern Sweden, Romania, Germany, Uzbekistan, and Kazakhistan (Ahlén 1997; Heinz and Braun 1997; Strelkov 1969, 1997). In contrast, a sedentary population of *Vespertilio murinus* in Denmark appears to be geographically isolated from a migratory population in eastern Europe (Masing 1989; Rydell and Baagøe 1994; Strelkov 1969).

In North America, the long-distance migrant *T. brasiliensis* is a well-known partial migrant (Cockrum 1969). Sedentary and migratory populations of this species are geographically isolated throughout the year (fig. 4.6). Sedentary populations occur along the Pacific coast of the United States and in the southeastern United States east of Texas; long-distance migrants occur from eastern Arizona to Texas and in Oklahoma, southeastern Utah, and southwestern Colorado. In some coastal areas of subtropical Australia, sedentary and migratory populations of *Pteropus poliocephalus* share communal roosts (Eby 1991). The existence of non-migratory populations within migratory species suggests that

Table 4.3. Bats for which there is evidence of partial or sex-biased migration

Family/Species	Maximum Migration Distance (km)	Partial Migration	Sex-Biased Migration	Region	References
Pteropodidae:					
Eidolon helvum	1,000	Yes	...	Africa	Thomas 1983
Myonycteris torquata	750	...	M	Africa	Thomas 1983
Pteropus alecto	u/k	Yes	...	Australasia	N. Markus, unpublished data; Palmer et al. 1999
Pteropus poliocephalus	978	Yes	...	Australia	Eby 1991; Webb and Tidemann 1996
Molossidae:					
Tadarida brasiliensis	>1,000	Yes	F	North America	Cockrum 1969
Phyllostomatidae:					
Leptonycteris curosoae	>1,000	Yes	F	North America	Ceballos et al. 1997; Fleming et al. 1993
Leptonycteris nivalis	>1,000	...	F	North America	Hensley and Wilkins 1988
Leptonycteris sanbornii	u/k	...	F	North America	Cockrum 1991
Vespertilionidae:					
Lasionycteris noctivagans	u/k	Yes	F	North America	Kunz 1982a; Barclay 1984
Lasiurus borealis	u/k	...	F	North & South America	Redford and Eisenberg 1992
Lasiurus cinereus	u/k	Yes	F	North & South America	Shump and Shump 1982
Miniopterus schreibersi	1,290	Yes	F	Australia/Europe/Africa	Dwyer 1966
Myotis grisescens	>500	...	F	North America	Tuttle 1976
Myotis lucifugus	>400	...	F	North America	Davis and Hitchcock 1965; Humphrey and Cope 1976
Myotis sodalis	500	...	F	North America	Thomson 1982
Nyctalus leiseri	975	Yes	F	Europe/Africa	Neri and Aulagnier 1996; Strelkov 1969
Nyctalus noctula	1,600	Yes	F	Europe/China	Strelkov 1969
Pipistrellus nathusii	2,000	Yes	F	Europe	Strelkov 1969
Pipistrellus pipistrellus	1,160	Yes	...	Europe	Strelkov 1969
Vespertilio murinus	1,440	Yes	...	Europe	Rydell and Baagoe 1994

Note. F = females more likely to migrate than males, M = males more likely to migrate than females. Ellipses dots indicate no data or unknown.

selection pressures for migratory behavior vary geographically and that this behavior will not evolve unless local selection pressures favor it. Clearly, migratory behavior is not a fixed characteristic of a species.

In addition to partial migration, sex-biased migration occurs in a variety of temperate and subtropical bats (table 4.3). This bias can take several forms: differential tendency to migrate, differential migration distances, and/or differential migration locations. In many species, females are more likely to migrate and/or migrate longer distances than are males. Many of these migratory movements involve females moving to summer roosts to form maternity colonies. Sex-biased migration is well documented in *Miniopterus schreibersii* in southeastern Australia and *M. schreibersi natalensis* in southern Africa; females move long distances in spring to congregate at maternity sites from which males are largely absent (Dwyer 1966, 1969; van der Merwe 1975). In eastern and central Europe, females of *P. nathusii* migrate longer distances to summer habitats than males, which do not reach the northernmost parts of the range (Strelkov 1969). There also is evidence of a lower tendency to migrate in males, which make up a high proportion of the sedentary population (Kapteyn and Lina 1994). In North America, most of the northern migrants in two species of *Leptonycteris* and *T. brasiliensis* are females that form maternity colonies in the Sonoran Desert (*L. curasoae*), northeastern Mexico (*L. nivalis*), and the south-central United States (*T. brasiliensis* [Cockrum 1969, 1991; A. Moreno, personal communication]). *Lasiurus cinereus,* which migrates into the United States and Canada from winter roosts in Mexico (and farther south?), is a species in which the sexes migrate to different geographic areas. Males spend the summer in the montane west, whereas females roost in the north-central and northeastern United States and Canada during the maternity period (Findley and Jones 1964). As discussed below, predominantly male migration in bats is known from *Myonycteris torquata* in the seasonal tropics of Africa (Thomas 1983).

Why are females more likely to be migratory or to migrate longer distances than males in temperate and subtropical species? There are at least two possible reasons for this sex-biased migration. The first can be explained by cost-benefit balance. Since females have high energetic requirements during pregnancy and lactation (Kunz et al. 1995; Kurta et al. 1989; Speakman and Racey 1987), the benefit they derive from moving to regions of seasonally high resource abundance may outweigh the costs of migration. However, males, with lower energetic requirements, may accept low resource density in order to reduce migratory costs. Although levels of flower and fruit resources during spring and summer have not been rigorously documented in the tropical dry forests of south-central Mexico (but see Bullock and Solís-Magallanes 1990), it is likely that they are much lower than resource levels in the Sonoran Desert when columnar cacti are flowering and fruiting (fig. 4.2). Thus, by migrating north, females of *L. curasoae* move into a region of high resource availability during the maternity period (Horner et al. 1998). Males remain to the south, perhaps to defend mating territories year-round as occurs in males of *Carollia perspicillata* in western Costa Rica (Fleming 1988). Likewise, females (and some males) of *T. brasiliensis* likely move along a resource gradient when they migrate from central Mexico to the south-central United States in the spring as insect populations begin to increase (G. McCracken, personal communication).

A second reason for sex-biased migration is that females have different roost requirements during the summer than do males. Females of some cave-dwelling species, such as *M. grisescens,* choose warm caves with air temperatures within their thermoneutral zones for rearing their young. These caves reduce the cost of thermoregulation in females (which, unlike most temperate vespertilionids, remain homeothermic during the day) and allow them to allocate more energy to milk production. In contrast, males of this species roost in cool summer caves and undergo daily bouts of torpor, thereby reducing their overall energy requirements (Tuttle 1975, 1976). The size of maternity populations affects the microclimates of cave roosts and may also influence the migration patterns of females. Metabolic heat produced by bats increases roost temperatures and assists in maintaining conditions necessary for optimum growth of embryos and dependent young (Dwyer 1969; Racey 1982). Females of *M. schreibersi,* which are widely distributed in Australia during the nonbreeding period, may migrate long distances and aggregate to achieve maternity colonies of sufficient size to sustain favorable roost temperatures (Baudinette et al. 1994). Females of other cave-dwelling species, such as *M. lucifugus,* use trees or buildings as maternity roosts and undergo daily torpor to conserve energy (Humphrey and Cope 1976). As in *M. grisescens,* males of *M. lucifugus* roost separately from females during the maternity season. Spatial segregation of males and females during the summer also reduces intraspecific competition for food and can select for differential migration in the sexes.

Tropical Bats

Not enough is known about the ecology of tropical bats to classify most species into the above three migration categories. Detailed, year-round data are needed to document seasonal migrations, and these data exist for only a handful of tropical species and/or for only a few tropical localities. Based on our current knowledge, however, it appears that many tropical and subtropical species are sedentary and non-migratory. This is likely to be true of insectivorous or animalivorous bats that live and feed in the understory of tropical forests. Such habitats tend to exhibit relatively low seasonal fluctuations in the availability of insect or plant resources (e.g., Karr and Braun 1990; Levey 1988; Smythe 1982), and forest understory birds and bats tend to have relatively stable populations (Greenberg 1980; Kalko et al. 1996; Loiselle and Blake 1990).

In contrast, bats (and birds) that feed on nectar and fruit, especially in the forest canopy, experience food supplies that undergo significant seasonal fluctuations (Fleming 1992; Levey and Stiles 1992; Loiselle and Blake 1990). We might expect many of these species to be migratory or, at least, to undergo seasonal habitat shifts. Little information is available on the annual movement patterns of tropical insectivorous species that feed above the canopy.

Sedentary, non-migratory lifestyles have been reported in a diverse array of tropical bats. In the Neotropics, many kinds of insectivores (e.g., most Emballonuridae, Furipteridae, Thyropteridae, Natalidae, Mormoopidae, phyllostomine phyllostomids, and lowland species of Vespertilionidae) as well as piscivores (Noctilionidae), carnivores (e.g., *Vampyrum spectrum*), certain omnivore/frugivores (e.g., *Phyllostomus hastatus*), and sanguinivores (Phyllostomidae: Desmodontinae) are sedentary (Bradbury and Emmons 1974; Bradbury and Vehrencamp 1976; Brooke 1997; McCracken and Bradbury 1981; Vehrencamp et al. 1977; Wilkinson 1985; Wilson 1971). Examples of sedentary Paleotropical bats include the aerial insectivore *Coelura afra* and the terrestrial insectivores *Lavia frons*, *Cardioderma cor*, and *Nycteris grandis* in Africa (McWilliam 1987; Vaughan 1976; Vaughan and Vaughan 1986); the carnivore *Macroderma gigas* and the aerial insectivore *Taphozous australis* in Australia (Churchill 1998); and the aerial insectivore *Chaerephon plicata* in Southeast Asia (Payne et al. 1985). O'Shea and Vaughan (1980) noted that 11 of 25 species (one each in the Pteropodidae, Emballonuridae, Nycteridae, Megadermatidae, and Rhinolophidae as well as six Vespertilionidae) occurring at a thorn-scrub site in central Kenya were year-round residents. In the Philippines, Heideman and Heaney (1989) reported that several species of subcanopy pteropodids, including *Haplonycteris fischeri*, *Ptenochirus jagori*, *Harpyionycteris whiteheadi*, *Macroglossus minimus*, and *Nyctimene rabori*, were sedentary. Finally, in Papua New Guinea, pteropodids such as *Dobsonia minor* and *Syconycteris australis* are highly sedentary and survive in very small patches of productive rainforest (Bonaccorso 1998; Winkelmann et al. 2000).

Short-distance migrants undoubtedly occur within the tropics, although we lack detailed knowledge of this because of a paucity of banding studies. In the Neotropics, Fleming (1988) reported that females of *Carollia perspicillata*, a frugivorous phyllostomid, migrate from lowland dry tropical forest, perhaps to moister upland forests, during the dry season (fig. 4.2). Distances between wet and dry season roosts are likely to be <200 km. Similar migration distances have been reported for at least eight species of insectivorous East African bats (four species each in the Rhinolophidae and Vespertilionidae) that move between coastal and inland habitats seasonally (O'Shea and Vaughan 1980; Vaughan 1977). In both of these examples, bats migrate out of habitats during dry periods when food levels are known (Costa Rica) or suspected (Africa) to be low and return with the resumption of rains when food levels increase.

A few cases of long-distance migration have also been documented in the tropics (and Australian subtropics). Except for the previously discussed arid-zone nectarivorous bats (*L. curasoae* and *C. mexicana*) and *T. brasiliensis*, which migrate from seasonal tropical habitats to subtropical or temperate habitats annually, current evidence for long-distance migrations in Neotropical bats is scant. One other phyllostomid nectarivorous bat, *Platalina genovensium*, which, like *L. curasoae*, is a cactus-visiting species, is suspected to undertake such migrations in arid parts of Peru (Sahley and Baraybar 1996). In a bat fauna as ecologically diverse as that of the Neotropics, it is likely that other examples of long-distance migrants will eventually be found, especially in southern South America.

In West Africa, three species of frugivorous pteropodid bats—*Eidolon helvum, Myonycteris torquata*, and *Nanonycteris veldkampi*—migrate from the forest zone north into the savanna during the major wet season, which begins in March. One-way migration distances range from about 400 km in *N. veldkampi* to at least 1,500 km in *E. helvum*. Individuals of both sexes migrate in *E. helvum* and *N. veldkampi*, but only immature males appear to migrate in *M. torquata*. Thomas (1983) argued that these migrations are driven by seasonal fluctuations in food supplies, but not in the same manner as those in *Leptonycteris*. Instead of migrating up a resource gradient, these species migrate down a resource gradient, that is, from forest habitats in which fruit levels are much higher year-round to savannas where fruit levels are always lower. Greater seasonal resource fluctuations (and lower per capita resource demands by resident bats at peak fruiting times) in the savanna habitat, however, make this habitat richer in food resources on a per capita basis than forest habitats during the wet season. A similar argument has been applied on a much larger geographic scale to many intercontinental avian migrants (e.g., Levey and Stiles 1992; Ricklefs 1980, 1992).

As we have described, most migratory animals move between regions of the world where the suitability of habitats and availability of resources vary in reliable annual cycles in association with shifts in temperature or rainfall. The migratory patterns of these species are regular and synchronized. Other animals, including some species of bats, however, have adapted their migrations to more irregular and ephemeral resource shifts (Dingle 1996; Ratcliffe 1931). In areas of subtropical and temperate Australia, for example, patterns of habitat suitability and resource availability for flower-feeding *Pteropus* are unreliable and ephemeral relative to the distinctly seasonal habitats of the temperate zone and wet/dry tropics. The eucalypt species that provide the main nectar and pollen resources for *P. scapulatus* and migrant populations of *P. poliocephalus* have supra-annual, noncyclical phenologies in which the bulk of nectar production occurs as eruptive, short-term flowering events that are ephemeral in time but that produce relatively high, short-term carrying capacities for bats. This variability becomes more pronounced with increasing latitude and increasing distances from the coast (P. Eby and D. Sommerville, unpublished manuscript). Sedentary populations of *P. poliocephalus* occupy floristically diverse coastal areas, while a nomadic population tracks patchy resources and migrates hundreds of kilometers between successive nectar pulses (fig. 4.7; Eby 1991; Parry-Jones and Augee 1992; Spencer et al. 1991).

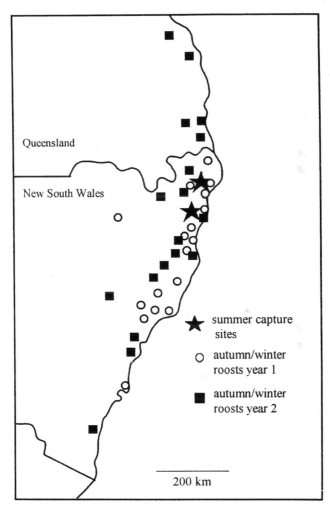

Figure 4.7. Annual variations in the autumn and winter migrations of 36 radio-collared *Pteropus poliocephalus*. Autumn/winter roosts used in year 1 were either empty or had substantially reduced populations in year 2. Similarly, autumn/winter roosts used in year 2 had been either empty or substantially smaller in year 1. (Sources of data: Eby 1991, 1996.)

No sedentary populations of *P. scapulatus* are known. However, the length and regularity of their movements also vary geographically, being more regular in coastal than inland areas (P. Birt, L. Hall, and M. Vardon, personal communication).

Long-distance migrants in tropical and subtropical regions tend to have similar wing characteristics and foraging traits as those of temperate migrants. However, no associations are apparent between their migration patterns and roost preferences. Migrant populations of the aerial hawking species *T. brasiliensis*, for example, live in tremendous maternity colonies (of up to about 20 million adults) in caves; its non-migrant populations tend to live in much smaller colonies in buildings (Cockrum 1969). Other North American, long-distance migrants (*Leptonycteris* spp., *C. mexicana*) are nectar feeders that forage in relatively open arid and semiarid habitats and also roost in caves. *Leptonycteris curasoae* is highly gregarious, with up to 100,000 adults living in a maternity roost, whereas *C. mexicana* usually lives in colonies of <25 individuals (Y. Petryszyn and E. L. Cockrum, unpublished manuscript; Wilkinson and Fleming 1996). Several long-distance migrants among Old World nectar and fruit bats forage away from clutter in canopy vegetation and roost in trees, either in large aggregations, as in Australian *P. poliocephalus* and *P. scapulatus* and African *Eidolon helvum*, or as solitary individuals or in small groups, as in African *Myonycteris torquata* and *Nanonycteris veldkampi* (Nelson 1965b; Ratcliffe 1931; Thomas 1983).

Physiological Consequences of Migration

Rather profound seasonal changes occur in the behavior and physiology of many kinds of migrant animals (Dingle 1996). In birds, these changes include cessation of reproduction, hyperphagia, fat deposition, and directionally nonrandom "premigratory" movements (zugenruhe). The adaptive value of such changes is easy to understand because migration can require a tremendous investment in time and energy. What physiological adjustments occur in migrant bats, and how energetically expensive is their migration?

The primary fuel that birds and bats use for migration is fat (Alerstam and Lindstrom 1990; Blem 1980; Klaassen 1996; Lindstrom and Alerstam 1992), and temperate species of both taxa typically increase their fat loads in the late summer and fall prior to migration. Fat deposition in temperate bats at this time can be rapid (Kunz et al. 1998). Females of *Myotis thysanodes*, for example, can increase their fat loads ninefold in only 11 d (Ewing et al. 1970). Use of daily bouts of torpor during the fall feeding period reduces energy needs by at least 40% and can help these bats maximize their rates of fat deposition (Blem 1980; McNab 1982). Unlike birds, however, fat deposition in most temperate bats must serve a duel purpose: it must provide fuel for hibernation as well as migration. Since hibernation lasts much longer than migration (up to 6

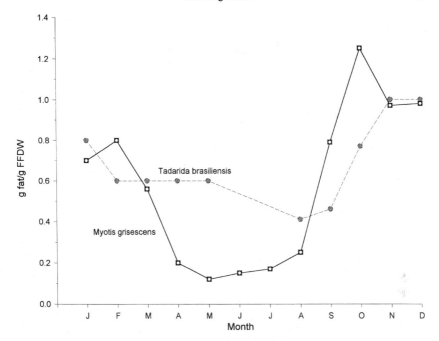

Figure **4.8.** Annual fat cycles of *Myotis grisescens* and *Tadarida brasiliensis* in the United States. *FFDW* = fat free dry weight. (Based on Krulin and Sealander 1972; O'Shea 1976; Pagels 1975.)

mo vs. a few days or weeks, respectively), and bats can refuel by feeding during migration, it is likely that fall fat deposition in most temperate bats is more strongly associated with hibernation than with migration. In contrast, because tropical bats do not hibernate, their fat stores can be used exclusively for migration (or for mating purposes in the case of males).

The annual fat cycles in two North American temperate bats are shown in figure 4.8. *Myotis grisescens* is a 15-g short-distance migrant that hibernates for about 6 mo. The fat content of both males and females increases by about an order of magnitude in September and October to about 26% of their wet mass (Krulin and Sealander 1972), but the proportion of this fat that is used directly for migration is poorly known. Based on changes in mean population masses between summer roosts and winter hibernacula, Tuttle (1976) estimated that females of this species use 1.5–4 g of fat during migration, with mass loss being proportional to migration distance. These indirect estimates suggest that a substantial fraction of the fat deposited in early fall could be depleted during migration. If this is true, then we can make three predictions: (1) some individuals will need to refatten after they have reached their winter roost before hibernation, (2) selection will favor individuals that minimize the distance that they migrate between summer and winter roosts to conserve fat for hibernation, and (3) hibernating bats will minimize fat loss during lengthy migrations by using stopovers to "refuel." In support of prediction 2, Tuttle (1976) suggested that the seasonal movements of *M. grisescens* generally are limited to those that minimize the distance between physiologically appropriate roost sites.

Individuals in both migrant and non-migrant populations of the 12.5-g *Tadarida brasiliensis* also undergo an annual fat cycle, but changes in fat levels are not as great as those in *M. grisescens* (fig. 4.8). O'Shea (1976) and Pagels (1975) reported 2–2.5-fold seasonal increases in fat levels in non-migrating and migrating populations of this non-hibernating species, respectively. Although *T. brasiliensis* neither migrates nor hibernates in southern Louisiana, Pagels (1975) demonstrated experimentally that fat deposited in the fall helps individuals resist hypothermia in winter; he also suggested that such deposits give these bats the option of migrating during especially cold winters.

Tropical and subtropical migrant bats are also known to deposit fat prior to migration. Both males and females of the New World nectarivorous bat *L. curasoae* deposit substantial amounts of subcutaneous fat in the fall, prior to mating and departure (migration?) for non-mating roosts (Ceballos et al. 1997). Based on changes in body mass, these bats put on at least 3 g of fat (ca. 12% of their wet mass) at this time of the year. Similarly, O'Shea and Vaughan (1980) reported large fat buildups in individuals of *Hipposideros commersoni, Miniopterus schreibersi, Scotophilus dinganii* (= *S. nigrita*), and *Scotoecus hindei* and four species of molossids prior to their disappearance (migration?) from a site in central Kenya.

How far can small microchiropteran bats migrate on a given amount of fat? Clearly, migration ranges will be positively correlated with fat load, but more fat is not necessarily better for migrating animals. Because increased body mass via fattening increases the absolute cost of transport and affects maneuverability and predation risk (Alerstam 1991; Klaassen 1996; see Speakman and Thomas, this volume), selection should favor an optimal fat load that minimizes this cost while maximizing net migration range (e.g., the distance between refueling stops in long-distance migrants [Lindstrom and Alerstam 1992]). To our knowledge, no one has yet addressed the topic of optimal fat loads in migrating bats.

Despite the absence of such studies, it is relatively straightforward to calculate how far bats can migrate on their fat loads. Ewing et al. (1970) proposed the following equation for this calculation:

$$\text{maximum migration range} = \frac{(\text{fat mass})(\text{kcal/g fat})(\text{flight speed kph})10^3}{(\text{BMR})(20)(\text{bat mass})(4.8 \text{ kcal/L } 0_2)},$$

in which kcal/g fat = 9.4, BMR = basal metabolic rate, and 20 = cost of flight relative to BMR (Winter and Helversen 1998; but see Speakman and Thomas,

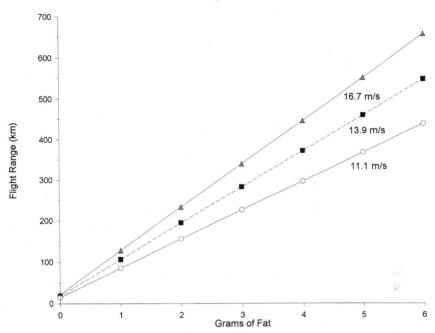

Figure 4.9. Potential migration distances in *Leptonycteris curasoae* based on amounts of stored fat and flight speed. See text for further explanation.

this volume). Examples of calculations for *L. curasoae*, which weighs about 23 g before fattening, are shown in figure 4.9. By putting on 3 g of fat, this bat will have a migratory range of 225–340 km, depending on flight speed. Sahley et al. (1993) reported that flight speed in this bat is about 8.3 m s^{-1} when it commutes between its day roost and feeding areas. Clearly, this bat can increase its flight range substantially either by flying faster and/or by increasing its fat load. At an average flight speed of 13.9 m $^{-1}$, a migration of 1,500 km will "cost" a female *L. curasoae* about 17 g of fat. Acquiring this amount of fat probably requires several stopovers for refueling during migration.

Using the above equation (but with the relative cost of flight set at 30 rather than 20), O'Shea (1976) estimated that migrant *T. brasiliensis* arriving in central Arizona in March had enough fat to continue flying for another 700 km, whereas individuals leaving Arizona in September had only enough fat to fly about 350 km before refueling. Observed fat levels on long-distance migrating microchiropterans suggest that they typically do not carry enough fat to last them the entire trip without stopping to refuel at least once. In contrast, certain long-distance avian migrants (e.g., ruby-throated hummingbirds and parulid warblers) carry twice as much fat and consequently have much longer flight ranges than do temperate bats (Blem 1980).

In addition to consuming fat, migration (and flight in general) results in significant rates of water loss. Carpenter (1969), for example, estimated that individuals of *L. curasoae* and *Eptesicus fuscus* lose about 3.9% and 3.1% of their mass via evaporation of water per hour of flight, respectively. These rates of loss suggest that bats can quickly become dehydrated during periods of prolonged flight and that they must regularly rehydrate to avoid serious physiological consequences. Migrating at high air temperatures and/or at high elevations leads to especially high rates of evaporative water loss. Most vespertilionids and molossids rehydrate by drinking freestanding water, whereas *L. curasoae* rehydrates by drinking nectar or eating water-rich fruits. Klaassen (1996) points out that rates of water loss can potentially constrain the speed of migration in birds, and a similar situation is likely to occur in bats. This topic deserves further study in bats.

How fast do bats migrate and are there seasonal and sexual differences in migration speeds? Evidence from radiotelemetry and banding studies suggests that some migratory bats reduce their rate of migration by breaking their journeys at several transient roosts, rather than completing lengthy migrations in a few sustained flights. Radio-collared *P. poliocephalus*, for example, stopped at transient roosts for periods of 1–8 d during migrations of hundreds of kilometers (Eby 1991, 1996). Distances traveled between these roosts were equivalent to distances traveled during nightly foraging bouts and ranged from about 50 to 130 km. Individuals traveled these nightly distances by making several flights of relatively short duration (30–90 min) and at ground speeds of 7.5–8.3 m s^{-1}; flights were interspersed with periods of roosting or foraging (P. Eby, D. Wahl, and C. Palmer, unpublished data). Thus, an overnight flight of 130 km took 12 h to complete, although actual flying time was approximately 5 h. These field data support predictions that, during sustained flight, free-flying *P. poliocephalus* travel at speeds that minimize power input while maximizing flight duration (6–8 m s^{-1} [Carpenter 1985]).

Precise data on speed of migration by North American long-distance migrants such as *T. brasiliensis*, *L. curasoae*, and *Lasiurus cinereus* are not yet available. It is likely, however, that each of these species also makes stopovers at several "transient" roosts during their northward and southward journeys. Brazilian free-tailed males apparently migrate north in the spring faster than females (Cockrum 1969). Migration at this time of the year is thought to take several weeks, with bats stopping at various transient roosts before arriving at their summer roosts. Cockrum speculated that both sexes migrate faster in the fall than in the spring, perhaps aided by the southward passage of cold fronts. Glass (1982) reported that when maternity roosts of *T. brasiliensis* in Oklahoma break up in late summer, some bats move north to transient caves before heading south into eastern Mexico in October. Banding returns indicate that some young-of-the-year move up to 360 km from their natal caves by early August (Glass 1982). Because they have not been the subjects of large banding studies, migration rates of the lesser long-nosed bat *L. curasoae* are less well-known than those of *T. brasiliensis*. Most females (and many males) leave a mating cave in coastal Jalisco, Mexico, in December but do not show up in their northern maternity roosts until late March or April (Ceballos et al. 1997; Cockrum 1991). Whether they stay in south-central Mexico or begin to migrate slowly north early in the year is not yet known. Finally, because it roosts solitarily in trees and relatively few have been banded, even less is known about migration rates of the hoary bat *L. cinereus*. Findley and Jones (1964) noted that females migrate through New Mexico earlier in the spring than do males. Zinn and Baker (1979) reported that hoary bats migrate through Florida in late October–late November and again in February–early April. These data suggest that northward migration is slower than southward migration in eastern populations of this species.

Studies of North American temperate bats indicate that female *M. lucifugus* and *M. grisescens* tend to leave their hibernacula earlier in the spring than males and fly directly to their summer maternity roosts at rates of 50–100 km per night (Davis and Hitchcock 1965; Humphrey and Cope 1976; Tuttle 1976). Females of *M. grisescens* probably make similarly rapid migrations back to their hibernacula in the fall (Tuttle 1976). Males and young-of-the-year in these species apparently migrate more slowly in both spring and fall. "Swarming" behavior, which can occur during both the spring and fall in *M. lucifugus* and other species (Barbour and Davis 1969; Humphrey and Cope 1976), can slow down rates of migration. This behavior involves the occurrence of large numbers of bats, sometimes of just one sex but at other times in mixed-sex or mixed-species groups, at the entrance of caves (often at hibernacula in the fall) from dusk to dawn on several days. Bats typically do not reside in these caves during the swarming period. Such behavior is thought to be a prelude to mating, which occurs in the fall in hibernating temperate bats. General "wandering" and short-term visits to transient caves during fall migration also occur in various temperate bats, especially in males and young-of-the-year (Davis and Hitchcock 1965; Dwyer 1966; Humphrey and Cope 1976).

Seasonal movements of long-distance European migrants, such as *N. noctula* and *P. nathusii*, have been studied in extensive banding programs (see reviews in Roer 1995; Strelkov 1969, 1997). Short-term, periodic occupations of areas from which migratory species are usually rare or absent indicate that these bats use several short flights rather than lengthy sustained flights during migration (Arnold et al. 1996; Heinz and Braun 1997; Strelkov 1969; Weidinger 1994). Autumn migrations occur over several weeks, and patterns of departure from maternity sites and passage through stopover habitats are staggered both within and between species (Ahlén 1997; Gerell-Lundberg and Gerell 1994; Strelkov 1969). Individuals begin to depart from breeding habitats in northern and central Europe in late July; *Vespertilio murinus* departs from central Russia earlier than *P. pipistrellus*, which departs prior to *N. noctula* and

P. nathusii (Strelkov 1969). Adult females of *P. nathusii* depart prior to males and young. These slow, staggered movements should reduce the density of animals and competition for resources along migration routes. In contrast, spring migrations in these bats occur over short periods (Strelkov 1969). Mass influxes that last 2–3 d have been recorded at stopover sites.

While these studies provide evidence for the use of relatively short flights by migrating bats, there is also evidence from movements over water that some species are capable of sustained flights of several hours. Field observations and band returns show that long-distance migrants such as *P. nathusii* and *N. noctula* fly 140–180 km over the Baltic Sea on routes from southern Sweden to Germany or Poland (Ahlén 1997; Gerell 1987). These movements represent sustained flights of 7–9 h, given flight speeds of 5.6–6.1 m s^{-1} (Ahlén 1997; Baagøe 1987; Jones 1995). Vagrants of several species of North American and European bats have been recorded several hundred kilometers from land on oil rigs and small vessels in the North Sea (Corbet 1970) and have also been recorded in Iceland (Petersen 1994). Although some of these records may reflect transportation by ships, there is also evidence of long, wind-assisted flights over the North Atlantic (Peterson 1994).

Our final topic in this section deals with group size in migrating bats. Current evidence suggests that many species of bats, including those that normally roost as solitary individuals during the summer, migrate in groups. These species include *L. cinereus, T. brasiliensis,* and *Lasionycteris noctivagans* in North America, *N. noctula, P. nathusii,* and *V. murinus* in Europe, *Eidolon helvum* in Africa, and various species of *Pteropus* in Australia. In each of these species, groups or "waves" of individuals have been seen in flight or have entered roosts during migration (Ahlén 1997; Cockrum 1969; Eby et al. 1999; Findley and Jones 1964; Gerell 1987; Glass 1982; Parry-Jones and Augee 1992; Ratcliffe 1931; Strelkov 1969; Thomas 1983). Cockrum (1969), for example, reported that about 100,000 females of *T. brasiliensis* arrived at Cueva del Tigre, Sonora, Mexico, on one night in March. Similarly, thousands of individuals of *E. helvum* arrive simultaneously at tree roosts in the savannas of Ivory Coast, Africa, early in the wet season (Thomas 1983). Overnight changes in roost populations of several hundred to several thousand animals have been repeatedly reported in Australian *Pteropus* (Eby et al. 1999; Nelson 1965b; Parry-Jones and Augee 1992; Ratcliffe 1931; C. Palmer, unpublished data). Swarming behavior, which is widespread in temperate hibernating bats, also implies that many individuals move among roosts in groups.

Group migration would appear to offer several advantages to bats, including reduced per capita predation risk, increased ease with which young bats learn migration routes and stopover sites, increased probability of the sexes finding each other for mating purposes, and the physiological advantages of group roosting during stopovers. Disadvantages of group migration include increased competition for food and mates. To our knowledge, however, none of these topics have yet been investigated in detail.

Life-History, Social, Population Genetic, and Community Consequences of Migration

Life-History Consequences

Because it is a risky, time- and energy-intensive activity, migration should have a strong effect on the evolution of life-history strategies of bats, including effects on fecundity and survivorship, as well as on the evolution of mating systems (see Barclay and Harder, this volume). Furthermore, because it involves the movement of large numbers of individuals across extensive landscapes, migration can have important population and community consequences, including seasonal changes in population sizes and in levels of interspecific competition between resident and migrant species, as well as possible effects on their genetic structure.

Data needed to assess the effect of migration on life-history strategies include life and fecundity tables for migratory and non-migratory species. If migration is risky, we expect rates of annual survivorship in juveniles and adults to be lower in migrants than in non-migrants. At first glance, this prediction might seem paradoxical because selection should never favor migratory behavior if it results in higher rates of mortality than sedentary behavior. As discussed previously, however, the proper test for this prediction involves intraspecific (ideally, intrapopulational) rather than interspecific comparisons. Such comparisons are currently impossible for bats because of a lack of data. Despite the absence of relevant intraspecific data, it still is logical to expect to find significant differences in annual survivorship in migratory versus non-migratory species. Furthermore, if migration involves a significant investment in time and energy, we also might expect annual fecundity to be lower in migrant than in non-migrant species. Finally, since migration "shuffles" populations and potentially breaks up social structures within roosts, we might expect mating systems and levels of behavioral cooperation among individuals to differ in migrant and non-migrant species. Specifically, we predict that monogamous mating systems and other temporally stable male-female associations should not exist in migrant species.

Unfortunately, the paucity of detailed life tables for bats prevents us from rigorously testing for significant differences in survivorship and fecundity among migrant and non-migrant species. As summarized in Tuttle and Stevenson (1982, table III), data for temperate hibernating bats indicate that annual survival rates in adults are 66%–77% in sedentary taxa ($n = 2$ species) and 64%–98% in regional migrants ($n = 7$ species). These authors pointed out that these estimates are fraught with technical problems and must be interpreted with caution. No data currently exist on annual adult survivorship in temperate, long-distance migrants and, except for *Carollia perspicillata* (Fleming 1988), *Artibeus jamaicensis* (Handley et al. 1991), and *Desmodus rotundus* (Tschapka and Wilkinson 1999), for tropical species. Annual adult survival rates in *C. perspicillata* in a Costa Rican dry tropical forest are about 62%; survival rates in *A. jamaicensis* in a Panamanian moist tropical forest are about 47% and 58% in males and females, respectively. Though scant, these data suggest that some tropical species have shorter life expectancies than certain temperate species that evade predators and harsh climatic conditions by hibernating. We suspect that hibernation is likely to have a stronger net effect on survivorship than migration in temperate bats. In contrast, several authors, as summarized in Ricklefs (1992), have noted that annual rates of survival in passerine birds tend to be higher in long-distance migrants than in sedentary species or short-distance migrants. Harsh winters apparently cause higher mortality rates in these birds than do the rigors of migration, which is contrary to our prediction for bats. This raises the question, Why are there no hibernating temperate birds (except for the 55-g poorwill [Csada and Brigham 1992])?

Ricklefs (1992, 1997) also reported that temperate North American resident songbirds produce more fledglings per individual than migrant songbirds and that production of offspring is negatively correlated with annual survival rates in the genus *Turdus* (thrushes). Does a similar trend hold for bats? Virtually all temperate bats, regardless of their migratory status, undergo a single pregnancy each year. Females in most species produce a single pup, but some species, including the long-distance migrants *P. nathusii* and *N. noctula* in Europe (Roer 1995) and *Lasiurus* spp. and *L. noctivagans* in North America, produce two or more pups per pregnancy. Larger litter sizes in certain long-distance migrants suggest that these species might have lower annual survival rates than more sedentary temperate bats. But differences in summer roost types (i.e., trees vs. caves or buildings), which could also affect juvenile survival, confound this comparison. Finally, litter sizes of two or more pups are not restricted to long-distance migrants. For example, females in the sedentary *Eptesicus fuscus* produce two pups per litter in the eastern United States as does *Nycticeius humeralis* in the southeastern United States (Barbour and Davis 1969).

Although monestry is also a common reproductive pattern in tropical bats, many species are polyestrous and undergo two or more pregnancies (invariably producing a single pup) per year (Hayssen and Kunz 1996; Kurta and Kunz 1987; Racey 1982; Wilson 1979). The distribution of monestry and polyestry, and hence annual female fecundity, among tropical bats appears to be more closely related to trophic position and/or phylogeny than it is to migratory behavior. Insectivorous and/or animalivorous bats other than the Vespertilionidae as well as *Pteropus* are monestrous, whereas certain vesper bats (e.g., *Myotis nigricans* in Central America) as well as most nectarivorous and frugivorous phyllostomids and pteropodids are polyestrous (Fleming 1988; Heideman 1988, 1995; Wilson 1971). Exceptions to this include the long-distance migrant nectarivores *Leptonycteris* spp. and *C. mexicana,* which are monestrous. While these species conform to our prediction, it could be argued that they really are "temperate" bats because of the northern locations of some of their maternity roosts. The only other phyllostomid with a temperate distribution, *Macrotus californicus,* which is a non-migratory insectivore, is also monestrous (Bleier 1975). Thus, it is not clear whether *Leptonycteris* and *Choeronycteris* are monestrous because they are migratory or because they produce their young under more or less temperate (= highly seasonal) conditions. More convincing support for our prediction of reduced fecundity in migrant species comes from O'Shea and Vaughan's (1980) study of an East African bat assemblage. All four migrant species produced one pup per pregnancy, whereas three of six resident species produced two to three young per pregnancy.

Regional migration in female *C. perspicillata* in Costa Rica has an interesting consequence for the reproductive potential of males (Fleming 1988). Although most females undergo two pregnancies a year, some males participate in only one of two annual mating periods. Females mate with males in lowland caves during the wet season and, after migrating, give birth in upland caves in the dry season. They then mate with males in the upland caves and, after again migrating, give birth in lowland caves. In both sets of caves, males apparently tolerate offspring within their harems that were fathered by distant males. Year-round defense of their mating territories, coupled with seasonal migration by females, causes most lowland males to miss out on the dry season mating period each year (Williams 1986).

Consequences for Bat Sociality

Migration in all bats and hibernation in temperate bats place significant constraints on the mating systems and social stability within species of bats. In general, migration reduces the likelihood that bats will form year-round stable associations within and between groups of males and females. This, in turn, prevents the evolution of monogamous mating systems, which occur in a variety of sedentary tropical bats, and year-round harems containing single males that defend relatively stable groups of females (McCracken and Wilkinson 2000). In the absence of stable associations of bats, socially cohesive or cooperative behaviors such as allogrooming, food sharing, and possibly communal nursing are unlikely to evolve (Wilkinson 1987). What kinds of mating systems are found in migrating bats and to what extent do these species exhibit socially cohesive behaviors?

Highly polygynous and sometimes promiscuous mating systems have been reported in most species of migrating bats (reviewed in McCracken and Wilkinson 2000). In many temperate species, regardless of migration status (e.g., table 4.1), mating occurs early in the fall, often in or near hibernacula. In cave-dwelling species such as *M. lucifugus*, both males and females mate with several partners without forming harems or mating territories. In other temperate species (e.g., *M. myotis*, *N. noctula*, *P. nathusii*, and *P. pipistrellus*), males set up mating territories, either in buildings or in hollow trees (and nest boxes), and mate with females as they pass through the area during migration. Again, females may mate with several males, and it is known that twins produced in *N. noctula* sometimes have different fathers (Mayer 1995). In other species (e.g., temperate *Nycticeius humeralis* and subtropical *Myotis adversus*), the mating system consists of seasonal harems in which single males defend and mate with groups of females. Regardless of the form of the mating system, females in most temperate species store sperm during the winter and ovulate in the spring before migrating back to their summer roosts.

The timing of mating and/or ovulation in non-hibernating North American migrant bats differs somewhat from that of hibernators. *Leptonycteris curasoae*, for example, mates and presumably ovulates in October or November. Gestation lasts 5–6 mo in this species (Ceballos et al. 1997), and many females are in advanced pregnancy as they migrate north. Although the mating system of this bat has not yet been studied, male-female associations are not likely to involve temporally stable harems, given the relatively short time (ca. 2 mo) both sexes roost together. In contrast, mating in *L. borealis* and *T. brasiliensis* is likely to occur in the early spring, just before or during migration (Cockrum 1969; Findley and Jones 1964; McCracken et al. 1994). Recent observations on the latter species indicate that males form mating territories within transient colonies; as in many temperate species, males and females have multiple sex partners (French and Lollar 1998; A. Keeley, personal communication).

In highly colonial *P. poliocephalus*, adult males and females share roosts throughout the year (Eby 1991, 1996; Ratcliffe 1931; Nelson 1965a). The process of mating is protracted, commencing in January with the establishment of male-defended mating territories (containing one or more females) and culminating in conception in March and April. Females produce a single pup in October-November. The weeks in which mating territories are defended and conception occurs coincide with late summer and autumn migrations (Eby 1991). In sedentary populations, mating groups are stable during this period (Nelson 1965a). However, it is unknown whether mating groups within migrating populations are also stable. Both adult males and females migrate during this time and roosts of migrating animals are known to contain mating groups. The composition of mating groups may change during migrations. Alternatively, it is possible that mating groups migrate together and male/female associations are maintained.

While socially cohesive or cooperative behaviors such as allogrooming and food sharing have been most often reported in sedentary tropical bats that form long-term associations (e.g., in the phyllostomids *Phyllostomus hastatus*,

Vampyrum spectrum, and *Desmodus rotundus* [Wilkinson 1987] but not in the fishing bat, *Noctilio leporinus* [Brooke 1997], high levels of female philopatry to natal roosts, which occurs in many species of bats (Kunz 1982b; Kunz and Lumsden, this volume; Lewis 1995; Tuttle 1976), mean that related females could potentially roost together during the maternity period. This, in turn, might favor the evolution of some form of cooperative behavior among migrant females via kin selection or reciprocity. In line with this hypothesis, Wilkinson (1992) reported that female evening bats *N. humeralis* nurse unrelated pups at the end of the lactation period. Non-descendent females were nursed significantly more frequently than were males. Neither kin selection nor reciprocity, however, appears to be involved in this behavior. Instead, Wilkinson (1992) suggested that communal nursing is a form of mutualistic behavior in which nursing females "dump" excess milk into young females that can help them find good foraging sites. Howell (1979) proposed that migrant individuals of *L. curasoae* also behave cooperatively by foraging at *Agave* inflorescences in orderly fashion and by allogrooming in night roosts. Fleming et al. (1998), however, disputed that this behavior is really cooperative or altruistic and reported that females in a northern maternity roost exhibited neither communal nursing nor allogrooming. Finally, although communal nursing was proposed to occur in maternity roosts of *T. brasiliensis* (Davis et al. 1962), McCracken's (1984a) genetic studies indicate that females nearly always nurse their own pups in large colonies. In summary, it appears that truly cooperative behavior has not yet been convincingly documented within summer colonies of migrant bats. Low average levels of relatedness (e.g., near zero in relatively small colonies of *N. humeralis* [Wilkinson 1992]) despite strong female philopatry may account for this.

Consequences for Genetic Structure

In addition to influencing the evolution of mating systems and social behavior, migration should have an important effect on genetic structure of bat populations. Specifically, we expect to see lower levels of genetic subdivision in migrant species than in sedentary species despite high levels of philopatry to summer and winter roosts in many species of migrants. Male dispersal and occasional natal and/or hibernaculum roost shifts by both sexes can lead to large panmictic populations with little genetic subdivision. This topic has been most thoroughly investigated in the Brazilian free-tailed bat *T. brasiliensis* (McCracken and Gassel 1997; McCracken et al. 1994; Svoboda et al. 1985). Based on the geographic distribution of banding returns, Cockrum (1969) predicted that North American populations of this species consist of four genetically distinct populations: two non-migratory populations (Pacific coast and southeastern United States) and two migratory populations (eastern Arizona–western New Mexico and eastern New Mexico–Texas–Oklahoma) that move into western and eastern Mexico, respectively. However, allozyme studies refute this scenario and indicate that *T. brasiliensis* has a large genetically effective population size (N_e) in Mexico that exhibits little geographic differentiation, as measured by Wright's (1978) fixation index F_{ST} (table 4.4). Genetic subdivision is also very low among non-migratory and migratory populations in the United States. These studies indicate that gene flow is substantial among all populations of this species. Bats may migrate along geographically distinct pathways, but rates of between-roost or between-pathway movement are high enough to prevent the formation of genetically distinct populations. High vagility translates into a more or less panmictic mating structure in this species.

Low levels of between-population genetic subdivision and large N_e's also occur in other species of long-distance migrants (table 4.4). Populations of *L. curasoae* in western Mexico, Baja California, and the southwestern United States, for example, extensively share mitochondrial DNA (mtDNA) haplotypes, and Wilkinson and Fleming (1996) estimated that its N_e is on the order of 50,000–100,000 individuals. Similarly, three migratory species of *Pteropus* experience little genetic differentiation over vast areas of Australia (table 4.4). Petit and Mayer (2000) used both mtDNA and microsatellite (nuclear) DNA to study the genetic structure of *N. noctula* in Europe. Their data indicate that this species has an unusually large breeding population and that its region of panmixia in central Europe has a diameter of about 3,000 km, which is the largest panmictic unit yet reported for terrestrial mammals.

Regional migrants also appear to exhibit low levels of genetic subdivision. Based on banding returns, Humphrey and Cope (1976) proposed that gene flow in *M. lucifugus* is restricted to a series of geographically distinct "demes" in eastern North America. More recent allozyme studies (summarized in Fenton and Thomas [1985]), however, suggest that gene flow is extensive among populations of this species and that the demic structure proposed by Humphrey and Cope (1976) does not exist. Low levels of genetic subdivision and extensive gene flow among populations also occur in *C. perspicillata* in western Costa Rica (Fleming 1988; table 4.4).

In contrast, high levels of genetic subdivision have been reported in a variety of sedentary bats. For example, most of the genetic variation in the Australian ghost bat *Macroderma gigas* resides within rather than between populations; 87% of its mtDNA variation is distributed among populations (Worthington Wilmer et al. 1994, 1999). High female philopatry and small N_e's are thought to contribute to this pattern. Non-migratory pteropodid and phyllostomid bats also have relatively high values of F_{ST} and low values of Nm (an estimate of the number of migrants per generation; table 4.4). Finally, the tropical emballonurid *Saccopteryx bilineata* and, in Europe, *M. myotis* and *Plecotus auritus* exhibit substantial genetic differentiation among colonies separated by only 43–56 km (Burland et al. 1999; table 4.4). Although we need more data before any firm conclusions regarding the relationship between migratory status

Table 4.4. Estimates of genetic subdivision among populations of migratory and nonmigratory bats

Family/Species	Method	No. of Populations	Max. Distance (km)	F_{ST}	N_m	References
Migratory species:						
Pteropodidae:						
Pteropus alecto	A	6	2,916	0.023	10.6	Webb and Tidemann 1996
Pteropus poliocephalus	A	6	744	0.014	17.6	Webb and Tidemann 1996
Pteropus scapulatus	A	6	2,700	0.028	4.2	Sinclair et al. 1996
Phyllostomidae:						
Leptonycteris curasoae	Mt	13	2,650	Very low (>0.01)	High	Wilkinson and Fleming 1996
Carollia perspicillata	A	5	70	0.007	35.5	Fleming 1988
Molossidae:						
Tadarida brasiliensis, SW U.S.	A	8	1,080	0.008	31.0	McCracken et al. 1994
Tadarida brasiliensis, Mexico (females)	A	4	720	0.017	15	McCracken et al. 1994
Tadarida brasiliensis, U.S.	A	6	3,780	0.158	1.33	McCracken and Gassel 1997
Vespertilionidae:						
Nyctalus noctula	Ms	13	...	0.006	ca. 41.4	Petit et al. 1999
Nonmigratory species:						
Pteropodidae:						
Cynopterus brachyotis	A	6	500	0.11	7.53	Peterson and Heaney 1993
Cynopterus musatenggara	A	11	1,080	0.17	1.22	Schmitt et al. 1995
Haplonycteris fischeri	A	6	500	0.61	0.05	Peterson and Heaney 1993
Emballonuridae:						
Saccopteryx bilineata males	A	4	43	0.104	2.2	McCracken 1984b
Megadermatidae:						
Macroderma gigas	M	4	3,700	Very high (ca. 0.87)	Very low	Worthington Wilmer et al. 1994
Vespertilionidae:						
Myotis myotis	M	2	56	0.028	...	Petri et al. 1997

Note. For method of analysis, A = allozymes, G = genomic DNA, Ms = microsatellites, and Mt = mitochondrial DNA. Maximum distance refers to the distance between populations or colonies. F_{ST} is Wright's (1978) fixation index indicating proportion of total heterozygosity that occurs among populations; this index ranges from 0 (total panmixia) to 1.0 (each population is genetically unique). N_m is the number of migrants per generation based on Wright's F_{ST} and calculated as $N_m = [(1/F_{ST}) - 1]/4$; values < 1.0 indicate that genetic drift can strongly influence the genetic structure of a species whereas values > 4.0 indicate that gene flow is the predominant force influencing genetic structure (Slatkin 1985).

and genetic structure can be reached, available data support the hypothesis that populations of mobile bats exhibit less genetic structure and have higher rates of between-population gene flow and larger effective breeding populations than do sedentary species.

Community Consequences

Seasonal movements of large numbers of bats from one location (community) to another can also have potentially important ecological consequences. These include seasonal changes in rates of resource consumption in different communities as well as changes in the demand for safe roost sites (see Kunz and Lumsden, this volume). If either food or roost sites are in limited supply, then the influx of migrant species could increase levels of interspecific competition within communities. These consequences of migration actually apply only to tropical communities because, as we have already discussed, temperate bat assemblages basically shut down during cold seasons as bats enter hibernation. Most temperate bat assemblages, therefore, are only seasonal phenomena. Tropical bat assemblages, in contrast, are active year-round and change in composition as migrant species enter and leave. It is in these assemblages that we might expect to see seasonal changes in the intensity of interspecific competition for food or roost sites (see Patterson et al., this volume). Unfortunately, as discussed at length by Findley (1993), it is difficult to conduct rigorous studies of interspecific competition in bats. As a result, we currently lack a quantitative understanding of the impact of migrant bats on their alternate assemblages. Instead, we can only describe changes in assemblage composition and speculate about what these changes mean.

To our knowledge, no European long-distance migrant species enter tropical regions during the winter, and only a handful of North American migrants do so. These include the aerial insectivores *L. cinereus* and *T. brasiliensis* and the nectarivores *L. curasoae, L. nivalis,* and *C. mexicana.* Of these species, only two, *T. brasiliensis* and *L. curasoae,* are common enough to significantly increase demands for food and/or shelter in their winter range. But whether they do so is currently unknown. Intratropical migrants, rather than extratropical migrants, are more likely to have a significant ecological impact on resource levels, at least in certain tropical regions. In West Africa, for example, three pteropodid bats (*E. helvum, M. torquata,* and *N. veldkampi*) migrate seasonally from the equatorial forest belt to savanna habitats 400–1000 km to the north. Around the Lamto field station in Ivory Coast, they enter an assemblage of six resident species of pteropodid bats when food levels (primarily fruit) are high relative to demand. Thomas (1982) reported that dietary overlap was low between migrant and resident bats as a result of differences in habitat selection. In central Kenya, 14 species of insectivorous bats join an assemblage of 10 insectivorous bats when food levels increase during the wet season. According to O'Shea and Vaughan (1980), factors that tend to reduce interspecific resource overlap among resident and migrant bats include differences in jaw width and body size, foraging heights, and echolocation call designs. As a result, although most species (18 of 21) of residents and immigrants are aerial insectivores, their feeding niches tend to be quite distinct. In both of these cases, potential for interspecific competition between resident and migrant bats appears to be low. Whether this is true in other tropical assemblages remains to be determined.

Conservation Consequences of Migration

Migration has been identified as a trait that can compound the detrimental impact of human activities on wildlife and increase extinction risk (Pagel and Payne 1996; Pimm et al. 1988). Migratory animals are disadvantaged by extensive and complex habitat requirements, tendencies to congregate into restricted areas, and reliance on broad-scale, integrated conservation programs (Brower and Malcolm 1991; Eby et al. 1999; Myers et al. 1987; Pagel and Payne 1996). Migrants are unlikely to be conserved incidentally within general programs of resource management. Rather, they require initiatives that target their specific needs.

Habitat Requirements

Migratory populations require a progression of spatially distinct, often apparently unrelated, habitats to complete their annual cycles. Loss or degradation of any one can create a point of vulnerability. Migrating bats generally require breeding sites (i.e., mating and maternity sites), overwintering sites, and migration corridors that link the two. For a range of conditions, reduction of either wintering or breeding habitat results in population reduction in bats, and preservation of these habitats is the foundation of most conservation programs (e.g., Hutson et al. 2000; Mickleburgh et al. 1992; Racey and Entwistle, this volume). However, the preservation of adequate stopover habitat is equally important, particularly for long-distance migrants (Hutto 1998).

Stopover habitats must provide adequate resources in precise sequence both in time and space. Broad-scale qualitative declines in habitat potentially interrupt the continuity of resources along migration corridors and increase the energy that individuals expend during migration (Fleming 2002). Studies of waterbirds show that distances between wetland stopovers affect the frequency and length of migrations (Farmer and Parent 1997). Birds moving within highly connected landscapes migrate longer distances and expend less energy in searching effort than birds that occupy highly fragmented habitats. In hummingbirds, exposure to poor-quality stopover habitat (i.e., habitat that is appropriately spaced but provides inadequate foraging resources) slows the progress of migration, exposes animals to deteriorating climatic conditions, intensifies competition for resources, reduces fuel stores, and ultimately affects abundance (Russell et al. 1994). The ability to alter migration routes in response to environmental change is essential to the viability of migratory populations, providing a mechanism for ameliorating the consequences of localized habitat loss. Various characteristics of population dynamics and genetics in birds have been linked to migration flexibility and may be useful in predicting resilience to change (Dolman and Sutherland 1995).

There is evidence that individual bats visit several stopover sites during long-distance movements. However, the habitat characteristics of migration corridors are largely unknown. Species such as *L. curasoae* migrate up a resource gradient, use relatively narrow corridors, and require a strict progression of stopover sites (fig. 4.5; Ceballos et al. 1997; Fleming et al. 1993). Although habitat options may be limited for this species, the identification and preservation of a small number of critical sites may provide an effective conservation strategy (Fleming 2002). Conversely, several European long-distance migrants such as *Noctula noctula* and *Pipistrellus nathusii* move in a diffuse fashion along wide migration corridors that are shared by a number of species (Strelkov 1969). Defining critical habitat for these bats poses substantial problems given our poor understanding of the size and configuration of stopover habitats, competitive relationships between species, and the characteristics that determine habitat quality. The problems associated with identifying and conserving critical habitat are further compounded in species with highly irregular migration paths, such as *Pteropus scapulatus* and *P. poliocephalus* (Fahse et al. 1998; Frederick et al. 1996; Richards and Hall 1998).

Periods of Concentration

Annual cycles of migration characteristically include periods when large proportions of a population congregate into a restricted space (Brower and Malcolm 1991). In migratory bats, periods of concentration are associated with mating, breeding, and hibernation or with the production of local resource flushes during times of general scarcity (e.g., Dwyer 1966; Eby et al. 1999; Tuttle 1976). Migratory animals are particularly vulnerable in these circumstances. Periods of concentration reduce independence in the fate of individuals and limit the capacity for large population size to reduce extinction risk (Brower and Malcolm 1991). High rates of mortality can occur from loss or deterioration of comparatively small areas or from localized exposure to chemicals, hunting pressure, and so on.

Periods of concentration occur at all phases of the migration cycle. For example, females of *Miniopteris schreibersii* that disperse over thousands of kilometers at other times of the year converge at a single maternity roost (Dwyer 1966), and both males and females of *M. grisescens* and *M. sodalis* aggregate in a small number of hibernation caves (Humphrey 1978; Tuttle 1976). Resource bottlenecks along migration corridors can also result in concentrations at stopover sites, as occurs in shorebirds (Myers 1983; Safina 1993), and resource pulses during general food scarcity can cause nectar-eating bats to concentrate. In many winters, *Pteropus poliocephalus* converges on restricted coastal habitats in the northern part of its range, exposing a high proportion of the population to the impacts of ongoing forest clearance in the area (Duncan et al. 1999; Eby et al. 1999).

Integrated Conservation Strategies for Migratory Bats

Protected lands form the basis of conservation strategies in many countries (Pressey et al. 1993). However, even reserve networks designed to sample regional species diversity are unlikely to preserve the temporally and spatially complex resource systems required by migratory organisms (Powell and Bjork 1995; Pressey 1994). Individual reserves are generally assumed to be functionally discrete (ecological interdependencies are rarely considered in the spatial configuration of reserves [Fahrig and Merriam 1994]) and temporally stable (Fahrig 1992). These considerations limit the suitability of reserve networks for migrants. Conservation programs for these species must augment reserved landscapes with careful management of areas outside formal reserves and must target public education and community involvement (Racey and Entwistle, this volume; Walker 1995).

During their annual cycles, migrating bats visit habitats that are managed or controlled by a wide range of governments, agencies, communities, and private individuals. The protection afforded bats varies between these jurisdictions, and conservation outcomes are only as effective as the protection provided at the most vulnerable point. Strategies for conserving migrants require a broad-scale, cooperative approach that engages all relevant parties (Pierson 1998; Walker 1995). Such programs are difficult to initiate and administer. They challenge participants to perceive their areas of jurisdiction as integral components of a wide-ranging system and to acknowledge that their management decisions have ramifications far beyond their holdings. Migratory bats in Europe and in North America benefit from international conservation agreements such as the Agreement on the Conservation of Bats in Europe 1992, which was formulated under the Bonn Convention on the Conservation of Migratory Species of Wild Animals 1979, and the Program for the Conservation of Migratory Bats of Mexico and the United States (Racey 1998; Racey and Entwistle, this volume; Walker 1995). These initiatives provide useful models for integrated programs in other regions of the world.

In countries such as the United States and Australia, land management and conservation programs are often state, rather than federal, responsibilities, and interstate migrants confront problems similar to those of international migrants (Pierson 1998). For example, in Australia the neighboring states of New South Wales and Queensland differ in policies for managing migratory *Pteropus* in commercial fruit orchards and also differ in their assessments of the threatened status of these bats (Eby 1995).

Many bats are partial migrants that have relatively large sedentary populations, for example, *V. murinus* (Rydell and Baagøe 1994), *L. curasoae* (Ceballos et al. 1997; Rojas-Martinez et al. 1999), and *P. alecto* (Palmer et al. 2000; Vardon and Tidemann 1999). The preservation of migration in these species will not necessarily be an outcome of programs designed to minimize extinction risk. It is likely that self-sustaining, viable populations could be preserved by effectively managing sedentary groups. However, there is a growing case for conserving migrations as specific phenomena and designating migratory populations as separate conservation units, particularly if conservation goals are set to preserve the functional roles of organisms (Brower and Malcolm 1991; Curio 1996; Smith et al. 1993). Migratory animals serve as identifiable linkages between distant habitat areas and demonstrate rhythms in ecological processes that occur over large spatial and temporal scales (Gilbert 1980). Bats such as *P. poliocephalus* and *L. curasoae* link diverse vegetation communities via shared pollinators and seed dispersers. Similarly, migrant populations of *P. nathusii* and *T. brasiliensis* create linkages between an array of habitats, including urban and agricultural systems, via shared insect predators. If, as predicted, the population and habitat thresholds necessary to sustain migration are substantially larger than minimum viability thresholds (Frederick et al. 1996; Pagel and Payne 1996), migratory populations and their important ecological interactions will become critically endangered before the viability of these species is of significant conservation concern.

The greatest conservation problem faced by all bats is lack of biological and ecological information (Racey and Entwistle, this volume). Our capacity to preserve viable populations of migratory bats and their ecological functions would be substantially enhanced by a better understanding of the complex resource systems they inhabit and their behavioral, physiological, and life-history adaptations to these systems.

Basic Differences between the Migratory Ecology of Birds and Bats

To complete our discussion of migration, we review some of the major similarities and differences in the migratory biology of birds and bats. For birds, we will limit most of our discussion to species of passerines—birds whose size and feeding ecology most closely match those of bats. Our overall conclusion from this comparison is that migration in birds and bats differs in a large number of quantitative and qualitative ways, probably as a result of significantly different selective pressures during its evolution in each group as well as the many profound differences in their life histories.

Similarities and differences in the migratory ecology of passerine birds and bats are summarized in table 4.5. In terms of quantitative differences, a much higher proportion of passerine species are long-distance, intercontinental migrants. From 31% to 42% of species of temperate passerines are tropical mi-

Table 4.5. Summary of some similarities and differences in the migratory ecology of passerine birds and bats

Parameter	Passerine Birds	Bats
Proportion of migrants that are "long-distance migrants"	High	Low
Maximum migration distance	≫1800 km	1800 km
Proportion of migrants that undergo intercontinental movements	High	Low
Long, sustained flights during migration?	Common	Uncommon
Amount of fat deposited prior to migration	Up to 50% of lean mass	≪50% of lean mass
Migration associated with hibernation?	No	Common in temperate bats
Is migration sex-biased?	No, both sexes migrate	Often yes; females more likely to migrate
Do stopovers occur during long-distance migration?	Yes	Yes
Mating occurs on summer grounds?	Yes	No
Females migrate while pregnant?	No	Yes
Geographic or habitat segregation between sexes on:		
Summer grounds	No	Yes
Winter grounds	Yes	No

grants (Mönkkönen et al. 1992); only New World *Lasiurus* bats are intercontinental migrants. On average, bats that are long-distance migrants fly much shorter distances between summer and winter habitats than do migrant passerines. Some passerines fly more than 4,000 km between summer and winter grounds, whereas few bats fly farther than 1,000 km during migration. Large fat buildups and long, sustained flights are common in migratory passerines. Despite the importance of fat for both migration and hibernation, bats deposit less fat prior to migration than do birds, and few species undergo long, sustained flights during migration.

Migration in passerines and bats also differs qualitatively (table 4.5). Migration is intimately related to hibernation in most species of temperate bats but not in passerine birds. A sex bias in migration distances is much more common in bats than in birds. Females are more likely to migrate and/or to migrate longer distances in several species of bats. Passerines are more likely to exhibit sex-based habitat segregation on their winter grounds, whereas bats are more likely to exhibit such segregation on their summer grounds. In most species of bats, mating and pregnancy/lactation generally occur in different geographic locations, whereas mating and brooding in birds are spatially concordant. Also, most species of passerines are ostensibly monogamous, and mating and incubation occur on the summer breeding grounds after migration. In contrast, all migrant bats are polygamous, and mating in most species occurs on or near the wintering grounds prior to or during migration.

Of all the differences we have described, perhaps the most profound is the large number of intercontinental migrants in passerine birds compared to bats. Many migrant passerines spend approximately half their lives in temperate habitats and half their lives in tropical habitats. In the New World, eastern migrants tend to live in late successional habitats during both the winter and summer, whereas western migrants tend to live in early successional habitats year round. Like their western North American counterparts, Palaearctic long-distance migrants typically inhabit early successional habitats year round. But unlike most Nearctic migrant passerines, which are sedentary and often defend feeding territories during the winter, Palaearctic passerines are less sedentary and track spatiotemporal changes in the availability of insects over large areas in winter (Mönkkönen et al. 1992). Finally, according to Mönkkönen et al. (1992), Nearctic and Palaearctic migrant passerines differ strongly in their evolutionary affinities. Most Nearctic tropical migrants have evolved from tropical ancestors, whereas most Palaearctic tropical migrants originally evolved in seasonal Palaearctic environments.

Except for species of *Lasiurus,* whose basal group is mostly Neotropical in distribution (Morales and Bickham 1995), no bats migrate from high temperate latitudes into the tropics annually. Instead, most temperate migrants remain within temperate latitudes year round. Rather than moving to the tropics, these bats migrate just far enough to encounter winter roost sites that provide suitable conditions for hibernation. Only non-hibernators among bats leave the temperate zone and migrate relatively long distances to the tropics. The presence of the ability to hibernate in bats, and its absence in birds, has resulted in the evolution of relatively short-distance migrations in the former group and long-distance migrations in the latter.

Conclusions

Findley (1993) and others (e.g., Altringham 1996) have dismissed migration in bats as being a "minor phenomenon." We do not agree with this view. Although a relatively small fraction of all bats are migratory, migration plays a potentially profound role in the lives of a substantial number of species, including those whose population sizes can be enormous. Some of the most gregarious bats in the world (e.g., *T. brasiliensis, E. helvum, M. schreibersi,* and various species of *Pteropus*) migrate substantial distances during their annual cycles. As we have documented in this chapter, such movements have numerous ecological, behavioral, and physiological effects on migrant species. And such movements place these species at considerable risk from a conservation viewpoint.

Throughout this chapter, we have identified topics that we believe to be important but understudied aspects of bat migration. These topics include the demographic consequences of migration, fat deposition and rehydration strategies of migrants, and conservation strategies. We have made considerable progress in understanding bat migration in the past 50–60 yr but much still remains to be learned. New technologies (e.g., stable isotope analysis, lighter and more powerful radio transmitters) and new tools (e.g., DNA and phylogenetic analyses) will aid us in gaining a much deeper understanding of "life on the move" in bats. The next 50 yr will be exciting ones for migration research in bats and other organisms, but only if we develop effective conservation strategies. Without such strategies, migration in bats and other organisms will continue to be an endangered phenomenon.

Acknowledgments

We thank the following people for discussions and help with the widely scattered literature on bat migration: R. Barclay, P. Birt, W. Bogdanowicz, F. Bonaccorso, M. Brigham, R. Coles, M. B. Fenton, L. Hall, S. Hand, J. N. Holland, S. Hoofer, A. Hutson, A. Keeley, S. Kerrison, L. Lumsden, N. Markus, G. McCracken, C. Palmer, J. Rydell, and M. Vardon. A. Hutson, T. Kunz, C. Palmer, E. Petit, and M. Vardon kindly sent us copies of unpublished manuscripts. We thank B. Fenton, J. Rydell, D. Thomas, and an anonymous reviewer for constructive comments on a draft of this chapter. Fleming acknowledges financial support for his migration studies from the following sources: Arizona Game and Fish Department, National Geographic Society, Ted Turner Endangered Species Fund, and U.S. National Science Foundation. Eby acknowledges financial support for her migration studies from New South Wales National Parks and Wildlife Service and the Australian National Rainforest Conservation Programme.

This chapter is dedicated to Martin Eisentraut, Donald Griffin, and Francis Ratcliffe for their pioneering efforts in the study of bat migration.

Literature Cited

Ahlén, I. 1997. Migratory behaviour of bats at south Swedish coasts. Zeitschrift für Säugetierkunde, 62:375–380.

Alerstam, T. 1991. Bird flight and optimal migration. Trends in Ecology and Evolution, 6:210–215.

Alerstam, T., and A. Lindstrom. 1990. Optimal bird migration: the relative importance of time, energy and safety. Pp. 331–351 in: Bird Migration: Physiology and Ecophysiology (E. Gwinner, ed.). Springer-Verlag, Berlin.

Altringham, J. D. 1996. Bats: Biology and Behaviour. Oxford University Press, Oxford.

Arnold, A., A. Scholz, V. Storch, and M. Braun. 1996. The Nathusius' bats in flood plain forests in Nordbaden (SW-Germany). Carolinea, 54:149–158.

Baagøe, H. J. 1987. Summer occurrence of Vespertilio murinus Linnaeus 1758 and Eptesicus serotinus Schreber 1780 Chiroptera Mammalia on Zealand Denmark. Annals of Natural History Museum Wien B, 88–89:281–292.

Barbour, R. W., and W. H. Davis. 1969. Bats of America. University Press of Kentucky, Lexington.

Barclay, R. M. R. 1984. Observations on the migration, ecology and behaviour of bats at Delta Marsh, Manitoba, Canada. Canadian Field Naturalist, 98:331–336.

Baudinette, R. V., R. T. Wells, K. J. Sanderson, and B. Clark. 1994. Microclimatic conditions in maternity caves of the bent-wing bat, *Miniopterus schreibersii:* an attempted restoration of a former maternity site. Wildlife Research, 21:607–619.

Bleier, W. J. 1975. Early embryology and implantation in the Californian leaf-nosed bat *Macrotus californicus.* Anatomical Records, 182:237–254.

Blem, C. R. 1980. The energetics of migration. Pp. 175–224 *in:* Animal Migration, Orientation and Navigation (S. A. Gauthreaux, ed.). Academic Press, New York.

Bonaccorso, F. J. 1979. Foraging and reproductive ecology in a Panamanian bat community. Bulletin of the Florida State Museum, Biological Sciences, 24:359–408.

Bonaccorso, F. J. 1998. Bats of Papua New Guinea. Conservation International, Washington, D.C.

Bradbury, J. W., and L. Emmons. 1974. Social organization of some Trinidad bats. 1. Emballonuridae. Zeitschrift für Tierpsychologie, 36:137–183.

Bradbury, J. W., and S. L. Vehrencamp. 1976. Social organization and foraging in emballonurid bats. I. Field studies. Behavioral Ecology and Sociobiology, 1:337–381.

Brooke, A. P. 1997. Social organization and foraging behaviour of the fishing bat, *Noctilio leporinus* (Chiroptera: Noctilionidae). Ethology, 103:421–436.

Brosset, A. 1990. The migrations of *Pipistrellus nathusii* in France: possible implication on the spreading of rabies. Mammalia, 54:207–212.

Brower, L. P., and S. B. Malcolm. 1991. Animal migrations: endangered phenomena. American Zoologist, 31:265–276.

Bullock, S. H., and J. A. Solís-Magallanes. 1990. Phenology of canopy trees of a tropical deciduous forest in Mexico. Biotropica, 22:22–35.

Burland, T. M, E. M. Barratt, M. A. Beaumont, and P. A. Racey. 1999. Population genetic structure and gene flow in a gleaning bat, *Plecotus auritus.* Proceedings of the Royal Society of London B, 266:975–980.

Carpenter, R. E. 1969. Structure and function of the kidney and the water balance of desert bats. Physiological Zoology, 42:288–302.

Carpenter, R. E. 1985. Flight physiology of flying foxes, *Pteropus poliocephalus.* Journal of Experimental Biology, 114:619–647.

Ceballos, G., T. H. Fleming, C. Chavez, and J. Nassar. 1997. Population dynamics of *Leptonycteris curasoae* (Chiroptera: Phyllostomidae) in Jalisco, Mexico. Journal of Mammalogy, 78:1220–1230.

Churchill, S. 1998. Australian Bats. Reed New Holland, Frenchs Forest, New South Wales.

Cockrum, E. L. 1969. Migration in the guano bat, *Tadarida brasiliensis.* Miscellaneous Publication, University of Kansas, Museum of Natural History, 51:303–336.

Cockrum, E. L. 1991. Seasonal distribution of northwestern populations of the long-nosed bats *Leptonycteris sanborni* Family Phyllostomidae. Anales del Instituto de Biologia Universidad Nacional Autonoma de Mexico, Serie Zoologia, 62:181–202.

Corbet, G. B. 1970. Vagrant bats in Shetland and the North Sea. Journal of Zoology (London), 161:281–282.

Csada, R. D., and R. M. Brigham. 1992. Common poorwill. Pp. 1–16 *in:* Birds of North America (A. Poole, P. Stettenheim, and F. Gill, eds.). Academy of Natural Sciences, Philadelphia.

Curio, E. 1996. Conservation needs ethology. Trends in Ecology and Evolution, 11: 260–263.

Davis, R. B., C. F. Herred II, and H. L. Short. 1962. Mexican free-tailed bats in Texas. Ecological Monographs, 32:311–346.

Davis, W. H., and H. B. Hitchcock. 1965. Biology and migration of the bat *Myotis lucifugus* in New England. Journal of Mammalogy, 46:296–313.

Dingle, H. 1980. Ecology and evolution of migration. Pp. 1–101 *in:* Animal Migration, Orientation and Navigation (S. A. Gautreaux, ed.). Academic Press, New York.

Dingle, H. 1996. Migration: The Biology of Life on the Move. Oxford University Press, New York.

Dolman, P. M., and W. J. Sutherland. 1995. The response of bird populations to habitat loss. Ibis, 137 (suppl. 1):S38–S46.

Duncan, A., G. B. Baker, and N. Montgomery. 1999. The Action Plan for Australian Bats. Environment Australia, Canberra.

Dwyer, P. D. 1966. The population pattern of *Miniopterus schreibersi* (Chiroptera) in north-eastern New South Wales. Australian Journal of Zoology, 14:1073–1137.

Dwyer, P. D. 1969. Population ranges of *Miniopterus schreibersii* (Chiroptera) in south-eastern Australia. Australian Journal of Zoology, 17:665–686.

Eby, P. 1991. Seasonal movements of grey-headed flying-foxes, *Pteropus poliocephalus* from two maternity camps in northern New South Wales. Wildlife Research, 18: 547–59.

Eby, P. 1996. Interactions between the grey-headed flying fox *Pteropus poliocephalus* and its diet plants—seasonal movements and seed dispersal. Ph.D. Thesis. University of New England, Armidale, Australia.

Eby, P., G. Richards, L. Collins, and K. Parry-Jones. 1999. The distribution, abundance and vulnerability to population reduction of the grey-headed flying-fox *Pteropus poliocephalus* in New South Wales. Australian Zoologist, 31:240–253.

Entwistle, A. C., P. A. Racey, and J. R. Speakman. 1997. Roost selection by the brown long-eared bat (*Plecotus auritus*). Journal of Applied Ecology, 34:399–408.

Entwistle, A. F., P. A. Racey, and J. R. Speakman. 2000. Social and population structure of a gleaning bat, *Plecotus auritus*. Journal of Zoology (London), 252:11–17.

Ewing, W. G., E. H. Studier, and M. J. O'Farrell. 1970. Autumn fat deposition and gross body composition in three species of *Myotis*. Comparative Biochemistry and Physiology A, 36:119–129.

Fahrig, L. 1992. Relative importance of spatial and temporal scales in a patchy environment. Theoretical Population Biology, 41:300–314.

Fahrig, L., and G. Merriam. 1994. Conservation of fragmented populations. Conservation Biology, 8:50–59.

Fahse, L., W. R. J. Dean, and C. Wissel. 1998. Modelling the size and distribution of protected areas for nomadic birds: Alaudidae in the Nama-Karoo, South Africa. Biological Conservation, 85:105–112.

Farmer, A. H., and A. H. Parent. 1997. Effects of the landscape on shorebird movements at spring migration stopovers. Condor, 99:698–707.

Fenton, M. B. 1990. The foraging behavior and ecology of animal-eating bats. Canadian Journal of Zoology, 86:411–422.

Fenton, M. B., and D. W. Thomas. 1985. Migrations and dispersal of bats (Chiroptera). Contributions to Marine Science (Special supplement: Migration: Mechanisms and Adaptive Significance, ed. M. A. Rankin), 27:409–424.

Findley, J. S. 1993. Bats: A Community Perspective. Cambridge University Press, Cambridge.

Findley, J. S., and C. Jones. 1964. Seasonal distribution of the hoary bat. Journal of Mammalogy, 45:461–470.

Fleming, T. H. 1988. The Short-Tailed Fruit Bat: A Study in Plant-Animal Interactions. University of Chicago Press, Chicago.

Fleming, T. H. 1992. How do fruit- and nectar-feeding birds and mammals track their food resources? Pp. 355–391 in: Effects of Resource Distribution on Animal-Plant Interactions (M. D. Hunter, T. Ohgushi, and P. W. Price, eds.). Academic Press, San Diego, Calif.

Fleming, T. H. 2003. Nectar corridors: migration and the annual cycle of lesser long-nosed bats. In: Conservation of Migratory Pollinators and Their Nectar Corridors in North America (G. Nabhan, ed.). University of Arizona Press, Tucson (in press).

Fleming, T. H., A. A. Nelson, and V. M. Dalton. 1998. Roosting behavior of the lesser long-nosed bat, Leptonycteris curasoae. Journal of Mammalogy, 79:147–155.

Fleming, T. H., R. A. Nuñez, and L. Sternberg. 1993. Seasonal changes in the diets of migrant and non-migrant nectarivorous bats as revealed by carbon stable isotope analysis. Oecologia (Berlin), 94:72–75.

Fleming, T. H., M. D. Tuttle, and M. A. Horner. 1996. Pollination biology and the relative importance of nocturnal and diurnal pollinators of three species of Sonoran Desert columnar cacti. Southwestern Naturalist, 41:357–369.

Frederick, P. C., K. L. Bildstein, B. Fleury, and J. Ogden. 1996. Conservation of large, nomadic populations of white ibis (Eudocimus albus) in the United States. Conservation Biology, 10:203–216.

French, B., and A. Lollar. 1998. Observations on the reproductive behavior of captive Tadarida brasiliensis mexicana (Chiroptera: Molossidae). Southwestern Naturalist, 43:484–490.

Gaisler, J. 1979. Ecology of bats. Pp. 281–342 in: Ecology of Small Mammals (D. M. Stoddard, ed.). Chapman & Hall, London.

Gerell-Lundberg, K., and R. Gerell. 1994. The mating behaviour of the Pipistrelle and the Nathusius' Pipistrelle (Chiroptera)—a comparison. Folia Zoologica, 43:315–324.

Gerell, R. 1987. Do Swedish bats migrate? Fauna och flora (Stockholm), 82:79–83.

Gilbert, L. E. 1980. Food web organization and conservation of Neotropical diversity. Pp. 11–34 in: Conservation Biology: An Evolutionary-Ecological Perspective (M. E. Soulé and B. A. Wilcox, eds.). Sinauer Associates, Sunderland, Mass.

Gill, F. B. 1995. Ornithology. 2d ed. W. H. Freeman & Co., New York.

Glass, B. P. 1982. Seasonal movements of Mexican freetail bats. Southwestern Naturalist, 27:127–133.

Graham, G. L. 1983. Changes in bat species diversity along an elevational gradient up the Peruvian Andes. Journal of Mammalogy, 64:559–571.

Greenberg, R. 1980. Demographic aspects of long-distance migration. Pp. 493–504 *in:* Migrant Birds in the Neotropics (A. Keast and E. S. Morton, eds.). Smithsonian Institution Press, Washington, D.C.

Greenwood, P. J. 1980. Mating systems, philopatry, and dispersal in birds and mammals. Animal Behaviour, 28:1140–1162.

Griffin, D. R. 1958. Listening in the Dark. Yale University Press, New Haven, Conn.

Griffin, D. R. 1970. Migrations and homing of bats. Pp. 233–264 *in:* Biology of Bats. (W. A. Wimsatt, ed.). Vol. 1. Academic Press, New York.

Hand, S. J., and J. A. W. Kirsch. 1998. A southern origin for the Hipposideridae (Microchiroptera)? evidence from the Australian fossil record. Pp. 72–90 *in:* Bat Biology and Conservation (T. H. Kunz and P. A. Racey, eds.). Smithsonian Institution Press, Washington D.C.

Handley, C. O., Jr., D. E. Wilson, and A. L. Gardner, eds. 1991. Demography and natural history of the common fruit bat, *Artibeus jamaicensis,* on Barro Colorado Island, Panama. Smithsonian Contributions to Zoology, 511:1–173.

Hayssen, V., and T. H. Kunz. 1996. Allometry of litter mass in bats: maternal size, wing morphology, and phylogeny. Journal of Mammalogy, 77:476–490.

Heideman, P. D. 1988. The timing of reproduction in the fruit bat *Haplonycteris fischeri* (Pteropodidae): geographic variation and delayed development. Journal of Zoology (London), 215:577–595.

Heideman, P. D. 1995. Synchrony and seasonality of reproduction in tropical bats. Symposia of the Zoological Society of London, no. 67:151–165.

Heidemann, P. D., and L. R. Heaney. 1989. Population biology and estimates of abundance of fruit bats (Pteropodidae) in Phillipine submontane rainforest. Journal of Zoology (London), 218:565–586.

Heinz, B., and M. Braun. 1997. Study on bats occurring in the park of the castle of Schwetzingen. Carolinea, 55:49–56.

Heithaus, E. R., T. H. Fleming, and P. A. Opler. 1975. Foraging patterns and resource utilization in seven species of bats in a seasonal tropical forest. Ecology, 56:841–854.

Hensley, A. P., and K. T. Wilkins. 1988. Leptonycteris nivalis. Mammalian Species, 307:1–4.

Hill, J. E., and J. D. Smith. 1984. Bats: A Natural History. University of Texas Press, Austin

Hooper, J. D. H., and W. M. Hooper. 1956. Habits and movements of cave-dwelling bats in Devonshire. Proceedings Zoological Society of London, 127:1–26.

Horner, M. A., T. H. Fleming, and C. T. Sahley. 1998. Foraging behaviour and energetics of a nectar-feeding bat, *Leptonycteris curasoae* (Chiroptera: Phyllostomidae). Journal of Zoology (London), 244:575–586.

Howell, D. J. 1979. Flock foraging in nectar-feeding bats: Advantages to the bats and the host plants. American Naturalist, 114:23–49.

Humphrey, S. R. 1978. Status, winter habitat, and management of the endangered Indiana bat, *Myotis sodalis.* Florida Scientist, 41:65–76.

Humphrey, S. R., and J. B. Cope. 1976. Population Ecology of the Little Brown Bat, *Myotis lucifugus,* in Indiana and North-central Kentucky. Special Publication of the American Society of Mammalogists, vol. 4. Allen Press, Lawrence, Kans., 81 pp.

Hutson, A. M., S. P. Mickleburgh, and P. A. Racey, eds. 2001. Global Action Plan for

Microchiropteran Bats. International Union for Conservation of Nature and Natural Resources, Gland, Switzerland.

Hutto, R. L. 1998. On the importance of stopover sites to migrating birds. Auk, 115: 823–824.

Johnson, W. E., W. L. Franklin, and J. A. Iriarte. 1992. The mammalian fauna of the Northern Chilean Patagonia: a biogeographical dilemma. Mammalia, 56:445–456.

Jones, G. 1995. Flight performance, echolocation and foraging behaviour in noctule bats *Nyctalus noctula*. Journal of Zoology (London), 237:303–312.

Kaitala, A., V. Kaitala, and P. Lundberg. 1993. A theory of partial migration. American Naturalist, 142:59–81.

Kalko, E. K. V., C. O. Handley, and D. Handley. 1996. Organization, diversity and long-term dynamics of a Neotropical bat community. Pp. 503–553 *in:* Long-Term Studies of Vertebrate Communities (M. Cody and J. Smallwood, eds.). Academic Press, Los Angeles

Kapteyn, K., and P. H. C. Lina. 1994. First record of a nursery roost of Nathusius' pipistrelle *Pipistrellus nathusii* in the Netherlands. Lutra, 37:106–109.

Karr, J. R., and J. D. Braun. 1990. Food resources of understory birds in central Panama, quantification and effects on avian populations. Studies in Avian Biology, 13:58–64.

Karr, J. R., and K. E. Freemark. 1983. Habitat selection and environmental gradients: dynamics in the "stable" tropics. Ecology, 64:1481–1494.

Kerth, K., and B. Koenig. 1999. Fission, fusion and non random associations in female Bechstein's bat (*Myotis bechsteinii*). Behaviour, 136:1187–1202.

Klaasen, M. 1996. Metabolic constraints on long-distance migration in birds. Journal of Experimental Biology, 199:57–64.

Krulin, G. S., and J. A. Sealander. 1972. Annual lipid cycle of the gray bat, *Myotis grisescens*. Comparative Biochemistry and Physiology A, 42A:537–549.

Kunz, T. H. 1982a. Lasionycteris noctivagans Mammalian Species, 172:1–5.

Kunz, T. H. 1982b. Roosting ecology of bats. Pp. 1–55 *in:* Ecology of bats (T. H. Kunz, ed.). Plenum Press, New York.

Kunz, T. H., J. O. Whitaker, Jr., and M. D. Wadanoli. 1995. Dietary energetics of the insectivorous Mexican free-tailed bat (*Tadarida brasiliensis*) during pregnancy and lactation. Oecologia, 101:407–415.

Kunz, T. H., J. A. Wrazen, and C. D. Burnett. 1998. Changes in body mass and fat reserves in pre-hibernating little brown bats (*Myotis lucifugus*). Ecoscience, 5:8–17.

Kurta, A., G. P. Bell, K. A. Nagy, and T. H. Kunz. 1989. Energetics of pregnancy and lactation in free-ranging little brown bats (*Myotis lucifugus*). Physiological Zoology, 62:804–818.

Kurta, A., and T. H. Kunz. 1987. Size of bats at birth and maternal investment during pregnancy. Symposia of the Zoological Society of London, no. 57:79–106.

Levey, D. J. 1988. Spatial and temporal variation in Costa Rican fruit and fruit-eating bird abundance. Ecological Monographs, 58:251–269.

Levey D. J., and F. G. Stiles. 1992. Evolutionary precursors of long-distance migration: resource availability and movement patterns in Neotropical landbirds. American Naturalist, 140:447–476.

Lewis, S. E. 1995. Roost fidelity of bats—a review. Journal of Mammalogy, 76:481–496.

Lindstrom, A., and T. Alerstam. 1992. Optimal fat loads in birds: A test of the time minimization hypothesis. American Naturalist, 140:477–491.

Loiselle, B. A., and J. G. Blake. 1990. Diets of understory fruit-eating birds in Costa Rica: seasonality and resource abundance. Studies in Avian Biology, 13:91–103.

Masing, M. 1989. A long-distance flight of *Vespertilio murinus* from Estonia. Myotis, 27:147–150.

Mayer, F. 1995. Multiple paternity and sperm competition in the noctule bat (*Nyctalus noctula*) revealed by DNA fingerprinting. Bat Research News, 36:88.

McCracken, G. F. 1984a. Communal nursing in Mexican free-tailed bat maternity colonies. Science, 223:1090–1091.

McCracken, G. F. 1984b. Social organization and genetic variation in two species of emballonurid bats. Zeitschrift für Tierpsychologie, 66:55–69.

McCracken, G. F., and J. W. Bradbury. 1981. Social organization and kinship in the polygynous bat *Phyllostomus hastatus*. Behavioral Ecology and Sociobiology, 8:11–34.

McCracken, G. F., and M. F. Gassel. 1997. Genetic structure in migratory and non-migratory populations of Brazilian free-tailed bats. Journal of Mammalogy, 78: 348–357.

McCracken, G. F., M. K. McCracken, and A. T. Vawter. 1994. Genetic structure in migratory populations of the bat *Tadarida brasiliensis mexicana*. Journal of Mammalogy, 75:500–514.

McCracken, G. F., and G. S. Wilkinson. 2000. Bat mating systems. Pp. 321–362 *in:* Reproductive Biology of Bats (E. G. Crichton and P. H. Krutzsch eds.). Academic Press, New York.

McNab, B. K. 1982. Evolutionary alternatives in the physiological ecology of bats. Pp. 151–200 *in:* Ecology of Bats (T. H. Kunz, ed.). Plenum Press, New York.

McWilliam, A. N. 1987. The reproductive and social biology of *Coelura afra* in a seasonal environment. Pp. 324–350 *in:* Recent Advances in the Study of Bats (M. B. Fenton, P. Racey, and J. M. V. Rayner, eds.). Cambridge University Press, Cambridge.

Mickleburgh, S. P., A. M. Hutson, and P. A. Racey, eds. 1992. Old World Fruit Bats: An Action Plan for Their Conservation. International Union for Conservation of Nature and Natural Resources, Gland, Switzerland.

Mitchell-Jones, A. J. 1999. Atlas of European Mammals. Academic Press, London.

Mönkkönen, M., P. Helle, and D. Welsh. 1992. Perspectives on Palaearctic and Nearctic bird migration: comparisons and overview of life-history and ecology of migrant passerines. Ibis, 134 (suppl.):7–13.

Morales, J. C., and J. W. Bickham. 1995. Molecular systematics of the genus *Lasiurus* (Chiroptera: Vespertilionidae) based on restriction-site maps of the mitochondrial ribosomal genes. Journal of Mammalogy, 76:730–749.

Myers, J. P. 1983. Conservation of migrating shorebirds: staging areas, geographic bottlenecks, and regional movements. American Birds, 37:23–25.

Myers, J. P., R. I. G. Morrison, P. Z. Antas, B. A. Harrington, T. E. Lovejoy, M. Sallaberry, S. E. Senner, and A. Tarak. 1987. Conservation strategy for migratory species. American Scientist, 75:19–26.

Nelson, J. E. 1965a. Behaviour of Australian Pteropodidae (Megachiroptera). Animal Behaviour, 8:544–557.

Nelson, J. E. 1965b. Movements of Australian flying foxes (Pteropodidae: Megachiroptera). Australian Journal of Zoology, 13:53–73.

Neri, F., and S. Aulagnier. 1996. First recapture of *Nyctalus leisleri* (Mammalia, Chiroptera) in France. Mammalia, 60:317–319.

Norberg, U. M., and J. M. Rayner. 1987. Ecological morphology and flight in bats (Mammalia; Chiroptera): wing adaptations, flight performance, foraging strategy and echolocation. Philosophical Transactions of the Royal Society London B, 316: 335–427.

O'Shea, T. J. 1976. Fat content in migratory central Arizona Brazilian free-tailed bats, *Tadarida brasiliensis* (Molossidae). Southwestern Naturalist, 21:321–326.

O'Shea, T. J., and T. A. Vaughan. 1980. Ecological observations on an East African bat community. Mammalia, 44:485–496.

Pagel, M., and R. J. H. Payne. 1996. How migration affects estimation of the extinction threshold. Oikos, 76:323–329.

Pagels, J. F. 1975. Temperature regulation, body weight, and changes in total body fat of the free-tailed bat, *Tadarida brasiliensis cynocephala* (Le Conte). Comparative Biochemistry and Physiology A, 50:237–246.

Palmer, C., O. Price, and C. Bach. 2000. Foraging ecology of the black flying fox *Pteropus alecto* in the seasonal tropics of the Northern Territory, Australia. Wildlife Research, 27:169–178.

Parry-Jones, K. A., and M. L. Augee. 1992. Movements of grey-headed flying foxes (*Pteropus poliocephalus*) to and from a colony site on the central coast of New South Wales. Wildlife Research, 19:331–340.

Payne, J., C. M. Francis, and K. Phillips. 1985. A Field Guide to the Mammals of Borneo. Sabah Society, Kota Kinabalu, Malaysia.

Petersen, A. 1994. The occurrence of bats (order Chiroptera) in Iceland. Natturufraedingurinn, 64:3–12.

Peterson, A. T., and L. R. Heaney. 1993. Genetic differentiation in Philippine bats of the genera *Cynopterus* and *Haplonycteris*. Biological Journal of the Linnean Society, 49:203–218.

Petit, E., L. Excoffier, and F. Mayer. 1999. No evidence of bottleneck in the post-glacial recolonization of Europe by the noctule bat (*Nyctalus noctula*). Evolution, 53: 1247–1258.

Petit, E., and F. Mayer. 2000. A population genetic analysis of migration: the case of the noctule bat (*Nyctalus noctula*). Molecular Ecology, 9:683–690.

Petri, B., S. Paabo, A. von Haeseler, and D. Tautz. 1997. Paternity assessment and population subdivision in a natural population of the larger mouse-eared bat *Myotis myotis*. Molecular Ecology, 6:235–242.

Pierson, E. D. 1998. Tall trees, deep holes and scarred landscapes: conservation biology of North American bats. Pp 309–325 *in:* Bat Biology and Conservation (T. H. Kunz and P. A. Racey, eds.). Smithsonian Institution Press, Washington, D.C.

Pimm, S. L., H. L. Jones, and J. Diamond. 1988. On the risk of extinction. American Naturalist, 132:757–785.

Powell, G. V. N., and R. Bjork. 1995. Implications of intratropical migration on reserve design: a case study using *Pharomachrus mocinno*. Conservation Biology, 9:354–362.

Pressey, R. L. 1994. Land classifications are necessary for conservation planning but what do they tell us about fauna? Pp. 31–41 *in:* Future of the Fauna of Western New South Wales (D. Lunney, S. Hand, P. Reed, and D. Butcher, eds.). Surrey Beatty & Sons, Chipping Norton, New South Wales.

Pressey, R. L., C. J. Humphries, C. R. Margules, R. I. Vane-Wright, and P. H. Williams.

1993. Beyond opportunism: key principles for systematic reserve selection. Trends in Ecology and Evolution, 8:124–128.

Racey, P. A. 1982. Ecology of Bat Reproduction. Pp. 57–104 *in:* Ecology of Bats (T. H. Kunz, ed.). Plenum Press, New York.

Racey, P. A. 1998. Ecology of European bats in relation to their conservation. Pp. 249–260 *in:* Bat Biology and Conservation (T. H. Kunz and P. A. Racey, eds.). Smithsonian Institution Press, Washington, D.C.

Raesly, R. L., and J. E. Gates. 1986. Winter habitat selection by north temperate cave bats. American Midland Naturalist, 118:15–31.

Ransome, R. D. 1971. The effect of ambient temperature on the arousal frequency of the hibernating greater horseshoe bat, *Rhinolophus ferrum-equinum,* in relation to site selection and the hibernation state. Journal of Zoology (London), 164:77–112.

Ratcliffe, F. N. 1931. The flying fox (*Pteropus*) in Australia. Commonwealth Scientific and Industrial Research Organization Bulletin, 53:1–133.

Redford, K. H., and J. F. Eisenberg. 1992. Mammals of the Neotropics. Vol. 2: The Southern Cone. University of Chicago Press, Chicago.

Richards, G. C., and L. S. Hall. 1998. The conservation biology of Australian bats: are recent advances solving our problems? Pp. 271–281 *in:* Bat Biology and Conservation (T. H. Kunz and P. A. Racey, eds.). Smithsonian Institution Press, Washington D.C.

Ricklefs, R. E. 1980. Geographical variation in clutch size among passerine birds: Ashmole's hypothesis. Auk, 97:38–49.

Ricklefs, R. E. 1992. The megapopulation: a model of demographic coupling between migrant and resident landbirds. Pp. 537–548 *in:* Ecology and Conservation of Neotropical Migrant Landbirds (J. M. Hagan III and D. W. Johnston, eds.). Smithsonian Institution Press, Washington, D.C.

Ricklefs, R. E. 1997. Comparative demography of New World populations of thrushes (*Turdus* spp). Ecological Monographs, 67:23–43.

Roer, H. 1995. 60 years of bat banding in Europe—results and tasks for future research. Myotis, 32–33:251–261.

Rojas-Martinez, A., A. Valiente-Banuet, M. del C. Arizmendi, A. Alcantara-Eguren, and H. T. Arita. 1999. Seasonal distribution of the lesser long-nosed bat (*Leptonycteris curasoae*) in North America: does a generalized migration pattern really exist? Journal of Biogeography, 26.1065–1077.

Russell, R. W., F. L. Carpenter, M. A. Hixon, and D. C. Paton. 1994. The impact of variation in stopover habitat quality on migrant rufous hummingbirds. Conservation Biology, 8:483–490.

Ryberg, O. 1947. Studies on Bats and Bat Parasites. Svensk Natur, Stockholm.

Rydell, J., and H. J. Baagøe. 1994. Vespertilio murinus. Mammalian Species, 467:1–6.

Rydell, J., K. B. Strann, and J. R. Speakman. 1994. First record of breeding bats above the Arctic Circle—northern bats at 68–70 degrees N in Norway. Journal of Zoology (London), 233:335–339.

Safina, C. 1993. Population trends, habitat utilization, and outlook for the future of the sandhill crane in North America: a review and synthesis. Bird Populations, 1:1–27.

Sahley, C. T., and L. Baraybar. 1996. The natural history of the long-snouted bat, *Platalina genovensium* (Phyllostomidae: Glossophaginae), in southwestern Peru. Vida Silvestre Neotropical, 5:101–109.

Sahley, C. T., M. A. Horner, and T. H. Fleming. 1993. Flight speeds and mechanical power outputs of the nectar-feeding bat *Leptonycteris curasoae* (Phyllostomidae: Glossophaginae). Journal of Mammalogy, 74:594–600.

Schmidt-Koenig, K. 1975. Migration and Homing in Animals. Springer-Verlag, Berlin, 99 pp.

Schmitt, I. H., D. J. Kitchener, and R. A. How. 1995. A genetic perspective on mammalian variation and evolution in the Indonesian Archipelago: biogeographic correlates in the fruit bat genus *Cynopterus*. Evolution, 49:399–412.

Schnitzler, H., and E. K. Kalko. 1998. How echolocating bats search and find food. Pp. 183–196 *in:* Bat Biology and Conservation (T. H. Kunz and P. A. Racey, eds.). Smithsonian Institution Press, Washington D.C.

Shump, K. A., and A. U. Shump. 1982. *Lasiurus cinereus*. Mammalian Species, 185:1–5.

Simmons, N. B. 1998. A reappraisal of interfamiliar relationships of bats. Pp. 3–26 *in:* Bat Biology and Conservation (T. H. Kunz and P. A. Racey, eds.). Smithsonian Institution Press, Washington, D.C.

Simmons, N. B., and A. L. Wetterer. 2002. Phylogeny and convergence in cactophilic bats. Pp. 87–121 *in:* Columnar Cacti and Their Mutualists: Evolution, Ecology, and Conservation (T. H. Fleming and A. Valiente-Banuet, eds.). University of Arizona Press, Tucson.

Sinclair, E. A., N. J. Webb, A. D. Marchant, and C. R. Tidemann. 1996. Genetic variation in the little red flying-fox *Pteropus scapulatus* (Chiroptera: Pteropodidae): implications for management. Biological Conservation, 76:45–50.

Slatkin, M. 1985. Rare alleles as indicators of gene flow. Evolution, 39:53–65.

Smith, T. B., M. W. Bruford, and R. K. Wayne. 1993. The preservation of process: the missing element of conservation programs. Biodiversity Letters, 1:164–167.

Smythe, N. M. 1982. The seasonal abundance of night-flying insects in a Neotropical rainforest. Pp. 309–318 *in:* The Ecology of a Tropical Rainforest: Seasonal Rhythms and Long-Term Changes (E. G. Leigh, A. S. Rand, and D. M. Windsor, eds.). Smithsonian Institution Press, Washington, D. C.

Speakman, J. R., and P. A. Racey. 1987. The energetics of pregnancy and lactation in the brown long-eared bat, *Plecotus auritus*. Pp. 368–393 *in:* Recent Advances in the Study of Bats (M. B. Fenton, P. Racey, and J. M. V. Rayner, eds.) Cambridge University Press, Cambridge.

Spencer, H. J., C. Palmer, and K. Parry-Jones. 1991. Movements of fruit-bats in eastern Australia, determined by using radio-tracking. Wildlife Research, 18:463–468.

Stebbings, R. E. 1988. Conservation of European Bats. Christopher Helm, London.

Strelkov, P. P. 1969. Migratory and stationary bats (Chiroptera) of the European part of the Soviet Union. Acta Zoologica, Cracow, 14:393–440.

Strelkov, P. P. 1997. Nursing area and its position within the range in migratory bats (Chiroptera: Vespertilionidae) from Eastern Europe and neighbouring regions. Russian Journal of Zoology, 1:545–553.

Svoboda, P. L., J. R. Choate, and R. K. Chesser. 1985. Genetic relationships among southwestern populations of the Brazilian free-tailed bat. Journal of Mammalogy, 66:444–450.

Thomas, D. W. 1982. Ecology of an African savanna fruit bat community: resource partitioning and role in seed dispersal. Ph.D. Thesis. University of Aberdeen, Aberdeen, Scotland.

Thomas, D. W. 1983. The annual migrations of three species of West African fruit bats (Chiroptera: Pteropodidae). Canadian Journal of Zoology, 61:2266–2272.

Thomson, C. E. 1982. *Myotis sodalis.* Mammalian Species, 163:1–5.

Tinkle, D. W., and I. G. Patterson. 1965. A study of hibernating populations of *Myotis velifer* in northwestern Texas. Journal of Mammalogy, 74:155–167.

Tschapka, M., and G. S. Wilkinson. 1999. Free-ranging vampire bats (*Desmodus rotundus*, Phyllostomidae) survive 15 years in the wild. Zeitschrift für Säugetierkunde, 64:239–240.

Tuttle, M. D. 1975. Population ecology of the gray bat (*Myotis grisescens*): factors influencing early growth and development. Occasional Papers, Museum of Natural History, University of Kansas, 36:1–24.

Tuttle, M. D. 1976. Population ecology of the gray bat (*Myotis grisescens*): philopatry, timing and patterns of movement, weight loss during migration, and seasonal adaptive strategies. Occasional Papers, Museum of Natural History, University of Kansas, 54:1–38.

Tuttle, M. D., and D. E. Stevenson. 1977. An analysis of migration as a mortality factor in the gray bat based on public recoveries of banded bats. American Midland Naturalist, 97:235–240.

Tuttle, M. D., and D. E. Stevenson. 1982. Growth and survival in bats. Pp. 105–150 *in:* Ecology of Bats (T. H. Kunz, ed.). Plenum Press, New York.

van der Merwe, M. 1975. Preliminary study on the annual movements of the Natal clinging bat. South African Journal of Science, 71:237–241.

Vardon, M. J., and C. R. Tidemann. 1999. Flying-foxes (*Pteropus alecto* and *P. scapulatus*) in the Darwin region, North Australia: patterns in camp size and structure. Australian Journal of Zoology, 47:411–423.

Vaughan, T. A. 1976. Nocturnal behavior of the African false vampire bat (*Cardioderma cor*). Journal of Mammalogy, 57:227–248.

Vaughan, T. A. 1977. Foraging behaviour of the giant leaf-nosed bat (*Hipposideros commersoni*). East African Wildlife Journal, 15:237–250.

Vaughan, T. A., and R. P. Vaughan. 1986. Seasonality and the behavior of the African yellow-winged bat. Journal of Mammalogy, 67:91–102.

Vehrencamp, S., F. G. Stiles, and J. W. Bradbury. 1977. Observations on the foraging behavior and avian prey of the Neotropical carnivorous bat, *Vampyrum spectrum.* Journal of Mammalogy, 58:469–478.

Volleth, M., and K.-G. Heller. 1994. Phylogenetic relationships of vespertilionid genera (Mammalia: Chiroptera) as revealed by karyological analysis. Zeitschrift für Zoologische Systematische Evolutionforschung, 32:11–34.

Walker, S. 1995. Mexico-U.S. partnership makes gains for migratory bats. Bats, 13:3–5.

Webb, N. J., and C. R. Tidemann. 1996. Mobility of Australian flying-foxes, *Pteropus* spp. (Megachiroptera): evidence from genetic variation. Proceedings of the Royal Society of London B, 263:497–502.

Weidensaul, S. 1999. Living on the Wind. North Point Press, New York.

Weidinger, K. 1994. Bat communities of three small pseudokarstic caves in eastern Bohemia (Czech Republic). Folia Zoologica, 43:455–464.

Wilkinson, G. S. 1985. The social organization of the common vampire bat. 1. Pattern and cause of association. Behavioral Ecology and Sociobiology, 17:111–121.

Wilkinson, G. S. 1987. Altruism and cooperation in bats. Pp. 299–323 *in:* Recent

Advances in the Study of Bats (M. B. Fenton, P. Racey, and J. M. V. Rayner, eds.). Cambridge University Press, Cambridge.

Wilkinson, G. S. 1992. Communal nursing in the evening bat, *Nycticeius humeralis*. Behavioral Ecology and Sociobiology, 31:225–235.

Wilkinson, G. S., and T. H. Fleming. 1996. Migration and evolution of lesser long-nosed bats *Leptonycteris curasoae*, inferred from mitochondrial DNA. Molecular Ecology, 5:329–339.

Williams, C. F. 1986. Social organization of the bat *Carollia perspicillata*. Ethology, 71: 265–282.

Wilson, D. E. 1971. Ecology of *Myotis nigricans* (Mammalia: Chiroptera) on Barro Colorado Island, Panama Canal Zone. Journal of Zoology (London), 163:1–13.

Wilson, D. E. 1979. Reproductive patterns. Pp. 317–378 *in:* Biology of Bats of the New World Family Phyllostomatidae (R.J. Baker, J.K. Jones Jr., and D.C. Carter, eds.). Pt. 3. Special Publications, the Museum, Texas Tech University, 16. Texas Tech Press, Lubbock.

Winkelmann, J. R., F. J. Bonaccorso, and T. L. Strickler. 2000. Home range of the southern blossom bat, *Syconycteris australis,* in Papua New Guinea. Journal of Mammalogy, 81:408–414.

Winkler, H., and B. Leisler. 1992. On the ecomorphology of migrants. Ibis, 134 (suppl.):21–28.

Winter, Y., and O. von Helversen. 1998. The energy cost of flight: do small bats fly more cheaply than birds? Journal of Comparative Physiology B, 168:105–111.

Wolda, H. 1992. Trends in abundance of tropical forest insects. Oecologia, 89:47–52.

Worthington Wilmer, J., L. Hall, E. Barratt, and C. Moritz. 1999. Genetic structure and male-mediated gene flow in the ghost bat (*Macroderma gigas*). Evolution, 53: 1582–1591.

Worthington Wilmer, J., C. Moritz, L. S. Hall, and J. Toop. 1994. Extreme population structuring in the threatened ghost bat, *Macroderma gigas:* evidence from mitochondrial DNA. Proceedings of the Royal Society of London B, 257:193–198.

Wright, S. 1978. Evolution and Genetics of Populations. Vol. 4: Variability Within and Among Natural Populations. University of Chicago Press, Chicago.

Zinn, T. L., and W. W. Baker. 1979. Seasonal migration of the hoary bat, *Lasiurus cinereus,* through Florida. Journal of Mammalogy, 60:634–635.

CHAPTER 5

Life Histories of Bats: Life in the Slow Lane

Robert M. R. Barclay and Lawrence D. Harder

Introduction

An individual's absolute fitness depends on the timing, quality, and quantity of reproduction. This life history incorporates fundamental components of reproductive output and survival, including size at birth and maturity, rate of growth, age at maturity, size and number of offspring, and senescence. In general, these life-history traits arise from an individual's schedule of investment in growth, maintenance, and reproduction. Because all organisms must cope with limited resources, investment in one function tends to reduce resources available for other functions, creating trade-offs between life-history traits. For example, allocation to current reproduction often reduces an individual's survival and/or future reproductive effort. Such trade-offs create correlations between life-history traits. Thus, in some species, individuals reach sexual maturity at a young age, produce many small offspring, and live short lives. Individuals of other species mature late, produce few large offspring, and live long lives (Holmes and Sherry 1997; Read and Harvey 1989; Roff 1992; Stearns 1992).

Variation in the life histories of mammals has received considerable attention (e.g., Harvey and Read 1988; Millar 1977; Millar and Hickling 1991; Promislow and Harvey 1990). These analyses have identified body size as an important influence on life-history traits, whereby small mammals typically mature early, have brief gestation and lactation periods, produce many small offspring per litter, and die at an early age. Larger mammals tend to exhibit the opposite traits (e.g., Western and Ssemakula 1982). Even after accounting for the effect of body size, however, species fall along a "fast-slow continuum" of life histories (Read and Harvey 1989). Thus, for their size, some mammals grow quickly, reproduce intensively, and live short lives (e.g., lagomorphs), whereas others grow and reproduce more slowly and live relatively long lives (e.g., primates).

Independent of body size, mammalian life histories vary with mortality risk (Harvey and Zammuto 1985; Promislow and Harvey 1990) in a manner consistent with evolutionary senescence theory (Holmes and Austad 1994; Partridge and Barton 1993). This theory predicts that low extrinsic mortality favors slow reproduction and delayed senescence (e.g., Medawar 1952). Indeed, juvenile mortality rate correlates especially strongly with many life-history traits over a broad range of mammalian taxa. Body size still comes into play, however, as small-bodied species tend to have high mortality rates (Promislow and Harvey 1990). Recent models (e.g., Charnov 1991; Kozlowski and Weiner 1997) have combined an optimality approach with aspects of allometry and the influence of mortality patterns to explain relationships among life-history traits. Natural selection is viewed as acting on age at maturity to maximize lifetime reproductive success. High juvenile mortality favors early maturation, whereas later-maturing animals are larger and can afford greater reproductive investment (Charnov 1991; Purvis and Harvey 1996).

Previous reviews of mammalian life histories have largely ignored the second largest order of mammals, Chiroptera. Bats include approximately 20% of all mammal species but represent from 0% to 6.3% of species examined in previous studies (e.g., Allaine et al. 1987; Millar 1977; Promislow and Harvey 1990; Purvis and Harvey 1995; Sacher and Staffeldt 1974; Sibly et al. 1997; Stearns 1983; Wooton 1987; but see Jones and Purvis 1997; Purvis and Harvey 1996). Notwithstanding, bats have been recognized as exhibiting unusual life histories. Despite their small size, bats lie at the slow end of the fast-slow continuum (Promislow and Harvey 1990) and live remarkably long lives (Austad and Fischer 1991; Holmes and Austad 1994). This set of characteristics has been related to reduced mortality associated with flight (Holmes and Austad 1994; Partridge and Barton 1993).

The unusual life-history traits of bats provide a unique opportunity to test predictions concerning variation in mammalian life histories and have stimulated recent interest (Jones and MacLarnon 2001; Jones and Purvis 1997; Racey and Entwistle 2000). We therefore consider the life histories of bats from two perspectives. We analyze life-history data for bats and shrews (another diverse group of small-bodied mammals) and assess hypotheses regarding the evolution of the general life history of bats in a mammalian context. We also examine life-history variation among bat species. We conclude that the adaptive radiation responsible for chiropteran diversity occurred despite significant constraints on life-history evolution.

Life-History Data

We compiled data on the ecology, body size, and life-history characteristics of bats and shrews (Soricidae) from the primary literature, *Mammalian Species* accounts (American Society of Mammalogists), and recent reviews (Hayssen et al. 1993; Innes 1994; Norberg and Rayner 1987). When more than one study provided data for a particular trait, we averaged the values or used values from the study providing the most data on other characteristics. For three species (*Eptesicus fuscus, Nyctalus noctula,* and *Myotis myotis*), our data set included separate values from two populations with substantially different life-history characteristics. For example, *Eptesicus fuscus* gives birth to twins in eastern North America but primarily to single pups in western locations (Kunz 1974).

We classified most species of bat as either tropical or temperate (i.e., occurring almost exclusively between the tropics or outside them). A few species ($n = 12$) occur widely in both zones and we classified them as such. For diet, we distinguished bats that feed primarily on fruit, nectar, insects, blood, animals other than insects, or a broad range of plant and animal prey. For some analyses we combined all plant feeders and all animal feeders. We excluded the three species of vampire bats, as their life histories are particularly unusual.

We characterized each species with respect to size in several ways. As a measure of adult body size, we used the body mass for nonpregnant females whenever it was available. Occasionally we found only an overall body mass for a species and, rarely, only male body mass. As measures of gestation effort, we used mass of individual pups at birth and the modal number of pups per litter, which we multiplied to give total mass of the litter at birth. We used mass at weaning and litter size to obtain total mass of the litter at weaning, assuming that all pups survived to weaning. For bats, our measure of weaning mass is an underestimate, as most studies provide only pup size at first flight (fledging), whereas weaning occurs later (e.g., Isaac and Marimuthu 1996; Koehler and Barclay 2000).

We used the maximum number of litters produced per female per year as a measure of reproductive output. Litter number, multiplied by the modal litter size, estimated the number of offspring produced by a female per year.

Our survey included several temporal characters. We present gestation and lactation duration in days (again, duration of lactation is likely underestimated), age at maturity for females in months, and maximum known longevity in years. We defined the age at sexual maturity for females as the age at first mating, thereby eliminating variation due to reproductive delays (e.g. delayed fertilization [Racey 1982]). When temporal traits were presented only as a range, we used the midpoint.

Our data have a number of limitations. Although we compiled at least partial data for 360 species of bats and 60 species of shrews, representing 38% and 20% of the total number of species, respectively, large gaps remain in the data (tables 5.1 and 5.2). Although litter size and number of litters per year are reasonably well known for bats, few data exist for other traits, particularly size at birth and weaning, longevity, and age at sexual maturity. Annual survival rates are virtually unknown, especially for tropical species, and are frequently based on banding studies that cannot distinguish dispersal from mortality or variation in mortality with age (e.g., Humphrey and Cope 1977; Keen and Hitchcock 1980). We thus had to use maximum longevity as a measure of survival, recognizing the limitations of this measure. For example, temperate species have been studied more intensively than tropical taxa, and measures of longevity depend to some extent on the duration and intensity of study.

Our data for bats also involve taxonomic inequities. For example, we have at least some data for approximately 58% of Phyllostomidae ($n = 141$) but for only 25% of Rhinolophidae ($n = 130$). The unevenness in the data is indicated by the fact that while there are considerable litter-size data for phyllostomids, data for almost all other traits involve ≤12 species (<9% of the total number of species). This poor representation of phyllostomids is unfortunate given that they are the most ecologically diverse bat family and could thus be particularly enlightening about the relationships between ecological and life-history characteristics. Because of the sparse data for some small families (table 5.1), for a few analyses we pooled families into larger taxa based on a recent phylogeny (Simmons and Geisler 1998).

Phylogenetic Analysis

When possible, we considered life-history variation in the context of chiropteran phylogeny. This served two purposes. First, the phylogeny enabled inference concerning the history of specific life-history traits, including identification of potential historical constraints. Second, the phylogeny allowed us to assess correlated evolution between life-history traits. Unfortunately, this approach was limited because studies reconstructing chiropteran phylogeny have considered only a fraction of bat species.

The phylogenetic hypothesis that we considered involved a composite phylogeny constructed by grafting published species-level phylogenies for chiropteran subgroups onto Simmons and Geisler's (1998) family-level phylogeny. Simmons and Geisler presented two phylogenies, one including fossil taxa and one without them. We used the one that excluded fossil taxa for our analyses and discuss the implications of the other version. To be included in the composite phylogeny, a species-level phylogeny had to have been constructed according to the principles of phylogenetic systematics. The final composite phylogeny includes only 174 taxa of the 360 species for which we have life-history data, with some particularly diverse taxa (e.g., Myotinae) being represented as single undifferentiated groups.

Based on the composite phylogeny, we used MacClade 3 (Maddison and Maddison 1992) to identify the most parsimonious optimization of the evolution of several quantitative traits (temperate vs. tropical range, diet, litter size and litters per year). Taxa that exhibit more than one character state (e.g., a species' geographic range includes both temperate and tropical areas) were coded as being polymorphic. Character states were unordered, allowing state change in any direction.

We also used the composite phylogeny to identify contrasts between related species with differing numbers of litters per year and global ranges (temperate versus tropical). Analysis of other life-history traits in the context of these contrasts assesses whether evolutionary transitions in litter number or global range were accompanied by correlated life-history responses. We supplemented the contrasts identified from the composite phylogeny with species pairs within polymorphic genera not included in the phylogeny. Some of these additional contrasts involved genera for which phylogenies could not be grafted onto the backbone tree (e.g., *Lasiurus* [Morales and Bickham 1995]). For genera that have not been studied phylogenetically (e.g., *Myotis* and *Pipistrellus*), we constructed contrasts between species within accepted subgeneric divisions.

Although we considered life-history traits in a phylogenetic context when possible, we did not employ the method of "phylogenetically independent contrasts" (reviewed by Harvey and Pagel 1991), which attempts to account for a lack of independence in the characteristics of different species that could arise from phylogenetic relatedness. We did not use this approach for three main reasons. First, whether the state of a species for a given trait is independent of that for a second species depends on the trait and question of interest, in addition to the relatedness of the species (Barrett et al. 1996). Because the traits of species can change more rapidly than implied by phylogenetic relatedness, independence need not correlate directly with relatedness. Second, the technique of phylogenetically independent contrasts infers ancestral character states based on an unverifiable model of evolutionary change, so that interpretation can depend on the model considered. Third, given the limited current knowledge of chiropteran phylogeny, implementation of phylogenetically independent contrasts would exclude over half the species for which we have life-history data.

Statistical Analyses

Our statistical analysis of life-history variation involved general linear models (Neter et al. 1990). Selection of statistical models employed a backward elimination procedure, which initially included all categorical main effects, continuous covariates, and interactions. Non-significant terms ($P > 0.05$) were excluded one at a time, unless they were involved in a significant interaction with another effect. Most analyses used log-transformed dependent variables to correct for heterogeneous variances and assure normality of residuals. Continuous independent variables (especially body mass) were also log-transformed when necessary to assure linear relations between dependent and independent variables. In an analysis that considers the relation between the logarithm of a life-history trait and log(body mass), the partial regression coefficient for log(body mass) is often referred to as the allometry coefficient. An allometry coefficient of 1 indicates that the life-history trait varies in direct proportion with body mass, whereas a value <1 indicates that changes in body mass have less effect on life history for large animals than for small ones.

Table 5.1. Mean, ±SE, and range for life history traits for families of bats

Family	Litter Size	Litters/Year	Birth Mass (as proportion of female mass)	Weaning Mass (as proportion of female mass)	Gestation (d)	Lactation (d)	Age at Maturity (mo)	Longevity (yr)
Antrozoidae (2)	2 (1)	1 (1)	0.36 (1)	1.49 (1)	63 (1)	...	...	9.1 (1)
Craseonycteridae (1)	1 (1)	1 (1)	...	...	...	...	...	...
Emballonuridae (48)	1 (17)	1.45 (16) +0.15/−0.14 1–3	0.24 (6) ±0.04 0.12–0.31	0.77 (1)	115.6 (6) +5.8/−5.5 98–135	61.6 (6) +11.2/−9.5 28–90	10.8 (7) +1.1/−1.0 7.5–16	...
Furipteridae (2)	1 (1)	...	...	...	...	...	...	...
Megadermatidae (5)	1 (5)	1.15 (5) +0.17/−0.15 1–2	0.28 (1)	...	104.6 (4) +13.3/−11.8 90–150	54.6 (2) +19.9/−14.6 40–75	18 (1)	14 (1)
Molossidae (86)	1 (33)	1.27 (18) +0.14/−0.13 1–5	0.26 (6) ±0.02 0.21–0.27	0.93 (1)	87.4 (10) +5.4/−5.1 67–120	47.0 (7) +8.5/−7.2 21–75	6.9 (3) +3.6/−2.4 3–12	9.7 (2) +3.4/−2.5 7.2–13
Mormoopidae (8)	1 (6)	1 (5)	0.19 (1)	...	105 (1)	...	...	...
Mystacinidae (1)	1 (1)	1 (1)	0.54 (1)	...	...	...	...	...
Natalidae (5)	1 (1)	1 (1)	...	...	...	...	...	...
Noctilionidae (2)	1 (2)	1 (2)	...	0.87 (1)	...	80.5 (1)	...	...
Nycteridae (13)	1 (4)	1.57 (4) +0.48/−0.37 1–3	...	...	98.5 (3) +15.3/−13.2 75–122.5	59.7 (3) +4.7/−4.3 52.5–67.5	...	...
Phyllostomidae (148)	1 (82)	1.54 (44) +0.08/−0.08 1–3	0.31 (15) ±0.02 0.15–0.47	0.78 (3) ±0.14 0.67–0.94	119.1 (6) +14.4/−12.8 90–205	95.6 (3) +73.8/−41.7 49–300	4.6 (3) +1.5/−1.2 3–8	10.9 (3) +3.4/−2.6 7–18
Pteropodidae (173)	1 (49)	1.54 (39) +0.08/−0.08 1–2	0.18 (16) ±0.01 0.10–0.28	0.54 (8) +0.07 0.37–0.74	137.0 (22) +6.0/−5.8 90–180	79.8 (21) +8.7/−7.8 37.5–165	8.0 (14) +1.0/−0.9 5–18	17.3 (5) +1.4/−1.3 14–21.8
Rhinolophidae (132)	1 (30)	1.03 (25) +0.03/−0.03 1–2	0.27 (14) ±0.02 0.13–0.36	0.89 (2) ±0.0 0.88–0.89	93.7 (21) +6.3/−5.9 42.5–161	60.3 (16) +6.4/−5.7 30–150	14.3 (11) +2.7/−2.3 6–36	23.3 (2) +6.8/−5.2 18–30

Rhinopomatidae (3)	1 (2)	...	0.19 (2) ±0.04 0.15–0.23	...	103.5 (2) +6.1/−5.7 97.5–109	43.8 (2) +16.0/−11.7 32–60	12.9 (2) +5.6/−3.9 9–18.5	...
Thyropteridae (2)	1 (1)	...	0.47 (1)	...	150 (1)	...	...	...
Vespertilionidae (354)	1.38 (110) 1–3	1.11 (91) +0.03/−0.04 1–3	0.21 (39) ±0.01 0.11–0.41	0.75 (22) ±0.03 0.57–1.0	69.4 (34) +3.8/−3.6 40–122.5	40.9 (36) +2.4/−2.3 21–90	5.6 (24) +0.9/−0.8 2–20	14.9 (35) +1.2/−1.1 5–34
Total litter	...	...	0.30 (38) ±0.02 0.18–0.50	1.22 (21) ±0.10 0.57–2.0	...	...	...	...

Sources. A comprehensive data base is available from authors.

Note. Numbers in parentheses are the number of species in each family, or the number of species for which there are data. Asymmetrical standard errors represent back-transformations of the SE calculated for the log-transformed observations used in statistical analyses. The absence of standard errors indicates no variation in the trait within the family. Mass values for Antrozooidae combine the masses of both twins of *Antrozous pallidus*. Ellipses dots indicate that no data were available for that trait.

For all statistical tests we used, α = 0.05 unless otherwise noted. We present means ± SE. When we used log-transformed data, back-transformation for presentation resulted in asymmetrical standard errors.

Life Histories of Bats

Compared to other eutherian mammals, bats have relatively invariable life histories, despite considerable variation in ecology, morphology, and physiology (see also Austad and Fischer 1991; Barclay 1994; Hayssen and Kunz 1996; Kihlstrom 1972; Kunz and Stern 1995; Kurta and Kunz 1987; Millar 1977; Promislow and Harvey 1990; Tuttle and Stevenson 1982; Wooton 1987). Bats develop and reproduce slowly and live long lives (table 5.1), placing them at the slow end of the fast-slow continuum of life-history traits (Read and Harvey 1989), an unusual situation given their small body size.

Bats uniformly bear only one young per litter except in the Vespertilionidae and Antrozoidae, the latter of which was recently elevated from the vespertilionids to its own family (Simmons and Geisler 1998). Although twinning has been reported in species from many families (Hayssen et al. 1993), the rate of twinning is extremely low and does not differ from that of vespertilionid species that typically produce single offspring. Among the Vespertilionidae, only six species in our sample have a modal litter size greater than two: five *Lasiurus* species and *Nycticeius schlieffeni* (Hayssen et al. 1993). However, litters of two or more are relatively common in vespertilionids (46 of 108 species, or 42.6%), especially in genera other than *Myotis* (in which only one of 30 species in our sample has a litter >1).

At birth, bat pups average 23.0 ± 8.0% of their mother's mass (*n* = 91). Relative birth mass varies considerably among species, with young *Nyctalus lasiopterus* averaging 11.1% of adult mass (Hayssen et al. 1993) and those of several species averaging over 35%. Total mass for litters of more than one young can exceed 50% of adult mass, as in *Pipistrellus subflavus* (Hoying and Kunz 1998).

Most bat species for which we have data have only one litter per year (180 of 257 species, or 70.0%). This partly reflects the lack of information for tropical species compared to temperate ones, which cannot produce multiple litters due to severe seasonality (see below). Females of several tropical species have up to three litters per year, including *Myotis adversus* (Strahan 1995), *Myotis nigricans* (Wilson 1971), and *Pipistrellus mimus* (Isaac et al. 1994). The latter species produces three litters of twins per year, the most offspring per year of any bat studied to date. *Chaerophon pumilus* may have up to five litters per year (Hayssen et al. 1993). Unfortunately, the proportion of females that actually reproduce during each breeding opportunity is unknown for most species.

The duration of both gestation and lactation vary widely among bats (table 5.1). The common vampire bat (*Desmodus rotundus*) is the most extreme, gestating for 205 d and lactating for up to 10 mo (Greenhall et al. 1983). The other extreme includes *Pipistrellus hesperus,* with a 40-d gestation (Schmidley 1991), and *Pipistrellus mimus,* with a 33-d lactation period and 21 d to fledging (Isaac and Marimuthu 1996).

At first flight, subadult bats average 75.0 ± 21.0% of adult mass ($n = 36$). Size at actual weaning is undoubtedly larger but rarely has been recorded. Male *Epomops buettikofori* fledge at only 28.9% of adult male mass (females fledge at 45.8% of adult female mass in this highly sexually dimorphic ptero-podid [Thomas and Marshall 1984]). In contrast, *Tadarida brasiliensis* fledge at 92.5% of adult mass (Kunz and Robson 1995). Female vespertilionids with lit-ters of two or more frequently fledge their young at a total litter mass signifi-cantly larger than their own mass, the most extreme being *Pipistrellus subflavus* at 160% (although preweaning mortality may be high [Hoying and Kunz 1998]) and *Tylonycteris pachypus* at 181% (Medway 1972).

Bats reach sexual maturity relatively slowly, although females of many spe-cies mate and give birth during their first year. Males frequently mature later than females (e.g., Racey 1982), and spermatogenesis takes longer than in other mammals (Racey and Entwistle 2000). In the tropics, without the delays caused by hibernation, females of some species produce young well before they are 1 yr old. For example, females of *Pipistrellus mimus* mate at 2 mo and bear twins at as young as 103 d of age (Isaac and Marimuthu 1996). At the other extreme, females of *Rhinolophus ferrumequinum* generally do not mature until they are 3 yr old (Corbet and Harris 1991).

Bats are long lived. Although data are scarce, individuals of several species have lived at least 30 yr in the wild (*Rhinolophus ferrumequinum* [Corbet and Harris 1991], *Plecotus auritus* [Lehmann et al. 1992]), with the record being a little brown bat (*Myotis lucifugus*) that lived at least 34 yr (Davis and Hitch-cock 1995).

Bats versus Shrews

To investigate whether bats differ from other small mammals in terms of their life-history characteristics, we compared bats to shrews (Soricidae: Insec-tivora), another group of small, primarily insectivorous mammals. We chose shrews for this comparison not only because they are similar to bats in size and diet but also because, like bats, they occupy both tropical and temperate regions of the Old and New World, have high metabolic rates, and include spe-cies that enter torpor (Feldhamer et al. 1998). Because diet influences some life-history traits of bats (see below), we only compared shrews ($n = 60$) to insec-tivorous bats ($n = 236$).

We analyzed differences in log-transformed life-history variables between bats and shrews using analysis of covariance (ANCOVA), with log(female body mass) as the covariate and order as the main effect. In no case did the in-teraction between order and log(female mass) significantly affect life-history

Table 5.2. Mean ± SE life-history traits for insectivorous bats and shrews (Soricidae)

Life-History Trait	Bats	Shrews
Female mass (g)	15.8 ± 0.98 (230)	11.5 ± 1.95 (50) ns
Litter size	1.2 ± 0.03 (221)	4.4 ± 0.23 (58)***
Litters per year	1.2 ± 0.04 (177)	1.9 ± 0.10 (5)**
Gestation duration (d)	88.5 ± 3.01 (82)	24.3 ± 1.02 (18)***
Lactation duration (d)	52.4 ± 2.55 (72)	22.0 ± 0.78 (21)***
Pup birth mass/female mass	0.23 ± 0.01 (62)	0.08 ± 0.01 (23)***
Litter birth mass/female mass	0.29 ± 0.01 (61)	0.32 ± 0.02 (22) ns
Pup weaning mass/female mass	0.76 ± 0.03 (25)	0.72 ± 0.04 (20) ns
Litter weaning mass/female mass	1.16 ± 0.08 (25)	3.45 ± 0.26 (20)***
Age at maturity (mo)	10.0 ± 1.0 (49)	4.8 ± 1.1 (10)**
Longevity (yr)	16.1 ± 1.06 (41)	1.4 ± 0.1 (19)***

Note. Numbers in parentheses are numbers of species for which there were data. Asterisks indicate significant differences between orders in ANCOVAs based on log-transformed data, with log(female mass) as the covariate. ns = not significant ($P > 0.05$). Although mass of pups relative to that of the adult female is presented here, ANCOVAs considered absolute masses.

**$P < 0.01$.

***$P < 0.001$.

variation, indicating that female mass affects life-history traits equally in the two orders. We thus removed the interaction from the final models. Adult female body mass positively affected birth and weaning mass of each offspring and the entire litter and negatively affected litter size. In contrast, duration of gestation and lactation, number of litters per year, age at maturity of females, and longevity varied independently of adult female mass. These results generally agree with Purvis and Harvey's (1996) analysis, which indicated that only birth and weaning mass scaled with body size for small mammals, including bats.

Bats and shrews occupy opposite ends of the fast-slow continuum of life-history traits found among mammals (Read and Harvey 1989; see also Purvis and Harvey 1996). Bats differ significantly from shrews in eight of the 10 life-history traits we tested ($P < 0.01$ in each case; table 5.2). Bats reproduce and develop more slowly and produce fewer young per litter and per year but live much longer than do shrews. After a gestation period that averages 3.6 times longer, bats produce individual offspring that are almost three times larger at birth than those of shrews. However, shrews average 3.7 times more offspring per litter, so the total masses of the litters at birth do not differ between the orders ($F = 0.85$, df = 1, 91, $P > 0.25$). This was also found when bats were compared to a more diverse group of small, nonvolant eutherians (Kurta and Kunz 1987). In other words, shrews and bats invest equally per litter, but shrews invest more rapidly and divide the total investment among several small pups, whereas bats typically invest slowly in a single large neonate (fig. 5.1).

Bats develop to independence slowly, and females invest less in lactation than shrews do. Bats lactate an average of 2.4 times longer than shrews, yet

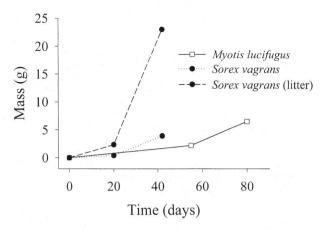

Figure 5.1. A comparison of development rates for individual young and entire litters of a similarly sized bat and shrew. The first point on each curve represents conception, the second birth mass, and the third weaning mass. Female little brown bats (*Myotis lucifugus*; 7.7 g) raise a single young to 6.5 g at weaning. In comparison, *Sorex vagrans* (6.6 g) raises an average of 5.9 young to 3.9 g each.

individual juveniles wean at the same mass (relative to that of the female) for the two taxa (approximately 75% of adult mass; table 5.2). If all juveniles survive to weaning, shrews wean litters three times heavier than those of insectivorous bats.

Based on limited data, shrews typically produce two litters per year, whereas most bats produce only one (table 5.2). Given the larger litters of shrews, they produce many more offspring per year than do bats (who typically produce only one pup per year), although we do not know the respective juvenile survival rates.

On average, most female bats reach sexual maturity within a year, but female shrews mature significantly faster (table 5.2). In many cases, female shrews produce at least one litter in the year of their birth, whereas this is rare among bats.

The maximum lifespan of bats averages over 10 times that of shrews (table 5.2). Estimates for both groups primarily stem from mark-recapture studies in the wild.

The Unique Biology of Bats

A striking feature of the life histories of bats is the general consistency among species (table 5.1). This homogeneity exists despite considerable variation among bats in other features often associated with variation in life-history traits. Bats range in body mass over almost three orders of magnitude, occupy tropical and temperate areas differing in the degree of seasonality, feed on various plants and animals, and exhibit monogamous to highly polygynous social systems. It seems reasonable to hypothesize that the homogeneity of bat life histories reflects an evolutionary consequence of the main feature that unifies all bats and that distinguishes them from other mammals: the ability to fly. Various morphological, anatomical, and physiological adaptations associated with flight make bats unique among mammals in many ways. Furthermore, flight influences the ecology of bats, via its effects on susceptibility to predation and access to resources. This unique biology may have influenced the evolution of bat life histories in two ways: it may have set limits on the range of feasible life histories available to bats, and it may have exposed bats to different selection regimes compared to those faced by other mammals, thereby establishing a different optimal life history for the Chiroptera as a whole.

Relevant Aspects of Anatomy, Morphology, and Energetics

Flight requires that an animal generate enough lift to overcome the force of gravity and enough thrust to overcome overall horizontal drag (Norberg 1985). Wing structure and body mass are thus critical traits that have undoubtedly experienced strong selection during the evolution of flight. For example, adaptations that reduce body mass should be evident. In addition, the maximum feasible size for flying animals is limited by geometric similarity because wing area does not keep pace with increases in body mass. Wing loading (the ratio of body weight to wing area) thus increases with body size, reducing the maneuverability of large flying animals (Norberg and Rayner 1987). Consequently, bats tend to be small compared to terrestrial or aquatic mammals.

Selection favoring reduced body mass has apparently resulted in considerable reduction of the female reproductive tract. In most bat species, one ovary and at least one uterine horn are reduced to the extent of being nonfunctional (Wimsatt 1979). Clearly, such restrictions on reproductive anatomy limit reproductive potential and may have important implications regarding life-history evolution.

Flight may also limit the increase in mass that occurs during pregnancy (Myers 1978). The additional mass of the litter increases the wing loading of pregnant females, thereby increasing minimum flight speed and the energetic cost of flight and decreasing maneuverability (Hayssen and Kunz 1996; Norberg and Rayner 1987). Foraging efficiency of female bats likely declines during pregnancy, a time when energy demands increase. This conflict could promote the evolution of different life-history traits, such as litter mass and number of young, compared to terrestrial mammals.

Bats are also unusual mammals in that young are not weaned until they approach adult size (Barclay 1994, 1995; Kunz 1987). Juvenile bats begin to fly at an average of 75% of adult mass (table 5.1), but as weaning may occur some weeks after initial flight (e.g., Koehler and Barclay 2000), they are even larger at independence. Length of forearm at fledging is even closer to that of the adult (>90% [Barclay 1994]), indicating that skeletal development is almost complete at weaning. This large size at fledging may be necessitated by anatomical constraints imposed by flight (Barclay 1994, 1995), as most young birds also fledge only when they approach or surpass adult size (O'Connor 1984; Ricklefs 1979). The wing bones of bats and birds experience twisting stresses (Swartz et al. 1992), which differ from the stresses placed on the limb bones of terrestrial mammals. This twisting may preclude flight until the wing bones are almost fully ossified and can resist torsional stresses (Barclay 1994). As well, the finger joints in bat wings may not be stable enough to maintain wing shape (e.g., camber) until ossification is almost complete (Kunz 1973; but see Papadimitriou et al. 1996). Thus, demands of flight may limit a bat's minimum age and relative size at independence.

Although flight is an efficient mode of locomotion per unit distance traveled, it is more energetically costly than terrestrial locomotion per unit time (Thomas 1987). This means that bats likely have relatively less energy to allocate to reproduction compared to other mammals, as they must spend a greater proportion of their total intake on maintenance (i.e., fuelling the high metabolic rate of flight [Kurta et al. 1989]). For example, female little brown bats spend about 66% of their daily energy while foraging, yet this activity occupies only about 15% of their time (Kurta et al. 1989). Such costs of foraging (see also Bell et al. 1986; Racey and Speakman 1987) far exceed those for terrestrial small mammals (Garland 1983). A low rate of allocation to reproduction likely manifests itself in different life-history traits than those found in mammals that have access to equivalent resources but that can allocate greater resources to reproduction.

Relevant Aspects of Ecology

The ability to fly is associated with a reduced risk of predation (Holmes and Austad 1994; Pomeroy 1990). This may result from an increased ability to escape predators and a more limited diversity of predators than faced by terrestrial animals. The nocturnal habits of bats further decrease their susceptibility to detection and capture (Speakman 1995). Indeed, predation on bats is rarely observed (e.g., Fenton et al. 1994), although it may vary depending on features such as roost location (Pavey et al. 1994). The generally low predation risk faced by bats increases longevity, reducing the imperative for intense, early reproduction, which can impose physiological and demographic costs.

Flight and nocturnality may also affect the availability of resources to bats. Flight provides access to resources that are inaccessible to terrestrial mammals and greatly extends an individual's foraging range. Flight also enables long distance migration, which allows some species to take advantage of seasonal resource abundance while avoiding periods of resource scarcity (Fleming and Eby, this volume). At the same time, flight limits access to some resources. Because of the incorporation of the hind limbs into the flight system, bats move awkwardly on the ground compared to terrestrial mammals, or even birds in which the hind limbs are separate from the wings. In addition, the nocturnal habits of bats limit resource availability, compared to shrews, for example, which frequently forage day and night (Rust 1978; Sorensen 1962).

Testing Hypotheses regarding Life-History Evolution

We now use the data on the life-history traits of bats and shrews to test various hypotheses regarding the evolution of bat life histories. We do this on two levels. Because certain features associated with flight are common among all bats, we assess explanations regarding the evolution of bat life histories in a mammalian context. However, we also test explanations for life-history variation among bats, given that bat species differ considerably in anatomy, morphology, energetics, and ecology, which may be associated with life-history variation. We first explore whether the low extrinsic mortality rate of bats is a key feature favoring their relatively "slow" life history (Promislow and Harvey 1990). We then examine potential ecological, physiological, and phylogenetic constraints on the life-history options available to bats and propose a scenario for life-history evolution during the transition from a nonvolant ancestor.

Low Extrinsic Mortality Hypothesis
General Considerations

The relation of mortality risk to age fundamentally influences the evolution of all aspects of an organism's life history (reviewed by Roff 1992; Stearns 1992). High adult mortality demands high reproductive effort to increase the chance of some reproductive output prior to death. In contrast, low adult mortality promotes reduced reproductive effort during each reproductive bout, particularly when current reproductive effort reduces survival or future effort.

An organism's chance of dying depends jointly on extrinsic and intrinsic factors (Medawar 1952). Extrinsic mortality involves fatal interactions with an organism's environment, including other organisms (e.g., predators, competitors, disease), extreme physical or chemical conditions, and accidents. Bats apparently experience a relatively low risk of extrinsic mortality, as data for 47 bat species indicate a mean longevity of 15.9 ± 0.96 yr (range 5–34 yr). This longevity greatly exceeds that of other mammals of equivalent size and active metabolic rate (e.g., shrews; table 5.2).

Intrinsic mortality involves physiological failure that could not have been avoided by an instantaneous change in behavior or environmental conditions. For example, a bat that starves because its severely worn teeth cause inefficient feeding suffers intrinsic mortality, whereas one that starves during prolonged food shortage does not. Intrinsic mortality differs from extrinsic mortality in being controlled genetically to some extent. This control takes two distinct forms: genetic disorders or senescence.

The incidence of genetic disorders depends on the genetic load of a species and the frequency of inbreeding. Unless bats carry an unusually heavy genetic load, genetic disorders likely represent a minor component of their intrinsic mortality because the mobility afforded by flight should promote outbreeding. Indeed, populations of three Australian flying foxes breed panmictically and the limited population subdivision in these species is more similar to the differentiation found in bird populations than to that of terrestrial mammals (Webb and Tidemann 1996). Such extensive outbreeding promotes heterozygosity and limits the expression of deleterious recessive alleles.

Bats may suffer considerable intrinsic mortality due to senescence among elderly adults, associated with physiological trade-offs between performance early versus late in life. Natural selection maintains such trade-offs because events early in life have greater fitness consequences than later events (Kirkwood and Rose 1991; Williams 1957). Selection thus favors individuals whose traits enable vigorous early reproduction, even though such effort increases later senescent mortality. However, the strength of such adaptive trade-offs depends on the risk of extrinsic mortality. Events late in life have greater value for animals such as bats that experience low extrinsic mortality and thus have a reasonable chance of living to an advanced age. As a result, selection should favor relatively less reproductive effort and greater somatic investment for bats than for organisms such as shrews that are subject to a high risk of extrinsic mortality. Unfortunately, studies of bats do not allow direct assessment of senescent mortality, which requires estimates of age-specific dynamics of mortality rates. Nevertheless, we suggest that the evolution of flight in mammals reduced mortality rate, thus establishing a "slow" life history as the optimal strategy

Variation among Bats

If life-history traits vary with mortality rate, then the reproductive effort of bats should be correlated with extrinsic mortality. For example, species that experience greater mortality due to predation or disease should have larger litters, more litters per year, or greater investment per offspring (i.e., larger mass at birth and/or weaning).

Among species of bats, reproductive output varies negatively with mortality rate. Species that produce more than one young per year (via larger litters and/or more than one litter per year) live shorter lives than those that produce one offspring per year (>1 young, mean $= 10.5 \pm 1.7$ yr, $n = 8$; 1 young, 17.7 ± 1.1 yr, $n = 33$; $t = 2.93$, df $= 39$, $P < 0.01$). This difference also occurs among vespertilionids (>1 young 10.8 ± 2.6 yr, $n = 5$; 1 young 17.8 ± 1.3 yr,

$n = 26$; $t = 2.24$, df = 29, $P < 0.05$). Although this negative association is consistent with the hypothesis that lower mortality favors reduced reproductive effort, it could also indicate a cost of reproduction whereby females of species that allocate more resources to reproduction, and thus less to maintenance, survive less well. Indeed, *Rhinolophus ferrumequinum* that breed early live shorter lives (Ransome 1995).

Bats that occupy temperate regions seem subject to lower annual mortality risks than those from tropical regions, based on their relative maximum longevity ($F = 4.70$, df = 1, 40, $P < 0.05$; see also Jurgens and Prothero 1987). On average, the seven tropical species for which longevity is known live only 55% as long (mean = 9.0, upper standard error [USE] = 2.01, lower standard error [LSE] = 1.61 yr) as the 39 temperate species (mean = 16.5, USE = 2.47, LSE = 2.22 yr). This geographic difference may reflect hibernation by many temperate species but not by tropical species. Hibernation may enhance survival by concealing bats for prolonged periods from many causes of extrinsic mortality and by postponing the somatic degradation responsible for senescent intrinsic mortality.

For the 46 species involved in the comparison among tropical and temperate regions, maximum longevity also varied significantly among families ($F = 3.22$, df = 4, 40, $P < 0.025$), but not with mean adult female mass ($F = 0.40$, df = 1,39, $P > 0.5$). The taxonomic difference largely represents a significant difference between three species of Molossoidea (mean = 7.0, USE = 2.21, LSE = 1.66 yr) and four species of Pteropodidae (mean = 22.6, USE = 6.33, LSE = 4.90 yr). This difference suggests that foliage roosting by pteropodids does not increase their vulnerability to predators. Perhaps the large size of many pteropodids limits the number of potential predators, although overall, body mass does not influence longevity.

The lower mortality of temperate species should promote lower reproductive effort. Indeed, geographic location significantly influences several life-history traits. Temperate bats have shorter gestation ($F = 10.59$, df = 1, 106, $P < 0.002$) and fewer litters per year (see "Resource Constraint Hypotheses," below). In addition, within the vespertilionids, temperate species have fewer offspring per litter than do tropical species and those that migrate from the subtropics to breed and are thus active year round (mean: temperate 1.30 ± 0.08; tropical/migratory 1.74 ± 0.09; $F = 14.34$, df = 1, 98, $P < 0.001$).

The difference in litter size between tropical and temperate bats can be viewed in several ways. It has been argued that the short growing season in temperate zones favors rapid growth (i.e., short gestation [Kunz and Stern 1995]) and that tropical/migratory species can simply afford larger litters than temperate zone species can because of prolonged growing conditions (Koehler 1991). More litters per year are also possible under such conditions. Although this may be true, we suggest that the smaller litter size of temperate-zone species supports the notion that lower mortality favors reduced reproductive effort. If having only one litter of one young per year and investing heavily in it were optimal, nothing prevents tropical species from doing so. Rather, we view the higher mortality in tropical and migratory species as favoring increased reproductive effort. Put another way, the reduced mortality suffered by temperate species allows species with reduced reproductive output to occupy temperate regions.

Resource Constraint Hypotheses
General Considerations

The unique biology of bats, especially features associated with flight, may restrict the evolution of bat life histories. In particular, the high costs that flight entails and the limited resources available to bats may limit the maximum reproductive effort they can afford.

Our data clearly indicate that shrews allocate more resources to reproduction than do insectivorous bats. Shrews achieve this high reproductive effort despite a basal metabolic rate several times higher than that of bats, at least among smaller species for either group (Hayssen and Lacy 1985). For example, the basal metabolic rate for the little brown bat (*Myotis lucifugus*, 6.5 g body mass) is 1.43 mL O_2 g-hr^{-1}, whereas that of the soricid *Cryptotis parva* (6.4 g) is 7.00 mL O_2 g-hr^{-1} (Hayssen and Lacy 1985). Although the metabolic rate of bats may increase up to 20 times during flight (Racey and Speakman 1987), flight typically occupies only a few hours per night (e.g., Swift 1980; Wilkinson and Barclay 1997). Given the variation among bats in time spent flying and the efficiency of flight, it seems unlikely that the high energetic cost of flight, by itself (the cost of flight hypothesis), can be a general explanation for differences in life-history patterns between bats and other small mammals.

Because of increases in body mass, and thus wing loading, reduced foraging efficiency and pregnancy may impose greater constraints on bats than on terrestrial mammals such as shrews (Myers 1978). Some studies support this hypothesis (the wing-loading constraint hypothesis), whereas others do not. In several species, reproductive females spend less time foraging per night during late pregnancy (Kunz 1974; Kunz et al. 1995; Rydell 1990), suggesting that the mass of the litter reduces foraging ability. In contrast, pregnant and lactating Yuma bats (*Myotis yumanensis*) forage in the same habitat (Brigham et al. 1992), despite different wing loadings. Furthermore, as they prepare for hibernation, female and male bats carry even greater added mass (in the form of fat) than do pregnant females (Kunz et al. 1998).

Data on the mass of litters of bats and terrestrial mammals also do not support the hypothesis that litter mass is constrained by flight. We found no significant difference in the relative mass of litters between insectivorous bats and shrews (table 5.2). Similarly, relative litter mass did not vary significantly over a broader range of bats and similarly sized terrestrial mammals (Kurta and Kunz 1987). Hence, female bats carry litters equal in mass to those of terrestrial mammals. Flight does not appear to limit life-history options in this way.

That bats wean their young at a large size means that each offspring is expensive to raise, which may limit the number of young a female can rear at one time (Barclay 1994, 1995). Under this hypothesis (the size-at-independence hypothesis), the high cost of each offspring results in the low, relatively invariable litter size of bats, and other life-history traits are then tied to this key character. While previous analyses using a broad range of species (Barclay 1994) indicated that bats wean young at a larger mass than terrestrial mammals of similar size, we found that bats and shrews wean individual young at an equivalent size, relative to maternal mass (table 5.2). Although this result may be confounded by our underestimate of weaning mass for bats, it emphasizes that shrews have the resources to invest more heavily in each litter than do bats.

Fledging a large pup requires considerable resources, but the limiting resource need not be energy. Instead, Barclay (1994, 1995) postulated that calcium availability limits reproductive output by bats. This calcium constraint hypothesis is based on the estimated calcium demand during pregnancy (Keeler and Studier 1992) and the relatively low calcium content of insects, fruit, and pollen (Herrera 1987; Stanley and Linskens 1974; Studier and Sevick 1992; Turner 1982). That females of several bat species experience temporary osteoporosis during lactation (Bernard and Davison 1996; Kwiecinski et al. 1987), and that calcium is the limiting mineral in the milk of *Tadarida brasiliensis* (Studier and Kunz 1995), supports this hypothesis.

The ability of birds and shrews to obtain calcium from inanimate sources (e.g., bones and mollusk shells) or invertebrates high in calcium (e.g., *Gammarus* spp.) may account for the greater variation in life-history traits in these groups than is evident among bats. For example, clutch/litter size of birds and shrews varies far more than litter size of bats. Bats are less able to forage for calcium-rich items on the ground due to their limb design, and microchiropterans cannot locate small inanimate objects on the ground using echolocation. Birds and shrews can thus augment their dietary intake of calcium whereas bats cannot (Barclay 1994, 1995). This hypothesis needs to be tested by analyzing the calcium content of shrew and bat diets.

Variation among Bats

Among bat species, variation in the effect of resource limitation on reproductive output could occur for individual reproductive bouts and/or over the entire year. In the first case, reproductive limitation should affect the size and number of young produced per litter and the time required for gestation and lactation. However, because a physiological constraint may preclude many bat species from producing multiple young per litter (polyovulation constraint hypothesis; see below), we only consider the duration of gestation and lactation. In contrast, variation in resource limitation on annual reproductive output should largely be reflected in the number of litters produced per year.

Duration of gestation depends on the geographic range of a species and the family to which it belongs. This conclusion arose whether we considered 11 pairs of related species with contrasting temperate or tropical ranges (location: $F = 13.2$, df = 1, 16, $P < 0.01$; family: $F = 24.6$, df = 4, 16, $P < 0.001$) or all 107 species for which we have relevant data (location: $F = 9.23$, df = 1, 100, $P < 0.005$; family: $F = 14.35$, df = 5, 100, $P < 0.001$). A similar analysis (Jones and MacLarnon 2001), using phylogenetically independent contrasts of a different data set and latitude as the geographic variable, found no effect of geographic location. In our analyses, the mean mass of adult females did not explain a significant proportion of the variation in gestation duration ($P > 0.05$ in both cases). However, log(litter mass) contributed to explaining log(gestation duration) for the 57 species for which these variables have been recorded ($F = 4.54$, df = 1, 49, $0.05 > P > 0.025$). Thus, we interpret variation in gestation duration in relation to geographical range, family affiliation, and litter mass.

Overall, tropical species gestate longer (mean = 98 d, USE = 4.3, LSE = 4.1 d) than temperate species (mean = 83 d, USE = 6.0, LSE = 5.6 d: $F = 4.84$, df = 1, 49, $0.05 > P > 0.025$). This difference contradicts the expectation that temperate species gestate longer than do tropical species because of greater costs of thermoregulation, shorter foraging times caused by brief nights, and delays in fetal growth caused by maternal use of torpor (Racey 1973). Instead, the longer gestation by tropical species could represent either less temporal limitation in tropical species, because of a longer active season, or greater resource limitation. As with postnatal growth (Kunz and Hood 2000; Kunz and Stern 1995), temperate species should benefit from rapid gestation to allow both females and their young to prepare for hibernation during the short growing season. All else being equal, the elevated reproductive effort needed to finance rapid development should demand a cost on females of temperate species, manifested as either a reduced proportion of females reproducing each year or reduced survival. The reproductive history of female bats has not been documented, so we cannot assess this cost. However, more rapid gestation does not seem to exact a survival cost as temperate species live longer than tropical species. Alternatively, the seasonal surge of productivity in temperate regions may provide more resources in the short term than is found in tropical regions, allowing more rapid development of fetuses for temperate species. Given the lack of evidence supporting temporal limitation, the possibility of resource limitation of fetal development in tropical species warrants further study.

The positive influence of litter mass (but not maternal mass) on gestation length indicates general resource limitation of fetal development. In addition, the partial regression coefficient of 0.103 ± 0.048 indicates a rapidly decelerating relation between gestation duration and litter mass. Thus, development of a gram of fetal tissue requires less time, on average, for species with heavy litters than for those with light litters.

Once the effects of geographic range and litter mass have been controlled, differences in gestation duration between families ($F = 3.78$, df $= 5$, 49, $P < 0.01$) are due to rapid development by vespertilionids relative to all other families ($F = 15.40$, df $= 1$, 49, $P < 0.001$). In particular, the gestation period of vespertilionids (mean $= 67$, USE $= 4.4$, LSE $= 4.1$ d) lasts about two-thirds as long as that of other species, on average (mean $= 96$, USE $= 5.2$, LSE $= 4.9$ d). In contrast, gestation period does not vary significantly among non-vespertilionid families ($F = 3.85$, df $= 4$, 49, $P > 0.05$).

The duration of lactation depends on some factors that also affect gestation time (family and litter mass), but it differs from gestation time in varying with mean adult female mass but not between temperate and tropical regions. Again, variation among families ($F = 3.44$, df $= 4$, 46, $P < 0.025$) arises solely because vespertilionids lactate for shorter periods (mean $= 39$, USE $= 3.5$, LSE $= 3.2$ d) than do non-vespertilionids (mean $= 55$, USE $= 4.5$, LSE $= 4.1$ d; $F = 8.57$, df $= 1$, 46, $P < 0.01$), whereas lactation period does not vary among non-vespertilionid families ($F = 1.74$, df $= 3$, 46, $P > 0.25$). In particular, the only significant contrast between vespertilionids and another family involves rhinolophids (mean $= 62$, USE $= 7.1$, LSE $= 6.4$ d), which lactate more than 50% longer than vespertilionids do.

Litter mass has a more complex influence on lactation time than on gestation, as its effect depends on an interaction with mean adult female mass ($F = 7.79$, df $= 1$, 46, $P < 0.01$; fig. 5.2). In general, log(lactation period) does not vary significantly with log(litter mass) for species with adult females weighing less than 185 g, whereas above this threshold lactation duration lengthens with increasing litter mass. For example, species with 10-g, 100-g, 200-g, or 500-g females have expected partial regression coefficients for litter mass of -0.180 ± 0.143, 0.076 ± 0.078, 0.153 ± 0.0716, and 0.255 ± 0.078, respectively. This suggests that raising a heavy litter relative to adult female size imposes a disproportionately large cost on large-bodied species.

The absence of differences in lactation time between species inhabiting temperate versus tropical regions ($P > 0.75$) suggests rather different influences on lactation than on gestation. For example, although gestation may depend on energy availability, lactation may be influenced more by calcium or protein availability.

In terms of annual productivity, the proportion of species that produce more than one litter per year varies significantly among families ($G = 48.58$, df $= 5$, $P < 0.001$), between temperate and tropical regions ($G = 11.86$, df $= 1$, $P < 0.001$), and with average adult female mass ($G = 20.46$, df $= 1$, $P < 0.001$: generalized linear model with binomial error structure; Proc Genmod, SAS V7.0 [McCullagh and Nelder 1989]). In general, larger-bodied species are less

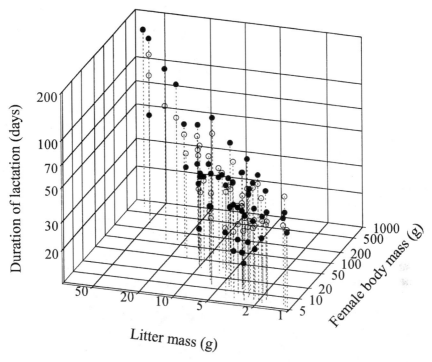

Figure 5.2. Relation of the duration of lactation to litter mass and female body mass of bats. *Open symbols* represent vespertilionids and *filled symbols* other bats.

likely to produce multiple litters than are small species, perhaps due to the increased time required to rear large offspring. Not surprisingly, the proportion of tropical species producing multiple litters per year (mean = 0.240, USE = 0.037, LSE = 0.034) greatly exceeds that of temperate species (mean = 0.021, USE = 0.037, LSE = 0.013). After the effects of body mass and geographic range have been isolated, the differences between families arise largely because pteropodids (mean = 0.461, USE = 0.169, LSE = 0.160) include a higher proportion of species with multiple litters per year than the five other families considered ($G = 35.88$, df = 1, $P < 0.001$). In addition, emballonurids (mean = 0.125, USE = 0.101, LSE = 0.060) have a higher incidence of multiple litters than rhinolophids do (mean = 0.016, USE = 0.019, LSE = 0.009). No other comparisons between families indicated significant differences in the incidence of multiple litters per year.

Among tropical species, diet significantly influenced the incidence of multiple litters per year ($G = 5.44$, df = 1, $P < 0.025$). A higher proportion of species that feed on plants (fruit and/or nectar) have multiple litters per year (mean = 0.474, USE = 0.143, LSE = 0.139) than species that feed on animals (mean = 0.154, USE = 0.047, LSE = 0.038). Once this effect of diet had been accounted for, the differences between families evident in the preceding analysis diminish considerably, so that pteropodids differ only from phyllostomids ($G = 15.04$, df = 1, $P < 0.001$). Hence, the relatively high incidence of multiple litters per year among pteropodids can be attributed largely to their plant-derived diets. A similar conclusion arises from the fact that within the Phyllostomidae, species feeding on plants (fruit and nectar) have more litters per year than do species feeding on animals (plants: mean = 1.78 ± 0.09, $n = 32$; animals: mean = 1.22 ± 0.15, $n = 9$; $t = $ df $= 39, 3.08$, $P < 0.005$). These results indicate that, in tropical environments, resources (especially energy) limit annual reproductive output less for species that consume plant material than for those that eat animals.

In addition to resource availability, variation in maintenance costs may also influence variation in the life-history traits of bats. Flight costs likely vary among bats due to differences in wing design and flight style (Norberg and Rayner 1987) or to differences in the amount of time spent flying. Among aerial insectivorous bats, for example, larger species may require long foraging periods due to features of their echolocation calls that reduce their ability to detect small prey and thus limit prey availability (Barclay and Brigham 1991).

Costs associated with thermoregulation also vary among bats, thereby influencing the energy available for reproduction. Indeed, gray bats (*Myotis grisescens*) grow faster in warmer caves (Tuttle 1975), presumably because both females and their young spend less energy thermoregulating. Colonial species can also reduce thermoregulatory costs by clustering (Roverud and Chappell 1991) and may thus have more energy to allocate to reproduction than solitary species. In this context, the hoary bat (*Lasiurus cinereus*), a large (30 g), solitary, foliage-roosting species, has one of the lowest growth rates of any temperate vespertilionid (Koehler and Barclay 2000). This slow growth may not only reflect high thermoregulatory costs but also the long foraging trips and thus high flight costs of lactating females (Barclay 1989).

Bats bear litters equivalent in mass to the litters of other mammals of the same size, indicating that the evolution of flight did not affect gross investment in newborns. However, differences among bat species in wing design or flight style may influence variation in maximum litter mass and, thereby, affect other life-history traits. For example, it has been proposed that bats with high wing loading cannot afford as heavy a litter as bats with low wing loading (Norberg and Rayner 1987; Hayssen and Kunz 1996). Under this component of the wing-loading hypothesis, log(litter mass) should increase with log(wing loading) less quickly than expected by geometric similarity alone (Hayssen and Kunz 1996).

In support of this hypothesis, Hayssen and Kunz (1996) reported that large bats bear relatively smaller neonates than do small-bodied species. They also found that megachiropterans, which have high wing loadings, bear smaller neonates for their body size than microchiropterans. Further, litter mass in megachiropterans scaled with wing loading, as expected by geometric similarity, whereas microchiropterans exhibited negative allometry. Thus, microchiropteran bats with high wing loading tend to have litters with relatively low mass (Hayssen and Kunz 1996), although considerable variation in litter mass remained unexplained by wing loading. Hayssen and Kunz (1996) concluded that selection for minimum increase in wing loading during pregnancy might be stronger among microchiropterans than megachiropterans due to different demands on flight ability.

In contrast to Hayssen and Kunz's (1996) analysis, we found marginally significant effects of log(wing loading) on log(litter mass) for 70 species, after accounting for variation associated with log(female mass), litter size, and taxonomic affiliation. Species with heavy females bear heavier litters than small-bodied species do ($F = 23.94$, df $= 1, 59$, $P < 0.001$), and species that typically bear one young have significantly lighter litters than those with two young ($F = 31.60$, df $= 1, 59$, $P < 0.001$: figs. 5.3 and 5.4). In addition, litter mass differed significantly among suborders ($F = 4.45$, df $= 1, 59$, $P < 0.05$) and among families within suborders ($F = 4.43$, df $= 4, 59$, $P < 0.005$). The latter difference reflects a difference between phyllostomids and vespertilionids ($P < 0.005$). No other contrast between microchiropteran families was significant. In addition to an overall difference in litter mass between megachiropterans and microchiropterans, the effect of wing loading differed weakly between suborders (suborder $\times$ log [wing loading] interaction, $F = 4.70$, df $= 1, 59$, $0.05 > P > 0.025$). The effect of wing loading was further complicated by a weak interaction with female body mass ($F = 4.14$, df $= 1, 59$, $0.05 > P > 0.025$). As figure 5.4 illustrates, variation in wing loading significantly affected litter mass for only small megachiropterans. Hence, wing loading has a limited effect on litter mass. Indeed, excluding wing loading and its interactions from the general linear model reduces the overall R^2 only slightly (0.936 to 0.930). Extension of this simplified analysis to all 98 species for which we had litter-mass data revealed significant, non-interacting effects of adult female mass, litter size, and family. Family differences ($F = 6.17$, df $= 5, 90$, $P < 0.001$) arose because phyllostomids bear significantly heavier litters than emballonurids, rhinolophids, and vespertilionids, after accounting for differences in female mass and litter size. In contrast to Hayssen and Kunz's (1996) conclusion, we find no evidence in our analysis that pteropodids produce unusually light litters.

Hayssen and Kunz's (1996) hypothesis likely fails to explain variation in litter mass because it overlooks variation in the flight requirements of bats that forage in different ways. Bats with low wing loading include species that forage for insects within the vegetation, glean prey from surfaces, hover while collecting nectar or fruit, or carry large prey (Norberg and Fenton 1988; Norberg and Rayner 1987). Such foraging behaviors require low wing loading, so that increased loading due to pregnancy may be just as costly as for species

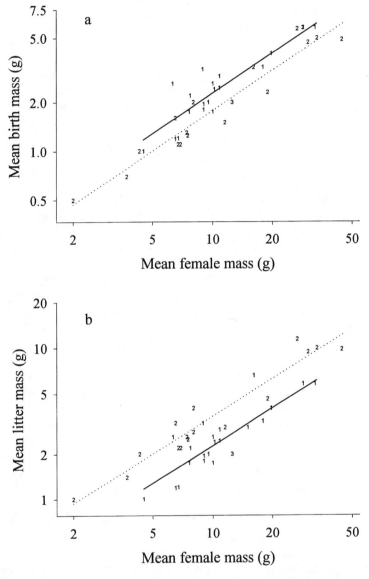

Figure 5.3. Relation between (*a*) mean birth mass of individual offspring or (*b*) mean litter mass and mean maternal mass, for vespertilionid species that produce one (*solid line*) or multiple young per litter (*dotted line*). Plotted numbers indicate litter sizes.

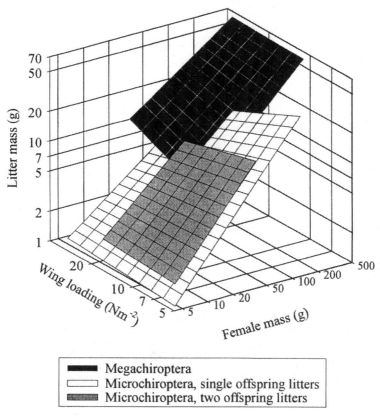

Figure 5.4. Dependence of mean litter mass of bat species on mean adult female mass and mean wing loading. The gray regression surface for vespertilionids with two offspring per litter lies above the white surface for the remaining Microchiroptera that produce single offspring. The black regression surface represents Megachiroptera. The regression surfaces are based on a general linear model that considered suborder, family within suborder, litter size, wing loading, female mass, and suborder × log [wing loading] and female mass × wing-loading interactions.

with high wing loading that forage in the open (e.g., molossids) or use flight simply to move from a roost to a feeding tree (pteropodids). Low wing loading need not indicate a larger buffer or safety margin enabling increased mass during pregnancy. Flight and foraging ability during pregnancy should directly influence selection during the evolution of wing design. Consequently, if producing young of a certain size or developmental stage benefits all bats, natural selection should produce females with the same ability to carry additional mass, regardless of their wing loading when not pregnant.

Sexual dimorphism in many bat species supports our argument. In many species, females are larger than males, and within vespertilionids, the degree of sexual dimorphism in body size correlates with litter size (Myers 1978); species with larger litters have the greatest size difference. In addition, females have larger wing areas than males do, even after controlling for body mass. Selection has thus modified traits other than litter mass to deal with increased wing loading during pregnancy (e.g., body size and wing area of females).

Offspring Size and Number Hypotheses

Our comparison of bats and shrews highlights the fact that bats produce very few large offspring per litter, with one being the modal number. These characteristics could represent either the production pattern that maximizes maternal fitness or physiological constraints on offspring production, particularly in species with single-offspring litters. The opportunity for evolutionary flexibility in characteristics of individual litters has widespread consequences for all other life-history characteristics. Therefore, we consider hypotheses regarding the trade-off between offspring size and number. We first review prevailing evolutionary explanations for offspring size and number of young per litter before addressing whether litter characteristics represent adaptive optima or imposed constraints.

Limited availability of resources for reproduction imposes a trade-off between the size and number of offspring produced during a reproductive bout. In particular, if a female invests all her reproductive resources (R) producing n offspring of size s, so that $R = ns$, then size and number must vary inversely, $n = R/s$. Given this resource constraint, natural selection favors the compromise offspring size (s^*) and number per litter (n^*) that maximizes a female's genetic contributions to subsequent generations (Lloyd 1987; Smith and Fretwell 1974).

In general, the optimal offspring size and number depend on the relation of a mother's fitness to her increasing investment in individual offspring (Lloyd 1987; Smith and Fretwell 1974). Some minimal resources (s_{min}) must be invested for an offspring to be viable. Allocation to an individual offspring in excess of s_{min} probably always enhances the offspring's survival and future reproductive prospects (performance). However, maternal fitness need not increase linearly with enhanced investment in an individual offspring. In general, the optimal offspring size from the parent's perspective maximizes offspring performance per unit of invested resources (i.e., slope of dotted line from origin in fig. 5.5 [Lloyd 1987]). If increased allocation enhances an offspring's performance in an accelerating manner, then investment of all reproductive resources in a single large offspring maximizes maternal fitness (policy of consolidation: fig. 5.5a). In contrast, decelerating increases in offspring performance with enhanced allocation favor division of reproductive resources among multiple offspring (policy of diversification: fig. 5.5b).

Given a policy of diversification, the influences on optimal offspring size (and number) depend on whether the available reproductive resources allow

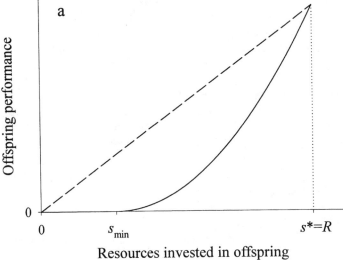

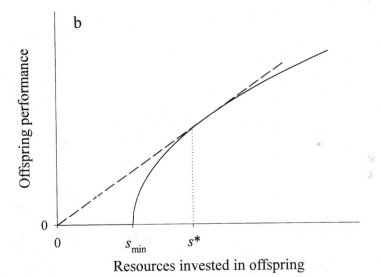

Figure 5.5. Consequences of the relation of offspring "performance" (survival and reproduction) to parental investment for the parental optimum offspring size. The solid curves illustrate that an offspring's performance generally increased with enhanced maternal investment in excess of the minimum viable offspring size (s_{min}). The optimal investment in a single offspring (s^*) maximizes performance per unit of investment, which is graphically represented by the slope of the dashed lines. a, A policy of consolidation (i.e., expend all reproductive resources [R] on a single offspring, $s^* = R$) maximizes parental fitness when offspring performance accelerates with increased investment. b, A policy of diversification (i.e., $s^* < R$) maximizes parental fitness when offspring performance decelerates with increased investment.

production of few or many viable offspring (Charnov et al. 1995; Ebert 1994). When many viable offspring can be produced, the optimal offspring size (s') depends on the minimum viable size and the deceleration of offspring performance but not on resource availability (Lloyd 1987; Smith and Fretwell 1974; although, see Venable 1992). However, when resources severely limit offspring production, the optimal offspring size also depends on resource availability because production of offspring of size s' typically leaves appreciable residual resources (i.e., given $n' = R/s'$, n' need not be an integer). Rather than waste this residual, the parent benefits from distributing it equally among offspring, causing the optimal offspring size (s^*) to deviate from s' (Charnov et al. 1995; Ebert 1994). For example, consider a female whose reproductive resources (R) exceed those needed to produce one offspring of size s' but would not allow production of two offspring of this size (fig. 5.6—i.e., $s'<R<2s'$). If the residual resources almost equal s', then maternal fitness is maximized by producing two offspring of size $s^* = R/2$ (fig. 5.6a). However, when the residual resources fall well below s', producing one large offspring maximizes maternal fitness, so $s^* = R$ (fig. 5.6b). Hence if reproductive resources allow production of only a few viable offspring, both a female's optimal offspring size and number should depend on her resource status.

Chiropteran families can be divided into two groups with respect to the number of young produced per litter: Vespertilionidae and Antrozoidae (sensu Simmons and Geisler 1998) include species with differing litter sizes, whereas species in all other families produce only one young per litter (fig. 5.7). Given that Vespertilionidae and Antrozoidae seem to be derived sister taxa (Simmons and Geisler 1998), this phylogenetic pattern suggests a significant evolutionary transition from the ancestral state of producing one large offspring, or two independent transitions, in the ancestor of vespertilionids and in the ancestor of *Antrozous pallidus*. Three explanations emerge for such a transition based on the preceding review of the influences on offspring size and number.

The policy-shift hypothesis proposes that all bat species produce the number of young that maximizes maternal fitness within their respective environment. In particular, the ancestral production of one young reflects an adaptive policy of consolidation (fig. 5.5a), whereas production of multiple young by derived taxa represents an adaptive shift to a policy of diversification (fig. 5.5b). According to this explanation, the relation of offspring performance to size changed from an ancestral accelerating pattern to a derived decelerating pattern in some vespertilionids and antrozoids.

Alternatively, two constraint hypotheses propose that conditions universally favor a policy of diversification among bats, but some ancestral physiological limitation (phylogenetic constraint) precludes production of more than one offspring in most species. The specific cause of this constraint determines whether production of a single offspring per litter is optimal. In the context of the resource constraint hypothesis, limited resources for reproduction and/or

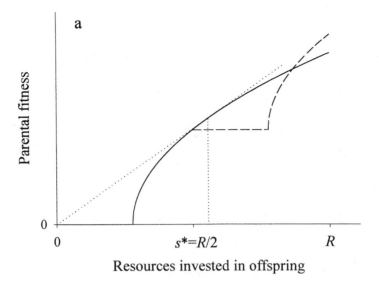

a

Parental fitness

0

0          $s^*=R/2$ R

Resources invested in offspring

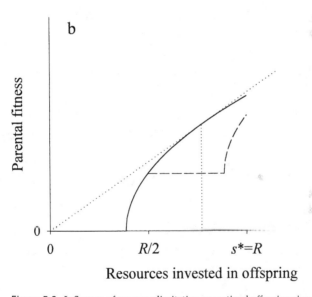

b

Parental fitness

0

0         $R/2$ $s^*=R$

Resources invested in offspring

Figure 5.6. Influence of resource limitation on optimal offspring size and number when available resources (R) allow production of only a few viable offspring. The solid curves illustrate the relation of maternal fitness to increased investment in a single offspring, with fitness increasing only when investment exceeds a fixed minimum viable offspring size (s_{min}). The dashed curves (hidden by solid curves for investment below $R/2$) depict the corresponding relation of fitness to increased allocation to two offspring. The point at which the dotted diagonal line is tangent to the fitness function for a single offspring identifies the optimal offspring size if resource availability allowed production of many offspring (s'). However, given very limited resources, the production of two offspring of size $s^* < s'$ maximizes parental fitness when $R/2$ is slightly less than s' (panel a). In contrast, panel b demonstrates that production of a single offspring of size $s^* < s'$ maximizes parental fitness when $R/2$ slightly exceeds s_{min}.

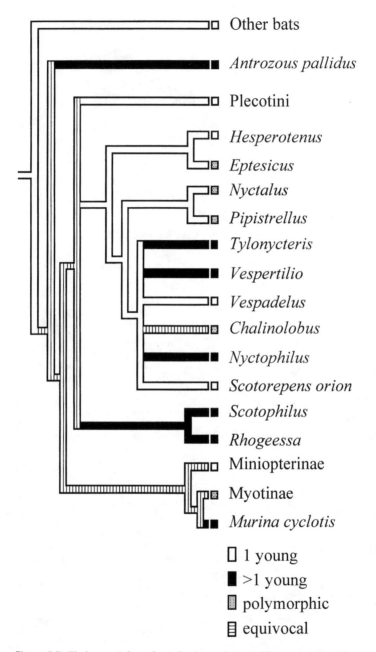

□ 1 young

■ >1 young

▦ polymorphic

▤ equivocal

Figure 5.7. Phylogenetic hypothesis for the evolution of litter size within the Chiroptera. See text for methods.

high reproduction costs select for the production of single young, despite decelerating offspring performance (e.g., fig. 5.6b). According to this explanation, vespertilionids and antrozoids that produce multiple young live in richer environments and/or are able to devote more resources to reproduction than other bats. Alternatively, the reproductive constraint hypothesis proposes that ancestral reproductive physiology in bats does not allow production of more than one offspring, regardless of resource availability, so that the widespread occurrence of single-young litters need not be adaptive. A fundamental change in the reproductive physiology of the ancestor of the antrozoid-vespertilionid clade enabled production of adaptive litter sizes, including multiple young when allowed by resource availability. We evaluate these competing hypotheses in light of available data on offspring size and number.

We consider only the policy-shift and resource constraint hypotheses for antrozoids and vespertilionids because variation in their litter size indicates the physiological ability to produce multiple young. Species in this group produce from one to four young per litter, even though they all eat insects and have relatively small bodies. This reproductive pattern implicates a policy of diversification, at least for species with multiple young, indicating that offspring performance decelerates with increasing investment (fig. 5.5b). Given the relatively similar ecologies within vespertilionids and antrozoids, species that produce one young per litter likely also experience decelerating offspring performance but cannot obtain or allocate sufficient resources to produce multiple young (e.g., fig. 5.6b). For example, *Eptesicus fuscus* (Holroyd 1993; Kunz 1974), *Nyctalus noctula* (Gaisler et al. 1979), and *Pipistrellus pipistrellus* (Rakhmatulina 1972; Stebbings 1968) typically produce one young in part of their geographic range and two, smaller young elsewhere. If this represented a shift in allocation policy, but not resource availability, then young from litters of two would be roughly half as large as those from litters of one (cf. figs. 5.5a and 5.6a). In contrast, if geographic variation involved resource availability, a consistent policy of diversification would result in young from litters of two having individual masses considerably exceeding half the mass of single-young litters (cf. fig. 5.5a and b). For *E. fuscus*, the only species of the three for which we have data on birth mass, the latter pattern is apparent. Birth mass in eastern North America, where twins are produced, is 3.3 g (Burnett and Kunz 1982), whereas it is 4.0 g in Alberta, where single offspring are the norm (Holroyd 1993). This indicates geographic variation in reproductive investment rather than a policy shift.

The resource constraint hypothesis seems unsatisfactory as an explanation for the limited number of offspring for species not included in the Antrozoidae or Vespertilionidae. All these species produce only one young per litter, even though they encompass the complete range of diet, duration of reproductive season, and body size exhibited by bats as a whole. This extensive ecological

diversity must surely cause greater interspecific variation in resource availability than that implicated in the geographic variation in litter size within
E. fuscus, N. noctula, and *P. pipistrellus.*

The policy-shift explanation can also be excluded for species not included
in the antrozoid-vespertilionid clade. These lineages include many taxa that
are ecologically similar to vespertilionids (e.g., Emballonuridae). Why then
would similar conditions cause the positive association between maternal investment and offspring performance to accelerate for ancestral lineages but
not for antrozoids or vespertilionids? It seems more reasonable to reject the
policy-shift explanation and instead examine the possibility that ancestral reproductive physiology precludes production of multiple young.

We propose that an incapacity for multiple simultaneous ovulations (the
polyovulation constraint hypothesis) limits litter size for species not in the vespertilionids or antrozoids. Females of most bat species have only one functional ovary, and even in many species with two, only one matures an ovum
during an estrous cycle. Indeed, production of a single ovum seems typical of
species that bear single offspring (e.g., Gopalakrishna et al. 1991; Pow and
Martin 1994; Rasweiler 1982; Wimsatt 1975). Although some of these species
occasionally produce two young per litter (see Hayssen et al. 1993), multiple
births could involve monozygotic twins (e.g., Bernard 1985) and so represent
the product of single ovulations. In contrast, vespertilionids that bear multiple
young possess symmetrical reproductive tracts (Wimsatt 1979) and ovulate
ova from each ovary during an estrous cycle, even producing more ova than
offspring (e.g., Hosken 1997; Krishna 1985; Wimsatt 1975). Hence, reproductive features that enable some vespertilionids (and presumably *A. pallidus*) to
produce several offspring per litter seem to represent a derived condition that
evolved from an ancestral constrained state of producing only one ovum per
estrous cycle (fig. 5.7).

Although the polyovulation constraint hypothesis presents a general proximate explanation for the life-history pattern of bats, it does not completely
explain the small litters of bats relative to shrews. Even vespertilionids and
A. pallidus, in which the reproductive constraint has been overcome, bear significantly fewer young (mean = 1.38 ± 0.05, $n = 110$) than do shrews (mean =
4.40 ± 0.23, $n = 58$). Thus, we view the ovulation constraint as only part of the
scenario for the evolution of bat life histories (see below).

The Consequences of Polyovulation for
Life-History Variation among Bats

The dichotomy between flexible litter characteristics for vespertilionids and
antrozoids and constrained production of a single offspring by all other bats
has diverse consequences for the life-history evolution of the two groups. In
particular, parallel variation in ecological characteristics need not promote

similar evolutionary responses by these two groups because of their differing scopes for reproductive adjustment. Consequently, a search for a single unifying explanation of life-history variation among bats will be both futile and misleading. Instead, we expect different patterns of covariation between life-history traits for vespertilionids and antrozoids than for other bats.

This dichotomy is evident for birth mass. Overall, mean birth mass of bats varies positively with mean female mass ($F = 568.1$, df $= 1, 90$, $P < 0.001$), but the offspring of large-bodied species (e.g., pteropodids) represent a smaller proportion of female mass than for small-bodied species (allometry coefficient $= 0.805 \pm 0.034$; comparison with slope $= 1$, $t = -5.76$, df $= 90$, $P < 0.001$). Once this effect has been accounted for, vespertilionids have significantly smaller offspring than do other bats ($F = 27.21$, df $= 1, 90$, $P < 0.001$), whereas the five other families for which we had sufficient data do not differ from each other ($F = 5.03$, df $= 4, 90$, $P > 0.05$). This pattern persists when vespertilionids that produce multiple young per litter are excluded, and it is not affected by interspecific differences in diet (animal vs. plant), location (temperate vs. tropical), or number of litters per year ($P > 0.25$ in all cases). Within vespertilionids, species producing one offspring per litter bear larger young than those producing two or three young ($F = 6.05$, df $= 2, 32$, $P < 0.01$; fig. 5.3).

Vespertilionids differ from other bats in several other life-history traits. For example, vespertilionids have significantly shorter periods of gestation and lactation than other families of bats (table 5.1), as discussed above. The relatively rapid pre- and postnatal development and smaller newborns of vespertilionids supports our suggestion that the evolution of polyovulation opened a greater range of life-history options to vespertilionids compared to that available to other species.

The Fast-Slow Continuum among Bats

There are further significant influences on life-history variation among bats, most of which involve differences among families such that some families appear to develop and reproduce more slowly than others do. For example, age at maturity for females varies with female size and family ($F = 6.17$, df $= 4, 35$, $P < 0.001$) but not with diet or geographic location. Age at maturity increases with log(female mass) ($F = 4.38$, df $= 1, 35$, $P < 0.05$; slope $= 0.17 \pm 0.08$). In addition, of the four families we could include in our analysis (emballonurids, pteropodids, rhinolophids, and vespertilionids), rhinolophids mature the latest and significantly later than either pteropodids or vespertilionids ($P < 0.002$ in each case).

The total number of young produced per female also varies significantly among bats, due to variation in both litter size and number of litters per year. Number of young per year declines significantly with female mass ($F = 11.22$, df $= 1, 220$, $P < 0.001$, slope $= -0.20 \pm 0.06$). Tropical species produce more offspring per year (1.84 ± 0.08) than do temperate species (0.98 ± 0.14; $F = 33.9$, df = 1, 220, $P < 0.001$), and plant feeders produce more (1.73 ± 0.18) than do animal feeders (1.08 ± 0.10; $F = 8.56$, df = 1, 220, $P < 0.005$). Families also differ, even when the other factors are controlled for ($F = 9.40$, df = 5, 220, $P < 0.001$). This variation results primarily from vespertilionids, which produce significantly more offspring per year than do emballonurids, phyllostomids, or rhinolophids ($P < 0.003$ in each case), but not more than pteropodids or molossids. None of the other family comparisons are significant.

In contrast to the above, body mass at weaning is not influenced by family, or by diet, or by geographical location ($P > 0.25$ in each case). The data are particularly scanty, however, and the only significant influence is a positive correlation with log(female body mass) ($F = 137.3$, df = 1, 27, $P < 0.001$).

Taking all the life-history traits together, among families for which we have sufficient data, vespertilionids and rhinolophids appear to fall at opposite ends of a fast-slow continuum of life-history traits. Vespertilionids produce significantly more offspring per year, each of which is significantly smaller and produced after a significantly shorter gestation than those of rhinolophids. Vespertilionids also mature significantly faster. These differences are intriguing because both families occupy both tropical and temperate areas, and both involve insectivorous species and species that can hibernate. The difference in life-history strategies between the two families supports our suggestion that the evolution of polyovulation in vespertilionids had major implications for other life-history traits. If mortality rate is also an important factor in the evolution of life-history variation within the Chiroptera, the fact that rhinolophids reproduce more slowly than vespertilionids implies that rhinolophids experience less extrinsic mortality. Unfortunately, only two longevity records are available for rhinolophids, so testing this prediction must await further data.

The particularly slow life history of rhinolophids is of interest in light of recent molecular evidence suggesting that they are more closely related to the Megachiroptera than to other Microchiroptera (Teeling et al. 2000). In many respects, pteropodids resemble rhinolophids in having "slow" traits (small litter size, large birth mass, long gestation, and long lives). However, pteropodids also wean their young at a relatively small mass and mature rapidly for their size. This unusual suite of life-history characteristics suggests a decoupling of life-history traits in this family. Given the placement of pteropodids at the base of the bat phylogeny (Simmons and Geisler 1998), it might be instructive to compare the life history of pteropodids to that of the nearest ancestor of the Chiroptera. The sister group to bats is under debate, but flying lemurs (Dermoptera) are often placed in that position (Simmons 1998). Unfortunately, little is known about the life history of dermopterans, other than that, like most bats, they produce single offspring (Hayssen et al. 1993).

The common vampire bat (*Desmodus rotundus*) stands out in having extreme life-history traits. It has exceptionally long gestation and lactation periods (Greenhall et al. 1983) and engages in a complex social system involving food sharing (Wilkinson 1984). Together, these features suggest that resource intake is especially limited and constrains reproduction. Such slow reproduction can only be advantageous for species with high survival; however, the maximum longevity of 18 yr (Lord et al. 1976) is not exceptional among bats. More data are required for this species, and the two other blood-feeding species (*Diphylla ecaudata* and *Diaemus youngi*), to determine whether sanguivory is indeed a low-return feeding niche, and if so, how mortality is concurrently reduced to allow flying mammals to occupy this niche.

Many life-history hypotheses cannot yet be tested on bats because of a lack of relevant data. In particular, the age-specific data on fecundity and survival required for cohort life tables are almost completely lacking for bats, and this underlies much of the poor representation of bats in some recent general analyses of mammalian life-history patterns (Purvis and Harvey 1995; Ricklefs 1998; Sibly et al. 1997). Data regarding fecundity, especially the proportion of females that reproduce during each reproductive season, and survival rates rather than just maximum longevity, are needed for a range of species from different families, occupying temperate and tropical areas, and with different diets and body sizes. In this respect, the Phyllostomidae should be particularly enlightening, given their ecological diversity.

Long-term data on known individuals are also required for bats. Only with such data could we address questions regarding costs of reproduction and their influence on subsequent reproduction and survival. Associated with such observations, experiments manipulating the costs of reproduction (e.g., Boutin 1990; Dijkstra et al. 1990) will ultimately be needed to rigorously test predictions stemming from the various hypotheses outlined in this chapter.

Scenario for the Evolution of Bat Life Histories

In summary, several hypotheses propose explanations for the difference in life-history traits of bats compared to shrews and other small mammals (table 5.3). Of these hypotheses, several are not supported by our analyses or by previous data. Of the others, several cannot be tested adequately because of insufficient data. In particular, although the low extrinsic mortality hypothesis seems to be supported by longevity data for bats and shrews, its specific predictions will be testable only with age-specific mortality data. Likewise, a test of the calcium constraint hypothesis requires assessment of the calcium demand by bats and shrews during reproduction and comparison of that demand to calcium availability in their diets. Finally, the polyovulation constraint hypothesis would be supported if taxa in basal lineages of the Chiroptera were shown to ovulate only one ovum at a time.

We propose that the combination of low extrinsic mortality, a constraint on the number of ova produced at one time, and the limited availability of resources explains the unusual life histories of Chiroptera relative to other

Table 5.3. Summary of hypotheses concerning
the evolution of bat life histories

Hypothesis	Status
Low extrinsic mortality	Supported
Resource constraint:	
Cost of flight	Not supported
Wing-loading constraint	Not supported
Size at independence:	
Calcium constraint	Lack of data
Offspring size/number:	
Policy-shift	Not supported
Reproductive constraint:	
Polyovulation constraint	Supported

mammals of similar size. In particular, we suggest the following evolutionary scenario.

In the transition from a terrestrial ancestor to a flying mammal, the unique stresses placed on the skeleton by flight required large size at independence, thus prolonging dependence on the mother. This, coupled with limited availability of resources for insectivorous mammals flying at night, meant that the ancestors of bats could only afford a low reproductive output, and they produced only one ovum at a time. This limited output per breeding attempt became possible because flight reduced extrinsic mortality, but the lack of polyovulation constrained selection for increased litter size during much of the chiropteran radiation. Once polyovulation evolved in the ancestors of vespertilionids and antrozoids, litter size remained constrained by the limited availability of calcium relative to the demands made by the need to produce large young at independence. Hence, the evolutionary options for bat life histories depend on low mortality. For example, migration, which likely involves considerable mortality risk, is far less common in bats than in birds, limiting the diversity of bats in temperate areas. In contrast, hibernation is relatively common in bats and may be possible because it further reduces mortality rates relative to non-hibernators.

Conclusions

Although bats have been relatively ignored in previous analyses of mammalian life-history variation, we believe that some of the same hypotheses applied to other orders, in combination with unique aspects of the biology of bats, can explain the evolution of bat life histories in general and at least some of the variation among bats. In particular, we suggest that while resource availability may be generally low for bats, thereby limiting reproductive output, flight reduces extrinsic mortality and thus permits reduced reproductive effort per breeding event. In addition, however, we view the evolution of polyovulation in the Vespertilionidae as a key adaptation, at least partially re-

a

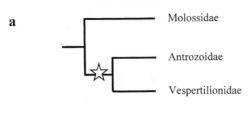

b

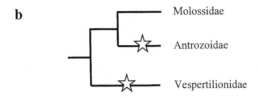

c

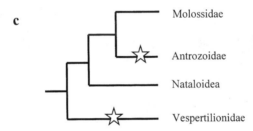

Figure 5.8. Three proposed relationships among bat taxa relevant to the polyovulation constraint hypothesis (*a* and *b* from Simmons and Geisler 1998; *c* from Simmons 1998). Stars represent transitions from litters of one offspring to more variable litter sizes. Figure 5.8*a* represents the phylogeny used in this chapter.

sponsible for the diversity of this family. Vespertilionids make up approximately one-third of all bat species and have the greatest geographical range of any of the families. This range and diversity is in spite of a uniformly small body size and an almost uniformly insectivorous diet. Although the ability to hibernate has allowed the family to radiate in temperate regions, the diversity is still greatest in tropical areas. The flexibility in life-history traits that polyovulation allows is a vital component of that diversity.

Our view of polyovulation is relevant to the phylogenies proposed by Simmons (1998) and by Simmons and Geisler (1998). While the phylogeny we have used requires only a single transition from an ancestral monotocous condition (fig. 5.8*a*), the other two proposed phylogenies require either two independent evolutionary events in the ancestors of antrozoids and vespertil-ionids or loss of polyovulation in one or more clades (fig. 5.8*b*, *c*). In terms of this trait, a phylogeny with Antrozoidae and Vespertilionidae as sister groups is most parsimonious. Determining whether polyovulation occurs in the Molossidae would test the validity of these phylogenies.

Acknowledgments

We thank F. Bonaccorso, K. Jones, C O'Donnell, P. Racey, R. Utzurrum, and M. Vonhof for providing unpublished data and manuscripts and M. Brigham, B. Fenton, L. Hollis, K. Jones, C. Lausen, K. Patriquin and two anonymous re-viewers for comments that significantly improved the chapter. Funding was provided from research grants to Barclay and Harder from the Natural Sci-ences and Engineering Research Council of Canada.

Literature Cited

Allaine, D., D. Pontier, J. M. Gaillard, J. D. Lebreton, J. Trouvilliez, and J. Clobert. 1987. The relationship between fecundity and adult body weight in homeotherms. Oe-cologia, 73:478–480.

Austad, S. N., and K. E. Fischer. 1991. Mammalian ageing, metabolism, and ecology: ev-idence from the bats and marsupials. Journal of Gerontology, 46:B47–B53.

Barclay, R. M. R. 1989. The effect of reproductive condition on the foraging strategy of female hoary bats (*Lasiurus cinereus*). Behavioral Ecology and Sociobiology, 24:31–37.

Barclay, R. M. R. 1994. Constraints on reproduction by flying vertebrates: energy and calcium. American Naturalist, 144:1021–1031.

Barclay, R. M. R. 1995. Does energy or calcium availability constrain reproduction by bats? Symposia of the Zoological Society of London, no. 67:245–258.

Barclay, R. M. R., and R. M. Brigham. 1991. Prey detection, dietary niche breadth, and body size in bats: why are aerial insectivorous bats so small? American Naturalist, 137:693–703.

Barrett, S. C. H., L. D. Harder, and A. C. Worley. 1996. The comparative biology of pol-lination and mating in flowering plants. Philosophical Transactions of the Royal Society of London B, 351:1271–1280.

Bell, G. P., G. A. Bartholomew, and K. A. Nagy. 1986. The role of energetics, water econ-omy, foraging behavior, and geothermal refugia in the distribution of the bat, *Macro-tus californicus*. Journal of Comparative Physiology B, 156:441–450.

Bernard, R. T. F. 1985. Reproduction in the Cape horseshoe bat (*Rhinolophus capensis*) from South Africa. South African Journal of Zoology, 20:129–135.

Bernard, R. T. F., and A. Davison. 1996. Does calcium constrain reproductive activity in insectivorous bats? some empirical evidence for Schreibers' long-fingered bat (*Miniopterus schreibersii*). South African Journal of Zoology, 31:218–220.

Boutin, S. 1990. Food supplementation experiments with terrestrial vertebrates: pat-terns, problems and the future. Canadian Journal of Zoology, 68:203–220.

Brigham, R. M., H. D. J. N. Aldridge, and R. L. Mackey. 1992. Variation in habitat use and prey selection by Yuma bats, *Myotis yumanensis*. Journal of Mammalogy, 73:640–645.

Burnett, C. D., and T. H. Kunz. 1982. Growth and age estimation in *Eptesicus fuscus* and comparison with *Myotis lucifugus*. Journal of Mammalogy, 63:33–41.

Charvov, E. L. 1991. Evolution of life history among female mammals. Proceedings of the National Academy of Sciences of the USA, 88:1134–1137.

Charnov, E. L., J. F. Downhower, and L. P. Brown. 1995. Optimal offspring sizes in small litters. Evolutionary Ecology, 9:57–63.

Corbet, G. B., and S. Harris, eds. 1991. Handbook of British Mammals. Blackwell Scientific Publications, Oxford.

Davis, W. H., and H. B. Hitchcock. 1995. A new longevity record for the bat *Myotis lucifugus*. Bat Research News, 36:6.

Dijkstra, C., A. Bult, S. Bijlsma, S. Daan, T. Meijer, and M. Zijlstra. 1990. Brood size manipulations in the kestrel (*Falco tinnunculus*): effects on offspring and parental survival. Journal of Animal Ecology, 59:269–285.

Ebert D. 1994. Fractional resource allocation into few eggs: *Daphnia* as an example. Ecology, 75:568–571.

Feldhamer, G. A., L. C. Drickamer, S. H. Vessey, and J. F. Merritt. 1998. Mammalogy: Adaptation, Diversity, and Ecology. WCB/McGraw-Hill, Boston.

Fenton, M. B., I. L. Rautenbach, S. M. Smith, C. M. Swanepoel, J. Grossell, and J. van Jaarsveld. 1994. Bats and raptors: threats and opportunities. Animal Behaviour, 48:9–18.

Gaisler, J., V. Hanak, and J. Dungel. 1979. A contribution to the population ecology of *Nyctalus noctula* (Mammalia, Chiroptera). Acta Scientiarum Naturalium Academiae Scientiarum Bohemoslovacae Brno, 13:1–38.

Garland, T., Jr. 1983. Scaling the ecological cost of transport to body mass in terrestrial mammals. American Naturalist, 121:571–587.

Gopalakrishna, A., A. Madhavan, and N. Badwaik. 1991. Breeding biology of the Indian leaf-nosed bat, *Hipposideros speoris* (Schneider) with notes on its ecology in Marathwada, Maharashtra state, India. Mammalia, 55:275–283.

Greenhall, A. M., G. Joermann, and U. Schmidt. 1983. Desmodus rotundus. Mammalian Species, 202:1–6.

Harvey, P. H., and M. D. Pagel. 1991. The Comparative Method in Evolutionary Biology. Oxford University Press, Oxford.

Harvey, P. H., and A. F. Read. 1988. How and why do mammalian life histories vary? Pp. 213–232 *in:* Evolution of Life Histories of Mammals (M. S. Boyce, ed.). Yale University Press, New Haven, Conn.

Harvey, P. H., and R. M. Zammuto. 1985. Patterns of mortality and age at first reproduction in natural populations of mammals. Nature, 315:319–320.

Hayssen, V., and T. H. Kunz. 1996. Allometry of litter mass in bats: maternal size, wing morphology, and phylogeny. Journal of Mammalogy, 77:476–490.

Hayssen, V., and R. C. Lacy. 1985. Basal metabolic rates in mammals: taxonomic differences in the allometry of BMR and body mass. Comparative Biochemistry and Physiology A, 81:741–754.

Hayssen, V., A. van Tienhoven, and A. van Tienhoven. 1993. Asdell's Patterns of Mammalian Reproduction. Cornell University Press, Ithaca, N.Y.

Herrera, C. M. 1987. Vertebrate-dispersed plants of the Iberian Peninsula: a study of fruit characteristics. Ecological Monographs, 57:305–331.

Holmes, D. J., and S. N. Austad. 1994. Fly now, die later: life-history correlates of gliding and flying in mammals. Journal of Mammalogy, 75:224–226.

Holmes, D. J., and D. Sherry. 1997. Selected approaches to using individual variation for understanding mammalian, life-history evolution. Journal of Mammalogy, 78: 311–319.

Holroyd, S. L. 1993. Influences of some extrinsic and intrinsic factors on reproduction by big brown bats (Eptesicus fuscus) in southeastern Alberta. M.Sc. Thesis, University of Calgary, Calgary.

Hosken, D. J. 1997. Reproduction and the female reproductive cycle of Nyctophilus geoffroyi and N. major (Chiroptera: Vespertilionidae) from south-western Australia. Australian Journal of Zoology, 45:489–504.

Hoying, K. M., and T. H. Kunz. 1998. Variation in size at birth and post-natal growth in the insectivorous bat Pipistrellus subflavus (Chiroptera: Vespertilionidae). Journal of Zoology (London), 245:15–27.

Humphrey, S. R., and J. B. Cope. 1977. Survival rates of the endangered Indiana bat, Myotis sodalis. Journal of Mammalogy, 58:32–36.

Innes, D. G. L. 1994. Life histories of the Soricidae: a review. Pp. 111–136 in: Advances in the Biology of Shrews (J. F. Merritt, G. L. Kirkland, Jr., and R. K. Rose, eds.). Special Publication of Carnegie Museum of Natural History, 18. Carnegie Museum of Natural History, Pittsburgh.

Isaac, S. S., and G. Marimuthu. 1996. Postnatal growth and age estimation in the Indian pygmy bat Pipistrellus mimus. Journal of Mammalogy, 77:199–204.

Isaac, S. S., G. Marimuthu, and M. K. Chandrashekaran. 1994. Fecundity in the Indian pygmy bat (Pipistrellus mimus). Journal of Zoology (London), 234:665–668.

Jones, K .E., and A. MacLarnon. 2001. Bat life histories: testing models of mammalian life-history evolution. Evolutionary Ecology Research, 3:465–476.

Jones, K. E., and A. Purvis. 1997. An optimum body size for mammals? comparative evidence from bats. Functional Ecology, 11:751–756.

Jurgens, K. D., and J. Prothero. 1987. Scaling of maximal lifespan in bats. Comparative Biochemistry and Physiology A, 88:361–367.

Keeler, J. O., and E. H. Studier. 1992. Nutrition in pregnant big brown bats (Eptesicus fuscus) feeding on June beetles. Journal of Mammalogy, 73:426–430.

Keen, R., and H. B. Hitchcock. 1980. Survival and longevity of the little brown bat (Myotis lucifugus) in southeastern Ontario. Journal of Mammalogy, 61:1–7.

Kihlstrom, J. E. 1972. Period of gestation and body weight in some placental mammals. Comparative Biochemistry and Physiology A, 43:673–679.

Kirkwood, T. B. L., and M. R. Rose. 1991. Evolution of senescence: late survival sacrificed for reproduction. Philosophical Transactions of the Royal Society of London B, 332:15–24.

Koehler, C. E. 1991. The reproductive ecology of the hoary bat (Lasiurus cinereus) and its relation to litter size variation in vespertilionid bats. M.Sc. Thesis, University of Calgary, Calgary.

Koehler, C. E., and R. M. R. Barclay. 2000. Postnatal growth and breeding biology of the hoary bat (Lasiurus cinereus). Journal of Mammalogy, 81:234–244.

Kozlowski, J., and J. Weiner. 1997. Interspecific allometries are by-products of body size optimization. American Naturalist, 149:352–380.

Krishna, A. 1985. Reproduction in the Indian pigmy pipistrelle bat, *Pipistrellus mimus.* Journal of Zoology (London), 206:41–51.

Kunz, T. H. 1973. Population studies of the cave bat (*Myotis velifer*): reproduction, growth and development. Occasional Papers, Museum of Natural History, University of Kansas, 15:1–43.

Kunz, T. H. 1974. Reproduction, growth, and mortality of the vespertilionid bat, *Eptesicus fuscus,* in Kansas. Journal of Mammalogy, 55:1–13.

Kunz, T. H. 1987. Postnatal growth and energetics of suckling bats. Pp. 395–420 *in:* Recent Advances in the Study of Bats (M. B. Fenton, P. Racey, and J. M. V. Rayner, eds.). Cambridge University Press, Cambridge.

Kunz, T. H., and W. H. Hood. 2000. Parental care and postnatal growth in the Chiroptera. Pp. 415–468 *in:* Reproductive biology of bats (E. G. Crichton and P. H. Krutzsch, eds.). Academic Press, New York.

Kunz, T. H., and S. R. Robson. 1995. Postnatal growth in the Mexican free-tailed bat (*Tadarida brasiliensis*): size at birth, growth rates and age-estimation. Journal of Mammalogy, 76:769–783.

Kunz, T. H., and A. A. Stern. 1995. Maternal investment and post-natal growth in bats. Symposia of the Zoological Society of London, no. 67:123–138.

Kunz, T. H., J. O. Whitaker, Jr., and M. D. Wadonoli. 1995. Dietary energetics of the insectivorous Mexican free-tailed bat (*Tadarida brasiliensis*) during pregnancy and lactation. Oecologia, 101:407–415.

Kunz, T. H., J. A. Wrazen, and C. D. Burnett. 1998. Changes in body mass and fat reserves in pre-hibernating little brown bats (*Myotis lucifugus*). Ecoscience, 5:8–17.

Kurta, A., G. P. Bell, K. A. Nagy, and T. H. Kunz. 1989. Energetics of pregnancy and lactation in free-ranging little brown bats (*Myotis lucifugus*). Physiological Zoology, 62:804–818.

Kurta, A., and T. H. Kunz. 1987. Size of bats at birth and maternal investment during pregnancy. Symposia of the Zoological Society of London, no. 57:79–106.

Kwiecinski, G. G., L. Krook, and W. A. Wimsatt. 1987. Annual skeletal changes in the little brown bat, *Myotis lucifugus lucifugus,* with particular reference to pregnancy and lactation. American Journal of Anatomy, 178:410–420.

Lehmann, J., L. Jenni, and L. Maumary. 1992. A new longevity record for the long-eared bat (*Plecotus auritus,* Chiroptera). Mammalia, 56:316–318.

Lloyd, D. G. 1987. Selection of offspring size at independence and other size-versus-number strategies. American Naturalist, 129:800–817.

Lord, R. D., F. Murdali, and L. Lazavo. 1976. Age composition of vampire bats (*Desmodus rotundus*) in northern Argentina and southern Brazil. Journal of Mammalogy, 57:573–575.

Maddison, W. P., and D. R. Maddison. 1992. MacClade 3: Interactive Analysis of Phylogeny and Character Evolution. Sinauer Associates, Sunderland, Mass.

McCullagh, P., and J. A. Nelder. 1989. Generalized Linear Models. 2d ed. Chapman & Hall, London.

Medawar, P. B. 1952. An Unsolved Problem of Biology. H. K. Lewis, London, 24 pp.

Medway, Lord. 1972. Reproductive cycles of the flat-headed bats *Tylonycteris pachypus* and *T. robustula* (Chiroptera: Vespertilionidae) in a humid equatorial environment. Zoological Journal of the Linnean Society, 51:33–61.

Millar, J. S. 1977. Adaptive features of mammalian reproduction. Evolution, 31: 370–386.

Millar, J. S., and G. J. Hickling. 1991. Body size and the evolution of mammalian life histories. Functional Ecology, 5:588–593.

Morales, J. C., and J. W. Bickham. 1995. Molecular systematics of the genus *Lasiurus* (Chiroptera: Vespertilionidae) based on restriction-site maps of the mitochondrial ribosomal genes. Journal of Mammalogy, 76:730–749.

Myers, P. 1978. Sexual dimorphism in size of vespertilionid bats. American Naturalist, 112:701–711.

Neter, J., W. Wassermann, and M. H. Kutner. 1990. Applied Linear Statistical Models. 3d ed. Irwin, Homewood, Ill.

Norberg, U. M. 1985. Flying, gliding and soaring. Pp. 129–158 *in:* Functional Vertebrate Morphology (M. Hildebrand, D. M. Bramble, K. F. Liem, and D. B. Wake, eds.). Harvard University Press, Cambridge, Mass.

Norberg, U. M., and M. B. Fenton. 1988. Carnivorous bats? Biological Journal of the Linnean Society, 33:383–394.

Norberg, U. M., and J. M. V. Rayner. 1987. Ecological morphology and flight in bats (Mammalia; Chiroptera): wing adaptations, flight performance, foraging strategy and echolocation. Philosophical Transactions of the Royal Society of London B, 316:335–427.

O'Connor, R. J. 1984. The Growth and Development of Birds. Wiley, New York.

Papadimitriou, H. M., S. M. Swartz, and T. H. Kunz. 1996. Ontogenetic and anatomic variation in mineralization of the wing skeleton of the Mexican free-tailed bat, *Tadarida brasiliensis*. Journal of Zoology (London), 240:411–426.

Partridge, L., and N. H. Barton. 1993. Optimality, mutation and the evolution of ageing. Nature, 362:305–311.

Pavey, C. R., A. K. Smith, and M. T. Mathieson. 1994. The breeding season diet of the powerful owl *Ninox strenua* at Brisbane, Queensland. Emu, 94:278–284.

Pomeroy, D. 1990. Why fly? the possible benefits for lower mortality. Biological Journal of the Linnean Society, 40:53–65.

Pow, C. S. T., and L. Martin. 1994. The ovarian-uterine vasculature in relation to unilateral endometrial growth in flying foxes (genus *Pteropus*, suborder Megachiroptera, Order Chiroptera). Journal of Reproduction and Fertility, 101:247–255.

Promislow, D. E. L., and P. H. Harvey. 1990. Living fast and dying young: a comparative analysis of life-history variation among mammals. Journal of Zoology (London), 220:417–437.

Purvis, A., and P. H. Harvey. 1995. Mammalian life-history evolution: a comparative test of Charnov's model. Journal of Zoology (London), 237:259–283.

Purvis, A., and P. H. Harvey. 1996. Miniature mammals: life-history strategies and macroevolution. Symposia of the Zoological Society of London, no. 69:159–174.

Racey, P. A. 1973. Environmental factors affecting the length of gestation in heterothermic bats. Journal of Reproduction and Fertility, Suppl., 19:175–189.

Racey, P. A. 1982. Ecology of bat reproduction. Pp. 57–104 *in:* Ecology of Bats (T. H. Kunz, ed.). Plenum Press, New York.

Racey, P. A., and A. C. Entwistle. 2000. Life history and reproductive strategies of bats. Pp. 363–414 *in:* Reproductive Biology of Bats (E. G. Crichton and P. H. Krutzsch, eds.). Academic Press, San Diego, Calif.

Racey, P. A., and J. R. Speakman. 1987. The energy costs of pregnancy and lactation in heterothermic bats. Symposia of the Zoological Society of London, no. 57: 107–125.

Rakhmatulina, I. K. 1972. The breeding, growth, and development of pipistrelles in Azerbaidzhan. Ekologiya (Moscow), 2:54–61.

Ransome, R. D. 1995. Earlier breeding shortens life in female greater horseshoe bats. Philosophical Transactions of the Royal Society of London B, 350:153–161.

Rasweiler, J. J. 1982. The contribution of observations on early pregnancy in the little sac-winged bat, *Peropteryx kappleri,* to an understanding of the evolution of reproductive mechanisms in monovular bats. Biology of Reproduction, 27:681–702.

Read, A. F., and P. H. Harvey. 1989. Life history differences among the eutherian radiations. Journal of Zoology (London), 219:329–353.

Ricklefs, R. E. 1979. Adaptation, constraint, and compromise, in avian postnatal development. Biological Reviews, 54:269–290.

Ricklefs, R. E. 1998. Evolutionary theories of aging: confirmation of a fundamental prediction, with implications for the basis and evolution of life span. American Naturalist, 152:24–44.

Roff, D. A. 1992. The evolution of life histories; theory and analysis. Chapman & Hall, New York.

Roverud, R. C., and M. A. Chappell. 1991. Energetic and thermoregulatory aspects of clustering behavior in the Neotropical bat *Noctilio albiventris.* Physiological Zoology, 64:1527–1541.

Rust, A. K. 1978. Activity rhythms in the shrews, *Sorex sinuosus* Grinnell and *Sorex trowbridgii* Baird. American Midland Naturalist, 99:369–382.

Rydell, J. 1990. Ecology of the northern bat *Eptesicus nilssoni* during pregnancy and lactation. Ph.D. Thesis, Lund University, Lund.

Sacher, G. A., and E. F. Staffeldt. 1974. Relation of gestation time to brain weight for placental mammals: implications for the theory of vertebrate growth. American Naturalist, 108:593–615.

Schmidly, D. J. 1991. Bats of Texas. Texas A&M University Press, College Station.

Sibly, R. M., D. Collett, D. E. L Promislow, D. J. Peacock and P. H. Harvey. 1997. Mortality rates of mammals. Journal of Zoology (London), 243:1–12.

Simmons, N. B. 1998. A reappraisal of interfamilial relationships of bats. Pp. 3–26 *in:* Bat Biology and Conservation (T. H. Kunz and P. A. Racey, eds.). Smithsonian Institution Press, Washington D.C.

Simmons, N. B., and J. H. Geisler. 1998. Phylogenetic relationships of *Icaronycteris, Archaeonycteris, Hassianycteris,* and *Palaeochiropteryx* to extant bat lineages, with comments on the evolution and foraging strategies in Microchiroptera. Bulletin of the American Museum of Natural History, 235:1–182.

Smith, C. C., and S. D. Fretwell. 1974. The optimal balance between size and number of offspring. American Naturalist, 108:499–506.

Sorensen, M. W. 1962. Some aspects of water shrew behavior. American Midland Naturalist, 68:445–462.

Speakman, J. R. 1995. Chiropteran nocturnality. Symposia of the Zoological Society of London, no. 67:187–201.

Stanley, R. G., and H. F. Linskens. 1974. Pollen Biology, Biochemistry, Management. Springer, Berlin.

Stearns, S. C. 1983. The impact of size and phylogeny on patterns of covariation in the life history traits of mammals. Oikos, 41:173–187.

Stearns, S. C. 1992. The Evolution of Life Histories. Oxford University Press, Oxford.

Stebbings, R. E. 1968. Measurements, composition and behaviour of a large colony of the bat *Pipistrellus pipistrellus*. Journal of Zoology (London), 156:15–33.

Strahan, R., ed. 1995. Mammals of Australia. Smithsonian Institution Press, Washington D.C.

Studier, E. H., and T. H. Kunz. 1995. Accretion of nitrogen and minerals in suckling bats, *Myotis velifer* and *Tadarida brasiliensis*. Journal of Mammalogy, 76:32–42.

Studier, E. H., and S. H. Sevick. 1992. Live mass, water content, nitrogen and mineral levels in some insects from south-central lower Michigan. Comparative Biochemistry and Physiology A, 103:579–595.

Swartz, S. M., M. B. Bennett, and D. R. Carrier. 1992. Wing bone stresses in free flying bats and the evolution of skeletal design for flight. Nature, 359:726–729.

Swift, S. M. 1980. Activity patterns of pipistrelle bats (*Pipistrellus pipistrellus*) in northeast Scotland. Journal of Zoology (London), 190:285–295.

Teeling, E. C., M. Scally, D. J. Kao, M. L. Romagnoli, M. S. Springer, and M. J. Stanhope. 2000. Molecular evidence regarding the origin of echolocation and flight in bats. Nature, 403:188–192.

Thomas, D. W., and A. G. Marshall. 1984. Reproduction and growth in three species of West African fruit bats. Journal of Zoology (London), 202:265–281.

Thomas, S. P. 1987. The physiology of bat flight. Pp. 75–99 *in:* Recent Advances in the Study of Bats (M. B. Fenton, P. Racey and J. M. V. Rayner, eds.). Cambridge University Press, Cambridge.

Turner, A. K. 1982. Timing of laying by swallows (*Hirundo rustica*) and sand martins (*Riparia riparia*). Journal of Animal Ecology, 51:29–46.

Tuttle, M. D. 1975. Population ecology of the gray bat (*Myotis grisescens*): factors influencing early growth and development. Occasional Papers, Museum of Natural History, University of Kansas, 36:1–24.

Tuttle, M. D., and D. Stevenson. 1982. Growth and survival of bats. Pp. 105–150 *in:* Ecology of Bats (T. H. Kunz, ed.). Plenum Press, New York.

Venable, D. L. 1992. Size-number trade-offs and the variation of seed size with plant resource status. American Naturalist, 140:287–304.

Webb, N. J., and C. R. Tidemann. 1996. Mobility of Australian flying-foxes, *Pteropus* spp. (Megachiroptera): evidence from genetic variation. Proceedings of the Royal Society of London B, 263:497–502.

Western, D., and J. Ssemakula. 1982. Life history patterns in birds and mammals and their evolutionary interpretation. Oecologia, 54:281–290.

Wilkinson, G. S. 1984. Reciprocal food sharing in the vampire bat. Nature, 308:181–184.

Wilkinson, L. C., and R. M. R. Barclay. 1997. Differences in the foraging behaviour of male and female big brown bats (*Eptesicus fuscus*) during the reproductive period. Ecoscience, 4:279–285.

Williams, G. C. 1957. Pleiotropy, natural selection and the evolution of senescence. Evolution, 11:398–411.

Wilson, D. E. 1971. Ecology of *Myotis nigricans* (Mammalia: Chiroptera) on Barro Colorado Island, Panama Canal Zone. Journal of Zoology (London), 163:1–13.

Wimsatt, W. A. 1975. Some comparative aspects of implantation. Biology of Reproduction, 12:1–40.

Wimsatt, W. A. 1979. Reproductive asymmetry and unilateral pregnancy in Chiroptera. Journal of Reproduction and Fertility, 56:345–357.

Wooton, J. T. 1987. The effects of body mass, phylogeny, habitat, and trophic level on mammalian age at first reproduction. Evolution, 41:732–749.

PART II

FUNCTIONAL ECOLOGY

Ecomorphology of Bats: Comparative and Experimental Approaches Relating Structural Design to Ecology

Sharon M. Swartz, Patricia W. Freeman, and Elizabeth F. Stockwell

Introduction

Interconnections between morphological design and function are central to biology; they underlie natural patterns in species distribution, phylogenetic diversification, and morphological specialization. At its core, ecomorphology explores the causal relationships between organismal design and behavioral performance and investigates how these relationships influence an organism's ability to exploit its environment. To the extent that we can understand mechanisms that dictate these relationships, we can gain broad insight into the ecology and evolution of species, higher-order clades, and ecological assemblages.

The past several decades have seen the beginning of integration of discoveries in the ecology and morphology of bats into an ecomorphological whole that promises to be greater than the sum of its parts. However, achieving this integration is challenging. Each of the elements of this interdisciplinary field is, in itself, a vast and complex subject. This book is a testament to the breadth and depth of ecological studies of bats. Morphological studies now extend beyond descriptive anatomy in both outlook and method, and incorporate, for example, aspects of genetics, physiology, solid and fluid mechanics, and developmental biology. There is also greater awareness that a particular morphology embodies not only the requirements of present life but also an organism's developmental and evolutionary history.

In ecomorphological analysis, one seeks to establish the nature and strength of the relationship between morphology and ecology. The relationship between an animal's structure and its interactions with its environment may itself be complicated and multifactorial. Morphology dictates an individual's performance limits and restricts its behavioral repertoires; regardless of habitat, a bat cannot fly faster or eat larger prey than its anatomy will allow. However, while morphology may constrain potential activities, the ecology of an animal is also strongly influenced by the local environment in which it functions day to day; flight speeds employed may be dictated primarily by the nature of the three-dimensional spatial environment, while prey selection may be driven by availability and abundance.

Part of the intrinsic challenge of the study of ecomorphology is that it is fundamentally synthetic. This complex subject draws on diverse conceptual approaches, integrates data from the field, the laboratory, and the museum, and takes into account both ecological and evolutionary timescales. Moreover, the volant, nocturnal lifestyles of bats present exceptional methodological challenges. Nonetheless, steadily increasing basic knowledge of bat biology along with a growing repertoire of conceptual and methodological approaches have set the stage for unprecedented innovations in the next few years.

Here, we examine recent advances in the study of the relationships among morphology, behavior, and ecology of bats. In this chapter, we provide one view of ecomorphology as a discipline, and highlight some case studies of particular relevance to understanding how the morphology of bats relates to their ecology, with special attention to the structure of the feeding apparatus, particularly the teeth and skull, in relation to dietary preferences, and the morphology of the wing in relation to flight performance. In bat ecomorphology, the feeding and flight apparatus have been the subjects of much study; we direct our focus to these areas, recognizing that other aspects of morphology influence behavior and ecology and equally merit further study.

We emphasize several themes: the central role of body size as a determinant of mechanical and ecological function; the importance of recognizing ontogenetic, intersexual, and other intraspecific variation; the need for integrative analyses that span field and laboratory; and the role of computer modeling in present and future studies. We offer our views concerning what kinds of information are most likely to lead to new insights and effective integration of the work of morphologists and ecologists in the future.

Correlational Approaches to Assessing Form and Its Ecological Significance
Background

One basic and widely applied approach to ecomorphological analysis assesses aspects of organismal structure as well as its ecological attributes and seeks to describe patterns of interrelationship among them. Comparisons of this kind may take many forms, depending on the manner in which one characterizes form, function, and/or their interrelationship. Assessment of morphology can range from a simple description of one or a few characters to exhaustive characterization of shape. In some cases, investigators evaluate structure with no preconceived notions about the relative importance of particular features (e.g., Birch 1997; Bookstein 1997; Bookstein et al. 1985; Strauss and Altig 1992). These studies seek general descriptors that may capture some critical aspect of form independent of functional analysis. Alternatively, theoretical or mechanical analyses and/or previous experimental studies may suggest that a given feature or suite of features are especially important.

Function or ecology may be quantified as categorizations of aspects of habitat, locomotion, food, and so on (e.g., Aldridge 1986b; Aldridge and Rautenbach 1987; Britton et al. 1997; Saunders and Barclay 1992). Studies of this kind then typically compare morphology and ecology among multiple taxa. Taxonomic sampling can focus specifically on phylogenetic lineages (e.g., Freeman 1981b; Saunders and Barclay 1992), sympatric communities (e.g., Bonaccorso 1979; Findley 1993), or dietary types (e.g., Fenton 1989, Norberg and Fenton 1988) or may seek to sample more broadly, even sampling the total diversity of bats (e.g., Norberg and Rayner 1987). Patterns of interrelationship among structural and ecological variables can then be described qualitatively or statistically.

Approaches that reveal patterns of correlation assess general patterns of relationship between morphology on the one hand and function or other ecological parameters on the other. Vaughan's (1959) research on wing anatomy and ecology in three bat species is one classic case in point. Analyses that extensively sample morphological and taxonomic diversity may reveal, for example, general associations among wing shape and habitat type (e.g., Findley et al. 1972; Norberg and Rayner 1987; Smith and Starrett 1979) or among robustness of jaws and types of foods eaten (Freeman 1981a, 1981b, 1998). These patterns, in and of themselves, give new insights into the relationship between structure and ecology. Additionally, patterns of strong correlation can be used to predict aspects of the ecology of rare or poorly studied taxa in those instances in which their anatomy is well described or is preserved in museum collections. These taxonomically comprehensive studies are particularly effective in identifying morphological and ecological extremes, in which relationships among form and ecology may be most readily described and interpreted. They also characterize subtle variation among more similar species. Correlational approaches can also generate specific and testable hypotheses that explore structural and behaviorial relationships more deeply. Results of basic descriptive and correlational analyses are long-lived and can be revisited repeatedly as new approaches and data become available.

Critical Issues in Correlational Analysis
Body Size Relationships in Ecomorphological Studies

Body size dictates most aspects of organismal ecology and physiology (see Speakman and Thomas, this volume, Willig et al., this volume). It is shaped by intra- and interspecific interactions and by the physical environment (e.g., La Barbera 1986; 1989; Peters 1983; Schmidt-Nielsen 1984). In bats, body size influences flight behavior, diet selection, roosting, reproductive behavior and physiology, and virtually all other aspects of biology. Understanding patterns of structure and behavior in relation to body size, then, is central to gaining an overall understanding of ecomorphology in the Chiroptera.

Numerous studies have demonstrated links among body size, basal metabolic rates, and the effectiveness of thermoregulation in mammals (Elgar and Harvey 1987; Hayssen 1984; Hayssen and Lacy 1985; McNab 1983, 1990; Nagy 1987). Among bats, similar patterns have been suggested by several studies, as reviewed by Speakman and Thomas (this volume). For example, populations of one species of blossom bats (*Macroglossus minimus*) in New Guinea vary in body mass along an altitudinal gradient. These metabolically flexible tropical bats are significantly larger at lower elevations (16.4 g, vs. 15.3 g at higher altitudes), and the lower but not the higher altitude populations can readily enter torpor; the same is true for *Syconycteris australis* (16.3 g in lowlands and 15.3 g at higher elevation (Bonaccorso and McNab 1997). Cotterill (1998) found that in two sympatric African insectivorous bats the larger *Rhinolophus hildebranti* has a longer prenatal development and a longer period of lactation than the smaller *R. simulator* (table 6.1). Because of the delayed but shortened embryonic development and shorter period of lactation in the latter species, pups are born closer to peak food availability in the wet season. However, body size interacts with many diverse influences on metabolic characteristics, including developmental patterns, life-history traits, environmental temperature variability, and diet. For example, *Melonycteris melanops*, the largest obligate nectarivore among bats (table 6.1), does not enter torpor at all body masses. As ongoing work extends studies to more species of diverse feeding habits and phylogenetic affinities, we may achieve a clearer view of the relative importance of body size as a determinant of metabolic traits.

Using body size to explain large-scale patterns in the ecomorphology of bats, Barclay and Brigham (1991) postulated that a combination of aerodynamic and sensory constraints limits body size in aerial insectivores. They proposed that increased body size decreases maneuverability and necessitates the detection of prey at greater distances. Hence, larger bats should employ relatively low-frequency echolocation pulses to reduce signal attenuation over long distances. As the frequency of echolocation signals decreases, however, the spatial resolution needed to detect small prey may be lost (see Jones and Rydell, this volume). The limited abundance of large prey, then, constrains dietary resources for large hawking bats (Barclay and Brigham 1991). This hypothesis has been tested in communities of insectivorous bats, and large species were found to take large prey, but small bats took only small food items (Aldridge and Rautenbach 1987, O'Neill and Taylor 1989). This positive relationship between body size and prey size is also consistent with Fenton's (1989) prediction that head length determines prey size for animal-eating bats. Additional evidence suggests that large, aerial insectivorous bats may not necessarily be limited to large prey. Aldridge and Rautenbach (1987) showed that large bats ate insects of a broad range of sizes and that there was a significant association between foraging habitat and the prey type. Consider-

Table 6.1. Body masses for species mentioned in text

Species	Body Mass (g)[a]
Antrozous pallidus	20.3
Artibeus jamaicensis	46.1
Artibeus lituratus	62.8
Carollia	10–25.0[b]
Carollia perspicillata	18.6
Carollia castanea	14.7
Corynorhinus townsendi	9.9
Desmodus rotundus	35.2
Eptesicus fuscus	17.6
Hipposideros commersoni	91.4
Lasiurus cinereus	25.6[c]
Lasiurus borealis	10.9
Leptonycteris curosoae	23.0
Macroglossus minimus	Lowlands 16.4
Macroglossus minimus	Highlands 15.3
Melonycteris melanops	53.3[d]
Myotis bechsteinii	8.4
Myotis evotis	4–11.0
Myotis lucifugus	7.9
Myotis volans	5–10.0
Myotis yumanensis	5.1
Nyctalus noctula	27.9
Phyllostomus hastatus	98.0
Pipistrellus pipistrellus	4.5
Plecotus auritus	7.5
Pteronotus parnellii	12–24.0
Pteropus giganteus	1175.0
Pteropus poliocephalus	800.0
Pteropus scapulatus	358.0
Rhinolophus ferrumequinum	19.0
Rhinolophus hildebranti	25.4
Rhinolophus simulator	6.0
Syconycteris australis	Lowlands 16.3
Syconycteris australis	Highlands 15.3
Tadarida brasiliensis	13.5

[a] Means or ranges derived from Silva and Downing (1995), from Freeman's unpublished database of diet in bats, and from references in text.

[b] Adult range for *Carollia*.

[c] Adult mass in non-Hawaiian *L. cinereus*.

[d] *Melonycteris melanops* adults can range to 63 g.

ation of the relationship between body size and flight performance highlights the need to develop approaches that allow reliable, consistent assessments of maneuverability.

In evaluating relationships between aspects of body size and prey choice, assessing the validity of the hypothesis that small prey are unavailable to large bats calling at relatively low frequencies remains important. Recent work shows that the strength of echoes from insects is independent of the frequency of the echolocation pulse (Waters et al. 1995). *Nyctalus noctula*, a large aerial insectivore (table 6.1), includes a substantial proportion of small dipterans in its diet (Jones 1995). Thus, body size may exert a broad influence on behavior and/or morphology of insectivorous bats that can be observed in large-scale, multi-species comparisons, but care must be exercised to isolate the effects of body size per se from those of other factors that may be correlated with body size to varying degrees.

Body size exerts a significant influence on foraging behavior in frugivores and nectarivores, as well as insectivores. Body size influences foraging behavior, albeit in somewhat different ways, in both frugivorous megachiropteran and phyllostomid microchiropteran lineages (Fleming 1993). Phyllostomids have relatively small foraging ranges, highly maneuverable flight, and highly selective diets; megachiropterans are typically less maneuverable, long-distance fliers, and dietary generalists. In phyllostomids, body size correlates with size of figs eaten, although large-bodied species will occasionally select small figs (Kalko et al. 1996). In contrast, there is no relationship between body size and size of fruit selected in megachiropteran frugivores (Dumont, this volume; Kalko et al. 1996). Body size also significantly influences diet and habitat use in bats that occur together in both wet and dry tropical communities in Costa Rica (Fleming 1991). Within the genus *Carollia*, as body size increases, the proportion of *Piper* fruits in the diet decreases, the proportion of large fruits in the diet increases, and time spent feeding decreases (table 6.1). In contrast, *Leptonycteris curasoae*, a large-bodied phyllostomid characterized by high wing loading, has wing morphology and foraging behavior convergent on that of megachiropterans (table 6.1). Unlike most phyllostomids that forage in mesic habitats, it traverses long distances across arid or semiarid environments to feed on flowers and fruits of three species of columnar cacti (Fleming 1993; Fleming and Eby, this volume, Sahley et al. 1993).

Richards (1995) teased out subtle results regarding body size and diet in 13 species of Australian megachiropterans. He grouped species into five feeding categories and found that there were large (>300 g) and small (<60 g) fruit and nectar specialists but only large-bodied generalists (table 6.1). Each group contains both an abundant species and one or more rare species. *Pteropus scapulatus*, with its reduced dentition, is indeed a specialized nectarivore (Freeman 1995). The large nectarivorous species are highly mobile and have wide distributions correlated with the distribution and diversity of eucalyptus. They are highly sensitive to olfactory cues and can respond quickly to mass flowering. These discoveries have implications for coevolution of bats and native forests and for the conservation of both (see Racey and Entwistle, this volume).

Intraspecific Patterns in Ecomorphological Studies

An important element in ecomorphological thinking is the recognition that if interspecific morphological variation is associated with variation in behavior and ecology, then morphological variation within species might also have behavioral consequences. For example, wing morphology varies not only with body size but also between sexes and among developmental stages within single species. This variation influences wing loading, aspect ratio, and mass distribution, which, in turn, affect flight performance characteristics such as turning ability, speed, and metabolic cost (e.g., Adams 1996, 1997; Hughes et al. 1989, 1995; Jones and Kokurewicz 1994). For example, wing loading in developing horseshoe bats, *Rhinolophus ferrumequinum,* decreases with age as length of the hand wing increases disproportionately relative to length of the arm wing. Ontogenetic change in the relative lengths of these two regions of the wing can thus modify flight efficiency with age (Hughes et al. 1995). As wing development progresses in *Pipistrellus pipistrellus,* wingbeat frequency decreases. This basic shift in flight kinematics may be driven by ontogenetic changes in wing moment of inertia as animals mature and increase in body mass and by the maturation of wing musculature (Hughes et al. 1989).

Ontogenetic variation in wing shape in *Myotis lucifugus* also appears to be ecologically significant (Adams 1996, 1997). During postnatal growth, as wing size increases, aspect ratio increases and wing loading decreases; simultaneously, growing bats forage in increasingly cluttered habitats. Interestingly, adult bats forage in a variety of habitats but shift to more cluttered habitats when juveniles became volant. Fecal samples also show a shift in diet with age, suggesting that adults may selectively limit competition with juveniles (Adams 1997). In this species, the greatly varying rates of muscle development may also be functionally or ecologically significant; the primary postural muscles of the hind limb are more mature at birth and reach adult fiber-type characteristics far more rapidly than do flight muscles (Powers et al. 1991).

In female bats, wing loading and flight kinematics can also vary with reproductive condition. For example, in *Phyllostomus hastatus,* a large phyllostomid, wing loading changes with seasonal changes in body mass in females (Stern et al. 1997; table 6.1). Wing loading and body mass also change in *Pipistrellus pipistrellus* in different reproductive conditions, and the highest wing loadings are seen in pregnant females (Webb et al. 1992). Lactating female *P. pipistrellus* that may be under metabolic stress and/or may have lower mass than nonreproductive females had lower wingbeat frequencies than either pregnant or postlactating bats (Hughes and Rayner 1993). These findings may have important implications for habitat use and prey selection, especially if pregnant females switch to less maneuverable prey types (Aldridge and Brigham 1988) or if power requirements for flight increase during pregnancy, as one might expect when wing loading increases (Hughes and Rayner 1991 1993).

Kalcounis and Brigham (1995) found that wing loading was a significant predictor of habitat use by *Myotis lucifugus,* regardless of age or reproductive status. Individuals with higher wing loading foraged in less cluttered habitat, suggesting that wing loading rather than dietary needs imposed by reproductive condition constrained foraging behavior. In *Myotis yumanensis*, Brigham et al. (1992) found no intraspecific differences in diet or habitat use among four reproductive and age classes. Clearly, we have only begun to assess the degree to which intraspecific morphological variation relates to ecology; this subject certainly merits further study.

Case Studies

Contemporary researchers have contributed to our understanding of the complex patterns of interrelationship among ecological and morphological characteristics of bats. These analyses have elucidated many anatomical systems in numerous taxa and continue to provide the primary data for broad generalizations. Here, we select two groups of analyses that illustrate some of the diversity of current study of bat ecomorphology. These case studies represent only a small fraction of the ongoing work in this area but highlight the kinds of results that derive from this approach.

Specialization of the Musculoskeletal System of the Desmodontine Hind Limb

It has long been recognized that the dimensions of bones vary in relationship to body size (Alexander et al. 1979; Galilei 1637; McMahon 1973; Schmidt-Nielsen 1984). Among mammals, and among subgroupings within mammals, linear dimensions such as length and diameter of the long bones typically change in proportion to (body mass)$^{0.33}$ (e.g., Biknevicius 1993; Demes and Jungers 1993; Alexander et al. 1979). Given that body mass is proportional to body volume, in turn a function of linear dimensions to the third power, these results demonstrate that, by and large, the shape of bones changes little in relation to body size, a pattern often designated as geometric similarity or isometry (Swartz and Biewener 1992). General patterns of scaling can serve as background descriptions against which specializations of bone shape deviations can be discerned.

Within bats, the dimensions of the long bones of the limbs generally demonstrate geometric similarity, but bats are distinctive in comparison to nonvolant mammals in several ways (Swartz 1997, 1998). If the evolution of flight is associated with reduced mechanical importance of the hind limbs for quadrupedal locomotion, and if the construction and transport of bone tissue requires a significant input of metabolic or nutritional resources, we predict that selection will favor reduction in mass of the hind limb skeleton. Indeed, hind limb bones of bats are clearly reduced in size and strength in comparison to those of nonvolant mammals (Howell and Pylka 1977; Swartz 1997, 1998; fig. 6. 1). Along with this morphological specialization for the ecology of flight, bats and their closest gliding relatives have acquired distinctive specializations of the tendons and tendon sheaths of the digits of the foot (Bennett 1993;

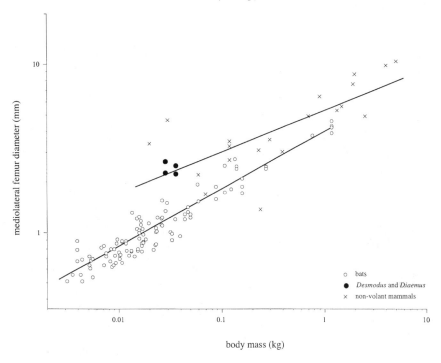

Figure 6.1. Log-log plot of femur diameter versus body mass of diverse bats and related and/or similarly sized nonvolant mammals. Vampire bats clearly cluster with nonvolant mammals and are distinctive in relation to bats of their body mass. The great robusticity of the desmodontine femur is likely related to their distinctive pattern of terrestrial locomotion (modified from Swartz 1997).

Quinn and Baumel 1993; Schutt 1993; Simmons and Quinn 1994). Structural modifications of adjacent surfaces of the long digital flexor tendons and their sheaths produce a ratchet-like, passive locking mechanism that may reduce or eliminate the need for activity in foot musculature in hanging postures. This interlocking of tendons and their sheaths, however, likely reduces the capacity for rapid ankle and toe flexion and extension.

The vampire bats, genera *Desmodus*, *Diaemus*, and *Diphylla*, composing the subfamily Desmodontinae, differ from all other bats in the structure of the hind limb and in their highly specialized feeding ecology (Schutt 1998). To approach and contact warm-blooded prey without detection, blood-feeding bats employ locomotor behaviors that are unusual in bats, moving stealthily across the ground and/or climbing rapidly. The evolution of this lineage clearly involved a fundamental ecological shift that has profoundly altered the mechanical demands placed on the hind limb. Terrestrial or arboreal locomotion, in contrast to flight, favor hind limb morphology in which digital tendons can slide freely in their sheaths throughout a large range of motion and in which the structural strength of the hind limb bones is increased relative to that of other bats. Indeed, recent work has directly documented the large forces experienced by the hind limbs of vampire bats during jumping (Schutt et al. 1997). Thus, although bats are unique among nonaquatic mammals in the degree of reduction of the hind limb bones, desmodontines alone among bats have reversed this trend and are distinctly characterized by remarkably robust femora and tibiae (fig. 6. 1), and the loss of the characteristic bat digital foot tendon-locking mechanism (Howell and Pylka 1977; Schutt 1993; Swartz 1997).

Chiropteran Craniodental Structure and Feeding Ecology

Bats exploit a wider range of food types than any other mammalian order. Teeth and jaws are the morphological locus of direct interaction with food, and the relationship between dietary diversity and craniodental structure has been the subject of a series of ecomorphological analyses (e.g., Dumont 1997a, 1997b, 1999; Freeman 1979, 1981a, 1981b, 1984, 1988, 1992, 1995, 1998, 2000). Studies of many species from diverse dietary and taxonomic groups have focused on a variety of components of the skull and dentition. Amid the great structural diversity in the teeth and jaws of bats, this approach has uncovered a number of important generalizations.

The structure of the teeth and their skeletal supports are relatively consistent among insectivorous bats but are distinctive in groups that have more specialized diets. For example, carnivorous phyllostomids, megadermatids, and nycterids have undergone evolutionary transitions from eating "a hard-covered package with soft insides [insects, especially beetles] to eating a soft-covered package with hard insides [vertebrates]" (Freeman, 1984). This distinctive feeding ecology, independently evolved several times, is correlated with a suite of morphological specializations. Carnivorous bats possess an elongated metastylar ridge on the upper molars and an enlargement of the reciprocally interlocking protoconids. These features have long been recognized as correlates of flesh eating and the ability to slice rather than crush food items in diverse mammalian groups. Carnivorous bats also possess lower molars with small talonid areas, relative to trigonid areas, and a relatively increased total area of the upper molar row. Both may be features related to improved slicing at the expense of crushing effectiveness (Freeman 1998, 2000).

Carnivorous bats also possess relatively thin mandibles, large brains, large pinnae, negatively tilted basicranial axes, and large body mass in comparison to their purely insectivorous relatives (Freeman 1984; fig. 6. 2). The morphological pattern observed in carnivorous species is also observed to lesser degrees among insectivorous bats that eat primarily soft-bodied insects, for example, moths, in contrast to those that specialize on relatively hard or tough prey, for example, beetles (Freeman 1979, 1981a). Bats that eat tough prey often

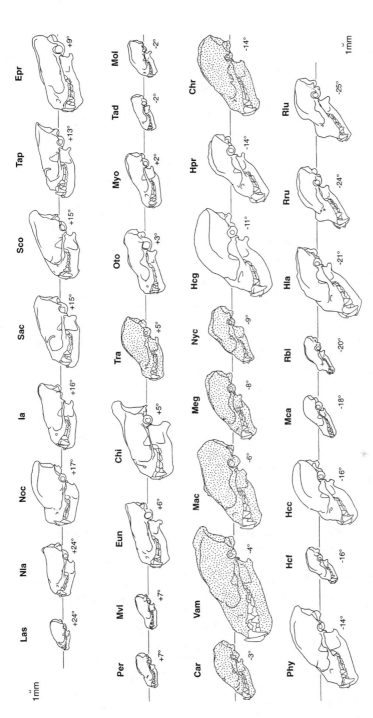

Figure 6.2. Bats are arranged so that the skull's basicranial axis is aligned horizontally (modified from Freeman 1984). Oral emitting bats have a positive or only slightly negative tilt of the head, and nasal-emitting bats have a negative tilt (Freeman 1984). Carnivorous bats from three families are shaded. *Trachops* may not be an obligate carnivore (Freeman 1984). Abbreviations of species are alphabetic within phylogenetic family order: *Peropteryx kappleri* (Per), *Saccolaimus peli* (Sac), *Taphozous nudiventris* (Tap); *Nycteris grandis* (Nyc); *Cardioderma cor* (Car), *Macroderma gigas* (Mac), *Megaderma lyra* (Meg); *Rhinolophus blasii* (Rbl), *R. luctus* (Rlu), *R. rufus* (Rru); *Hipposideros c. commersoni* (Hcc), *H. c. gigas* (Hcg), *H. lankadiva* (Hla), *H. pratti* (Hpr), *H. ruber* (= *H. caffer*, Hcf); *Noctilio leporinus* (Noc); *Chrotopterus auritus* (Chr), *Macrotus californicus* (Mca), *Phyllostomus hastatus* (Phy), *Trachops cirrhosus* (Tra), *Vampyrum spectrum* (Vam); *Ia io* (Ia), *Lasiurus borealis* (Las), *Myotis myotis* (Myo), *M. velifer* (Mvl), *Nyctalus lasiopterus* (Nla), *Scotophilus nigrita gigas* (Sco); *Chetromeles torquatus* (Chi), *Eumops perotis* (Epr), *E. underwoodi* (Eun), *Molossus molossus* (Mol), *Otomops martiensseni* (Oto), and *Tadarida brasiliensis* (Tad).

have robust skulls with thick dentaries, elongated canines, and short, wide faces (fig. 6. 3).

Skulls of oral-emitting insectivorous bats that eat tough prey show a convergence toward felids and hyaenids (Freeman 1984, 2000). These independent lineages share short, wide faces that bring the canines close to the fulcrum of the jaw and allow increased volume of the masseter muscle and, in some cases, the temporalis as well; the two large jaw-closing muscles are the primary determinants of bite force magnitude. In contrast, the nasal-emitting beetle specialist, *Hipposideros commersoni,* has a vertically expanded but thin dentary and taller sagittal crest that expands the skull vertically but not in breadth (Freeman 2000; fig. 6.3).

Fruit-eating microchiropteran bats have wide palates and faces—associated, potentially, with the ability to remove large chunks effectively from relatively large fruits and/or the ability to transport large fruits—and small stylar shelf areas that create a rim on the labial aspect of basin-like molars (Freeman 1988). Rimming the molars produces an efficient cutting edge that surrounds the entire perimeter of palate, with the rim of the flattened lower molars nesting inside that of the upper teeth (Freeman 1988). Molars with shallow basins surrounded by a rim are also typical of other mammalian frugivores, particularly megachiropterans, marsupials, and primates (Freeman 1988, 1995, Lucas 1979; Rosenberger and Kinzey 1976; Slaughter 1970). Relatively flat molars in the rear of the mouth of frugivores crush in a mortar-and-pestle-like arrangement; rims on the lower molars make up the "pestles" that are driven into the basins interior to the surrounding rim on the upper molars to shear as "mortars." In this way, bats extract nutritious juice and separate it from nonnutritive pulp or indigestible seeds (Freeman 1988, Lucas and Luke 1984).

The most derived phyllostomid frugivores have small canines, possibly related to burying the face in the fruit and not having to transport fruit (e.g., *Centurio senex*). Other micro- and megachiropteran frugivores have longer canines that are likely useful in gripping fruit during transport (Freeman 1988). Frugivorous bats also differ from non-fruit-eating species by possessing greater tooth area in the anterior part of the tooth row. In contrast, more omnivorous bats have more equal allocation of occlusal area among the different kinds of teeth (Freeman 1988, 1998, 2000).

Employing a functional perspective to understand and interpret tooth and skull design of frugivores is not a trivial exercise. Fruits are a very broad class of food items and may vary substantially in the mechanical challenges they present to the feeding apparatus. The outside covering may be tough or brittle, soft or hard, and the inside of the fruit can range between hard and extremely soft, sometimes even in a single species, depending on the degree of ripeness (see Dumont, this volume). Overall fruit size relative to animal's jaw apparatus is also critical. Although it may be most appropriate botanically to

Gracile bat skulls   # Robust bat skulls

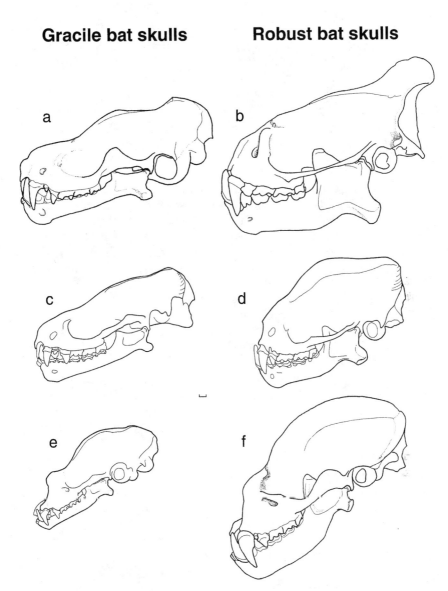

Figure 6.3. Bat skulls on the *left* are delicately built or gracile, and ones on the *right* are robust. Robust skulls have large cranial crests, fewer but larger teeth, and thicker dentaries (Freeman 1979, 1981a). Robustness or lack thereof is likely, in part, mediated by softness or hardness of items in the diet. *a, Otomops martiensseni*, an Old World molossid bat, has one of the most gracile skulls in its family. Recent data indicate it is a moth specialist (Rydell and Yalden 1997). *b, Cheiromeles torquatus*, an Old World bat and robust extreme, eats harder prey, such as grasshoppers and beetles (Freeman 1979, 1981b). *c, Nyctinomops macrotis*, a New World molossid with an extreme gracile skull, is a moth specialist (Freeman 1979, 1981a). *d, Molossus ater* is a New World molossid with an extreme skull that eats harder items, such as beetles (Freeman 1979, 1981a). *e, Rhinolophus blasii* is a known moth specialist (Freeman 1981a). *f, Hipposideros commersoni* is a known beetle specialist (Vaughan 1977).

classify fruits taxonomically or by their morphology (i.e., drupes, berries, pomes, aggregate and multiple fruits), these classification schemes are far less relevant to the frugivore than the size and mechanical nature of the fruit (Dumont 1999, this volume; Freeman 1988; Lucas 1979; Lucas and Luke 1984).

Just as the ecology of nectar- and pollen-feeding bats is distinctive, so is their craniodental morphology. These taxa possess, to varying degrees, elongate, narrow rostra and greatly reduced postcanine teeth, although the degree to which these specializations are seen differs between microchiropteran and megachiropteran nectarivores (Freeman 1995). Nectarivore canines remain relatively large, perhaps for structural support as the tongue works to obtain nectar. In mammals other than bats, dental reduction is most common among insectivorous, particularly ant-eating, taxa, in which the tongue has taken on a central role in food processing. Similarly, the tongues of nectar-feeding bats appear to play a central role in food acquisition and processing and are as specialized in their morphology as the mineralized portions of the feeding apparatus (Griffiths 1982). Secondarily, the distinctive skull and palate shape in chiropteran nectarivores necessarily limits the attachment area and available space for jaw musculature. This reorganization of musculature may be associated with a substantial reduction in the magnitude of bite forces these species can generate; however, the function of the enlarged canines, which are long and sharp, has yet to be explored in species that specialize on nectar and pollen.

From a broad comparative perspective, species within nine microchiropteran families (Phyllostomidae, Molossidae, Vespertilionidae, Emballonuridae, Rhinolophidae, Megadermatidae, Nycteridae, Noctilionidae, and Mormoopidae) are unevenly distributed throughout the relatively large morphospace representing ecologically significant craniodental features (Freeman 2000; figs. 6.3 and 6.4). Insectivorous and carnivorous bats from many lineages fall within a circumscribed area that can be designated as "insectivore morphospace." However, although numerous phyllostomids have retained insectivory and its associated morphologies, four unique feeding ecologies have evolved from the primitive insectivorous condition within this group. Concomitantly, four divergent suites of morphological specializations appear within subgroups of this diverse family (fig. 6.4). Within the Vampyrinae, a tendency toward increased carnivory is associated with changes in tooth form and an increase in the size of the teeth relative to the palate and appears, secondarily, to have facilitated the evolution of increased body size (Freeman 1984, 1988, 2000). Within the Glossophaginae, nectarivory is associated with rostral lengthening, reduction of tooth size relative to the palate, and specialization of the structure and function of the tongue (Freeman 1995, 1998, 2000; Griffiths 1982). Many bats in the Stenodermatinae possess shortened rostra and diverse patterns of loss of the insectivorous dilambdodont molar morphology and pursue a primarily fruit-eating ecology (Freeman 1988, 1998,

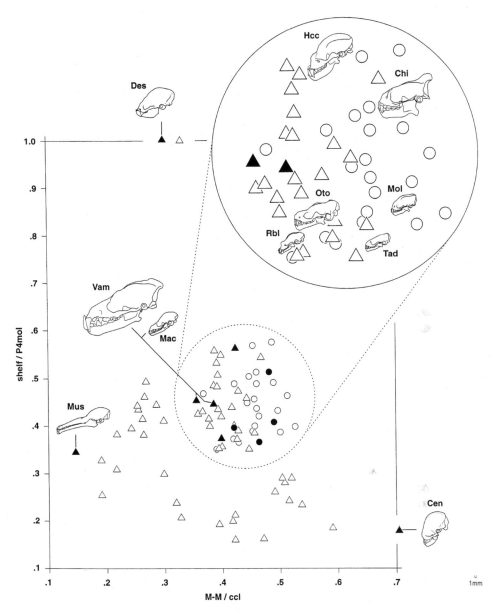

Figure 6.4. Bivariate plot of the relative area of the stylar shelf on the upper molariform tooth row (shelf/P4mol) versus breadth of the palate at the molars relative to the length of skull (M-M/ccl) showing the comparative scatter of insectivorous species from nine microchiropteran families (*encircled*) and phyllostomids, which fall largely outside the circle (modified from Freeman 2000). Diversity in Phyllostomidae is greater than that of all other families put together. *Circles* represent oral emitters and *triangles*, nasal emitters. Skull profiles in the insert replace the *filled symbols* on the graph. Diversity of insectivores within the circle includes large and small, oral and nasal emitters, and bats with robust and gracile skulls. Most bats from figure 6.3 are included for comparison of diversity within insectivorous bats (*Rhinolophus blasii* [*Rbl*], *Otomops martiensseni* [*Oto*], and *Tadarida brasiliensis* [*Tad*] are gracile; and *Hipposideros c. commersoni* [*Hcc*], *Cheiromeles torguatus* [*Chi*], and *Molossus molossus* [*Mol*] are robust). Extreme phyllostomids are pictured for comparison of diversity among Microchiroptera (*Musonycteris harrisoni* [*Mus*], a long-faced nectarivore; *Centurio senex* [*Cen*], a short-faced frugivore; *Desmodus rotundus* [*Des*], a sanguinivore; *Macrotus californicus* [*Mac*], a small-sized insectivore; and *Vampyrum spectrum* [*Vam*], a carnivorous giant).

2000). In association with the evolution of blood feeding, the Desmodontinae display reduction of many elements of the dentition.

The phyllostomid pattern of interrelationships among feeding ecology and morphology of the teeth and jaws may shed light on ecological patterns at higher levels and contribute to a better understanding of bat community ecology in Neotropical forests (Freeman 2000). The evolutionary acquisition of novel feeding modes appears to have influenced not only the range of phyllostomid diversity but also the structure of bat assemblages (Patterson et al., this volume). Although non-insectivorous phyllostomids make up only approximately 40% of the bat species diversity in La Selva in Costa Rica and Barro Colorado Island in Panama, four genera—*Glossophaga, Sturnira, Carollia, and Artibeus* —account for between 86% and 89% of bat biomass, respectively (Freeman 2000 from data in Bonaccorso 1979; LaVal and Fitch 1977; Timm et al. 1989). Future studies may be able to gain deeper insight into the interspecific interactions that determine this pattern and help to provide a clear causal linkage between morphological novelty in the feeding apparatus and ecological and evolutionary success.

Challenges to Correlational Analyses: Some Caveats and Limitations

Correlational approaches provide meaningful insight into the ecological significance of patterns of morphological variation, but the power of these approaches depends on a number of factors. Ideally, morphological and ecological variables employed in ecomorphological analyses are causally related. However, given our limited understanding of biological complexity, we face a significant probability of spurious results, and therefore, sometimes it may be virtually impossible to link conclusively the morphology in question to the ecological parameters of interest. Organismal morphology is highly integrated (e.g., Voss et al. 1990; Shubin and Wake 1996), and the characters of interest to us in ecomorphological studies are rarely if ever discrete, autonomous features controlled independently by the genetic system, free of covariation with characters unrelated to the function or ecological attribute under study. Changes in one part of an organism's structure necessarily bring at least some correlated changes, seriously confounding ecomorphological analysis. Independent of genetic correlations, there may be correlations among structural features because of morphological or physiological requirements not related to those aspects of ecology that are the foci of a particular analysis. To reveal causal links between structure and function, investigators must therefore sometimes employ additional, noncorrelational approaches. The more direct the connection between morphology and function, the greater the investigator's ability to discern important aspects of ecomorphology and avoid misleading correlations. Moreover, in some cases, very small differences in morphology can profoundly affect function and, in others, even relatively large morphological divergence may have little functional or ecological significance.

Correlational approaches to ecomorphology also depend critically on the characterization of the relevant ecology. For example, analysis of the relationship between craniodental structures and feeding ecology requires dietary description. But, by virtue of availability or choice, the type of food processed by the jaws and teeth may vary from year to year and from season to season in any given year and, in species with multiple feeding bouts in a single night, depending on the time of night. Within a single species characterized by a particular morphology, there may also be considerable geographic variation in diet. "Menus" at different times may have vastly different mechanical properties and thus impose disparate demands on the feeding apparatus. Patterns of correlation among craniodental characteristics and diet may be meaningful only when one considers the entire dietary repertoire of the species of interest, rather than predominant or preferred food items.

Recent research emphasizes the need for detailed fieldwork to obtain a realistic and biologically meaningful view of a species' diet. To obtain a clear picture of relationships among diet and morphology, sampling effort, prey availability, nightly, seasonal, and even year-to-year, variation must be taken into account. For example, lactating females of *Tadarida brasiliensis* employ two distinctive feeding pulses: a bout between early evening and midnight and an early morning bout (Whitaker et al. 1996). During the first bout, these bats fed largely on beetles and lygaied bugs, prey of relatively high strength and toughness, and in the early morning, most prey were soft-bodied and more easily processed moths. Clearly, if dietary data were collected from only one of these feeding bouts, the reliability of patterns of correlation between this species' morphology and feeding ecology would be limited.

Bats can take advantage of seasonally abundant prey and may opportunistically select prey at high densities (Brigham et al. 1992; Jones 1995; Waters et al. 1995). Under these conditions, animals may forage in habitats not predicted by the general patterns of relationship between wing morphology and flight mode. For example, frugivorous phyllostomid species such as *Artibeus lituratus* or *A. jamaicensis* may take advantage of seasonal nectar resources at flowering trees (Handley et al. 1991), although they lack the small body size and relatively low wing loading, high wing aspect ratio, and long wing tips often seen in nectarivorous phyllostomids (Norberg and Rayner 1987; table 6.1). In insectivorous bats, it may be especially difficult to establish a clear relationship between flight morphology and diet when it is necessary to rely on indirect analysis of feeding ecology. Fecal analysis or examination of culled insect parts may not reliably indicate whether prey have been captured in midair or gleaned from a substrate.

Variation in diet for a given species is also important over longer timescales and over its geographical distribution. Indeed, it may be the foods eaten in times of extreme environmental or population stress that exert the greatest selection pressures and are therefore most critical for determining morphological structure (Grant 1986; Grant and Grant 1989). Studies of the feeding ecology of *Lasiurus cinereus* in New Mexico show that this species has strong preferences for soft-bodied moths. However, in Canada, at the height of the insect season, odonates, beetles, and other hard prey dominate its diet (Barclay 1985). Further, changing distributions of prey or fruit abundance can influence bat diets directly or in more subtle ways, through interactions among consumers of local food items. Competition among sympatric species for specific foods will change from year to year in association with changes in population structure of either prey or predator species. For example, dietary overlap between *L. cinereus* and *L. borealis* varies from year to year, with the greatest overlap in a year of minimum food abundance (Hickey et al. 1996).

Limited knowledge of the full range of food items for a species can, in such instances, have a large impact on our ability to relate ecology to structure. Based on morphology of the skull and teeth, in the absence of dietary evidence, Freeman (1981a) suggested that *L. cinereus* should be readily able to consume hard prey such as beetles. This apparent mismatch of structure and function might well contribute to a view that the relationship between feeding performance and craniodental morphology is not always strong. But how closely the morphology of the *Lasiurus* feeding apparatus relates to diet depends on whether one considers the southern or northern populations. The more complete picture of the diet in this species suggests that the ability to process hard prey is, at least periodically, a critical functional element of its skull and dentition.

Analyses of fecal or stomach contents must account for the timing and/or geography of sample collection to avoid misleading interpretations. To date, however, there are few species for which we have chronicled dietary information in sufficient detail to address these diel, seasonal, annual, and geographic variations. Some of the most comprehensive studies of dietary variation have been carried out on endangered temperate species (Best et al. 1997; Lacki et al. 1995; Sample and Whitmore 1993). Such analyses consistently demonstrate hourly, daily, and seasonal variation in dietary composition and often show that distribution of prey types differs significantly from their availability in the environment. For tropical bats, seasonal changes often mean dramatic dietary changes; these remain largely undocumented. As a consequence, inferences concerning diet based on distribution of available resources may not accurately reflect the diet of a given species.

In general, correlational analyses are most meaningful when specific confounding factors are identified. In particular, virtually all aspects of organismal form and behavior are strongly affected by the overall size of the organism; thus, comparisons among taxa require explicit consideration of the effects of body size. Likewise, it is well established that groups of species may share certain morphologies or behaviors by virtue of their inheritance from a common ancestor instead of through the selective pressures imposed by the contemporary ecological setting. It is thus necessary to account for the effects of phylogenetic relationship to discern ecomorphological patterning (e.g., Felsenstein 1985; Garland et al. 1999; Losos 1996; Losos and Miles 1994; Perry 1999; see below, "Phylogenetic Considerations in Ecomorphology").

Comparative/correlational approaches have produced the core of our understanding of the relationships between morphology and ecology of bats. Through research over the past several decades, broad, general patterns of structure/function relationship have emerged (e.g., Aldridge and Rautenbach 1987; Findley 1993; Freeman 1998; Norberg and Rayner 1987). Finer-scale patterns have also been uncovered, particularly within relatively diverse and speciose genera or families (e.g., Findley [1972] for *Myotis*, and Freeman [1981b] for Molossidae). This approach simultaneously improves our understanding of the meaning of structural diversity and points to new directions for functional morphological and ecological research. Strong patterns of correlation among ecological and morphological features can be used to generate robust hypotheses about function or ecology for species about which little is known. In species accessible to further ecological study, these predictions can, ultimately, be tested by new data, further refining our understanding of the basic ecomorphological relationships.

Of equal importance, deviations from robust patterns can be identified. "Outliers" from otherwise highly regular patterns can direct our attention to interesting and important biological phenomena and/or point to critical features of the structure or ecology of the group that have been underappreciated or ignored. Discovery of patterns and deviations from these patterns describe diversity and generate hypotheses that can then be further developed and tested in the field or laboratory.

Function-Focused Approaches to Morphological Analysis

Ecomorphological analysis builds on the rich databases of observations of field ecology and comparative morphology and on the interpretations of these data within their primary disciplinary contexts. Correlative comparative approaches identify patterns of covariation among ecological and morphological characteristics of organisms by synthesizing diverse primary data; to further interpret these patterns, physiological and experimental approaches can often explore why particular aspects of morphology relate to certain ecological characteristics. Explicitly functional analyses can also offer opportunities to confirm that proposed relationships of structure to ecology are biologically critical and to distinguish causation from secondary correlations.

Probing causal relationships between organismal design and behavioral performance may require that one quantify morphology and/or performance in a very specific manner. This approach often requires analysis of how design controls or constrains behavior. A focus on the biological roles of specific structures, and on how multiple structures interact in a functioning whole, may be particularly well suited to achieving the goals of ecomorphological analysis.

Morphologists can choose general descriptors of organismal structure by attention to those aspects of morphology most likely to affect performance. For example, even in the absence of a detailed understanding of the aerodynamics of flapping flight of bats, one can feel confident that wing size and shape influence flight performance in some way and that skull form does not. However, this intuitive approach can sometimes be imprecise or misleading, particularly when comparisons are made among taxa of differing body sizes. Some functional parameters change as linear functions of body mass, and some as various power functions; hence functional equivalence at differing body sizes may require substantial modifications of structural design (Koehl 1995; LaBarbera 1989; McMahon 1984; Schmidt-Nielsen 1984). Conversely, similar shape at differing body sizes often implies differences in performance capabilities.

Even when structural descriptors are based on biomechanical analyses, function cannot be inferred unambiguously from morphology alone. No functional analysis can encompass all relevant structural and physiological complexity, and a number of critical variables in all real-world functional analyses are estimated, not measured directly. We illustrate this by considering the analysis of the form of a muscle that one believes is functionally important to a particular behavior.

The mass of a given muscle, or interspecific variation in mass, are often construed as indicative of functional and mechanical importance. However, it is not mass but physiological cross-sectional area and three-dimensional anatomical location relative to bony levers and joints that are required to estimate the largest force a muscle could produce under maximal stimulation and for a particular posture (Gans 1982). The internal architecture of the muscle, including its fiber length, degree of pinnation, and location of its origin and insertion, provide better guides to a muscle's functional potential than its size. Moreover, the intensity with which a muscle is activated and the timing of the onset and offset of muscle contraction with respect to a particular activity cannot be predicted a priori and often are not consistent with classic descriptions of muscle function based on anatomy alone. What appears to be a flexor, as determined by anatomical location, may remain electrically silent during primary flexing motions of the joint it crosses and instead may generate significant forces only during controlled extension. The positions of bones at a joint, and thus a muscle's leverage, also change continuously during movement (e.g., Loeb and Gans 1986). Moreover, multiple muscles that may act either synergistically or antagonistically to the muscle of interest cross many joints. Hence, a particular motion generated by the application of a specific force moment about a joint can be achieved by a virtually infinite number of combinations of force magnitudes contributed by each of the anatomically relevant muscles.

Although direct measurement of muscle activity patterns, or electromyography (EMG), presents considerable technical challenges, this technique provides invaluable information that can contribute directly to the interpretation of musculoskeletal anatomy. Recently, numerous studies have employed electromyography to assess directly the timing and relative intensity of muscle activity during particular movements in humans and a great diversity of animals, including bats (e.g., Altenbach and Hermanson 1987; De Gueldre and De Vree 1984, 1988, 1990; Hermanson and Altenbach 1981; Lancaster et al. 1995). These analyses can be employed to develop an understanding of how muscle activation patterns vary among behaviors. For example, the pectoralis muscle of *Artibeus jamaicensis, Antrozous pallidus,* and *Eptesicus fuscus* does not simply adduct the wings during the downstroke. Instead, during slow flight, the pectoralis is activated midway through the upstroke and ceases activity before the mid-downstroke (Altenbach and Hermanson 1987).

Electromyography, like other experimental methods, including in vivo bone strain measurement (Biewener 1992; Swartz 1991, 1998; Swartz et al. 1992) and force platform analysis (Schutt et al. 1997), requires detailed information concerning movements of specific anatomical structures during the behaviors of interest; this kind of information is also extremely informative in and of itself. In addition, outputs of many functional analyses are very sensitive to aspects of kinematics. This is particularly true in analyses of flight because of the complex, nonlinear nature of aerodynamic forces and flight energetics. Photographic methods, particularly multicamera, short-exposure still photography, and high-speed cinematography and videography provide considerable insight into mechanically and aerodynamically important aspects of the three-dimensional conformation of wings (Aldridge 1986a, 1987; Altenbach 1979; Norberg 1970, 1972, 1976a, 1976b, 1976c; Rayner et al. 1986). For example, recognition of multiple distinctive kinematic patterns or gaits in bat flight required accurate records of multiple wing landmarks over a range of flight speeds (Aldridge 1986a, 1986b; Norberg 1976a, 1976b, 1976c; Rayner 1986). Visualization techniques can, ultimately, provide information that is critical to functional interpretations of morphology, and fortunately, it is increasingly possible to attain high resolution, high-speed sequences at relatively low cost. Imaging equipment for low light environments and compact, portable instrumentation will also greatly expand opportunities to see and record flight and feeding behaviors in great detail, both in the field and in the lab.

The techniques of direct functional study all share certain limitations. They are typically time, labor, and equipment intensive, and necessitate that investigators have access to live animals in relatively controlled laboratory conditions. These features constrain sample sizes and restrict the number of taxa that can be studied. However, direct functional study of a limited nature can serve as a key intermediate step in analysis that can thereafter guide the selection of measurements that can be made more easily from large numbers of individuals and/or species.

Experimental Approaches to Understanding Ecomorphology of the Feeding Apparatus

Coupled with ecological data, measures of functional performance can provide important insight into the constraints that affect behavior in the wild. To understand the functional role of a particular morphology requires that one identify the mechanisms behind correlations between morphological traits and behavior. In studies of tooth morphology and diet in bats, experimental studies that examine the mechanical limitations of tooth and jaw morphology with respect to the mechanical properties of prey are crucial to establishing a link between morphology and diet (see Dumont, this volume). The physical nature of food provides a selective influence on the shapes and configurations of the teeth (Lucas 1979). Experimental approaches that identify how the morphology of particular elements of the masticatory apparatus affect food processing can help elucidate the selective pressures influencing the evolution of tooth shape.

This general approach assumes that the morphology of the skull and/or teeth can constrain diet, for example, that there are detectable, available, palatable, and nutritious food items that an animal is unable to acquire or process because of limitations to its structural design. Few studies have rigorously tested this assumption for any vertebrates (but see Kiltie 1982, Moore and Sanson 1995; Wainwright 1987). *Pteronotus parnellii*, for example, can readily capture tethered beetles whose exoskeletons it is unable to puncture (Goldman and Henson 1977). Independent measures of the hardness of prey and limitations imposed by tooth morphology would be extremely useful to determine the range of prey types functionally available to various bats.

To date, there have been few studies of any kind that empirically quantify the mechanical properties of bat prey. In an important first step, Freeman (1981a) qualitatively partitioned various insect orders commonly eaten by bats into five hardness categories, ranging from the softest (e.g., Ephemeroptera) to hardest (Coleoptera). Further refinements of this approach might also account for the possibility that hardness varies within orders, at least in part in relation to body size, among different anatomical regions (e.g., odonate heads would be harder than abdominal segments) and among developmental stages. Understanding the mechanical nature of insect cuticle, a tissue commonly consumed by bats, is complicated, however, by the complex nature of its physical properties. Cuticle is a fibrous composite material like plywood and varies in its stiffness and toughness depending on its moisture content (Vincent 1980). The mechanics of hard tissues in noninsect arthropods may be even more complex, as they often incorporate mineral salts into the cuticular tissue, thereby increasing hardness and strength (Vincent 1980).

Only a few studies have addressed the issue of how teeth process different types of foods, although biomechanical approaches hold much promise in this arena. For example, reducing the contact area between teeth and prey increases the local stresses at the contact interface and, thereby, decreases the force required to fracture exoskeletons of prey (Evans and Sanson 1998; Freeman and Weins 1997; Popowics and Fortelius 1997). Strait (1993) predicted that species that feed on hard prey should have relatively short shearing surfaces and found that comparisons of the length of shearing crests on second molars from diverse insectivorous mammals, including bats, supported this hypothesis (but see Evans and Sanson 1998). Such mechanically advantageous localization of tooth/cuticle contact may thus have influenced the structural design of the teeth of diverse insectivorous mammals, including bats.

Determining whether variation in tooth morphology leads to differences in food processing mechanics can effectively test theoretical analyses. In studies of the relationship between the shape of bat teeth and their ability to puncture an apple, teeth with sharper tips required lower puncture force than those with blunt tips (Freeman and Weins 1997). These results have been further refined with large two-dimensional, Plexiglas scale models of teeth with either sharp or blunt tips and a beam-shaped model substrate. For a given load, stresses were more highly concentrated at the point of contact for a smaller than for a larger apical radius—that is, stresses are more concentrated by sharp tips (Freeman and Wiens 1997). This study represents some of the first experimental work to quantify puncture performance relative to apical sharpness of teeth.

Evans and Sanson (1998) also used physical models to test directly the efficiency of different tooth shapes in breaking down foods of varying physical properties. With models that varied in tip sharpness (radius of apical curvature) and cusp sharpness (volume or surface area of tooth per unit distance from apex of cusp), force, and energy required to puncture the cuticular surfaces of beetles decreased with decreasing contact area. These results also highlight the importance of size considerations in comparative studies of tooth morphology; similarly shaped teeth of different sizes will not puncture foods with equal efficiency.

Studies of mastication (e.g., Storch 1968) and analytical predictions of bite forces (e.g., Reduker 1983) have also provided important bases from which to form hypotheses about functional consequences of morphology. For example, estimates of mechanical advantage for major muscle complexes involved in jaw adduction based on skull and dentary measurements, coupled with estimates of adductive muscle force, suggest that *Myotis evotis*, a substrate gleaner, has a more forceful and quicker bite than *M. volans*, an aerial insectivore (table 6.1). Direct measurement of bite forces of live *M. evotis* and *M. volans*, coupled with analyses of jaw motion during insect capture, would be particularly useful in elucidating this case.

Feeding analyses using electromyography, cineradiography, or measurement of associated forces applied by the jaw muscles during mastication have been limited to only a few studies in bats (e.g., De Gueldre and De Vree 1984, 1990; Kallen and Gans 1972). De Gueldre and De Vree (1990) created a three-dimensional model to estimate applied and reaction forces at the bite point and temporomandibular joints in response to foods of different consistencies in *Pteropus giganteus*. Their kinematic and biomechanical analysis showed that food consistency affected both the magnitude and orientation of the bite force. In particular, the differences between the magnitude of masseter and temporalis activities influenced both the orientation of the bite force and the mechanically optimal position of food. Their conclusions could not have been reached without experimental determination of the sequence in which the muscles fire and their anatomical placement with respect to the jaw joint (De Gueldre and De Vree 1988).

Experimental Approaches to Understanding Ecomorphology of Flight

Experimental approaches are critical to a better understanding of the relationship between flight performance and morphology. Conventional aerodynamic theory has provided a starting point from which to generate hypotheses about the functional significance of the diversity of wing morphologies among species. However, bat wings are structurally complex compared to wings of conventional aircraft. In comparison with aircraft materials, the constituent tissues of wings, skin, muscles, ligaments, tendons and bones are highly variable, nonlinear, and anisotropic in mechanical properties (Papadimitriou et al. 1996; Swartz 1998; Swartz et al. 1996). Moreover, to date, studies of wing morphology are based on fully outstretched wings that represent the minimum wing loading achieved during the wingbeat. Three-dimensional wing conformation, wing mass distribution, and mechanical characteristics of wing tissues also change dynamically throughout even a single wingbeat cycle (fig. 6.5) of an animal flying horizontally at constant velocity (Swartz 1998). The functional significance of wing morphologies of bats and other flying animals will likely best be defined in the context of flapping kinematics.

To the extent that one can test the limits of performance abilities associated with particular morphologies, it will be possible to gain deeper insight into

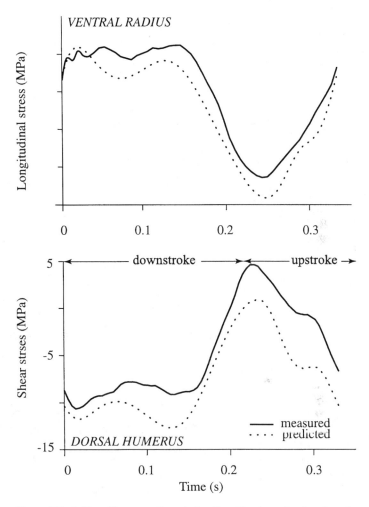

Figure 6.5. *A,* Dorsal humerus; *B,* ventral radius. The stress developed on the surface of major wing bones of *Pteropus poliocephalus* during a complete wing-beat cycle (downstroke from $t = 0$ s to $t = 0.22$ s, upstroke from $t = 0.23$ s to $t = 0.34$ s). Empirical values (*heavy black lines*) are computed from data measured directly from bone surfaces using in vivo implantation of strain gauges onto wild-caught animals; model values are estimated from an abstract computer model of bat flight (see Watts et al. 2001).

mechanical or aerodynamic factors that may constrain habitat use or exploitation of food resources. For example, because some habitats, such as the forest understory, can be spatially complex, differences among bat species in maneuverability differences could directly affect habitat use (Aldridge and Rautenbach 1987).

Flight Maneuvers and Maneuverability

Flight performance through an obstacle course is one useful experimental metric of maneuverability (e.g., Aldridge 1986a, Aldridge and Rautenbach 1987; Stockwell 2001). Among several species of insectivorous British bats, those species best able to negotiate the most tightly spaced string-array obstacles also foraged in the most cluttered habitats (Aldridge 1986b). Similarly, in several species of African microchiropterans, wing loading and body mass were negatively correlated with obstacle course maneuverability, and foraging in habitats of great vegetation density was positively related to maneuverability (Aldridge and Rautenbach 1987). In Neotropical phyllostomids, morphological variables associated with body size (e.g., mass, length of forearm, and wing span) were negatively correlated with maneuverability in an obstacle course in which obstacle spacing was scaled to wing span—that is, larger species were less maneuverable than smaller species, even when performance tasks accounted for absolute variation in wing span (Stockwell 2001). Morphological variables associated with depth of maximum wing camber were positively correlated with maneuverability. The most maneuverable species in the obstacle course were relatively small understory frugivores (Carollia perspicillata and C. castanea) with wings that could be cambered deeply. In contrast, less maneuverable large canopy frugivores, Artibeus jamaicensis and A. lituratus, had wings of shallow maximum camber (Stockwell 2001; table 6.1).

Insight into the morphological and behavioral determinants of maneuverability is central to understanding the variation among species in performing maneuvers. At least two strategies for maneuverable flight have been proposed (Thollesson and Norberg 1991). Mass moments of inertia—measures of the distribution of mass in the body and wings with respect to a particular axis of rotation, such as the body's midline—are a major influence on turning ability, at least for rolling turns while wings are maximally extended. For maneuverability at high flight speeds, small roll moments of inertia—wing mass concentrated proximally with minimal mass in the hand wing—are favored to achieve high roll accelerations. Turn radius can also be decreased if body mass and hence wing loading are small. Bats flying at slow speeds, however, can achieve high roll accelerations in spite of high roll moments of inertia when wing mass is concentrated more distally, with relatively heavier dactylopatagia and relatively lighter plagiopatagia. Several slow-flying taxa possess broad wings and wide wing tips that generate extra lift necessary to produce the

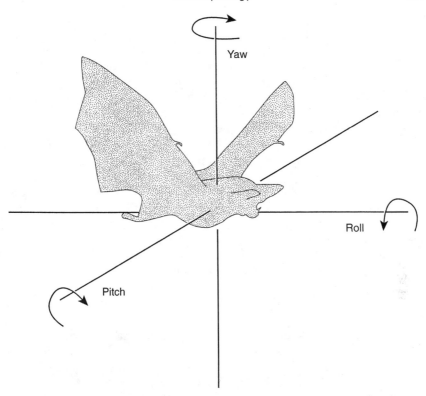

Figure 6.6. Schematic of a bat in flight demonstrating the three rotational degrees of freedom that must be controlled during flight. *Roll* refers to rotations about an axis passing through the animal's center of mass from head to tail in a horizontal plane, *pitch* to rotations about an axis that passes mediolaterally in a horizontal plane, and *yaw* to rotations about a vertical axis. (Adapted from Stockwell 2001.)

aerodynamic moment about the roll axis (Thollesson and Norberg 1991). Thus, a number of aspects of wing shape, including mass distribution along the wingspan and the chordwise dimensions of the wing, interact to influence turning performance. As we gain increasingly detailed information on wing kinematics in both straight and maneuvering flight, the relative importance of these and other yet unidentified design criteria will become clearer.

Many maneuvers by bats are not simple rolls but, rather, also involve pitching and yawing moments (Norberg 1976a; Stockwell 2001; fig. 6.6). Moreover, flapping affects aerodynamic stability because wing movements constantly change the spatial location of the center of aerodynamic force relative to center of mass (Stockwell 2001). Realistic and biologically meaningful comparisons of maneuverability among species must ultimately examine variation in stability about roll, pitch, and yaw axes to describe the functionally significant aspects of bat flight adequately.

It is increasingly evident that both wing morphology and flight kinematics are important determinants of maneuverability, although, to date, we have little real knowledge of their relative importance. It is clear, for example, that modulation of wingbeat kinematics may allow for the exploitation of a novel or seasonal resource that would not otherwise be predicted based on fixed wing morphology alone. Whether a bat flaps its wings or glides though a turn affects turning performance, and one simple kinematic variable that affects turning performance is the relative proportions of the turn through which wings flap or are held relatively stationary. Among six species of British bats, only *Rhinolophus ferrumequinum* flapped its wings while turning and, as a result, achieved tighter turns than predicted based on morphology alone (Aldridge 1987). Thus, it appears that adjustments of flight kinematics can enable a typically less maneuverable species to negotiate a cluttered habitat or turn tightly in at least some circumstances. The energetic costs and mechanical demands of such short-term adjustments to kinematics are unexplored to date and must be weighed against any gains in maneuverability. We advocate expanding views of what determines flight maneuverability and propose that a more complete understanding of wing movements in relation to flight will also require consideration of important biological functions distinct from flight per se. For example, emission of echolocation pulses, especially in aerial insectivores is coupled with wingbeat (Jones 1994; Kalko and Schnitzler 1989; Lancaster et al. 1995). The coordination of these two major functions must produce as yet poorly understood and complex interactions among aerodynamics, wing kinematics, echolocation, and energetic costs of flight behaviors.

Experimental Manipulation of Body Mass

A number of investigators have explored the limits of flight performance by experimentally manipulating body mass and hence wing loading. Body mass changes substantially on a daily or seasonal basis due to fluctuations in stomach contents, transport of food and young, and so forth. The influence of body mass on flight mechanics and energetics is thus particularly pertinent (Hughes and Rayner 1991; Hughes et al. 1995; Norberg and Fenton 1988; Norberg and Rayner 1987; Schutt et al. 1997; Webb et al. 1992). In a pioneering study of flight performance in five species of North American bats, Davis and Cockrum (1964) found that *Tadarida brasiliensis* took off with loads of no more than 9% of its body mass, in contrast to *Plecotus townsendi,* which readily took off with loads up to 70% of unloaded body mass (table 6.1). More recent work demonstrates that loads as small as 5% of unloaded body mass can result in a substantial loss of maneuverability (Aldridge and Brigham 1988). The decreased maneuverability with increasing wing loading in individuals mirrors the overall pattern seen in broad interspecific comparisons over a large range of body sizes (Aldridge and Rautenbach 1987).

Artificial loading also affects flight kinematics (Hughes and Rayner 1991). Wingbeat frequency and amplitude are greater in artificially loaded than in unloaded *Plecotus auritus* during steady, forward flight, in keeping with theoretical predictions (Hughes and Rayner 1991). In contrast to predictions, flight speed decreases, approximately doubling the predicted power requirements for flight. Such results illustrate the potential impact of changes in total body mass on flight energetics as mediated by subtle changes in wing loading and kinematics.

Computer Modeling Approaches in Ecomorphology

Computer modeling approaches made more accessible by increasing computer power and ease of use can effectively address the functional significance of morphology and kinematics in bats. In particular, when a biologically important behavior can be characterized as resulting from the mechanical workings of a morphological system, computer programs can reproduce organismal function in simplified fashion, sometimes with a high degree of accuracy and precision. This approach offers several important strengths. One can rarely isolate or manipulate single morphological features for functional analysis; it is impossible to alter experimentally the stiffness or strength of structural tissues such as teeth or bone or to induce an individual to activate only a single muscle. However, computer models built on detailed information from living organisms, which thereby reflect biological reality, are not limited in this way. Individual elements of models can be manipulated in ways that the investigator hypothesizes will significantly influence performance. Kinematic patterns that have not been observed in real animals can be imposed on model systems, force magnitudes and orientations can be altered, the size and shape of support elements can be controlled. Computer modeling approaches also enable one to construct potential evolutionary intermediates or extreme forms not represented in extant faunas. For example, one could postulate that the maximum body size of bats is limited by the ability of the bones of the wing to withstand bending stresses imposed by flapping flight or that the diet of a species is limited because it is mechanically unable to process certain food items. Computer models make it possible to assess whether such hypotheses are reasonable and can identify morphological characteristics that dictate performance.

Computer Modeling of the Mechanics and Aerodynamics of Bat Flight

Many components of wing structure have the potential to affect the range of velocities that a species can employ, the energetic cost of flight behaviors, and the aerodynamic forces experienced by a bat. Particularly important design elements include the distribution of mass and lifting surface area along the length of the wing, placement of the bones and muscles within the wing

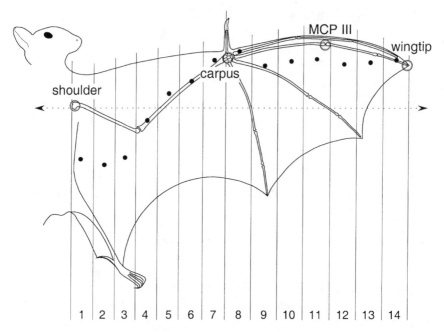

Figure 6.7. Plan-view of the ventral surface of the wing of a *Pteropus poliocephalus* held in a position similar to that of mid-downstroke, showing the subdivision of the wing into 14 chordwise strips for the purpose of computer modeling. The *filled circles* indicate the locations of the centers of masses of these strips relative to a reference line connecting the two shoulders. The *open circles* indicate anatomical locations used as landmarks for collecting kinematic data.

membrane, the relative lengths and orientations of various skeletal elements, and the structure of the wing membrane skin. The three-dimensional movements of wing elements interact with this complex structural organization to determine flight performance. To date, our understanding of the specific roles of various aspects of morphology or of kinematics on flight mechanics and energetics remains quite limited, in part because of the near impossibility of experimentally manipulating single parameters of interest, separating them from their normal network of biological interactions.

   Computer modeling of the details of the mechanics and aerodynamics of the wingbeat can probe relationships among morphology, kinematics, and wing structural mechanics (Watts et al. 2001). One recent model is based on the morphology and wing kinematics of *Pteropus poliocephalus*, a species whose flight is well studied (Carpenter 1985; Swartz et al. 1992; Thomas 1975, 1981). The model comprises an abstraction of a bat composed of numerous interconnected wing and body segments that reflect anatomy with high precision (fig. 6.7). In particular, model segments accurately reflect the distribution of wing mass and surface area. The model mathematically imposes movements of the wing landmarks through the three-dimensional space based on empirically measured patterns of wing motion based on films of wind tunnel flights.

Building on morphology—a characterization of wing form designed from features likely critical for flight performance—and kinematics, this model computes the magnitude and orientation of each of the forces (gravity, inertia, lift, drag, internal force carried by wing structures, added mass force) acting on each wing segment at small increments of the wingbeat. From the resulting force estimates, it is possible to test the model's validity in two independent ways. First, the model calculates the rise and fall of the bat's center of mass during the downstroke and upstroke respectively. Second, it computes stresses developed in the proximal wing bones. These results can then be compared directly to (1) whole body movements measured directly from films and (2) stress magnitude and orientations measured in vivo from strain gauge recordings from the same wing bones in the same species (Swartz et al. 1992; Watts et al. 2001; figs. 6.5, 6.8, and 6.9). For horizontal flight at moderate speed, the model and empirical data match extremely well, providing good evidence that the model captures many of the most important aspects of flight mechanics and aerodynamics in this species. Once the accuracy and precision of the model are validated in this way, the model can be employed to compute measures of energetics, maneuverability, joint forces, and so on with confidence. Future analyses that employ realistic models such as this one, which can be extended to diverse species, will allow exploration of many questions of interest to the ecomorphology of bats: functional significance of wing mass and area distributions, aerodynamic and/or mechanical limits on body size, energetic consequences of load carrying, and many others.

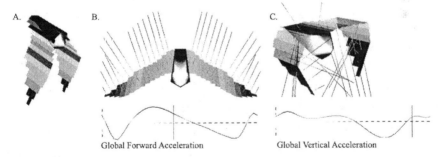

Figure 6.8. Single frame images from dynamic visualization of computer model of bat flight. *A,* Animation allows the user to view the bat from any position and at any degree of zoom (mouse controlled) as the wings beat. *B,* Distribution of lift on the wing: users may select a single wing force and display its vector representation on each segment from any view; these vectors change dynamically in concert with a display of the forward or vertical acceleration of the center of mass (*bottom:* vertical bar moves horizontally with wingbeat cycle). *C,* Total force on the wing: any combination of forces may be selected (here portions of the vectors are off the figure).

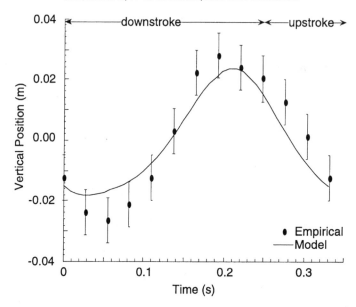

Figure 6.9. Comparison of the vertical oscillations of the bat's center of mass over one wingbeat cycle as computed from the model (*solid line*) and as measured directly from wind-tunnel film footage (*filled circles; bars* represent 2 SDs).

Computer Models and Analysis of Tooth Morphology

Teeth must successfully process different foods, and, for bats, teeth take the form of slicing blades, crushing devices, juice extractors, et cetera (Lucas 1979). The relationship between tooth form and diet in bats is well exemplified by an example of a dietary type not observed among any other mammals: blood feeding in the vampire bats. Desmodontines, obligate sanguinivores, have no need to mechanically process their food and possess small, razor-sharp, blade-like incisors that can readily inflict superficial wounds, and yet their teeth appear unable to withstand the stresses normally generated during biting or chewing (Freeman 1992; Van Valkenburgh and Ruff 1987).

For most bats, however, the interplay between tooth form and diet is complex. Effectiveness of a particular tooth morphology in puncturing hard-bodied prey items will depend on the size and shape of the puncturing teeth, the size and hardness of prey items, and the magnitude of muscle force that can be recruited for biting as amplified by the lever mechanics of the jaws. These design considerations must then be balanced against competing design constraints. For a given amount of bite force applied to a jaw, long, very sharp blades or cusps can generate very high, localized stresses in whatever an animal bites but have a greatly increased risk of breakage.

How can one realistically assess the role of shape in tooth performance? The morphology of mammalian teeth is more complex than the human-manufactured structures to which classic engineering design theory can be readily applied. A relatively new engineering analysis technique, finite element analysis, is, however, well suited to this problem (Beaupré and Carter 1992; Huiskes and Chao 1983; Zienkiewicz and Taylor 1989). In finite element analysis (FEA), complex geometries are redefined as composites of a large number of simple shapes, each of which can be analyzed using conventional beam theory. The mechanical response of each of these simple shapes to applied forces can be readily calculated, and mathematical algorithms employed to link results among the many component elements, thereby estimating mechanical behavior of the whole structure (Zienkiewicz and Taylor 1989). How accurately a combination of many simple forms reproduces a complicated and irregular geometry depends largely on the number of elements employed; partitioning of a complex shape into 100 relatively large elements will not provide the same precision of analysis as dividing the same shape into 10,000 elements that are each 100-fold smaller. However, it takes far more computational time and power to analyze models with many thousands of elements than those with a few. This approach provides a powerful way to gain insight into the mechanics of structures of complex shape and holds great promise for the study of biological structure.

Freeman (1998) used FEA to examine the effects of tooth shape on the propagation of cracks through a food substance as the tooth penetrated the food and compared the effects of an edged versus a nonedged canine tooth as it penetrated a uniform substance. Stresses were highest at the margins of the edged tooth model and, in contrast, were less but were uniformly distributed around the nonedged tooth model. To validate results from the finite element model, Freeman (1998) constructed scale models representing edged and nonedged teeth and applied them to a pressure-sensitive, photoelastic material to simulate biting into food. Preliminary results of this physical modeling are consistent with and lend strength to the FEA simulation results. Both approaches document a substantial increase of surface energy at the edge of a puncture initiated by an edged tooth, and no such build-up with a nonedged or circular tooth. Further experiments will examine more subtle shapes of teeth and will extend this analysis to three dimensions.

Phylogenetic Considerations in Ecomorphology

Several recent studies emphasize the importance of interpreting interspecific variation in morphology in the context of well-defined phylogenies (reviewed in Losos and Miles 1994). Unless clear cases of convergence in morphology are demonstrated among distantly related taxa, the mechanism underlying a particular morphological trait will remain obscure. Variation in degree of phylogenetic relationship among taxa will, in and of itself, produce patterned morphological variation, and explicit efforts are needed to distinguish morphological similarity due to similar selective pressures from that due to shared ancestry alone. Multiple independent evolutions of particular characteristics, or convergences, can be especially illuminating in this context.

A clear case of convergence in both tooth and flight morphologies associated with the evolution of carnivory (feeding on vertebrates as opposed to invertebrates) has been demonstrated among species of Nycteridae, Megadermatidae, Vespertilionidae, and Phyllostomidae (Freeman 1984; Norberg and Fenton 1988; fig. 6.2). Because these families are relatively distantly related (Baker et al. 1989; Koopman 1984; Pierson 1986; Simmons 1998; Smith 1976), low wing loadings and low aspect ratios of carnivorous members of these four families appear to have evolved independently under similar selection pressures for slow, maneuverable flight and the ability to carry heavy prey. The independent acquisition of particular morphological features in the distinct lineages provides more compelling evidence for a functional relationship between these morphological traits and the flight behavior associated carnivorous bats than would be possible in the absence of phylogenetic context.

Phylogenetic mapping of morphological traits in bats is exemplified by recent work superimposing a phylogeny of the family Phyllostomidae onto ecologically important craniodental characteristics (Baker et al. 1989; Freeman 2000; fig. 6.4). This coupling of morphology and phylogeny leads to a better understanding of the adaptive radiation of the phyllostomids, in particular, and the evolution of Microchiroptera, in general. Microchiropteran bats are overwhelmingly insectivorous and constraints on morphology imposed by insectivory have produced a dynamic equilibrium in bat morphologies that has persisted for 60 million years. The morphological diversification within phyllostomids is greater than in all other families of microchiropteran bats, and although many phyllostomid lineages have undergone substantial change, those that have maintained insectivory have changed little (Freeman 2000). This mapping approach suggests that the ability to eat fruit may be the key synapomorphy that has facilitated phyllostomid ecological and morphological diversity. In addition, this study demonstrated that morphology, ecology, and phylogenetics can be effectively integrated to deepen our understanding of bat evolution.

Conclusions

Studies that have identified patterns of association between morphology and behavior (e.g., Freeman 1981a; Norberg and Rayner 1987) have provided a big-picture view of the ecomorphology of flight and feeding and serve as important foundations from which to generate new hypotheses about the function of morphological traits and factors that might have influenced bat evolution. There are many gaps in current knowledge about bat ecomorphological relationships, and many of the patterns of morphological differences and associated behaviors have yet to be experimentally tested or quantified. It is our hope that the approaches we have outlined in this chapter will aid in the design of future studies. Future field studies will continue to expand our knowledge of the ecology and behavior of bats in their natural environments, and experimental work and detailed kinematic studies can help test theories based on mechanics or fixed-wing aerodynamics. In cases where experiments are difficult to conduct on live animals, computer modeling provides a new tool with which to understand better the mechanical limitations imposed by morphology and to help identify functionally important morphological characters.

How, then, can we best study the ecomorphology of bats? Integration of field and laboratory methods and collaborations among ecologists and morphologists are integral to future progress. Time-consuming and labor-intensive research will be of greatest value when it addresses questions that are relevant to better understanding of both the morphology of bats and the ecological significance of morphology. Morphological and experimental studies that focus on traits of known ecological importance are particularly valuable, as are ecological studies that highlight behaviors whose mechanistic basis is well understood. The effects of body size on interspecific patterns, intraspecific variation in ecology and morphology, and phylogenetic effects on observed patterns of structural and behavioral variation have yet to be fully addressed. In all, we believe this field will advance rapidly in coming years, as new insights into the ecomorphology of bats arise from increasingly successful synthesis of morphological and ecological study.

Acknowledgments

We gratefully acknowledge the reviewers who thoughtfully and carefully improved this chapter. Swartz and Stockwell were partially supported by a grant from the National Science Foundation to Swartz during the preparation of this chapter. Freeman thanks the following museums, which loaned specimens for data analysis: American Museum of Natural History; Field Museum of Natural History; Royal Ontario Museum; Texas A&M University, Texas Cooperative Wildlife Collection; Texas Tech University, the Museum; U.S. National Museum, Fish and Wildlife Labs; University of California, Berkeley, Museum of Vertebrate Zoology; University of Michigan, Museum of Zoology; University of Kansas, Museum of Natural History; and University of Nebraska State Museum (UNSM). Angie Fox, UNSM artist, assisted with several figures.

Literature Cited

Adams, R. A. 1996. Size-specific resource use in juvenile little brown bats, *Myotis lucifugus* (Chiroptera: Vespertilionidae): is there an ontogenetic shift? Canadian Journal of Zoology, 74:1204–1210.

Adams, R. A. 1997. Onset of volancy and foraging patterns of juvenile little brown bats, *Myotis lucifugus.* Journal of Mammalogy, 78:239–246.

Aldridge, H. D. J. N. 1986a. Kinematics and aerodynamics of the greater horseshoe bat, *Rhinolophus ferrumequinum,* in horizontal flight at various flight speeds. Journal of Experimental Biology, 126:479–497.

Aldridge, H. D. J. N. 1986b. Manoeuvrability and ecology in British bats. Myotis, 23–24:157–160.

Aldridge, H. D. J. N. 1987. Turning flight of bats. Journal of Experimental Biology, 128:419–425.

Aldridge, H. D. J. N., and R. M. Brigham. 1988. Load carrying and maneuverability in an insectivorous bat: a test of the 5% "rule" of radio-telemetry. Journal of Mammalogy, 69:380–382.

Aldridge, H. D. J. N., and I. L. Rautenbach. 1987. Morphology, echolocation and resource partitioning in insectivorous bats. Journal of Animal Ecology, 56:763–778.

Alexander, R. M., A. S. Jayes, G. M. O. Maloiy, and E. M. Wathuta. 1979. Allometry of the limb bones of mammals from shrews (*Sorex*) to elephants (*Loxodonta*). Journal of Zoology (London), 89:305–314.

Altenbach, J. S. 1979. Locomotor morphology of the vampire bat, *Desmodus rotundus.* American Society of Mammalogists, Special Publication no. 6:1–137.

Altenbach, J. S., and J. W. Hermanson. 1987. Bat flight muscle function and the scapulo-humeral lock. Pp. 100–118 *in:* Recent Advances in the Study of Bats (M. B. Fenton, P. Racey, and J. M. V. Rayner, eds.). Cambridge University Press, Cambridge.

Baker, R. J., C. S. Hood, and R. L. Honeycutt. 1989. Phylogenetic relationships and classification of the higher categories of the New World bat family Phyllostomidae. Systematic Zoology, 38:228–238.

Barclay, R. M. R. 1985. Long- versus short-range foraging strategies of hoary (*Lasiurus cinereus*) and silver-haired (*Lasionycteris noctivagans*) bats and the consequences for prey selection. Canadian Journal of Zoology, 63:2507–2515.

Barclay, R. M. R., and R. M. Brigham. 1991. Prey detection, dietary niche breadth, and body size in bats: why are aerial insectivorous bats so small? American Naturalist, 137:693–703.

Beaupré, G. S., and D. R. Carter. 1992. Finite element analysis in biomechanics. Pp. 149–174 *in:* Biomechanics: A Practical Approach. Vol. 2. Structures (A. A. Biewener, ed.). Oxford University Press, Oxford.

Bennett, M. 1993. Structural modifications involved in the forelimb and hindlimb grip of some flying foxes (Chiroptera, Pteropodidae). Journal of Zoology (London), 229:237–248.

Best, T. L., B. A. Milam, T. D. Haas, W. S. Cvilikas, and L. R. Saidak. 1997. Variation in diet of the gray bat (*Myotis grisescens*). Journal of Mammalogy, 78:569–583.

Biewener, A. A. 1992. In vivo measurement of bone strain and tendon force. Pp. 123–147 *in:* Biomechanics: A Practical Approach. Vol. 2. Structures (A.A. Biewener, ed.). Oxford University Press, Oxford.

Biknevicius, A. R. 1993. Biomechanical scaling of limb bones and differential limb use in caviomorph rodents. Journal of Mammalogy, 74:95–107.

Birch, J. M. 1997. Comparing wing shape of bats: the merits of principal-components analysis and relative-warp analysis. Journal of Mammalogy, 78:1187–1198.

Bonacorrso, F. J. 1979. Foraging and reproductive ecology in a Panamanian bat community. Bulletin of the Florida State Museum, Biological Science, 24:359–408.

Bonaccorso, F. J., and B. K. McNab. 1997. Plasticity of energetics in blossom bats (Pteropodidae): impact on distribution. Journal of Mammalogy, 78:1073–1088.

Bookstein, F. L. 1997. Morphometric Tools for Landmark Data: Geometry and Biology. Cambridge University Press, Cambridge.

Bookstein, F. L., B. Chernoff, R. Elder, J. Humphries, G. Smith, R. Strauss. 1985. Morphometrics in Evolutionary Biology. Academy of Natural Sciences of Philadephia, Philadelphia.

Brigham, R. M., H. D. J. N. Aldridge, and R. L. Mackey. 1992. Variation in habitat use and prey selection by Yuma bats, Myotis yumanensis. Journal of Mammalogy, 73:640–645.

Britton, A., G. Jones, J. M. V. Rayner, A. M. Boonman, and B. Verboom. 1997. Flight performance, echolocation and foraging behaviour in pond bats, Myotis dasycneme (Chiroptera: Vespertilionidae). Journal of Zoology (London), 241:503–522.

Carpenter, R. E. 1985. Flight physiology of flying foxes, Pteropus poliocephalus. Journal of Experimental Biology, 114:619–647.

Cotterill, F. P. D. 1998. Female reproduction in two species of tropical horseshoe bats (Rhinolophidae) in Zimbabwe. Journal of Mammalogy, 79:1306–1316.

Davis, R., and E. L. Cockrum. 1964. Experimentally determined weight lifting capacity in individuals of five species of western bats. Journal of Mammalogy, 45:643–644.

De Gueldre, G., and F. De Vree. 1984. Movements of the mandibles and tongue during mastication and swallowing in Pteropus giganteus (Megachiroptera): a cineradiographical study. Journal of Morphology, 179:95–114.

De Gueldre, G., and F. De Vree. 1988. Quantitative electromyography of masticatory muscles of Pteropus giganteus (Megachiroptera). Journal of Morphology, 196:73–106.

De Gueldre, G., and F. De Vree. 1990. Biomechanics of the masticatory apparatus of Pteropus giganteus (Megachiroptera). Journal of Zoology (London), 220:311–332.

Demes, B., and W. L. Jungers. 1993. Long bone cross-sectional dimensions, locomotor adaptations and body size in prosimian primates. Journal of Human Evolution, 25:57–74.

Dumont, E. R. 1997a. Cranial shape in fruit, nectar and exudate feeding mammals: implications for interpreting the fossil record. American Journal of Physical Anthropology, 76:1127–1136.

Dumont, E. R. 1997b. Salivary pH and buffering capacity in frugivorous and insectivorous bats. Journal of Mammalogy, 78:1210–1219.

Dumont, E. R. 1999. The effect of food hardness on feeding behaviour in frugivorous bats (Family Phyllostomidae): an experimental study. Journal of Zoology (London), 248:219–229.

Elgar, M. A., and P. H. Harvey. 1987. Basal metabolic rates in mammals: allometry, phylogeny and ecology. Functional Ecology, 1:25–26.

Evans, A. R., and G. D. Sanson. 1998. The effect of tooth shape on the breakdown of insects. Journal of Zoology (London), 46:391–400.

Felsenstein, J. 1985. Phylogenies and the comparative method. American Naturalist, 125:1–15.

Fenton, M. B. 1989. Head size and foraging behaviour of animal-eating bats. Canadian Journal of Zoology, 67:2029–2035.

Findley, J. S. 1972. Phenetic relationships among bats of the genus *Myotis*. Systematic Zoology, 21:31–52.

Findley, J. S. 1993. Bats: A Community Perspective. Cambridge University Press, Cambridge

Findley, J. S., E. H. Studier, and D. E. Wilson. 1972. Morphologic properties of bat wings. Journal of Mammalogy, 53:429–444.

Fleming, T. H. 1991. The relationship between body size, diet, and habitat use in frugivorous bats, genus *Carollia* (Phyllostomidae). Journal of Mammalogy, 72:493–501.

Fleming, T. H. 1993. Plant-visiting bats. American Scientist, 81:460–467.

Freeman, P. W. 1979. Specialized insectivory: Beetle-eating and moth-eating molossid bats. Journal of Mammalogy, 60:467–479.

Freeman, P. W. 1981a. Correspondence of food habits and morphology in insectivorous bats. Journal of Mammalogy, 62:166–173.

Freeman, P. W. 1981b. A multivariate study of the family Molossidae (Mammalia, Chiroptera): morphology, ecology, evolution. Fieldiana (Zoology), 7:1–173.

Freeman, P. W. 1984. Functional cranial analysis of large animalivorous bats (Microchiroptera). Biological Journal of the Linnean Society, 21:387–408.

Freeman, P. W. 1988. Frugivorous and animalivorous bats (Microchiroptera): dental and cranial adaptations. Biological Journal of the Linnean Society, 33:249–272.

Freeman, P. W. 1992. Canine teeth of bats (Microchiroptera): size, shape, and role in crack propagation. Biological Biological Journal of the Linnean Society, 45:97–115.

Freeman, P. W. 1995. Nectarivorous feeding mechanisms in bats. Biological Journal of the Linnean Society, 56:439–463.

Freeman, P. W. 1998. Form, function, and evolution in skulls and teeth of bats. Pp. 140–156 *in:* Bat Biology and Conservation (T. H. Kunz and P. A. Racey, eds). Smithsonian Institution Press, Washington, D.C.

Freeman, P. W. 2000. Macroevolution in Microchiroptera: Recoupling morphology and ecology with phylogeny. Evolutionary Ecology Research, 2:317–335.

Freeman, P. W., and W. N. Weins. 1997. Puncturing ability of bat canine teeth: the tip. Pp. 225–232 *in:* Life among the Muses: Papers in Honor of James S. Findley (T. L. Yates, W. L. Gannon, and D. E. Wilson, eds.). University of New Mexico Press, Albuquerque.

Galilei, G. 1637. Proposition VIII. Pp. 1–108 *in:* Dialogues concerning the Two New Sciences (H. Crew and A.T. De Salvio, eds.). Macmillan, New York.

Gans, C. 1982. Fiber architecture and muscle function. Exercise Sport Science Review, 10:160–207.

Garland, T., Jr., P. E. Midford, and A. R. Ives. 1999. An introduction to phylogenetically based statistical methods, with a new method for confidence intervals on ancestral values. American Zoologist, 39:374–388.

Goldman, L. J., and O. W. Henson, Jr. 1977. Prey recognition and selection by the constant frequency bat, *Pteronotus p. parnellii*. Behavioral Ecology and Sociobiology, 2:411–419.

Grant, B. R., and P. R. Grant. 1989. Evolutionary Dynamics of a Natural Population: The Large Cactus Finch of the Galapagos. University of Chicago Press, Chicago.

Grant, P. R. 1986. Ecology and Evolution of Darwin's Finches. Princeton University Press, Princeton, N.J.

Griffiths, T. A. 1982. Systematics of the New World nectar-feeding bats (Mammalia, Phyllostomidae), based on the morphology of the hyoid and lingual regions. American Museum Novitates, 2742:1–45.

Handley, C. O., Jr., D. E. Wilson, and A. L. Gardner. 1991. Demography and natural history of the common fruit bat, *Artibeus jamaicensis*, on Barro Colorado Island, Panama. Smithsonian Contributions to Zoology, no. 511. Smithsonian Institution Press, Washington, D.C.

Hayssen, V. 1984. Basal metabolic rate and the intrinsic rate of increase: an empirical and theoretical reexamination. Oecologia, 64:419–421.

Hayssen, V., and R. C. Lacy. 1985. Basal metabolic rates in mammals: taxonomic differences in the allometry of BMR and body mass. Comparative Biochemistry and Physiology A, 81:741–754.

Hermanson, J. W., and J. S. Altenbach. 1981. Functional anatomy of the primary downstroke muscles in the pallid bat, *Antrozous pallidus*. Journal of Mammalogy, 62:801–805.

Hickey, M. B. C., L. Acharya, and S. Pennington. 1996. Resource partitioning by two species of vespertilionid bats (*Lasiurus cinereus* and *Lasiurus borealis*) feeding around street lights. Journal of Mammalogy, 77:325–334.

Howell, D. J., and J. Pylka. 1977. Why bats hang upside-down: a biomechanical hypothesis. Journal of Theoretical Biology, 69:625–631.

Hughes, P., R. D. Ransome, and G. Jones. 1989. Aerodynamic constraints on flight ontogeny in free-living greater horseshoe bats, *Rhinolophus ferrumequinum*. Pp. 255–262 *in:* European Bat Research 1987. (V. Hanak, I. Horacek, and J. Gaisler, eds.). Proceedings of the Fourth European Bat Conference, Prague 1987. Charles University Press, Prague, Czechoslovakia.

Hughes, P.M., and J. M. V. Rayner. 1991. Addition of artificial loads to long-eared bats *Plecotus auritus*: handicapping flight performance. Journal of Experimental Biology, 161:285–298.

Hughes, P, and J. M. V. Rayner. 1993. The flight of pipistrelle bats *Pipistrellus pipistrellus* during pregnancy and lactation. Journal of Zoology (London), 230:541–555.

Hughes, P., J. M. V. Rayner, and G. Jones. 1995. Ontogeny of "true" flight and other aspects of growth in the bat *Pipistrellus pipistrellus*. Journal of Zoology (London), 235:291–318.

Huiskes, R., and E. Y. S. Chao. 1983. A survey of finite element analysis in orthopedic biomechanics: the first decade. Journal of Biomechanics, 16:385–409.

Jones, G. 1994. Scaling of wingbeat and echolocation pulse emission rates in bats: why are aerial insectivorous bats so small? Functional Ecology, 8:450–457.

Jones, G. 1995. Flight performance, echolocation and foraging behaviour in noctule bats *Nyctalus noctula*. Journal of Zoology (London), 237:303–312.

Jones, G., and T. Kokurewicz. 1994. Sex and age variation in echolocation calls and flight morphology of Daubenton's bats *Myotis daubentonii*. Mammalia, 58:41–50.

Kalcounis, M. C., and R. M. Brigham. 1995. Intraspecific variation in wing loading affects habitat use by little brown bats (*Myotis lucifugus*). Canadian Journal of Zoology, 73:89–95.

Kalko, E. K. V., E. A. Herre, and C. O. Handley, Jr. 1996. Relation of fig fruit characteristics to fruit-eating bats in the New and Old World tropics. Journal of Biogeography, 23:565–576.

Kalko, E. K. V., and H.-U. Schnitzler. 1989. The echolocation and hunting behavior of Daubenton's bat, *Myotis daubentonii*. Behavioral Ecology and Sociobiology 24: 225–238.

Kallen, F. C., and C. Gans. 1972. Mastication in the little brown bat, *Myotis lucifugus*. Journal of Morphology, 136:385–420.

Kiltie, R. A. 1982. Bite force as a basis for niche differentiation between rain forest peccaries (*Tayassu tajacu* and *T. pecari*). Biotropica, 14:188–195.

Koehl, M. A. R. 1995. When does morphology matter? Annual Review of Ecology and Systematics, 27:501–542.

Koopman, K. F. 1994. Chiroptera: Systematics. Pt. 60. Handbook of Zoology. Vol. 8. Mammalia. Walter de Gruyter, New York.

LaBarbera, M. 1986. The evolution and ecology of body size. Pp. 60–98 *in:* Patterns and Processes in the History of Life (D. M. Raup and D. Jablonski, eds.). Springer-Verlag, Berlin.

LaBarbera, M. 1989. Analyzing body size as a factor in ecology and evolution. Annual Review of Ecology and Systematics, 20:97–118.

Lacki, M. J., L. S. Burford, and J. O. Whitaker, Jr. 1995. Food habits of gray bats in Kentucky. Journal of Mammalogy, 76:1256–1259.

Lancaster, W. C., O. W. Henson, Jr., and A. W. Keating. 1995. Respiratory muscle activity in relation to vocalization in flying bats. Journal of Experimental Biology, 198:175–191.

LaVal, R. K., and H. S. Fitch. 1977. Structure, movements, and reproduction in three Costa Rican bat communities. Occasional Papers, Museum of Natural History, University of Kansas, 69:1–28.

Loeb, G. E., and C. Gans. 1986. Electromyography for Experimentalists. University of Chicago Press, Chicago.

Losos, J. B. 1996. Phylogenies and comparative biology. 2. Testing causal hypotheses derived from phylogenies with data from extant taxa. Systematic Biology, 45:259–270.

Losos, J. B., and D. B. Miles. 1994. Adaptation, constraint, and the comparative method: phylogenetic issues and methods. Pp. 60–98 *in:* Ecological Morphology, Integrative Organismal Biology (P. C. Wainwright and S. M. Reilly, eds.). University of Chicago Press, Chicago.

Lucas, P. W. 1979. The dental-dietary adaptations of mammals. Neues Jahrbuch für Geologie und Palaeontologie Monatshefte, 8:486–512.

Lucas, P. W., and D. A. Luke. 1984. Chewing it over: basic principles of food breakdown. Pp. 283–301 *in:* Food Acquisition and Processing in Primates (D. J. Chivers, B. A. Wood, and A. Bilsborough, eds.). Plenum Press, London.

McMahon, T. A. 1973. Size and shape in biology. Science, 179:1201–1204.

McMahon, T. A. 1984. Muscles, Reflexes, and Locomotion. Princeton University Press, Princeton, N.J.

McNab, B. K. 1983. Energetics, body size, and the limits to endothermy. Journal of Zoology (London), 199:1–29.

McNab, B. K. 1990. The physiological significance of body size. Pp. 11–21 *in:* Body Size in Mammalian Paleobiology: Estimation and Biological Implications (J. Damuth and B. J. McFadden, eds.). Cambridge University Press, Cambridge.

Moore, S. J., and G. D. Sanson. 1995. A comparison of the molar efficiency of two insect-eating mammals. Journal of Zoology (London), 235:175–192.

Nagy, K. A. 1987. Field metabolic rate and food requirement scaling in mammals and birds. Ecological Monographs, 57:111–128.

Norberg, U. M. 1970. Functional osteology and myology of the wing of *Plecotus auritus* Linnaeus (Chiroptera). Arkiv för Zoologi, 22:483–543.

Norberg, U. M. 1972. Functional osteology and myology of the wing of the dog-faced bat, *Rousettus aegypticus* (E. Geoffroy) (Pteropidae). Zeitschrift für Morphologie der Tiere, 73:1–44.

Norberg, U. M. 1976a. Some advanced flight maneuvers of bats. Journal of Experimental Biology, 64:489–495.

Norberg, U. M. 1976b. Aerodynamics, kinematics and energetics of horizontal flapping flight in the long-eared bat *Plecotus auritus*. Journal of Experimental Biology, 65:179–212.

Norberg, U. M. 1976c. Aerodynamics of hovering flight in the long-eared bat *Plecotus auritus*. Journal of Experimental Biology, 65:459–470.

Norberg, U. M., and M. B. Fenton. 1988. Carnivorous bats? Biological Journal of the Linnean Society, 33:383–394.

Norberg, U. M., and J. M. V. Rayner. 1987. Ecological morphology and flight in bats: wing adaptations, flight performance, foraging strategy and echolocation. Philosophical Transactions of the Royal Society of London B, 316:335–427.

O'Neill, M. G., and R. J. Taylor. 1989. Feeding ecology of Tasmanian bat assemblages. Australian Journal of Ecology, 14:19–31.

Papadimitriou, H. M., S. M. Swartz, and T. H. Kunz. 1996. Ontogenetic and anatomic variation in mineralization of the wing skeleton of the Mexican free-tailed bat, *Tadarida brasiliensis*. Journal of Zoology (London), 240:411–426.

Perry, G. 1999. The evolution of search modes: ecological versus phylogenetic perspectives. American Naturalist, 153:98–109.

Peters, R. H. 1983. The Ecological Implications of Body Size. Cambridge University Press, Cambridge.

Pierson, E. D. 1986. Molecular systematics of the Microchiroptera: higher taxon relationships and biogeography. Ph.D. Dissertation. University of California, Berkeley.

Popowics, T. E., and M. Fortelius. 1997. On the cutting edge: tooth blade sharpness in herbivorous and faunivorous mammals. Annales Zoologica Fennici, 34:73–88.

Powers, L. V., S. C. Kandarian, and T. H. Kunz. 1991. Ontogeny of flight in the little brown bat, *Myotis lucifugus:* behavior, morphology, and muscle histochemistry. Journal of Comparative Physiology A, 186:675–681.

Quinn, T. H., and J. J. Baumel. 1993. Chiropteran tendon locking mechanism. Journal of Morphology, 216:197–208.

Rayner, J. V. M. 1986. Vertebrate flapping flight mechanics and aerodynamics and the evolution of flight in bats. Pp. 27–74, Bat Flight—Fledermausflug. Biona (W. Nachtigall, ed.). Report 5. Gustav Fischer, Stuttgart.

Rayner, J. V. M., G. Jones, and A. Thomas. 1986. Vortex flow visualizations reveal change in upstroke function with flight speed in bats. Nature, 321:162–164.

Reduker, D. W. 1983. Functional analysis of the masticatory apparatus in two species of *Myotis*. Journal of Mammalogy, 64:277–286.

Richards, G. C. 1995. A review of ecological interactions of fruit bats in Australian ecosystems. Symposia of the Zoological Society of London, no. 67:79–96.

Rosenberger, A. L., and W. G. Kinzey. 1976. Functional patterns of molar occlusion in platyrrhine primates. American Journal of Physical Anthropology, 45:281–298.

Rydell, J., and D. W. Yalden. 1997. The diets of two high-flying bats from Africa. Journal of Zoology (London), 242:69–76.

Sahley, C. T., M. A. Horner, and T. H. Fleming. 1993. Flight speeds and mechanical power outputs of the nectar-feeding bat, *Leptonycteris curasoae* (Phyllostomidae: Glossophaginae). Journal of Mammalogy, 74:594–600.

Sample, B. E., and R. C. Whitmore. 1993. Food habits of the endangered Virginia big-eared bat in West Virginia. Journal of Mammalogy, 74:428–435.

Saunders, M. B., and R. M. R. Barclay. 1992. Ecomorphology of insectivorous bats: a test of predictions using two morphologically similar species. Ecology, 73:1335–1345.

Schmidt-Nielsen, K. 1984. Scaling: Why Is Animal Size So Important? Cambridge University Press, Cambridge.

Schutt, W. A., Jr. 1993. Digital morphology in the Chiroptera—the passive digital lock. Acta Anatomica, 148:219–227.

Schutt, W. A., Jr. 1998. Chiropteran hindlimb morphology and the origin of blood-feeding in bats. Pp. 157–168 *in:* Bat Biology and Conservation (T. H. Kunz and P. A. Racey, eds.). Smithsonian Institution Press, Washington, D.C.

Schutt, W. A., Jr., J. S. Altenbach, Y. H. Chang, D. M. Cullinane, J. W. Hermanson, F. Murdali, and J. E. A. Bertram. 1997. The dynamics of flight-initiating jumps in the common vampire bat *Desmodus rotundus*. Journal of Experimental Biology, 200:3003–3012.

Shubin, N., and D. Wake. 1996. Phylogeny, variation, and morphological integration. American Zoologist, 36:51–60.

Silva, M., and J. A. Downing. 1995. CRC Handbook of Mammalian Body Masses. CRC Press, Boca Raton, Fla.

Simmons, N. B. 1998. A reappraisal of interfamilial relationships of bats. Pp. 3–26 *in:* Bat Biology and Conservation (T. H. Kunz and P. A. Racey, eds.). Smithsonian Institution Press, Washington, D.C.

Simmons, N. B., and T. H. Quinn. 1994. Evolution of the digital tendon locking mechanism in bats and dermopterans: a phylogenetic perspective. Journal of Mammalian Evolution, 2:231–254.

Slaughter, B. H. 1970. Evolutionary trends of chiropteran dentitions. Pp. 51–83 *in:* About Bats (B. H. Slaughter and D. W. Walton, eds.). Southern Methodist University Press, Dallas.

Smith, J. D. 1976. Chiropteran evolution. Pp. 49–69 *in:* Biology of Bats of the New World Family Phyllostomatidae (R. J. Baker, J. K. Jones, Jr., and D. C. Carter, eds.). Pt. 1. Special Publications, the Museum, Texas Tech University, 10. Texas Tech Press, Lubbock.

Smith, J. D., and A. Starrett. 1979. Morphometric analysis of chiropteran wings. Pp. 229–316 *in:* Biology of Bats of the New World Family Phyllostomatidae. (R. J. Baker, J. K. Jones, and D. C. Carter, eds.). Pt. 3. Special Publications the Museum, Texas Tech University, 16. Texas Tech Press, Lubbock.

Stern, A. A., T. H. Kunz, and S. S. Bhatt. 1997. Seasonal wing loading and the ontogeny of flight in *Phyllostomus hastatus* (Chiroptera: Phyllostomidae). Journal of Mammalogy, 78:1199–1209.

Stockwell, E. F. 2001. Morphology and flight manoeuvrability in New World leaf-nosed bats (Chiroptera: Phyllostomidae). Journal of Zoology (London), 254:505–514.

Storch G. 1968. Funktionsmorphologische Untersuchungen an der Kaumuskulatur und an korrelierten Schadelstrukturen der Chiropteren. Abhandlungen der senckenbergischen naturforschenden Gesellschaft, 517:1–92.

Strait, S. G. 1993. Molar morphology and food texture among small-bodied insectivorous mammals. Journal of Mammalogy, 74:391–402.

Strauss, R. E., and R. Altig. 1992. Ontogenetic body form changes in three ecological morphotypes of anuran tadpoles. Growth, Development, and Aging, 56:3–16.

Swartz, S. M. 1991. Strain analysis as a tool for functional morphology. American Zoologist, 31:655–669.

Swartz, S. M. 1997. Allometric patterning in the limb skeleton of bats: implications for the mechanics and energetics of powered flight. Journal of Morphology, 234:277–294.

Swartz, S. M. 1998. Skin and bones: functional, architectural, and mechanical differentiation in the bat wing. Pp. 109–126, *in:* Bat Biology and Conservation (T. H. Kunz and P. A. Racey, eds.). Smithsonian Institution Press, Washington, D.C.

Swartz, S. M., M. B. Bennett, and D. R. Carrier. 1992. Wing bone stresses in free flying bats and the evolution of skeletal design for flight. Nature, 359:726–729.

Swartz, S. M., and A. A. Biewener. 1992. Shape and scaling. Pp. 20–43 *in:* Biomechanics: A Practical Approach. Vol. 2. Structures (A. A. Biewener, ed.). Oxford University Press, Oxford.

Swartz, S. M., M. S. Groves, H. D. Kim, and W. R. Walsh. 1996. Mechanical properties of bat wing membrane skin. Journal of Zoology (London), 239:357–378.

Thollesson, M., and U. M. Norberg. 1991. Moments of inertia of bat wings and body. Journal of Experimental Biology, 158:19–35.

Thomas, S. P. 1975. Metabolism during flight in two species of bats, *Phyllostomus hastatus* and *Pteropus gouldii*. Journal of Experimental Biology, 63:273–293.

Thomas, S. P. 1981. Ventilation and oxygen extraction in the bat *Pteropus gouldii* during rest and steady flight. Journal of Experimental Biology, 94:231–250.

Timm, R. M., D. E. Wilson, B. L. Clauson, R. K. LaVal, and C. S. Vaughan. 1989. Mammals of the La Selva-Braulio Carrillo complex. North American Fauna, 75:1–162.

Van Valkenburgh, B., and C. B. Ruff. 1987. Canine tooth strength and killing behavior in large carnivores. Journal of Zoology (London), 212:379–397.

Vaughan, T. A. 1959. Functional morphology of three bats: Eumops, Myotis, Macrotus. University of Kansas Publications, Museum of Natural History, 12:1–153.

Vaughan, T. A. 1977. Foraging behaviour of the giant leaf-nosed bat (*Hipposideros commersoni*). East African Wildlife Journal, 15:237–249.

Vincent, J. F. V. 1980. Insect cuticle: a paradigm for natural composites. Pp. 183–210 *in:* The Mechanical Properties of Biological Materials (J. F. V. Vincent and J. D. Currey, eds.). Cambridge University Press, Cambridge.

Voss, R. S., L. F. Marcus, and P. Escalante P. 1990. Morphological evolution in muroid rodents. I. Conservative patterns of craniometric covariance and their ontogenetic basis in the Neotropical genus *Zygodontomys*. Evolution, 44:1568–1587.

Wainwright, P. C. 1987. Biomechanical limits to ecological performance: mollusc-crushing by the Caribbean hogfish, *Lachnolaimus maximus* (Labridae). Journal of Zoology (London), 213:283–297.

Waters, D. A., J. Rydell, and G. Jones. 1995. Echolocation call design and limits on prey size: a case study using aerial-hawking bat *Nyctalus leisleri*. Behavioral Ecology and Sociobiology, 37:321–328.

Watts, P., E. J. Mitchell, and S. M. Swartz. 2001. A computer model for estimating mechanics of horizontal flapping flight in bats: model description and validation. Journal of Experimental Biology, 204:2873–2898.

Webb, P. I., J. R. Speakman, and P. A. Racey. 1992. Inter- and intra-individual variation in wing loading and body mass in female pipistrelle bats: theoretical implications for flight performance. Journal of Zoology (London), 228:669–673.

Whitaker, J. O., Jr., C. Neefus, and T. H. Kunz. 1996. Dietary variation in the Mexican free-tailed bat (*Tadarida brasiliensis mexicana*). Journal of Mammalogy, 77:716–724.

Zienkiewicz, O. C., and R. L. Taylor. 1989. The Finite Element Method. Vol. 1. McGraw-Hill, London.

CHAPTER 7

Attack and Defense: Interactions between Echolocating Bats and Their Insect Prey

Gareth Jones and Jens Rydell

Introduction

Insects and other small arthropods are the most widespread food of bats. About 70% of the world's approximately 1,100 or so extant bat species are insectivorous. Most bats and all insectivorous ones are in the suborder Microchiroptera (Hill and Smith 1984). Because all microchiropterans use echolocation, insectivory is intimately linked to the natural history of echolocation. Frugivorous bats (including some Megachiroptera) may also eat insects occasionally, perhaps as a supplementary source of protein (Courts 1998).

Over evolutionary time, bats have influenced the evolution of many traits in nocturnal insects, and features of insects have also influenced how bats use echolocation. Here we review how bats detect and capture insects and how insects evade echolocating bats. We ask questions that we think are of general interest from an evolutionary point of view.

Echolocation and Insectivory in Early Bats

The earliest remains of echolocating bats were found in sediments from the middle of the Eocene Epoch (beginning of the Tertiary Period, ca. 50 million years ago) in Wyoming, USA (Green River Formation [Jepsen 1970; Novacek 1985]) and southern Germany (Messel Formation [Habersetzer et al. 1994]). In the Messel quarry, bats of four species from two genera (*Palaeochiropteryx* and *Hassianycteris*) have fossilized stomach contents showing that they fed mainly on small moths (Microlepidoptera), caddis-flies (Trichoptera), and true flies (Diptera) (Habersetzer et al. 1994). Thus, insects and bats have coexisted for most of the Tertiary, or at least for 50 million years.

The moderately enlarged cochlea of *Icaronycteris index*, the only bat found in the Green River Formation, suggests that Eocene bats echolocated, even if not with the sophistication shown by extant aerial-hawking bats (Novacek 1985; Simmons and Geisler 1998). A similar claim can be made for the Messel bats (Habersetzer et al. 1994).

When bats first appeared in the fossil record, several taxa of diurnal predatory birds such as hawks and falcons were present (Peters 1992). Predation by these may have restricted bats to nocturnality during most of their early history (Rydell and Speakman 1995).

Before insectivorous bats evolved, nocturnal flying insects may have been relatively unexploited by predators except when they were resting. Although it is difficult to speculate about environmental conditions when the first bats evolved, today insects are an abundant source of food during the night. Indeed, in temperate ecosystems there are few other suitable food resources for bats.

Vision is of limited use for tracking small, mobile, aerial prey in the dark or in unpredictable lighting (see Altringham and Fenton, this volume). In contrast, echolocation is effective for this, although bats need to produce intense ultrasonic pulses in order to receive audible echoes from targets as small as insects (Pye 1993). Flying bats tend to correlate call emission with wing flapping (review in Schnitzler and Henson 1980) and use their flight muscles to increase air pressure behind the larynx (Lancaster et al. 1995). The coupling of calling and flapping probably permits the production of intense echolocation calls during flight at little extra energetic expenditure on top of the flight cost (Speakman and Racey 1991). Conversely, the added energetic cost of echolocation when bats are stationary may be very high (Speakman et al. 1989).

The evolution and refinement of flight and echolocation was obviously of great fitness value for the exploitation of nocturnal insects. However, the short range for ultrasonic echolocation of small targets in the air (see below) requires slow flight, and this seems to constrain the performance of bats because it renders them susceptible to predation by raptorial birds during the day (Speakman 1991, 1995).

Which Insects Are Eaten by Bats and When?

Bats eat a diverse range of insects. Among the smallest prey are biting and nonbiting midges (Diptera; Ceratopogonidae and Chironomidae), which have wingspans of only a few millimeters. More than half of the insects eaten by European pipistrelles, *Pipistrellus* spp. (body mass, ca. 5 g) in Scotland, for example, have a body length of only 4–12 mm (Swift et al. 1985). At the other extreme, Commerson's leaf-nosed bat, *Hipposideros commersoni* (74–180 g), feeds on beetles with body lengths up to 5 cm (Vaughan 1977).

What sorts of insects do bats eat, and what selective forces do these insects place on bats? Although most bat species seem to be flexible in their diets, some species apparently specialize on certain insect taxa. Although bats can eat large numbers of insects each night, we still do not know whether bats regulate populations of prey species.

Diptera are among the most abundant nocturnal insects, at least in temperate ecosystems, and are important prey for many bats. Many Diptera are small and, in theory, should be best detected by high-frequency calls with short wavelengths. Dipteran flight activity peaks early in the night (Racey and Swift 1985; Rydell et al. 1996a), so bats feeding on Diptera may be selected to leave their roosts early. Bats that feed on aerial insects are relatively fast flying (for bats), a trait that helps them to avoid avian predators, which may still be active at dusk (Jones and Rydell 1994). Bats that eat large numbers of aerial Diptera include *Pipistrellus* spp. in Europe (reviewed by Vaughan 1997) and *Myotis lucifugus* and *Lasionycteris noctivagans* in North America (Barclay 1985; Belwood and Fenton 1976; reviewed by Freeman 1981). Some aerial-hawking species emit echolocation calls of relatively high frequencies (e.g., *Pipistrellus* uses terminal frequencies of 35–60 kHz, often sweeping from 80–90 kHz [Kalko and Schnitzler 1993]). However, several larger bats whose call frequencies are 20–25 kHz (wavelengths 14–17 mm), such as *Nyctalus noctula* (Jones 1995), *N. leisleri* (Waters et al. 1995), and *Vespertilio murinus* (Rydell 1992a), eat many small dipterans. Some bats feed mainly by gleaning Diptera that are either resting or that are active close to vegetation. For example, *Myotis nattereri* eats many dipterans that fly by day and rest at night (reviewed by Vaughan 1997). This species has relatively broad wings conferring slow, manoeuvrable flight suitable for the close vicinity of vegetation (Norberg and Rayner 1987). *Myotis nattereri* can detect prey very close to surfaces by using broadband echolocation calls (Siemers and Schnitzler 2000).

Coleoptera, particularly dung beetles (Scarabaeidae), are eaten in large numbers by *Eptesicus serotinus* in Europe (Vaughan 1997), by *E. fuscus* in North America (Freeman 1981), and by *Hipposideros commersoni* in Africa (Vaughan 1977). Some molossids that eat beetles, such as *Molossus ater* (Fenton et al. 1998c) have relatively few but large teeth and powerful jaws compared with related species that eat more soft-bodied insects (Freeman 1979; Schwartz et al., this volume).

Many bats eat large numbers of moths (Lepidoptera). Apparent moth specialists include the gleaning *Plecotus auritus* and *P. austriacus* in Europe (Vaughan 1997), *Corynorhinus rafinesquii* in North America (Hurst and Lacki 1997), and *Cloeotis percivali* in Africa (Whitaker and Black 1976). Some large (25–40 g), fast flying molossids, including *Tadarida teniotis* and *Otomops martiensseni* (Rydell and Arlettaz 1994, Rydell and Yalden 1997), also eat many moths. Other bats specializing in moth predation include *Barbastella barbastellus* (Rydell et al. 1996a; Sierro and Arlettaz 1997) and *Euderma maculatum* (Ross 1967). *Euderma* is probably not a gleaner, and it may catch moths by aerial hawking in open habitats (Woodsworth et al. 1981). Aerial-hawking *Lasiurus* species capture many moths around streetlights in North America (Acharya and Fenton 1992, 1999).

Bats such as gleaning *Myotis blythii* (Arlettaz and Perrin 1995) and *Otonycteris hemprichi* (Arlettaz et al. 1995) often eat Orthoptera such as crickets and katydids, as do some Neotropical phyllostomids such as *Tonatia silvicola* and *Micronycteris hirsuta* (Belwood and Morris 1987). Bats also eat Hemiptera. For example, water striders (Gerridae) are sometimes captured by *Myotis daubentonii* by trawling over the water surface (A. M. Svensson, I. Danielsson, and J. Rydell, unpublished data), and *Nycticeius humeralis* eats stinkbugs (Pentatomidae) and chinchbugs (Lygaeidae [Whitaker and Clem 1992]). Barclay (1985) found remains of Odonata (dragonflies) under day roosts of *Lasiurus cinereus*.

Nocturnal swarms of Hymenoptera such as ants, and Isoptera (termites) are often exploited by bats (Gould 1978; Kunz et al. 1995). This also applies to Ephemeroptera (mayflies), Trichoptera (caddis flies [e.g., Belwood and Fenton 1976; Brigham et al. 1992]) and some Neuroptera (e.g., alder flies), insects that typically form crepuscular or nocturnal swarms over water (e.g., Sweeney and Vannote 1982). Exploitation of swarms of aquatic insects seems particularly important for bats at high latitudes in Europe and North America, areas where lakes and rivers are prominent landscape features (e.g., Fenton et al. 1983b; Racey et al. 1998; Rydell 1986, 1992a).

There may be geographical variation in the relative importance of different insect taxa in the diets of bats (Vaughan 1997). Differences in natural vegetation, land use, geology, or climate may affect prey availability. British bats eat mainly Diptera, which are the most abundant insects captured in light traps (Williams 1939). In North America, fewer studies find Diptera to be a major component of bat diets, even in aerial-hawking species (Barclay 1985; Freeman 1981; Ross 1967; but see Anthony and Kunz 1977). In southern Africa, Coleoptera and Lepidoptera are recorded most frequently in dietary studies (Aldridge and Rautenbach 1987; Fenton et al. 1998a; Freeman 1981), although this could reflect the high diversity of bats there.

In many insects, males may be more vulnerable to bat predation than females as a consequence of sexual selection. Singing male orthopterans may be especially vulnerable to predation because bats often locate these insects by listening to their songs (Bell 1985; Belwood and Morris 1987; Hosken et al. 1994), though nonstridulating females make up a significant proportion of the diet of some gleaning bats (Bell 1982; Belwood and Morris 1987). Males perhaps are at greater risk of predation than females in many Diptera and other swarming insects because males dominate swarms (Downes 1969). In insects that use pheromones for mate attraction (many beetles and moths), females are the signaling sex. Nevertheless, males pay the price of a higher predation risk because they must spend considerable time in flight searching for females (Acharya 1995; Acharya and McNeil 1998; Jones 1990).

How Bats Detect and Capture Insects

All microchiropteran bats studied to date use echolocation, although not always for detection, classification, and localization of insects. Before describing

how bats capture insects, we first describe some major features of bat echolocation. Echolocation can be considered in terms of "duty cycles," that is, how
much of the time the signal is "on" (Fenton et al. 1995). Low duty-cycle (less
than ca. 20% [Jones 1999]) bats use short (0.5–30 ms) echolocation pulses separated by much longer listening intervals and, hence, separate the outgoing
pulse and returning echo in the time domain. In contrast, high-to-medium
duty-cycle (greater than ca. 30% [Jones 1999]) species use longer pulses (sometimes 50 ms or more) separated by shorter intervals (mainly for breathing).
They separate the outgoing pulse and the returning echo in the frequency domain and, therefore, tolerate a temporal pulse-echo overlap. These two approaches to echolocation thus operate in very different ways.

Low Duty-Cycle Echolocation

Bat echolocation calls are composed of frequency-modulated (FM) or constant
frequency (CF) components or a combination of both (CF-FM, FM-CF-FM).
Frequency-modulated components may be either steep (broadband) or shallow (narrowband), and in the latter case they are sometimes called "quasi-CF"
components (QCF [Schnitzler and Kalko 1998]). Broadband FM signals typically sweep downward through about an octave in a short duration (typically
0.5–5 ms). Such signals provide accurate measures of the range and angle
of a target (Simmons and Stein 1980) and are therefore well suited for prey localization (Schnitzler and Kalko 1998). As they also encode spectral information, FM signals may be useful for classifying targets according to texture
(Schmidt 1988).

Bats that search for prey within vegetation often use broadband FM calls
(Neuweiler 1989), but those that do so in open areas or in edge habitats often
add a longer (typically 5–10 ms) narrowband component to the signal. The
narrowband component is usually at the end of the call, where frequencies are
lowest (e.g., *Pipistrellus* spp. [Kalko and Schnitzler 1993]) although often at the
start of the call in emballonurids (e.g., Barclay 1983) and noctilionids (Kalko
et al. 1998; Schnitzler et al. 1994). Rather than spreading energy broadly over a
range of frequencies, most energy is thus focused in a relatively narrowband
component, and this may facilitate detection of distant objects (Neuweiler
1989). The low frequency and high amplitude (typically >100 dB peak-
equivalent sound pressure level at 10 cm from the bat's mouth [Griffin 1950;
Surlykke et al. 1993; Waters and Jones 1995]) of the narrowband component
probably maximizes the bat's detection range. Narrow neuronal filters may facilitate the detection of "acoustic glints," that is, modulations in the amplitude
and frequency of echoes from fluttering insect wings (Schnitzler et al. 1983).

Bats that use pure FM echolocation calls, and species that emit FM calls that
are followed by narrowband tails, keep outgoing pulses and echoes returning
separate in time because they cannot detect echoes while calling and, thus,
are said to be intolerant of pulse-echo overlap (Kalko and Schnitzler 1993).

Likewise, weak echoes from insects are harder to detect if they overlap temporally with echoes returning from the background (i.e., unwanted echoes are called clutter). As bats approach an object they shorten the pulses, so that pulse-echo overlap is avoided or minimized (Kalko and Schnitzler 1993). However, some bats that emit FM calls at low duty cycles and feed close to vegetation (e.g., *Myotis nattereri*) may be partly tolerant of pulse-echo overlap (Siemers and Schnitzler 2000). Bats also vary the duration of the interpulse (listening) interval depending on the distance to the background. Because bats using this type of echolocation must separate pulses and echoes temporally, short pulses and long listening intervals are employed—that is, they operate with low duty cycles.

Because atmospheric attenuation is particularly severe at high frequencies (Griffin 1971; Lawrence and Simmons 1982), ultrasonic calls have short ranges. Bats must therefore be close to small targets before they can detect echoes from them (typically 1–5 m) for frequencies near 20 kHz, and closer for higher frequencies (Kick 1982; Kick and Simmons 1984; Simmons 1973; Waters et al. 1995).

Ultimately it is the short effective range of echolocation that puts such high demand on the maneuverability and flight speed, and hence on the body mass, of insectivorous bats (Barclay and Brigham 1991; Jones 1996). For example, at a flight speed of 5 m s^{-1}, typical for many small bats (Baagøe 1987), and a detection range of 1 m, typical for small insect targets (Waters et al. 1995), the bat has only 0.2 s between detection and prey contact. It is probably this time constraint that has shaped much of bat feeding behavior, including prey selection (see below [Barclay and Brigham 1994]).

High Duty-Cycle Echolocation

Some bats (rhinolophid and hipposiderid species in the Old World, *Pteronotus parnellii* [Henson et al. 1980] in the New World) use long CF components in their calls. Bats in the family Hipposideridae typically emit calls of shorter duration (<20 ms) than do bats in the family Rhinolophidae (often >50 ms), but both groups search for prey by using medium or high duty cycles (e.g., 35%–70% [Jones and Rayner 1989; Jones et al. 1993]). Rhinolophids echolocate by using higher duty cycles than hipposiderids (Jones 1999). A high duty cycle increases the chance of detecting glints from insect wingbeats (Schnitzler et al. 1983; Kober and Schnitzler 1990). In high duty-cycle echolocation, outgoing pulses and returning echoes are separated in frequency rather than time (Fenton et al. 1995). To achieve this, bats make use of Doppler shifts that are induced by their own movement. Lowering the emitted frequency compensates for Doppler shifts caused by the bat's flight speed (Schnitzler 1968) so that echoes return in a narrow frequency window of heightened sensitivity (the "acoustic fovea" [Schuller and Pollak 1979]) that is slightly higher in frequency than the emitted calls. The emitted sound can then be "filtered out" because the bats have reduced auditory sensitivity at these frequencies, thanks to specializations in the inner ear and the brain (reviewed in Pollak and Casseday 1989).

Bats that use high duty-cycle echolocation can detect the echoes of fluttering insects among background echo clutter. Rhinolophid and hipposiderid bats generally echolocate at much higher frequencies compared to similar-sized species that use low duty-cycle echolocation, and most rhinolophoid bats emphasize the second harmonic of the calls (Heller and Helversen 1989; Jones 1999). The reason for this is not understood. Rhinolophids also combine long durations and high frequencies in their calls because they can tolerate temporal overlap between outgoing pulse and returning echo. Conversely, low duty-cycle bat species emit short pulses if they call at high frequencies. This is because extreme atmospheric attenuation limits detection to close ranges, and short pulses are then needed to avoid overlap between the outgoing pulse and the echo, which will return soon after pulse emission (Waters et al. 1995).

Bats generally trade off accuracy in localization (best achieved with broadband FM signals) against detection (best achieved with CF, or narrowband signals) by emphasizing different components in various echolocation pulses (Schnitzler and Kalko 1998). For example, all bat species that produce terminal buzzes (see below) exaggerate the FM components of the calls after prey detection to facilitate target localization. Including harmonics, hence increasing bandwidth, may further enhance target ranging (Simmons and Stein 1980; Surlykke 1992) and localization (Beuter 1980; Zbinden 1988).

Foraging Techniques Used by Bats

We will now outline the major ways in which bats capture insects and use echolocation. Many species use several different foraging techniques, so it is unwise to classify any particular bat into one category (Fenton 1990). The long-eared bat *Myotis evotis*, for example, captures insects both by gleaning and by aerial hawking and uses different sensory cues for the capture of resting and flying prey (Faure and Barclay 1994; fig. 7. 2).

Aerial prey are often caught in flight by hawking and typically captured in the bat's wing or tail membrane and then transferred to the mouth while in flight (Griffin and Webster 1962; Kalko 1995; fig. 7.1). Aerial prey are located in space by a series of echolocation pulses, ultimately ending in a terminal "feeding" buzz (fig. 7.1). When searching for prey, aerial-hawking bats produce search phase pulses (Griffin et al. 1960), often at a rate of one per wingbeat (Britton et al. 1997; Jones 1994; Kalko 1994). In open habitats, pulses may correspond to one every second or third wingbeat (Jones 1999; Rydell 1990).

After a prey item has been detected, the interpulse interval and the pulse duration decrease as the bat moves closer to the target (the approach phase), so that pulse-echo overlap is avoided, at least until the very end of the capture

(a)

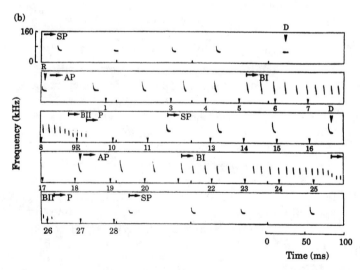

(b)

Figure 7.1. Simultaneous multiflash image (*a*) and echolocation call sequence (*b*) of an aerial hawking bat (*Pipistrellus pipistrellus*) showing how echolocation calls change during pursuit and capture of prey by an aerial hawking bat. Numbers show sequential order of flashes and their corresponding synchronization pulses in the sonogram. The bat makes two capture attempts, but the first insect captured is not visible on the image (capture between images 10 and 11). The second insect is captured at about flash 27. A terminal buzz is emitted at each capture attempt. The phases in echolocation call sequences are indicated with arrows in the sonogram. *D* = detection, *R* = reaction, *SP* = search phase, *AP* = approach phase, *BI* = buzz I, *BII* = buzz II, *BI*+*BII* = terminal phase, *P* = postbuzz pause. From Kalko (1995), with permission from Academic Press.

attempt (Britton et al. 1997; Kalko and Schnitzler 1989, 1993; Kalko et al. 1998). Several pulses are then produced per wingbeat, and the approach phase is followed by a terminal phase, when pulse interval and pulse duration is shortest. Approach and terminal phase pulses are typically broadband or contain broadband components presumably to optimize ranging performance (Parsons et al. 1997; Simmons and Stein 1980; Surlykke 1992). In many aerial-hawking and trawling bats, the terminal buzz may be divided into two parts, buzz I and buzz II (fig. 7.1). Buzz II is characterized by a sudden drop in frequency and a continuous high repetition rate (Kalko and Schnitzler 1989a). It is not known whether the change in pulse characters between buzzes I and II is adaptive or the consequence of physiological constraints associated with a repetition rate that may exceed 200 Hz (Surlykke et al. 1993). Buzz II is not usually present in gleaning bats or in some aerial-hawking species. Many bats that hunt predominantly by aerial hawking emit pulses containing a narrow-band component when searching for prey and switch to pure broadband signals during approach (e.g., *Pipistrellus* spp. [Kalko and Schnitzler 1993]). Their moderate-to-high wing loadings confer relatively fast flight and limited maneuverability (Norberg and Rayner 1987).

When detecting prey in complex habitats such as within foliage, background (clutter) echoes (Neuweiler 1989, 1990) often mask echoes from insects. Broadband, low duty-cycle echolocation may be of limited use for detection of insects in these situations. Indeed, bats that use high and low duty-cycle echolocation use different approaches to detect prey in clutter. As we have seen, high duty-cycle bats can separate echoes of fluttering insects from both the emitted signal and from background clutter. Such flutter-detecting bats detect prey using the acoustic glints in their CF echoes, produced by prey wing movements (Schnitzler et al. 1983). Bats that emit broadband signals at low duty cycle also feed on prey moving in clutter, but in their case it is probably the sounds produced by prey that are the cues by which prey are detected and located. It is not known if flutter-detecting bats also exploit prey-generated sounds in prey detection. Taking prey from surfaces is called gleaning.

Some gleaning bats can make fine discriminations between targets of different texture by means of broadband echolocation calls of low intensity (Habersetzer and Vogler 1983; Schmidt 1988). Some species, however, do not distinguish prey echoes from clutter echoes (Schnitzler and Kalko 1998), and this discrepancy is hard to explain.

Gleaning bats that use FM echolocation for orientation typically use other cues to locate prey on surfaces, and some stop echolocating when they search for and approach prey (Anderson and Racey 1991; Fiedler 1979; Marimuthu 1997; fig. 7.2). Localization occurs by listening for prey-generated sounds by *Myotis evotis* (Faure and Barclay 1992), *M. septentrionalis* (Faure et al. 1993), *Antrozous pallidus* (Bell 1982; Fuzessery et al. 1993), by vision if sufficient light

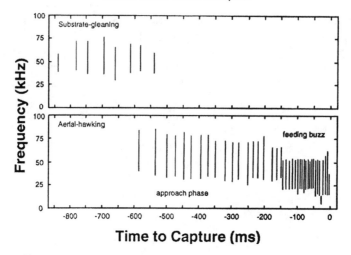

Figure 7.2. Echolocation calls of *Myotis evotis* capturing aerial prey (*above*) and gleaning (*below*) in a flight cage. Note the lack of a terminal buzz during the gleaning sequence. (From Faure and Barclay [1994], with permission from Springer-Verlag.)

is available, as in *Macrotus californicus* (Bell 1985), or by olfaction and passive listening, as in *Mystacina tuberculata* (G. Jones, personal observation). Gleaning bats have low wing loading that facilitates take off from the ground and allows maneuverable flight in obstacle-rich woodland habitats (Norberg and Rayner 1987). Gleaning bats usually eat larger prey than do aerial-feeding species of similar size (Fenton 1990). They are often sensitive to the low frequencies of the rustling sounds produced by prey in clutter. For example *Plecotus auritus* and *Megaderma lyra* are both most sensitive to frequencies around 20 kHz, which is below the frequency range used in their echolocation (Coles et al. 1989; Schmidt et al. 1984).

Several bats glean insects from the water surface by using their large feet and/or their tail membrane (trawling). Trawling insectivorous bats include *Myotis daubentonii* (Jones and Rayner 1988; Kalko and Schnitzler 1989a), *M. adversus* (Jones and Rayner 1991), *M. dasycneme* (Britton et al. 1997), and *Noctilio albiventris* (Kalko et al. 1998). Trawling might be considered a type of gleaning, but the acoustic challenges facing trawling and those facing most other gleaning bats are quite different. In gleaning, prey are often hidden in clutter, while water acts as an acoustic mirror, provided that the bat flies low over the surface and emits the sound beam forward (Rydell et al. 1999), so that prey items on or near the surface become acoustically conspicuous. In contrast to true gleaners, trawling bats, like aerial-hawking species, use high-intensity echolocation calls to detect, track, and evaluate prey, and they also emit feeding buzzes (fig. 7.3). Echolocation calls recorded from trawling bats often contain

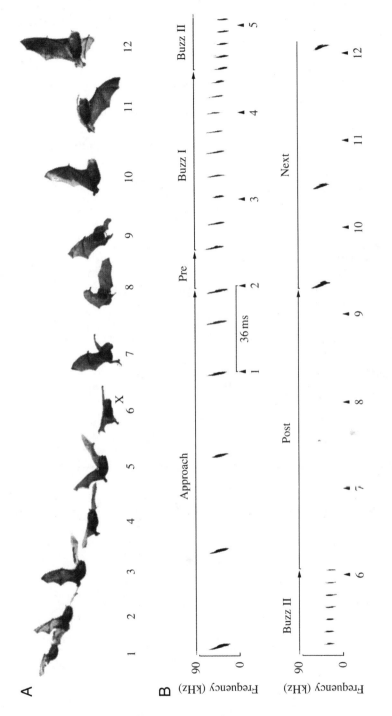

Figure 7.3. Simultaneous multiflash image (A) and echolocation call sequence (B) of a trawling, insectivorous bat, *Myotis daubentonii*. There are 38 ms between images, and X marks the point of prey capture. In (B) *APPROACH* = approach phase, *PRE* = prebuzz pause, *BUZZ I* = buzz I of terminal phase, *BUZZ II* = buzz II of terminal phase, *POST* = postbuzz pause, *NEXT* = resumption of search phase echolocation after prey capture. (From Britton and Jones [1999], with permission from the Company of Biologists.)

interference patterns, seen as notches in sonograms, waveforms, or power spectra. Such patterns are the consequence of a microphone detecting sound from the bat's mouth and reflected sound from the water surface simultaneously (Kalko and Schnitzler 1989b).

Interestingly, trawling *Myotis* bats use broadband echolocation signals even though they hunt essentially in open spaces. In contrast, *Noctilio albiventris* captures insects from the water surface and emits echolocation calls that begin with a narrowband component and terminate with a sweep. The CF components may permit detection of glints from fluttering prey both during aerial hawking and when trawling (Kalko et al. 1998). In *Noctilio albiventris*, the CF portion activates a gating mechanism to process FM pulse-echo pairs for ranging (Roverud and Grinnell 1985). Trawling bats often have relatively low wing loadings that permit slow flight, and their wings may be long (Norberg and Rayner 1987). The species that are most similar in morphology are often not sister taxa in trawling and gleaning bats, and gleaning and trawling appear to have evolved many times (Ruedi and Mayer 2001), perhaps because targets on water reflect stronger echoes than do airborne targets (Siemers et al. 2001). Some bats hang on perches and detect insects that fly past or move on the ground. This behavior is termed "flycatching" or "perch hunting." We restrict the term "flycatching" to the capture of aerial prey, detected from a perch. Flycatching is seen in several rhinolophoid species, which detect prey by echolocation, as in *Rhinolophus rouxi* (Neuweiler et al. 1987; Schnitzler et al. 1985; fig. 7.4), *R. ferrumequinum* (Duvergé and Jones 1994, in press), and *Hipposideros commersoni* (Vaughan 1977). Perch hunting is also used by many gleaning species, which take ground-dwelling prey, presumably detected by listening

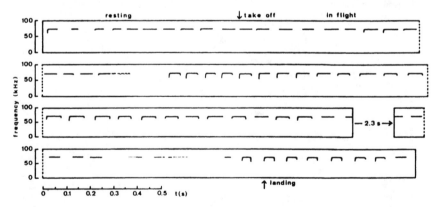

Figure 7.4. Echolocation calls during flycatching in *Rhinolophus rouxi*. (From Schnitzler et al. [1985], with permission from Springer-Verlag.)

for prey-generated sounds, as in *Cardioderma cor* (Vaughan 1976), *Megaderma lyra* (Fiedler 1979), and several species in the genus *Nycteris* (Fenton et al. 1983a, 1993).

Presumably, a bat that uses a sit-and-wait strategy encounters prey at a lower rate than it would if searching for mobile prey. Many flycatching species use high duty-cycle echolocation, perhaps increasing their chance of detecting glints from the wingbeats of sparsely distributed prey. *Rhinolophus rouxi* uses flycatching more than aerial hawking when insects are scarce, thereby saving energy (Neuweiler et al. 1987).

Flycatching and perch-hunting bats are usually large and often return to perches to eat large prey such as beetles and moths, presumably to minimize energy used in flight. Flycatching species typically have large wingtips and a low aspect ratio, conferring maneuverable flight in confined habitats (Norberg and Rayner 1987).

Capture Success and Prey Selection

Capture Success

Although echolocating bats emit terminal buzzes during the pursuit of prey in the air or on water surfaces, prey capture is not always successful. Thus, feeding buzzes do not necessarily indicate that capture has occurred. Capture success is a necessary component for calculation of the profitability of various available prey types, though it is difficult to quantify in nature. Direct visual observations of capture success by the northern bat *Eptesicus nilssonii* vary from 100% for small dung beetles (including *Aphodius* spp.) pursued over grassland to 36% for moths flying around streetlights (Rydell 1992b) and 41% for displaying ghost moths (*Hepialus humuli*) pursued over a field (Rydell 1998). *Lasiurus* bats catch 90% of moths they pursue around streetlights (Acharya and Fenton 1992). Capture success is expected to vary greatly, often because of the sophisticated defenses employed by insects (see below).

Finding methods to establish the outcome of bat-insect interactions for foraging in the field remains an important challenge, because attempts by bats to capture insects are not usually visible to human observers. Capture success may be estimated by acoustic methods. The postbuzz pause after a "feeding buzz" is longer after successful capture attempts in *M. daubentonii* trawling prey in the laboratory, although there is no evidence that it is a valid predictor of capture success in the field. A better indicator of capture success is the lowered repetition rate of echolocation pulses immediately after prey capture (Britton and Jones 1999). Acharya and Fenton (1992) found significantly longer postbuzz pauses after successful attacks by *Lasiurus* species hawking moths in the field, but there was no threshold that distinguished successful from unsuccessful attacks.

Choice of Frequencies for Echolocation

Some bats emit search phase calls with frequencies as low as 11 kHz, as in *Euderma maculatum* (Fullard and Dawson 1997) and *Tadarida teniotis* (Zbinden and Zingg 1986), and as high as 212 kHz in *Cloeotis percivali* (Fenton and Bell 1981). These are rare exceptions, however, as the great majority of insectivorous bats use echolocation calls dominated by frequencies between 20 and 60 kHz (Fenton et al. 1998b). The reason for this probably lies in the frequency-dependent effects of atmospheric attenuation and target strength. Atmospheric attenuation undoubtedly sets an upper limit on useful call frequencies. The absorption of sound energy increases with frequency and is especially pronounced in the ultrasonic range (Lawrence and Simmons 1982). Because sound must travel both to and from a target, and only weak echoes are reflected from small targets, atmospheric attenuation limits the effective range of echolocation at high frequencies.

The relationship between the proportion of acoustic energy reflected from a target (target strength) and call frequency is complex but is probably vital for understanding why most bats avoid echolocation calls dominated by low frequencies. The wavelength of a sound is inversely proportional to frequency (e.g., 10 kHz translates to a wavelength of 34 mm in air, while 100 kHz means 3.4 mm). Target strength is theoretically greatest when incident wavelengths are near the same size or smaller than the target (Pye 1993). Target strength falls rapidly once the ratio of target circumference to wavelength falls below one, the so-called Rayleigh region (the theoretical relationship between target strength and frequency was developed for spheres by Lord Rayleigh [see Pye 1993; Sales and Pye 1974]).

Echolocation Call Structure Influences Detectability of Small Prey

Recent studies provide clear evidence that the theoretical relationship between prey size and call frequency, as predicted by the Rayleigh effect, also holds for insects (Houston et al. 2003). Reflectivity decreases sharply when the wavelength exceeds the wing length of the ensonified insect, and low-frequency ultrasound (20–30 kHz) reflects poorly from small insects (2.5–5.0 mm wing length).

Echolocating bats hence use call frequencies that are compromises between maximizing echo strength from insects and minimizing energy loss by atmospheric attenuation. This has important ecological implications; while small insects may be eaten by bats using higher frequencies, they may be practically unavailable to bats that emit low-frequency calls because the insects are not detectable (Barclay and Brigham 1991). Such sensory constraints must always be considered when attempting to measure the abundance of insects available to echolocating bats (Barclay 1985).

However, the relationship between prey size and echolocation call fre-

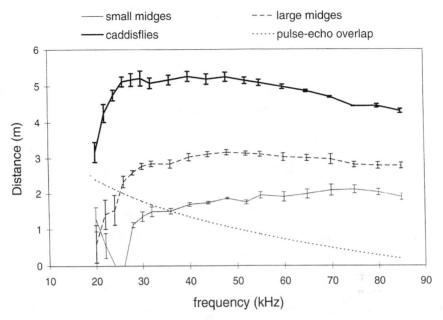

Figure 7.5. How Rayleigh scattering and atmospheric attenuation constrain the detection of prey size for echolocating bats. The *solid lines* represent the estimated maximum detection distances of a fictive bat as a function of call frequency, based on measurements of the target strengths of insects. The estimates assume a source level of 100 dB, an echo detection threshold of 0 dB, and atmospheric attenuation at 12°C and 88% relative humidity. The *dashed line* represents distances where pulse-echo overlap would occur derived from the relationship between frequency and duration in aerial hawking vespertilionids (Waters et al. 1995). Detection distances fall rapidly at low frequencies because of Rayleigh scattering, and fall more slowly at high frequencies because of atmospheric attenuation. Thus, bats using calls < ca. 35 kHz should not detect small midges, and bats calling < ca. 25 kHz should not detect large midges because echoes would return from the insects during pulse emission. Standard errors of target strengths are shown for small midges (wing length 2.6–3.1 mm, $n = 5$), large midges (wing length 4.0–5.0 mm), and caddis flies (wing length 8.0–9.0 mm). (From Houston et al. [2003], with permission from the University of Chicago Press.)

quency is complicated. For example, call frequency and body size are negatively correlated, even within families of bats (Barclay and Brigham 1991; Jones 1996), so that body size and echolocation call frequency are confounding variables. Nevertheless, Houston et al. (2003) predicted what the smallest insects eaten by European aerial-hawking bats would be with reasonable accuracy using a model that included pulse duration (an indirect measure of the minimum detection range to avoid pulse-echo overlap [Kalko and Schnitzler 1993]) and wavelength (Rayleigh scattering limits the detectability of small prey; fig. 7.5).

Many bats that emit low-frequency calls often feed on surprisingly small prey (Jones 1995; Rydell 1986, 1992a; Waters et al. 1995). Although frequency constraints may explain whether single targets are detectable by echolocation, other factors need to be considered. Small prey often swarm, and this may increase their conspicuousness both visually and acoustically. Target strength of a swarm may be related to the number of individual insects in the swarm, as with fish in shoals (Mitson 1983). Swarms may then be detected easily by bats that emit low-frequency calls, that travel long distances, and that provide wide search beams. Alternatively, bats may not be as constrained as we think; those that emit low-frequency calls often have surprisingly flexible echolocation behavior and sometimes make use of higher frequencies (Jones 1995).

Active Prey Selection and Optimal Foraging Theory

Although factors other than energy content may be important in prey selection, such as calcium (Barclay 1995) or essential fatty acids (Schalk and Brigham 1995), bats should be sensitive to prey size when evaluating whether a prey item should be eaten or ignored. Optimal foraging theory predicts that animals should select prey that maximize the rate of net energy return (Stephens and Krebs 1986).

However, bats may not always be able to evaluate the profitably of prey in their environment. The quality of information that a bat receives about a prey item depends on the structure of the bat's echolocation calls and, ultimately, on the duty cycle used in echolocation. Echoes from long CF calls will encode many glints from insect wingbeats. Because the wingbeat rate of insects is negatively correlated with body size (Greenewalt 1962), bats that use long CF calls may be able to evaluate prey characteristics (including size and, hence, profitability) independently of echo strength. Under laboratory conditions, bats that use CF calls can resolve differences in wingbeat rates better than those that rely on FM pulses (Roverud et al. 1991). Bats that use high-duty-cycle echolocation will probably be able to evaluate prey most effectively and will therefore be most likely to conform to predictions of optimal foraging models.

Because glint rate depends on wingbeat frequency, time/frequency features in echoes from CF calls can be signatures of a particular prey item (Kober and Schnitzler 1990; Schnitzler 1987). Long CF components can therefore be used to classify prey items that could be used in foraging decisions. In the laboratory, *Rhinolophus ferrumequinum* can discriminate between glint frequencies that differ by less than 8% (von der Emde and Menne 1989) and can even discriminate between wingbeat echoes of different species fluttering at the same rate (von der Emde and Schnitzler 1990).

Jones (1990) studied the diet of *R. ferrumequinum* in the field and assessed prey abundance simultaneously. Large moths were eaten in proportion to their abundance, but small Diptera and Hymenoptera were not; these insects were important in the diet only when large (more energetically profitable) moths were scarce. These relationships between the incorporation of profitable and unprofitable prey in relation to prey abundance follow predictions based on optimal foraging theory (Charnov 1976) and are similar to trends found in visually hunting insectivorous birds (Davies 1977; Turner 1982). Thus, greater horseshoe bats seem to be sufficiently informed of prey characteristics to select profitable prey actively.

Can bats that use low duty-cycle echolocation make similarly informed choices? Laboratory studies show that such bats make remarkable discriminations between targets at different ranges and of different textures (see reviews by Moss and Schnitzler 1995; Simmons et al. 1995). However, evidence that they use echolocation to select certain prey from those available under natural conditions is lacking (Arlettaz and Perrin 1995; Barclay and Brigham 1994).

Barclay and Brigham (1994) presented free-living *Myotis lucifugus* and *M. yumanensis* with a series of alternative prey types. Bats preferred moving prey over immobile items, so movement is a cue as to whether an item is edible. However, these *Myotis* also attacked moving, inedible prey such as small leaves as frequently as moving, edible items such as moths of similar size. Barclay and Brigham (1994) concluded that the short-range operation of echolocation and the rapid flight of bats make it difficult to discriminate and thus select certain types of prey under natural conditions. However *Lasiurus cinereus* and *L. borealis* rarely made contact with stones thrown in the air, after pursuing the stones and echolocating at them. Some bats may have learned that stones were inedible and seemed to use information during the terminal phase of echolocation for evaluation (Acharya and Fenton 1992).

Recent experiments with trawling *Myotis daubentonii* suggest that this bat mistakes patches of duckweed (Lemnacaea) for prey (Boonman et al. 1998), and it apparently even has problems distinguishing insect echoes from noise pulses produced by moving water (Rydell et al. 1999). *Myotis daubentonii* detects insects at an average distance of 128 cm (Kalko and Schnitzler 1989a), so that even if unprofitable prey are detected, a bat may have little time to reject them. However *M. daubentonii* chooses large over small prey in choice experiments both in the laboratory and field (Houston and Jones 2003). It is not clear whether the bats reacted to the more conspicuous of two acoustic targets (i.e., the larger amplitude echo) or whether they were making discriminations based on fine spectral differences. Thus, choice may have been either passive, with the bats taking the first prey item detected, or active, based on prey evaluation. There is evidence for active selection by the FM bat *Myotis nattereri* foraging in a flight cage (Siemers and Schnitzler 2000).

Clearly, the topic of prey choice by bats is marked by controversies and is a fruitful area for future research. Measurements of prey availability in nature are difficult to make, and controlled-choice experiments in the field and in field-like laboratory conditions may offer the best ways forward.

Insect Defenses against Bat Predation

Echolocation performance of insectivorous bats has probably evolved considerably since its first appearance more than 50 million years ago and has become more sophisticated and diverse (Fenton et al. 1995). However, insects can probably evolve faster than bats because they have shorter generation times and perhaps also because they are the prey and not the predator (the "life-dinner" principle [Dawkins and Krebs 1979]). Thus, it comes as no surprise that insects possess a suite of defenses against bats. The performance of bats as insect predators is therefore strongly influenced by the defensive behavior of the prey, and this is another complicating factor that must be considered when attempting to test foraging theory.

Insect behavior may, in fact, be one of the most important determinants of what bats are able to catch and eat, and for this reason we believe that an understanding of insect defensive behavior is a prerequisite for a proper understanding of the feeding biology of insectivorous bats. This also means that much of what nocturnal insects do, or rather how they do it, is affected by bat predation. J. David Pye (1968) summarized the evolution of countermeasures by insects to bat sonar in a poem.

> In days of old and insects bold
> (Before bats were invented),
> No sonar cries disturbed the skies—
> Moths flew uninstrumented.
>
> The Eocene brought mammals mean
> And bats began to sing;
> Their food they found by ultrasound
> And chased it on the wing.
>
> Now deafness was unsafe because
> The loud high-pitched vibration
> Came in advance and gave a chance
> To beat echolocation.
>
> Some found a place on wings of lace
> To make an ear in haste;
> Some thought it best upon the chest
> And some below the waist.
>
> Then Roeder's keys upon the breeze
> Made Sphingids show their paces.
> He found the ear by which they hear
> In palps upon their faces.
>
> Of all unlikely places!

Nocturnality in insects may have first evolved because it protected them against heavy daytime predation from dinosaurs, early birds, and pterosaurs during the Cretaceous period or earlier. Later, when bats evolved and radiated during the early Tertiary period, nocturnal insects were confronted with predators that hunted by sound instead of vision. This would have resulted in strong selection pressure favoring insects that could hear the calls of bats and take evasive action or that could avoid them by other means.

The diversity of insects, particularly in the tropics, argues in favor of diversity of defenses against bats. However, our knowledge about this subject is strongly biased in favor of the relatively low-diversity bat and insect faunas of the north temperate regions. It is also strongly biased from studies that have focused on ultrasonic hearing and its associated evasive behavior of some groups of nocturnal insects, particularly moths. Many nocturnal insects are probably deaf, and presumably they use nonauditory methods to reduce predation by bats. Thus, predation pressure from bats has almost certainly affected the morphology, physiology, and behavior of nocturnal insects in many different ways.

Ultrasonic Hearing in Insects

Ultrasonic hearing and the associated defensive behavior of certain Lepidoptera provide the best-known examples of bat defense in insects. The ears of moths were discovered and described in some detail long ago (Eggers 1919; Kennell and Eggers 1933), but their function in bat defense did not become clear until the 1950s (reviewed by Roeder 1967a), following the discovery and early studies of echolocation in bats (Griffin 1958; references listed in Sales and Pye 1974). In Lepidoptera, ultrasonic hearing is known from at least seven superfamilies. The ears of moths are located on different parts of the body (Fullard and Yack 1993) and show important differences in fine structure among superfamilies. Thus, ultrasonic hearing evolved at least once within each superfamily. For example, the Pyraloidea, Drepanoidea, and Geometroidea have tympanic organs situated on the first abdominal segment, whereas the Noctuoidea have them on the metathorax (Scoble 1992). Ultrasonic hearing also occurs in some Sphingidae (Roeder et al. 1970), as well as in some butterflies (Papilionoidea, family Nymphalidae [Swihart 1967]) and the tropical Uranoidea (Sick 1937). The latter two groups are predominantly diurnal, so it remains uncertain if their hearing systems are involved in bat defense.

Although the morphology of tympanic ears varies among different clades of Lepidoptera, all possess a thin membrane, backed by an air-filled sac, to which are attached a small number (one in the family Notodontidae, two in other Noctuoidea, and four in the Geometroidea) of acoustic sensory "A cells," which send primary afferent spikes to the central nervous system. The A cells show a broad frequency response that varies somewhat from species to

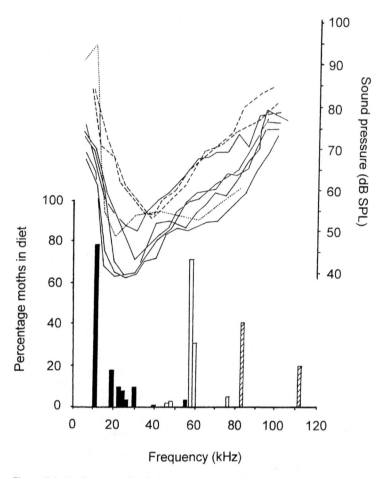

Figure 7.6. Audiograms of eight European species of moths and the incorporation of moths in the diets of European bats in relation to the dominant frequencies of their echolocation calls. Note how bat species that call at frequencies between 20 and 50 kHz eat few moths, and moth sensitivity to bat calls is greatest at these frequencies. Bats that hunt by gleaning, or that call at frequencies <15 or >60 kHz often eat large numbers of moths. Audiograms (*top*) are from *Agrotis segetum, Diarsia mendica, Cerapteryx graminis, Apamea crenata* and *A. maillardi* (Noctuidae—*continuous lines*), *Phalera bucephala* and *Phoesia tremula* (Notodontidae—*dashed lines*), and *Ephestia kuehniella* (Pyralidae—*dotted line*). The importance (% volume or % frequency) of moths in the diets of European bats in relation to the frequency of most energy in echolocation calls normally emitted in search flight is shown at bottom. *Black bars* are aerial-hawking species (from left to right, *Tadarida teniotis, Nyctalus noctula, N. leisleri, Vespertilio murinus, Eptesicus serotinus, E. nilssonii, Myotis daubentonii,* and *Pipistrellus pipistrellus*); *white bars* are gleaners (from left to right, *Myotis myotis, M. nattereri, Plecotus austriacus, P. auritus,* and *Myotis emarginatus*); and *hatched bars* are flutter detectors (*Rhinolophus ferrumequinum* and *R. hipposideros*). (Sources are detailed in Rydell et al. [1995], from which the figure is taken with the permission of the Nordic Ecological Society.)

species. However, the best frequency range is usually between 20 and 60 kHz, which corresponds to the bandwidth used by most species of aerial-hawking bats (Fenton and Fullard 1979; Fullard 1987, 1988; Fullard and Thomas 1981; fig. 7.6). Despite variation in the best frequency of hearing among moths, the tuning curves are similar enough to suggest the same selective agent (aerial-hawking bats [Fullard 1987]). Slight differences in the response between moths of different geographic areas may reflect differences in the frequencies emitted by different bat faunas (Fullard 1984, 1987). Higher hearing sensitivity in larger moth species reflects the fact that the detection distance in bats is likely to increase with increasing size of the target (Surlykke et al. 1999).

The A cells in a moth ear normally differ in sensitivity. For example, the A1 cell of the noctuid ear is 15–30 dB more sensitive than the A2 cell (Roeder 1974). The response by one or both A cells in each of the two ears may therefore encode information about the intensity and direction of an incoming sound. In this way the bat's position can be determined, providing the basis for a graded evasive response shown by many moths (Roeder 1967a, 1967b). However, Surlykke (1984) inferred a graded response even in notodontids, which have only a single sensory cell in each ear.

The tympanic ears of moths evolved from specialized mechanoreceptors, called chordotonal organs, that normally link across joints in the cuticle and respond to stretching (Scoble 1992). In contrast, the hearing organs in hawk-moths (Sphingidae) are not tympanate but consist of modified mouthparts. This pilifer hearing organ, while sensitive, is not very directional (Göpfert et al. 1998; Roeder et al. 1970). Ultrasonic hearing may have evolved independently in the two subfamilies of hawkmoths, the Sphinginae and Macro-glossinae (Göpfert et al. 1998). Hawkmoths alter their flight speed on hearing ultrasound, and some species even respond by emitting acoustic signals that may startle bats (Göpfert and Wasserthal 1999a, 1999b; Göpfert et al. 1998).

Most moth ears are used exclusively for defense against bats, and the hearing organs must have evolved in that context (Spangler 1988b). There are examples of moths that use ultrasound for intraspecific communication (Conner 1999; Spangler 1985), but this behavior most likely evolved secondarily (Spangler 1988b).

At least six other insect orders have ultrasonic hearing in one form or another (Fullard and Yack 1993; Hoy 1992; Hoy and Robert 1996; Hoy et al. 1989). Tympanal hearing organs have been described in nocturnal beetles (Coleoptera), including tiger beetles (Cincindellidae [Spangler 1988a; Yager and Spangler 1995, 1997]) and scarab beetles (Scarabaeidae [Forrest et al. 1995, 1997]). Scarab ears are located behind the head and consist of a thin tympanic membrane backed by an air sac and a chordotonal sensory structure. In contrast, tiger beetles have abdominal ears, and the difference in location between various beetle families suggests that ultrasonic hearing evolved more than once in beetles. However, in both cases, the ears are tuned to frequencies used by bats, and stimulation with batlike ultrasound results in evasive flight responses in scarabs and the emission of click trains in combination with distasteful secretions in tiger beetles. Thus, the ultrasonic hearing and associated behavioral responses function as a defense against echolocating bats in both groups.

Ultrasonic hearing is also known in green lacewings (Neuroptera [Miller and MacLeod 1966]), where the receptor and tympanic membrane are located in a small swelling on the radial vein of each forewing (Miller 1970). Flying green lacewings react to batlike sounds by folding their wings and dropping to the ground (Miller and Olesen 1979). The cyclopian ear of preying mantids (Dictyoptera) is another example in which ultrasonic hearing serves primarily as a defense against bats (Yager and Hoy 1986; Yager and May 1990). Behavioral observations in the field suggest that ultrasonic hearing promotes survival in mantids (Cumming 1996; Yager et al. 1990).

Many orthopterans, such as locusts, crickets, and katydids, hear ultrasound, and some react evasively to batlike sounds when in flight, either by closing the wings and dropping to the ground or by flying away from the sound source (Libersat and Hoy 1991; Moiseff et al. 1978; Nolen and Hoy 1986; Robert 1989). In contrast to most other insect orders, orthopterans communicate intraspecifically with sound, so that in this case hearing could first have evolved for communication and only later became involved in defense against bats (Hoy and Robert 1996).

Ultrasonic hearing sensitivity has recently been discovered in some parasitoid Diptera (Robert et al. 1992). However, their best hearing frequencies do not match those used by most echolocating bats, so their ears are probably not involved in defense. Most nocturnal flies are relatively slow and, thus, may be incapable of the powerful evasive flight responses that may be a necessary component of an efficient hearing-based bat defense system. In the latter case, possession of ears alone may not lead to lower predation rate, and this may not evolve toward a bat defense function.

Defense against Bats by Hearing Moths

Most defensive responses of insects to batlike ultrasound result in changes in flight speed, direction, or cessation of flight altogether. The sensitivity of moth ears suggests they can detect an approaching bat at least at 20–30 m (Roeder 1967a), provided that bat's calls are of high amplitude (Waters and Jones 1995). A bat is unlikely to detect even a big moth (such as an average sized noctuid) >5 m or so away (Kick 1982; Waters et al. 1995). The insect will typically be at an advantage compared with the bat. Difference in detection range between the predator, which relies on echoes reflecting from a small target, and the prey, which detects the sound directly, provides the basis for the graded evasive response shown by hearing insects (Roeder 1962).

From the insect's point of view, the risk of being caught is presumably related to the distance to the bat, and "decisions" on which evasive response to

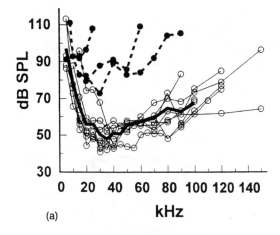

(a)   **kHz**

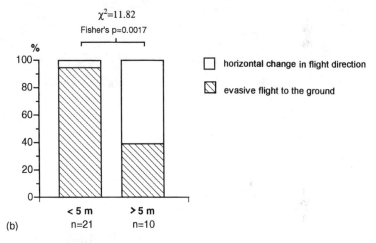

(b)

Figure 7.7. Bat defense in the geometrid moth *Agriopis marginaria* from Scandinavia, a species with fully winged males but flightless females. *a*, Audiograms of individual males (*open circles* and *solid lines; thick line* indicate mean threshold) and females (*filled circles* and *dotted lines*). *b*, Behavioral reactions of free-flying (wild) males to pulses of ultrasound (26 kHz, 110 dB SPL at 1 m) at various distances (<5 m and 5–12 m, respectively) from the sound source. (Modified from Rydell et al. [1997], with permission from the Royal Society.)

use may be based on the perceived risk, at least if the evasive behavior is costly. For example moths typically react to simulated bat calls of low intensity, indicating a distant bat, by simply turning away from the sound source or by increasing the unpredictability of the flight path (Roeder 1962; fig. 7.7). This response seems appropriate when the bat is far away and the insect is unlikely to have been detected. In contrast, bat calls of high intensity, indicating a close bat, induce more dramatic and erratic flight responses from insects (figs. 7.7 and 7.8). In such cases the evasive behavior may consist of a passive drop, a

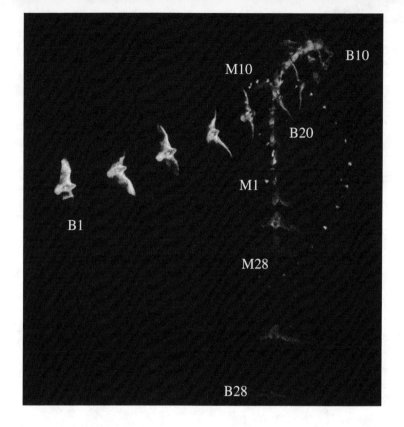

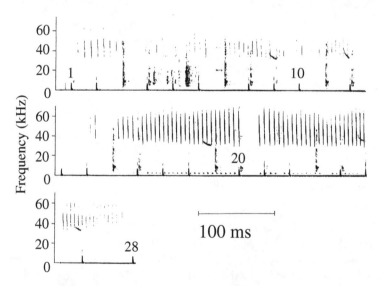

Figure 7.8. Photo of evasive response of a moth pursued by a red bat *Lasiurus borealis* in the field. The bat flies toward the moth, overshoots it, and by *B10* loops back toward the moth. At about *M12* (not shown in figure), the moth enters a power dive, and although it heads toward the bat's descending flight path, the bat again overshoots it and fails to capture the moth. Throughout the pursuit, the bat emits a prolonged terminal buzz (sonogram) in an attempt to track the moth. Images of the bat (*B*) and moth (*M*) are labeled chronologically, so the bat was at *B10* when the moth was at *M10*, 333 ms into the sequence. Flash rate was 30 Hz. The low frequency clicks on the sonogram are flash synchronization pulses. The search phase echolocation calls of a second bat can sometimes be seen in the background on the sonogram.

powered dive (fig. 7.8), a downward spiral flight, or a loop, followed by a vertical dive—maneuvers that carry the moth toward the ground or into protective vegetation (Miller and Olesen 1979; Roeder 1967a; Rydell et al. 1997; fig. 7.7). Variation in the evasive behavior may in itself be an important part of the defensive strategy, particularly if it includes a random component, because unpredictability counteracts learning by predators (Roeder 1975).

Predation on moths by aerial-hawking bats usually occurs in flight. Moths sitting on substrates also respond to batlike calls, however, usually by freezing (Werner 1981) or by stopping the release of their pheromones (Acharya and McNeil 1998). However, the hearing capacity of flightless female moths in some species is reduced considerably compared to that of the flying males (Cardone and Fullard 1988; Rydell et al. 1997). This suggests that predation from aerial-hawking echolocating bats is of relatively little importance for the nonflying females compared with the flying males. Alternatively, bat predation on the females still occurs, but there is little or no selection that maintains the hearing because the ears no longer trigger efficient evasive maneuvers and hence do not increase survival.

Defensive Sound Production

Aposematism (warning about dangerous attributes) is a defense strategy commonly employed by insects, but most known examples are based on color and are thus directed at diurnal, visual predators (Edmunds 1974). Many tiger moths (Arctiidae) are nocturnal and tympanate, and some are distasteful and aposematically colored (Bissett et al. 1960). Their coloration presumably protects them during the day, and the trains of ultrasonic clicks emitted by some when encountering bats at night may function in the same way (Carpenter and Eltringham 1938; Dunning 1968; Dunning and Roeder 1965).

The click trains of arctiids are elicited by stimulation of the tympanic organs with ultrasound and are produced by special tymbal organs situated on the metathorax (Blest et al. 1963; Fullard and Fenton 1977). When arctiid click trains are played back to a bat approaching an insect, the bat usually aborts the attack (Acharya and Fenton 1992; Dunning and Roeder 1965; Dunning et al. 1992). It thus seems clear that the clicks protect the moth from echolocating bats, although functions other than aposematism are also possible. The clicks could startle an attacking bat (Bates and Fenton 1990) or jam the reception and interpretation of the echoes and thus provide the moth with acoustic camouflage (Fullard et al. 1979). The clicks could interfere with the bat's ranging performance in the moments shortly before capture (Fullard et al. 1994; Miller 1991). Clicking seems adaptive because arctiids are caught and eaten less frequently than other moths of comparable size (Dunning et al. 1992).

The ultrasonic clicks produced by some nymphalid butterflies, such as *Inachis io*, may be aposematic signals aimed against bats, with which the butterflies often share hibernacula in winter (Møhl and Miller 1976). As nymphalids are generally aposematically colored and distasteful, it seems possible that the clicks may warn bats of a noxious prey, just as in the Arctiidae.

Considering how common Batesian mimicry is among diurnal Lepidoptera, it is surprising that there is yet no described example of acoustic Batesian mimicry among nocturnal species. Mimics of the aposematic arctiid clicks would be expected among palatable nocturnal moths.

Defense in Earless Insects

Most nocturnal or crepuscular insects, including Diptera, Trichoptera, Ephemeroptera, Plecoptera, Isoptera, Hymenoptera, and Heteroptera, most of the small Lepidoptera, and most Coleoptera, are probably deaf to bat calls. An adequate defense against predation is essential for every animal, and other anti-bat adaptations may therefore be expected in these insects. Bat predation is likely to affect insects in many different ways and result in many morphological and behavioral adaptations as alternatives and additions to hearing-based defense systems. Predator defense is usually complex (Edmunds 1974). Traits that result in lower predation rates are favored by natural selection regardless of why they evolved or whether other defenses already exist.

Size, Body Temperature, and Flight Style in Insects

Among larger nocturnal moths (Macrolepidoptera and Pyralidae), an overwhelming majority (ca. 94%) possess ultrasonic organs (Rydell and Lancaster 2000), illustrating the importance and efficiency of hearing-based defense systems in these insects. At the same time, more than half of the extant moth species are small (Microlepidoptera), most of which seem to be earless and therefore deaf.

Insects with wingspans of only a few millimeters may be too small to be detected by most echolocating bats or deemed unprofitable (Houston et al. 2003). At the other extreme, large insects that approach the size and mass of small bats, may be too big for aerial-hawking bats to handle (Roeder 1974). It is therefore interesting that many earless moths seem to be either very small or very large. Among the Macrolepidoptera, earless species and families have significantly larger wingspans and higher body masses than forms with ears (Rydell and Lancaster 2000). Bat predation could therefore have influenced the size of moths. In contrast, the presence of bat-detecting ears in some hawkmoths suggests that large size alone does not guarantee protection from bat predation.

A general increase of body size in insects may not only provide protection by size directly but also because it facilitates faster flight. Big moths should fly faster than smaller moths because they generally have higher wing loadings (Casey and Joos 1983), and, presumably, fast-flying moths are harder to catch than moths that fly slowly (Roeder 1974). Some medium-sized Lasiocampidae (e.g., *Malacosoma* spp.), which are earless and strictly nocturnal, tend to fly more erratically than tympanate moths do (Lewis et al. 1993), and they also

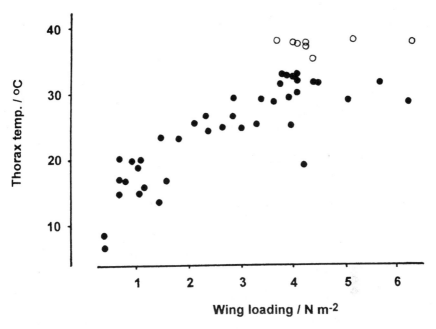

Figure 7.9. Thorax temperature in flight in relation to the wing loading of earless (*open circles*) and tympanate (*filled circles*) moths from Scandinavia. Each circle represents one species (one to 12 individuals). The earless species are of the Hepialoidea (1), Bombycoidea (4), and Sphingoidea (3), and the tympanate ones are of the Geometroidea (10), Drepanoidea (4), and Noctuoidea (26). The body temperatures were measured in the thorax, using a thermoprobe, of moths coming to a mercury light, and at ambient temperatures between 8° and 23°C. (From Rydell and Lancaster [2000], with permission from the Nordic Ecological Society.)

have a significantly higher flight body temperature (near 40°C) than do tympanate species of the same size (Rydell and Lancaster 2000; fig. 7.9). This suggests that a higher body temperature in earless moths may facilitate increased energy output during flight (Heinrich 1993). Fast or erratic flight could thus be an alternative defense against bats in earless moths, but one that is energetically demanding, as it requires a higher metabolic rate (Bartholomew and Casey 1978).

Predictability in Bat Feeding Activity and Its Possible Exploitation by Insects

Predictable variation in bat activity may be exploited by insects that have become partly or entirely diurnal (Fullard and Dawson 1999; Fullard et al. 1998), crepuscular (Andersson et al. 1998), or that emerge and fly particularly early or late in the season (Fullard and Barclay 1980; Yack 1988).

Partial degeneration of the ultrasonic hearing in some of the early spring and late autumn flying noctuid moths suggests that they may face less predation from bats compared with moths that fly during the summer (Surlykke and Treat 1995). However, more dramatic degeneration of the hearing occurs in those that have become diurnal (Fullard et al. 1998; Surlykke et al. 1998).

The winter moths *Operophtera* spp., which emerge in late October and November, seem to maintain a sophisticated ultrasound-based bat avoidance mechanism despite the unusual flight period. However, in contrast to the noctuid "winter moths" studied by Surlykke and Treat (1995), *Operophtera* spp. fly slowly (Svensson and Rydell 1998; Svensson et al. 1999) and may rely more on hearing for defense than the noctuids, which presumably are protected by their fast flight speed (Rydell et al. 1997). Although predation from bats on winter moths may be relatively low, this could be a consequence of the unusual flight period rather than the cause. By flying and mating in late autumn or early spring, moths may benefit by allocating the larval stage to late spring and early summer, when the leaves are most nutritious.

Acoustic Concealment and Crypsis

Roeder (1974) suggested that earless moths could reduce the risk of being detected by bats by flying close to the ground, thus taking advantage of the fact that prey detection by echolocation is complicated by acoustic clutter. Lewis et al. (1993) tested this acoustic concealment hypothesis and found that two species of lasiocampid moths flew significantly closer to the ground than sympatric tympanate species. Likewise, acoustic concealment may explain why males of the ghost moth *Hepialus humuli* usually stay <0.5 m from the vegetation when they are visually signaling to attract females. Moths within this zone are relatively safe from attacks by patrolling *Eptesicus nilssonii*, but remaining there seems to compromise the quality of the visual display used to attract females (Rydell 1998; fig. 7.10).

Insects that emit high intensity mate calling songs that are conspicuous to bats, such as many Orthoptera (e.g., crickets and katydids), may signal from within bushes or other hiding places. These are examples of physical or visual rather than acoustic concealment.

Most insects are probably concealed from aerial-hawking bats as long as they remain motionless on the vegetation and keep quiet, as some flightless female moths seem to do (Cardone and Fullard 1988). Acoustic concealment is presumably further enhanced if the wing area, that is the target strength, is reduced (Rydell et al. 1997). However, bat predation is not necessarily the primary reason why some female moths have become flightless. By abandoning flight, they can allocate more energy to egg production (Marden 1995).

Predator Swamping

Mayflies, caddis flies, stone flies, some lacewings (Neuroptera), and many two-winged flies (Diptera), including chironomids, typically emerge at dusk and swarm over bodies of water or near the surrounding vegetation. These insects generally lack ultrasonic hearing, as far as is known, and are also rela-

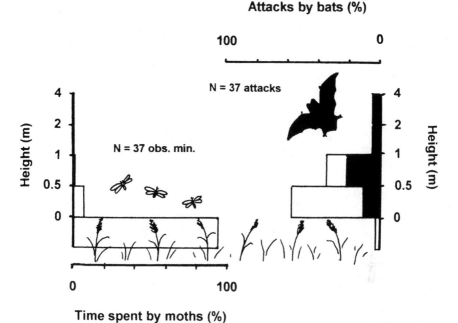

Figure 7.10. The flight height of displaying male ghost swifts, *Hepialus humuli,* over a hayfield in relation to the top of the panicle-bearing grass stems (*left bars*) and the height where the displaying moths were attacked (*right bars*) and caught (*black sections*) by northern bats, *Eptesicus nilssonii.* The focal observations of bats and moths were made separately. (From Rydell [1998], with permission from the Royal Society.)

tively slow flyers compared with moths and beetles. Thus, it is not surprising that many aerial-hawking and trawling bats preferentially feed on swarms of insects over water rather than searching for less conspicuous prey in terrestrial habitats (Anthony and Kunz 1977; Brigham and Fenton 1991; Rydell 1986). Because bats that hunt in aquatic and riparian habitats heavily exploit swarming insects, which usually lack recognized bat-defense systems, it is reasonable to ask whether swarming has an antipredator function. Swarming behavior of some mayflies is apparently affected by predation from birds and insects (Sweeney and Vannote 1982) but, surprisingly, has not been examined with respect to bat predation.

The short crepuscular flight period of many slow-flying and earless insects in temperate areas, such as mayflies and chironomids, may minimize their exposure to bats. This hypothesis requires, of course, that swarming has already started to decline by the time bats emerge to feed. Limited evidence suggests that this may be the case (Rautenbach et al. 1988; Rydell et al. 1996a).

Defense against Bats—Does It Work?

It is clear that tympanate moths gain a substantial selective advantage by having ears, as they are less frequently consumed by bats compared to earless species (Pavey and Burwell 1998). Based on field observations, it has been estimated that moths that react to bat calls (hearing ones) have at least a 40% better
chance of surviving a bat attack compared with moths that do not react
(Acharya and Fenton 1992; Miller 1980; Roeder 1967a). This is most likely an
underestimate because hearing moths probably avoid most attacks entirely by
leaving the zone of danger long before they are detected. In contrast, small
dung beetles (Scarabaeidae) pursued by *Eptesicus nilssonii* are normally captured, so these insects seem quite helpless once detected (Rydell 1992b).

Coevolution?

It seems as if most evasive responses of moths are induced in flight, and such
behavior presumably evolved in response to predation pressure from aerial-
hawking bats (Fenton and Fullard 1979; Fullard 1987, 1988). However, there
are also gleaning bats, some of which feed extensively on moths. Moths respond to ultrasound even when not in flight, either by becoming motionless
(Werner 1981; Acharya and McNeil 1998) or by falling to the ground (Rydell
et al. 1997). However, gleaning bats normally use low-intensity echolocation
calls (Faure et al. 1990; Miller and Treat 1993) or rely on listening for prey-
generated sounds (Anderson and Racey 1991, Bell 1982; Faure and Barclay
1992) or vision (Bell 1985). The weak echolocation calls of gleaning bats are
usually not easily detected by moths (Faure and Barclay 1992; Faure et al. 1990,
1993), and at present it seems as if moths generally lack efficient defense
against gleaning bats. Gleaning may represent an example of a counteradaptation by bats, coevolved in response to the defense systems of moths. However, gleaning could also have evolved because it facilitates detection of prey
among clutter, and the increased availability of moths and other tympanate insects may just be a fortunate side effect of this.

Bats that use echolocation call frequencies outside the range of the best
hearing in tympanate insects could gain the advantage of increased availability of prey. This may be another counteradaptation in bats that bypasses the
defensive adaptations in insects (i.e. "the allotonic frequency hypothesis"
[Fenton and Fullard 1979]). Bats that use very low frequencies (ca. 10 kHz) and
feed almost exclusively on tympanate insects include the large molossids
Tadarida teniotis and *Otomops martiensseni* (Rydell and Arlettaz 1994; Rydell
and Yalden 1997) as well as *Euderma maculatum*, a long-eared vespertilionid
(Fullard and Dawson 1997). At the other end of the spectrum are the Rhinolophidae and the Hipposideridae, some of which use extremely high frequencies (>100 kHz) and also feed extensively on tympanate prey (Jones 1992;
Pavey and Burwell 1998; Whitaker and Black 1976). However, it is again

difficult to argue with any confidence that the unusual echolocation strategies used by these bats evolved as counteradaptations against insect defenses. Low frequencies also improve the detection distance and high frequencies facilitate the detection of small prey. Nevertheless, regardless of its origin, the use of very low or particularly high frequencies by some bats increases predation on tympanate prey.

Although it is clear that insects have evolved many adaptations as defenses against bats, it is not equally obvious that bats have evolved in response to the antipredator adaptations in insects. Thus, it is unclear if bat-insect coevolution in its strictest sense (Janzen 1980) has occurred, although this depends largely on how coevolution is defined.

Streetlights

Moths and other insects are attracted to lights, and this in turn attracts aerial-hawking bats that come to feed on the insects (Acharya and Fenton 1992; Belwood and Fullard 1984; Dunning et al. 1992; Griffin 1958; Hickey and Fenton 1990). At these lights, bats seem to capture more moths than they normally do in other habitats (Rydell 1992b), probably because the moths become concentrated into predictable and easily exploitable patches. A recent study (Svensson and Rydell 1998) suggests that the hearing-based defense system of winter moths (*Operophtera* spp.) is inhibited near mercury-vapor lamps, presumably further facilitating the hunting success of bats near streetlights. Other moth species also seem to have a selective disadvantage around streetlights (Acharya and Fenton 1999).

The situation at streetlights illustrates that the outcome of predator-prey interactions is never fixed but varies depending on prevailing circumstances. Streetlights have only recently become part of the evolutionary setting of bats and insects, but this example illustrates how the selection pressure may shift as habitats change in space and time, with or without the influence of humans. Such habitat heterogeneity is presumably one reason why antipredator adaptations never become perfect and why so many tympanate moths continue to turn up in bat stomachs.

Acknowledgments

The Royal Society, Biotechnology and Biological Sciences Research Council (BBSRC), and Natural Environment Research Council (NERC) have supported Jones's research. Rydell was funded by the Swedish Natural Science Research Council (NFR), Lunds Djurskyddsfond, and SAAB Dynamics in Göteborg. We thank Brock Fenton, Stuart Parsons, and three referees for comments on earlier drafts.

Literature Cited

Acharya, L. 1995. Sex-biased predation on moths by insectivorous bats. Animal Behaviour, 73:1461–1468.

Acharya, L., and M. B. Fenton. 1992. Echolocation behaviour of vespertilionid bats (*Lasiurus cinereus* and *L. borealis*) attacking airborne targets including arctiid moths. Canadian Journal of Zoology, 70:1292–1298.

Acharya, L., and M. B. Fenton. 1999. Bat attacks and moth defensive behaviour around street lights. Canadian Journal of Zoology, 77:27–33.

Acharya, L., and J. N. McNeil. 1998. Predation risk and mating behavior: the responses of moths to bat-like ultrasound. Behavioral Ecology, 9:552–558.

Aldridge, H. D. J. N., and I. L Rautenbach. 1987. Morphology, echolocation and resource partitioning in insectivorous bats. Journal of Animal Ecology, 56:763–778.

Anderson, M. E., and P. A. Racey. 1991. Feeding behavior of captive brown long-eared bats *Plecotus auritus*. Animal Behaviour, 42:489–493.

Andersson, S., J. Rydell, and M. G. E. Svensson. 1998. Light, predation and the lekking behavior of the ghost swift *Hepialus humuli* (L.) (Lepidoptera: Hepialidae). Proceedings the Royal Society of London B, 265:1345–1351.

Anthony, E. L. P., and T. H. Kunz. 1977. Feeding strategies of the little brown bat, *Myotis lucifugus*, in southern New Hampshire. Ecology, 58:775–786.

Arlettaz, R., G. Dändliker, E. Kasybekov, J.-M. Pillet, S. Rybin, and J. Zima. 1995. Feeding habits of the long-eared desert bat, *Otonycteris hemprichi* (Chiroptera: Vespertilionidae). Journal of Mammalogy, 76:873–876.

Arlettaz, R., and N. Perrin. 1995. The trophic niches of sympatric sibling *Myotis myotis* and *M. blythii*: do mouse-eared bats select prey? Symposia of the Zoological Society of London, no. 67:361–376.

Baagøe, H. J. 1987. The Scandinavian bat fauna: adaptive wing morphology and free flight in the field. Pp. 57–74 *in*: Recent Advances in the Study of Bats (M. B. Fenton, P. Racey, and J. M. V. Rayner, eds.). Cambridge University Press, Cambridge.

Barclay, R. M. R. 1983. Echolocation calls of emballonurid bats from Panama. Journal of Comparative Physiology A, 151:515–520.

Barclay, R. M. R. 1985. Long- versus short-range foraging strategies of hoary (*Lasiurus cinereus*) and silver-haired (*Lasionycteris noctivagans*) bats and the consequences for prey selection. Canadian Journal of Zoology, 63:2507–2515.

Barclay, R. M. R. 1995. Does energy or calcium availability constrain reproduction by bats? Symposia of the Zoological Society of London, no. 67:245–258.

Barclay, R. M. R., and R. M. Brigham. 1991. Prey detection, dietary niche breadth, and body size in bats: why are aerial insectivorous bats so small? American Naturalist, 137:693–703.

Barclay, R. M. R., and R. M. Brigham. 1994. Constraints on optimal foraging: a field test of prey discrimination by echolocating insectivorous bats. Animal Behaviour, 48:1013–1021.

Bartholomew, G. A., and T. M. Casey. 1978. Oxygen consumption of moths during rest, pre-flight warm up, and flight in relation to body size and wing morphology. Journal of Experimental Biology, 76:11–25.

Bates, D. L., and M. B. Fenton. 1990. Aposematism or startle? predators learn their responses to the defences of prey. Canadian Journal of Zoology, 68:49–52.

Bell, G. P. 1982. Behavioral and ecological aspects of gleaning by a desert insectivorous bat, *Antrozous pallidus* (Chiroptera: Vespertilionidae). Behavioral Ecology and Sociobiology, 10:217–223.

Bell, G. P. 1985. The sensory basis of prey location by the California leaf-nosed bat *Macrotus californicus*. Chiroptera: Phyllostomidae. Behavioral Ecology and Sociobiology, 16:343–347.

Belwood, J. J., and M. B. Fenton. 1976. Variation in the diet of *Myotis lucifugus* (Chiroptera: Vespertilionidae). Canadian Journal of Zoology, 54:1674–1678.

Belwood, J. J., and J. H. Fullard. 1984. Echolocation and foraging behaviour in the Hawaiian hoary bat *Lasiurus cinereus semotus*. Canadian Journal of Zoology, 62:2113–2120.

Belwood, J. J., and G. K. Morris. 1987. Bat predation and its influence on calling behavior in Neotropical katydids. Science, 238:64–67.

Beuter, K. J. 1980. A new concept of echo evaluation in the auditory system of bats. Pp. 747–761 *in:* Animal Sonar Systems (R.-G. Busnel and J. F. Fish, eds.). Plenum Press, New York.

Bisset, G. W., J. F. D. Frazer, M. Rothschild, and M. Schachter. 1960. A pharmacologically active choline ester and other substances in the garden tiger moth *Arctia caja* (L.). Proceedings of the Royal Society of London B, 152:255–262.

Blest, A. D., T. S. Collett, and J. D. Pye. 1963. The generation of ultrasonic signals by a New World arctiid moth. Proceedings of the Royal Society of London B, 158:196–207.

Boonman, A. M., M. Boonman, F. Bretschneider, and W. A. van de Grind. 1998. Prey detection in trawling insectivorous bats: duckweed affects hunting behavior in Daubenton's bat, *Myotis daubentonii*. Behavioral Ecology and Sociobiology, 44:99–107.

Brigham, R. M., H. D. J. N. Aldridge, and R. L. Mackey. 1992. Variation in habitat use and prey selection by Yuma bats, *Myotis yumanensis*. Journal of Mammalogy, 73:640–645.

Brigham, R. M., and M. B. Fenton. 1991. Convergence in foraging strategies by two morphologically and phylogenetically distinct nocturnal aerial insectivores. Journal of Zoology (London), 223:475–489.

Britton, A. R. C., and G. Jones. 1999. Echolocation behaviour and prey capture success in foraging bats: laboratory and field experiments on *Myotis daubentonii*. Journal of Experimental Biology, 202:1793–1801.

Britton, A. R. C., G. Jones, J. M. V. Rayner, A. M. Boonman, and B. Verboom. 1997. Flight performance, echolocation and foraging behaviour in pond bats, *Myotis dasycneme* (Chiroptera: Vespertilionidae). Journal of Zoology (London), 241:503–522.

Cardone, B., and J. H. Fullard. 1988. Auditory characteristics and sexual dimorphism in the gypsy moth. Physiological Entomology, 13:9–14.

Carpenter, G. D. H., and H. Eltringham. 1938. Audible emission of defensive froth by insects; with an appendix on the anatomical structures concerned in a moth. Proceedings of the Royal Society of London A, 108:243–252.

Casey, T. M., and B. A. Joos. 1983. Morphometrics, conductance, thoracic temperature, and flight energetics of noctuid and geometrid moths. Physiological Zoology, 56:160–173.

Charnov, E. L. 1976. Optimal foraging: attack strategy of a mantid. American Naturalist, 110:141–151.

Coles, R. B., A. Guppy, M. E. Anderson, and P. Schlegel. 1989. Frequency sensitivity and directional hearing in the gleaning bat, *Plecotus auritus* (Linnaeus 1758). Journal of Comparative Physiology A, 165:269–280.

Conner, W. E. 1999. "Un chant d'appel amoureau": acoustic communication in moths. Journal of Experimental Biology, 202:1711–1723.

Courts, S. E. 1998. Dietary strategies of Old World fruit bats (Megachiroptera, Pteropodidae): how do they obtain sufficient protein? Mammal Review, 28:185–194.

Cumming, G. S. 1996. Mantis movements by night and the interactions of sympatric bats and mantises. Canadian Journal of Zoology, 74:1771–1774.

Davies, N. B. 1977. Prey selection and the search strategy of the spotted flycatcher (*Muscicapa striata*): a field study on optimal foraging. Animal Behaviour, 25: 1016–1033.

Dawkins, R., and J. R. Krebs. 1979. Arms races between and within species. Proceedings of the Royal Society of London B, 205:489–511.

Downes, J. A. 1969. The swarming and mating flight of Diptera. Annual Review of Entomology, 14:271–298.

Dunning, D. C. 1968. Warning sounds of moths. Zeitschrift für Tierpsychologie, 25:129–138.

Dunning, D. C., L. Acharya, C. B. Merriman, and L. Dal Ferro. 1992. Interactions between bats and arctiid moths. Canadian Journal of Zoology, 70:2218–2223.

Dunning, D. C., and K. D. Roeder. 1965. Moth sounds and the insect catching behavior of bats. Science, 147:173–174.

Duvergé, P. L., and G. Jones. 1994. Greater horseshoe bats—activity, foraging behaviour and habitat use. British Wildlife, 6:69–77.

Duvergé, P. L. and G. Jones. In press. Use of farmland habitats by greater horseshoe bats. *In:* Conservation and Conflict: Mammals and Farming in Britain (F. H. Tattersall and W. J. Manleym, eds.). Linnean Society Occasional Publication, Westbury Publishing, Yorkshire.

Edmunds, M. 1974. Defence in Animals. Longman Group. Ltd., Harlow.

Eggers, F. 1919. Das thoracale bitympanale Organ einer Gruppe der Lepidoptera. Zoologische Jahrbücher, Abteilung für Anatomie und Ontogenie der Tiere, 41:273–376.

Faure, P. A., and R. M. R. Barclay. 1992. The sensory basis for prey detection by the long-eared bat, *Myotis evotis,* and the consequences for prey detection. Animal Behaviour, 44:31–39.

Faure, P. A., and R. M. R. Barclay. 1994. Substrate-gleaning versus aerial-hawking: plasticity in the foraging and echolocation behavior of the long-eared bat, *Myotis evotis.* Journal of Comparative Physiology A, 174:651–660.

Faure, P. A., J. H. Fullard, and R. M. R. Barclay. 1990. The response of tympanate moths to the echolocation calls of a substrate gleaning bat, *Myotis evotis.* Journal of Comparative Physiology A, 166:843–849.

Faure, P. A., J. H. Fullard, and J. W. Dawson. 1993. The gleaning attacks of the northern long-eared bat, *Myotis septentrionalis,* are relatively inaudible to moths. Journal of Experimental Biology, 178:173–189.

Fenton, M. B. 1990. The foraging behaviour and ecology of animal-eating bats. Canadian Journal of Zoology, 68:411–422.

Fenton, M. B., D. Audet, M. K. Obrist, and J. Rydell. 1995. Signal strength, timing, and self-deafening: the evolution of echolocation in bats. Paleobiology, 21:229–242.

Fenton, M. B., and G. P. Bell. 1981. Recognition of species of insectivorous bat by their echolocation calls. Journal of Mammalogy, 62:133–243.

Fenton, M. B., D. H. M. Cumming, I. L. Rautenbach, G. S. Cumming, M. S. Cumming, J. Dunlop, M. D. Hovorka, et al. 1998a. Bats and the loss of tree canopy in African woodlands. Conservation Biology, 12:399–407.

Fenton, M. B., and J. H. Fullard. 1979. The influence of moths' hearing on bat echolocation strategies. Journal of Comparative Physiology A, 132:77–86.

Fenton, M. B., C. L. Gaudet, and M. L. Leonard. 1983a. Feeding behaviour of the bats *Nycteris grandis* and *Nycteris thebaica* (Nycteridae) in captivity. Journal of Zoology (London), 200:347–354.

Fenton, M. B., H. G. Merriam, and G. L. Holroyd. 1983b. Bats of Kootenay, Glacier, and Mount Revelstoke national parks in Canada: identification by echolocation calls, distribution, and biology. Canadian Journal of Zoology, 61:2503–2508.

Fenton, M. B., C. V. Portfors, I. L. Rautenbach, and J. M. Waterman. 1998b. Compromises: sound frequencies used in echolocation by aerial-feeding bats. Canadian Journal of Zoology, 76:1174–1182.

Fenton, M. B., I. L. Rautenbach, D. Chipese, M. B. Cumming, M. K. Musgrave, J. S. Taylor, and T. Volpers. 1993. Variation in foraging behaviour, habitat use, and diet of large slit-faced bats (*Nycteris grandis*). Zeitschrift für Säugetierkunde, 58:65–74.

Fenton, M. B., I. L. Rautenbach, J. Rydell, H. T. Arita, J. Ortega, S. Bouchard, M. D. Hovorka, et al. 1998c. Emergence, echolocation, diet and foraging behavior of *Molossus ater* (Chiroptera: Molossidae). Biotropica, 30:314–320.

Fiedler, J. 1979. Prey catching with and without echolocation in the Indian false vampire *Megaderma lyra*. Behavioral Ecology and Sociobiology, 6:155–160.

Forrest, T. G., H. E. Farris, and R. R. Hoy 1995. Ultrasound acoustic startle response in scarab beetles. Journal of Experimental Biology, 198:2593–2598.

Forrest, T. G., M. P. Read, H. E. Farris, and R. R. Hoy. 1997. A tympanal hearing organ in scarab beetles. Journal of Experimental Biology, 200:601–606.

Freeman, P. W. 1979. Specialized insectivory: beetle-eating and moth-eating molossid bats. Journal of Mammalogy, 60:467–479.

Freeman, P. W. 1981. Correspondence of food habits and morphology in insectivorous bats. Journal of Mammalogy, 62:166–173.

Fullard, J. H. 1984. Acoustic relationships between tympanate moths and the Hawaiian hoary bat (*Lasiurus cinereus semotus*). Journal of Comparative Physiology A, 155:795–801.

Fullard, J. H. 1987. Sensory ecology and neuroethology of moths and bats: interactions in a global perspective. Pp. 244–272 *in:* Recent Advances in the Study of Bats (M. B. Fenton, P. Racey, and J. M. V. Rayner, eds.). Cambridge University Press, Cambridge.

Fullard, J. H. 1988. The tuning of moth ears. Experientia, 44:423–428.

Fullard, J. H., and R. M. R. Barclay. 1980. Audition in spring species of arctiid moths as a possible response to differential levels of bat predation. Canadian Journal of Zoology, 58:1745–1750.

Fullard, J. H., and J. W. Dawson. 1997. The echolocation calls of the spotted bat *Euderma maculatum* are relatively inaudible to moths. Journal of Experimental Biology, 200:129–137.

Fullard, J. H., and J. W. Dawson. 1999. Why do diurnal moths have ears? Naturwissenschaften, 86:276–279.

Fullard, J. H., J. W. Dawson, L. D. Otero, and A. Surlykke. 1998. Bat-deafness in day-flying moths (Lepidoptera, Notodontidae, Dioptinae). Journal of Comparative Physiology A, 181:477–483.

Fullard,. J. H., and M. B. Fenton. 1977. Acoustic behavioural analyses of the sound produced by some species of Nearctic Arctiidae (Lepidoptera). Canadian Journal of Zoology, 55:1213–1224.

Fullard, J. H., M. B. Fenton, and J. A. Simmons. 1979. Jamming bat echolocation: the clicks of arctiid moths. Canadian Journal of Zoology, 57:647–649.

Fullard, J. H., J. A. Simmons, and P. A. Saillant. 1994. Jamming bat echolocation: the dogbane tiger moth Cycnia tenera times its clicks to the terminal attack calls of the big brown bat Eptesicus fuscus. Journal of Experimental Biology, 194:285–298.

Fullard, J. H., and D. W. Thomas. 1981. Detection of certain African, insectivorous bats by sympatric, tympanate moths. Journal of Comparative Physiology A, 143:363–368.

Fullard, J. H., and J. E. Yack. 1993. The evolutionary biology of insect hearing. Trends in Ecology and Evolution, 8:248–252.

Fuzessery, Z. M., P. Buttenhoff, B. Andrews, and J. M. Kennedy. 1993. Passive sound localization of prey by the pallid bat (Antrozous p. pallidus). Journal of Comparative Physiology A, 171:767–777.

Göpfert, M. C., A. Surlykke, and L. T. Wasserthal. 1998. Hearing in hawkmoths—scales replace the tympanum in some species. P. 282 in: Göttingen Neurobiology Report, 1998 (N. Elsner and R. Wehner, eds.). Georg Thieme Verlag, Stuttgart.

Göpfert, M. C., and L. T. Wasserthal. 1999a. Hearing with the mouthparts: behavioural responses and the structural basis of ultrasound perception in acherontiine hawkmoths. Journal of Experimental Biology, 202:909–918.

Göpfert, M. C. and L. T. Wasserthal. 1999b. Auditory sensory cells in hawkmoths: identification, physiology and structure. Journal of Experimental Biology, 202:1579–1587.

Gould, E. 1978. Opportunistic feeding by tropical bats. Biotropica, 10:75–76.

Greenewalt, C. H. 1962. Dimensional Relationships for Flying Animals. Smithsonian Miscellaneous Collections, vol. 144, no. 2. Smithsonian Institution, Washington, D.C., 46 pp.

Griffin, D. R. 1950. Measurements of the ultrasonic cries of bats. Journal of the Acoustical Society of America, 22:247–255.

Griffin, D. R. 1958. Listening in the Dark. Yale University Press, New Haven, Conn.

Griffin, D. R. 1971. The importance of atmospheric attenuation for the echolocation of bats (Chiroptera). Animal Behaviour, 19:55–61.

Griffin, D. R., and F. A. Webster. 1962. The role of the flight membranes in insect capture by bats. Animal Behaviour, 10:332–340.

Griffin, D. R., F. A. Webster, and C. R. Michael. 1960. The echolocation of flying insects by bats. Animal Behaviour, 8:141–154.

Habersetzer, J., G. Richter, and G. Storch. 1994. Paleoecology of early Middle Eocene bats from Messel, FRG: aspects of flight, feeding and echolocation. Historical Biology, 8:235–260.

Habersetzer, J., and B. Vogler. 1983. Discrimination of surface-structures targets by the echolocating bat *Myotis myotis* during flight. Journal of Comparative Physiology A, 152:275–282.

Heinrich, B. 1993. The Hot-Blooded Insects. Springer-Verlag, Berlin.

Heinrich, B., and T. P. Mommsen. 1985. Flight of winter moths near 0°C. Science, 228:177–179.

Heller, K.-G., and O. von Helversen. 1989. Resource partitioning of sonar frequency bands in rhinolophid bats. Oecologia, 80:178–186.

Henson, O.W., Jr., M. M. Henson, J. B. Kobler, and G. D. Pollak. 1980. The constant frequency component of the biosonar signals of the bat, *Pteronotus p. parnellii*. Pp. 913–916 *in:* Animal Sonar Systems (R.-G. Busnel and J. F. Fish, eds.). Plenum Press, New York.

Hickey, B., and M. B. Fenton. 1990. Foraging by red bats (*Lasiurus borealis*): do interspecific chases mean territoriality? Canadian Journal of Zoology, 68:2477–2482.

Hill, J. E., and J. D. Smith. 1984. Bats: A Natural History. British Museum (Natural History), London.

Hosken, D. J., W. J. Bailey, J. E. O'Shea, and J. D. Roberts. 1994. Localization of insect calls by the bat *Nyctophilus geoffroyi* (Chiroptera: Vespertilionidae): a laboratory study. Australian Journal of Zoology, 42:177–184.

Houston, R. D., A. M. Boonman, and G. Jones. 2003. Do echolocation signal parameters restrict bats' choice of prey? *In:* Echolocation in Bats and Dolphins (J. Thomas, C. F. Moss, and M. Vater, eds.). University of Chicago Press, Chicago (in press).

Houston, R. D., and G. Jones. 2003. Discrimination of prey during trawling by the insectivorous bat, *Myotis daubentonii. In:* Echolocation in Bats and Dolphins (J. A. Thomas, C. F. Moss, and M. Vater, eds.). University of Chicago Press, Chicago (in press).

Hoy, R. R. 1992. The evolution of hearing in insects as an adaptation to predation from bats. Pp. 115–129 *in:* The Evolutionary Biology of Hearing (D. B. Webster, R. R. Fay, and A. N. Popper, eds.). Springer-Verlag, Berlin.

Hoy, R. R., T. G. Nolen, and, P. D. Brodfuehrer. 1989. The neuroethology of acoustic startle and escape in flying insects. Journal of Experimental Biology, 146:287–306.

Hoy, R. R., and D. Robert. 1996. Tympanal hearing in insects. Annual Review of Entomology, 41:433–450.

Hurst, T. E., and M. J. Lacki. 1997. Food habits of Rafinesque's big-eared bat in southeastern Kentucky. Journal of Mammalogy, 78:525–528.

Janzen, D. H. 1980. When is it coevolution? Evolution, 34:611–612.

Jepsen, G. L. 1970. Bat origins and evolution. Pp. 1–64 *in:* Biology of Bats (W. A. Wimsatt, ed.). Vol. 1. Academic Press, New York.

Jones, G. 1990. Prey selection by the greater horseshoe bat *Rhinolophus ferrumequinum:* optimal foraging by echolocation? Journal of Animal Ecology 59:587–602.

Jones, G. 1992. Bats vs. moths: studies on the diets of rhinolophid and hipposiderid bats support the allotonic frequency hypothesis. Pp. 87–92 *in:* Prague Studies in Mammalogy (I. Horáček and V. Vohralík, eds.). Charles University Press, Prague.

Jones, G. 1994. Scaling of wingbeat and echolocation pulse emission rates in bats: why are aerial insectivorous bats so small? Functional Ecology, 8:450–457.

Jones, G. 1995. Flight performance, echolocation and foraging behaviour in noctule bats *Nyctalus noctula.* Journal of Zoology (London), 237, 303–312.

Jones, G. 1996. Does echolocation constrain the evolution of body size in bats? Symposia of the Zoological Society of London, no. 69:111–128.

Jones, G. 1999. Scaling of echolocation call parameters in bats. Journal of Experimental Biology, 202:3359–3367.

Jones, G., M. Morton, P. M. Hughes, and R. M. Budder. 1993. Echolocation, flight morphology and foraging strategies of some West African hipposiderid bats. Journal of Zoology (London), 230:385–400.

Jones, G., and J. M. V. Rayner. 1988. Flight performance, foraging tactics, and echolocation in free-living Daubenton's bats Myotis daubentoni (Chiroptera: Vespertilionidae). Journal of Zoology (London), 215:113–132.

Jones, G., and J. M. V. Rayner. 1989. Foraging behavior and echolocation of wild horseshoe bats Rhinolophus ferrumequinum and R. hipposideros (Chiroptera, Rhinolophidae). Behavioral Ecology and Sociobiology, 25:183–191.

Jones, G., and J. M. V. Rayner. 1991. Flight performance, foraging tactics, and echolocation in the trawling insectivorous bat Myotis adversus (Chiroptera: Vespertilionidae). Journal of Zoology (London), 225:393–412.

Jones, G., and J. Rydell. 1994. Foraging strategy and predation risk as factors influencing emergence time in echolocating bats. Philosophical Transactions of the Royal Society of London B, 346:445–455.

Kalko, E. K. V. 1994. Coupling of sound emission and wingbeat in naturally foraging European pipistrelle bats (Microchiroptera: Vespertilionidae). Folia Zoologica, 43:363–376.

Kalko, E. K. V. 1995. Insect pursuit, prey capture and echolocation in pipistrelle bats (Microchiroptera). Animal Behaviour, 50:861–880.

Kalko, E. K. V., and H.-U. Schnitzler. 1989a. The echolocation and hunting behavior of Daubenton's bat, Myotis daubentoni. Behavioral Ecology and Sociobiology, 24:225–238.

Kalko, E. K. V., and H.-U. Schnitzler. 1989b. Two-wave-front interference patterns in frequency-modulated echolocation signals of bats flying low over water. Journal of the Acoustical Society of America, 85:961–962.

Kalko, E. K. V., and H.-U. Schnitzler. 1993. Plasticity in echolocation signals of European pipistrelle bats in search flight: implications for habitat use and prey detection. Behavioral Ecology and Sociobiology, 33:415–428.

Kalko, E. K. V., H.-U. Schnitzler, I. Kaipf, and A. D. Grinnell. 1998. Echolocation and foraging behavior of the lesser bulldog bat, Noctilio albiventris: preadaptations for piscivory? Behavioral Ecology and Sociobiology, 42:305–319.

Kennell, J., and F. Eggers. 1933. Die abdominalen Tympanalorgane der Lepidopteren. Zoologische Jahrbücher, Abteilung für Anatomie und Ontogenie der Tiere, 57:1–104.

Kick, S. A. 1982. Target detection by the echolocating bat, Eptesicus fuscus. Journal of Comparative Physiology A, 145:431–435.

Kick, S. A., and J. A. Simmons. 1984. Automatic gain control in the bat's sonar receiver and the neuroethology of echolocation. Journal of Neurosciences, 4:2725–2737.

Kober, R., and H.-U. Schnitzler. 1990. Information in sonar echoes of fluttering insects available for echolocating bats. Journal of the Acoustical Society of America, 87:882–896.

Kunz, T. H., J. O. Whitaker, Jr., and M. D. Wadanoli. 1995. Dietary energetics of the insectivorous Mexican free-tailed bat (*Tadarida brasiliensis*) during pregnancy and lactation. Oecologia, 101:407–415.

Lancaster, W. C., O. W. Henson, and A. W. Keating. 1995. Respiratory muscle activity in relation to vocalization in flying bats. Journal of Experimental Biology, 198: 175–191.

Lawrence, B. D., and J. A. Simmons. 1982. Measurements of atmospheric attenuation at ultrasonic frequencies and the significance for echolocation by bats. Journal of the Acoustical Society of America, 71:585–590.

Lewis, F., J. H. Fullard, and S. B. Morrill. 1993. Auditory influences on the flight behaviour of moths in a Nearctic site. II. Flight times, heights and erraticism. Canadian Journal of Zoology, 71:1562–1568.

Libersat, F., and R. R. Hoy. 1991. Ultrasonic startle behavior in bushcrickets (Orthoptera; Tettigonidae). Journal of Comparative Physiology A, 169:507–514.

Marden, J. 1995. Evolutionary adaptation of contractile performance in muscle of ectothermic winter-flying moths. Journal of Experimental Biology, 19:2087–2094.

Marimuthu, G. 1997. Stationary prey insures life and moving prey ensures death during the hunting flight of gleaning bats. Current Science, 72:928–931.

Miller, L. A. 1970. Structure of the green lacewing tympanal organ (*Chrysopa carnea*, Neuroptera). Journal of Morphology, 131:359–382.

Miller, L. A. 1980. How the green lacewing avoids bats: behavior and physiology. Pp. 941–943 *in:* Animal Sonar Systems (R.-G. Busnel and J. F. Fish, eds.). Plenum Press, New York.

Miller, L. A. 1991. Arctiid moth clicks can degrade the accuracy of range difference discrimination in echolocating big brown bats, *Eptesicus fuscus*. Journal of Comparative Physiology A, 168:571–579.

Miller, L. A., and E. G. MacLeod. 1966. Ultrasonic sensitivity: a tympanal receptor in the green lace wing *Chrysopa carnea*. Science, 154:891–893.

Miller, L. A., and J. Olesen. 1979. Avoidance behavior in green lacewings, *Chrysopa carnea*, in the presence of ultrasound. Journal of Comparative Physiology, 131: 113–120.

Miller, L. A., and A. E. Treat. 1993. Field recordings of echolocation and social signals from the gleaning bat *Myotis septentrionalis*. Bioacoustics, 5:67–87.

Mitson, R. B. 1983. Fisheries Sonar. Fishing News Books, Ltd., Farnham.

Møhl, B., and L. A. Miller. 1976. Ultrasonic clicks produced by the peacock butterfly: a possible bat-repellent mechanism. Journal of Experimental Biology, 64:639–644.

Moiseff, A., G. S. Pollock, and R. R. Hoy. 1978. Steering responses of flying crickets to sound and ultrasound: mate attraction and predator avoidance. Proceedings of the National Academy of Sciences of the USA, 75:4052–4056.

Moss, C. F., and H.-U. Schnitzler. 1995. Behavioral studies of auditory information processing. Pp. 87–145 *in:* Hearing by Bats (A. N. Popper and R. R. Fay, eds.). Springer-Verlag, New York.

Neuweiler, G. 1989. Foraging ecology and audition in echolocating bats. Trends in Ecology and Evolution, 4:160–166.

Neuweiler, G. 1990. Auditory adaptations for prey capture in echolocating bats. Physiological Reviews, 70:615–641.

Neuweiler, G., W. Metzner, U. Heilmann, R. Rübsamen, M. Eckrich, and H. H. Costa.

1987. Foraging behavior and echolocation in the rufous horseshoe bat *Rhinolophus rouxi* of Sri Lanka. Behavioral Ecology and Sociobiology, 20:53–67.

Nolen, T. G., and R. R. Hoy. 1986. Phonotaxis in flying crickets. I. Attraction to the calling song and avoidance of bat-like ultrasound are discrete behaviors. Journal of Comparative Physiology A, 159:423–439.

Norberg, U. M. 1986. Evolutionary convergence in foraging niche and flight morphology in insectivorous aerial-hawking birds and bats. Ornis Scandinavica, 17:153–160.

Norberg, U. M., and J. M. V. Rayner. 1987. Ecological morphology and flight in bats (Mammalia; Chiroptera): wing adaptations, flight performance, foraging strategy and echolocation. Philosophical Transactions of the Royal Society of London B, 316:335–427.

Novacek, M. J. 1985. Evidence for echolocation in the oldest known bats. Nature, 315:140–141.

Parsons, S., C. W. Thorpe, and S. M. Dawson. 1997. Echolocation calls of the long-tailed bat: a quantitative analysis of types of calls. Journal of Mammalogy, 78:964–976.

Pavey, C., and C. Burwell. 1998. Bat predation on eared moths: a test of the allotonic frequency hypothesis. Oikos, 81:143–151.

Peters, D. S. 1992. Messel birds: a land based assemblage. Pp. 135–151 *in:* Messel: An Insight into the History of Life on Earth (S. Schaal and W. Ziegler, eds.). Clarendon Press, Oxford.

Pollak, G. D., and J. H. Casseday. 1989. The neural basis of echolocation in bats. Springer-Verlag, Berlin.

Pye, J. D. 1968. How insects hear. Nature, 218:797.

Pye, J. D. 1993. Is fidelity futile? the "true" signal is illusory, especially with ultrasound. Bioacoustics, 4:271–286.

Racey, P. A., and S. M. Swift. 1985. Feeding ecology of *Pipistrellus pipistrellus* (Chiroptera: Vespertilionidae) during pregnancy and lactation. I. Foraging behaviour. Journal of Animal Ecology, 54: 205–215.

Racey, P. A., S. M. Swift, J. Rydell, and L. Brodie. 1998. Bats and insects over two Scottish rivers with contrasting nitrate status. Animal Conservation, 1:195–202.

Rautenbach, I. L., A. C. Kemp, and C. H. Scholtz. 1988. Fluctuations in availability of arthropods correlated with microchiropteran and avian predator activities. Koedoe, 31:77–90.

Robert, D. 1989. The auditory behavior of flying locusts. Journal of Experimental Biology, 147:279–310.

Robert, D., J. Amoroso, and R. R. Hoy 1992. The evolutionary convergence of hearing in a parasitoid fly and its cricket host. Science, 258:1135–1137.

Roeder, K. D. 1962. The behaviour of free flying moths in the presence of artificial ultrasonic pulses. Animal Behaviour, 10:300–304.

Roeder, K. D. 1967a. Nerve Cells and Insect Behavior. 2d ed. Harvard University Press, Cambridge, Mass.

Roeder, K. D. 1967b. Turning tendency of moths exposed to ultrasound while in stationary flight. Journal of Insect Physiology 13:873–888.

Roeder, K. D. 1974. Acoustic sensory responses and possible bat evasion tactics of certain moths. Pp. 71–74 *in:* Proceedings of the Canadian Society of Zoologists Annual Meeting, 1974 (M. D. B. Burt, ed.). University of New Brunswick, Fredericton.

Roeder, K. D. 1975. Neural factors and evitability in insect behavior. Journal of Experimental Zoology, 194:75–88.

Roeder, K. D., A. E. Treat, and J. S. Vandeberg. 1970. Distal lobe of the pilifer: an ultrasonic receptor in Choerocampine hawkmoths. Science, 170:1098–1099.

Ross, A. 1967. Ecological aspects of the food habits of insectivorous bats. Proceedings of the Western Foundation of Vertebrate Zoology, 1:205–263.

Roverud, R. C., and A. D. Grinnell. 1985. Echolocation sound features processed to provide distance information in the CF/FM bat, *Noctilio albiventris*: evidence for a gated time window utilizing both CF and FM components. Journal of Comparative Physiology A, 156:457–469.

Roverud, R. C., V. Nitsche, and G. Neuweiler. 1991. Discrimination of wingbeat motion by bats, correlated with echolocation sound pattern. Journal of Comparative Physiology A, 168: 259–263.

Ruedi, M., and F. Mayer. 2001. Molecular systematics of bats of the genus *Myotis* (Vespertilionidae) suggests deterministic ecomorphological convergences. Molecular Phylogentics and Evolution, 21:436–448.

Rydell, J. 1986. Foraging and diet of the northern bat *Eptesicus nilssoni* in Sweden. Holarctic Ecology 9:272–276.

Rydell, J. 1990. Behavioural variation in echolocation pulses of the northern bat *Eptesicus nilssoni*. Ethology, 85:103–113.

Rydell, J. 1992a. The diet of the parti-coloured bat *Vespertilio murinus* in Sweden. Holarctic Ecology, 15:195–198.

Rydell, J. 1992b. Exploitation of insects around streetlamps by bats in Sweden. Functional Ecology, 6:744–750.

Rydell, J. 1998. Bat defence in lekking ghost swifts (*Hepialus humuli*), a moth without ultrasonic hearing. Proceedings of the Royal Society of London B, 265: 1373–1376.

Rydell, J., and R. Arlettaz. 1994. Low frequency echolocation enables the bat *Tadarida teniotis* to feed on tympanate insects. Proceedings of the Royal Society of London B, 257:175–178.

Rydell, J., and W. C. Lancaster. 2000. Flight and thermoregulation in moths have been shaped by predation from bats. Oikos, 88:13–18.

Rydell, J., A. Entwistle, and P. A. Racey. 1996a. Timing of foraging flights of three species of bats in relation to insect activity and predation risk. Oikos, 76:243–252.

Rydell, J., G. Jones, and D. A. Waters. 1995. Echolocating bats and hearing moths: who are the winners? Oikos, 73:419–424.

Rydell, J., L. A. Miller, and M. Jensen. 1999. Echolocation constraints of Daubenton's bat foraging over water. Functional Ecology, 13:247–255.

Rydell, J., G. Natuschke, A. Theiler, and P. E. Zingg. 1996b. Food habits of the barbastelle bat *Barbastella barbastellus*. Ecography, 19:62–66.

Rydell, J., N. Skals, A. Surlykke, and M. G. E. Svensson. 1997. Hearing and bat defence in geometrid winter moths. Proceedings of the Royal Society of London B, 264:83–88.

Rydell, J., and J. R. Speakman. 1995. Evolution of nocturnality in bats: potential competitors and predators during their early history. Biological Journal of the Linnean Society, 54:183–191.

Rydell, J., and D. W. Yalden. 1997. The diets of two high-flying bats from Africa. Journal of Zoology (London), 242:69–76.

Sales, G., and J. D. Pye. 1974. Ultrasonic Communication by Animals. Chapman & Hall, London.

Schalk, G., and R. M. Brigham. 1995. Prey selection by insectivorous bats: are essential fatty acids important? Canadian Journal of Zoology, 73:1855–1859.

Schmidt, S. 1988. Evidence for a spectral basis of texture perception in bat sonar. Nature, 331:617–619.

Schmidt, S., B. Türke, and B. Vogler. 1984. Behavioral audiogram from the bat, *Megaderma lyra* (Geoffroy, 1810; Microchiroptera). Myotis, 21–22:62–66.

Schnitzler, H.-U. 1968. Die Ultraschall-Ortungslaute der Hufeisen-Fledermäuse (Chiroptera-Rhinolophidae) in verschiedenen Orientierungssituationen. Zeitschrift für Vergleichende Physiologie, 57:376–408.

Schnitzler, H.-U.. 1987. Echoes of fluttering insects: information for echolocating bats. Pp. 226–243 *in:* Recent Advances in the Study of Bats (M. B. Fenton, P. Racey, and J. M. V. Rayner, eds.). Cambridge University Press, Cambridge.

Schnitzler, H.-U., H. Hackbarth, U. Heilmann, and H. Herbert. 1985. Echolocation behavior of rufous horseshoe bats hunting for insects in the flycatcher-style. Journal of Comparative Physiology A, 157:39–64.

Schnitzler, H.-U., and O. W. Henson, Jr. 1980. Performance of airborne animal sonar systems. I. Microchiroptera. Pp.109–181 *in:* Animal Sonar Systems (R.-G. Busnel and J. F. Fish, eds.). Plenum Press, New York.

Schnitzler, H.-U., and E. K. V. Kalko. 1998. How echolocating bats search and find food. Pp. 183–196 *in:* Bat Biology and Conservation (T. H. Kunz and P. A. Racey, eds.). Smithsonian Institution Press, Washington, D.C.

Schnitzler, H.-U., E. K. V. Kalko, I. Kaipf, and A. D. Grinnell. 1994. Fishing and echolocation behavior of the greater bulldog bat, *Noctilio leporinus,* in the field. Behavioral Ecology and Sociobiology, 35:327–345.

Schnitzler, H.-U., D. Menne, R. Kober, and K. Heblich. 1983. The acoustical image of fluttering insects in echolocating bats. Pp. 235–250 *in:* Neuroethology and Behavioral Physiology (F. Huber and H. Markl, eds.). Springer-Verlag, Berlin.

Schuller, G. and G. D. Pollak. 1979. Disproportionate frequency representation in the inferior colliculus of horseshoe bats: evidence for an "acoustic fovea." Journal of Comparative Physiology A, 132:47–54.

Scoble, M. J. 1992. The Lepidoptera: Form, Function and Diversity. Natural History Museum in association with Oxford University Press, Oxford.

Sick, H. 1937. Die Tympanalorgane der Uraniden und Epiplemiden. Zoologische Jahrbucher, Abteilung für Anatomie und Ontogenie der Tiere, 63:351–398.

Siemers, B. M., and H.-U. Schnitzler. 2000. Natterer's bat (*Myotis nattereri* Kuhl, 1818) hawks for prey close to vegetation using echolocation signals of very broad bandwidth. Behavioral Ecology and Sociobiology, 47:400–412.

Siemers, B. M., P. Stilz, and H.-U. Schnitzler. 2001. The acoustic advantage of hunting at low heights above water: behavioural experiments on the European trawling bats *Myotis capaccinii, M. dasycneme* and *M. daubentonii.* Journal of Experimental Biology, 204:3843–3854.

Sierro, A., and R. Arlettaz. 1997. Barbastelle bats (*Barbastella* spp.) specialize in the predation of moths: implications for foraging tactics and conservation. Acta Oecologica, 18:91–106.

Simmons, J. A. 1973. The resolution of target range by echolocating bats. Journal of the Acoustical Society of America, 54:157–173.

Simmons, J. A., and R. A. Stein. 1980. Acoustic imaging in bat sonar: echolocation signals and the evolution of echolocation. Journal of Comparative Physiology A, 135:61–84.

Simmons, J. A, M. J. Ferragamo, P. A. Saillant, T. Haresign, J. M. Wotton, S. P. Dear, and D. N. Lee. 1995. Auditory dimensions of acoustic images in echolocation. Pp. 146–190 in: Hearing by Bats (A. N. Popper and R. R. Fay, eds.). Springer-Verlag, New York.

Simmons, N. B., and J. H. Geisler. 1998. Phylogenetic relationships of Icaronycteris, Archaeonycteris, Hassianycteris, and Palaeochiropteryx to extant bat lineages, with comments on the evolution of echolocation and foraging strategies in Microchiroptera. Bulletin of the American Museum of Natural History, 235:1–182.

Spangler, H. G. 1985. Sound production and communication by the greater wax moth (Lepidoptera: Pyralidae). Annals of the Entomological Society of America, 78:54–61.

Spangler, H. G. 1988a. Hearing in tiger beetles (Cicindelidae). Physiological Entomology, 13:447–452.

Spangler, H. G. 1988b. Moth hearing, defence, and communication. Annual Review of Entomology, 33:59–81.

Speakman, J. R. 1991. Why do insectivorous bats in Britain not fly in daylight more frequently? Functional Ecology, 5:518–524.

Speakman, J. R. 1995. Chiropteran nocturnality. Symposia of the Zoological Society of London, no. 67:187–201.

Speakman, J. R., M. E. Anderson, and P. A. Racey. 1989. The energy cost of echolocation in pipistrelle bats (Pipistrellus pipistrellus). Journal of Comparative Physiology A, 165:679–685.

Speakman, J. R., and P. A. Racey. 1991. No cost of echolocation for bats in flight. Nature, 350:421–423.

Stephens, D. W., and J. R. Krebs. 1986. Foraging Theory. Princeton University Press, Princeton, N.J.

Surlykke, A. 1984. Hearing in notodontid moths: a tympanate organ with a single auditory neurone. Journal of Experimental Biology, 113:323–335.

Surlykke, A. 1992. Target ranging and the role of time-frequency structure of synthetic echoes in big brown bats, Eptesicus fuscus. Journal of Comparative Physiology A, 170: 83–92.

Surlykke, A., M. Filskov, J. F. Fullard, and E. Forrest. 1999. Auditory relationships to size in noctuid moths: bigger is better. Naturwissenschaften, 86:238–241.

Surlykke, A., L. A. Miller, B. Møhl, B. B. Andersen, J. Christensen-Dalsgaard, and M. B. Jørgensen. 1993. Echolocation in two very small bats from Thailand: Craseonycteris thonglongyai and Myotis siligorensis. Behavioral Ecology and Sociobiology, 33:1–12.

Surlykke, A., N. Skals, J. Rydell, and M. G. E. Svensson. 1998. Sonic hearing in a diurnal geometrid moth Archiearis parthenias (L.) temporally isolated from bats. Naturwissenschaften, 85:36–37.

Surlykke, A., and A. E. Treat. 1995. Hearing in wintermoths. Naturwissenschaften, 82:382–384.

Svensson, A. M., and J. Rydell. 1998. Mercury vapour lamps interfere with moths' bat defence. Animal Behaviour, 55:223–226.

Svensson, M. G. E., J. Rydell, and R. Brown. 1999. Bat predation and the flight timing of winter moths, *Epirrita* and *Operophtera* spp. (Lepidoptera, Geometridae). Oikos, 84:193–198.

Sweeney, B. W., and R. L. Vannote. 1982. Population synchrony in mayflies (*Dolania americana*): a predator satiation hypothesis. Evolution, 36:810–821.

Swift, S. M., P. A. Racey, and M. I. Avery. 1985. Feeding ecology of *Pipistrellus pipistrellus* (Chiroptera: Vespertilionidae) during pregnancy and lactation. II. Diet. Journal of Animal Ecology, 54:217–254.

Swihart, S. L. 1967. Hearing in butterflies (Nymphalidae: *Heliconius, Ageronia*). Journal of Insect Physiology, 13:469–476.

Turner, A. K. 1982. Optimal foraging by the swallow (*Hirundo rustica,* L.): prey size selection. Animal Behaviour, 30:862–872.

Vaughan, N. 1997. The diets of British bats (Chiroptera). Mammal Review, 27:77–94.

Vaughan, T. A. 1976. Nocturnal behavior of the African false vampire bat (*Cardioderma cor*). Journal of Mammalogy, 57:227–248.

Vaughan, T. A. 1977. Foraging behavior of the giant leaf-nosed bat (*Hipposideros commersoni*). East African Wildlife Journal, 15:237–249.

von der Emde, G., and D. Menne. 1989. Discrimination of insect wingbeat-frequencies by the bat *Rhinolophus ferrumequinum*. Journal of Comparative Physiology A, 164:663–671.

von der Emde, G., and H.-U. Schnitzler. 1990. Classification of insects by echolocating greater horseshoe bats. Journal of Comparative Physiology A, 167:423–430.

Waters, D. A., and G. Jones. 1995. Echolocation call structure and intensity in five species of insectivorous bats. Journal of Experimental Biology, 198:475–489.

Waters, D. A., J. Rydell, and G. Jones. 1995. Echolocation call design and limits on prey size: a case study using the aerial-hawking bat *Nyctalus leisleri*. Behavioral Ecology and Sociobiology, 37:321–328.

Werner, T. 1981. Responses of nonflying moths to ultrasound: the threat of gleaning bats. Canadian Journal of Zoology, 59:525–529.

Whitaker, J. O., Jr., and H. L. Black. 1976. Food habits of cave bats from Zambia. Journal of Mammalogy, 57:199–205.

Whitaker, J. O., Jr., and P. Clem. 1992. Food of the evening bat *Nycticeius humeralis* from Indiana. American Midland Naturalist, 127:211–214.

Williams, C. B. 1939. An analysis of four years' captures of insects in a light trap. 1. General survey; sex proportion; phenology; and time of flight. Transactions of the Royal Entomological Society of London, 89:79–131.

Woodsworth, G. C., G. P. Bell, and M. B. Fenton. 1981. Observations on the echolocation, feeding behaviour and habitat use of *Euderma maculatum* in southcentral British Columbia. Canadian Journal of Zoology, 59:1099–1102.

Yack, J. E . 1988. Seasonal partitioning of atympanate moths in relation to bat activity. Canadian Journal of Zoology, 66:753–755.

Yager, D. D., and R. R. Hoy. 1986. The cyclopian ear: a new sense for the praying mantis. Science, 231:727–729.

Yager, D. D., and M. L. May. 1990. Ultrasound-triggered, flight-gated evasive maneuvers in the praying mantis, *Parasphendale agrionina*. II. Tethered flight. Journal of Experimental Biology, 152:41–58.

Yager, D. D., M. L. May, and M. B. Fenton. 1990. Ultrasound-triggered, flight-gated evasive maneuvers in the praying mantis, *Parasphendale agrionina*. I. Free flight. Journal
of Experimental Biology, 152:17–39.

Yager, D. D., and H. G. Spangler. 1995. Characterization of auditory afferents in the tiger
beetle, *Cincindela marutha* Dow. Journal of Comparative Physiology A, 176:587–599.

Yager, D. D., and H. G. Spangler. 1997. Behavioral response to ultrasound by the tiger
beetle, *Cincindela marutha* Dow combines aerodynamic changes and sound production. Journal of Experimental Biology, 200:649–659.

Zbinden, K. 1988. Harmonic structure of bat echolocation signals. Pp. 581–587 *in:* Animal Sonar: Processes and Performance (P. E. Nachtigall and P. W. B. Moore, eds.).
Plenum Press, New York.

Zbinden, K., and P. E. Zingg. 1986. Search and hunting signals of echolocating European
free-tailed bats, *Tadarida teniotis*, in southern Switzerland. Mammalia, 50:9–25.

CHAPTER 8

Glossophagine Bats and Their Flowers: Costs and Benefits for Plants and Pollinators

Otto von Helversen and York Winter

Introduction

Birds and bats share feeding niches that include flowers in both the Old and New World tropics. In the process of feeding on nectar and pollen, these plant-visiting vertebrates serve as important pollinators. In the Old World tropics, flower-visiting bats belong to the group commonly referred to as flying foxes or megabats, of which the most specialized are members of the subfamily Macroglossinae. In the New World, flower-visiting bats are members of the family Phyllostomidae, of which the most specialized are the subfamilies Phyllonycterinae and Glossophaginae. Worldwide, there are approximately 50 flower-visiting bat species and many more plant species that are pollinated by these bats. Dobat (1985) listed about 750 species in 270 genera of plants visited by bats, but since then many more species have been identified.

Flowers pollinated by bats exhibit a number of characteristic adaptations, which are collectively described by the "syndrome of chiropterophily" (e.g., Dobat 1985; Faegri and van der Pijl 1971; Heithaus 1982; Porsch 1931b; Vogel 1968, 1969a, 1969b). "Pollination syndromes" are the reflection of the pollinator's requirements (see table 8.3); accordingly, they are expressed to various degrees, depending on whether a plant is more of a generalist or specialist with respect to its pollination biology. Major characteristics of bat-pollinated flowers include nocturnal anthesis, often limited to a single night; shape and sturdiness (often brushlike flowers or bells, which sometimes are only the size of a head mask for a bat; fig. 8.1); a large production of nectar and pollen; an intense, typical scent; inconspicuous colors; and a freely exposed position on the plant. The adaptive nature of these and other traits will be discussed below.

Bat pollination is a phenomenon restricted to the tropics and subtropics. This can be seen in the distribution of glossophagine bats in the New World in comparison to that of hummingbirds (fig. 8.2). The species richness of nectarivorous bats and birds is highest in the circumequatorial regions. One of the reasons for this limitation to the tropics presumably is that pollinators need a

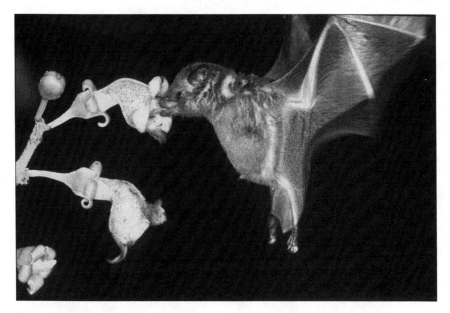

Figure 8.1. The flower of the glossophagine-pollinated *Paliavana prasinata* (Gesneriaceae) is no larger than a head mask for the visiting bat.

reliable year-round supply of flower nectar (Fleming et al. 1993). Similarly, frugivorous species are also restricted to the tropics and subtropics. In this case, the reasons are not as easy to understand, since many species of frugivorous birds live at temperate latitudes. Many species of bats are capable of hibernation (see Thomas and Speakman, this volume), which in principle should enable plant-visiting bats to sustain themselves by eating nectar and fruits during the warm months of the year. Presumably it is their inability to hibernate that has generally restricted the Phyllostomidae and the Megachiroptera to the tropics and, thus, has prevented the evolution of bat frugivory and nectarivory in temperate zones.

In this chapter we focus on New World glossophagine bats (and glossophagine-pollinated flowers, as the ecology, behavior, physiology, and energetics of these bats have been studied in considerable detail during the past decade.

For flower visitors, in particular, the energetics of foraging constitutes a crucial point for specialization (Heinrich 1979; Heinrich and Raven 1972). In a guild of flower visitors, individuals and species with the greatest efficiency should be the most superior competitors. Here efficiency refers to the gain (measurable as intake of sugar or the energy equivalent per unit time) diminished by the costs (measurable as the energy consumption during the flight

Hummingbirds **Glossophagine Bats**

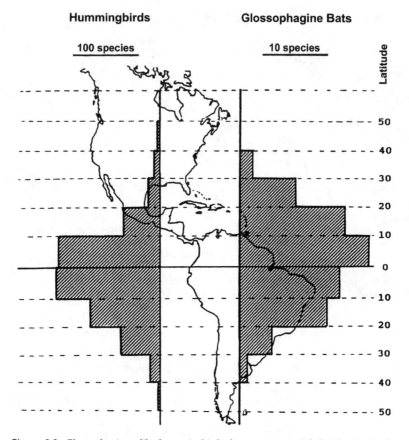

Figure 8.2. Glossophagines, like hummingbirds, have an equatorial distribution in the Americas. However, they comprise only about 35 species compared to the roughly 330 species of hummingbirds—a ratio of 1:10—similar to the overall ratio of bat-to-bird species richness (from Helversen 1993). Bars show number of species per 10 degrees latitude according to scale at top of the figure.

from flower to flower and the extraction of food). The cost-benefit ratio of foraging is therefore presumably the decisive parameter for evolution within the ecological niche of a flower visitor.

During periods of nectar limitation, pollinators have to compete for flower nectar (Law 1995), but the various plant species must also compete for the visits of their pollinators. Given that the reproductive fitness of a plant depends substantially on the number of pollen grains transferred, the number of visits by pollinators is a critical factor in competition. In some cases pollen limitation for seed set could be demonstrated in the field (*Agave* [Howell and Schropfer-

Roth 1981] and *Puya* [H. Schmid, personal observation]). But, even when experimental evidence for pollen limitation is lacking, it is important to understand that plants also evolved traits making it possible for individuals to compete for chances of pollen deposition. Genotypes that are able to deposit pollen early and in high quantities on the stigmata of other plants will have a selective advantage. The amount of nectar produced by a plant that is needed to effectively attract a pollinator creates a dilemma for plants. If the reward obtainable from one kind of flower is less than that offered by other plants blooming at the same time, the pollinators will select the other plants, and as a consequence the visitation rate to some may be low. However, if a highly competitive plant were to produce an increased volume of nectar, the pollinators might not visit so often because they would obtain enough energy in only one or a few visits. Thus, the reward should be offered in the smallest possible portions. One potential solution to this dilemma is for a plant competing with other plants to use the pollinator's efficiency of foraging (the cost-benefit ratio) and not necessarily raise the amount of energy that it provides. Thus, a successful competitor for a pollinator does not necessarily need to increase the amount of reward but may instead reduce the cost of flower exploitation for the pollinator. These costs are mainly incurred in the energy expended to find a flower and to feed from it but also, perhaps, in the risk of predation at the flower.

In the following discussion we consider the characteristics of flowers pollinated by glossophagine bats by asking the following questions: (1) How do bat flowers make themselves easier to find and thus save the pollinator energetic costs and time? (2) How have flower bats reduced their costs for flower visiting and increased their efficiency? How, in the process, have various species avoided or reduced competition with one another, and how are species assemblages of nectarivorous bats structured? (3) What are the adaptations of the bat flowers, and how has the evolution of bat flowers been affected by pollinators?

How Flowers Are Made Conspicuous for Pollinating Bats

If a plant can increase its fitness by reducing the cost of visitation by potential pollinators, the reduction in the time spent searching for a flower is one of the most important consequences for bats. For a given distribution of flowers, the search time of the pollinator depends primarily on the sensory stimuli that the plant can send out to signal the flower's position. Olfaction, echolocation, and vision are the most important sensory modalities that bats can employ during foraging flight (see Altringham and Fenton, this volume). These sensory abilities, complemented by a well-developed spatial memory, are exploited by bat flowers.

Flower Scent

The intense scent of most bat-pollinated flowers is markedly different from the scent of flowers visited by other pollinators (e.g., bees, moths, or flies). Even in the early studies of bat pollination, a characteristic scent was considered to be one of the most important indications of chiropterophily (Porsch 1931b; van der Pijl 1936; Vogel 1968, 1969a, 1969b). The following examples illustrate the difficulty of describing these odors verbally because most of them are perceived as unpleasant by humans. Descriptions of these odors include "strong, unpleasantly stale" (*Hippeastrum calyptratum* [Vogel 1969a]); "nasty" (*Vriesea longicaulis* [Vogel 1969a]); "like freshly cut kohlrabi" (*Crescentia cujete* [Porsch 1931a]); "peculiar, somewhat reminiscent of cabbage" (*Musa* [van der Pijl 1936]); "scent of kohlrabi and garlic" (*Amphitecna macrophylla* [Porsch 1938]); "like damp decaying leaves" (*Myrciaria cauliflora* [Porsch 1942]); "like sour milk, nasty sweet, causing a feeling of nausea" (*Durio zibethinus* [van der Pijl 1936]); "like rotten urine" (*Crescentia cujete* [Winkler 1907]); "smell like opossums" (*Vriesea jonghii* [Müller 1897]); "unpleasant like mouse urine" (*Bassia latifolia* [van der Pijl 1936]); "mild but skunk-like" (*Lemaireocereus thurberi* [Alcorn and Olin 1962]); "smell of mustard oil, smell of carcass" (*Adansonia digitata* [Porsch 1935]); and "smell of corpse" (*Weberocereus tunilla* [Bauer 1991]).

Chemical analysis of odor is now sufficiently developed for a precise evaluation of the constituents of flower scent. Scented air around a flower is passed through a filter that absorbs volatile molecules ("headspace analysis"). The molecules that accumulate on the filter are then separated by gas chromatography and analyzed by mass spectrometry. To date, the odor spectrum of 22 different bat flowers has been analyzed (Bestmann et al. 1997; Kaiser and Tollsten 1995; Knudsen and Tollsten 1995). The odor components thus far measured belong to four classes of substances: aliphatics, aromatics, terpenoids, and sulphur-containing compounds (fig. 8.3).

The most striking result of these studies has been the identification of a large concentration of sulphur-containing compounds in most of the scents that have been analyzed. These sulphur compounds (particulary dimethyl disulfide, dimethyl trisulfide, and dimethyl tetrasulfide) are produced by many phylogenetically unrelated bat-pollinated plant species (Knudsen and Tollsten 1995). The rare occurrence of sulphur-containing scents in other pollination systems (Knudsen et al. 1993) suggests convergent evolution as an adaptation to bat pollination.

In fact, dimethyl disulfide—which smells extremely unpleasant to humans—is a strong attractant for glossophagines. Field experiments in the tropical lowland rainforest of Costa Rica have shown that free-ranging *Glossophaga commissarisi* were attracted to dummy flowers when these contained either dimethyl disulfide or 2,4-dithiapentane as an odor source. Other com-

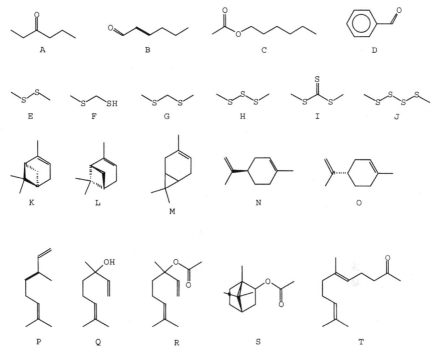

Figure 8.3. Chemical compounds found in the headspace of bat-pollinated flowers. *A*, 3-hexanone; *B*, (E)-2-hexenal; *C*, hexyl acetate; *D*, benzaldehyde; *E*, dimethyl disulphide; *F*, 2-thiapropane-1-thiol; *G*, 2,4 dithiapentane; *H*, dimethyl trisulphide; *I*, dimethyl trithiocarbonate; *J*, dimethyl tetra-sulphide; *K*, (+)-alpha-pinene; *L*, (−)-alpha-Pinene; *M*, delta-3-carene; *N*, (R)-limonene, *O*, (S)-limonene; *P*, ocimene; *Q*, linalool; *R*, linalyl acetate; *S*, bornyl acetate; *T*, geranylacetone (from Helversen et al. 2000).

ponents of bat-flower odors were distinctly less effective (Helversen et al. 2000; fig. 8.4).

Flower Color

Bat flowers are mostly inconspicuously colored, being either green or brownish to brownish-red, and a few are whitish, which yields a high contrast at night. Specific spectral colors appear to have no important function for the Glossophaginae because bats have so far been regarded as colorblind (Jacobs 1992, 1993). Recently, however, U. Schmidt (personal communication) found evidence that *Carollia perspicillata* (Carolliinae) can, in fact, see colors. This species, which is primarily frugivorous but occasionally visits flowers, was able to distinguish between the colors red and green in a training experiment. In contrast, discrimination experiments with the flower specialist *Glossophaga*

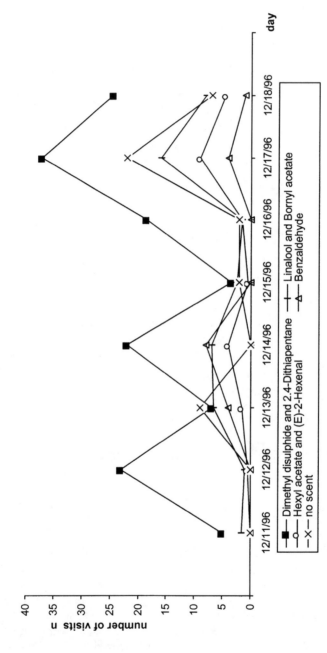

Figure 8.4. Visits of flower bats (mainly *Glossophaga commissarisi*) to artificial flowers scented with various odor substances in natural surroundings in the tropical lowland rainforest of Costa Rica. The artificial flowers were placed at the edge of a forest and the approach flights were counted by means of photoelectric barriers. Sulphur-containing substances were clearly preferred.

provided no evidence of color vision (J. López, Y. Winter, and O. von Helversen, unpublished data).

Glossophagines have a well-developed form of vision (Suthers et al. 1969). They can detect the contrast of dark structures against a light night sky, as well as that of light flowers compared to a dark background. Interestingly, Barthlott and coworkers showed that several chiropterophilous flowers clearly reflect ultraviolet (UV) light (e.g., Burr and Barthlott 1993; Burr et al. 1995). This UV reflection is especially striking, for example, in flowers of certain columnar cacti, such as *Neobuxbaumia* and *Stenocereus*. Our experiments with *Glossophaga soricina* have demonstrated that this species is highly sensitive in the ultraviolet region (fig. 8.5). Ultraviolet light is presumably important in the twilight, when the spectrum of the light is determined primarily by the scattering in the atmosphere and is thus shifted toward shorter wavelengths in comparison to direct sunlight (Smith 1982). Ultraviolet sensitivity is lacking in almost all mammals so far investigated because UV is absorbed by the media in the eye; only a few rodents such as the house mouse and gerbil are likewise UV-sensitive (Jacobs 1992, 1994).

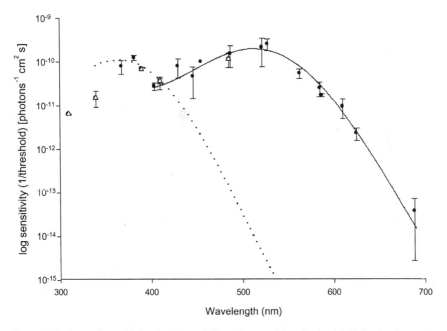

Figure 8.5. Spectral sensitivity function of *Glossophaga soricina* obtained in behavioral experiments. Data are the mean sensitivity values for three animals (± 1 SD) at the different spectral wavelengths. The solid curves represent the best fit Dartnall nomograms with a λ_{max} of 514 and 360 nm, respectively. (From J. López, Y. Winter, and O. von Helversen, unpublished data.)

Light-colored, UV-reflecting flowers could thus serve as visual signals for flower-visiting bats, especially in open habitats. Flowers that tend to be generalists may also attract sphingid moths. The brownish-red colors of some bat flowers appear to be either relics of former ornithophily or a means of "defense" against sphingids in the case of flowers exclusively specialized for pollination by bats, as such colors make flowers less conspicuous to moths (Helversen 1993).

Acoustic Properties: Reflection of Echolocation Calls

Glossophagines, like most Microchiroptera, navigate primarily by echolocation. As members of the Phyllostomidae, glossophagines are considered to be "whispering bats," producing rather faint echolocation calls (Schnitzler and Kalko 1998). Only a few phyllostomatids have relatively loud calls, such as *Choeronycteris mexicana* (K. G. Heller and O. von Helversen, unpublished observation) and *Leptonycteris curasoae* (U. Schnitzler, E. Kalko, and T. Fleming, in preparation). Thus, if a flower is to be acoustically conspicuous to glossophagines, it should have structures that reflect echolocation calls in a conspicuous manner. Here "conspicuous" could mean that the echoes are especially loud or that they are easily identifiable because of a specific modulation of frequency.

One bat-pollinated flower that attracts its pollinators by conspicuous echoes is *Mucuna holtonii*, a member of the Fabaceae that is widely distributed in Central America. *Mucuna holtonii* is a liana that grows chiefly on trees near the banks of brooks and small rivers, reaching into the canopy. From there the inflorescences, which comprise many blossoms at the end of peduncles, hang suspended in the open. The large, pale, night-blooming flowers of *M. holtonii* are of the papilionaceous type (fig. 8.6): two petals form the keel and two lateral petals form the short "wings" (alae). The fifth, upper, petal forms the so called standard or vexillum, which is shaped like a small concave mirror measuring about 19 × 19 mm.

Uniquely for bat pollinated flowers (though it is not atypical of the Papilionaceae), the deposition of pollen in *Mucuna* is released by an "explosion mechanism" when the flower is visited for the first time (Sazima in Dobat 1985; Vogel 1969a, 188–190; Helversen and Helversen 1999). To gain access to the nectar, the bat must grasp the flower with its thumbs and hind legs and press its snout into the slit between the two alae. This causes the keel to burst and the staminal column, which is under tension, to catapult most of the pollen load onto the bat's rump directly above the tail membrane. This part of the body is struck because the bat must grip the flower with its feet to brace itself for pushing the snout between the alae and thus trigger the explosion.

At the first visit to the flower, the bat receives up to 100 μL nectar, which is by far the largest reward obtainable. Once exploded, further nectar accumulation results only in smaller amounts of nectar (0–20 μL) to hovering bats.

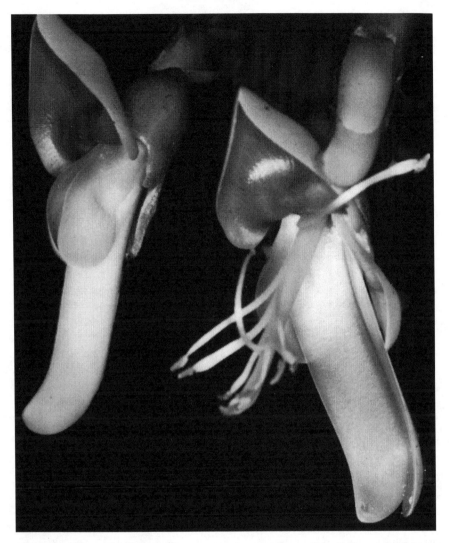

Figure 8.6. A virgin (*left*) and an exploded (*right*) flower of *Mucuna holtonii*. The upper petals are the raised vexilla that function as echo-acoustic reflectors.

However, the plant opens new flowers successively from dusk to about midnight, and thus the inflorescence remains attractive to bats. When a bud becomes mature, its vexillum ("standard") is raised (fig. 8.6). Thus, a bat should mostly search for newly opened flowers where the raised standard provides a signal that a flower is "ripe for cracking."

A series of experiments confirmed that the raised vexillum is a critical signal indicating the presence of mature flowers to bats and that exploded

Table 8.1. Visits of glossophagine bats (mostly *Glossophaga commissarisi*) at *Mucuna holtonii* flowers during a field experiment

	Experiment A		Experiment B	
	Intact Controls	Vexillum Removed	Intact Controls	Pad of Cotton Wool in Cavity of Vexillum
N (flowers offered)	321	203	143	172
Exploded after 2h	282	42***	94	34***
	(88)	(21)	(66)	(17)

Note. Numbers in parentheses are percentages.

*** $P < 0.001$.

flowers can be distinguished from virgin flowers (Helversen and Helversen 1999). First, a number of inflorescences were individually marked. Before the beginning of the experiment, up to about 2100 h, these inflorescences were covered with plastic bags to prevent bats from visiting mature flowers, so that more virgin flowers would be available for manipulations. The vexillum was removed from one-half of all virgin flowers in each inflorescence; the other half were maintained intact as controls. Removal of the vexillum caused a marked reduction in the number of visits (experiment A, table 8.1). A possible role of olfactory cues emanating from the vexillum was excluded by inserting a small pad of cotton into the cavity of the vexillum. Thus, only the echo-reflecting properties of the vexillum were altered while retaining possible sources of odor and also retaining the visual shape of the flower. The results of this experiment were similar to the first: the probability of explosion of manipulated flowers was significantly reduced compared to that of intact flowers (Helversen and Helversen 1999; experiment B, table 8.1). In addition, microscopic examination of the vexillum revealed no gland cells (D. von Helversen, unpublished observations).

The ability of bats to distinguish between virgin and exploded flowers could not be demonstrated by simply counting the number of cracked flowers but needed to be observed directly. We manipulated entire inflorescences so that each inflorescence consisted exclusively of either intact or exploded flowers. Visits of bats to these inflorescences were observed and videotaped using an infrared camera. At exploded flowers, the bats never tried to land but always hovered. We concluded that the bats had no difficulty discriminating between virgin and exploded flowers (D. von Helversen and O. von Helversen, unpublished observations).

The echoes reflected from *Mucuna* flowers exposed to artificial sounds that imitated natural echolocation calls of *Glossophaga commissarisi*, a principal pollinator of *Mucuna holtonii* at La Selva (Tschapka 1998), are shown in figure 8.7. We used short sweeps (1 ms duration) that were modulated downward, from 120 to 80 kHz, produced by an electronic ultrasound generator. The echoes were recorded by a microphone mounted close to the signal-producing loud-speaker to simulate the natural position of a bat's ears being close to its mouth. In a comparison of the echoes produced by virgin flowers, buds and flowers in which the vexillum had been filled with pads of cotton revealed that the echo of the entire flower was strongly dominated by the echo of the vexillum (fig. 8.7). The echoes reflected from an isolated vexillum mounted on a thin wire had an astonishingly high amplitude, considering the small size of the flower (target strength: -15.5 ± 1.2 dB at a distance of 10 cm, compared to the intensity of the sound signal at the position of the flower). The spectral composition of the echo was dependent on the angle of sound incidence, but the amplitude was high within a large cone of incidence angles (about $-40°$ to $-50°$, to $+40°$ to $+50°$; fig. 8.7).

The echoes of such a concave vexillum should be acoustically conspicuous to a bat because of the long persistence during several calls emitted by the passing bat. Thus, the vexillum should be easily detectible among many other perceived loud echoes, that is, from leaves, which reflect echoes in one direction only. The vexillum of the flower of *M. holtonii* appears to act like a cat's eye or a triple mirror in the optical domain, reflecting most of the energy toward the direction of incidence (Helversen and Helversen 1999).

The peculiar concave geometry and the upright position of the *M. holtonii* vexillum may be an adaptation to the echolocation system of glossophagine pollinators, as its sound-reflecting properties facilitate detection and recognition by the bat's sonar system. This hypothesis is corroborated by the fact that neither Paleotropic bat-pollinated species of *Mucuna*, visited by small Megachiroptera (which cannot echolocate), nor bird-pollinated species of *Mucuna* possess vexilla with the specialized shape and stiffness. This is demonstrated by illustrations of the African *Megaloglossus*-pollinated *Mucuna flagellipes* (Grünmeier 1993) and of the Australian *Syconycteris*-pollinated *Mucuna macropoda* (Hopkins and Hopkins 1993). In both species the vexillum is not erect enough to reflect sound back to the emitter.

Similar adaptations may be found in other glossophagine-pollinated flowers. As one promising example, the flowers of several columnar cacti are produced within a hairy zone, the "cephalium," of the cactus. While cactus cephalia were formerly considered to serve only basic functions such as protection against low air temperature and also — in bat-pollinated species — protecting the wings of the hovering bat from the spines (Helversen 1993), they may also enhance the echoacoustic contrast of the flowers by absorbing sound energy.

Spatial Memory and "Steady State" Flowering

Once a plant that is in bloom has been located by a bat, the effort of searching during the following nights—as long as the plant continues to open new flowers each night—is much reduced. Glossophagines have a highly developed

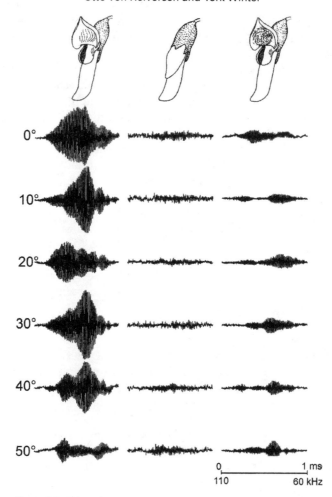

Figure 8.7. Echoes from a virgin, intact *Mucuna* flower (*left*), from a bud (*middle*), and from a flower, the vexillum of which was filled with a pad of cotton wool (*right*). The incident signal was a 1-ms sound pulse, linearly downward frequency modulated from 110 to 70 kHz, with a triangular envelope. The microphone was mounted side by side with the loudspeaker, simulating ears and mouth of a bat. The flower was exposed from different angles in the horizontal plane (from Helversen and Helversen 1999).

spatial memory, as do other bats (Altringham and Fenton, this volume; Neuweiler and Möhres 1967). In the laboratory, even after days have passed they often search precisely, down to the centimeter, at sites where they were previously rewarded (O. von Helversen and Y. Winter, unpublished observation). The use of spatial memory for the localization of known flowers was demonstrated by *Glossophaga* during a spatial orientation task (J. Thiele and Y. Winter, unpublished data). Bats were trained to feed from a single feeder among an array of four feeders, all four of which differed in geometric shape and thus echo-acoustic reflectivity. To elucidate the feeder localization strategy used by the bats, we evaluated the bats' behavior after rearranging feeders. Bats showed a distinct hierarchy of search strategies during tests with unrewarded feeders; they initially preferred the feeder at the former location and thus chose on the basis of spatial memory. Only during the second phase of search did they orient toward the learned geometric pattern and thus echoacoustic property.

Most of the plants pollinated by glossophagines exploit the behavior of bats by opening new flowers each night for an extended period but often opening relatively few flowers per night ("steady state" type of flowering [Gentry 1974]). Examples of this flowering behavior include the Bromeliaceae, such as *Vriesea gladioliflora* (22 ± 7 flowers per inflorescence, 0.8 flowers open per night, mean blooming period therefore about 1 mo) and *Vriesea kupperiana* (ca. 220 flowers per inflorescence and two to three flowers per night, mean blooming period 2–3 mo [M. Tschapka and O. von Helversen, unpublished observation]), the Capparidaceae *Cleome moritziana* (6–8 mo [O. von Helversen, unpublished observation]), and the Solanaceae *Markea neurantha* (up to 10 mo [Tschapka 1998; Voss et al. 1979]).

While most glossophagine-pollinated plants exhibit a steady state flowering strategy, other plants have shorter blooming periods, with a massive production of flowers ("big bang" type [Gentry 1974]), such as *Ceiba pentandra*, *Agave* sp., and *Hymenaea courbaril*. These plants are visited mainly by the larger Glossophaginae (e.g., *Leptonycteris*) and by less specialized, larger flower bats such as *Phyllostomus* spp. or fruit-eating bats like the Stenodermatinae and Carolliinae and, also, by birds and insects during the day.

Energy and Food Resource Requirements

Although there are only about 35 species of glossophagines in the Neotropics (Koopman 1981), many localities have several syntopic species. This raises the question of how guilds or ensembles of these nocturnal nectar feeders are structured and how the nectar resources are divided among the constituent members. Unlike hummingbirds and many nectar-feeding insects, mutualistic coevolution between single species of glossophagine bats or subgroups of bat species, on the one hand, and the floral morphology of specific groups of plants, on the other hand, appears to be rare or absent. In general, flowers are visited by several bat species if syntopic flower visitors occur in an area (see Dobat 1985). Thus, interspecific competition for nectar resources may play an important role in shaping the organization of glossophagine assemblages (see Patterson et al., this volume).

An examination of hummingbird assemblages illuminates some of the general principles involved in the competition between species for nectar food. Critical parameters for the structuring of hummingbird assemblages are the efficiency of nectar feeding and the body size–related daily energy requirements (Brown et al. 1978; Feinsinger 1987). These energetic parameters appear to determine the minimum nectar content of a flower for which exploitation is still economic. Only bee-sized hummingbirds can profitably exploit even the small amounts of nectar in insect-visited flowers (Feinsinger 1987). Knowledge of basic energetic parameters seems therefore essential for a mechanistic, bottom-up analysis of assemblage structure in glossophagine bats.

Energetic Resource Requirements and Efficiency of Foraging

How can we determine densities of food energy required by individuals of a particular species? Obviously, one minimum requirement is that available food energy must minimally match an animal's daily energy expenditures (see Speakman and Thomas, this volume). Food energy must be distributed in portions that are large enough for economic harvesting—that is, the energy of an average food portion must exceed the cost of acquiring it. Habitats are suitable only if they allow an energy gain that is significantly higher than the energy cost of foraging. In order to arrive at quantitative predictions of the cost-benefit ratio, the relevant energetic parameters must be known.

Daily Energy Expenditure

The daily energy expenditure (DEE; also, "field metabolic rate" or FMR—see Speakman and Thomas, this volume) of a nectar-feeding bat determines the overall amount of nectar energy that must be available to an individual bat on a daily basis. Daily energy expenditures of glossophagines have been estimated both in the field and in flight cages, using methods ranging from doubly labeled water and feeding trials to energy budget estimates derived from time budgets obtained by radiotelemetry of free-ranging bats (Voigt 1998; Helversen et al. 1986; Helversen and Reyer 1984; Winter 1998b; Winter and Tschapka, in preparation). Larger glossophagine species have higher DEEs than do smaller species (see allometric relations in table 8.2), and the slope of this log-linear relationship coincides with values derived for other endothermic vertebrates (Nagy 1987, 1994). However, the average DEE of glossophagines (the offset of the regression line) is 60%–70% higher than the allometric average calculated for other eutherian mammals of their size class on the basis of DEE estimates with the doubly labeled water method (Nagy 1994). Thus, the high rates of DEE in glossophagines coincide with the range typical of birds (fig. 8.8) and to date have the highest estimated daily energy turnover in mammals (see Speakman and Thomas, this volume).

Table 8.2. Energy relations and foraging parameters in glossophagine bats and their scaling with body mass

Variable	Unit of Measure	Energy Relation	References
DEE	kJ d^{-1}	1,555 M$^{0.755}$	Winter and von Helversen, unpublished
BMR	W	2.98 M$^{0.713}$	Arends et al. 1995; McNab 1988
RMR	W	1.25 BMR day, 1.5 BMR night	Aschoff and Pohl 1970, von Helversen and Reyer 1984 (and references cited therein)
Minimum thermal conductance (MTC)	W°C^{-1}	0.218 M$^{0.5}$	Arends et al. 1995; Herreid and Kessel 1967
Thermoregulatory energy	W°C^{-1}	1 MTC day, 1.5 MTC night	Aschoff 1981
Lower critical temperature	°C	30	Arends et al. 1995 (and references cited therein)
Forward flight cost	W	50.2 M$^{0.771}$	Winter and von Helversen 1998
Hovering flight cost	W	128 M$^{0.95}$	Voigt and Winter 1999; Winter 1998a
Foraging flight speed	m s^{-1}	20 M$^{0.23}$	Winter 1999
Flight acceleration	m s^{-2}	4.4 (for G.c.)	Winter 1999
Mean hovering duration	s	0.5	This study
Nectar sugar (for 26% di- and 74% monosaccharides)	kJ g^{-1}	15.92	. . .
Sugar content of 17% sugar (wt/wt) nectar	g l^{-1}	181.5	Baker and Baker 1990; Bolten et al. 1979; Wolf et al. 1983
Sugar assimilation efficiency	. . .	0.99	Winter 1998b

Note. Equations based on body mass M in kg. W = J s^{-1}. To convert from W to daily rates in kJ multiply by (3600 × 24)/1000. Formula conversions: $y = a M[g]b = a 1000^b M[kg]^b$; $y = a M[kg]^b = (a/1000^b) M[g]^b$; if animal specific y [kJ] $= a M^b$ then mass specific y [kJ kg^{-1}] $= a M^{b-1}$. For the conversion of ml O$_2$ to energy rates it was assumed that metabolism in nectar-feeding bats during the inactive period (daytime) is fueled by combustion of fat (oxicaloric equivalent 19.59 J/mL O$_2$) and during the active, absorptive period (nighttime) by combustion of sugars (oxicaloric equivalent 21.10 J/mL O$_2$). BMR/RMR measurements were performed with fasting animals and hence fat catabolism was assumed. The factor of 2.98 for glossophagine BMR is 1.15 times the chiropteran slope (2.59 [McNab 1988]) as suggested by the data in Arends et al. (1995). Estimates of daily variation in RMR and thermoregulatory costs are preliminary current best estimates in need of verification by field measurements. DEE = daily energy expenditure (also field metabolic rate, or FMR), BMR = basal metabolic rate, RMR = resting metabolic rate, MTC = minimum thermal conductance, and *G.c.* = *Glossophaga commissarisi*. Flower search efficiency is an unknown parameter depending on bat/flower system and habitat.

Daily energy turnover can also be viewed from a different perspective: in glossophagines, it equals two-thirds of the total caloric content of the bat's body (assuming 22 kJ/g dry mass [Masman et al. 1986]). This illustrates the magnitude of energy turnover in these small mammals. If we compare glossophagines with other nectar feeders, it is interesting to note that DEEs of the nectar-feeding pteropodid *Syconycteris australis* (Geiser and Coburn 1999) and also of 6–15-g sunbirds (Nectariniidae [Peaker 1990]) are within the range of values of the glossophagines. Only the DEE of hummingbirds appears on average to be about 30% higher than that of glossphagines (Lopez Calleja et al. 1997; Tiebout and Nagy 1991).

One result of our DEE estimates was surprising to us. Even under ad lib. feeding conditions in the laboratory, glossophagine bats maintained flight activity and daily energy turnover corresponding to the levels of free-ranging

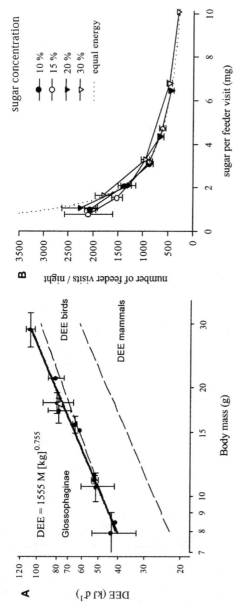

Figure 8.8. *A*, Daily energy expenditure (DEE) of nectar-feeding glossophagine bats as a function of body mass (data based on over 450 24-h measurements in 58 individuals of 11 species). Least squares regression of *DEE* on body mass *M* yields DEE [kJ d^{-1}] = 1555 M [kg]$^{0.755}$ (from Y. Winter and O. von Helversen, unpublished data). The *open triangle* is a measurement from nectar-feeding *Sycomycteris australis* (Pterodopidae, Macroglossinae [Geiser and Coburn 1999]; data points not included for calculating the regression). *Dashed lines* are regressions of DEE based on doubly labeled water estimates for birds (Nagy 1987) and eutherian mammals (Nagy 1994). *B*, Effects of the nectar volume obtained during single feeder visits and of sugar concentration (percent sucrose weight/weight) on the number of feeder visits by *Glossophaga soricina* (experimental setup as in fig. 8.9*B*). Food was offered from two feeders 14 m apart. It was delivered by computer-controlled syringe pumps that were programmed such that a bat had to visit the feeders alternately in order to receive food. The *dashed line* is the line of equal energy intake calculated from the overall mean nightly intake of 2.7 g sucrose equivalents for the 9.5-g bats. The bats adjusted their feeding effort (number of feeder visits) to the availability of food (volume per feeder visit) or the quality of the food (sugar concentration) during most experimental conditions and thus kept daily energy intake at around 48 kJ d^{-1}. Bats became unable to balance their daily energy budget at rewards of less than 1 mg of sugar. Legend shows sugar concentration in % mass/mass (data in *B* from O. von Helversen, Y. Winter, and H.-J. Rübbelke, unpublished data).

bats. An extensive series of feeding trials with *Glossophaga soricina* demonstrated that DEE was regulated within a narrow range that was only slightly affected by food resource availability (O. von Helversen, Y. Winter, and H.-J. Rübbelke, in preparation; fig. 8.8B). For these experiments, we manipulated foraging effort by programming computer-operated nectar feeders to deliver a range of volumes from small to large during single visits (or delivering solution of low or high sugar concentration). At the same time we required bats to fly a fixed minimum flight distance to obtain a food reward (by forcing them to alternate between opposite feeders). As a result, the number of visits to feeders per night varied from 300 to more than 2,000 on "high energy reward" nights and "low energy reward" nights, respectively. In spite of this large variation, the total amount of sugar consumed was nearly the same (fig. 8.8). Despite the large difference in number of visits to feeders, the DEE remained similar, not reflecting the nearly tenfold difference in required foraging effort. The constancy in DEE was achieved by a flight activity that remained at an average of about 4–5 h per night, irrespective of resource availability. During nights of high energy reward, bats alternated between the opposite ends of the flight cage in their regular fashion but visited the feeders at the ends only occasionally. Conversely, during nights of low energy reward, most of the flight activity was devoted to flying between feeders.

In the experiments mentioned above, we tested single bats, but tests with group-living bats confirmed this pattern. Communally housed individuals of *Glossophaga commissarisi* marked with passive integrated transponder (PIT) tags for automatic identification at a computer-controlled nectar feeder had DEEs within the same range as single individuals (Winter and Tschapka, in preparation). One last finding in this context is that even lactating females of *G. commissarisi* in the field did not deviate significantly from non-reproductive bats with respect to DEE (Voigt 1998; Winter and Tschapka, in preparation). In these females, a reduction in nightly flight activity compensated for the increased cost of maintaining a higher body mass in flight. A similar reduction in flight activity was observed using telemetry for free-ranging, pregnant, and lactating *Carollia perspicillata* (Charles-Dominique 1991).

One conclusion from these observations of daily energy turnover in glossophagine bats is that DEE appears to be actively regulated within a genetically determined range. An examination of other vertebrates indicates that the regulation of DEE within a particular range is not unusual. Hummingbirds maintain high flight activity irrespective of sugar concentration and foraging effort required (Lopez Calleja et al. 1997; Tamm and Gass 1986). Similarly, house mice maintain high constancy in their DEE by partitioning energy between locomotor activity (in a running wheel) and thermoregulation (Perrigo and Bronson 1983, 1985). Just as in glossophagines, the requirements of foraging (manipulated in the mice by requiring a minimum number of wheel revolutions per food reward) have only a small effect on total locomotor effort for a given ambient temperature.

One mechanism that may confer some flexibility in daily energy partitioning is the lowering of body temperature during periods of rest (Speakman and Thomas, this volume). In general, deep torpor is not an energy-saving strategy in tropical phyllostomids, including glossophagines (McNab 1969). Our own observations confirm that glossophagines at moderate temperatures (above 15°–20°C) are always alert and immediately responsive and do not enter torpor as long as they have access to food. However, during instances of food withdrawal, we have observed bats to become torpid and less mobile (Y. Winter, personal observation). Thus, shallow torpor may be an energy-saving strategy during times of food shortage, although this has not been quantified in glossophagines. Similar observations of moderate thermolability have been made for the phyllostomids *Carollia perspicillata* and *Sturnira lilium* (Audet and Thomas 1997; McManus 1977). It would be interesting to know if occurrence at higher altitudes (i.e., in *Anoura geoffroyi* or *A. cultrata*) is associated with a pronounced capacity for torpor.

While many questions regarding the regulation of DEE remain open, the present data (fig. 8.8) provide a first estimate of the total amount of energy that must, on average, be available to an individual glossophagine on a daily basis. A correspondingly larger amount of total energy is needed to support the number of animals that constitute the minimum viable population size. In the next section, we explore how food energy must be distributed within the habitat in order to permit economic harvesting.

Energy Cost of Flight

Although flight is one of the most energetically expensive forms of locomotion (Speakman and Thomas, this volume), its evolutionary success reflects the fact that it is not only fast but is also an economical means of moving forward. The high speed of flight compensates for the high power requirement so that the cost of transport (the cost per unit of distance moved) is lower than for all other terrestrial forms of locomotion (only swimming is less expensive [Schmidt-Nielsen 1972; Tucker 1970]). In addition, flight is the only possible solution to economically exploit a large three-dimensional terrestrial resource space.

Horizontal Forward Flight

The energy expended for flight is the major energetic cost factor for a foraging bat. Thus, the net energy gained from a flower visit by a glossophagine bat is the difference between the energy content of the imbibed nectar sugar and the flight cost incurred during the commute from a previous flower, plus the hovering expenditures during feeding. Measurements of the energy cost of horizontal forward flight in small bats have so far been obtained by indirect energy balance methods. Using this protocol, the total energy turnover of an animal is ascertained over a period of time consisting of periods of both flight and rest. Flight cost is determined by subtracting the estimated cost entailed during the nonflight period from the total energy turnover (Speakman and Racey 1991; Speakman and Thomas, this volume; Winter and Helversen 1998; fig. 8.9). The results of measurements carried out to date for small bats are about 20%–25% lower than predicted by the majority of allometric equations for bird flight (fig. 8.9A). This indicates a metabolically less costly forward flight by small bats at their slower speeds.

Flight Speed

The cost of flight for a certain distance is a function of flight power (W or $J\,s^{-1}$) and flight speed ($m\,s^{-1}$). Knowledge of flight speeds is thus necessary to estimate the cost of flight for covering the distance between two flowers. Speeds of commuting flight have been estimated for several species of glossophagine bats in the field (Sahley et al. 1993; Tschapka 1998; Winter 1999). In addition, general scaling factors, derived from aerodynamic models, allow predictions of flight speed from the body mass of a bat (Norberg and Rayner 1987). Presently available data suggest that commuting flight speeds of glossophagines scale with body mass M as $V\,[ms^{-1}] = 20\,M\,[kg]^{0.23}$ (Winter 1999). When commuting between neighboring flowers that are close to one another, acceleration and deceleration phases at reduced speeds constitute a significant portion of a flight interval. To take this into account, travel times of *Glossophaga commissarisi* were estimated while the bats were foraging between flowers of the chiropterophilous bromeliad *Vriesea gladioliflora* in Costa Rica. The presence of a bat hovering and feeding at a flower was detected with photoelectric sensors at the flowers and timed by a computer (Winter 1999). Data on flight speeds obtained from this arrangement can be used for estimating flight costs over short distances (see table 8.2).

Hovering Flight

Hovering flight permits a glossophagine bat to remain airborne while imbibing the nectar-solution. As the brevity of the hovering visits allows a bat to move rapidly between flowers, it can visit many flowers per unit time, which reduces the amount of energy that is required from each single flower visit in the form of nectar sugar (and pollen, Howell 1973).

The hovering flight of glossophagines is certainly the most highly developed among bats. Glossophagines have a kinematic feature of wing movement during hovering flight that seems to be unique. Three-directional high-speed films (900 frames s^{-1}, Helversen 1986; Helversen and Helversen 1975a) revealed that during the backstroke of the flapping wing the morphological underside of the hand wing is turned upward (supinated), thus forming a distal wing triangle that generates lift even during a short phase of the wing's backstroke (fig. 8.10). This pronounced "wing tip reversal" is not known for other

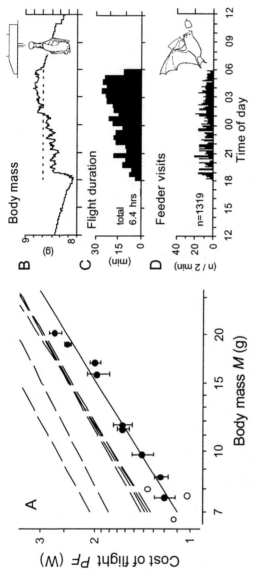

Figure 8.9. *A*, Energy cost of horizontal flight in small bats and birds as a function of body mass. *Filled circles* are estimates of forward flight cost in glossophagine bats and *open symbols* are data from vespertilionids (Speakman and Racey 1991). The best fit equation relating flight energy expenditure P_F to body mass for small bats (<35 g) is P_F [W] = 50.2 M [kg]$^{0.771}$. *Dashed lines* are allometric functions relating flight cost with body mass in birds (references in Winter and Helversen 1998). *B–D*, Experimental setup used for determining 24-h time and energy budgets in nectar-feeding bats (here, *Glossophaga commissarisi*). *B*, Body mass and flight activity are monitored automatically as the only roost available to a bat is suspended from an electronic balance connected to a computer. The *dashed line* indicates mean body mass during flight activity. *C*, Time spent in flight during consecutive 30-min intervals. Durations of flights are calculated from the times of departure and arrival at the balance. *D*, Number of feeder visits during consecutive 2-min intervals. During each hovering visit the bat receives a fixed amount of sugar water from a computer-controlled pump (from Winter and Helversen 1998).

hovering bats. During hovering flight, the wing tips, as viewed from the side, move along a path that does not correspond to a horizontal figure eight, as in hummingbirds but, rather, to a "bent" eight, so that the wings are brought close together in front of and above the body (Helversen 1986; Helversen and Helversen 1975; Dudley and Winter 2002). The glossophagine flowers from which the bats feed while hovering must therefore leave room for the wings at these two places (see below and fig. 8.10).

We measured the energy cost of hovering flight after training glossophagine bats to an artificial feeder that also served as a respirometric mask (fig. 8.11). When inserting the head into the mask for feeding, respiratory gases were withdrawn to determine oxygen and carbon dioxide concentrations to estimate metabolic rates. The metabolic cost (energy input) of hovering flight turned out to be only slightly higher than the cost of horizontal forward flight, which was contrary to the expectations of a previous quasi-steady aerodynamic analysis of hovering flight in glossophagines (Norberg et al. 1993). In 12-g *Glossophaga*, the energy cost of hovering flight was only 1.1–1.2 times that of horizontal forward flight, and in 17-g *Choeronycteris*, the factor was about 1.4 (Voigt and Winter 1999; Winter 1998a). While this small difference in energy requirements for hovering and horizontal forward flight seemed surprising to us at first, there is parallel evidence for this in hummingbirds as well (Berger 1985; Ellington 1991). Based on currently available data, the cost of hovering in glossophagines scales with mass M as P_h [W] $= 128\ M$ [kg] $^{0.95}$ (Voigt and Winter 1999).

It is interesting to compare the solution of hovering flight in glossophagines with other groups of hovering nectar feeders. Comparable cost estimates of hovering flight are available for three groups that hover as part of their flower-feeding habits: sphingid moths, hummingbirds, and glossophagine bats. Between these three groups, glossophagine bats have the lowest mass-specific cost of hovering flight. In comparison, glossophagines need 40% less energy than a hummingbird and 60% less energy than a sphingid moth to counterbalance their body mass during hovering flight. The overlap in individual flight power among these taxa illustrates these differences (fig. 8.11). At a power of 1.1 W, a glossophagine bat can support a body mass of 7 g, a hummingbird one of 4 g, and a sphingid moth a mass of 3 g. Thus, with the same energy input, a glossophagine can maintain twice the mass aloft as a sphingid moth in hovering flight.

Knowing the duration of hovering events allows the calculation of feeding energy expenditures, which is needed for a cost-benefit analysis of foraging. A large number of measurements with infrared photoelectric devices installed at flowers both in the field and in the laboratory have revealed that glossophagines will normally hover for durations of less than 1 s, more typically around 0.3–0.6 s (Fleming et al. 1996; O. von Helversen, unpublished data;

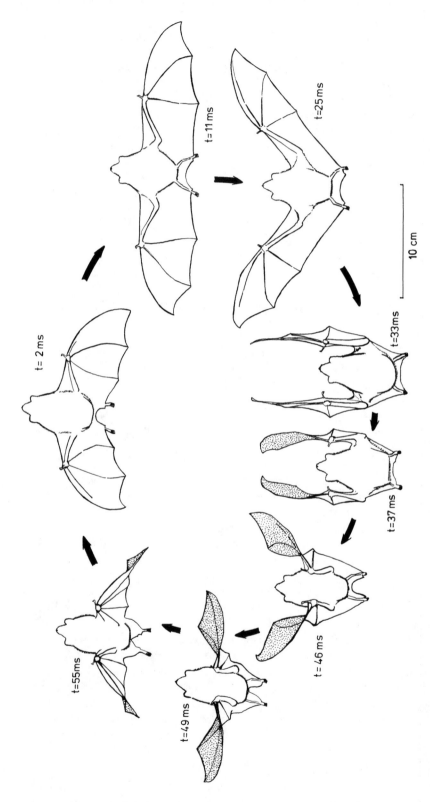

Figure 8.10. A hovering *Glossophaga soricina* seen from above. During the backstroke ($t = 46$ and 49 ms) the characteristic wing tip reversal can be seen (from Helversen 1986).

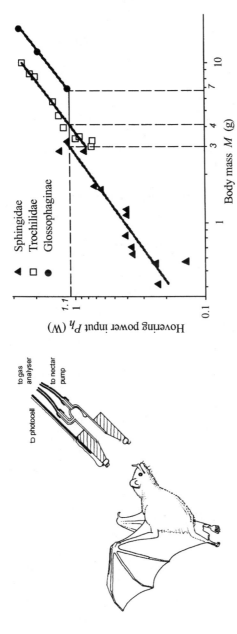

Figure 8.11. The energy cost of hovering flight. *Left*, experimental setup for determining rates of respiratory gas exchange during hovering flight in a glossophagine bat. *Right*, energy cost of hovering flight and body mass in sphingid moths, hummingbirds (Trochilidae), and glossophagine bats (Phyllostomidae). *Solid lines* are regressions relating hovering power input P_H to body mass M. For glossophagines, this follows the relation P_H [W] = 128 M[kg]$^{0.95}$. At a power input of 1.1 W the three groups of flower specialists overlap in energy expenditure for hovering but support very different body masses (*left*, from Winter et al. 1998; *right*, from Voigt and Winter 1999).

only under special training conditions in the laboratory did we record hovering durations for 7–17 s [Voigt and Winter 1999; Winter 1998a]). Thus, hovering accounts for only a small fraction of the foraging time budget of glossophagines. The flight energy budget of a glossophagine is dominated by the expenditure for commuting and for moving between flowers in horizontal forward flight. This may be the reason why these bats have evolved a relatively high wing loading and wing length, which is shorter than predicted for optimal hovering (Norberg and Rayner 1987).

Scaling of Energy Expenditures and Community Assembly Rules

The energy relations summarized in table 8.2 provide a basis from which minimum food requirements can be estimated. For many animals, estimating resource densities in units that are relevant is difficult, as many cost factors (i.e. search effort, effort of obtaining food or capturing prey, caloric value of food items) usually are unknown. Attempting this for nectar-feeding bats, however, is more straightforward than for other groups of foragers. The energy content of nectar can be quantified easily, and steady state flowering plants are highly predictable resources at fixed spatial locations. We assume that the bats have detailed knowledge of the locations and spatial distribution of these food sources from their previous encounters. The food resource space for a glossophagine bat can then be simplified as an area with predictable locations of renewable food sources from which nectar is collected by repeatedly visiting profitable flowers during the night. Thus, food energy density of the habitat from a bat's perspective is a function of the mean distance between neighboring flowers and the mean caloric value of the nectar obtainable during a feeding visit at a blossom. To estimate the minimum amount of nectar energy E_{fl} that a glossophagine bat requires from an average flower visit to balance its daily energy budget, we derived a linear model detailed in Winter and Helversen (2001), which resulted in the following equation:

$$E_{fl} = DEE \, \frac{eff_s(S_{ff}/V_f+1+t_h)}{T_f}.$$

The minimum amount of nectar energy required from an average feeder visit E_{fl} is a function of daily energy expenditure DEE, the mean distance between flowers S_{ff}, the commuting flight speed of the bat V_f, the mean duration of hover-feeding visits at flowers t_h, and the total time spent for foraging flights T_f (for estimates of these parameters, see table 8.2). The constant with a value of 1 (second) approximates the time loss due to reduced flight speed during flower approach and departure. The degree to which a bat is able to commute along the shortest routes within its habitat is modeled by the factor eff_s, which has a value of 1 for a perfect solution of the "traveling salesman problem" but drops below 1 if bats travel along detours or skip the nearest flowers.

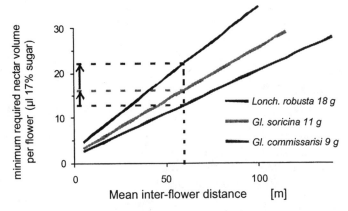

Figure 8.12. Boundary values of minimum nectar energy density for three differently sized species of glossophagine bats. Data were calculated from equation 1 and energy relations given in table 8.2, assuming that all nightly flight activity (4.5 h) was devoted to foraging and that successive flowers were visited along minimal routes.

This equation provides a first approximation of the minimum nectar energy that must be available during an average flower visit. Minimum nectar levels can be estimated from DEE (fig. 8.9A, table 8.2) and an average flight time budget of typically 4–5 h during a night (Horner et al. 1998; Helversen and Reyer 1984).

Equation 1 can be used to compare minimum nectar energy requirements of different-sized species, as shown in figure 8.12. One major result of this model calculation is the prediction that individuals of different-sized species differ in minimum resource requirements (fig. 8.12). Individuals of smaller species can subsist on lower nectar energy densities (kJ per unit of foraging distance) than those of larger species. This size-dependent difference in minimum resource requirements is important for discussing the consequences of interspecific competition in the next section.

If many individuals together depend on the nectar pool in a given area, the rate of nectar flow acts at the level of population density. If feeding niches of different species largely overlap and food is limited, then species will "interact" competitively. The consumption of shared resources by bat A will directly influence resource availability to bat B. This situation, in which competing individuals do not actually have to meet, is termed "exploitation competition" (e.g., Begon et al. 1996). One consequence of consumption is the depletion of resources. If two species differ in their energetic foraging efficiencies or in their overall energy requirements (reflected, e.g., in the DEE), then such a reduction in resource levels can lead to the exclusion of the species with the higher requirements. This relation between (1) daily energy requirement and energetic efficiency during foraging and (2) required density of food resources within the habitat can, in the case of resource limitation, fundamentally influence the processes that determine community assembly (see Patterson et al., this volume).

Several ecological predictions follow from the proposed differences in resource requirements between different-sized glossophagine species (fig. 8.12). During a period of diminishing rates of nectar production (e.g., as a seasonal effect), the model predicts that a habitat will become unprofitable for individuals of larger species first. In similar fashion, increases in population abundance without simultaneous increase in primary food production and carrying capacity will result in diminishing resource shares per individual. Again, the model predicts that larger species will be the first to give up the habitat.

Bats are long lived, highly mobile, and relatively K-selected (see Barclay and Harder, this volume). A general assumption is that bat communities with their low beta diversities are influenced by deterministic interactions at least at intermediate organizational levels (Findley 1993; Willig and Moulton 1989; Willig et al., this volume). This general contention finds support from statistical evaluations of Antillean island communities. The analysis of species presence-absence matrices showed that the pattern of species co-occurrence on the Antilles is nonrandom with many species co-occurring much more often than is predicted by a null model distribution (McFarlane 1989; see also McNab 1971). Although two South American assemblages of glossophagines were found not to deviate in their species' morphospace from random assembly (Willig and Moulton 1989), this negative result should be treated with caution. Only two sites were evaluated that were not species-rich in glossophagines (two species at one site and three at the other).

The predictions of the above model are valid only within the framework of shared food resources and competition on the basis of exploitation. Glossophagine bats may be one of the few species groups in which both conditions are largely met, at least in lowland rainforest communities. There are no records of interspecific territoriality in nectar-feeding glossophagines. Conversely, numerous records of observing and mist netting bats at flowers have indicated that single flowers may be visited by different individuals and species in short succession (O. von Helversen, M. Tschapka, and Y. Winter, personal observation). The only documented case of *intra*specific food resource defense in *Glossophaga* is consistent with this argument. Here, individuals defended only a few square meters around a single, nectar-rich *Agave* plant, and even this only during a few hours of the night (Lemke 1984). There is a large overlap in resource use of flowers between different species of bats (see below and fig. 8.17). At least in rainforest habitats, it does not appear that any bat flowers provide an advantage to the larger species of glossophagines. This is in contrast to the more typical situation in fruit- or seed-eating assemblages, where larger fruits or seeds are used predominantly by the larger species (e.g., Darwin's finches [Grant 1986]).

The large overlap in the use of flower resources between nectar-feeding glossophagines resembles the pattern of "shared preference habitat (or resource) allocation" (Chase and Belovsky 1993; Rosenzweig 1991), which has been advocated as the main mechanism accounting for included-niche patterns (Rosenzweig 1991; Rosenzweig and Lomolino 1997). Field investigations have often revealed patterns of fundamental niches, which are called included niches (Chase and Belovsky 1993; Colwell and Fuentes 1975). Here, several species in a guild overlap with their niche positions in one particular habitat or resource use type. While some species can use more than one type, others—the intolerant species—cannot. The most tolerant species have the potential to use many habitat or resource types. "Shared preference habitat allocation" is one mechanism that may account for such included-niche patterns. If the habitat that every species within an assemblage can use is also the best habitat (i.e., the preferred) for all species, then a mechanism of tolerance variation can mediate coexistence. This mechanism depends on the trade-off between niche-breadth and the ability to outcompete others in the commonly preferred habitat. Intolerant or specialist species, without the ability to use many habitat or resource types, must be able to dominate the preferred habitat (Rosenzweig and Lomolino 1997). *Hylonycteris underwoodi* and *Glossophaga commissarisi* in lowland Atlantic Costa Rica are an example for such a specialist-tolerant species pair (see below).

In summary, both requirements of our model—competition by exploitation and large overlap between species in the use of nectar resources—seem to be fulfilled in assemblages of glossophagine bats.

An Experiment on Interspecific Exploitation Competition

The prediction that a larger species should be inferior to a smaller one during pure exploitation competition was tested in a laboratory experiment (Y. Winter, unpublished data). One individual each of a small and a large glossophagine species were maintained in separate flight cages so that competition by interference was not possible. Each cage was equipped with a separate, computer-controlled nectar feeder. An important feature of the experimental design was that these two feeders were treated by the computer as if they were two separate corollas leading to the same, single nectary. The computer secreted virtual nectar units into a nectary account. On a bat's visit to its feeder, the computer "paid out" from this common nectar account if the current balance was above zero. Thus, the feeding activity of one individual directly affected resource availability to the other individual, in a purely exploitative fashion.

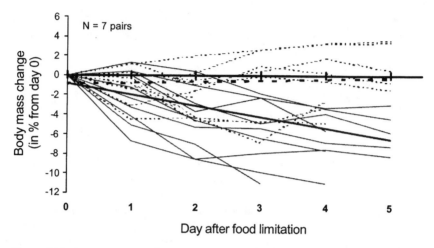

Figure 8.13. Experiment on exploitation competition for limited nectar between *Glossophaga soricina* (10 g, *dashed line*) and *Choeronycteris mexicana* (17 g, *continuous line*). See text for experimental procedure. The effect of food limitation was evaluated by determining the rate of body mass loss during the experimental phase based on the day before the onset of food limitation (day 0). On average, the small species could maintain body mass while the larger *Choeronycteris* lost body mass (from Y. Winter, unpublished data).

After initial ad lib. food provisioning, nectar availability was reduced during a 5-d food limitation phase of the experiment to about 70% of ad lib. consumption rates. The results of this experiment between individuals of 17-g *Choeronycteris* and 10-g *Glossophaga* (n = 7 pairs) were unequivocal (fig. 8.13). The immediate behavioral response of both species to diminishing nectar portions at their feeders was to step up foraging effort by increasing feeder visitation rates. This further reduced portion size. For the smaller *Glossophaga*, even the reduced nectar rewards, however, were still marginally profitable, whereas for the larger *Choeronycteris* they were not. As a consequence, in all seven replications of the experiment, the larger *Choeronycteris* lost body mass at a faster rate than did individuals of *Glossophaga*. On average, *Glossophaga* was even able to maintain body mass during the food limitation period (fig. 8.13). This experiment supports the model's prediction that a large species is inferior to a small one during exploitative competition for limited nectar.

The outcome of this experiment raises a new question: if large species are competitively inferior to smaller species in such a restricted laboratory setup, then what factors allow coexistence under the wider range of conditions available in a natural habitat? We think that an important aspect in answering this question lies in a different type of energetic advantage of larger species: larger glossophagines live more parsimoniously relative to their energy reserves (fig. 8.14). The relative body fat reserves of nectar-feeding bats are in the range

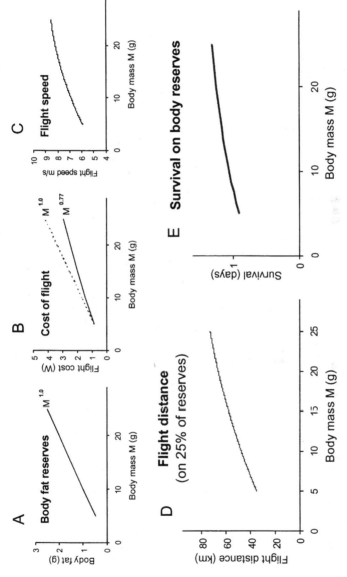

Figure 8.14. Body fat reserves, flight range, and starvation endurance in glossophagine bats. Allometric relations from A to C (see table 8.2) were used to compute the flight range (D) and the duration of starvation endurance assuming regular DEE as given in table 8.2 (E), assuming a consumption of 25% and 100% of total body fat reserves, respectively. Body fat reserves were assumed to total 10% of body mass (Y. Winter and P. Koteja, unpublished data).

of 10%–15% of body mass (measured by Soxhlet extraction [Y. Winter and P. Koteja, unpublished data]). Given in exponential form, body fat thus scales in isometric proportion to body mass M as M^1 (because $M = M^1$). In contrast, all major energy-consuming processes scale to body mass with an exponent smaller than 1, often between 0.5 and 0.8 (table 8.2). Therefore, in terms of relative units of energy in fat reserves, the maintenance of major energy expenditures requires an increasingly smaller percentage of fat reserves per unit of time as an animal becomes larger. Larger animals, therefore, can subsist for longer on their fat reserves than can smaller animals.

The major expenditure during foraging in nectar-feeding glossophagine bats is the energy cost of flight, which scales to body mass with an exponent of about 0.77. Thus, while the absolute cost of flight is higher for larger species, the energy required for each gram of body mass decreases with increasing body mass (mass specific flight cost approximately $M^{-0.23}$). Larger species also fly faster than smaller species (mean foraging flight speed scales as 20 $M^{0.23}$ [Winter 1999]). Using this information we can calculate the distance that an individual bat can cover by flight if it fuels metabolism solely with a fixed fraction of its body fat reserves. This calculation shows that, within the body mass range of glossophagines, large species can fly farther than can smaller species (fig. 8.14). A second relationship is similar: as DEE increases with a 0.75 exponent to body mass (i.e., smaller than 1), larger species can also maintain routine activities and survive for longer on their fat reserves than can smaller species (fig. 8.14, table 8.2).

The combination of these two effects may yield the energetic prerequisite for a fundamentally different foraging strategy in larger species (fig. 8.15). Their greater flight range may predispose them to explore and gain knowledge of wide rainforest areas by flight. The investments for potentially unsuccessful exploration and search flights can be covered by energy reserves. The knowledge of a larger area may enable larger species to exploit selectively temporary "hot spots" of nectar superabundance. While such resource clumps may also be utilized by local populations of smaller species, they will remain attractive as long as the standing crop of nectar is not depleted below the profitability threshold of the larger species. When large and small species meet while food resources are not limited, competitive exclusion mechanisms do not come into effect. Whether such a foraging strategy may also be associated with social organization that facilitates learning and information transfer is currently unknown.

An Assemblage of Glossophagine Bats in the Lowland Rainforest of Costa Rica

How closely do the predictions of this deterministic model fit the pattern of community assembly and resource use in the field? The best studied assemblage of glossophagine bats is at the La Selva biological station in the tropical

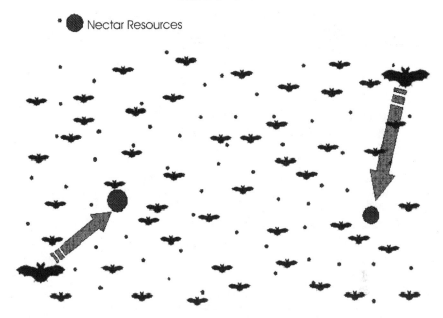

Figure 8.15. Hypothetical model of coexistence between small and large species of glossophagine bats. Large species are characterized by higher absolute cost of flight and a higher absolute daily energy expenditure; inferior exploitation competition with smaller species; and farther flight range than smaller species and a longer fasting endurance. The possible coexistence of large and small species might be explained by the following. The individual distribution of small species is spatially more restricted and the population density is limited by locally available resources. Large species, in contrast, explore and know a wide rainforest area and can exploit temporary sources of nectar surplus that are not exhausted by local populations of small species (from Y. Winter and O. von Helversen, unpublished data).

lowland rainforest on the Atlantic side of Costa Rica, which has been examined in great detail by Tschapka (Tschapka 1998; Tschapka, in preparation; figs. 8.16, 8.17). Five species of glossophagine bats occur sympatrically in the lowlands region of the Atlantic side of Costa Rica.

Of these five species, only two small species, *Glossophaga commissarisi* (8.8 g) and *Hylonycteris underwoodi* (7.6 g), are able to maintain widely distributed, resident populations in the lowland rainforest. Neither of the two larger species, *Lonchophylla robusta* (15.7 g) or *Glossophaga soricina* (11–12 g), have widely distributed, permanent resident populations within the continuous lowland rainforest. This pattern conforms to the predicted competitive advantage of smaller species.

During the months of April–August, nectar density is reduced, and it is during this period that only the smaller of the two resident species, *Hylonycteris*, remains in the nectar niche, whereas the larger and morphologically less specialized *G. commissarisi* includes fruit in its diet. This would be expected in

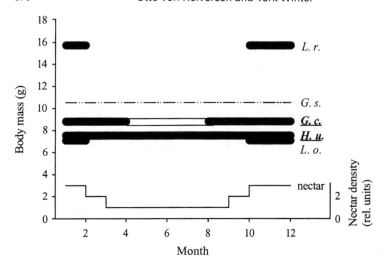

Figure 8.16. Use of nectar resources during the annual cycle by differently sized species of a nectar-feeding bat assemblage in the lowland Atlantic rainforest of Costa Rica. *Horizontal bars* indicate the periods of species presence at the study site. *Black bars* indicate periods of predominant nectar consumption and *white bars* those of fruit consumption at the study site. Species abbreviations are Lr = *Lonchophylla robusta*, Gs = *Glossophaga soricina*, Gc = *G. commissarisi*, Hu = *Hylonycteris underwoodi*; and Lo = *Lichonycteris obscura*. (Data are from Tschapka [1998 and unpublished manuscript].) Gs did not occur on the old-growth rainforest study site but was found in secondary habitats in the surroundings (M. Tschapka, personal communication).

view of the lower energy requirements of *Hylonycteris* and a possibly higher feeding efficiency on small nectar volumes due to its more extensible tongue (Winter and Helversen, submitted).

Between October and February, nectar is abundant and two seasonal species move into the area: the larger *L. robusta* and the small *Lichonycteris obscura*. *Lonchophylla* appears to follow the depicted strategy of a wide-ranging opportunist that selectively exploits temporary hot spots of nectar superabundance (fig. 8.15). Its nightly times of arrival at the La Selva feeding site suggests that these bats have commuted from a longer (presently still unknown) distance (Tschapka 1998). The 6–8-wk flowering period of big-bang (Gentry 1974) *Matisia cordata* (a cultivated tree at La Selva), which has copious amounts of nectar between January and February each year, repeatedly attracted flocks of *Lonchophylla* during 3 yr of observation (Tschapka 1998; Tschapka, in preparation). Only the status of the other seasonal species, *L. obscura*, alludes to an explanation at present. As the smallest species (7 g), it should, from an energetic point of view, not be inferior to resident *Hylonycteris* or *G. commissarisi*. All the same, this small species occurs only seasonally, and even then is not common (Tschapka 1998). It currently remains unknown as to whether its rarity results

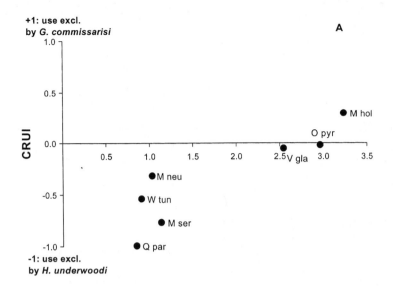

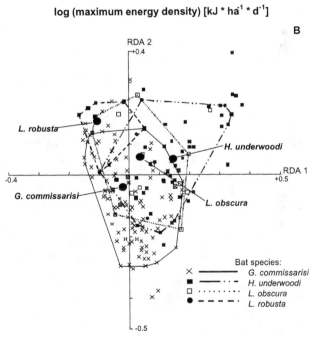

Figure 8.17. Differential use of floral resources by four species of glossophagine bats in the lowland rainforest of Atlantic Costa Rica. *A*, Plant species providing low quantities of nectar energy per forest area were visited less by *Glossophaga commissarisi* than by *Hylonycteris underwoodi*. Utilization of higher-yield species was approximately equal by these two resident nectarivores. *CRUI* is the comparative resource use index defined for *Gs* and *Hu*. The index ranges from −1 (exclusive use by *Hu*) to +1 (exclusive use by *Gc*). *B*, Reduced major axes representation of the 18-dimensional floral resource space that was determined from presence/absence percentages of pollen on bat fur from 18 plant species. Note the large overlap between all four species. The two resident species, *G. commissarisi* and *H. underwoodi*, differed most, and this was largely due to the exploitation pattern shown in *A*. Of the two seasonal species, *Lichonycteris obscura* had an intermediate position. The more outward position of *Lonchophylla robusta* was influenced by heavy visitation of planted *Matisia cordata* during its "big-bang" flowering period. *Large filled circles* are the polygon centers of the four species. Note that while *H. underwoodi* is predominantly nectarivorous, the species *G. commissarisi*, *L. robusta*, and *L. obscura* will seasonally consume fruit and insects. This is not included in the food resource space analyzed here (both figures from M. Tschapka, unpublished manuscript).

from special behavioral or morphological adaptations that may restrict this species to a specialized foraging niche. Its social system and center of distribution are not known.

In summary, despite the many unresolved questions, it can be concluded that nectar-feeding glossophagines in the Atlantic lowland region of Costa Rica bear significant features of a deterministically organized assemblage. The bottom-up analysis presented here demonstrates the value of quantitative physiology for addressing ecological questions related to community diversity. While much less information is available on Southeast Asian communities of nectar-feeding megachiropterans, their gross pattern of resource use also seems to comply with our predictions. In Malaysia, the two smaller *Macroglossus* species are resident nectar specialists roosting singly or in small groups in trees, while the larger *Eonycteris* roost in large colonies from which single individuals commute for up to at least 40 km to their foraging ranges (Gould 1978; Mickleburgh et al. 1992; Start 1974; Start and Marshall 1976; Watzke 1999).

Evolution of Bat Pollination

Although plants certainly enlisted "megabats" in the Old World and "microbats" in the New World independently as pollinators, the two bat-pollination systems have presumably evolved from a common ancestor, since both probably derive from a pollination system involving ancient nonflying mammals (Sussman and Raven 1978). Remnants of this line persist today, many in the Old World and at least a few in the New World (e.g., Lumer [1980] for rodents, Janson et al. [1981] and Ferrari and Strier [1992] for primates, Vieira and Carvalho-Okano [1996] and Tschapka and Helversen [1999] for marsupials as pollinators of otherwise large bat–pollinated flowers). Because reciprocal adaptation syndromes that have arisen by coevolution stabilize one another, they can be maintained in evolution even if all individual partner species on both sides have been exchanged in the course of time. Thus, there may well be an unbroken connection between Old World and New World bat flowers given by their common origin. That interpretation would also help resolve the question of "monophyly" or "diphyly" of the Old and New World bat pollination syndrome. This question has been discussed especially with reference to *Ceiba* and *Parkia*. Both genera include trees pollinated by large bats and occur naturally in the Old and New World (i.e., not having been introduced by man [Baker and Harris 1957; Dobat 1985; Vogel 1969a, 1969b, 1980]).

"Glossophagine Bats" and "Large-Bat Flowers"

Not all Neotropical bat flowers are typical glossophagine-flowers. For some flowers, large bats and arboreal mammals are the main pollinators, and glossophagine bats are not important pollinators (e.g., Tschapka and Helversen 1999; Vieira and Carvalho-Okano 1996). An example is the balsa tree, *Ochroma pyramidale*. The big cuplike flowers are visited by *Phyllostomus discolor, P. hastatus*, a number of stenodermatine bats (Fleming 1982; O. von Helversen, unpublished observation—northeast Colombia, December 1975, and Maracay, Venezuela, March 1982), *Carollia perspicillata*, and also by marsupials like *Caluromys* (La Selva, Costa Rica, O. von Helversen, personal observation) and monkeys (M. Fogden, personal observation). While smaller stenodermatine bats "jump" into the *Ochroma* flowers with a head-first dive from the air, larger bats like *P. hastatus* and *P. discolor* hold the petals with their hind legs (leaving many claw marks) and dip their heads into the nectar, becoming completely covered with pollen and nectar on the head and back. Small glossophagine bats, in contrast, mostly pilfer nectar from the side, sticking their heads between two petals while hovering, without any contact with pollen or the pistil (O. von Helversen, unpublished observation). In the case of *Mabaea* and *Marcgravia nepenthoides*, glossophagines are also only "nectar thieves"—the main pollinators are arboreal mammals, such as opossums (Tschapka and Helversen 1999; Vieira and Carvalho-Okano 1996), and possibly large bats. It is likely that the large flower heads of *Parkia* and inflorescences like those of *Hymenaea courbaril* or *Ceiba pentandra*, in which many flowers are open simultaneously, are also pollinated mainly by *Phyllostomus* and stenodermatine bats and not by glossophagines (Fleming 1983; Gribel et al. 1990; Heithaus et al. 1975; Hopkins 1984, 1998; Vogel 1968). In contrast, many typical glossophagine flowers cannot be easily exploited by large, nonhovering bats. To this group belong many small pendulous bat flowers (like *Cobaea, Mucuna, Markea*, or *Weberocereus* [Tschapka et al. 1999]), more delicate, herbaceous plants (*Irlbachia* [Machado et al. 1998], and many others), or the slender flowers of *Quararibea* (e.g., *Q. pendula* and *Q. parvifolia*) or *Helicteres baruensis* (Helversen and Voigt 2002).

Other flowers are visited by glossophagines as well as by smaller Stenodermatinae and Carolliinae. Examples are *Lafoensia pacari* (Sazima and Sazima 1975), *L. glyptocarpa* (Sazima et al. 1994), *L. punicifolia* (O. von Helversen, personal observation, northeast Colombia, November 1973), *Passiflora mucronata* and *Tetrastylis ovalis* (Buzato and Franco 1992; Sazima and Sazima 1978), *Mabea* sp. (Steiner 1983; Vieira and Carvalho-Okano 1996), and others. Still others are visited by *P. discolor* together with glossophagines (e.g., *Bauhinia* sp. [Heithaus et al. 1974; Ramirez et al. 1984; Sazima and Sazima 1977]). The primarily omnivorous and frugivorous species turn to bat flowers particularly during the dry seasons, when nectar serves as a source of drinking water as well as an energy supply (Heithaus et al. 1975).

The major reasons for the observed differences in flower exploitation between glossophagines and less specialized, often larger flower bats lie in the body size differences, different capacities for tongue extension (Helversen and Helversen 1975b; Nicolay 1999; Nicolay and Winter, in press; Winter and Helversen 2003), and the glossophagines' specialized hovering ability.

The Floral Syndrome: Adaptations of Glossophagine Flowers to Their Pollinators

If energy is limited and the cost-benefit ratio of foraging becomes critical, a pollinator increasingly must select which flowers to visit, especially to exclude flowers for which combined foraging costs are larger than is energy gain. Thus, the pollinator and its requirements may become one of the important selective forces for a flowering plant. The demands made by the bats are reflected in the adaptations to the pollinators that the flowers evolve. Because such demands are multiple and include general energetic, sensory, and cognitive abilities of the pollinators, the flowers would also be expected to develop a number of general adaptations to bat pollination. Together, these are termed the "syndrome of chiropterophily" (reviewed by Dobat 1985). Moreover, many plant species have evolved special adaptations that are needed for or are advantageous only under the specific boundary conditions for these plants (see table 8.3).

The degree to which a particular pollination syndrome is developed in a given plant species naturally depends on how highly specialized that species is for a particular group of pollinators. If a plant fares better when it attracts a variety of visitors—if in this respect it is a generalist—the syndrome characteristics will, of course, be less pronounced or will combine the adaptations to different pollinators and strive for a compromise, as Waser et al. (1996) discuss in detail. The advantages and disadvantages of specialists and generalists, which terms obviously mark only the extremes of a continuous distribution, are related to the pollinator spectrum in basically the same way as in the case of any other kind of ecological niche. The advantage of a specialist, from the standpoint of the male function of the flower, lies mainly in the more specific transfer of pollen to the stigmata of conspecific flowers and thus in a lower wastage of pollen (and nectar). From the perspective of female function, the advantages lie, in part, on ensuring the availability of sufficient pollen grains for every ovule but probably in many cases also in the mating isolation from related species with which hybridization would be possible. Preventing this is advantageous if it would lead to a loss of energy due to ovules yielding inferior offspring.

The generalists are evidently at an advantage when pollinators are limited or when their availability in space and time varies (Fleming et al. 1996; Groman and Pellmyr 1999; Waser et al. 1996) because a lesser degree of specialization allows a wider range of pollinators to be used. For bat-pollinated flowers, as for other flowers, this is mainly the case in the more extreme biotopes, such as on the paramo or in deserts. The greater abundance of specialized

Table 8.3. Characteristic traits of glossophagine flowers (right) and their possible causes (left)

Possible Origin of Selection Pressure	"Syndrome" Character of Flower
Properties of the pollinator:	
Large size (7–35 g) compared to insects	Rather big and sturdy flowers
Night active	Blooming at night
High energy turnover; minimum reward for a 10-g bat approximately 1-mg sugar per visit	High rate of nectar secretion (0.15 to >2 mL/night); large nectaries; devices that prevent the dropping out of nectar in pendulous flowers
Regular grooming of the fur; pollen as nitrogen-source	High pollen production (extremely pollen-rich anthers, many anthers or additional anthers from female sterile flowers)
Head with elongated rostrum that can be inserted into the corolla	Often bell-shaped type of flower, never narrow tubes or spurs
Bats approach flowers in flight	Free exposure of the flowers; inflorescences often unilateral (secund)
Excellent hovering ability	Smaller size compared to big-bat flowers (flowers sometimes only a nectar cup or head-mask for the bat); no landing devices (claw marks rare); free space for wing movement; protecting "wool" (when spines might damage wings)
Highly developed sense of olfaction	Characteristic and intense smell
Innate preference of sulphur-containing odors	Often disulphides or other sulphur-containing components in flower scent
No color vision	No conspicuous colors usable for attraction of bats; white flowers may offer a brightness contrast
Good visual pattern recognition	Free exposure to give visibility against background
Spectral sensitivity extended to near UV	Spectral remission sometimes contains near UV
Echolocation	Adaptations of flower shape
Excellent spatial memory	Flowering time of individual plants long, often longer than several months
Fear of predation at visiting flowers	Flowers with reduced accessibility to predators (e.g., flagelliflorous inflorescences)
"Rudiments" from precedent adaptations to other pollinators:	
Transition from bird flowers	Reddish to red-brown or purplish
Transition from sphingid flowers	Whitish or pale
Origin from pollination by ancient nonflying mammalian pollinators?	Characteristic scent? Cauliflory?
Defense against other visitors:	
Hummingbirds and other flower-visiting birds	Flower blooming only between dusk and dawn; nectar low in sucrose
Bees and wasps	Low nectar sugar concentration?
Sphingid moths	Dark colors; high nectar viscosity?
Ants	Timing of nectar secretion; hard-walled buds; circlets of spines or sticky hairs; smooth, waxy surface; bud embedded in mucous secretion

niches that accompanies the much greater diversity of tropical forests is also evident in the fact that flowers with a specialized pollination system are more common and conspicuous in tropical forests than in those of temperate latitudes.

In the seminal publication by Waser et al. (1996), little attention was paid to the difference between specialization to single or a few species within a pollinator group and specialization to larger groups of pollinators. The bat-pollination syndrome is an interesting example in this regard because, as discussed above, there are essentially no flowers specialized to single bat species, whereas a particularly strict distinction is made between bats and other pollinator groups. From our field observations (mostly from Costa Rica), we estimate that at least about 60% of all the plant species visited by glossophagines are pollinated almost exclusively by these bats, because either no other visitors are attracted or other visitors are incapable of appreciable pollen transfer purely on mechanical grounds. Only about 30% are more or less generalists, which also make use of birds and/or insects (and then mostly sphingid moths) as pollinators. In any case, the chiropterophilous flowers in particular seem to be an especially good example of the predictive value of floral syndromes. That the syndrome of chiropterophily is neither some kind of mystical entity nor a mere construct is demonstrated, above all, by the many cases in which, when authors such as Vogel (1958a, 1958b, 1968, 1969a, 1969b) inferred bat pollination of plants from herbarium studies, their predictions have been subsequently proven correct in the field.

Table 8.3 summarizes the most important and most characteristic traits of glossophagine-pollinated flowers, along with corresponding demands of their pollinators. For more details, see Helversen (1993) and Winter and Helversen (2001).

Transitions between Pollination Systems

Megabats and the larger frugivorous and omnivorous phyllostomid bats land on flowers to access nectar, whereas glossophagines have evolved the ability to hover and extract nectar with protrusible tongues. Because their body mass is much lower, down to about 7 g, glossophagines can manage with much smaller amounts of nectar per visit, as discussed above. This opened new opportunities for Neotropical flowers to exploit bats as pollinators: glossophagine flowers need not provide any sort of landing platform and many hummingbird- or sphingid-pollinated flowers already produce enough nectar to attract a glossophagine bat occasionally.

It must have been a relatively frequent event during the evolution of flowering plants to change pollinator adaptations and switch from one type to another. The best indication of this phenomenon is provided by the pattern of pollination types within related groups of plants. Whereas pollinators normally form monophyletic groups and their adaptations are mostly homologous within a given group, the corresponding flower types are normally scattered across many unrelated taxa and their adaptations are acquired by convergent evolution (Vogel 1990). Glossophagine pollination appears to have developed independently from other systems more than 100 times (see Dobat 1985).

The driving force for a plant to switch from one pollinator system to another may be (1) the occurrence of a plant in an area where the original pollinator is rare or missing, so that another occasional visitor becomes the pollen carrier more frequently, (2) use of the pollinator as a premating isolating mechanism to avoid hybridization with related plant species as long as hybrid seeds can be produced but hybrid viability is inferior (a process that possibly is limited to the phase of speciation when two populations that were separated meet again, so that the question of speciation or amalgamation of the gene pools arises), and (3) special ecological needs of a plant that can be met better or only by a certain pollinator. Glossophagine bats, as an example, offer to the plants an extreme long-distance pollination capability compared to most other pollinators. An 11-g glossophagine bat such as *Anoura caudifer* will cover up to about 60 km per night during foraging flight (Helversen 1986), and Horner et al. (1998), using radiotelemetry, estimated foraging flights ranging from 80 to 100 km for a 25-g *Leptonycteris* in the Sonoran Desert. Of course, this is not the distance between two visited flowers, but these figures give at least an impression of how large foraging areas of glossophagine bats may be. Long-distance transfer of pollen may be especially important in plant communities of high species diversity (because a greater number of species per area correlates with a greater mean distance between individual plants) or in other areas, as in deserts, where plant species have low population densities.

If transitions between pollinator systems are relatively frequent evolutionary events, it might be possible to make a prediction of the chances of transitions between particular systems. Of course, such a transition should be more probable if morphological and physiological changes do not have to be too great. Table 8.4 compares the main features of several important pollinator systems and suggests predictions of likely transitions. Obviously, the most probable candidates for transition to and from bat flowers are bird flowers and sphingid flowers. (Both have at least five important traits in common with glossophagine flowers.) Very improbable candidates are bee-, fly-, and wind-pollinated flowers. (They have hardly any traits in common.)

This consideration is corroborated by many cases where related bird- and bat-pollinated (or sphingid- and bat-pollinated) plant species exist within the same genus or even subgenus. Vogel (1990) gives a list of examples. In some species, bird- and bat-pollination systems appear to exist even in different

Table 8.4. Relationships between adaptations to pollination by bats and to some other pollen vectors

	Bats	Birds, Especially Hummingbirds	Sphingid Moths	Bees	Carrion Flies	Beetles	Wind
Shape and size	**Bell-shaped or brush-type, large and robust**	**Tubular, sometimes brush-type, medium size and robust**	Long and narrow tubes or brush-type, medium or large size	Variable shape, small and delicate	Variable	**Variable**	Variable, inconspicuous
Amount of nectar	**Very high**	**High**	High	Small	Normally small (if any)	Small (if any)	None
Nectar concentration	**Low**	**Low to medium**	**Low to medium**	High	N.A.	N.A.	N.A.
Amount of pollen	**High**	**Medium**	**Medium**	Variable	Small	High	High
Flowering time	**Night**	**Day**	**Night**	Day	Day/night	Day/night	Day/night
Exposure	**Well exposed**	**Relatively well exposed**	**Relatively well exposed**	Exposure of lesser importance	Exposure of lesser importance	Exposure of lesser importance	Well exposed
Scent	**Characteristic bat flower odor**	No scent	Characteristic perfume like scent	Variety of aromatic scents	Often carrion	Variable	None
Color	Inconspicuous, greenish, brownish or purple	Conspicuous often red or red and yellow	White	Conspicuous colors, often UV-free nectar guides	Often brown or whitish	Variable, but normally inconspicuous	Variable, but inconspicuous
Heat production	No	No	No	No	Yes	Yes	No
Flowering phenology	**Long flowering duration of individual plants**	Variable	Variable	Variable	Variable	Variable	Variable

Note. Boldface type indicates specific characters from flowers in nonbat pollination syndromes that are especially likely to have attracted bats, which in turn may have led to evolutionary transitions between floral syndromes.

populations of the same plant species, at least according to their present taxonomic definition. Examples are *Cirsium subcoriaceum*, Asteraceae (S. Vogel personal communication, O. von Helversen and Y. Winter, personal observation); *Vriesea platynema*, Bromeliaceae; *Erythrina glauca*, Fabaceae (O. von Helversen, unpublished observation); and *Symbolanthus calygonus*, Gentianaceae (P. Vogel personal communication, O. von Helversen and Y. Winter, unpublished observation).

Bat-pollinated plants that are generalists are mostly visited by bats and birds and/or bats and sphingid moths. Examples of bat/bird generalists are, in the Andean paramo, *Purpurella grossa* (K. Schuchmann, personal communication, Vogel 1968), *Puya*, and *Macrocarpaea* (F. Matt, H. Schmid, and O. von Helversen, unpublished observation); in desert and arid zones of South America and Mexico, *Agave*, columnar cacti (Alcorn and Olin 1961; Fleming et al. 1996; Howell and Schropfer-Roth 1981; Sahley 1996), *Ceiba* (Baker et al. 1971), *Abutilon* sp. (Buzato et al. 1994; Gottsberger 1972), *Siphocampylus sulfureus* (Sazima et al. 1994); and others. Examples of flowers that are visited and pollinated by bats and sphingids are also relatively common. *Capparis hastata* was visited by many sphingids and by *Glossophaga* in Venezuela (Ocumare, April 1992, O. von Helversen, unpublished observation), *Condaminea corymbosa*, Rubiaceae, by *Anoura* and several species of sphingids in Ecuador [F. Matt, M. Tschapka, and O. von Helversen, unpublished observation (from Zamorra, March 1998]), and *Cleome spinosa* in western Mexico (M. Tschapka, personal communication), while *Luehea* and *Bauhinia* are visited by bats and sphingids in the dry forests of Costa Rica (Haber and Frankie 1989; O. von Helversen, personal observation).

Thus, once the possibility of visits by the small and hovering glossophagines arose, bird- and sphingid-pollinated plants must have changed to bat pollination, with the result that bat-pollinated flowers in the Neotropics became more numerous and diversified than in the Old World. As a conservative estimate, there could be a total of 800–1,000 species of Neotropical bat flowers. The ratio of pollinator species to flower species would then be about 1:20 (see Vogel and Westerkamp 1991). Bat flowers would account for about 0.7%–1% of the Neotropical angiosperms. About the same relation was found in a more detailed examination of the well-known Costa Rican angiosperm flora. At least 80 of about 9,000 angiosperm species are bat pollinated in Costa Rica (Tschapka and Helversen, in preparation), which amounts to about 0.9%.

Loose Ends and Directions for Future Research
Spatial Memory

In order to balance its energy budget, a bat needs to obtain a minimum amount of nectar energy from a flower visit. The size of this minimum reward critically depends on a bat's ability to move during foraging along optimal routes between flowers and food patches (see eq. 1). What determines search efficiency in glossophagine bats? The majority of flower visits are revisits to previously encountered plants. Ongoing nectar secretion may lead to 10–30 visits or more to an individual flower during a night (O. von Helversen and M. Tschapka, unpublished observation), and steady state flowering plants present their flowers at the same location over many nights. Spatial memory for food location is therefore likely to be the most important mechanism enabling glossophagines to relocate flowers. If different bat species have equal abilities in this respect, then our model's predictions for relative differences in species requirements will not be affected. But this need not be the case. Spatial memory forms part of an animal's mental and cognitive abilities. Idiothetic orientation and sensory information from external cues are combined to infer a relative and absolute knowledge of spatial relations. The central nervous system resides at the base of this process, not only for the perception of the outside world but also for the generation of biologically adequate responses. If the evolutionary differentiation of neural systems responsible for orientation is influenced by ecological factors such as the spatial dimensions of foraging areas or seasonal differences in foraging strategy, we may well expect differences in spatial memory for food locations between species.

Metabolic Rate and Thermoregulation

Specialized adaptations for inhabiting higher elevations or using them as seasonal refuges have not been investigated in glossophagines (but see Bartels et al. [1998] for macroglossines). Clearly, more field data on DEE are necessary, including for higher altitudes and different seasons. At higher elevations, thermoregulatory expenditures increase as temperatures decrease. Increased thermoregulatory cost will lead to a reduction of energy available for flight and foraging activity and, therefore, an increase in required resource densities, unless energy demands for increased thermoregulation are compensated for. The use of hypothermia under natural conditions remains undocumented for glossophagines. For understanding species distributions along altitudinal gradients, field measurements of body temperatures and metabolic rates will be necessary.

Foraging Strategies and Population Genetics

The postulated dichotomy of foraging strategies between small resident species, on the one hand, and larger species with widely spaced feeding areas, on the other hand, should lead to differences in the genetic differentiation of the populations involved. Populations of larger species should be more homogenous over significantly larger areas. This prediction was tested for *Leptonycteris curasoae* by Wilkinson and Fleming (1996) using molecular methods of population genetics, but additional studies are needed for comparative analysis.

Brachyphyllinae/Phyllonycterinae

The flower-visiting bats of the Antilles, the Brachyphyllinae and (often included) Phyllonycterinae, are a relict-like, isolated group of phyllostomids. Extinction of this group on mainland America may have been due to competitive displacement by the more advanced and energetically more efficient glossophagines. The likelihood of this interpretation could be evaluated by an energetic comparison of the two groups.

Is Opportunistic Frugivory Suboptimal?

Glossophagines are clearly specialized for imbibing nectar from flowers. Nonetheless, many species are (seasonal) generalists that also take fruit and insects. Do more generalized genera such as *Glossophaga* fare less well during times of frugivory? Is there a general difference in the cost-benefit ratio of fruit versus nectar and pollen feeding? In Costa Rica, for example, *Glossophaga commissarisi* raises its October cohort of young on a nectar and pollen diet while the April cohort is raised predominantly on fruit (Tschapka 1998). It might be illuminating to search for fitness differences between the two seasons that might reflect differences in foraging efficiency and diet.

Generalists/Specialists and Evolutionary Changes in Pollination Systems

Quantitative data on the pollination efficiency of different visitors are difficult to obtain but urgently needed. At first, pollination success of day-active and night-active pollinators should be quantified in more species and correlated with local abundance of pollinators and their fluctuations in time. Another approach might be the phylogenetic analysis of related groups of plant species and a comparison with their pollination demands. Bat pollination, which is one of the best-defined pollination syndromes, and its relation to bat/bird and bat/moth generalists may be especially promising subjects to study.

Acknowledgments

We are grateful for the comments of Marco Tschapka, Tom Kunz, Ted Fleming, and an anonymous reviewer who helped improve and clarify this manuscript. Dagmar von Helversen participated in discussion, cooperation, and companionship during our field studies over many years.

Literature Cited

Alcorn, S., and G. Olin. 1961. Pollination of the Saguaro cactus by doves, nectar-feeding bats, and honey bees. Science, 133:1594–1595.

Alcorn, S., and G. Olin. 1962. Pollination requirements of the organpipe cactus. Cactus and Succulent Journal, 34:134–138.

Arends, A., F. J. Bonaccorso, and M. Genoud. 1995. Basal rates of metabolism of nectarivorous bats (Phyllostomidae) from a semiarid thorn forest in Venezuela. Journal of Mammalogy, 76:947–956.

Aschoff, J. 1981. Thermal conductance in mammals and birds: its dependence on body size and circadian phase. Comparative Biochemistry and Physiology A, 69:611–619.

Aschoff, J., and H. Pohl. 1970. Der Ruheumsatz von Vögeln als Funktion der Tageszeit und Körpergröße. Journal für Ornithologie, 111:38–47.

Audet, D., and D. W. Thomas. 1997. Facultative hypothermia as a thermoregulatory strategy in the phyllostomid bats, Carollia perspicillata and Sturnira lilium. Journal of Comparative Physiology B, 167:146–152.

Baker, H. G., and I. Baker. 1990. The predictive value of nectar chemistry to the recognition of pollinator types. Israel Journal of Botany, 39:157–166.

Baker, H. G., I. Baker, and S. A. Hodges. 1998. Sugar composition of nectars and fruits consumed by birds and bats in the tropics and subtropics. Biotropica, 30:559–586

Baker, H. G., R. W. Cruden, and I. Baker. 1971. Minor parasitism in pollination biology and its community function: the case of Ceiba acuminata. BioScience, 21:1127–1129.

Baker, H. G., and B. J. Harris. 1957. The pollination of Parkia by bats and its attendant evolutionary problem. Evolution, 11:449–460.

Bartels, W., B. S. Law, and F. Geiser. 1998. Daily torpor and energetics in a tropical mammal, the northern blossom-bat Macroglossus minimus (Megachiroptera). Journal of Comparative Physiology B, 168:233–239.

Bauer, R. 1991. Weberocereus tunilla, eine übelriechende Schönheit. Cactus and Succulent Journal, 42:6–8.

Begon, M., J. L. Harper, and C. R. Townsend. 1996. Ecology: Individuals, Populations, and Communities. 3d ed. Blackwell Science, Malden.

Berger, M. 1985. Sauerstoffverbrauch von Kolibris (Colibri coruscans und C. thalassinus) beim Horizontalflug. Pp. 307–314 in: Bird Flight (W. Nachtigall, ed.). Biona Report 3. Fischer Verlag, Stuttgart.

Bestmann, H. J., L. Winkler, and O. von Helversen. 1997. Headspace analysis of volatile flower scent constituents of bat-pollinated plants. Phytochemistry, 46:1169–1172.

Bolten, A. B., P. Feinsinger, H. G. Baker, and I. Baker. 1979. On the calculation of sugar concentration in flower nectar. Oecologia, 61:301–304.

Brown, J. H., W. A. Calder III, and A. Kodric-Brown. 1978. Correlates and consequences of body size in nectar feeding birds. American Zoologist, 18:687–700.

Burr, B., and W. Barthlott. 1993. Untersuchungen zur Ultraviolettreflexion von Angiospermenblüten II. Tropische und subtropische Pflanzenwelt, 87:1–193.

Burr B., D. Rosen, and W. Barthlott. 1995. Untersuchungen zur Ultraviolettreflexion von Angiospermenblüten. III. Dillenidae und Asteridae s.l. Tropische und Subtropische Pflanzenwelt, 95:1–185.

Buzato, S., and A. L. M. Franco. 1992. Tetrastylis ovalis: a second case of bat-pollinated passionflower (Passifloraceae). Plant Systematics and Evolution, 181:261–267.

Buzato, S., M. Sazima, and I. Sazima. 1994. Pollination of three species of Abutilon (Malvaceae) intermediate between bat and hummingbird flower syndromes. Flora, 189:327–334.

Charles-Dominique, P. 1991. Feeding strategy and activity budget of a frugivorous bat Carollia perspicillata (Chiroptera, Phyllostomatidae) in French Guiana. Journal of Tropical Ecology, 7:243–256.

Chase, J. M., and G. E. Belovsky. 1993. Experimental evidence for the included niche. American Naturalist, 143:514–527.

Colwell, R. K., and E. R. Fuentes. 1975. Experimental studies of the niche. Annual Review of Ecology and Systematics, 6:281–310.

Dobat, K. 1985. Blüten und Fledermäuse. Waldemar Kramer, Frankfurt am Main.

Dudley, R., and Y. Winter. 2002. Hovering performance of flower bats. Journal of Experimental Biology (in press).

Ellington, C. P. 1991. Limitations on animal flight performance. Journal of Experimental Biology, 160:71–91.

Faegri, K., and L. van der Pijl. 1971. The Principles of Pollination Ecology. 2d ed. Pergamon Press, Oxford.

Feinsinger, P. 1987. Approaches to nectarivore-plant interactions in the New World. Revista Chilena Historia Natural, 60:285–319.

Ferrari, S. F., and K. B. Strier. 1992. Exploitation of *Mabea fistulifera* nectar by marmosets (*Callithrix flaviceps*) and muriquis (*Brachyteles arachnoides*) in Southeast Brazil. Journal of Tropical Ecology, 8:225–239.

Findley, J. S. 1993. Bats: A Community Perspective. Cambridge University Press, Cambridge.

Fleming, T. H. 1982. Foraging strategies of plant-visiting bats. Pp. 287–325 *in:* Ecology of Bats (T. H. Kunz, ed.). Plenum Press, New York.

Fleming, T. H. 1983. *Carollia perspicillata.* Pp. 457–458 *in:* Costa Rican Natural History (D. H. Janzen, ed.). University of Chicago Press, Chicago.

Fleming, T. H., R. A. Nuñez, and L. da Silveira Lobo Sternberg. 1993. Seasonal changes in the diets of migrant and non-migrant nectarivorous bats as revealed by carbon stable isotope analysis. Oecologia, 94:72–75.

Fleming, T. H., M. D. Tuttle, and M. A. Horner. 1996. Pollination biology and the relative importance of nocturnal and diurnal pollinators in three species of Sonoran Desert columnar cacti. Southwestern Naturalist, 41:257–269.

Geiser, F., and D. K. Coburn. 1999. Field metabolic rates and water uptake in the blossom-bat *Syconycteris australis* (Megachiroptera). Journal of Comparative Physiology B, 169:133–138.

Gentry, A. H. 1974. Flowering phenology and diversity in tropical Bignoniaceae. Biotropica, 6:64–68.

Gottsberger, G. 1972. Blütenbiologische Beobachtungen an brasilianischen Malvaceen. II. Österreichische Botanische Zeitschrift, 120:439–509.

Gould, E. 1978. Foraging behavior of Malaysian nectar-feeding bats. Biotropica, 10:184–193.

Grant, P. 1986. The Ecology and Evolution of Darwin's Finches. Princeton University Press, Princeton, N.J.

Gribel, R., I. Sazia and M. Sazia. 1990. Flores pedem morcegos. Ciência Hoje, 11:22–28.

Groman, J. D., and O. Pellmyr. 1999. The pollination biology of *Manfreda virginica* (Agavaceae): relative contribution of diurnal and nocturnal visitors. Oikos, 87:373–381.

Grünmeier, R. 1993. Bestäubung der Fabaceae *Mucuna flagellipes* durch Flughunde in Kamerun. Pp. 29–39 *in:* Animal-Plant Interactions in Tropical Environments (W. Barthlott et al., eds.). Museum Koenig, Bonn.

Haber, W. A., and G. W. Frankie. 1989. A tropical hawkmoth community: Costa Rica dry forest Sphingidae. Biotropica, 21:155–172.

Heinrich, B. 1979. Bumblebee Economics. Harvard University Press, Cambridge, Mass.

Heinrich, B., and P. H. Raven. 1972. Energetics and pollination ecology. Science, 176:597–602.

Heithaus, E. R. 1982. Coevolution between bats and plants. Pp. 327–367 in: Ecology of Bats (T. H. Kunz, ed.). Plenum Press, New York.

Heithaus, E. R., T. H. Fleming, and P. A. Opler. 1975. Foraging patterns and resource utilization in seven species of bats in a seasonal tropical forest. Ecology, 56:841–854.

Heithaus, E. R., P. R. Opler, and H. G. Baker. 1974. Bat activity and pollination of Bauhinia pauletia: plant-pollinator coevolution. Ecology, 55:412–419.

Helversen, D. von, and O. von Helversen. 1975a. Glossophaga soricina (Phyllostomatidae)—Flug auf der Stelle [hovering flight] [text and 7-min high speed film, 900 fr/s]. Pp. 3–17 in: Encyclopaedia Cinematographica. (G. Wolf, ed.). Vol. E 1838. Institut für den wissenschaftlichen Film/Institute for the Scientific Film, Göttingen.

Helversen, D. von, and O. von Helversen. 1975b. Glossophaga soricina (Phyllostomatidae) Nahrungsaufnahme (Lecken) [feeding, high speed film]. Pp. 3–17 in: Encyclopaedia Cinematographica. (G. Wolf, ed.). Vol. E 1837. Institut für den wissenschaftlichen Film/Institute for the Scientific Film, Göttingen.

Helversen, D. von, and O. von Helversen. 1999. Acoustic guide in bat-pollinated flower. Nature, 398:759–760.

Helversen, O. von. 1986. Blütenbesuch bei Blumenfledermäusen: Kinematik des Schwirrfluges und Energiebudget im Freiland. Pp. 107–126 in: Fledermausflug—Bat Flight. (W. Nachtigall, ed.). Biona-Report 5. G. Fischer, Stuttgart.

Helversen, O. von. 1993. Adaptations of flowers to pollination by glossophagine bats. Pp. 41–59 in: Animal-Plant Interactions in Tropical Environments (W. Barthlott et al., eds.). Museum Alexander Koenig, Bonn.

Helversen, O. von, and H-.U. Reyer. 1984. Nectar intake and energy expenditure in a flower-visiting bat. Oecologia, 63:178–184.

Helversen, O. von, and C. C. Voigt. 2002. Glossophagine bat pollination in Helicteres baruensis (Sterculiaceae). Ecotropica, 8:23–30.

Helversen, O. von, M. Volleth, and J. Núñez. 1986. A new method for obtaining blood from a small mammal without injuring the animal: use of triatomid bugs. Experientia, 42:809.

Helversen, O. von, L. Winkler, and H. J. Bestmann. 2000. Sulphur containing "perfumes" attract flower-visiting bats. Journal of Comparative Physiology A, 186:143–153.

Herreid, C. F., and B. Kessel. 1967. Thermal conductance in birds and mammals. Comparative Biochemistry and Physiology, 21:405–414.

Hopkins, H. C. 1984. Floral biology and pollination ecology of the Neotropical species of Parkia. Journal of Ecology, 72:1–23.

Hopkins, H. C. F. 1998. Bat pollination and taxonomy in Parkia (Leguminosae: Mimosoideae). Pp. 31–55 in: The Biological Monograph (H. C. F. Hopkins, C. R. Huxley, C. M. Pannell, G. T. Prance, and F. White, eds.). Royal Botanic Gardens, Kew.

Hopkins, H. C. F., and M. G. Hopkins. 1993. Rediscovery of Mucuna macropoda (Leguminosae: Papilionoideae), and its pollination by bats in Papau New Guinea. Kew Bulletin, 48:297–306.

Horner, M. A., T. H. Fleming, and C. T. Sahley. 1998. Foraging behaviour and energetics of a nectar-feeding bat, *Leptonycteris curasoae* (Chiroptera: Phyllostomidae). Journal of Zoology (London), 244:575–586.

Howell, D. J. 1973. Bats and pollen: physiological aspects of the syndrome of chiropterophily. Comparative Biochemistry and Physiology A, 48:263–276.

Howell, D. J., and B. Schropfer-Roth. 1981. Sexual reproduction in agaves: the benefits of bats; the cost of semelparous advertising. Ecology, 62:1–7.

Jacobs, G. H. 1992. Ultraviolet vision in vertebrates. American Zoologist, 32:544–554.

Jacobs, G. H. 1993. The distribution and nature of colour vision among the mammals. Biological Reviews, 68:413–471.

Janson, C. H., J. Terborgh, and L. Emmons. 1981. Non-flying mammals as pollinating agents in the Amazonian forest. Biotropica (Supplement: Reproductive Botany), 13:1–6.

Kaiser, R., and L. Tollsten. 1995. An introduction to the scent of cacti. Flavour and Fragrance Journal, 10:153–164.

Knudsen, J. T., and L. Tollsten. 1995. Floral scent in bat-pollinated plants: a case of convergent evolution. Botanical Journal of the Linnean Society, 119:45–57.

Knudsen, J. T., L. Tollsten, and G. Bergström. 1993. Floral scents—a checklist of volatile compounds isolated by head-space techniques. Phytochemistry 33:253–280.

Koopman, K. F. 1981. The distributional patterns of New World nectar-feeding bats. Annals of the Missouri Botanical Garden, 68:352–369.

Law, B. S. 1995. The effect of energy supplementation on the local abundance of the common blossom bat, *Syconycteris australis,* in south-eastern Australia. Oikos, 72:42–50.

Lemke, T. O. 1984. Foraging ecology of the long-nosed bat, *Glossophaga soricina,* with respect to resource availability. Ecology, 65:538–548.

Lopez Calleja, M. V., F. Bozinovic, and C. Martinez Del Rio. 1997. Effects of sugar concentration on hummingbird feeding and energy use. Comparative Biochemistry and Physiology A, 118:1291–1299.

Lumer, C. 1980. Rodent pollination of *Blakea* (Melanostomaceae) in a Costa Rican cloud forest. Brittonia, 32:512–517.

Machado, I. C. S., I. Sazima, and M. Sazima. 1998. Bat pollination of the terrestrial herb *Irlbachia alata* (Gentianaceae) in northeastern Brazil. Plant Systematics and Evolution, 209: 231–237.

Masman, D., M. Gordijn, S. Daan, and C. Dijkstra. 1986. Ecological energetics of the kestrel: field estimates of energy intake throughout the year. Ardea, 74:24–39.

McFarlane, D. A. 1989. Patterns of species co-occurance in the Antillean bat fauna. Mammalia, 53:59–65.

McManus, J. J. 1977. Thermoregulation. Pp 281–292 *in:* Biology of Bats of the New World Family Phyllostomatidae. (R. J. Baker, J. K. Jones, Jr., and D. C. Carter, eds.). Pt. 2. Special Publications, the Museum, Texas Tech University, 13. Texas Tech Press, Lubbock.

McNab, B. K. 1969. The economics of temperature regulation in Neotropical bats. Comparative Biochemistry and Physiology, 31:227–268.

McNab, B. K. 1971. The structure of tropical bat faunas. Ecology, 52:352–358.

McNab, B. K. 1988. Complications inherent in scaling the basal rate of metabolism in mammals. Quarterly Review of Biology, 63:25–54.

Mickleburgh, S. P., A. M. Hutson and P. A. Racey, eds. 1992. Old World Fruit Bats: An Action Plan for Their Conservation. International Union for Conservation of Nature and Natural Resources, Gland, Switzerland.

Müller, F. 1897. Einige Bemerkungen über Bromeliaceen. Flora, 83:454–474.

Nagy, K. A. 1987. Field metabolic rate and food requirement scaling in mammals and birds. Ecological Monographs, 57:111–128.

Nagy, K. A. 1994. Field bioenergetics of mammals: what determines field metabolic rates. Australian Journal of Zoology, 42:43–53.

Neuweiler, G., and F. P. Möhres. 1967. Die Rolle des Ortsgedächtnisses bei der Orientierung der Großblatt-Fledermaus *Megaderma lyra*. Zeitschrift für vergleichende Physiologie, 57:147–171.

Nicolay, C. W. 2001. Ecological morphology and nectar-feeding performance in flower-visiting bats. Ph.D. Dissertation. Kent State University, Kent, Ohio.

Nicolay, C. W., and Y. Winter. In press. Performance analysis as a tool for understanding the ecological morphology of flower-visiting bats. *In:* Functional and Evolutionary Ecology of Bats (Z. Akbar, G. F. McCracken, and T. H. Kunz, eds.). Oxford University Press, Oxford.

Norberg, U. M., T. H. Kunz, J. Steffensen, Y. Winter, and O. von Helversen. 1993. The cost of hovering and forward flight in a nectar-feeding bat, *Glossophaga soricina*, estimated from aerodynamic theory. Journal of Experimental Biology, 182:207–227.

Norberg, U. M., and J. M. V. Rayner. 1987. Ecological morphology and flight in bats (Mammalia: Chiroptera): wing adaptations, flight performance, foraging strategy and echolocation. Philosophical Transactions of the Royal Society of London B, 316:337–419.

Peaker, M. 1990. Nutritional requirements and diets for hummingbirds and sunbirds. International Zoo Yearbook, 29:109–118.

Perrigo, G., and F. H. Bronson. 1983. Foraging effort, food intake, fat deposition, and puberty in female mice. Biology of Reproduction, 29:455–463.

Perrigo, G., and F. H. Bronson. 1985. Behavioral and physiological responses of female house mice to foraging variation. Physiology and Behavior, 34:437–440.

Porsch, O. 1931a. Crescentia-eine Fledermausblume. Österreichische Botanische Zeitschrift, 80:31–44.

Porsch, O. 1931b. Das Problem Fledermausblume. Anzeiger der Akademie der Wissenschaften Wien, Mathematisch-Naturwissenschaftliche Klasse, 69:27–28.

Porsch, O. 1935. Säugetiere als Blumenausbeuter und die Frage der Säugetierblume. II. Biologica Generalis, 11:171–188.

Porsch, O. 1938. Das Bestäubungsleben der Kakteenblüte: Cactaceae. Jahrbuch der Deutschen Kakteen-Gesellschaft, 39:1–142.

Porsch, O. 1942. Ein neuer Typus Fledermausblumen. Biologica Generalis, 15:283–294.

Ramirez, N., C. Sobrevila, N. X. De Enrech, and T. Ruiz-Zapata. 1984. Floral biology and breeding system of *Bauhinia ungulata*, Leguminosae: a bat pollinated tree in Venezuelan Llanos. American Journal of Botany, 71:273–280.

Rosenzweig, M. L. 1991. Habitat selection and population interactions: the search for mechanisms. American Naturalist, 137:S5.

Rosenzweig, M. L., and M. V. Lomolino. 1997. Who gets the short bits of the broken stick? Pp. 63–90 *in:* The Biology of Rarity (W. E. Kunin, and K. J. Gaston, eds.). Chapman & Hall, London.

Sahley, C. T. 1996. Bat and hummingbird pollination of an autotetraploid columnar cactus, *Weberbauerocereus weberbaueri* (Cactaceae). American Journal of Botany, 83: 1329–1336.

Sahley, C. T., M. A. Horner, and T. H. Fleming. 1993. Flight speeds and mechanical power outputs of the nectar-feeding bat, *Leptonycteris curasoae* (Phyllostomatidae: Glossophaginae). Journal of Mammalogy, 74:594–600.

Sazima, I., and M. Sazima. 1975. Quiropterophilia em *Lafoensia pacari* St. HIL. (Lythraceae), na Serra do Cipo, Minas Gerais. Revista Ciencia e Cultura, 27: 405–416.

Sazima, I., and M. Sazima. 1977. Solitary and group foraging: two flower-visiting patterns of the lesser spear-nosed bat *Phyllostomus discolor.* Biotropica, 9:213–215.

Sazima, M., and I. Sazima. 1978. Bat pollination of the passion flower *Passiflora mucronata,* in Southeastern Brazil. Biotropica 10:100–109.

Sazima, M., I. Sazima, and S. Buzato. 1994. Nectar by day and night: *Siphocampylus sulfureus* (Lobeliaceae) pollinated by hummingbirds and bats. Plant Systematics and Evolution, 191:237–246.

Sazima, I., Wagner A. Fischer, M. Sazima, and E. A. Fischer. 1994. The fruit bat *Artibeus lituratus* as a forest and city dweller. Ciencia e Cultura (Journal of the Brazilian Association for the Advancement of Science), 46:164–168.

Schmidt-Nielsen, K. 1972. Locomotion: energy cost of swimming, flying, and running, Science, 177:222–228.

Schnitzler, H. U., and E. K. V. Kalko. 1998. How echolocating bats search and find food. Pp. 183–196 *in:* Bat Biology and Conservation (T. H. Kunz and P. A. Racey, eds.). Smithsonian Institution Press, Washington, D.C.

Smith, H. 1982. Light quality, photoreception and plant strategy. Annual Review of Plant Physiology, 33: 481–518.

Speakman, J. R., and P. A. Racey. 1991. No cost of echolocation for bats in flight. Nature, 350:421–423.

Start, A. N. 1974. The feeding biology in relation to food resources of nectarivorous bats in Malaysia. Ph.D. Thesis, University of Aberdeen, Aberdeen.

Start, A. N., and A. G. Marshall. 1976. Nectarivorous bats as pollinators of trees in West Malaysia. Pp. 141–150 *in:* Tropical Trees: Variation, Breeding, and Conservation (J. Burley and B. T. Styles, eds.). Symposium of the Linnean Society, London.

Steiner, K. E. 1983. Pollination of *Mabea occidentalis* (Euphorbiaceae) in Panama. Systematic Botany, 8:105–117.

Sussman, R. W., and P. H. Raven. 1978. Pollination by lemurs and marsupials: an archaic coevolutionary system. Science, 200:731–736.

Suthers, R. A., J. Chase, and B. Bradford. 1969. Visual form discrimination by echolocating bats. Biological Bulletin, 137:535–546.

Tamm, S., and C. L. Gass. 1986. Energy intake rates and nectar concentration preferences by hummingbirds. Oecologia, 70:20–23.

Tiebout, H. M., and K. A. Nagy. 1991. Validation of the doubly labeled water method ($^3HH^{18}O$) for measuring water flux and CO_2 production in the tropical hummingbird *Amazilia saucerottei.* Physiological Zoology, 64:362–374.

Tschapka, M. 1998. Koexistenz und Ressourcennutzung in einer Artengemeinschaft von Blumenfledermäusen (Phyllostomidae: Glossophaginae) im atlantischen Tieflandregenwald Costa Ricas. Doctoral Thesis, Erlangen University, Erlangen.

Tschapka, M., and O. von Helversen 1999. Pollinators of syntopic *Marcgravia* species in a Costa Rican lowland rain forest: bats and opossums. Plant Biology, 1:382–388.

Tschapka, M., O. von Helversen, and W. Barthlott. 1999. Bat pollination of *Weberocereus tunilla*, an epiphytic rain forest cactus with functional flagelliflory. Plant Biology, 1:554–559.

Tucker, V. A. 1970. Energetic cost of locomotion in animals. Comparative Biochemistry and Physiology, 34:841–846.

van der Pijl, L. 1936. Fledermäuse und Blumen. Flora, 31:1–40.

Vieira, M. F., and R. M. Carvalho-Okano. 1996. Pollination biology of *Mabea fistulifera* (Euphorbiaceae) in Southeastern Brazil. Biotropica, 28:61–68.

Vogel, S. 1958. Fledermausblumen in Südamerika. Österreichische Botanische Zeitschrift, 104:491–530.

Vogel, S. 1968. Chiropterophilie in der neotropischen Flora: Neue Mitteilungen I. Flora, 157:562–602.

Vogel, S. 1969a. Chiropterophilie in der neotropischen Flora: Neue Mitteilungen II. Flora, 158:185–222.

Vogel, S. 1969b. Chiropterophilie in der neotropischen Flora: Neue Mitteilungen III. Flora, 158:289–323.

Vogel, S. 1980. Florengeschichte im Spiegel blütenökologischer Erkenntnisse. Vorträge der rheinisch-westfälischen. Akademie der Wissenschaften, 291:7–48.

Vogel, S. 1990. Radiacion adaptativa del sindrome floral en las familias neotropicales. Boletin de la Academia Nacional de Ciencias, 59:5–30.

Vogel, S., and C. Westerkamp. 1991. Pollination: an integrating factor of biocenoses. Pp. 159–170 *in:* Species Conservation: A Population-Biological Approach. Birkauser Verlag, Basel.

Voigt, C. C. 1998. Die energetischen Kosten des Fluges und der Reproduktion bei Blumenfledermäusen (Phyllostomidae; Glossophaginae). Doctoral Thesis, Erlangen University, Erlangen.

Voigt, C. C., and Y. Winter. 1999. The energetic cost of hovering flight in nectar-feeding bats (Phyllostomidae: Glossophaginae) and its scaling in moths, birds and bats. Journal of Comparative Physiology B, 169:38–48.

Voss R., M. Turner R. Inouye, M. Fisher, and R. Cort. 1979. Floral biology of *Markea neurantha* Hemsley (Solanaceae), a bat-pollinated epiphyte. American Midland Naturalist, 103:262–268.

Waser, N. M., L. Chittka, M. V. Price, N. Williams, and J. Ollerton. 1996. Generalization in pollination systems, and why it matters. Ecology, 77:1043–1060.

Watzke, S. 1999. An investigation of the ecology of nectarivorous pteropodid bats (Pteropodidae: Macroglossinae) in West-Malaysia. Thesis (in German), University of Suttgart-Hohenheim, Hohenheim.

Wilkinson, G. S., and T. H. Fleming. 1996. Migration and evolution of lesser long-nosed bats *Leptonycteris curasoae*, inferred from mitochondrial DNA. Molecular Ecology, 5:329–339.

Willig, M. R., and M. P. Moulton. 1989. The role of stochastic and deterministic process in structuring Neotropical bat communities. Journal of Mammalogy, 70:323–329.

Winkler, H. 1907. Beiträge zur Morphologie und Biologie tropischer Blüten und Früchte. Engler's botanisches Jahrbuch, 38:233–271.

Winter, Y. 1998a. *In vivo* measurement of near maximal rates of nutrient absorption in a mammal. Comparative Biochemistry and Physiology A, 119:853–859.

Winter, Y. 1998b. Energetic cost of hovering flight in a nectar-feeding bat measured with fast-response respirometry. Journal of Comparative Physiology B, 168:434–444.

Winter, Y. 1999. Flight speed and body mass of nectar-feeding bats (Glossophaginae) during foraging. Journal of Experimental Biology, 202:1917–1930.

Winter, Y., and O. von Helversen. 1998. The energy cost of flight: do small bats fly more cheaply than birds? Journal of Comparative Physiology B, 168:105–111.

Winter, Y., and O. von Helversen. 2001. Bats as pollinators: foraging energetics and floral adaptations. Pp. 148–170 *in:* Cognitive Ecology of Pollination (L. Chittka and J. Thomson, eds.). Cambridge University Press, Cambridge.

Winter, Y., and O. von Helversen. 2003. Operational tongue length in phyllostomid nectar-feeding bats. Journal of Mammalogy (in press).

Winter, Y., Voigt, C., and O. von Helversen. 1998. Gas exchange during hovering flight in a nectar-feeding bat, *Glossophaga soricina.* Journal of Experimental Biology, 201:237–244.

Wolf, A. V., M. G. Brown, and P. G. Prentiss. 1983. Concentrative Properties of Aqueous Solutions: Conversion Tables. Pp. D223–D272 *in:* CRC Handbook of Chemistry and Physics. 64th ed. CRC Press, Boca Raton, Fla.

CHAPTER 9

Bats and Fruit:
An Ecomorphological Approach

Elizabeth R. Dumont

Introduction

Bats are among the most ecologically and taxonomically diverse group of living mammals. Like many other groups of organisms (Huston 1994), bats reach their highest diversity in the tropics, where more than 100 species may coexist in lowland Amazonian forests (Simmons and Voss 1998). Driving this diversity is the expansion of bats out of insectivory and into a broad range of specialized trophic niches. In particular, the large number of plants that rely on vertebrates for seed dispersal and pollination supports an array of plant-visiting bats that focus their feeding and foraging efforts on fruit, nectar, and pollen. The trend toward increasing frugivore diversity in tropical latitudes is not unique to bats but is repeated in many vertebrate groups (e.g., Fleming 1973; Karr 1971; Terborgh 1986a).

The important contribution of bats to the diversity of vertebrate frugivore communities and their ecological roles in tropical ecosystems is increasingly recognized. Field studies of bats, particularly in the Neotropics, illustrate that frugivores often make up a large part of local species assemblages in tropical lowland forests, in numbers of species and individuals (e.g., Findley 1993; Fleming 1986; Kalko et al. 1996a; Simmons and Voss 1998). For instance, Barro Colorado Island, Panama, harbors 20 species of frugivorous bats with an estimated population of more than 10,000 individuals (Bonaccorso 1979; Handley et al. 1991; Kalko et al. 1996a). Likewise, Old World fruit bats often roost in huge colonies of thousands of individuals and may disperse great distances in their nightly foraging (see Fleming and Eby, this volume).

Because of their species richness, high biomass, feeding habits, and mobility, fruit-eating bats play important roles as seed dispersers in tropical ecosystems (see Patterson and Willig, this volume). Many species of trees and shrubs rely heavily on bats as seed dispersal agents. In contrast to other frugivores such as monkeys, birds, and rodents, which may be seed predators or consume unripe fruit (e.g., Norconk et al. 1997; Terborgh et al. 1993), bats eat primarily ripe fruits, and the seeds are often swallowed and excreted unharmed or dropped during fruit processing (e.g., Banack 1998; Bonaccorso and Gush 1987; Dumont and Irvine 1998). (The report of *Carollia perspicillata* destroying the seeds of *Anacardium excelsum* [Bonaccorso 1979] appears to be an exception.) The passage of intact seeds though a bat's digestive system reportedly improves or, at least, does not negatively affect germination rates (Bizerril and Raw 1998; Iudica and Bonaccorso 1997; Utzurrum 1995; Utzurrum and Heideman 1991). Many fruit-eating bats depend heavily on plant resources throughout the year (e.g., Banack 1998; Fleming 1988; Handley et al. 1991; Tan et al. 1998). These bats are reliable seed dispersers that must certainly have played, and continue to play, significant roles in the evolution and maintenance of tropical ecosystems.

Throughout the course of chiropteran evolution, two clades of bats have converged toward frugivory. Members of the family Phyllostomidae (New World leaf-nosed bats) represent a diverse radiation that is endemic to the New World. Among phyllostomids, frugivory occurs in 27 genera and approximately 96 species from the subfamilies Stenodermatinae (sensu Wilson and Reeder 2003), Carollinae, Brachyphyllinae, and Glossophaginae. New World fruit bats are a relatively derived clade whose fossil record extends back to the Miocene (Czaplewski 1997; Simmons and Geisler 1998). In the Old World, members of the family Pteropodidae represent a much more ancient frugivore radiation based on their basal location within the Chiroptera (Simmons and Geisler 1998). Pteropodids are found throughout the Old World from West Africa to the Cook Islands in the South Pacific and from southern Australia to the eastern Mediterranean (Mickleburgh et al. 1992). The family includes 171 frugivorous species within 35 genera (Mickleburgh et al. 1992; Wilson and Reeder 2003); the remaining 14 species focus on nectar as their primary source of food.

New and Old World fruit bats exhibit many similarities in the organization of their frugivore assemblages. However, despite the many convergences between the two fruit bat families, phyllostomids and pteropodids are very different kinds of animals (Fleming 1993). Pteropodids may be relatively large (up to 1.2 kg), lack complex echolocation, and are often viewed as morphologically and ecologically generalized. In contrast, phyllostomid frugivores are relatively small (<100 g), are competent echolocators, and are more likely to be dietary specialists (Swartz et al., this volume). Some of the most extreme variation in craniodental anatomy within any group of mammals is found among phyllostomids (Freeman 2000). In addition, species diversity in phyllostomid assemblages may be as much as three times higher than that of pteropodid assemblages (Fleming 1993; Francis 1990).

The goal of this chapter is first to provide a brief review of fruits as a food source and then to investigate three interrelated mechanisms of resource partitioning within fruit bat assemblages: the physical properties of fruit, morphological variation in the masticatory systems of bats, and variation among bats in fruit-processing strategies. Strategies of resource partitioning are driven by influences extending from individual to landscape levels of the ecological hierarchy. While the complexity of the resource-partitioning system prohibits explanation from a single subset of variables, there are associations among groups of variables that speak to specific aspects of the puzzle of resource partitioning and provide a glimpse of underlying themes in patterns of resource use.

Much of the data presented here draws from two geographically and historically distinct sets of fruit bats: an assemblage of Old World fruit bats studied in Madang Province, Papua New Guinea, and an assemblage of New World fruit bats from Costa Rica. The ecological, morphological, and behavioral convergences among members of these two groups make them an attractive model with which to investigate mechanisms of resource partitioning in chiropteran frugivores.

Fruits as Food
Distribution and Abundance

The distribution and abundance of fruits is a fundamental factor in determining the size and complexity of any given frugivore community (Fleming et al. 1987). The simplest indication of this relationship is found in the increase in frugivorous species in equatorial forests where plants with vertebrate-dispersed fruits also reach their peak diversity and abundance (Willig et al., this volume). The importance of fruit resources is underscored by the fact that, even at low latitudes, geographical regions that lack large fruit crops also support relatively limited frugivore assemblages (Goodman and Ganzhorn 1997; Goodman et al. 1997). Differences in fruit diversity and abundance have been cited as the possible cause of high species diversity and dietary specialization among phyllostomids and lower species diversity and more generalized feeding strategies among pteropodids. In contrast to the New World, fruits in Paleotropical forests are both more patchy in their distribution and produced by far fewer plant species (Fleming 1993).

On a more local scale, fruit distribution and abundance vary among plant species within forests. Some plants attract dispersers by offering huge crops of fruit. The most common example of this is figs (*Ficus*) and other members of the family Moraceae. These trees produce hundreds or thousands of fruits over a limited time and attract many different frugivores. Plants that use this fruiting strategy are often patchily distributed within forests and individual trees fruit asynchronously, thus providing a locally abundant but widely dispersed fruit supply. At the opposite extreme are plants that produce small but steady crops. Examples of these plants include *Piper*, *Solanum*, *Cecropia* and wild gingers, which often grow in patches and produce only a few ripe fruits a day but are productive over longer periods of time. This fruiting strategy provides a consistent low-volume supply of fruit dispersed over a larger area. Based on Gentry's (1974) study of flowers, these fruiting strategies are often termed "big bang" and "steady state," respectively. Foraging strategies that capitalize on both big bang and steady state fruit crops have been documented among phyllostomid bats, and a similar picture of variation in foraging strategies among pteropodids is beginning to emerge as well.

In the New World, many species from the subfamily Stenodermatinae specialize on big bang canopy fruits such as figs. These animals commute (sometimes long distances) directly to a fruiting tree, and once in the area make short trips between the fruiting tree and feeding roost(s) (Handley et al. 1991; Morrison 1978). This is probably a means of avoiding predators that also congregate where many frugivores are drawn to the abundant fruit crop (Howe 1979). In contrast to fig-feeers, other bats may focus their efforts on steady state crops (i.e., *Piper, Solanum,* and *Cecropia*). These species employ a foraging strategy in which an often predictable nightly foraging route passes a number of plants in the search for widely distributed ripe fruits (Fleming 1982, 1988; Fleming et al. 1977; Heithaus et al. 1975; Heithaus and Fleming 1978). These species often forage in light gaps and disturbed areas where pioneer understory species thrive. Not unexpectedly, these bats experience population booms in disturbed areas (Brosset et al. 1996; Schulze et al. 2000; Wilson et al. 1996).

Although the foraging strategies of Old World fruit bats are not as well documented as those of New World species, studies of pteropodid assemblages are beginning to reveal significant variation in foraging behavior. Many *Pteropus* species rely heavily on figs in their diet (e.g., Mickleburgh et al. 1992) and may commute long distances each day to reach fruiting trees (Marshall 1983). Several smaller pteropodids also use figs extensively. For example, in Papua New Guinea, *Dobsonia minor* feeds on both figs and introduced *Piper*. Alterations in individual home ranges appear to track the availability of ripe figs, suggesting that these fruits are a pivotal resource (Bonaccorso et al. 2000). *Nyctimene,* an even smaller pteropodid, also relies heavily on figs (Bonaccorso 1998; Richards 1986) and the foraging movements of this bat are similar to those of New World fig specialists from the subfamily Stenodermatinae (Spencer and Fleming 1989). At the other extreme of foraging style is *Syconycteris australis,* a macroglossine that is a facultative frugivore over the northern part of its range (Bonaccorso 1998; Law and Spencer 1995). *Syconycteris* in New Guinea feeds heavily on introduced *Piper*, and individuals have been documented returning to the same set of *Piper* patches on consecutive nights (Winkelmann et al. 2000). Similarly, *Cynopterus sphinx* in India forages for steady state crops early in the evening, then switches to big bang resources later in the night (Elangovan et al. 1999). Like the New World *Carollia, Syconyc-teris* and *Cynopterus* take advantage of reliable yet low-volume fruit supplies.

Fruit Color, Display, and Odor

Many studies of plant-animal interactions have focused on characteristics of plants that are associated with specific seed dispersal or pollination agents. Among vertebrate frugivores, fruit color, display, and odor have been of particular interest. For example, birds tend to favor small, dark-colored fruits that are displayed for either perching or reaching birds (e.g., Moermond et al. 1986; Wheelwright and Janson 1985; Willson et al. 1989), whereas primates generally favor larger, more brightly colored fruits that are accessible to animals of relatively large body size (Gauthier-Hion et al. 1985). The classic view of bat fruits is that they are drab or light in color, displayed openly by plants so that they are easily accessed by bats in flight, and have a distinct odor (Van Der Pijl 1957). Importantly, these associations between frugivores and fruit characteristics are not absolute but point to general patterns of variation in resource use within frugivore assemblages. For bats, worldwide variation in features of bat fruits reflect not only the varying characteristics of New and Old World forests but also the different sensory and locomotor systems of phyllostomid and pteropodid bats. While many fruit bats appear to use olfactory cues to locate fruit (Bloss 1999; Laska 1990; Oldfield et al. 1993; Reiger and Jakob 1988), the color and display of bat-dispersed fruits often vary significantly between the Old and New World.

New World bat fruits generally follow the classic pattern of having relatively dull coloration (green or brown [Fleming 1988; Fleming and Williams 1990; Kalko et al. 1996b]). For figs, coloration (along with fruiting phenology) is an important feature differentiating bat figs (green) from bird figs (red [Korine et al. 2000]). While New World plants such as *Piper* and *Cecropia* present fruits in an open and accessible location, Neotropical bat-dispersed figs are often hidden within surrounding foliage (Kalko et al. 1996b). The extent to which phyllostomids use echolocation in foraging for fruit varies among species. An experimental study of the use of echolocation and olfaction by *Carollia* demonstrated that *Piper* fruits are initially detected and localized using olfaction, then echolocation is used to pinpoint fruit location (Thies et al. 1998). In contrast, *Phyllostomus hastatus* uses echolocation over olfaction to detect openly displayed, pendant fruits (Kalko and Condon 1998).

Old World fruit bats do not echolocate (with the exception of *Rousettus*) but rely on olfaction and vision when foraging. Kalko et al. (1996b) predicted that Old World fruits are displayed more openly and are of more contrasting colors than New World fruits. In support of this prediction, a recent survey of bat fruits in Northeastern Australia and Papua New Guinea (Dumont and Irvine 1998) illustrates that of a total of 63 plant species known to be used by bats,

40% were light in color (i.e. yellow or pale green/yellow/white) and an additional 36% were red to orange. Each of these colorations promotes visibility by providing contrast with surrounding foliage. The remaining fruits in the survey were darker and with more cryptic colors, including dull green, purple, and blue. Also in support of Kalko et al.'s (1996b) prediction, 78% of the fruits in the survey are clearly displayed in relatively uncluttered spaces. These displays include not only the typical pendant and terminal branch locations often seen in New World fruits but also the caulicarpous (trunk-fruiting) pattern that is common among Old World fruits such as *Ficus* and *Syzygium*. In the 22% of cases in which fruits are more hidden, they are often located along branches, sometimes deep within the canopy of a tree. Overall, the presence of fruits that are hidden from view by either their location or coloration in this survey suggests the importance of olfaction in pteropodid foraging.

Variation in fruit display has been linked to differences between pteropodids and phyllostomids in wing design and flight characteristics. As a group, New World fruit bats have relatively short, broad wings that allow them to fly slowly and with good maneuverability (Norberg and Rayner 1987). These characteristics are useful both in cluttered understory spaces and in situations where canopy fruits (such as figs) need to be identified and plucked from surrounding foliage. In general, pteropodids have relatively longer and narrower wings than do phyllostomids (Norberg and Rayner 1987). This is especially true of the larger species (e.g., *Pteropus*), whose wing morphology supports efficient long-distance flight (Fleming and Eby, this volume). In part, the less maneuverable flight of large pteropodids led Kalko et al. (1996b) to predict that Old World fruits would be displayed in easily accessible, low-clutter environments. At least with respect to fruits from Australia and Papau New Guinea, 20% of fruits are found in cluttered spaces. This percentage would be even higher if caulicarpous fruits were included as bats must navigate under (and perhaps through) the canopy to reach them. There are two explanations of how these more hidden fruits are accessed by pteropodids. First, although many large pteropodids are not maneuverable fliers, many smaller species have the short, broad wings that provide the maneuverability needed to navigate in cluttered subcanopy spaces (Norberg and Rayner 1987). Second, some of the larger pteropodids land among the terminal branches of trees and subsequently climb quite effectively inside the canopy to search out fruits (E. R. Dumont, personal observation). In this case, flight is not used during the last phase of retrieving a fruit.

In summary, general differences in the color and display of New and Old World fruits appear to be broadly associated with differences in the sensory and locomotor capabilities of the bats that eat them. Nevertheless, there is a wide range of fruit colors and presentations in each region. Both pteropodids and phyllostomids use a variety of alternative solutions to resolve the problem of locating and retrieving fruits. Two particularly important variables that deserve more thorough investigation are the role of olfaction in foraging (currently known for only a few species) and the variation in flight capabilities among pteropodids.

Nutritional Qualities

Fruits vary widely in their nutritional properties. In particular, proportions of protein, fiber, lipids, and minerals can differ by orders of magnitude (e.g., Dinerstein 1986; Fleming 1988; Herbst 1986; Herrera 1987; Jordano 1992; Mattson 1980; O'Brien et al. 1998; Wendeln et al. 2000). In general, figs contain a relatively large proportion of indigestible fiber but are low in nitrogen (protein) and lipids. In contrast, many steady state fruits found in the understory, such as *Piper* and *Solanum*, tend to have low fiber content, high nitrogen content, and, in the case of *Piper*, a high concentration of lipids. Within fruit bat assemblages, there is a tendency for smaller bats to focus on high-quality fruits, whereas large bats tend to utilize fruits with less nitrogen and more fiber. Small-bodied phyllostomids that use figs extensively have been a perplexing exception to this pattern. A more inclusive perspective on the nutrients available from all the figs in a bat's diet may offer an explanation. A recent analysis of figs from Barro Colorado Island, Panama, suggests that bats can acquire sufficient carbohydrates, minerals, amino acids, and lipids solely from figs by feeding on a combination of different fig species (Wendeln et al. 2000).

Because fruits provide little protein relative to many other types of foods, it has been suggested that fruit bats avoid overingesting fruit by supplementing their diet with nitrogen-rich foods (Thomas 1984). Behaviors including the ingestion of leaves, pollen, and insects may reduce the need for fruit bats to consume large quantities of fruit in order to meet nitrogen requirements (e.g., Courts 1998; Fleming 1986; Kunz and Diaz 1995). Although many fruit-feeding bats do supplement their fruit diets, several studies suggest that both pteropodids and phyllostomids are particularly efficient at retaining nitrogen or have lower than predicted nitrogen needs (Delorme and Thomas 1996, 1999; Korine et al. 1996). This characteristic is just one of a suite of related specializations of fruit bat digestive systems (e.g., Dumont 1997b; Dumont et al. 1999; Forman 1972; Martinez del Rio 1994; Martinez del Rio and Restrepo 1993; Nagato et al. 1998; Tandler et al. 1997; Tedman and Hall 1985).

Mineral content is increasingly recognized as an important factor in evaluating the nutritional qualities of fruits. Rather than energy resources being a factor limiting food choice in bats, Barclay (1994, 1995) suggested that calcium might be a more critical constraint on reproduction. In addition, calcium is an essential resource for basic cellular processes, bone maintenance, and lactation (e.g., Favus 1996; Kwiecinski et al. 1987; Nordin and Frankel 1989). Several studies support the prediction that foods commonly eaten by bats are high in calcium. In particular, a survey covering several geographic regions reported that figs are higher in calcium than other sympatric fruits (O'Brien et al. 1998). Moreover, a recent survey of figs in Panama indicates that a diet of mixed fig species can provide bats with all of the nutrients they require (Wendeln et al. 2000). Not only do many native fruits provide higher contents of calcium and other minerals than cultivars (Nelson et al. 2000a), but several studies indicate that leaves also constitute a significant source of both minerals and protein for many plant-visiting bats (e.g., Kunz and Diaz 1995; Ruby et al. 2000).

Patterns of Resource Use

Terborgh's (1986a, 371–372) comment that "fruit availability is extremely variable in both space and time" and that, during periods of fruit scarcity, "frugivores must pursue alternative lifestyles in order to survive" summarizes the results of many studies of tropical frugivores. Although many species use fruit year-round by turning to keystone resources such as Neotropical figs (O'Brien et al. 1998; Terborgh 1986b), most supplement their diet of fruits with other resources. A recent study using stable isotopes found that many phyllostomid frugivores ingest insects and that only a few rely exclusively on fruit (Herrera et al. 1998). Free-ranging pteropodids are not generally thought to pursue insects actively (Bonaccorso 1998; but see Courts 1997, 1998), although many supplement their diets with leaves, pollen, unripened fruits, and nectar during periods of low fruit availability (e.g., Kunz and Diaz 1995; Kunz and Ingalls 1994; Mickleburgh et al. 1992; Nelson et al. 2000b; Richards 1995; Ruby et al. 2000; Tan et al. 1998).

Certainly, seasonal variation in fruit abundance makes a simple description of the diet of a fruit bat species a difficult and in some cases an almost impossible task. Dietary flexibility appears to be most marked among pteropodid frugivores, where strong seasonality, periods of limited fruit availability, and frequent cyclones may markedly influence diet composition (e.g., Banack 1998; Funakoshi et al. 1993; Nelson et al. 2000b; Palmer et al. 2000). Despite their reliance on more clearly delimited sets of fruits (Fleming 1986), phyllostomids also exhibit important seasonal variation in diet, as demonstrated by analyses of feeding habits and stable isotopes (e.g., Galetti and Morellato 1994; Herrera et al. 1998; Willig et al. 1993). As pointed out by Swartz et al. (this volume), our knowledge of the diets of most bats is extremely limited, and developing an understanding of the nutritional, biomechanical, and ecological implications of variation in diet will no doubt be a difficult endeavor.

Nevertheless, despite the dietary breadth of many bats and the overlapping diets of many species, we do know that fruit bats are not simply opportunistic feeders (e.g., Fleming 1986; Kitchener et al. 1990; Utzurrum 1995; Willig et al. 1993). Many species focus their feeding and foraging efforts on specific suites of fruits. For example, Fleming (1986) noted that phyllostomid frugivores focus their foraging efforts on a few "core" plant species and supplement this core with other plant (or insect) resources to meet nutritional requirements. The dietary emphasis on separate core fruits is supported by many studies of phyllostomids (e.g., Bonaccorso 1979; Fleming 1988; Handley et al. 1991; Heithaus et al. 1975; Hernandez-Conrique et al. 1997; Iudica 1995; Willig et al. 1993).

The focus on core fruits among pteropodid species is less clear. Many studies document broad overlap in fruit resource-use within pteropodid frugivore assemblages and add to the perception that pteropodids are more generalized than are phyllostomids. However, there is also ample evidence that different species focus their feeding and foraging efforts on specific suites of fruits (e.g., Bhat 1994; Bonaccorso 1998; Juste and Perez Del Val 1995; Marshall 1983; Richards 1995; Tan et al. 1998; Utzurrum 1995). In a recent study of *Pteropus* on the Samoan archipelago, Banack (1998) found that while two sympatric species used a nonrandom subset of plants, there was a high degree of dietary overlap between them due, in part, to seasonal variation in fruit supply. Instead these species were more accurately classified as "sequential specialists," focusing on specific resources as they become available. Variation in patterns of resource use among pteropodids is less clear and certainly less well documented than for phyllostomids. Although many pteropodids use a broad array of resources, there is a discernible trend toward a trophic dichotomy between species that rely heavily on canopy resources (especially figs) and other, often smaller, species that use a higher proportion of steady state understory resources.

Given that fruits vary in their distribution, abundance, and nutritional composition and that fruit bats are not opportunistic feeders, the remaining sections of this chapter investigate a mechanism of resource partitioning that may be at work in assemblages of fruit bats. The central theme of the mechanism discussed here is the direct interaction between bats and fruits that occurs during feeding. In these bat-fruit interactions, both characteristics of fruits and the morphological and behavioral attributes of bats are equally important factors mediating resource partitioning.

Vertical Stratification

The separation of sympatric species in three-dimensional space is a common mechanism of niche partitioning and serves to promote species diversity. For example, both birds and primates use vertical stratification to create many smaller niches within a single area of forest (e.g., Erard 1990; Fleagle and Mittermeier 1980). Many studies demonstrate that bats also utilize vertical stratification (Ascorra et al. 1996; Bernard 1998; Francis 1990, 1994; Kalko and Handley 2001; Simmons and Voss 1998). While some bats feed and forage in the understory, others focus their feeding and foraging efforts on canopy resources. Still other species show no clear preference for any particular forest level.

Bonaccorso (1979) described three frugivore ensembles (guilds) among New World fruit bats based on food preferences and foraging height. "Ground-story frugivores" are small-bodied bats that feed on high-quality, widely spaced, understory fruits. In contrast, "canopy frugivores" eat low-quality, locally abundant canopy fruits and may (or may not) be relatively large-bodied. Many canopy frugivores are considered to be fig specialists. Finally, "frugivore/omnivores" use fruit resources, but supplement their diet with insects and/or nectar. Bats in this latter category use the broadest range of food items of any of the frugivore guilds.

Similar variation in food preferences, foraging height, and even specialization on figs has been reported among pteropodid frugivore assemblages (Bonaccorso 1998; Fleming 1993; Francis 1990, 1994; Ingle 1993; Juste and Perez Del Val 1995; McKenzie et al. 1995; Richards 1995; Utzurrum 1995). However, the tight association between vertical stratification, fruit preferences, and foraging styles seen among phyllostomids is looser among pteropodids. Differences in the distribution of fruits in Neotropical and Paleotropical forests may be partly responsible. First, in contrast to Neotropical figs, which are canopy fruits, many caulicarpus (trunk-fruiting) figs produce fruits within a few meters of the ground (E. R. Dumont, personal observation; Utzurrum 1995). Thus, medium-sized fig-eaters such as *Nyctimene* are often captured in ground-level mist nets (Bonaccorso 1998; Spencer and Fleming 1989). Second, in comparison to Neotropical forests, some Old World forests exhibit a relative dirth of understory fruit (e.g., Francis 1990). The scarcity of understory fruits in the Old World may prevent the degree of specialization on these resources seen in the Neotropics. Instead, the pteropodid analog of the ground-story frugivore guild appears to include relatively small-bodied taxa (e.g., *Syconycteris, Balionycteris, Haplonycteris, Cynopterus*) that capitalize on understory fruits when they are available but frequently utilize big bang fruit and/or nectar resources.

In sum, vertical stratification appears to exist as a mechanism of resource partitioning in all fruit bat assemblages. However, the match between vertical stratification and foraging strategies described for canopy and ground-story frugivores in the Neotropics is not directly applicable to those of Old World fruit bats. Rather, the main contrast in Old World fruit bats is between fig-eating bats that focus their feeding and foraging efforts on figs (and other big bang crops) and ground-story bats that utilize more dispersed steady state resources usually found in the understory.

Fruit Processing

Based on observations of phyllostomids, Bonaccorso and Gush (1987) proposed a link between feeding behavior and the spatiotemporal distribution and nutritional quality of fruit; bats that eat locally abundant fruits (such as figs) that are high in fiber spend more time processing individual fruits than

Table 9.1. Mean fruit processing times (hours:minutes:seconds) for several pteropodid frugivores

Species	N	Mass (g)	Fruit	Processing Time	Category
Pteropus conspicillatus	6	814.3	Apple	00:02:15	Figs
Dobsonia minor	7	77.0	Apple	00:10:58	Figs
Paranyctimene raptor	3	24.8	Apple	01:12:25	Figs
Nyctimene albiventer	1	26.3	Apple	00:31:23	Figs
Syconycteris australis	2	18.5	*Muntingia*	00:01:06	Ground-story

Note. Mass is based on male and (non-pregnant) female midpoints reported in Bonaccorso (1998). All "apples" are spherical pieces of skinned apple of approximately 20 mm diameter. *N* indicates the number of individuals sampled, and "category" refers to the type of fruits emphasized in the bat's diet.

bats that consume higher-quality fruits with more patchy distributions. As illustrated in table 9.1, a similar pattern of variation was observed in fruit-processing times within an assemblage of pteropodids. These data reflect the time spent processing "fruits" (spherical pieces of skinned apple) of approximately 20-mm diameter. In general, fig-feeders spend more time processing fruits than the understory frugivore *Syconycteris*. The exception is *Pteropus conspicillatus*, which eats quickly. That *P. conspicillatus* feeds rapidly is probably a function of its large body size—breaking apart and chewing a 20-mm fruit is a less challenging task for *Pteropus* than for smaller bats.

Another way of examining food-processing time is to compare the number of chews that different species use to process mouthfuls of fruit. Figure 9.1 illustrates the number of chews used by fig-feeders and ground-story frugivores to process mouthfuls of papaya. These species in each category include both phyllostomids (Dumont 1999) and pteropodids (Dumont 1998). Fig-feeders as a group use significantly more chews to process each mouthful of fruit than do ground-story frugivores ($P < 0.001$, Mann-Whitney rank sum test). This pattern also applies within each family; fig-feeders consistently use more chews per mouthful than do ground-story frugivores.

All fig-feeders described here produce "spats," a bolus of seeds, fruit skin, and fiber that has been thoroughly chewed and the juice squeezed out. Presumably, this avoids the consumption of indigestible fibrous material (Martinez del Rio and Restrepo 1993; Morrison 1980). For these species, eating slowly and continuously over a long period of time, chewing thoroughly, and spitting out less digestible material increases the potential for extracting the maximum nutrition from each fruit. In contrast, understory frugivores use significantly fewer chews to process each mouthful of fruit and tend not to produce spats. The exception is *Sturnira*, which makes spats despite the fact that it is not a fig specialist. This appears to reflect the close phylogenetic link between *Sturnira* and other stenodermatines (*Dermanura* and *Artibeus*) and, conversely, the greater distance between *Sturnira* and both carollines (*Carollia*) and glossophagines (*Glossophaga*).

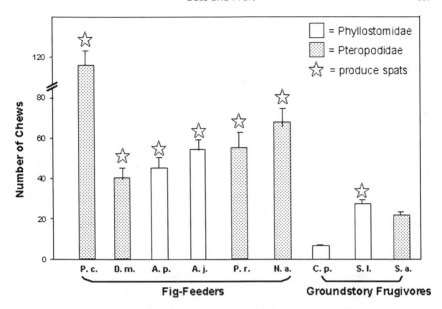

Figure 9.1. The mean (*bars*) and standard error (*lines*) of number of chews used to process mouthfuls of food by fig-feeders and ground-story frugivores. Statistics summarize data from two to six individuals. As a group, fig-feeders use significantly more chews than ground-story frugivores. All fig-feeders and one ground-story frugivore produce spats (*stars*). Species are *Pteropus conspicillatus* (*P. c.*), *Dobsonia minor* (*D. m.*), *Artibeus phaeotis* (*A. p.*), *Artibeus jamaicensis* (*A. j.*), *Paranyctimene raptor* (*P. r.*), *Nyctimene albiventer* (*N. a.*), *Carollia perspicillata* (*C. p.*), *Sturnira lilium* (*S. l.*), and *Syconycteris australis* (*S. a.*).

These differences between fig-feeders and understory frugivores likely reflect a balance between nutritional composition of fruits and the energetic cost of carrying excess mass in the form of indigestible material during flight. Although a diet of various fig species can provide adequate nutrients and minerals (Wendeln et al. 2000), figs are higher in fiber and often lower in protein and lipid content than many understory fruits (Dinerstein 1986; Fleming 1988; Herbst 1986). Fig-feeders may benefit from extracting nutrients from figs by oral processing rather than swallowing large quantities of indigestible fiber. That the fig-feeders studied thus far make spats when they eat fruits with a variety of textures and water contents (i.e., fig, apple, banana, papaya) suggests that spat production is part of a fixed behavioral program rather than a response to fruits with certain physical or compositional characteristics.

In contrast to fig-feeders, the decreased emphasis on chewing among ground-story frugivores is associated with the low fiber content, high nutritional content, and distribution of their primary foods. Ground-story frugivores (sensu Bonaccorso 1979) feed on low abundance, high-quality fruits that are patchy in their distribution (Dinerstein 1986; Fleming 1988; Herbst 1986). Even when eating a fig, *Carollia* does not make spats but drops relatively large pieces of unchewed fruit. The failure of *Carollia* and *Syconycteris* to produce spats is presumably linked to their foraging styles and choice of fruits, as there is nothing about their anatomy that precludes making spats.

Fruit Size and Hardness

Not only are the density, distribution, and nutritional content of fruits important elements of resource partitioning among frugivores, but physical characteristics of fruits may also be critical (Lucas 1979, 1994; Lucas and Luke 1984; van Roosmalen 1984). The physical properties of foods are factors in resource partitioning within vertebrate communities ranging from amphibians to fish, birds, and mammals (e.g., Freeman 1979, 1981; Freese and Oppenheimer 1981; Herring 1985; Kiltie 1982; Moermond et al. 1986; Toft 1980; Van Valkenburgh 1996; Wainwright 1987; Wheelwright 1985). Recent interest in identifying correlations between the physical properties of fruit and fruit selection by mammals has resulted in a growing dataset summarizing the physical properties of fruits. Through these studies, fruit size and hardness have emerged as particularly important variables in resource partitioning (e.g., August 1981; Bonaccorso 1979; Kinzey and Norconk 1990; Strait and Overdorff 1996; Ungar 1995; Yamashita 1996). The broader significance of resource partitioning among frugivores on the basis of fruit hardness and size is underscored by the likelihood that plants capitalize on the relationship between fruit properties and fruit preferences to select among potential seed dispersal agents (August 1981; Janson 1983; Janzen 1979).

Two aspects of fruit size—mass and dimension—may determine which frugivores can use them. At the lower end of the range of bat size, fruit mass is a limiting factor for bats that carry fruits to feeding roosts. This probably reflects a balance between the absolute size of a fruit and its nutritional value. That is, in commuting situations, the value of a fruit must be greater than or equal to the energetic cost incurred through carrying it. Bats may also be limited by the dimensions of fruits that will fit into their mouths. Gape—how widely animals can open the space between their upper and lower jaws—clearly limits the size of fruits eaten by birds (Wheelwright 1985), although it is more limiting for birds that swallow fruits whole than for those that crush them first (Moermond and Denslow 1985; Moermond et al. 1986). Fruit bats arguably possess a broader array of food-processing behaviors than do birds by virtue of their more complex dental morphologies and their ability to use a variety of biting styles and chewing behaviors. Although the horizontally wide mouths of fruit bats may augment their ability to grip and carry fruits (Freeman 1988), vertical gape may also limit the size of fruits that bats can carry and bite effectively. Although small bats are known to feed on large fruits in situ (e.g., mango, papaya, *Dipteryx*), they may use mechanically inefficient, and thus costly, processing behaviors that are offset by the energy savings gained by eating these fruits without transporting them.

Among some phyllostomids, fruit size is often positively correlated with frugivore size (Bonaccorso 1979; Heithaus et al. 1975; Kalko et al. 1996b; Wendeln et al. 2000). The bat-fruit size relationship is mediated by both physical limitations and energetic demands. With respect to their physical abilities to process fruits, large bats can carry/handle both large and small fruits, while small bats are limited to small fruits. However, large bats may be limited to large fruits that offer sufficient energy to cover the expenses of their foraging flights. The exception may be that some fruits contain essential nutrients or minerals that the bats require. Although the limitations imposed by bat size on fruit size may not apply in instances where bats feed from fruits in situ (i.e., small bats may feed on large fruits at the tree or on the ground), they do apply when bats pluck and carry fruits to feeding roosts.

The association between fruit size and bat size among pteropodids is more equivocal. Studies of Philippine fruit bats reveal a lack of association between fruit size and bat size (Utzurrum 1995). In contrast, a survey of bat fruits in Australia and Papua New Guinea suggests a loose association in which small bats consume small fruits, whereas larger bats consume both large and small fruits (Dumont and Irvine 1998; fig. 9.2A). If this pattern holds across all fruit bats, it may explain why the relatively small phyllostomids tend to exhibit a tighter correlation between bat size and fruit size than the larger pteropodids (Kalko et al. 1996b).

Regardless of whether fruits are transported to feeding roosts or eaten in situ, the hardness of fruits is another physical property that may influence patterns of resource partitioning. Hardness is a general description of food texture that refers to a suite of specific physical properties including toughness, strength, stiffness, and plasticity (see review in Strait 1997). Essentially, hard foods are more resistant to permanent deformation and therefore are more difficult for animals to break apart and chew than are soft foods. Indeed, a close relationship between food texture and craniodental design has recently been suggested for frugivorous bats (Freeman 1988, 1992). Several developing lines of evidence suggest that bats are limited by fruit hardness in much the same way that they are limited by fruit size. The mean hardness of fruits eaten by three pteropodid genera is illustrated in figure 9.2B. Members of the genus *Syconycteris* (ca. 14.5–23.5 g) consume only the softest fruits whereas members of the genus *Pteropus* (180–1,400 g), consume fruits with a wide range of hardness values (Dumont and Irvine 1988). A similar trend is evident among phyllostomids, though hardness data are currently only available for a limited array of species of fruit (E. R. Dumont, unpublished data). As in the case of fruit size, associations between fruit hardness and bat size are probably loose. Many other factors, including the distribution, abundance, and nutritional qualities of fruits, influence patterns of resource use.

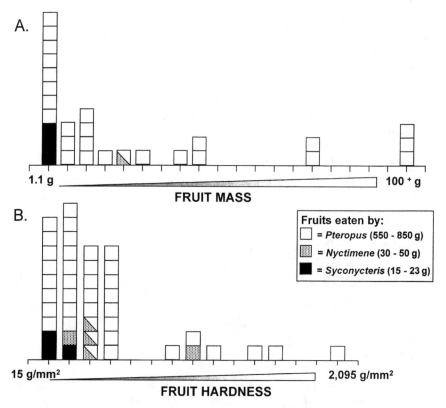

Figure 9.2. *A*, The wet mass (g) of Old World fruits and the size of bat consumers. *B*, The hardness (g mm^{-2} of puncture resistance) of Old World fruits and the size of bat consumers. In these histograms, each square represents the mean value of one fruit species.

Interestingly, among 28 species of bat fruits from Australia and Papua New Guinea, size and hardness are significantly and positively correlated ($r^2 = 0.79$, $P < 0.0001$; Dumont and Irvine 1998). Within this set of data, fruit size and hardness also vary according to their vertical distribution. Overall, ground-story fruits (most often produced by shrubs) tend to be small and soft while fruits from higher in the canopy tend to be relatively larger and harder. This same general pattern applies to a smaller sample of New World fruits including several species of *Ficus*, *Piper*, and *Solanum*. Based on their resistance to being punctured (measured as pressure in g mm^{-2}), canopy fruits are often five to 10 times harder than ground-story fruits. This is not only true for fruits that are considered to be hard by human standards (e.g., palm nuts, *Terminalia*, *Barringtonia*), but even ripe figs are commonly many times harder than ripe understory fruits. Although we may consider a ripe fig to be a soft fruit, it may pose a significant challenge to many small-bodied fruit bats.

The association between vertical stratification, fruit size, and fruit hardness provides circumstantial evidence that these fruit characteristics play a role in resource partitioning. The link between physical properties of fruit and patterns of resource use within fruit bat assemblages is further demonstrated by experimental studies that clearly indicate sympatric bat species from ground-story and fig-feeding (i.e., canopy or big bang) guilds or ensembles differ significantly in both craniofacial anatomy and their approaches to processing fruits of different hardness.

Morphological Diversity in Bats and Patterns of Resource Use

Many aspects of morphology are associated with partitioning of food resources among frugivorous bats (Norberg 1994, 1998; Norberg and Rayner 1987; Swartz et al., this volume). Both phyllostomids and pteropodids exhibit morphological correlates of feeding-guild membership in both their cranial and postcranial skeletons. With respect to the flight apparatus, several impressive studies illustrate that wing morphology reflects flight capabilities and that different flight dynamics are required for traversing different distances and maneuvering in different habitats. Not surprisingly, the wings of ground-story frugivores are designed to maneuver efficiently in cluttered, subcanopy spaces whereas bats that utilize canopy resources have wings that are designed to navigate in more open habitats and cover larger distances.

Similarly, many studies have documented that morphological divergence in the skulls and teeth of closely related mammals reflects specialization for eating foods of different size and hardness (e.g., Daegling 1992; Dumont 1995; Freeman 1979, 1981; Kiltie 1982; Kinzey 1992; Strait 1993). Similar patterns of morphological divergence in cranial form are beginning to emerge from studies of fruit bats. Although only two studies have focused on detailed functional analyses of frugivores (Dumont 1997a; Freeman 1988), both concluded that the traditional view of frugivore craniodental anatomy as generalized is inaccurate. Rather, the category "frugivore" is extremely broad and contains species with many unique dietary adaptations. Given the broad range of variation in the size and hardness of bat fruits, it is not surprising that morphological variation in the skulls and teeth of fruit bats suggests specialization for fruits with different physical properties.

Analyses of cranial-shape variables that represent mechanically relevant variation is a well-established means of looking for evidence of functional divergence among fruit bats (e.g., Dumont 1995; Freeman 1988, 1992). In figure 9.3, a series of 17 shape variables (i.e., size-adjusted; Falsetti et al. 1993) were used to summarize the morphology of the skull as a whole as well as specific aspects of the palate and the dentary using principal components analysis (Tabachnick and Fidell 1996). The sample includes 10 species of fig-eaters and ground-story frugivores representing both pteropodids and phyllostomids. This limited set of species was selected because several aspects of

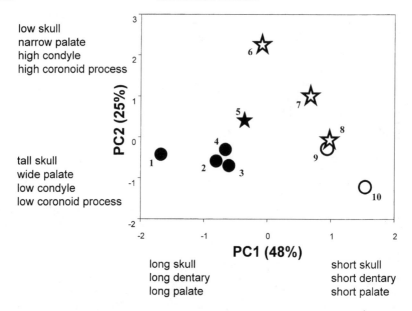

Figure 9.3. Plot of the first two principal components (73% total variation) describing the distribution of ground-story frugivores (*stars*) and fig-feeders (*circles*) on the basis of size-adjusted functional cranial variables. Pteropodids are indicated by *solid symbols* while phyllostomids are indicated by *open symbols*. Descriptions of significant morphological characters are provided along each axis. Species are *Pteropus conspicillatus* (*1*), *Paranyctimene raptor* (*2*), *Nyctimene albiventer* (*3*), *Dobsonia minor* (*4*), *Syconycteris australis* (*5*), *Glossophaga soricina* (*6*), *Carollia perspicillata* (*7*), *Sturnira lilium* (*8*), *Artibeus phaeotis* (*9*), and *Artibeus jamaicensis* (*10*).

their food-processing behaviors (i.e., biting and chewing) have been documented (Dumont 1999). Together, the first two principal components explain 73% of the variation within the sample. The first principal component (PC1) separates species with short skulls, dentaries, and palates (high PC scores) from those with relatively longer skulls and longer, wider dentaries. High scores along the second component (PC2) indicate increasingly low skulls, narrow palates, high condyles, and high coronoid processes.

   Two important trends are visible in the distribution of species across these principal components. First, as reported in previous analyses (Fleming 1993; Freeman 1995), phyllostomids (*open symbols*) and pteropodids (*solid symbols*) are distinct. Here, members of the two families are separated along the first component, which largely reflects skull length. From a mechanical perspective, these features suggest that pteropodids may have relatively wider gapes (Herring 1975; Herring and Herring 1974; Wall 1995), while phyllostomids may have dentaries that are designed to resist higher antero-posterior bending forces (Biknivicius and Ruff 1992; Daegling 1992, 1993; Hylander 1979a, 1979b).

A second trend in these data is that, regardless of family affiliation, fig-feeders and ground-story frugivores (fig. 9.3, *circles* and *stars,* respectively) fall into two distinct groups. Differences between fig-feeders and ground-story frugivores are primarily expressed along PC2. As a group, the fig-feeders included in this study exhibit relatively taller skulls, wider palates, and lower condyles and coronoid processes than do ground-story frugivores. Phyllostomid fig-feeders also have shorter palates and skulls. Overall, these features suggest crania with robust attachments for masticatory muscles and, especially for phyllostomids, crania that are designed to be more resistant to torsional strains than the crania of ground-story frugivores (Covey and Greaves 1994; Freeman 1984). Freeman (1988) has also suggested that the wide, shallow palates of many frugivorous bats function to increase the number of teeth that can be used in biting and/or carrying fruits. The apparent structural strength of fig-feeder skulls fits well with their dietary emphasis on relatively large, hard fruits. It is likely that the greater specialization observed in the skulls of phyllostomid fig-feeders, relative to pteropodid fig-feeders, is the result of focusing on relatively hard fruits while retaining small body size.

This brief analysis illustrates that at least a subset of frugivorous pteropodids and phyllostomids clearly differ in functional aspects of cranial morphology. It is often argued that pteropodid species are less distinct from one another in terms of their anatomy and behavior than are phyllostomid frugivores, which tend to be more clearly delineated specialists. The relatively large numbers of phyllostomid frugivores that occur in sympatry in contrast to the smaller numbers of sympatric pteropodids support this interpretation. However, this analysis also suggests that significant variation in cranial shape does exist among pteropodids. Assessing the functional implications of this variation is an important avenue for further research.

Behavioral Diversity, Plasticity, and Patterns of Resource Use

Within the context of ecomorphological approaches to comparative biology, functional morphologists are looking more closely at behavior as a means of understanding the relatiosnship between ecology and anatomy (e.g., Arnold 1983; Swartz et al., this volume; Wainwright 1991, 1996; Wainwright and Reilly 1994). Not only do body size and morphology circumscribe the range of an animal's mechanical abilities, but behavior, or how animals use their anatomy, can also influence mechanical efficiency. A long history of detailed analyses of the biomechanics of mastication (e.g., Biknivicius and Ruff 1992; Czarnecki and Kallen 1980; De Gueldre and De Vree 1990; Freeman 1988; Greaves 1978; Herring 1985, 1993; Kallen and Gans 1972; Kiltie 1982; Radinsky 1981a, 1981b; Reduker 1983; Thomason et al. 1990) has established a basic understanding of form-function relationships in mammalian skulls. However, only a few studies have approached the relationship among feeding behavior, morphological diversity, and ensemble structure in mammals (Dumont 1999; Van Valkenburgh 1996).

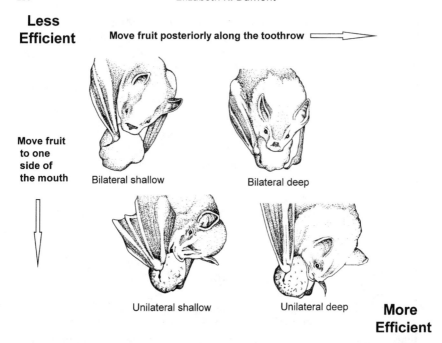

Less Efficient   **Move fruit posteriorly along the toothrow** ⟹

Move fruit to one side of the mouth

Bilateral shallow   Bilateral deep

Unilateral shallow   Unilateral deep   **More Efficient**

Figure 9.4. Bats use four different strategies for biting into fruits. From a mechanical perspective, shallow bilateral bites are least efficient while deep unilateral bites are most efficient for any individual bat. (Line drawings from Dumont [1999].)

Feeding behavior among both phyllostomids and pteropodids has been studied in the context of resource partitioning and feeding-guild structure (Dumont 1998, 1999). In these studies, individuals from sympatric species were subjected to feeding experiments in which fruit size was held constant and fruit hardness was varied. The hypothesis underlying these experiments was that feeding behavior of bats reflects the hardness (i.e., resistance to puncture) of the foods that bats frequently consume. Bats that naturally consume hard fruits (fig-feeders, including bats that focus on big bang canopy fruits) were expected to exhibit behaviors that increase the efficiency of mastication, while bats that naturally consume softer fruits (ground-story frugivores) were expected to lack specialized food-processing styles. In addition, because they are likely to encounter fruits with a wider range of hardness values, fig-feeders should exhibit a broader range of behavioral responses to changing food hardness than ground-story frugivores.

When biting into fruits to remove mouthfuls of food, bats use four basic behaviors that vary in mechanical efficiency (fig. 9.4). Here, estimates of mechanical efficiency are based on a simple model of the jaw as a class 3 lever in which forces applied to it are positioned between the temporo-mandibular

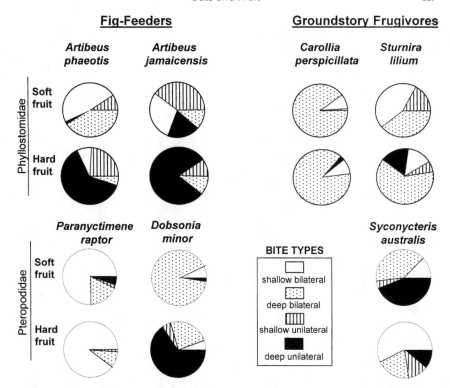

Figure 9.5. Pie charts illustrating the proportions of different bite types used by seven bat species during soft and hard fruit feeding. The biting behavior of fig-feeders and ground-story frugivores differ within both Phyllostomidae and Pteropodidae. With the exception of *Paranyctimene*, fig-feeders shift to more efficient biting styles (i.e., deep unilateral bites) when confronted with hard fruits.

joint (fulcrum) and the bite point (Herring 1993). By taking bites with the posterior teeth on only one side of the mouth (deep unilateral bites), an individual bat achieves a highly efficient transfer of muscle force to the fruit. This is because the distances from the biting teeth and chewing muscles to the jaw joint are roughly equal (i.e., lever and load arms are roughly equal), and relatively few teeth are in contact with the food, thus concentrating bite forces over a small area. At the other extreme, a bite taken with both left and right anterior teeth simultaneously (a shallow bilateral bite) is less efficient because of the disparity in lever and load arm lengths and the larger number of teeth involved.

The few species of phyllostomids studied thus far confirm the predicted differences between fig-eaters and ground-story frugivores (fig. 9.5). The fig-feeding *Artibeus jamaicensis* and *Dermanura phaeotis* employ higher proportions of mechanically efficient biting than do the ground-story frugivores

Carollia perspicillata and *Sturnira lilium,* regardless of whether they are eating soft or hard fruits. Moreover, the fig-eaters exhibit a more pronounced shift to a higher proportion of mechanically efficient biting styles when confronted with a hard fruit than do ground-story frugivores. Although the differences between phyllostomid ground-story and fig-eaters are clear, it is interesting that *Sturnira* differs from *Carollia* in having a broader repertoire and stronger shift in biting styles (Dumont 1999). This similarity between *Sturnira* and canopy frugivores may reflect the close phylogenetic relationship among stenodermatine species. Alternatively, it may also indicate that *Sturnira* regularly includes some harder foods in its diet. Despite these subtle variations, the biting behavior of the phyllostomid fig-feeders studied here is specialized for hard (i.e., resistant) fruit while the behavior of the ground-story frugivores sampled here is not.

The biting styles of pteropodids also differ from one another (Dumont 1998). For example, the biting style of the fig-feeding *Dobsonia minor* is similar to that of phyllostomid fig-feeders; *Dobsonia* increases the mechanical efficiency of its biting during hard fruit feeding by shifting to unilateral biting. A larger species that also consumes figs, *Pteropus conspicillatus,* displays a very similar pattern. Curiously, the fig-feeding *Paranyctimene raptor* emphasizes relatively inefficient shallow bilateral biting during both soft and hard fruit feeding. This same feeding style is shared by *Nyctimene albiventer.* Unlike other fig-feeders, the lack of a shift in biting styles with changing fruit hardness in *Paranyctimene* suggests that it is not behaviorally specialized for processing hard or resistant fruits. For the latter species (and for *Nyctimene*), specializations for processing hard fruits probably lie in craniodental morphology rather than behavior (Dumont 1998). Like New World ground-story frugivores, *Syconycteris australis* exhibits biting behaviors that include a broad range of biting styles and does not shift to more efficient biting strategies when processing hard fruits.

Despite the fundamental differences between phyllostomids and pteropodids, variation in food processing strategies within each assemblage is influenced by the physical properties of fruits. Although there are similarities among fig-eaters and ground-story frugivores from both families, it is clear that patterns of variation in biting behavior are not absolute. Whether differences in biting styles between phyllostomids and pteropodids reflect phylogenetic distance and/or aspects of morphological variation (including body size) is not clear at this time. Nevertheless, studies that combine anatomy, behavior, and ecology are beginning to identify associations between anatomical variation and ecological diversity. Additional controlled studies are underway that incorporate native fruits and identify feeding behaviors among outgroup taxa and more basal members of the two frugivore clades. Eventually, these data will provide a clearer picture of the evolution of anatomical and behavioral diversity among fruit bats.

Directions for Future Research

Detailed studies of bats and plants have contributed significantly to our understanding of the roles played by bats in tropical ecosystems. However, there is still much to learn about the ecology and evolution of fruit bats. Ultimately, a firm understanding of community-level processes is essential for the development of effective conservation policies. In order to achieve that goal, more data from a number of different sources is crucial.

We need more fundamental natural history data for fruit bats and the plants on which they feed. For example, the indication that the physical and nutritional properties of fruits are important factors in resource partitioning within bat assemblages is tantalizing, but the existing data are scarce. Knowing more about fruit size, hardness, spatiotemporal distribution, and nutritional composition would set the stage for identifying variables among plants that influence patterns of resource use. Much of that data could be generated through surveys of known bat fruits. By combining plant data with additional detailed studies of diet, foraging movements, migration, and energetics of sympatric fruit bat species, we may gain significant insights into subtle but perhaps significant variations in patterns of resource use on which larger assemblages have evolved.

As more basic information about bats and fruits becomes available, ecomorphological analyses will be an increasingly effective means of bridging the gap between the morphology and behavior of individuals and interactions among species. Studies of the intersection of the craniodental morphology of fruit bats, feeding behavior, and variation in the physical properties of fruits are still in their infancy. Nevertheless, the results thus far indicate that integrating morphology and behavior can shed light on the ways in which even apparently similar species coexist. Eventually, ecomorphological analyses of a variety of anatomical systems (e.g., feeding, flight, and sensory systems) will provide the foundation on which we can build a more complete picture of species interactions.

Finally, the independent evolution of frugivory in the New and Old World offers the unique opportunity to identify the overriding themes in the evolution of frugivory in bats. An important area for further research is to evaluate in more detail the similarities and differences in the ecology, morphology, and behavior of Neotropical and Paleotropical fruit bat assemblages. Evaluating the results of ecomorphological analyses against a backdrop of phylogenetic relationships among genera within both pteropodid and phyllostomid lineages will provide valuable insights into correlations between morphology, behavior, and ecology. These studies ultimately will identify fundamental morphological and behavioral adaptations that have defined the shifts to chiropteran frugivory and set the stage for the evolution of these influential members of tropical communities.

Acknowledgments

I extend special thanks to F. Bonaccorso and E. K. V. Kalko for their many thoughtful discussions about bats and fruit. This research was supported by National Science Foundation grants IBN-9905404 and IBN-9507488.

Literature Cited

Arnold, S. J. 1983. Morphology, performance and fitness. American Zoologist, 23: 347–361.

Ascorra, C. F., S. Solari, and D. E. Wilson. 1996. Diversidad y ecología de los quirópteros en Pakitza. Pp. 593–612 in: Manu: The Biodiversity of Southern Peru (D. E. Wilson and A. Sandoval, eds.). Smithsonian Institution Press, Washington, D.C.

August, P. V. 1981. Fig fruit consumption and seed dispersal by Artibeus jamaicensis in the Llanos of Venezuela. Biotropica, 13:70–76.

Banack, S. A. 1998. Diet selection and resource use by flying foxes (genus Pteropus). Ecology, 79:1949–1967.

Barclay, R. M. R. 1994. Constraints on reproduction by flying vertebrates: energy and calcium. American Naturalist, 144:1021–1031.

Barclay, R. M. R. 1995. Does energy or calcium availability constrain reproduction by bats? Symposia of the Zoological Society of London, no. 67:245–258.

Bernard, E. 1998. Vertical stratification of bat communities in tree-fall gaps in central Amazonian primary forests. Selbyana, 19:268–269.

Bhat, H. R. 1994. Observations on the food and feeding behavior of Cynopterus sphinx Vahl (Chiroptera, Pteropodidae) at Pune, India. Mammalia, 58:363–370.

Biknivicius, A. R., and C. B. Ruff. 1992. The structure of the mandibular corpus and its relationship to feeding behaviours in extant carnivorans. Journal of Zoology (London), 228:479–507.

Bizerril, M. X., and A. Raw. 1998. Feeding behavior of bats and dispersal of Piper arboreum seeds in Brazil. Journal of Tropical Ecology, 14:109–114.

Bloss, J. 1999. Olfaction and the use of chemical signals in bats. Acta Chiropterologica, 1:31–45.

Bonaccorso, F. J. 1979. Foraging and reproduction in a Panamanian bat community. Bulletin of the Florida State Museum, Biological Sciences, 24:359–408.

Bonaccorso, F. J. 1998. Bats of Papua New Guinea. Conservation International, Washington, D.C.

Bonaccorso, F. J., and T. J. Gush. 1987. Feeding behaviour and foraging strategies of captive phyllostomid fruit bats: an experimental study. Journal of Animal Ecology, 56:907–920.

Bonaccorso, F. J., J. R. Winkelmann, E. R. Dumont, and K. Thibault. 2002. Home range of Dobsonia minor (pteropadidae): a solitary, foliage-roosting fruit bat in Papua New Guinea. Biotropica, 34:138–146.

Brossett, A., P. Charles-Dominique, A. Cockle, J.-F. Cosson, and D. Masson. 1996. Bat communities and deforestation in French Guiana. Canadian Journal of Zoology, 74:1974–1982.

Courts, S. E. 1997. Insectivory in captive Livingstone's and Rodrigues fruit bats Pteropus livingstonii and P. rodricensis (Chiroptera: Pteropodidae): a behavioural adaptation for obtaining protein. Journal of Zoology (London), 242:404–410.

Courts, S. E. 1998. Dietary strategies of Old World fruit bats (Megachiroptera, Pteropo-didae): how do they obtain sufficient protein? Mammal Review, 28:185–193.

Covey, D. S. G., and W. S. Greaves. 1994. Jaw dimensions and torsion resistance during canine biting in the Carnivora. Canadian Journal of Zoology, 72:1055–60.

Czaplewski, N. J. 1997. Chiroptera. Pp. 410–431 in: Vertebate Paleontology in the Neotropics: The Miocene Fauna of La Venta, Colombia (R. F. Kay, R. H. Madden, R. L. Cifelli, and J. J. Flynn, eds.). Smithsonian Institution Press, Washington, D.C.

Czarnecki, R. T., and F. C. Kallen. 1980. Craniofacial, occlusal, and masticatory anatomy in bats. Anatomical Record, 198:87–105.

Daegling, D. J. 1992. Mandibular morphology and diet in the genus *Cebus*. International Journal of Primatology, 13:545–570.

Daegling, D. J. 1993. The relationship of in vivo bone strain to mandibular corpus mor-phology in *Macaca fascicularis*. Journal of Human Evolution, 25:247–269.

De Gueldre, G., and F. De Vree. 1990. Biomechanics of the masticatory apparatus of *Pteropus giganteus* (Megachiroptera), Journal of Zoology (London), 220:311–332.

Delorme, M., and D. W. Thomas. 1996. Nitrogen and energy requirements of the short-tailed fruit bat (*Carollia perspicillata*): fruit bats are not nitrogen constrained. Journal of Comparative Physiology, 166:427–434.

Delorme, M., and D. W. Thomas. 1999. Comparative analysis of digestive efficiency and nitrogen and energy requirements of the phyllostomid fruit-bat (*Artibeus jamaicen-sis*) and the pteropodid fruit-bat (*Rousettus aegyptiacus*). Journal of Comparative Physiology B, 169:123–132.

Dinerstein, E. 1986. Reproductive ecology of fruit bats and the seasonality of fruit pro-duction in a Costa Rican cloud forest. Biotropica, 18:307–318.

Dumont, E. R. 1995. Correlations between enamel thickness and dietary adaptation among extant primates and chiropterans. Journal of Mammalogy, 76:1127–1136.

Dumont, E. R. 1997a. Cranial shape in fruit, nectar and exudate feeding mammals: im-plications for interpreting the fossil record. American Journal of Physical Anthro-pology, 102:187–202.

Dumont, E. R. 1997b. Salivary pH and buffering capacity in frugivorous and insectivo-rous bats. Journal of Mammalogy, 78:1210–1219.

Dumont, E. R. 1998. The effect of fruit hardness on feeding behavior in pteropodid bats. Bat Research News, 39:165.

Dumont, E. R. 1999. The effect of food hardness on feeding behaviour in frugivorous bats (Family Phyllostomidae): an experimental study. Journal of Zoology (London), 248:219–229.

Dumont, E. R., K. Etzel, and J. Hempel. 1999. Bat salivary proteins segregate according to diet. Mammalia, 63:159–166.

Dumont, E. R., and A. K. Irvine. 1998. Old World bat fruits: diversity and implications for pteropodid ecology. Bat Research News, 39:166.

Elangovan, V., G. Marimuthu, and T. H. Kunz. 1999. Temporal patterns of individual and group foraging behaviour in the short-nosed fruit bat, *Cynopterus sphinx*, in south India. Journal of Tropical Ecology, 15:681–687.

Erard, C. 1990. Écologie et comportment des gobe-mouches (Aves: Muscicapinae, Platysteirinae, Monarchinae) du nord-est du Gabon. 2. Organisation sociale et écolo-gie de la reproduction des Muscicapinae. Mémoires Musee Nacional Histoire Na-turelle, 146:1–233.

Falsetti, A. B., W. L. Jungers, and T. M. Cole. 1993. Morphometrics of the callitrichid forelimb: a case study in size and shape. International Journal of Primatology, 14: 551–572.

Favus, M. J. 1996. Primer on the Metabolic Bone Diseases and Disorders of Mineral Metabolism. Lippincott-Raven Publishers, Philadelphia.

Findley, J. S. 1993. Bats: A Community Perspective. Cambridge University Press, Cambridge.

Fleagle, J. G., and R. A. Mittermeier. 1980. Locomotor behavior, body size and comparative ecology of seven Surinam monkeys. American Journal of Physical Anthropology, 52:301–322.

Fleming, T. H. 1973. The number of mammal species in several North and Central American forests. Ecology, 54:555–563.

Fleming, T. H. 1982. Foraging strategies of plant-visiting bats. Pp. 287–325 in: Ecology of Bats (T. H. Kunz, ed.). Plenum Press, New York

Fleming, T. H. 1986. Opportunism versus specialization: the evolution of feeding strategies in frugivorous bats. Pp. 105–118 in: Frugivores and Seed Dispersal (A. Estrada and T. H. Fleming, eds.). Dr. W. Junk Publishers, Dordrecht.

Fleming, T. H. 1988. The Short-Tailed Fruit Bat: A Study in Plant-Animal Interactions. University of Chicago Press, Chicago.

Fleming, T. H. 1993. Plant-visiting bats. American Scientist, 81:460–467.

Fleming, T. H., R. Breitwisch, and G. H. Whitesides. 1987. Patterns of vertebrate frugivore diversity. Annual Review of Ecology and Systematics, 18:91–109.

Fleming, T. H., E. R. Heithaus, and E. B. Sawyer. 1977. An experimental analysis of the food location behavior of frugivorous bats. Ecology, 58:619–627.

Fleming, T. H., and C. F. Williams. 1990. Phenology, seed dispersal, and recruitment in Cecropia peltata (Moraceae) in Costa Rican tropical dry forest. Journal of Tropical Ecology, 6:163–178.

Forman, G. L. 1972. Comparative morphological and histochemical studies of stomachs of selected American bats. University of Kansas Science Bulletin, 159:591–729.

Francis, C. M. 1990. Trophic structure of bat communities in the understory of lowland dipterocarp rain forest in Malaysia. Journal of Tropical Ecology, 6:421–431.

Francis, C. M. 1994. Vertical stratification of fruit bats (Pteropodidae) in lowland dipterocarp rain forest in Malaysia. Journal of Tropical Ecology, 10:523–530.

Freeman, P. W. 1979. Specialized insectivory: beetle-eating and moth-eating molossid bats. Journal of Mammalogy, 60:467–479.

Freeman, P. W. 1981. Correspondence of food habits and morphology in insectivorous bats. Journal of Mammalogy, 62:166–173.

Freeman, P. W. 1984. Functional cranial analysis of large animalivorous bats (Microchiroptera). Biological Journal of the Linnean Society, 21:387–408.

Freeman, P. W. 1988. Frugivorous and animalivorous bats (Microchiroptera): dental and cranial adaptations. Biological Journal of the Linnean Society, 33:249–272.

Freeman, P. W. 1992. Canine teeth of bats (Microchiroptera): size, shape, and role in crack propagation. Biological Journal of the Linnean Society, 45:97–115.

Freeman, P. W. 1995. Nectarivorous feeding mechanisms in bats. Biological Journal of the Linnean Society, 56:439–463.

Freeman, P. W. 2000. Macroevolution in Microchiroptera: recoupling morphology and ecology with phylogeny. Evolutionary Ecology Research, 2:317–335.

Freese, C., and J. R. Oppenheimer. 1981. The capuchin monkeys, genus Cebus. Pp 331–

390 *in:* Ecology and Behavior of Neotropical Primates (A. F. Coimbra-Filho and R. A. Mittermeier, eds.). Vol. 1. Academia Brasileira de Ciências. Rio de Janiero.

Funakoshi, K., H. Wantanabe, and T. Kunisaki. 1993. Feeding ecology of the northern Ryukyu fruit bat, *Pteropus dasymallus,* in a warm-temperate region. Journal of Zoology (London), 230:221–230.

Galetti, M. and L. P. C.Morellato. 1994. Diet of the large fruit-eating bat *Artibeus literatus* in a forest fragment in Brasil. Mammalia, 58:661–665.

Gauthier-Hion, A., J.-M. Duplantier, R. Quris, F. Freer, C. Sourd, J.-P. Decoux, G. Dubost, et al. 1985. Fruit characters as a basis of fruit choice and seed dispersal in a tropical vertebrate community. Oecologia, 65:324–337.

Gentry, A. H. 1974. Coevolutionary patterns in Central American Bignoniaceae. Annals of the Missouri Botanical Garden, 69:728–759.

Goodman, S. M., and J. U. Ganzhorn. 1997. Rarity of figs (*Ficus*) on Madagascar and its relationship to a depauperate frugivore community. Revue Ecologie (Terre et Vie), 52:321–329.

Goodman, S. M., J. U. Ganzhorn, and L. Wilmé. 1997. Observation at a *Ficus* tree in Malagasy humid forest. Biotropica, 29:480–488.

Greaves, W. S. 1978. The jaw lever system in ungulates: a new model. Journal of Zoology (London), 184:271–285.

Handley, C. O., Jr., D. E. Wilson, and A. L. Gardner, eds. 1991. Demography and Natural History of the Common Fruit Bat *Artibeus jamaicensis* on Barro Colorado Island, Panama. Smithsonian Institution Press, Washington, D.C.

Heithaus, E. R., and T. H. Fleming. 1978. Foraging movements of a frugivorous bat, *Carollia perspicillata* (Phyllostomidae). Ecological Monographs, 48:127–142.

Heithaus, E. R., T. H. Fleming, and P. A. Opler. 1975. Patterns of foraging and resource utilization in seven species of bats in a seasonal tropical forest. Ecology, 56:127–143.

Herbst, L. H. 1986. The role of nitrogen from fruit pulp in the nutrition of the frugivorous bat *Carollia perspicillata.* Biotropica, 18:39–44.

Hernandez-Conrique, D., L. I. Iñiguez-Dávalos, and J. F. Storz. 1997. Selective feeding by phyllostomid fruit bats in a subtropical montane cloud forest, Biotropica, 29: 376–379.

Herrera, C. M. 1987. Vertebrate-dispersed plants of the Iberian peninsula: a study of fruit characteristics. Ecological Monographs, 57:305–331.

Herrera, L. G., T. H. Fleming, and L. S. Sternberg. 1998. Trophic relationships in a Neotropical bat community: a preliminary study using carbon and nitrogen isotope signatures. Tropical Ecology, 39:187–191.

Herring, S.W. 1975. Adaptations for gape in the hippopotamus and its relatives. Forma et Functio, 8:85–100.

Herring, S. W. 1985. Morphological correlates of masticatory patterns in peccaries and pigs. Journal of Mammalogy, 66:603–617.

Herring, S. W. 1993. Functional morphology of mammalian mastication. American Zoologist, 33:289–299.

Herring, S. W., and S. E. Herring. 1974. The superficial masseter and gape in mammals. American Naturalist, 108:561–576.

Howe, H. F. 1979. Fear and frugivory. American Naturalist, 114:925–931.

Huston, M. A. 1994. Biological Diversity—the Coexistence of Species on Changing Landscapes. Cambridge University Press, Cambridge.

Hylander, W. L. 1979a. The functional significance of primate mandibular form. American Journal of Physical Anthropology, 106:223–240.

Hylander, W. L. 1979b. Mandibular function in *Galago crassicaudatus* and *Macaca fasicularis:* an in vivo approach to stress analysis of the mandible. Journal of Morphology, 159:253–296.

Ingle, N. R. 1993. Vertical stratification of bats in a Philippine rainforest. Asia Life Sciences, 2:215–222.

Iudica, C. A. 1995. Frugivoria en murcielagos: el futero comun (*Sturnira lilium*) en las Jungas de Jujuy, Argentina. Pp. 123–128 *in:* Investigacion, Conservacion y Desarrollo en Selvas Subtropicales de Montana (A. D. Brown and H. R. Grau, eds.). Proyecto de Desarrolla Agroforestal, Laboratorio de Investigaciones Ecologicas de las Yungas, Tucuman.

Iudica, C. A., and F. J. Bonaccorso. 1997. Feeding of the bat, *Sturnira lilium,* on fruits of *Solanium riparium* influences dispersal of this pioneer tree in forests of northwestern Argentina. Studies in Neotropical Fauna and Environment, 32:1–3.

Janson, C. H. 1983. Adaptation of fruit morphology for dispersal agents in a Neotropical forest. Science, 219:187–189.

Janzen, D. H. 1979. How to be a fig. Annual Review of Ecology and Systematics, 10:13–51.

Jordano, P. 1992. Fruits and frugivory. Pp. 105–156 *in:* Seeds: The Ecology of Regeneration in Plant Communities (M. Fenner, ed.). Redwood Press, Melksham.

Juste, J. B. and J. Perez Del Val. 1995. Altitudinal variation in the subcanopy fruit bat guild in Bioko Island, Equatorial Guinea, Central Africa. Journal of Tropical Ecology, 11:141–146.

Kalko, E. K. V., and M. A. Condon. 1998. Echolocation, olfaction and fruit display: how bats find fruits of flagellichorous cucurbits. Functional Ecology, 12:364–372.

Kalko, E. K. V., and C. O. Handley, Jr. 2001. Neotropical bats in the canopy: diversity, community structure, and implications for conservation. Plant Ecology, 153:319–333.

Kalko, E. K. V., C. O. Handley, Jr., and D. Handley. 1996a. Organization, diversity, and long-term dynamics of a Neotropical bat community. Pp. 503–553 *in:* Long-Term Studies in Vertebrate Communities (M. Cody and J. Smallwood, eds.). Academic Press, Los Angeles.

Kalko, E. K. V., E. A. Herre, and C. O. Handley, Jr. 1996b. Relation of fig fruit characteristics to fruit-eating bats in the New and Old World tropics. Journal of Biogeography, 23:565–576.

Kallen, F. C., and C. Gans. 1972. Mastication in the little brown bat, *Myotis lucifugus.* Journal of Morphology, 136:385–420.

Karr, J. R. 1971. Structure of avian communities in selected Panama and Illinois habitats. Ecological Monographs, 41:207–233.

Kiltie, R. A. 1982. Bite force as a basis for niche differentiation between rain forest peccaries (*Tayassu tajacu* and *T. peccari*). Biotropica, 14:183–195.

Kinzey, W. G. 1992. Dietary and dental adaptations in the Pitheciinae. American Journal of Physical Anthropology, 88:499–514.

Kinzey, W. G., and M. A. Norconk. 1990. Hardness as a basis of fruit choice in two sympatric primates. American Journal of Physical Anthropology, 81:5–15.

Kitchener, D. J., A. Gunnell, and Maharadatunkamsi. 1990. Aspects of the feeding biology of fruit bats (Pteropodidae) on Lombok Island, Nusa Tenggara, Indonesia. Mammalia, 54:561–578.

Korine, C. A., Z. Arad, and A. Arieli, and Z. A. Amichai. 1996. Nitrogen and energy balance of the fruit bat *Rousettus aegyptiacus* on natural fruit diets. Physiological Zoology, 69:618–634.

Korine, C. A., E. K. V. Kalko, and E. A. Herre. 2000. Fruit characteristics and factors affecting fruit removal in a Panamanian community of strangler figs. Oecologia 123:560–568.

Kunz, T. H., and C. A. Diaz. 1995. Folivory in fruit-eating bats, with new evidence from *Artibeus jamaicensis* (Chiroptera: Phyllostomidae). Biotropica, 27:106–120.

Kunz, T. H., and K. A. Ingalls. 1994. Folivory in bats: an adaptation derived from frugivory. Functional Ecology, 8:665–668.

Kwiecinski, G. G., L. Krook, and W. A. Wimsatt. 1987. Annual skeletal changes in the little brown bat, *Myotis lucifugus*, with particular reference to pregnancy and lactation. American Journal of Anatomy, 178:410–420.

Laska, M. 1990. Olfactory sensitivity to food odor components in the short-tailed fruit bat *Carollia perspicillata* (Phyllostomidae, Chiroptera). Journal of Comparative Physiology, 166:395–399.

Law, B. S., and H. J. Spencer. 1995. Common blossom-bat, *Syconycteris australis*. Pp. 423–425 *in:* Mammals of Australia (R. Strahan, ed.). Smithsonian Institution Press, Washington, D.C.

Lucas, P. W. 1979. The dental-dietary adaptations of mammals. Neus Jaubuch Geologie Palaontologie 8:486–512.

Lucas, P. W. 1994. Categorization of food items relevant to oral processing. Pp. 197–218 *in:* The Digestive Systems in Mammals: Food, Form and Function (D. J. Chivers and P. Langer, eds.). Cambridge University Press, Cambridge.

Lucas, P. W., and D. A. Luke. 1984. Chewing it over: basic principles of food breakdown. Pp. 283–301 *in:* Food Acquisition and Processing in Primates (D. J. Chivers, B. A. Wood, and A. Bilsborough, eds.). Plenum Press, New York.

Marshall, A. G. 1983. Bats, flowers and fruit: evolutionary relationships in the Old World. Biological Journal of the Linnean Society, 20:115–135.

Martinez del Rio, C. 1994. Nutritional ecology of fruit-eating and flower visiting birds and bats. Pp. 103–127 *in:* The Digestive Systems in Mammals: Food, Form and Function (D. J. Chivers and P. Langer, eds.). Cambridge University Press, Cambridge.

Martinez del Rio, C., and C. Restrepo. 1993. Ecological and behavioral consequences of digestion in frugivorous animals. Vegetatio, 197/108:205–216.

Mattson, W. J. 1980. Herbivory in relation to plant nitrogen content. Annual Review of Ecology and Systematics, 11:119–161.

McKenzie, N. L., A. C. Gunnell, M. Yani, and M. R. Williams. 1995. Correspondence between flight morphology and foraging ecology in some Palaeotropical bats. Australian Journal of Zoology, 43:241–257.

Mickleburgh, S. P., A. M. Hutson, and P. A. Racey, eds. 1992. Old World Fruit Bats: An Action Plan for Their Conservation. International Union for Conservation of Nature and Natural Resources, World Conservation Union, Gland.

Moermond T. C. and J. S. Denslow. 1985. Neotropical frugivores: patterns of behavior, morphology and nutrition with consequences for fruit selection. Pp. 865–897 *in:* Neotropical Ornithology (P. A. Buckley, M. S. Foster, E. S. Morton, R. S. Ridgely and N. G. Buckley, eds.). Ornithological Monographs, 36. American Ornithologists' Union, Washington, D.C.

Moermond, T. C., J. S. Denslow, D. J. Levey, and E. Santana. 1986. The influence of morphology on fruit choice in Neotropical birds. Pp. 137–146 *in:* Frugivores and Seed Dispersal (A. Estrada and T. H. Fleming, eds.). Dr. W. Junk Publishers, Dordrecht.

Morrison, D. W. 1978. Foraging ecology and energetics of the frugivorous bat *Artibeus jamaicensis.* Ecology, 59:716–723.

Morrison, D. W. 1980. Efficiency of food utilization by fruit bats. Oecologia, 45: 270–273.

Nagato, T., B. Tandler, and C. J. Phillips. 1998. An unusual parotid gland in the tent-building bat, *Uroderma bilobatum:* possible correlation of interspecific ultrastructural differences with differences in salivary pH and buffering capacity. Anatomical Record, 252:290–300.

Nelson, S. L., M. A. Miller, E. J. Heske, and E. J. Fahey. 2000a. Nutritional consequences of a change in diet from native to agricultural fruits for the Samoan fruit bat. Ecography, 23:393–401.

Nelson, S. L., M. A. Miller, E. J. Heske, and E. J. Fahey. 2000b. Nutritional qualities of leaves and unripe fruit consumed as famine foods by the flying foxes of Samoa. Pacific Science, 54:301–311.

Norberg, U. M. 1994. Wing design, flight morphology, and habitat use in bats. Pp. 205–239 *in:* Ecological Morphology (P.C. Wainwright and S.M. Reilly, eds.). University of Chicago Press, Chicago.

Norberg, U. M. 1998. Morphological adaptations for flight in bats. Pp. 93–108 *in:* Bat Biology and Conservation (T. H. Kunz and P. A. Racey, eds.). Smithsonian Institution Press, Washington, D.C.

Norberg, U. M., and J. M. V. Rayner. 1987. Ecological morphology and flight in bats: wing adaptations, flight performance, foraging strategy, and echolocation. Philosophical Transactions of the Royal Society of London B, 316:335–427.

Norconk, M. A., C. Wertis, and W. G. Kinzey. 1997. Seed predation by monkeys and macaws in Eastern Venezuela: preliminary findings. Primates, 38:177–184.

Nordin, M., and V. H. Frankel. 1989. Biomechanics of bone. Pp. 3–29 *in:* Basic Biomechanics of the Musculoskeletal System (M. Nordin and V. H. Frankel, eds.). Lea & Febiger, Philadelphia.

O'Brien, T. G., M. Kinnaird, E. S. Dierenfeld, N. L. Concklin-Brittain, R. W. Wrangham, and S. C. Silver. 1998. What's so special about figs? Nature, 392:668.

Oldfield, A. C., C. R. Tideman, and A. P. Robinson. 1993. Olfactory discrimination in the Australian flying foxes, *Pteropus poliocephalus* and *P. scapulatus.* Bat Research News, 34:33.

Palmer, C., O. Price, and C. Bach. 2000. Foraging ecology of the black flying fox (*Pteropus alecto*) in the seasonal tropics of the Northern Territory, Australia. Wildlife Research, 27:169–178.

Radinsky, L. B. 1981a. Evolution of skull shape in carnivores, 1. Representative modern carnivores. Biological Journal of the Linnean Society, 15:369–388.

Radinsky, L. B. 1981b. Evolution of skull shape in carnivores, 2. Additional modern carnivores. Biological Journal of the Linnean Society, 16:337–355.

Reduker, D. W. 1983. Functional analysis of the masticatory apparatus in two species of *Myotis*. Journal of Mammalogy, 64:277–286.

Reiger, J. F., and E. M. Jakob. 1988. The use of olfaction in food location by frugivorous bats. Biotropica, 20:161–164.

Richards, G. C. 1986. Notes on the natural history of the Queensland tube-nosed bat, *Nyctimene robinsoni*. Macroderma, 2:64–67.

Richards, G. C. 1995. A review of ecological interactions of fruit bats in Australian ecosystems. Symposia of the Zoological Society of London, 67:79–96.

Ruby, J., P. T. Nathan, J. Balasingh, and T. H. Kunz. 2000. Chemical composition of fruits and leaves eaten by short-nosed fruit bat, *Cynopterus sphinx*. Journal of Chemical Ecology, 26:2825–2841.

Schulze, M. D., N. E. Seavy, and D. F. Whitacre. 2000. A comparison of the phyllostomid bat assemblages in undisturbed Neotropical forest and in forest fragments of a slash-and-burn farming mosaic in Petén, Guatemala. Biotropica, 32:174–184.

Simmons, N. B., and J. H. Geisler. 1998. Phylogenetic relationships of *Icaronycteris*, *Archaeonycteris*, *Hassionycteris*, and Palaeochiropteryx to extant bat lineages, with comments on the evolution of echolocation and foraging strategies in Microchiroptera. Bulletin of the American Museum of Natural History, 235:1–182.

Simmons, N. B., and R. S. Voss. 1998. The mammals of Paracou, French Guiana: a Neotropical lowland rainforest fauna. Pt. 1. Bulletin of the American Museum of Natural History, 237:1–219.

Spencer, H. J., and T. H. Fleming. 1989. Roosting and foraging behaviour of the Queensland tube-nosed bat, *Nyctimene robinsoni* (Pteropodidae): preliminary radio-tracking observations. Australian Wildlife Research, 16:413–420.

Strait, S. G. 1993. Molar morphology and food texture among small-bodied insectivorous mammals. Journal of Mammalogy, 74:391–402.

Strait, S. G. 1997. Tooth use and the physical properties of food. Evolutionary Anthropology, 5:199–211.

Strait, S. G., and D. J. Overdorff. 1996. Physical properties of fruits eaten by Malagasy primates. American Journal of Physical Anthropology, 22:224.

Tabachnick, B. G., and L. S. Fidell. 1996. Using Multivariate Statistics. 2d ed. Harper & Row, New York.

Tan, K. I I., A. Zubaid, and T. I I. Kunz. 1998. Food habits of *Cynopterus brachyotis* (Muller) (Chiroptera: Pteropodidae) in Peninsular Malaysia. Journal of Tropical Ecology, 14:299–307.

Tandler, B., T. Nagato, and C. J. Phillips. 1997. Ultrastructure of the parotid salivary glands in seven species of fruit bats in the genus *Artibeus*. Anatomical Record, 248:176–188.

Tedman, R. A., and L. S. Hall, 1985. The morphology of the gastrointestinal tract and food transit times in the fruit bats *Pteropus alecto* and *P. poliocephalus* (Megachiroptera). Australian Journal of Zoology, 33:625–640.

Terborgh, J. 1986a. Community aspects of frugivory in tropical forests. Pp 371–384 *in*: Frugivores and Seed Dispersal (A. Estrada and T. H. Fleming eds.). Dr. W. Junk Publishers, Dordrecht.

Terborgh J. 1986b. Keystone plant resources in the tropical forest. Pp. 330–344 *in*: Conservation Biology (M. E. Soulé, ed.). Academic Press, New York.

Terborgh, J., E. Losos, P. Riley, and M. Balanos Riley. 1993. Predation by vertebrates and invertebrates on the seeds of five canopy tree species of an Amazonian forest. Vegetatio, 107/108:375–386.

Thies, W., E. K. V. Kalko, and H. U. Schnitzler. 1998. The roles of echolocation and olfaction in two Neotropical fruit-eating bats, *Carollia perspicillata* and *C. castanae,* feeding on *Piper.* Behavioral Ecology and Sociobiology, 42:397–409.

Thomas, D. W. 1984. Fruit intake and energy budgets of frugivorous bats. Physiological Zoology, 57:457–467.

Thomason, J. J., A. P. Russell, and M. Morgeli. 1990. Forces of biting, body size and masticatory muscle tension in the opossum *Didelphis virginiana.* Canadian Journal of Zoology, 68:318–324.

Toft, C. A. 1980. Feeding ecology of thirteen syntopic species of anurans in a seasonal tropical environment. Oecologia, 45:131–141.

Ungar, P. S. 1995. Fruit preferences of four sympatric primate species at Ketambe, Northern Sumatra, Indonesia. International Journal of Primatology, 16:221–245.

Utzurrum, R. C. B. 1995. Feeding ecology of Philippine fruit bats: patterns of resource use and seed dispersal. Symposia of the Zoological Society of London, no. 67: 63–77.

Utzurrum, R. B., and P. D. Heidemen. 1991. Differential ingestion of viable vs. non-viable *Ficus* seeds by fruit bats. Biotropica, 23:311–312.

Van Der Pijl, L. 1957. The dispersal of plants by bats (Chiropterochory). Acta Botanica Neerlandia, 6:291–315.

Van Roosmalen, M. G. M. 1984. Subcategorizing foods in primates. Pp. 167–175 *in:* Food Acquisition and Processing in Primates (D. J. Chivers, B. A. Wood, and A. Bilsborough, eds.). Plenum Press, New York.

Van Valkenburgh, B. 1996. Feeding behavior in free-ranging, large African carnivores. Journal of Mammalogy, 77:240–254.

Wainwright, P. C. 1987. Biomechanical limits to ecological performance: mollusc-crushing by the Caribbean hogfish, *Lachnolaimus maximus* (Labridae). Journal of Zoology (London), 213:283–297.

Wainwright, P. C. 1991. Ecomorphology: experimental functional anatomy for ecological problems. American Zoologist, 31:680–693.

Wainwright, P. C. 1996. Ecological explanation through functional morphology: the feeding biology of sunfishes. Ecology, 77:1336–1343.

Wainwright, P. C., and S. M. Reilly. 1994. Ecological Morphology: Integrative Organismal Biology. University of Chicago Press, Chicago.

Wall, C. E. 1995. Form and function of the temporomandibular joint in anthropoid primates. Ph.D. Dissertation. State University of New York, Stony Brook.

Wendeln, M. C., J. R. Rankle, and E. K. V. Kalko. 2000. Nutritional values of 14 species of figs and bat feeding preferences in Panama. Biotropica, 32:489–501.

Wheelwright, N. T. 1985. Fruit size, gape width and the diets of fruit-eating birds. Ecology, 66:808–818.

Wheelwright, N. T., and C. H. Janson. 1985. Colors of fruit displays of bird-dispersed plants in two tropical forests. American Naturalist, 126:777–799.

Willig, M. R., G. R. Camino, and S. J. Noble. 1993. Dietary overlap in frugivorous and insectivorous bats from edaphic cerrado habitats of Brazil. Journal of Mammalogy, 74: 117–128.

Willson, M. F., A. K. Irvine, and N. G. Walsh. 1989. Vertebrate dispersal syndromes in some Australian and New Zealand plant communities, with geographic comparisons. Biotropica, 21:133–147.

Wilson, D. M., and D. E. Reeder. 1993. Mammal Species of the World: A Taxonomic and Geographic Reference. Smithsonian Institution Press, Washington, D.C.

Wilson, D. M., C. F. Ascorra, and S. Solari. 1996. Bats as indicators of habitat disturbance. Pp. 613–625 *in:* Manu: The Biodiversity of Southern Peru (D. E. Wilson and A. Sandoval, eds.). Smithsonian Institution Press, Washington, D.C.

Winkelmann, J. R., F. J. Bonaccorso, and T. L. Strickler. 2000. Home range of the southern blossom bat, *Syconycteris australis,* in Papua New Guinea. Journal of Mammalogy, 81:408–418.

Yamashita, N. 1996. Seasonality and site specificity of mechanical dietary patterns in two Malagasy lemur families (Lemuridae and Indriidae). International Journal of Primatology, 17:355–387.

CHAPTER 10

Physiological Ecology and Energetics of Bats

John R. Speakman and Donald W. Thomas

Introduction

No two factors other than body temperature and body size have a more profound impact on how animals function. Temperature affects the rate of all metabolic processes and hence how rapidly nerves conduct, muscles contract, enzymes attack substrates, nutrients are absorbed, and tissues are built. As a consequence, temperature affects how animals perceive and react to predators or prey, how fast they can capture prey to digest and extract needed nutrients, and how rapidly they grow and reproduce. The effect of temperature on all of these processes is exponential, so that even a small increment in body temperature (T_b) can have an important effect. It is no wonder that homeothermic birds and mammals, and many poikilotherms, seek to maintain high T_b through the production of metabolic heat or the absorption of solar radiation.

Although high T_b confers advantages, it also entails a cost. For homeotherms, maintaining a high and stable T_b requires a constant allocation of energy that provides no direct material benefit. Sometimes this energetic cost of homeothermy is small compared to the benefits. At other times the cost is prohibitive, so that animals may be forced to abandon strict homeothermy to bring energy expenditures into line, either with rates at which they are able to harvest energy or by varying the size of their energy reserves.

Size affects the amount of food required to build bodies and sustain cellular integrity: large animals obviously require more food than do small animals. However, the importance of size in biology lies not just in its effect on the absolute requirements of energy and nutrients but, even more so, in its all-pervading effect on scaling. As size increases, volume and mass change more rapidly than does area. This change in proportions with increasing or decreasing body size, termed "allometry," dictates many patterns of body structure, metabolic rate, and heat flow. For example, large animals must allocate proportionally more of their ingested nutrients to building bone and support structures than small animals do because body mass increases more rapidly than the cross-sectional area of bone and, hence, strength. Due to the allometry of bone strength and body mass (M_b), a greater proportion of M_b is tied up in metabolically inactive bone. This is one of the reasons why large animals require fewer nutrients and metabolize energy at a slower rate on a mass specific basis than small animals do. Heat is lost to the surrounding environment through the body surface, while the capacity to produce heat is a function of the mass of organs and muscle tissue. Consequently, big animals are better equipped to deal with low ambient temperatures (T_a) than are small animals because they offer smaller surface areas through which heat is lost, relative to their capacity to produce it.

The cost-benefit trade-off of maintaining a high T_b dictates thermoregulatory patterns and energetic strategies and thus underpins much of physiological ecology. Bats, however, are particularly interesting in the context of physiological ecology for several of reasons. They are extremely thermolabile and exhibit a wide diversity of patterns in temperature regulation. Although bats can be classed as small, body masses span almost three orders of magnitude, ranging from less than 2 g for *Crasseonycteris thonglonglyai* to more than 1,000 g for the largest *Pteropus* species. Bats are well represented in almost all climatic zones and so adapt to an impressive array of environmental temperature and rainfall conditions. They exhibit a remarkable diversity of feeding and foraging strategies and include frugivores, nectarivores, carnivores, piscivores, sanguinivores, and aerial and gleaning insectivores. And, of course, because bats fly, this imposes specific constraints on locomotory costs, exploitable prey abundances, and fat storage. Together, these and other features explain the long-standing interest in bat evolutionary physiological ecology.

In this chapter, we begin by explaining the fundamental principles governing heat flow and temperature regulation. We then examine the energetic implications of different thermoregulatory strategies and metabolic rates and of different life-history traits of bats, including diet, torpor and hibernation, flight as a mode of locomotion, and reproductive investment. Our purpose is to expose underlying principles and show how they help explain the patterns that emerge so vividly within the Chiroptera.

A Primer on Heat Flow, Metabolic Rate, and the Regulation of Body Temperature

Euthermy or the Defense of an Elevated Body Temperature

The first law of thermodynamics states that heat flows down a temperature gradient ($\Delta T°$) from warm to cold. Consequently, all homeotherms are confronted with a problem of heat, and hence energy, management. The rate of heat flux is dominated by three factors: the size of $\Delta T°$, the resistance that any barriers offer to heat flow, and the surface area across which heat is exchanged. The only way to maintain a stable "high" T_b in the face of "low" T_a is to produce heat at the same rate that it is being lost.

Let's first consider the three factors that determine heat flux sequentially. Under still conditions, the rate of heat flow between two reference points, in our case the body core and a point just outside the layer of fur, is a linear function of $\Delta T°$ (fig. 10.1A). An animal that maintains T_b at 37°C when T_a is 17°C (a 20°C gradient) has to produce heat at twice the rate that it would if T_a were 27°C (a 10°C gradient). Thermodynamic theory predicts that $\Delta T°$ determines heat flux and not the precise values at the two reference points. Thus, in theory, an animal should have to produce heat at the same rate whether it regulates T_b at 37°C against a 10°C gradient ($T_a = 27°C$) or if it regulates T_b at 27°C against a 10°C gradient ($T_a = 17°C$).

The effect of $\Delta T°$ allows one to compare the relative heat flux under different temperature conditions, but it does not allow calculation of the absolute rate of heat flux. To know the absolute rate of heat flow out from the body core, one needs to know the resistance to heat as it flows out from the body core. The absolute rate of heat flow (H) is determined by the slope of the line in figure 1A. Biologists refer to this as the thermal conductance (C, measured as watts °C^{-1} or mL O_2 °C^{-1}) and think of it as the ease with which heat flows between the body core and the surrounding air. Objects and animals that have a steep slope, indicating a high thermal conductance, lose heat more readily than those having a lower thermal conductance. For inanimate objects, thermal conductance is constant, so the relationship between heat flow and $\Delta T°$ is linear, with an origin zero heat flow and zero $\Delta T°$ (or $T_a = T_b$). Animals, however, may adjust thermal conductance or depress T_b as T_a declines, which has the effect of curving the line relating heat flux to $\Delta T°$ or T_a (fig. 1B). In this case, a linear regression may not project back through a $\Delta T°$ of zero, thus apparently contravening thermodynamic laws. McNab (1980b) discusses the biological implications of varying thermal conductance and T_b in detail.

While the notion of thermal conductance may be intuitively satisfying, it masks the underlying processes that limit the flow of heat away from an animal's core. It is conceptually easier to think in terms of resistance to heat flow, where mathematically the resistance is the inverse of thermal conductance (R: equivalent to C^{-1}), which is expressed as (W °C^{-1})$^{-1}$. To visualize the flow of heat from an animal's core to the surrounding air, imagine that heat is produced at a single point within the core. Heat flows first through the tissues to the skin surface, carried partly by convective blood flow through the arterial and capillary network and partly by conduction through the cells and tissues themselves. Heat then flows through the fur barrier primarily by conduction through the hairs themselves, but also by conduction through the trapped air and by small convection eddies within the fur, and finally dissipates from the thin boundary air layer overlying the outer surface of the fur by convective air movement and net thermal radiation losses. The resistance that each layer offers to the flow of heat is determined by both the type of heat transport and the path length along which heat must travel. Heat is transferred more rapidly by convection and radiation than by conduction through tissues or air, and a thick fur layer offers more resistance to heat flow than a thin layer. Because each layer offers resistance to heat flow, temperature drops progressively be-

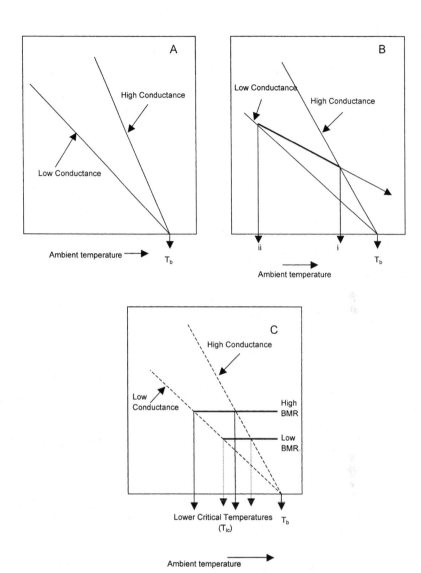

Figure 10.1. Theoretical patterns of resting metabolic rate (MR) (and hence energy expenditure) as a function of ambient temperature (T_a) for an animal regulating its body temperature (T_b) at euthermic levels. *A*, The theoretical pattern of resting MR as a function of T_a. The slope of the lines relating resting MR to T_a is the thermal conductance (C). For any given thermal gradient ($\Delta T°$) or T_a, resting MR can be predicted by MR = $C \times \Delta T°$. When conductance is constant this line extrapolates to T_b with steep slopes reflecting high conductance and shallow slopes low conductance. *B*, The pattern of resting MR as a function of T_a (*bold line*) when thermal conductance decreases with declining T_a. Note that the regression line indicating the average thermal conductance does not extrapolate to $T_a =$ T_b. *C*, The horizontal lines indicate two different levels of basal metabolic rate (BMR). High BMR results in a decrease in lower critical temperature (T_{lc}) and a decrease in the upper critical temperature (T_{uc}) but a wider thermoneutral zone.

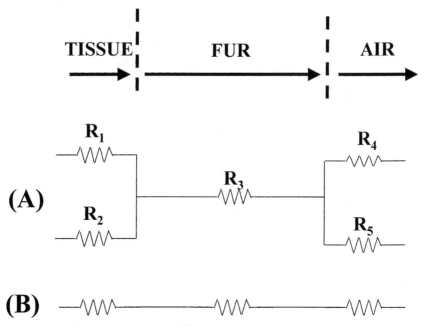

Figure 10.2. A schematic diagram showing the path for heat flux between the body core and the surrounding air. A, Heat flux is limited by resistances R_1–R_5 in series and in parallel. R_1 and R_2 represent conduction and convection through the tissues, R_3 conduction through the fur, and R_4 and R_5 convection and radiation away from the fur surface. B, Resistors in parallel can be replaced by equivalent resistors (e.g., $R_{equiv} = [(R_1 \times R_2)/(R_1 + R_2)]$). The total resistance to heat flux is set by the sum of the equivalent resistances in series.

tween the core and the skin, through the fur, and finally into the surrounding air. How much the temperature drops across a given layer is determined by the proportional contribution of that layer to the total resistance. A thick and efficient fur layer, which offers a high resistance to heat flow, will exhibit a large $\Delta T°$ between its inner and outer surfaces, and the temperature at the hair tips will approach T_a.

It is convenient to think of heat as flowing through a number of resistances that may be arranged in parallel within a given layer and in series between layers, as shown in figure 10.2. The value of this approach is that it allows us to apply the same formulas that predict current and voltage in electrical circuits: $I = V/R_{total}$ where I is the current, V is the voltage potential and R_{total} is the circuit resistance. When thinking in terms of heat rather than electricity, we simply substitute heat flow (H) for I, temperature gradient ($\Delta T°$) for V, and the sum of the resistances in the different layers for R_{total}. Thus, the overall flow of heat away from the core is calculated by $H = \Delta T°/R_{total}$, and since $R = 1/C$, this is the same as $H = \Delta T° \times C$ in figure 10.1A.

There are several benefits to thinking in terms of resistances and $\Delta T°$. One is that the overall resistance, R_{total}, can simply be found from the sum of the equivalent resistances for each layer, remembering that an equivalent resistance refers to a single resistance that could replace two or more resistances in parallel. A second is that the resistance offered by a given layer, x, can be calculated from the heat flow and the temperature drop across that layer if the total heat flow is known ($R_x = \Delta T_x°/H$). A third is that the relative importance of each layer in limiting heat flow is easy to visualize. Just as the voltage drop (ΔV_x) across a single resistor of a series can be predicted by the ratio of that resistor (R_x) to the total circuit resistance ($\Delta V_x = (R_x/R_{total}) \times V$), the temperature drop across a given layer is determined by the proportional contribution of the resistance in that layer to the total resistance offered by all layers ($\Delta T_x° = (R_x/R_{total}) \times \Delta T°$).

Let's consider two practical examples. First, consider the temperature drop between the core and the skin surface. When peripheral blood flow is high, much heat is transported to the skin surface and the ratio of tissue resistance to the total resistance (R_{tissue}/R_{total}) is small. As a result, there is only a small temperature difference between core and skin, and we see the skin as "hot." Because the skin is "hot," $\Delta T°$ between the base of the fur and its external surface is large, driving heat flow through the fur. As T_a declines and animals seek to conserve heat by reducing peripheral circulation, convective heat transport to the skin declines toward zero. Most of the tissue resistance is now determined by the slower conduction of heat through the tissue matrix. We see the skin as "cool" due to this increase in tissue resistance, which results in a higher R_{tissue}/R_{total} ratio. As a consequence, less heat is delivered to the base of the fur, $\Delta T°$ through the fur is smaller, and heat flow through the fur is reduced.

The ability to depress skin temperature is affected by body size. By definition, small animals have short path lengths between the organs (where much of body heat is produced) and the skin surface, while large animals necessarily have longer path lengths. Thus, small animals have relatively low tissue resistance and hence high skin temperatures compared with large animals. This is the reason why temperature-sensitive radio transmitters glued to the skin surface accurately predict T_b for small animals (Audet and Thomas 1997; Barclay et al. 1996) but should not do so for large animals. This is also one reason why small animals tend to have higher mass-specific thermal conductances than large animals do.

Next, let's consider whether fat can serve as an effective insulation for small terrestrial animals thermoregulating in air. It is well known that aquatic seals and whales rely on a thick layer of subcutaneous blubber as insulation, but terrestrial hibernating mammals also lay down substantial deposits of subcutaneous fat. Although fat serves primarily for energy storage, by analogy with aquatic mammals, fat might also have a secondary role as an insulation layer.

To analyze this problem, imagine a small lean animal with a tissue path length of 4 mm and a fur thickness of 6 mm and then consider what effect the deposition of a 1-mm thick subcutaneous fat deposit would have on heat flow. From Schmidt-Nielsen's (1990) table 7.3 (and converting thermal conductance to its inverse, R), we find that muscle tissue, fat, and fur offer resistances of about 21.7, 46.9, and 262.9 $(W \, °C^{-1} \, mm^{-1})^{-1}$, respectively. For a lean animal having no fat deposit, the tissue resistance is 4 mm $\times$ 21.7 $(W \, °C^{-1} \, mm^{-1})^{-1}$, the fur resistance is 6 mm $\times$ 262.9 $(W \, °C^{-1} \, mm^{-1})^{-1}$, and the overall resistance is the sum of the two or 1,664.2 $(W \, °C^{-1})^{-1}$. Adding a 1-mm subcutaneous layer of fat would increase the overall resistance by 1 mm $\times$ 46.9 $(W \, °C^{-1} \, mm^{-1})^{-1}$, raising the overall resistance to 1,711.1 $(W \, °C^{-1})^{-1}$, only 2.8% higher than the lean condition. Hence, 1 mm of fat is not an effective barrier to heat loss for small terrestrial animals.

Surface area also affects heat loss. Thus, large animals lose more heat in absolute terms than do small animals. However, due to the effects of allometry, small animals lose more heat per unit volume or unit mass than do large animals. One can appreciate the effect of even small changes in body size on mass-specific heat flux by simply comparing the surface areas and volumes of small spherical "animals." A sphere of 1 cm radius has a surface area of 12.6 cm^2, and a volume of 4.2 cm^3, giving a surface area to volume ratio of 3.0 (= 12.6/4.2). In contrast, a 3-cm radius sphere has a surface area of 113.1 cm^2 and a volume of 113.1 cm^3, for a surface area to volume ratio of 1.0. The larger animal has a surface area and hence a total heat flux approximately nine times greater than the smaller animal. However, if both animals have the same fur thickness and face the same T_a, then the smaller one will have to produce heat at 3.3 times the rate (on a mass-specific basis) of the larger animal if it is to maintain T_b constant.

Because heat flow (and hence the metabolic cost of homeothermy) is determined by both the thermal conductance and $\Delta T°$, animals can regulate their metabolic cost to some degree by both behavioral and physiological adjustments. Vasoconstriction and piloerection, the selection of areas sheltered from the convective effect of wind, and clustering all act to increase the thermal resistance of tissue, fur, and boundary air layers and so determine the rate of heat loss for a given $\Delta T°$. By selecting high T_a, animals can directly reduce $\Delta T°$ and again benefit from a reduced metabolic cost. As T_a increases toward T_b, $\Delta T°$ decreases and the thermoregulatory cost declines. Resting metabolic rate (MR), however, does not fall off to zero but, rather, stabilizes at a plateau that spans a limited range of T_a, called the thermoneutral zone (fig. 10.1C). The lower inflection point, termed the "lower critical temperature" (T_{lc}) defines the point at which resting MR is at its lowest, and a further reduction in $\Delta T°$ has no effect. This minimal resting MR, variously referred to as basal, resting, or standard MR depending on the exact criteria used to define it, represents the summed costs of maintaining organismal integrity or homeostasis. We will refer to resting MR within the thermoneutral zone as BMR. The level at which BMR is set for a given individual or species determines not only the lowest rate of energy expenditure but also the position of T_{lc} and the width of the thermoneutral zone (fig. 10.1C).

Regardless of the underlying processes that set the level of BMR, the energy produced ultimately dissipates as heat within the tissues. As T_a approaches T_b and $\Delta T°$ falls toward zero, it becomes progressively more difficult to unload heat through conduction, convection, and radiation. Although animals can rely on vasodilation to pump heat to the skin surface, flatten their fur to increase thermal conductance, and facilitate the passive conduction of heat out to the boundary air layer, there comes a point where they must actively facilitate heat loss. This transition from passive to active regulation of heat flow generates an upper inflection point in the metabolic curve, the upper critical temperature (T_{uc}). At T_{uc} and above, animals must increase either convection (e.g., fanning) or evaporative water loss. Because a stable T_b can only be achieved when heat production and loss are balanced, in the absence of any $\Delta T°$ (i.e., when $T_a = T_b$) or a negative $\Delta T°$ (i.e., when $T_a > T_b$), animals must actively regulate evaporative water loss by panting, sweating, or salivating. In hot, dry climates where water is limited, there may be a cost associated with a high BMR in terms of its implications for the management of body water.

When coupled with T_b, the two primary reference points in figure 10.1C, namely, BMR and thermal conductance, set the metabolic response of a resting homeotherm to a wide range of T_a. It is important to note that T_b, BMR, and thermal conductance are free to vary independently of each other. This can be a source of confusion since it is often noted that animals with low BMR also have low T_b (e.g., marsupials). However, this apparent link is a biological phenomenon and not an intrinsic aspect of the physical laws governing heat flow. Because BMR, T_b, and thermal conductance describe the metabolic responses of homeotherms, considerable attention has been paid to how these parameters vary with body size, taxonomic affiliation, and ecology (Elgar and Harvey 1987; Lasiewski and Dawson 1967; McNab 1980a, 1982, 1986, 1987, 1988, 1992; Reynolds and Lee 1996). We will examine these relations for bats in some detail later.

Heterothermy or the Reversible Depression of Body Temperature

Homeothermy implies the defense of a stable and elevated T_b and requires the production of metabolic heat to offset heat loss. When energy supply is limited the cost of homeothermy may be prohibitive. To escape this energy constraint, some mammals (well exemplified by the Chiroptera) can regulate T_b at a set point well below the normal euthermic levels of 30°–40°C. This is characterized by a complete shift in the thermoregulatory response curve to lower temperatures and lower MR (Geiser 1988). Animals that regulate T_b at 20° and 5°C rather than at 35°C would have response curves similar to those illustrated in

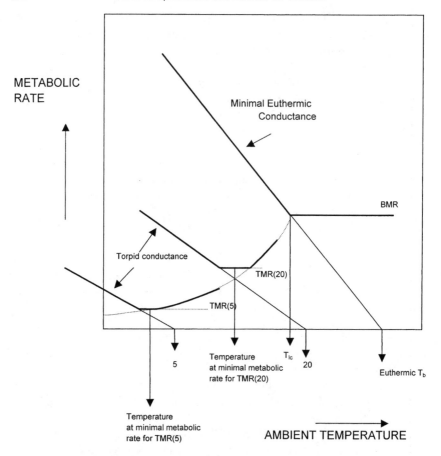

METABOLIC
RATE

Minimal Euthermic
Conductance

BMR

Torpid conductance

TMR(20)

TMR(5)

5

Temperature
at minimal metabolic
rate for TMR(20)

T_{lc}

20

Euthermic T_b

Temperature
at minimal metabolic
rate for TMR(5)

AMBIENT TEMPERATURE

Figure 10.3. Theoretical patterns of energy expenditure as functions of ambient temperature in animals regulating their body temperatures at euthermic temperatures (i.e., at BMR), 20° and 5°C. In all cases the patterns show increases in energy expenditure as ambient temperature exceeds or declines significantly below the regulated temperature. The level of expenditure at the base of the curves is dependent on the regulated temperature, being lower when body temperature is regulated at a lower level. TMR = torpid metabolic rate.

figure 10.3. When T_a is above the T_b set point (generally called $T_{b\text{-min}}$), animals appear to have abandoned homeothermy completely because T_b fluctuates more or less in parallel with T_a. This hypometabolic state or torpor, however, still involves the precise regulation of T_b, albeit at a lower level. As T_a drops to the set point, $T_{b\text{-min}}$, torpor MR (TMR) declines to its lowest level (termed "TMR$_{\text{min}}$"). But as T_a continues to fall below $T_{b\text{-min}}$, animals increase MR to defend a stable T_b, and they exhibit the same thermoregulatory characteristics as homeotherms. As with euthermic animals, the slope of the line relating MR to

T_a projects back through $T_{b\text{-min}}$, the regulated set point. The slope of this line, the thermal conductance in torpor, however, is lower than euthermic thermal conductance, probably because of changes in peripheral blood circulation.

Because $\Delta T°$ is not defended when T_a is above $T_{b\text{-min}}$, by definition thermoregulatory cost should be zero. The only metabolic cost is that determined by the direct effects of temperature on metabolic rate. As T_b falls from the euthermic level, the rate of all metabolic processes also declines exponentially, giving rise to the Q_{10} effect. A 10°C drop in T_b results in roughly a halving in metabolic rate ($Q_{10} = 2$), which is readily observable as a decline in oxygen consumption or CO_2 production. There has been considerable debate over factors that contribute to the reduced MR of animals when they are in torpor (reviewed in Geiser 1988; Heldmaier et al. 1993). Three factors may act in concert to depress MR in torpor. First, the animal abandons thermogenesis and down-regulates metabolism to the level set by BMR. This "down-regulation" reduces T_b because heat loss exceeds production. The reduction in T_b has a direct effect on MR due to the Q_{10} effect, decreasing MR by a factor of two to three for each 10°C drop in T_b (Geiser 1988; Heldmaier and Ruf 1992; Heldmaier et al. 1993; Snyder and Nestler 1990). Finally, MR may be further depressed by the altered acid-base status due to CO_2 retention (Malan 1980, 1986, 1988). Regardless of the relative importance of each of these effects, the combined effect is a marked reduction in MR in torpor. Hock (1951) and more recently Thomas et al. (1990a), for example, showed that MR for *Myotis lucifugus* at T_b of 5°C was reduced to roughly 1% of the euthermic level.

The ability to regulate T_b at a lower set point without lethal consequences, such as loss of contractile properties in heart muscle or the ability to regulate intra- and extracellular ion gradients, opens the possibility for intermediate homeothermic strategies, whereby T_b (and resting MR) is regulated at a lower set point, but above T_a, with the consequent effect of reducing, but not eliminating, $\Delta T°$. It is worth questioning whether the depression of T_b by, say, 10°C is equivalent to raising T_a by the same 10°C, thus maintaining an equivalent $\Delta T°$. The laws of thermodynamics state that heat flow is determined by $\Delta T°$ and thermal conductance, so varying either T_b or T_a should result in similar heat flows. However, lowering T_b depresses the oxygen requirement of peripheral tissues and thus permits a reduction in heart rate and peripheral blood flow without the consequent risk of anoxia. A reduction in peripheral blood flow acts to reduce convective heat transfer through the tissues, increasing tissue resistance and decreasing heat flux for a given $\Delta T°$. For this reason, "torpid" thermal conductance is lower than "euthermic" thermal conductance (fig. 10.3). Lowering T_b reduces the cost of homeothermy more than would be anticipated based on the reduction in $\Delta T°$ alone. For *M. lucifugus*, Studier (1981) showed that a 4°C depression in T_b from 37°C to 33°C resulted in a reduction in MR of 13.0 mW g^{-1}, whereas a further depression of 4°C from

$33\,^{\circ}C$ to $29\,^{\circ}C$ decreased MR by only $9.5\ \text{mW g}^{-1}$. The consequence of this non-linearity is that slight depressions of T_b (shallow torpor) result in dispropor-tionate savings (Studier 1981; Webb et al. 1993).

Resting Energy Expenditure of Euthermic Bats

Patterns in Basal Metabolic Rate with the Chiroptera: Diet, Phylogeny, or Environmental Temperature

The first analysis of the patterns of BMR within the Chiroptera was presented by McNab (1982), but since then a number of studies have provided new data allowing us to reexamine these trends and question the factors that are asso-ciated with variation in BMR. We compiled measures of BMR for 80 species of bats published between 1964 and 1999. Dr. A. Cruz-Neto (Sao Paulo State Uni-versity, Rio Claro) kindly provided data for an additional four species bring-ing the total to 84, or nearly 8% of known bats species (table 10.1). Where mul-tiple measurements were available, we averaged the measures to generate a single estimate per species. The data included representatives from 11 of the 19 currently recognized bat families (Simmons 1998), species that specialize on five diet types (insects, nectar, fruit, blood, and other vertebrate animals) and are distributed from the equator to $58\,^{\circ}$ N. No data were found for the Rhino-pomatidae, Craseonycteridae, Nycteridae, Mystacinidae, Myzapodidae, Thy-ropteridae, and Furipteridae, and studies on the energetics of representatives of these groups are sorely needed.

The major factor affecting BMR in bats is M_b (fig. 10.4A). The least squares regression

$$\log_e \text{BMR (mL O}_2\text{ h}^{-1}) = 1.0895 + 0.744 \log_e M_b\text{ (g)} \qquad (1)$$

explained 92.4% of the variation in BMR between species ($F = 989.93$, df $= 1$, 82, $P < 0.001$). The slope of this relationship is not significantly different from the slope derived by Kleiber (1961) over a much wider range of M_b for mam-mals (0.756). Despite the strong effect of M_b on BMR, substantial residual vari-ation exists. Previous studies have suggested that insectivorous bats typically have below-average BMR, whereas frugivores typically have above-average BMR. This same diet effect generally applies to both mammals and birds, so McNab (1969, 1980a, 1982, 1986, 1987) has argued that diet is the key determi-nant of BMR. However, diet and phylogeny covary, making it difficult to sep-arate an ecological (dietary) effect from a phylogenetic effect. In any analyses relating to bats, most insectivores are vespertilionids, whereas frugivores are phyllostomids or megachiropterans. Elgar and Harvey (1987) and Harvey and Elgar (1987) argued that when the effects of phylogeny are removed by the use of phylogenetically independent contrasts analysis (Felsenstein 1985; Harvey and Pagel 1991), then the dietary effect disappears. In their opinion, BMR reflects phylogenetic affiliation rather than diet.

Table 10.1. Basal metabolic rates of 84 species of bats

Species	Latitude	Family	Diet	T_{roost}	M_b	BMR	References
Paranyctimene raptor	5	1	3	...	23.6	35.7	McNab and Bonaccorso 1995
Nyctimene albiventer	5	1	3	...	30.9	26.3	McNab and Bonaccorso 1995
Melonycteris melanops	5	1	2	...	16.4	14.4	McNab and Bonaccorso 1995
Rousettus amplicaudatus	5	1	3	...	91.5	102.5	McNab and Bonaccorso 1995
Dobsonia praedatrix	5	1	3	...	179.5	141.8	McNab and Bonaccorso 1995
Pteropus pumilus	...	1	3	...	194.2	126.2	McNab and Bonaccorso 1995
Dobsonia pannietensis	5	1	3	...	241.4	173.8	McNab and Bonaccorso 1995
Pteropus rodriguensis	...	1	3	...	254.5	134.9	McNab and Bonaccorso 1995
Dobsonia moluccense	5	1	3	...	463.8	394.2	McNab and Bonaccorso 1995
Pteropus hypomelansus	...	1	3	...	470.1	216.2	McNab and Bonaccorso 1995
Pteropus giganteus	...	1	3	...	739.2	377.0	McNab and Bonaccorso 1995
Megaloglossus woermeri	20	1	2	...	12.4	21.7	McNab 1969
Macroglossus minimus	20	1	2	...	16.3	21.0	Bartels et al. 1998
Syconycteris australis[a]	22	1	2	...	18.0	30.7	Geiser et al. 1996
Cynopterus brachyotis[a]	12	1	3	30.0	37.4	47.5	McNab 1969
Eonycteris spelea[a]	0	1	2	25.0	51.6	48.0	McNab 1969
Dobsonia minor[a]	15	1	3	...	87.0	109.6	McNab 1969
Rousettus aegyptiacus	30	1	3	30.0	146.0	122.6	Noll 1979
Pteropus scapulatus	27	1	2	27.0	362.0	242.5	Bartholomew et al. 1964
Pteropus poliocephalus	27	1	3	27.0	598.0	316.9	Bartholomew et al. 1964
Saccopteryx bilineata	10	2	1	26.5	8.2	15.3	Genoud and Bonaccorso 1986
Saccopteryx leptura	10	2	1	...	4.2	9.5	McNab 1982
Peropteryx macrotis	10	2	1	27.5	5.1	11.8	Genoud et al. 1990
Macroderma gigas	23	3	4	...	148.0	139.1	Leitner and Nelson 1967
Rhinolophus ferrumequinum	52	4	1	22.0	28.0	46.5	Speakman et al. 2003
Hipposideros galeritus	5	4	1	30.0	8.5	9.4	McNab 1989
Noctilio labialis	22	5	4	30.0	27.0	23.8	McNab 1969
Noctilio albiventris	9	5	4	...	40.0	39.6	Chappell and Roverud 1990
Noctilio leporinus	22	5	1	26.0	61.0	47.0	McNab 1969
Pteronotus quadridens	18	6	1	27.0	4.8	6.0	Rodríguez-Durán 1998
Mormoops blainvilli	18	6	1	29.0	8.9	8.7	Rodríguez-Durán 1998
Pteronotus davii	10	6	1	30.0	9.4	15.3	Bonaccorso et al. 1992
Pteronotus personata	10	6	1	30.0	14.0	23.0	Bonaccorso et al. 1992
Mormoops megalophylla	10	6	2	30.0	16.5	24.4	Bonaccorso et al. 1992
Pteronotus parnellii	10	6	1	30.5	19.2	30.7	Bonaccorso et al. 1992
Monophyllus redmani	18	7	2	30.0	8.8	11.3	Rodríguez-Durán 1998
Vampyressa pusilla	23	7	2	27.3	8.8	18.7	Cruz-Neto, personal communication
Rhinophylla pumilio	0	7	3	...	9.5	16.2	McNab 1969
Glossophaga soricina	22	7	2	26.0	9.6	21.6	McNab 1969
Vampyressa nymphae	22	7	3	...	10.5	21.2	McNab 1969
Artibeus cinereus	10	7	3	...	10.5	21.2	McNab 1969
Anoura caudifer	20	7	2	25.4	11.5	35.1	McNab 1969
Macrotus californicus	34	7	1	29.0	13.0	22.5	Bell et al. 1986
Carollia perspicillata[a]	22	7	3	24.3	14.9	31.4	McNab 1969
Uroderma bilobatum	0	7	3	...	16.2	26.6	McNab 1969
Erophylla bombifrons	18	7	2	26.0	16.4	18.0	Rodríguez-Durán 1998
Chiroderma doriae	23	7	3	26.4	19.9	31.2	Cruz-Neto, personal communication
Artibeus concolor	0	7	3	...	19.7	32.9	McNab 1969
Sturnira lilium	22	7	3	28.4	21.0	37.6	McNab 1969

(continued on next page)

Table 10.1. *(continued)*

Species	Latitude	Family	Diet	T_{roost}	M_b	BMR	References
Sturnira tildae	23	7	3	. . .	20.9	41.0	Cruz-Neto, personal communication
Leptonycteris sanborni[a]	36	7	2	. . .	22.0	44.0	Carpenter and Graham 1967
Vampyrops lineatus	22	7	3	. . .	21.9	2.3	McNab 1969
Tonatia bidens	0	7	4	21.0	27.4	39.2	McNab 1969
Diphylla ecaudata	20	7	5	22.0	27.8	33.9	McNab 1969
Desmodus rotundus[a]	10	7	5	27.0	29.0	26.4	McNab 1969
Phyllostomus elongatus	0	7	1	. . .	35.6	38.8	McNab 1969
Phyllostomus discolor[a]	20	7	1	25.5	33.5	34.5	McNab 1969
Diameus youngi	0	7	5	24.0	36.6	34.0	McNab 1969
Artibeus jamaicensis[a]	10	7	3	. . .	45.0	56.3	McNab 1969
Artibeus hirsuitus	20	7	3	. . .	47.0	110.9	Carpenter and Graham 1967
Tonatia sylvicola	0	7	4	. . .	55.0	66.0	McNab 1969
Artibeus fimbriatus	23	7	3	. . .	63.9	78.0	Cruz-Neto, personal communication
Artibeus lituratus	10	7	3	19.0	70.1	84.8	McNab 1969
Phyllostomus hastatus[a]	10	7	1	24.0	84.2	70.7	McNab 1969
Chrotopterus auritus	22	7	4	17.0	96.1	101.8	McNab 1969
Antrozous pallidus	37	8	1	34.0	22.0	21.2	Licht and Leitner 1967b
Tadarida brasiliensis[a]	37	9	1	31.4	11.0	15.3	Licht and Leitner 1967b
Molossus molossus	30	9	1	27.1	16.5	20.1	McNab 1969
Eumops perotis	34	9	1	35.0	56.0	39.8	Leitner 1966
Myotis yumanensis	36	10	1	31.0	5.5	12.3	O'Farrell and Studier 1970
Pipistrellus pipistrellus[a]	58	10	1	17.0	6.6	12.1	Speakman 1993
Myotis austroriparius	45	10	1	. . .	7.5	10.9	McNab 1969
Myotis lucifugus	42	10	1	24.0	7.7	18.7	Kurta and Kunz 1988
Nyctophilus geoffroyi	35	10	1	. . .	8.0	11.4	Hosken and Withers 1999
Myotis thysanodes	36	10	1	32.0	8.1	17.4	O'Farrell and Studier 1970
Plecotus auritus[a]	58	10	1	24.5	11.1	16.0	McLean and Speakman 1999
Histiotus velatus	20	10	1	. . .	11.2	10.0	McNab 1969
Chalinolobus gouldii	32	10	1	. . .	17.5	25.2	Hosken and Wither 1997
Pizonyx vivesi	35	10	4	. . .	25.0	35.8	Carpenter 1968
Myotis myotis	45	10	1	25.0	25.0	25.0	Hanus 1959
Eptesicus serotinus	52	10	1	20.0	27.0	43.1	Speakman et al. 2003
Nyctophilus major	32	10	1	17.0	13.6	20.4	Hosken 1997
Natalus tumidirostris	10	11	1	27.5	5.4	8.3	Genoud et al. 1990

Note. "Latitude" represents the approximate latitude where the bats were captured prior to measurement (° north or south); roosting site temperatures are in °C; body mass is in grams; and basal metabolism (BMR) is oxygen consumption (mL h^{-1}). Families are as follows: 1 = Pteropodidae, 2 = Emballonuridae, 3 = Megadermatidae, 4 = Rhinolophidae, 5 = Noctilionidae, 6 = Mormoopidae, 7 = Phyllostomidae, 8 = Antrozoidae, 9 = Molossidae, 10 = Vespertilionidae, and 11 = Natalidae (after Simmons 1998). For diets, 1 = insectivory, 2 = nectarivory, 3 = frugivory, 4 = carnivory, and 5 = sanguinivory. Ellipses dots represent items for which no data were provided.

[a] Several measurements were available in the literature for this species, and the value is an average of these several measurements. Species are listed first by family and then by body mass. Only one reference (generally the largest sample size contributing to the pooled data) is cited.

Log$_e$ Basal Metabolic Rate
(mls O$_2$/h)

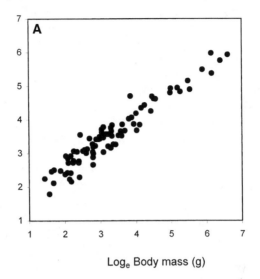

Log$_e$ Body mass (g)

Log$_e$ Basal Metabolic Rate
(mls O$_2$/h)

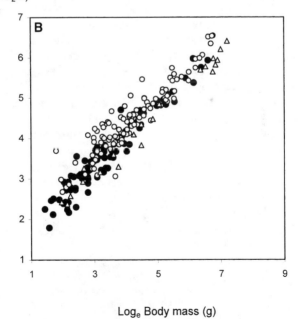

Log$_e$ Body mass (g)

Figure 10.4. *A,* The relationship between basal metabolic rate and body mass in 84 bat species. Each point represents a separate species. *B,* Basal metabolic rates of bats (*filled circles*) compared with rodents (*open circles*) and marsupials (*open triangles*) of the same approximate body size range. Bats have metabolic rates intermediate between those of marsupials and rodents. The order effects are significant. Each point represents a different species.

The debate over the relative importance of diet and phylogeny in setting BMR in bats is particularly confused. On the one hand, papers by both McNab (1982, 1986, 1987) and Elgar and Harvey (1987) drew heavily on data from a single classic paper by McNab (1969) on the MR of Neotropical species. Unfortunately, these data included an erroneous correction for standard temperature and pressure, which McNab (1989) later corrected, decreasing these estimates of BMR by an average of 20%. Since frugivorous bats are over-represented in this sample, the decrease in estimates of their BMR reduces the supposed diet effect. On the other hand, analyses involving phylogenetic contrasts remain problematic due to the lack of a reliable and uncontested phylogeny for the Chiroptera at the level of order, family, and even species. Over the past 2 decades opinion over the higher cladistic relationships of bats has oscillated from monophyly to diphyly and back to monophyly (Adkins and Honeycutt 1991; Allard et al. 1992; Ammerman and Hillis 1992; Baker et al. 1991; Pettigrew 1986, 1991; Pettigrew et al. 1989; Simmons 1998). There is also considerable debate over the phylogeny of the different families where, for example, the Vespertilionidae have been variously positioned as close to the Natalidae and Molossidae (van Valen 1979), the Myzopodidae (Eisenberg 1981), the Emballonuridae and Molossidae (Pierson 1986), somewhere between the Phyllostomidae and a Natalidae/Myzopodidae/Molossidae grouping (Fenton 1992), or within a superfamily Vespertilionoidea between the superfamilies Nataloidea and Molossidea (Simmons 1998; Simmons and Geisler 1998). Uncertainties concerning phylogenetic affiliations are critically important to the debate over the effect of diet and phylogeny on BMR because analyses based on phylogenetically independent contrasts require a completely known phylogeny (Harvey and Pagel 1991; Pagel and Harvey 1988). While nobody disputes the importance of known phylogenies at higher taxonomic levels, recent simulations show that uncertainties at the generic and species levels may bias the results of phylogenetically independent contrasts as much as errors at higher levels (Symonds 1999).

In our data set, which includes a substantially greater number of species than were originally included in previous analyses, there was a significant association between diet and residual BMR (BMR with the effect of M_b removed [ANOVA: $F = 3.03$, df $= 4, 79$, $P = 0.022$]). Because two diets are poorly represented in the sample (blood feeding with three species and carnivory with seven species), these groups may exert an undue effect on the ANOVA. Nevertheless, the diet effect remains significant if the analysis is restricted to the three dietary groups that are well sampled (ANOVA: $F = 4.06$, df $= 2, 69$, $P = 0.022$). As noted by McNab (1982, 1986, 1989), above-average BMR is associated with frugivory while below-average BMR is associated with insectivory (Tukey pairwise comparison: $P < 0.05$; mean residual BMR: insectivores $= -0.0924$, $n = 32$; frugivores $= +0.089$, $n = 28$). Residual BMR is also significantly affected by taxonomic affiliation at the family level (ANOVA: $F =$

2.14, df = 10, 73, P = 0.031), and this effect remains even when only those families represented by over 10 species (Pteropodidae—20 species; Phyllostomidae—31 species; Vespertilionidae—13 species) are included in the analysis (F = 3.44, df = 2, 61, P = 0.039). However, when both taxonomic affiliation and diet are included as independent factors in an ANCOVA model, with M_b as covariate, the effect of taxonomy loses significance (F = 1.8, df = 10, 63, P = 0.075), but the effect of diet is strengthened (F = 3.1, df = 4, 63, P = 0.021). This analysis supports McNab's (1987, 1989, 1992) contention that diet is the key factor setting residual BMR.

Although the slope of the regression of BMR on M_b for bats is similar to that derived by Kleiber (1961) for mammals in general, the mean BMR of bats is only 79.4% of the Kleiber line (median = 78.8%, SD = 20.8%, n = 84). However, values of BMR for bats range from a low of 42.9% (*Histiotus velatus* [McNab 1969 as revised in McNab 1989]) to 163% (*Artibeus hirsuitus* [Carpenter and Graham 1967]) of values predicted by the Kleiber line. The Kleiber line, however, is generated by a heterogeneous data set with unequal sampling across sizes and taxonomic groups, so it is more appropriate to compare bats with taxonomically homogeneous groups in the same size range. We obtained data on BMR for 117 species of rodents (mean M_b = 125.7 g; range 4–1,200 g) from McNab (1986), Elgar and Harvey (1987), Bozinovic (1992), McMillan and Hinds (1992) and for 31 marsupials in the same mass range (mean M_b = 415.6 g) from Bennett and Harvey (1987) and Dawson and Hulbert (1970). Comparing BMR across these groups using each species as an independent datum (fig. 10.4B) reveals that bats (mean residual BMR = −0.119) have significantly lower BMR than rodents do (mean residual BMR = +0.162) but that both bats and rodents have elevated BMR compared with marsupials (mean residual BMR = −0.291; ANCOVA: interaction F = 2.11, df = 1, 226, P = 0.123; M_b F = 2482.73, df = 1, 226, P < 0.0001; order F = 11.26, df = 1, 226, P < 0.0001). Post hoc group comparisons using residual BMR were all significant (P < 0.01).

Why do bats have lower BMR than similar-sized rodents? Although this difference may seem linked to taxonomic affiliation, other factors may be involved. One factor that has not previously been considered in analyses of BMR in bats (Elgar and Harvey 1987; McNab 1969, 1982, 1986, 1987, 1989, 1992) is the fact that most of the data are for tropical species. In contrast, most data for rodents are for temperate species, suggesting at least part of the difference might be attributable to climate rather than taxonomy. Basal metabolic rate is affected by the seasonal exposure of animals to low T_a (Casey et al. 1979; Chaffee and Roberts 1971; Hart 1953a, 1953b; Scholander et al. 1950), so the thermoregulatory challenge associated with climatic region may affect BMR (Haim and Izhaki 1993; Speakman 1995). Animals that need a high thermogenic capacity to cope with low T_a may require a higher BMR, while those that routinely live at temperatures above T_{lc} may suppress BMR to avoid heat stress

Residual Log$_e$ BMR

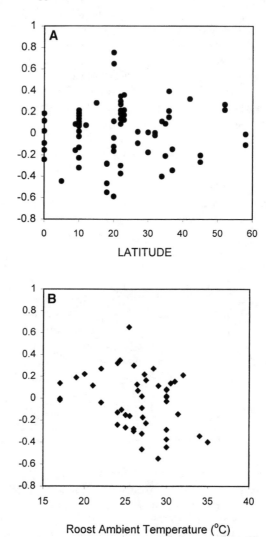

Roost Ambient Temperature (^{o}C)

Figure 10.5. Relationship between residual basal metabolic rate (BMR) and latitude (*A*) and between residual basal BMR and ambient roost temperature for bats (*B*).

(Bartholomew et al. 1962; Rodríguez-Durán 1994). Consequently, BMR is positively correlated with latitude in birds (Hails 1979), notably seabirds (Ellis 1984), but also in small mammals (principally rodents [Speakman 2000]).

Surprisingly, for bats, residual BMR is not significantly correlated with latitude (fig. 10.5A: $F = 0.20$, df $= 1, 44$, $P = 0.653$, where latitude was known). Since latitude has a consistent effect on BMR in other animal groups, it is worth considering why bats do not conform to this pattern. The level of BMR probably reflects the size of the metabolic and thermogenic machinery required for thermoregulation in cold climates. Animals living at high latitudes maintain larger organ masses and pay a maintenance cost, which we see as an elevated BMR. Bats, however, may not require the same metabolic capacity at high latitude because during most of the daytime they occupy roosting sites where internal temperatures may bear only a poor relation to the external environment (Kunz 1982). At night, when bats do go out, they spend most of their time flying, which generates enough heat to sustain T_b with no additional thermoregulatory costs. Due to the buffering capacity of their roosts, bats may never experience the thermoregulatory challenge offered by the local environment and so may escape the selection that latitude imposes on BMR.

To test this hypothesis, we reviewed the temperatures reported inside day roosts (T_{roost}) occupied by bats during the summer and compared these with the summer maximum T_a reported for 554 meteorological sites as a function of latitude (Oliver and Fairchild 1984). During the summer, T_{roost} does indeed show far less variation as a function of latitude than does T_a (fig. 10.6). At high latitudes, T_{roost} is generally close to that of the local environment. Bats probably select roosts so that T_{roost} is as close as possible to T_{lc}, thus minimizing their daily roosting costs. Many temperate bats avoid the thermal constraints imposed by latitude and climate by clustering during the day-roosting period, generating and trapping their body heat locally (Roverud and Chappell 1991), such that they routinely experience temperatures in the thermoneutral zone. In contrast, at latitudes below 20°–30° N and 20°–30° S, T_{roost} is generally cooler than T_a. At low latitudes and in hot climates, bats appear to select their roosts primarily to avoid overheating.

If T_{roost} rather than T_a is the environmental condition that bats most commonly encounter, we might anticipate that BMR would be more closely correlated with mean T_{roost} than latitude. Data for T_{roost} at the locations where bats had been captured and subsequently measured by respirometry were available for 47 of the 84 species. Residual BMR was a significantly and negatively correlated with T_{roost} (fig. 10.5B: $F_{1,45} = 4.26$, $P = 0.045$, $r^2 = 0.09$). The best-fit relationship including both M_b and T_{roost} as predictors explained 91.1% of the variation in BMR:

$$\log_e \text{BMR (mL O}_2 \text{ h}^{-1}) = 1.6317 + 0.719 \log_e M_b - 0.0187 \, T_{roost}. \qquad (2)$$

We suggest that although bats are found worldwide, throughout their geographic range they use day roosts that buffer the local climate, thus escaping

Temperature (°C)

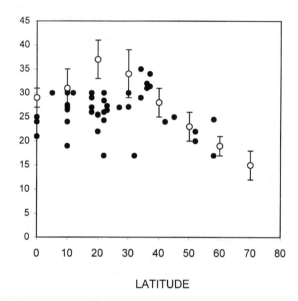

LATITUDE

Figure 10.6. Mean temperatures inside day roosts occupied by bats during summer as a function of latitude (degree north or south) compared with mean maximum midsummer (July or December) air temperature for 554 meteorological sites across the globe (*open symbols* ± SD).

temperatures that would require high thermogenic capacity and high BMR. In contrast, most rodents measured to date inhabit temperate and arctic regions, where they would routinely have to cope with low T_a requiring high BMR. Thus, at least part of the difference in BMR between bats and rodents may reflect these factors rather than taxonomic influences.

The importance of T_{roost} in setting the level of BMR in bats has been suggested previously but usually as a response to heat rather than cold (Bonaccorso et al. 1992; Genoud and Bonaccorso 1986; Licht and Leitner 1967b; Rodríguez-Durán 1994). At high T_{roost}, where $\Delta T°$ is insufficient to drive convective losses, bats must rely on evaporation to regulate T_b for periods of up to 12 h, normally without access to drinking water. This may favor a reduction in BMR to reduce dehydration risk. This same argument has been invoked to explain reduced BMR in desert rodents (Haim and Izhaki 1993; McNab 1966) and some birds (Bartholomew et al. 1962).

The importance of T_{roost} as a factor influencing the residual BMR of bats undermines previous attempts to explain residual BMR solely in terms of phylogeny and diet because both of these traits are linked to some extent with variations in T_{roost}. For example, the Pteropodidae and Phyllostomidae represent the most important frugivore and nectarivore groups but are also exclu-

sively tropical and subtropical. This makes it difficult to separate a phylogenetic effect from either a climatic or a diet effect. In this context, it is interesting that removing the variation in BMR that can be explained by M_b and T_{roost}
leaves a residual variability that is not explained by either dietary differences
($F = 1.93$, df $= 4, 33$, $P = 0.128$) or taxonomic affiliation ($F = 0.66$, df $= 9, 33$, P
$= 0.735$). This remains the case even when we restrict the analysis to well-
sampled categories (diets [insectivory, frugivory, nectar/pollen feeding]: $F =$
1.41,df $= 2, 19$, $P = 0.268$; families [Vespertilionidae, Phyllostomidae, pteropo-
dids]: $F = 0.75$, df $= 2, 19$, $P = 0.485$). On this basis we suggest that there is little
evidence linking variation in the BMR of bats to their dietary habits or to broad
phylogenetic groupings.

For a given M_b, residual BMR is also positively correlated with T_b ($F = 33.3$,
df $= 1$, 61, $r^2 = 0.34$; fig. 10.7). As mentioned earlier, this correlation between BMR and T_b is not determined by the laws governing heat flow but,
rather, reflects a biological or evolutionary adaptation. By definition, within
the thermoneutral zone, BMR generates more than enough heat to sustain T_b
(fig. 10.1C), thus low T_b is not simply a reflection of an incapacity to bring heat

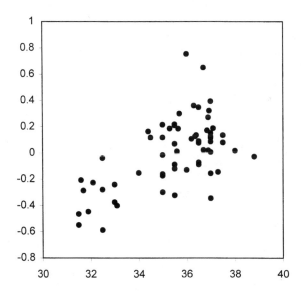

Residual Log$_e$ BMR

Body Temperature (°C)

Figure 10.7. Relationship between residual basal metabolic rate (BMR)
and euthermic body temperature in bats.

production (BMR) into balance with heat loss. The correlation between BMR and T_b reflects a more fundamental effect of temperature on the rate of the chemical processes that sustain life. The relationship between residual $\log_e$ BMR and T_b within the range of 31°–39°C has a slope of 0.086, which translates to a Q_{10} of approximately 2.4. The effect of T_b on BMR in bats is consistent with that found for most biological processes, where Q_{10} values lie in the range of 2–3 (Blaxter 1989). Animals therefore regulate their BMR in part by regulating their T_b set points.

Effects of Reproduction on Basal Metabolism

One of the criteria used for defining BMR is the absence of growth or reproduction (Kleiber 1961), so quantifying the effects of reproduction on BMR is technically a contradiction of terms. However, if all other conditions for measuring BMR are met, it is possible to measure the increment in resting MR over and above BMR, an increment that that can be attributed to reproductive effort (Mclean and Speakman 2000; Speakman and McQueenie 1996). For this reason, it is not entirely illogical to retain the term "BMR" when referring to resting MR during the reproductive period.

Reproduction increases BMR in several different ways. Reproductively active bats need to elevate biosynthesis to support processes such as pregnancy and lactation in females and spermatogenesis in males. Both pregnancy and lactation, but particularly lactation, involve the direct transfer of energy from the parent to offspring in the form of sugars, protein, and fats for growth. These materials must be synthesized, which increases BMR due to metabolic inefficiencies. Many animals also experience large changes in body composition that serve to support their reproductive effort. An obvious example is the development of mammary tissue to enable milk production, but the gut (Cripps and Williams 1975; Fell et al. 1963) and liver (Kennedy et al. 1958; Speakman and McQueenie 1996) also increase in size during lactation to allow females to process the extra food intake. In addition to costs associated with the synthesis of new tissue, the increase in maintenance costs of metabolically active tissues may affect BMR. These costs remain unquantified in most species of bats. However, Kurta and Kunz (1987) found that BMR in pregnant *Myotis lucifugus* was higher than that of nonreproductive individuals. Similarly, McLean and Speakman (2000) found that BMR of brown long-eared bats (*Plecotus auritus*) nearly doubled during pregnancy and lactation compared to that observed in nonreproducing individuals.

Metabolic and Behavioral Responses to High Ambient Temperatures

Upper critical temperatures were available for 50 of the 84 species. Mean T_{uc} was 36.26°C (median = 36.37, SD = 2.30, max = 41.0, min = 30.0). There was no significant relationship between T_{uc} and M_b ($F = 0.32$, df = 1, 49, $P = 0.573$) or minimum thermal conductance ($F = 0.05$, df = 1, 49, $P = 0.816$). However, as would be expected based on heat production and heat loss, T_{uc} was correlated positively with T_b and negatively with BMR (multiple regression r^2 = 0.212, F = 6.18, df = 2, 46, P = 0.004; BMR: t = −2.54, P = 0.015; T_b: t = 3.35, P =0.002). High T_b has the effect of increasing $\Delta T°$ at a given T_a, thus shifting T_{uc} to higher temperatures. Increasing BMR produces more heat, which requires a larger $\Delta T°$ to unload, thus shifting T_{uc}.

Several studies have documented the behavior of bats as T_a increases above T_{uc}. Licht and Leitner (1967a) reported that three species of bats (*Myotis yumanensis*, *Antrozous pallidus*, and *Tadarida brasiliensis*) in the wild became restless and moved to avoid areas of their roost where T_a approached 40°C but tolerated lower temperatures in the region of 35°–38°C. When bats were exposed to high T_a in captivity, they showed heat stress responses. The wings became engorged with blood at 40°C, bats panted with an open mouth at 41.5°–42°C, and finally, bats salivated profusely and self-anointed at 42°C. Similar responses at slightly lower T_a were reported in *Macroderma gigas*, in which ventilation rates increased from 55 to 154 breaths per minute at 35°–38°C and salivation and self-anointing occurred above 38°C (Leitner and Nelson 1967). Megachiropterans have similar responses. Egyptian fruit bats, *Rousettus aegyptiacus*, initiated self-anointing with saliva at 38°C, whereas the larger pteropodids started to pant at 39.5°C and self-anointed at higher T_a (Bartholomew et al. 1964). Pteropodids also fan their wings at high T_a to facilitate heat dissipation by convection and by promoting evaporative losses (Bartholomew et al. 1964; Ochoa and Kunz 1999). Apart from a report by Reeder and Cowles (1951) of wing fanning by *M. yumanensis*, no other studies have reported wing-fanning behavior in the Microchiroptera, and other studies have commented on its absence (Herreid 1963; Licht and Leitner 1967a).

The T_b that bats can tolerate before they die are not exceptional when compared with other small mammals. Speakman et al. (1994) reviewed the available data on lethal T_b across a range of bat species and found that the lowest lethal T_b (after 1-h exposure) averaged 42.1°C (SD = 1.98, n = 10) and the highest nonlethal T_b (1-h exposure) averaged 41.3°C (SD = 1.3, n = 29). Despite the fact that lethal T_b is not exceptional when compared with other small mammals (Adolph 1947), many studies have remarked on the amazing tolerance of bats to high T_a (e.g., Bronner et al. 1999). Tolerance to high T_a appears to result from the ability of bats to tolerate hyperthermia and thus delay the onset of evaporative cooling until T_b has risen close to the lethal limit, thereby retarding dehydration (Licht and Leitner 1967a). This is likely to be important, considering that bats normally do not have access to water in the day roosts.

Thermal Conductance and Metabolic Responses to Low Ambient Temperatures

Lower critical temperatures were available for 60 species in our database (fig. 10.8A). Mean T_{lc} was 28.96°C (median = 29.0, SD = 3.02, min = 19, max = 35).

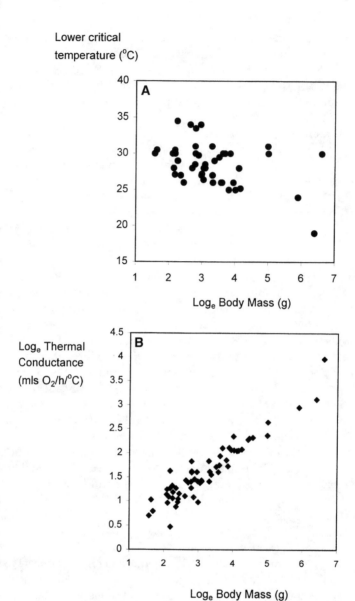

Figure 10.8. Relationship between body mass and lower critical temperature (*A*) and thermal conductance below lower critical temperature (*B*) in bats.

The position of T_{lc}, representing the lower boundary of the thermoneutral zone, is set by the interaction T_b and thermal conductance (C), which together determine $\Delta T°$ and the rate of heat loss, and BMR (the rate of heat production). Because thermal conductance and BMR are both strongly influenced by M_b (fig. 10.4 and 10.8B), T_{lc} is also correlated with M_b ($T_{lc} = 32.58 - 1.147 \log_e M_b$; $r^2 = 0.178$, $F = 12.54$, df = 1, 58, $P < 0.0001$). For the larger pteropodids, the thermoneutral zone spans almost 15°C, whereas in smaller bats it can be as narrow as 1°C. Treated independently, T_{lc} was also significantly negatively related to $\log_e$ BMR ($F = 18.08$, df = 1, 52, $P < 0.001$, $r^2 = 0.258$), positively related to thermal conductance ($F = 7.48$, df = 1, 52, $P = 0.009$, $r^2 = 0.158$), but not related to T_b ($F = 0.0$, df = 1, 52, $P = 0.965$).

Below T_{lc}, MR increases to compensate for the rate of heat loss set by the joint effects of thermal conductance and $\Delta T°$. While for some bats the regression line through data below T_{lc} projects back to intersect the x-axis at T_b, for others it intersects well above T_b. This latter pattern suggests that bats either continue to adjust thermal conductance below T_{lc} (e.g., *Macroderma gigas* [Leitner and Nelson 1967] and *Nyctophilus geoffroyi* [Hosken and Withers 1999]) or depress T_b with declining T_a (e.g., *Eumops perotis* [Leitner 1966]). The effect of changes in thermal conductance and T_b on metabolic response at low T_a has been discussed in detail by McNab (1980b). Below T_{lc}, mammals usually produce heat initially through nonshivering thermogenesis, followed by shivering thermogenesis. Leitner (1966) showed that increased muscle electrical activity (shivering) in *E. perotis* began at around 23°C, or 8°C below T_{lc}, suggesting that nonshivering thermogenesis alone could offset heat loss through a span of about 8°C. More direct evidence for nonshivering thermogenesis comes from observations that noradrenaline stimulated the energy expenditure of *Rousettus aegyptiacus* (Noll 1979), that the beta-blocker, propranonol, reduced the capability of *Myotis myotis* to arouse from torpor (Heldmaier 1969), and by the immunological identification of the mitochondrial uncoupling protein, UCP-1, in pipistrelle bats (*Pipistrellus pipistrellus* = *pygmaeus* [Trayhurn et al. 1995]).

Thermal conductance below T_{lc} is strongly related to M_b (fig. 10.8B). The least squares regression equation,

$$\log_e \text{conductance (mL } O_2 \text{ h}^{-1} °C^{-1}) = -0.1037 + 0.534 \log_e M_b \text{ (g)}, \quad (3)$$

explains 88.2% of the variation in thermal conductance ($F = 441.8$, df = 1, 59, $P < 0.0001$). The slope of 0.53 is similar to the slopes derived for other eutherian mammals (Aschoff 1981 [0.48]; Hart 1971 [0.50]; Herreid and Kessel 1967 [0.50]; Morrison and Ryser 1951 [0.50]). The intercept is equivalent to 0.201 watts °C^{-1} at $M_b = 1$ kg, which is higher than the intercepts reported for eutherian mammals [Aschoff 1981 (0.158); Hart 1971 (0.168); Herreid and Kessel 1967 (0.174); Morrison and Ryser 1951 (0.177)]. As McNab (1969, 1982) previously noted, bats have higher thermal conductances than other mammals. It has been argued that because of their naked wings, bats have much greater surface areas across which heat might be exchanged than mammals of equivalent body size (Licht and Leitner 1967b; Peters 1983). However, this is unlikely to explain high thermal conductance because at rest the wings are folded and because bats control the flow of blood into the wings (Cowles 1947; Kluger and Heath 1970), allowing them to cool to almost T_a (Lancaster et al. 1997). Indeed, Bartholomew et al. (1964) showed that by wrapping their wings around themselves, flying foxes lowered their thermal conductance, probably due to the layer of still air trapped beneath the wings. The high thermal conductance of bats therefore probably reflects their pelage.

Bats may have poor insulation to allow heat dissipation during flight. In flapping flight, endogenous heat production increases by an order of magnitude compared with resting rates (see below). Bats have a poor capacity to modulate heat loss via the respiratory tract (Carpenter 1986), so they must unload excess heat through convection. Thermal images of flying bats (Lancaster et al. 1997) indicate that most of this heat is lost from the body surface rather than by the wings as was previously thought (Cowles 1947; Kluger and Heath 1970). Bats allow T_b to rise during flight (Thomas 1987), which when coupled with a high thermal conductance facilitates the unloading of excess heat. McNab (1974) suggested that bats may have poor insulation to facilitate heat dissipation and allow them to attain low T_b when hibernating, but this seems unlikely because bats molt in the fall, and winter pelage has better insulative properties than the summer pelage (Shump and Shump 1980).

Poor summer pelage insulation (high thermal conductance) may be linked with other aspects of bat behavior. To minimize their energy requirements, bats should select day roosts offering T_a approaching or even exceeding T_{lc}. Comparing figure 10.5, which shows the temperatures of roosting sites selected by bats during summer, and figure 10.7A, which shows T_{lc}, reveals that bats appear to select roosts that offer temperatures close to T_{lc}, thereby minimizing their daytime energy requirements. Low thermal conductance therefore does not impose an energetic cost to roosting bats, but it does facilitate convective rather than evaporative heat loss during flight.

Bats may not always be able to find roosts that offer temperatures in their thermoneutral zone throughout the day, so they commonly cluster when they regulate T_b at euthermic levels (Trune and Slobodchikoff 1976). Clustering provides several thermal benefits. First, it reduces the surface area through which heat is lost and, second, the heat generated may be sufficient to elevate T_{roost} (Canals et al. 1989; Contreras 1984). Although it has been suggested that the local build-up of CO_2 suppresses metabolic rate (Herreid and Schlenker 1980), studies of clustering in nonbat species, which have attempted to partition these effects, suggest that the effects of local CO_2 concentrations are negligible (Contreras 1984; Speakman and Rossi 1999). In some species, clustering may also be linked with an increased propensity to regulate T_b at lower levels, thus depressing BMR (Roverud and Chappell 1991).

Regulation of Body Temperature and the Use of Torpor by Bats

Inconsistencies and Contradictions between Studies of Thermoregulatory Patterns in Bats

Bats exhibit a remarkable array of thermoregulatory patterns that span the range from the strict defense of a stable T_b almost to complete heterothermy. In this respect, greater variability can be found in the Chiroptera than in any other mammalian group, giving rise to a large (and often contradictory) literature on T_b in bats. Understanding how bats regulate T_b under natural conditions is confounded by their apparent extreme sensitivity to study conditions. Thus, studies performed under different conditions or at different times of the year often produce dramatically different results. It is not our intention to perform a comprehensive review of the literature on T_b regulation in bats, but we will detail three case studies to demonstrate the importance of context.

In Brazil, McNab (1969) found that *Artibeus jamaicensis*, *A. lituratus*, *Carollia perspicillata*, and *Glossophaga soricina* were able to maintain stable and elevated T_b across a wide range of T_a. Based on these data, the Phyllostomidae appeared to be competent thermoregulators. In contrast, Studier and Wilson (1970) found that *A. jamaicensis*, *C. perspicillata*, *Micronycteris hirsuta*, *Pteronotus suapurensis*, and *Vampyrodes caraccioli* from Panama all exhibited a pronounced heterothermic response in the face of declining T_a. These bats barely defended a stable T_b at T_a above 30°C, and they allowed T_b to fall by 20°–30°C as T_a declined from 30° to 5°C. Based on Studier and Wilson's (1970) data, the Phyllostomidae would appear to be weak thermoregulators at best. However, when *A. jamaicensis* was held in captivity, its thermoregulatory ability improved over a period of days (Studier and Wilson 1979), suggesting that its poor ability in Panama was either related to the initial stress of capture or to body condition at the time of capture.

By manipulating time in captivity and feeding opportunity independently, Audet and Thomas (1997) showed that thermoregulatory ability of both *C. perspicillata* and *Sturnira lilium* was strongly affected by feeding success and M_b. Most bats that received fruit ad lib. remained normothermic ($T_b \geq 33$°C) on all days, whereas most bats that received a restricted ration became hypothermic ($T_b \leq 32$°C) on the day following food restriction regardless of the time in captivity. Body temperature was highly correlated with small changes in M_b in both species. At masses ≥ 20 g for *C. perspicillata* and ≥ 18 g for *S. lilium*, individuals remained normothermic, whereas the proportion of individuals depressing T_b below 32°C increased as M_b decreased. One cannot conclude from these studies that hypothermia in the Phyllostomidae is an artifact induced by starvation or poor adaptation to captive conditions; some free-ranging *C. perspicillata* fitted with temperature-sensitive radio transmitters also depressed T_b, albeit less than did food-deprived individuals (Audet and Thomas 1996). The fact that not all individuals became hypothermic following food restriction in the laboratory and that the level at which bats regulated T_b differed in the laboratory and in the field suggest that the regulation of T_b is strongly dependent on both food intake and body condition, with individuals in better condition defending higher T_b than individuals in poorer condition.

A second example underlines the importance of context in studies of the regulation of T_b. Numerous authors (e.g., Hitchcock et al. 1984; Stones and Wiebers 1967; Thomas 1988) have suggested that males of temperate vespertilionid species resort to torpor regularly as an energy conservation strategy during the summer months because they are exempt from the reproductive constraints imposed on females (see Racey 1982). To test this, Kurta and Kunz (1988) measured resting MR of solitary male *Myotis lucifugus* roosting in the laboratory under simulated natural roost conditions. Based on daily patterns of resting MR and T_b, they concluded that male *M. lucifugus* did not resort to torpor but, rather, defended an elevated and stable T_b throughout the day (mean $T_b = 35.7°C$). They argued that males are not exempt from reproductive constraints and are forced to maintain high T_b to ensure testicular development and spermatogenesis. However, in their field study of *Eptesicus fuscus*, using temperature-sensitive radio transmitters to monitor patterns of T_b, Grinevitch et al. (1995) found the opposite pattern. During the May–June period when females were pregnant, males resorted to shallow torpor (depression of T_b by only a few degrees) on >95% of days and exhibited deep torpor (T_b approaching T_a) on >80% of days. The use of torpor persisted into mid-summer, when males used shallow torpor on >75% of days and deep torpor on >35% of days. Similarly, 96% of male *M. lucifugus* captured as solitary day-roosters in bat houses near Sherbrooke, Québec, in June and July ($n = 26$) either had $T_b < 30°C$ or showed clear behavioral signs of torpor (D. W. Thomas, unpublished data). The fact that males of the genus *Myotis* penetrate into higher altitude zones where T_a is low and insect abundance irregular (Thomas 1988; Thomas and Bell 1986) suggests that torpor is in fact a common thermoregulatory strategy for males.

Finally, studying T_b and resting MR of female *M. lucifugus* and *M. thysanodes*, Studier and O'Farrell (1972, 1976) found that females of both species exhibited great variation in T_b with respect to T_a. Individuals of both species could be roughly classed as "regulators" and "conformers" based on the difference between T_b and T_a that they defended. Regulators maintained $\Delta T° > 8°C$ and conformers $\Delta T° < 2°C$ at T_a's in the range of 16°–20°C. Although *M. thysanodes* tended to be regulators more often than *M. lucifugus*, 30%–50% of pregnant and lactating females of both species were conformers and allowed T_b to fall below 22°C. This resulted in a dramatic decrease in resting MR from >2 mL O_2 g^{-1} h^{-1} to well below 0.5 mL O_2 g^{-1} h^{-1}. Even regulating individuals entered torpor, allowing T_b to fall from 35°C to <32°C as T_a fell from 30°C to <25°C. Surprisingly, postlactating females of both species showed the greatest tendency to maintain high T_b, with over 85% of individuals classed as regulators. These results contrast markedly with those obtained by Kurta et al. (1987) and Kurta and Kunz (1988), who studied the metabolic rates and T_b of pregnant, lactating, and postlactating *M. lucifugus* roosting under conditions that more closely simulated the natural roost (a carved-out wooden beam). When in clusters of about eight individuals (Kurta et al. 1987), pregnant, lactating, and postlactating females all maintained resting MR in excess of 2.4 mL O_2 g^{-1} h^{-1}, indicating that they defended high T_b. Cluster temperatures exceeded 30°C and individual rectal temperatures all exceeded 35°C, providing no evidence of daily torpor. Even when forced to roost alone (Kurta and Kunz 1988), lactating females maintained high resting MR (>3.5 mL O_2 g^{-1} h^{-1}), indicating that they too regulated a high T_b.

How, then, can we reconcile conflicting results emerging from different studies of the same species or explain the extreme variation found between species within the Chiroptera? One often has the feeling that the implicit objective of laboratory-based studies of thermoregulation in bats is to identify *the* thermoregulatory pattern that characterizes the species or group. However, as the case studies that we have presented above so clearly show, one must be cautious before accepting any single pattern as characteristic or representative of a given species, genus, or family. The value of laboratory studies may not be to identify the true thermoregulatory pattern but, rather, to demonstrate the thermoregulatory flexibility of a given species or group and to identify the contexts that lead to euthermy or torpor and the consequences of regulating T_b at a given level. The challenge is then for field studies to provide the data showing what bats really do and when they do it. Only then can we really accept laboratory-based data as reliable.

Variation in the Use of Daily Torpor by Bats

The earliest studies of bat thermoregulation were made on temperate-zone species (e.g., Eisentraut 1934, 1960; Hock 1951). Since these bats regularly exhibited periods of depressed T_b they were invariably classed as heterotherms and considered to be primitive in their thermoregulatory ability. However, we now view torpor not as an indication of thermoregulatory incompetence but, rather, as a derived trait where T_b is regulated at a lower set point. While probably all temperate-zone species, dominated by the insectivorous Vespertilionidae, are capable of entering regulated torpor on a regular basis (Stones and Weibers 1967), for many small tropical species the use of torpor appears to be a last resort linked with energy shortfalls. These species generally will defend their euthermic T_b if they are in good nutritional state but will allow T_b to fall when in poor condition (e.g., Audet and Thomas 1997; Studier and Wilson

1970). This complete abandoning of regulation is distinguishable from regulated torpor in that spontaneous arousal is not possible, and if prolonged, the animals invariably die. However, if the animals are passively warmed they may recover (e.g., Genoud and Bonaccorso 1986).

Initial studies of the Megachiroptera suggested that they were obligate eutherms (e.g., Bartholomew et al. 1964; Kulzer et al. 1970; Morrison 1959). Reduced T_b occurred only after defense of the euthermic temperature and the animals appeared unable to recover spontaneously. However, more recent studies of smaller megachiropterans have demonstrated that they, too, are capable of torpor (Bartels et al. 1998; Bartholomew et al. 1970; Geiser et al. 1996). Bonaccorso and McNab (1997), and McNab and Bonaccorso (1995) noted that some large pteropodids defend a core T_b at euthermic levels while allowing the extremities to cool substantially. For some individuals, resting MR doesn't rise above basal levels, and it may even decline as T_a falls below the normal T_{lc}. This suggests a thermoregulatory strategy approaching torpor, but whether this is simply an indication of extreme peripheral vasoconstriction or a true torpor response remains to be determined.

Racey (1973) was the first to demonstrate a cost to torpor. In captivity, pregnant female *Pipistrellus pipistrellus* that were forced to express torpor by a combination of food deprivation and low T_a gave birth later than females that maintained high T_b, indicating that torpor and growth (or production) are not compatible. This cost to torpor has since been confirmed for free-ranging *P. pipistrellus*, where inclement weather and low insect availability result in a delay in parturition date (Racey and Swift 1981). If torpor and production are incompatible, then milk production will also be slowed and possibly stopped by torpor (Wilde et al. 1995, 1999), resulting in decreased growth rates of nursing young. For temperate-zone bats in general, early parturition and rapid growth appear to be important in providing juveniles the time needed to complete growth and acquire adequate fat reserves prior to hibernation (Kunz et al. 1998b; Pagels 1975). For *Rhinilophus ferrumequinum*, overwinter survival is lower for late-born young, providing the only direct evidence for a fitness cost to the use of torpor (Ransome 1990). As a result, it is now commonly assumed that, for reproductive females, torpor is a last-ditch solution to an energetic constraint. This view is increasingly confirmed by field and carefully controlled laboratory studies. Using temperature-sensitive radio transmitters to track daily patterns in T_b of *Eptesicus fuscus*, Audet and Fenton (1988), Grinevich et al. (1995), and Hamilton and Barclay (1995) have all shown that females normally maintain high T_b throughout pregnancy and lactation. Similarly, Kurta et al. (1987) showed that pregnant *M. lucifugus* also maintain high resting MR and T_b throughout the day. Pregnant *Myotis lucifugus* will, however, resort to torpor when food intake is restricted for a single night (Kurta 1991) as do phyllostomids when facing food restriction (Audet and Thomas 1997).

While the fitness benefit for maintaining a high T_b is relatively clear for pregnant and lactating females, the benefit for males and non-reproductive fe-males is less clear. During at least the period when females are lactating, males, too, must invest in production, witnessed by testicular development and spermatogenesis. Kurta and Kunz (1988) found that at least some males defended high T_b when well fed, and they suggested that males might face the same reproductive constraints as reproductive females. It is interesting to note that, at high latitudes, the number of male *Plecotus auritus* found in maternity colonies increases, and this may reflect their attempt to reduce thermoregula-tory costs while maintaining high T_b (Entwistle et al. 1996). However, the only field studies of thermoregulatory patterns of male and non-reproductive fe-males (*E. fuscus*) indicate that they normally select cooler roosts (Hamilton and Barclay 1995) and regularly use torpor (Grinevitch et al. 1995).

Hibernation and the Torpor-Arousal Cycle

During winter in the temperate zone, the decline in both food supply and T_a combine to make the defense of an elevated T_b problematic. Some bats escape this energetic bottleneck by migrating (Christensen 1947; Davis and Hitchcock 1965; Fenton and Thomas 1985; Fleming and Eby, this volume; Strelkov 1969). However, the most common overwintering strategy for temperate-zone spe-cies is hibernation. In autumn, bats undergo a marked and rapid fattening and generally begin hibernation with fat reserves amounting to $\geq 25\%$ of M_b (Baker et al. 1968; Ewing et al. 1970; Krulin and Sealander 1972; Kunz et al. 1998b; Polskey and Sealander 1979). Depending on latitude and local climates, these fat reserves must fuel metabolism for periods up to 200 d.

Superficially, hibernation differs from shallow daily torpor in the duration of torpor bouts and the depth of T_b depression (Geiser 1994; Geiser and Ruf 1995; Heldmaier et al. 1993). While bats in daily torpor remain torpid for peri-ods of 2–15 h, hibernating bats may remain torpid for 2–15 d, with continuous torpor up to 75 d having been reported (e.g., Brack and Twente 1985; Menaker 1964; Twente et al. 1985). During daily torpor, T_h seldom falls below $10°C$, but in hibernation T_b may fall to $2°C$. Supercooling ($T_b < 0°C$) has been reported (Davis and Reite 1967; Hock 1951), but considering that T_b is normally regu-lated at $T_{b\text{-}min}$, these reports need confirmation. True hibernation and the shal-low daily torpor that bats show during the summer might be viewed as sim-ply reflecting two levels of expression of the same thermoregulatory process. However, there is evidence that torpor during deep hibernation involves ad-ditional metabolic suppression. Analyzing data for mammals that express various degrees of torpor, Geiser (1988) found the Q_{10} associated with torpor during deep hibernation was higher than that for shallow daily torpor. He in-terpreted this as indicating a regulated suppression of TMR over and above a simple temperature effect during hibernation. Unfortunately no data were available to allow the comparison of Q_{10} for shallow and deep torpor within the same species, thus it is impossible to know whether the different Q_{10} are due to interspecific differences in the effect of T_b on TMR or a fundamental difference between shallow daily torpor and deep torpor during hibernation. Because many bats exhibit both shallow daily torpor and torpor during deep hibernation, they may be the ideal models for future studies of the regulation of MR in torpor. The depression of T_b during hibernation is also accompanied by changes in lipid composition that act to maintain membrane function (Geiser and McMurchie 1984; Geiser et al. 1984).

During hibernation, TMR is so low that even the smallest fat reserves could sustain animals throughout the winter if they remained in continuous torpor. Thomas et al. (1990a, 1990b) estimated that roughly 308 mg of fat could sustain *Myotis lucifugus* during 193 d of torpor if animals remained in continuous torpor. However, all hibernators arouse periodically, and the return to euthermic T_b is metabolically expensive. For *M. lucifugus*, each arousal of several hours duration costs 108 mg of fat or the equivalent of 67 d of torpor. At a rhythm of roughly one arousal every 12–15 d (Thomas 1995b; Thomas et al. 1990b), arousals account for about 85% of fat depletion through the winter for *M. lucifugus*, and this same pattern has been found for other hibernators (Kayser 1961, 1965; Wang 1978). If overwinter survival depends on the judicious meting out of energy from a limited fat reserve, then it is worth questioning just why animals expend so much energy on arousals. Many bats do die during winter, presumably due to exhaustion of their fat reserves (Davis and Hitchcock 1965; Ransome 1990), thus arousals probably satisfy some fundamental physiological requirement. Theories explaining arousals abound (see Thomas and Geiser 1997).

Until recently, the principal explanation for arousals was for feeding. White (1786) noted that *Nyctalus noctula* aroused more frequently on warmer nights when winter-active insects were flying, and observations confirmed that bats did feed. For *P. pipistrellus* in England, Avery (1985) formulated a cost-benefit model, which associated arousal decisions with energy balance during the arousal period. This model predicted that pipistrelles should not arouse at T_a <8°C due to the combination of high foraging costs and low insect availability. Most observations of winter activity in this species do indeed correspond with predictions. Brigham (1987) also suggested feeding might prompt arousals of *E. fuscus* because winter-active individuals had depleted fat reserves. However, low fat reserves may be the result rather than the cause of arousals, especially if little food is available. Based on laboratory studies of *E. fuscus*, Twente and Twente (1987) argued that arousals were timed by an internal "biological alarm clock" to coincide with evening feeding opportunities. Thomas (1993, 1995b), however, could find no field evidence for such a temporal pattern to arousals of *M. lucifugus*. Park et al. (1999), though, found that arousals in greater horseshoe bats (*Rhinolophus ferrumequinum*) in southern England were closely timed to coincide with sunset. These apparently conflicting data are not irreconcilable. In the severe winters experienced in Canada, feeding opportunities are rare or usually absent so that timing arousal to coincide with sunset is probably unimportant. However, in other areas, including the temperate climate of southern England, feeding opportunities occur throughout winter (e.g., Avery 1985), so timing arousals to coincide with sunset may be important. An important point, however, is that even in the complete absence of feeding opportunities bats continue to arouse. They may time these arousals so that they can emerge and feed (e.g., Avery 1985), but on occasion they may leave hibernacula without feeding (Whitaker and Rissler 1993) or they may arouse without leaving the hibernaculum (Thomas 1995b). Feeding may occur during arousals in some species, but it does not appear to be the underlying cause for arousals.

There is no question that T_a and T_b affect the duration of torpor bouts and the frequency of arousals (e.g., Brack and Twente 1985; Twente et al. 1985), but by itself this correlation is not a satisfactory explanation for the existence of arousals. A number of authors have argued that T_a and T_b determine the rate of depletion of energy substrates, the accumulation of metabolic wastes, the loss of membrane potentials, and even the accumulation of a sleep deficit (see Lyman et al. [1981] and Thomas and Geiser [1997] for reviews; see also Daan et al. 1991; Strijkstra and Daan 1997, 1998; Trachsel et al. 1991). However, the fundamental obstacle to identifying the cause of arousals is that all of these processes vary in the same way with T_a and T_b. Arousals may also be driven by thirst. Fisher and Manery (1967), Speakman and Racey (1989), and Thomas and Cloutier (1992) all suggested that low but finite rates of evaporation during prolonged torpor bouts lead to dehydration, forcing hibernators to arouse to replenish depleted water stores. Because evaporation is temperature dependent, a water-balance theory also predicts a correlation between T_b and arousal frequency. Thomas and Cloutier (1992) argued that metabolic water production could not cover evaporative losses for *M. lucifugus,* and, indeed, numerous authors have reported drinking in cave bats (e.g., Davis 1970; Shimoizumi 1959). Analyzing energy and water budgets of hibernating and arousing *P. pipistrellus,* Speakman and Racey (1989) concluded that bats fed to obtain water rather than energy, which adds a twist to Avery's (1985) foraging model. The most recent evidence in support of a water-balance theory comes from Thomas and Geiser (1997), who drew on data for *M. lucifugus* to model cutaneous and pulmonary water losses for hibernating *Spermophilus lateralis.* They found that the rate of evaporative water loss predicts torpor bout duration (and, hence, arousal frequency) far more accurately than did either T_b or TMR for T_a ranging from $-2°C$ to $+8°C$ (normal hibernation temperatures). While this study cannot establish a causal link between arousal and water balance, the strength of the correlation indicates that evaporative water loss warrants further study.

If arousals are an energetically expensive, but necessary, consequence of hibernation, then bats are presumably under strong selection pressure to optimize the size of the prehibernation fat store and its use during the winter. The fact that survival by hibernating bats is high (e.g., Keen and Hitchcock 1980) argues that bats normally are able to balance their winter energy budget successfully. However, by provoking additional arousals and thus dramatically altering overwinter energy expenditures, human intervention can shift the net hibernation energy balance from positive to negative and thereby affect survival. This reasoning has led to several studies on the effect of human disturbance on the frequency of arousals.

In a laboratory study on the effects of tactile (physical contact) and nontactile (temperature, light, and sound) stimuli on *P. pipistrellus,* Speakman et al. (1991) found that, while tactile stimuli provoked a clear and energetically costly arousal response, nontactile stimulation generally did not. In contrast, however, Hall (1832) noted that hibernating bats under natural conditions aroused when they were subjected to noise and light. Using infrared detectors and temperature-sensitive radio transmitters to study the effect of human presence on the arousals of hibernating *M. lucifugus,* Thomas (1995a, 1995b) found that even brief visits to a cave where no bats were handled resulted in a burst of activity that lasted up to 8.5 h. This suggests that bats are, in fact, sensitive to human disturbance. The difference between Thomas's and Speakman's results appears to stem from the conditions under which bats were studied. In the laboratory, bats are typically tested individually and soon after entry into torpor when they are less sensitive to disturbance. In the field, large numbers of bats are exposed to the stimulus, and the bats have been in torpor for various lengths of time. The probability of finding at least a few arousals is thus increased. Even nontactile events precipitated arousals in approximately 5% of bats in Speakman et al.'s (1991) study. If, on entering a cave containing approximately 1,300 bats (Thomas 1995a), arousals were precipitated, at this rate, 65 bats would arouse. By flying, attempting to mate, and finally reintegrating into hibernating clusters, these active bats would arouse others, resulting in a "cascade effect" (Thomas 1995b). Thus, human disturbance clearly does affect arousal frequencies of bats and may have a negative impact on their energy balance.

Metabolic Rate and Gas Exchange during Torpor

Considering the importance of torpor and hibernation in the Chiroptera, surprisingly few measurements of TMR (metabolic rate during torpor) have been made. We found measures of TMR for only 18 species, although several of these were each the subject of several independent studies. Data were available for three megachiropterans, 12 vespertilionids, and single representatives of the Natalidae, Molossidae, and the Emballonuidae (table 10.2). For each species for which data were available, we determined T_b at which TMR reached a minimum ($T_{b\text{-min}}$ and TMR_{min}). Mean $T_{b\text{-min}}$ was 11.7°C (median 9°C, SE = 1.68°C), but $T_{b\text{-min}}$ showed extreme variation, ranging from 2°C for *M. lucifugus*

Table 10.2. Torpid metabolic rates of 18 species of bats

Species	FA	Diet	M_b	TMR	$T_{a\text{-min}}$	References
Macroglossus minimus	1	2	15.5	10.9	20.0	Bartels et al. 1998
Syconycteris australis[a]	1	2	18.4	10.7	18.1	Geiser et al. 1996
Nyctimene albiventer	1	2	28.2	19.2	22.0	Bartholomew et al. 1970
Peropteryx macrotis	2	1	5.0	4.0	24.2	Genoud et al. 1990
Tadarida brasiliensis	9	1	16.9	1.7	8.0	Herreid and Schmidt-Nielsen 1966
Myotis lucifugus[a]	10	1	5.2	0.2	2.0	Hock 1951
Pipistrellus pipistrellus[a]	10	1	5.8	0.6	6.0	Speakman et al. 1991
Barbastella barbastellus	10	1	7.0	0.3	4.0	Cited in Speakman et al. 1991
Nyctophilus geoffroyi	10	1	8.0	2.1	15.0	Hosken and Withers 1999
Myotis nattereri	10	1	8.9	0.5	8.0	Speakman et al. 1991
Myotis daubentoni	10	1	9.5	0.4	8.0	Speakman et al. 1991
Plecotus auritus	10	1	10.2	0.7	6.0	Speakman et al. 1991
Nyctolphilus major	10	1	13.6	2.6	15.0	Hosken 1997
Chalinolobus gouldii	10	1	17.5	1.8	10.0	Hosken and Withers 1997
Eptesicus serotinus	10	1	22.7	0.6	6.0	Speakman et al. 1991
Nyctalus noctula[a]	10	1	26.5	0.8	6.0	Speakman et al. 1991
Myotis myotis[a]	10	1	30.0	6.0	10.0	Hanus 1959
Natalus tumidirostris	11	1	4.9	3.7	23.6	Genoud et al. 1990

Note. FA is family affiliation; temperature ($T_{a\text{-min}}$) is that at which the minimum metabolic rate occurred (°C); body mass (M_b) is in grams; and torpid metabolism (TMR) is oxygen consumption (mL h^{-1}).

[a]Several measurements were available in the literature for this species, and the value is an average of these several measurements. Only one reference is cited. Species are listed first by family and then by body mass. Refer to table 1 for key to family and diets.

(Hock 1951) to 24.2°C for the emballonurid, *Peropteryx macrotis* (Genoud et al. 1990). Bats appear to vary enormously in $T_{b\text{-min}}$, which might reflect differing physiological abilities to tolerate low T_b. The few data that are available for tropical species suggest that they defend much higher $T_{b\text{-min}}$ than temperate vespertilionids. This apparent inability to allow T_b to fall below approximately 15°C may have important implications for the biogeography of small tropical species (Bonaccorso and McNab 1997; Geiser et al. 1996; Law 1994).

As would be expected from the Q_{10} effect, TMR_{min} was positively correlated with $T_{b\text{-min}}$ (fig. 10.9A: $r^2 = 0.715$, $F_{1,\ 16} = 40.15$, $P < 0.0001$), and much of the residual variation was explained by M_b (fig. 10.9B). The multiple regression equation

$$\log_e TMR\ (mL\ O_2\ h^{-1}) = -3.87 + 0.163\ T_{b\text{-min}}\ (°C) + 0.988\ \log_e M_b\ (g) \quad (4)$$

explained 91.8% of the variation in TMR (multiple regression: $F = 84.3$, df = 2, 15, $P < 0.0001$). The slope of the mass effect was almost 1.0 (fig. 10.9B), as has been observed in other hibernating animals during torpor (e.g., Geiser 1988).

Because bats spend protracted periods in torpor when they are hibernating, one might predict that the optimal T_a that a hibernating bat should select would be equivalent to $T_{b\text{-min}}$, thus permitting bats to metabolize at TMR_{min} and draw down their fat reserves at the lowest possible rate. In their review of hibernacula temperatures, Webb et al. (1996) noted that hibernating bats

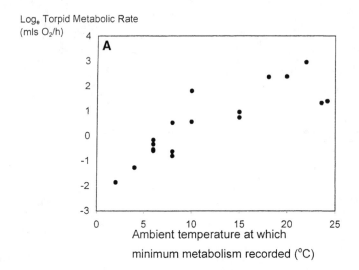

Log$_e$ Torpid Metabolic Rate (mls O$_2$/h)

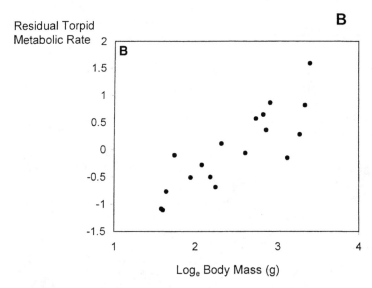

Residual Torpid Metabolic Rate

Log$_e$ Body Mass (g)

Figure 10.9. *A,* Torpid metabolic rate in relation to the ambient temperature at which metabolism reached a minimum; *B,* torpid metabolic rate (with the effects of ambient temperature removed) as a function of body mass (refer to table 10.2 for original data).

appear to select (or at least tolerate) a very wide range of T_a, spanning almost 15°–20°C depending on the species. The lack of any clear mode at $T_{b\text{-min}}$ suggests that bats either do not (or are unable to) select "optimal" hibernation temperatures. However, bats do alter their roosting locations through the hibernation period, and Thomas and Cloutier (1992) likened this to an evaporation/condensation process driven by arousals. Bats in suboptimal areas arouse more frequently and tend to converge on areas where the arousal frequency is reduced. Bats clearly move up the T_a gradient to avoid freezing temperatures, but it is unclear whether bats move down the gradient to avoid areas of high T_a. If TMR were the sole consideration, we might expect bats to converge on areas where T_a approaches $T_{b\text{-min}}$; however, the existence of arousals and their high cost complicate this simple prediction. The benefit of moving to areas where $T_a = T_{b\text{-min}}$ must integrate the costs of torpor and the costs of rewarming and euthermy during the arousal cycle. We will develop a quantitative model elsewhere but briefly outline our reasoning here. At $T_a <$ $T_{b\text{-min}}$, TMR increases due to the metabolic defense of $T_{b\text{-min}}$. At the same time, low T_a increases the cost of rewarming and the cost of euthermy during an arousal, thus bats should move to reduce these costs through the next torpor-arousal cycle. At $T_a > T_{b\text{-min}}$, TMR increases slowly due to the Q_{10} effect, but the increase in the cost of torpor with increasing T_a is offset by the reduced cost of rewarming and euthermy. Thus, there may be a strong advantage to move up a T_a gradient to $T_{b\text{-min}}$ but little advantage to move down the T_a gradient. This would explain the large variance in hibernacula temperatures, and this same reasoning may explain how some species of bats manage to hibernate successfully across large latitudinal gradients.

Regardless of the T_a that bats either select or tolerate when in torpor, the net effect of the reduction in T_b is a dramatic decrease in resting MR. The extreme reduction in MR with changes in torpor results in shifts in oxygen demand that must be tracked by changes in the uptake of oxygen at the lungs. Animals can alter the delivery of oxygen into their circulatory system by changing the ventilation rate, the tidal volume, or the oxygen extraction efficiency. Like other mammals, but unlike birds, bats adjust all three parameters to match the supply to the demand for oxygen (Chappell and Roverud 1990; Studier and O'Farrell 1976). Webb et al. (1992), for example, observed the changes in ventilation frequency of euthermic *P. auritus* as a function of changes in oxygen consumption as T_a declined. Although there was a linear relation between oxygen consumption and ventilation rate, the slope of this relation meant that bats simultaneously adjusted either extraction efficiency or tidal volume.

During torpor, oxygen demand falls to extremely low levels and this gives rise to an intermittent or apnaeic respiratory pattern. Bouts of rapid ventilation are punctuated by periods of apnea that may vary from 40 s to well over 60 min, depending on species and T_b (Hays et al. 1991; Shimoizumi 1959; Szewczak and Jackson 1992; Thomas and Cloutier 1992; Thomas et al. 1990a;

Webb et al. 1992). Similar breathing patterns are common in other hibernators and in reptiles (e.g., Milsom et al. 1993). There has been some debate over the processes that generate these periods of apnea. Some authors have suggested that the glottis remains open during apnea and gas exchange continues at a low level by diffusion (Malan [1980] for hedgehogs, *Erinaceus europaeus,* and Hays et al. [1991] and Szewczak and Jackson [1992] for bats). By partitioning oxygen uptake during the ventilation-apnea cycle, Thomas et al. (1990a) concluded the glottis remained closed during apnea for *M. lucifugus.* Computed tomography scans confirm the glottis remains closed in ground squirrels, *Spermophilus lateralis* (Garland and Milsom 1994), but these are substantially larger animals and generalizing this observation to bats is dubious.

Whether or not the glottis remains open during apnea, discontinuous breathing leads to wide fluctuations in blood-gas pressures, P_{O_2} and P_{CO_2} (Steffen and Riedesel 1982). Tähti and Soivio (1975) suggested that the progressive increase in P_{CO_2} through apnea may stimulate the onset of a breathing episode, while the subsequent drop in P_{CO_2} during the breathing bout might cause breathing to stop. If this were the case, one would expect that the duration of apnea to be shorter at higher T_a when oxygen demand and CO_2 production are greater and that at any given T_a the duration of apnea and the following ventilation bout would be correlated. Both patterns are observed in torpid pipistrelle bats (Hays et al. 1991), but direct measurements of end tidal gas compositions in hibernating ground squirrels at the beginning and end of breathing bouts does not suggest the existence of a fixed trigger or threshold P_{CO_2} level (Garland and Milsom 1994). Curiously, a large proportion of the oxygen demand during breathing bouts in ground squirrels comes from the cost of breathing itself, with the balance fueling the oxygen debt accumulated during the previous period of apnea. Whether this is also the case in bats remains to be established.

Flight and the Energetic Costs of Locomotion

Movement costs energy and the magnitude of the cost depends on the mode of locomotion and the speed at which animals travel. For the major modes of locomotion (swimming, walking, running, and flying), the elevation and shape of the curve relating energy cost to speed varies, thus locomotory mode has an important impact on the costs of transport. Although some species of bats are accomplished at quadrupedal terrestrial locomotion (in particular the Desmodontidae and the Mystacinidae [Dwyer 1962]), no bats rely on terrestrial locomotion as their primary mode. In consequence, the energetic costs associated with terrestrial locomotion do not form a large portion of the energy budget.

Because bats are morphologically highly specialized for flight and because flight style varies considerably with diet and foraging mode, considerable attention has been paid to the costs and kinematics of flight in bats. Terrestrial animals move by exerting force against the ground and using the reaction force to drive the body forward and upward until gravity pulls it forward and downward (McNeil Alexander 1999). To fly, animals not only need to generate the forces necessary for forward motion, but they must also generate the lift to offset gravitational attraction (Norberg 1990). And they must do so by exerting a force against a deformable fluid (air) that produces a more diffuse reaction force than does the solid ground. As a result, the energy cost of flight is considerably greater than the energy costs of terrestrial locomotion (Norberg 1990; Schmidt-Nielsen 1990).

The costs associated with flight can be resolved into three primary components: lift, body drag (parasite drag), and wing drag (profile drag [Norberg 1990; Pennycuick 1969, 1989; Rayner 1979]). Although the inertial costs of flapping the wings have been assumed to be negligible (Norberg 1990), recent studies suggest that they may be significant, possibly adding a fourth consideration to the energy cost of flight (Norberg et al. 1993). The three primary costs (lift, parasite drag, and profile drag) vary in different ways with speed (fig. 10.10). The cost of generating lift is inversely related to flight speed because it is the forward motion that forces air over the wing surface that generates lift; the faster the body moves, the more air passes over the wing, and the greater the lift generated. As forward air speed declines, the animal must alter the flapping motion of its wings to generate the airflow over the wings (the induced velocity), and this costs energy. Drag, however, increases with airspeed so that animals are forced to expend energy to maintain forward velocity. Combining the three aerodynamic costs results in a U-shaped curve of cost against speed (fig. 10.10). Because the curve is U-shaped rather than linear, the speed at which animals minimize their energy expenditure is not zero velocity (as it is for terrestrial locomotion) but, rather, at some fixed forward velocity set by the base of the U. This is called the minimum power speed, and the aerodynamic cost of flying at this speed, normally expressed in watts, is called the minimum power. By flying at minimum power speed, bats fly as cheaply as possible per unit time. However, animals probably fly to cover distance rather than to fill time. To cover as much distance as possible for a given energy cost, animals can do better by flying a little faster than the minimum power speed. Due to the relatively flat shape at the bottom of the curve, the extra speed results in only a slight increase in cost. The maximum-range speed occurs where a tangent intersects the curve (fig. 10.10). Aerodynamic mechanical power requirements at this speed are called maximum-range power. Aerodynamic models show that the parameters that determine the shape of the power-velocity curve are dependent on M_b and wing shape. In theory, long narrow wings, typified by many *Tadarida* species, reduce power costs but increase the minimum power and maximum-range speeds. Short broad wings, typified by hovering bats such as *Glossophaga*, should result in

POWER
REQUIREMENT

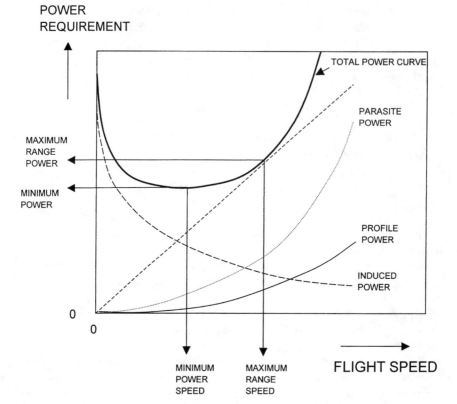

Figure 10.10. Components of the energy costs of flight used by aerodynamicists and the resultant total mechanical power cost of flight in relation to flight speed. The speed at the base of the U-shaped power-speed curve is the minimum power speed and the speed at a point where a tangent from the origin intersects the curve is the maximum range speed.

higher power costs but offer the lower minimum power and maximum-range speeds compatible with maneuverable hovering flight. Aerodynamic modeling therefore suggests that there is a trade-off between the flight cost, speed, and maneuverability.

Because animals are not 100% efficient at converting chemical energy to mechanical energy in their muscles, the mechanical costs of flight indicated by aerodynamic models represent only part of the total metabolic cost. Muscular efficiency lies in the range of 20%, thus animals face flight costs roughly five times those predicted by aerodynamic models, losing roughly 80% of this energy as heat. Aerodynamic modelers generally assume that the efficiency of mechanical power generation is constant across species, individuals within a species, and speeds within individuals. No data are available for bats, but for birds this assumption appears to be erroneous (Bishop 1997; Ward et al. 1999). For this reason, attempts to estimate flight energy costs based on morphology and assuming a fixed efficiency provide only a poor evaluation of the true metabolic costs of flight.

The standard method for measuring metabolic flight costs directly is based on mask respirometry (Tucker 1966), whereby birds and bats are trained to fly in wind tunnels while wearing a mask to draw off their respiratory gases. By varying wind and hence flight speed in the tunnel, the effects of flight speed on cost can be determined. The technique is difficult to apply because of the time required to train bats to fly for the protracted periods necessary for them to reach steady state oxygen consumption (Rothe et al. 1987). A total of only 10 individual bats across three species have been trained in this way, but these data suggest that bats broadly conform to the expectations of the aerodynamic models in that there is a U-shaped curve of cost against speed. Criticisms of these measurements include the potential problems of circulating vortices in enclosed spaces (Rayner and Thomas 1991), the costs of carrying the mask (Tucker 1972), and stress. A novel application of the respirometry approach was made by Winter (1998), who integrated a fixed-position mask with an automated feeder, allowing him to measure flight costs of nectarivorous bats as they hovered at the feeder. This approach overcomes the problems of vortices, carrying the mask, and stress but provides measures only for the costs of hovering (flight at speed 0), which in theory (fig. 10.10) substantially exceed the minimal power and maximum-range costs.

An alternative method of measuring the energy costs of flight is to use the doubly-labeled water method (DLW [Lifson and McClintock 1966; Nagy 1980; Speakman 1997]), which depends on the differential elimination of oxygen and hydrogen isotopes. Although the DLW method allows an estimate of energy demands for unrestrained animals, it is far less precise than indirect calorimetry and does not allow the determination of a power curve in relation to flight speed because the animals select the speed at which they will fly and inevitably include maneuvers such as turning and rolling that are absent in wind tunnel flight. Moreover, because a certain degree of isotope turnover needs to occur to obtain a precise and accurate estimate of expenditure (see Speakman [1997] for details), bats need to fly for considerable periods to derive an estimate of flight cost (one to several hours). This can be overcome to some extent by pooling measurements across individuals, plotting the total cost as a function of the percent time spent flying, and then extrapolating to estimate the cost at 100% time in flight (e.g., Racey and Speakman 1987) or by measuring the costs during the period the animals are not flying directly by respirometry and subtracting this from the total cost by DLW (Speakman and Racey 1991). The disadvantages may outweigh the advantages, however, and in total the DLW method has been applied to only a few species of bats

Table 10.3. Metabolic rates of 13 bat species during flight

Species	M_b	E_{fly}	Method	References
Pipistrellus pipistrellus	6.7	1.12	DLW	Speakman and Racey 1991
Plecotus auritus [a]	7.85	1.17	DLW	Speakman and Racey 1991
Glossophaga commissarisi	8.6	1.23	Mass bal	Winter and von Helversen 1998
Hylonycteris underwoodi	8.7	1.3	Mass bal	Winter and von Helversen 1998
Glossophaga soricina [a]	11.6	1.63	Mass bal/mask	Winter and von Helversen 1998
Glossophaga longirostris	16.3	1.97	Mass bal	Winter and von Helversen 1998
Choeronycteris mexicana	19.6	2.57	Mass bal	Winter and von Helversen 1998
Leptonycteris curasoae	28.2	3.16	Mass bal	Winter and von Helversen 1998
Phyllostomus hastatus	93	8.83	Mask	Thomas 1975
Hypsignathus monstrosus	258	23.3	Mask	Carpenter 1986
Eidolon helvum	315	22.4	Mask	Carpenter 1986
Pteropus polioeophalus	628	34.8	Mask	Carpenter 1985
Pteropus gouldii	779	43.7	Mask	Thomas 1981

Note. Body mass (M_b) is in grams and the flight cost (E_{fly}) is in watts. Species are ordered by their body masses. Method refers to the measurement approach. DLW = doubly-labeled water, mass bal = mass balance, and mask is by mask respirometry.

[a] Subjects of multiple studies — only one reference is cited.

(table 10.3). Although no studies have cross-calibrated DLW measures with indirect calorimetry for bats, one study on starlings (*Sturnus vulgaris*) indicated that the two estimates did not differ significantly (Ward et al. 1999).

Most recently, Winter and Helversen (1998) developed a method that combines measures of energy intake and mass balance for nectarivorous bats with their time budgets. By considering large numbers of observation periods in which the bats vary the proportions of time spent in flight and at rest, they were able to determine the energy costs of resting and flight of six species of nectarivorous bat (summarized in Winter and Helversen [1998]). As with the DLW method, this approach provides an estimate of flight cost at speeds that the bats select themselves and includes the cost of maneuvers within the total cost.

From mask respirometry studies that were made at a range of flight speeds, we calculated the flight costs at the minimum power speed. Combining these values with estimates based on the different methods, and averaging the measured costs where more than one estimate was available for a given species, flight cost is strongly dependent on M_b. The relationship

$$\log_e \text{flight cost (W)} = -1.46 + 0.794\, M_b\ (g) \tag{5}$$

explained 99.7 % of the variation in flight cost (fig. 10.11). The residual variability is insufficient at present to refine this prediction by incorporating other aspects of bat morphology. Indeed, the very high proportion of variation explained by M_b alone is remarkable because the estimates include costs derived using three different methods, at a range of different speeds from those where bats should have minimal flight costs, and those where they should be maximal (hovering).

The slope of this relationship is approximately the same as that for BMR in bats (0.794 vs 0.744). Flight cost thus appears to be set at a constant 16.5 times BMR across the mass range from 5 g up to 1.1 kg, in close agreement with the value of 15 times BMR found by Winter and Helversen (1998). The small sample size precludes any robust analysis of the residuals to test for an effect of variation in BMR or flight mode, thus we suggest that flight costs should be predicted directly from the allometric equation based on M_b (eq. 5).

Aerodynamic models that assume a constant efficiency of translating metabolic to mechanical power do not predict a fixed ratio between BMR and flight cost. Using published morphometric data for the species for which metabolic flight cost estimates are available, Norberg and Rayner (1987) and Rayner (1990) found that predicted mechanical costs of flight rise far more steeply than the observed metabolic costs, with an exponent of approximately 1.2 rather than 0.75. This means that flight efficiency is probably not constant

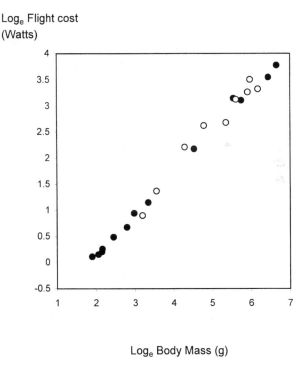

Figure 10.11. Energy costs of flight in relation to body mass for bats (*filled circles*) and from wind-tunnel studies of birds (*open circles*).

across species but scales positively with M_b. Small bats appear to fly with efficiencies of only 4% (i.e., 96% of the cost of flight is lost as heat), while larger bats fly at efficiencies of about 20%–25%. Similar positive scaling relationships for efficiency have recently been suggested for birds (Bishop 1997).

Winter and Helversen (1998) noted that the flight costs estimated from the mass balance approach were approximately 25% lower than allometric predictions of flight costs for birds (Masman and Klaasen 1987; Rayner 1990). The available data on flight costs for birds, however, represent an extremely heterogeneous data set, including a wide range of wing morphologies, M_b, and methods. Most estimates are for birds that are substantially larger than the bats measured by Winter and Helversen (1998), thus extrapolating predictions from the bird data set to bats at a smaller M_b is problematic. To refine this comparison, we selected measures of flight costs for birds to include estimates derived only by wind-tunnel respirometry (table 10.3). We then combined the estimated flight costs for bats (M_b: 7–700 g) with birds (M_b: 24–480g; fig. 10.11). This comparison is probably more valid than that made by Winter and Helversen (1998) because it compares costs over similar body masses and it permits more rigorous statistical testing based on the individual values. We found a strong effect of mass on flight cost ($F = 1,838.8$, df = 1, 19, $P < 0.001$), but no effect of taxon ($F = 0.77$, df = 1, 19, $P = 0.393$) and no significant interaction between mass and taxon. The residual values for bats averaged -0.018 and those for birds $+0.02$, in the same direction as that found by Winter and Helversen (1998), but this was far from significant. We found no evidence supporting the suggestion that bats fly more cheaply than birds do, but it is clear that more estimates are required for small birds.

One should note that the effect of M_b on flight cost is an interspecific relationship, and there is no reason to believe that the same relationship will hold within species. Species that differ in M_b also differ in wing area (Norberg and Rayner 1987), but the intraspecific variation in wing area is much reduced, and within individuals, wing area is constant. Norberg and Rayner (1987) calculated that the mechanical power requirements should scale with an exponent of 1.56 against M_b when wing dimensions remain fixed. Supporting this estimate, data on wing kinematics suggest that mechanical costs increased by 15% when a serotine bat (*Eptesicus serotinus*) was forced to carry an additional mass amounting to 11% of its body mass (Rayner et al. 1989). However, because mechanical costs make up only 4%–10% of metabolic costs and because mechanical requirements may only poorly match metabolic demands, it is difficult to predict how changes in M_b associated with foraging, fattening, pregnancy, or transporting fruits or radio transmitters will affect flight costs. Because mass changes are a regular component of the annual and breeding cycles of bats, solid data on flight costs are dearly needed. It is interesting to note that the maximum size of prehibernation fat reserves (ca. 30% of M_b [Kunz et al. 1998b]) closely matches fetal masses for hibernating bats like *M. lucifugus*

(Kurta and Kunz 1987). This may be coincidental, but it may also indicate the maximum mass that these bats can economically transport. Speakman and Racey (1991) were unable to detect an effect of individual variation in M_b on flight costs for *P. pipistrellus* when costs were estimated by a combination of respirometry and doubly-labeled water. This may indicate poor coupling between mechanical and metabolic flight costs, but it could also reflect measurement imprecision. More accurate measures of flight costs set the intraspecific scaling exponent of hovering costs in *Choeronycteris mexicana* as 1.76, compared with a theoretical value of 1.56 (Voigt and Winter 1999). For *G. soricina* in forward flight, Voigt (2000) found an intraindividual scaling exponent of 1.69 (compared with the theoretical value of 1.56) and an interindividual scaling exponent of 0.97 (compared with a theoretical expectation of 1.19). These comparisons broadly support the predictions of aerodynamic models, but for the time being the small database precludes any rigorous testing of aerodynamic models with empirical measures.

Estimating and Measuring Field Metabolic Rate in Bats

Bats differ from most other mammals by occupying thermally buffered roosts where they are at rest and near thermoneutrality for extended periods, while using an expensive locomotory mode (flight) when active. This daily shift between metabolic extremes clearly has an impact on total energy expenditures and a variety of decisions concerning where to roost, where to forage, and what and how much to eat. Because total daily energy expenditure (now usually called field metabolic rate, or FMR) sets the goal for the regulation of energy intake at the individual level and because it may even determine carrying capacity and community structure (Findley 1992), estimating or directly measuring FMR has been the focus of considerable effort.

The DLW method provides a direct measure of CO_2 production and hence FMR in free-ranging animals, but an alternate method of estimating the FMR is to construct a time-energy budget (TEB). If the energy demands associated with each activity are accurately known, one can then construct a total energy budget that should be equivalent to the FMR. Bat researchers have attempted to integrate knowledge of energy demands derived from the study of animals in the laboratory into estimates of FMR for at least 20 yr (e.g., Kunz 1980; Studier and O'Farrell 1980; Thomas 1984).

The equations that we have derived thus far allow predictions of the energy costs associated with resting (eq. 1), torpor (eq. 2) and flight (eq. 5), if various input parameters (such as M_b and T_a) are known. In combination with a known time-budget for the animal in question, this model should also be capable of generating estimates of daily energy expenditures (FMR) for bat species. To illustrate the capabilities of the equations, consider the following example. Pipistrelle bats (*P. pipistellus* = *pygmaeus*) live in roosts near Aberdeen with temperatures around 15°C in summer (Swift 1980). Within the roosts, however, they select specific sites and form clusters. Measurements in the vicinity of clusters (Speakman and Racey, unpublished data) suggest that, throughout the daytime roost period, the bats are normally within the thermoneutral zone ($T_a > 32$) and, therefore, metabolize at BMR. Individuals fly for an average of 225 min nightly (Swift 1980), and we have no evidence that they use night roosts. Non-reproducing bats weigh approximately 5.5 g (Speakman and Racey 1987). Using equation (1) the predicted BMR is 10.56 mL O_2 h^{-1}. Using an energy equivalent of 20.09 J mL O_2^{-1} gives an estimated metabolism of 212.3 J h^{-1}, or 4.2 kJ for the 20.25-h roosting period. From equation (5), the predicted flight cost is 1.0 W, giving a total flight cost of 13.5 kJ for the 3.75 h of flight. The total daily energy expenditure should be close to 17.7 kJ d^{-1}.

For many years, this type of TEB estimate for daily energy expenditures remained unvalidated because measuring the energy expenditures of free-living bats directly was technically difficult. Despite the fact that the DLW method now provides direct access to measures of FMR (see Speakman 1998), few studies have focused specifically on bats. The first measurements using DLW on a free-living bat were for the nectarivore *Anoura caudifer* (Helversen and Reyer 1984), and in the 15 yr since this first field application, there have been only eight further applications of the DLW method to bat species in the field (table 10.4). Bats remain one of the most poorly sampled of all small mammal groups (Nagy et al. 1999; Speakman 2000), with less than 1% of all bat species measured. Using the data for different species and subgroups (e.g., males and females, or pregnant and lactating individuals) as independent data yields 12 different estimates of daily energy expenditures.

The most significant factor influencing the energy demands of bats by DLW was M_b. The least squares regression

$$\log_e \text{FMR (kJ } d^{-1}) = 1.87 + 0.732 \log_e M_b \text{ (g)} \qquad (6)$$

explained 80% of the variation in FMR ($F = 40$, df $= 1, 10, P < 0.001$; fig. 10.12). The mean unsigned deviation of individual measures from equation (9) was 23.8% ($n = 12$, range $= -47\%$ to $+46\%$).

Despite the small sample size available for comparison, we can use these direct DLW measurements to test the precision and accuracy of our estimates of daily energy expenditure using the TEB approach illustrated above for *P. pipistrellus*. To derive a TEB model for each of the species for which measurements are available, we scoured the literature for information on the flight times of the species in question, the time spent roosting either in day or night roosts, and T_{roost}. In some cases, these measurements were available for the particular bats that had been measured (e.g., Kunz et al. 1998a), but in other cases we had to use data from the same population or even from other popu-

Table 10.4. Field metabolic rates (FMR) of nine bat species measured using the doubly-labeled water method

Species	M_b	FMR	Model 1	Model 2	Model 3	Model 4	References
Pipistrellus pipistrellus	7.6	29.3	21.1	23.8	21.1	23.8	Racey and Speakman 1987
Myotis lucifugus (L)	7.9	25.3	22.8	25.5	22.8	25.5	Kurta et al. 1989
Plecotus auritus	8.5	27	17.7	20.8	30.1	39.4	Speakman and Racey 1987
Myotis lucifugus (P)	9	29.9	25.7	28.7	25.7	28.7	Kurta et al. 1989
Anoura caudifer	11.5	51.9	36	39.5	36.3	39.9	Helversen and Reyer 1984
Macrotus californicus	12.9	22.8	27.5	31.7	34.4	42	Bell et al. 1986
Syconycteris australis	17.4	76.9	34.7	39.9	57.9	74.8	Coburn and Geiser 1999
Eptesicus fuscus (L)	17.4	75.3	34.7	39.9	50.9	64.3	Kurta et al. 1990
Carollia perspicillata	19.5	79.3	54.5	59.7	54.5	59.7	Thomas 1988
Eptesicus fuscus (P)	20.8	47.6	39.9	45.9	57.4	72.1	Kurta et al. 1990
Phyllostomus hastatus (F)	74.4	132.2	95.6	111.5	95.6	111.5	Kunz et al. 1998a
Phyllostomus hastatus (M)	87.1	159.7	84.9	103.4	84.9	103.4	Kunz et al. 1998a

Note. Some species are subdivided into distinct groups labeled "L" and "P" for lactating and pregnant or "M" and "F" for males and females, respectively. The data represent the mean metabolic rates in kJ d^{-1} and mean body masses (M_b, grams) measured across a group of animals for each species. In addition to the observed field metabolic rate (FMR), the estimated metabolic rates from four different time and energy budget models are presented. Model 1 assumes resting animals always metabolise at BMR. Model 2 assumes bats at rest metabolize at 1.5 times BMR to account for the costs of nonflight activity. Models 3 and 4 are the same as 1 and 2, respectively, except the temperature profiles of the roost sites are used to assess thermoregulatory demands as well as the costs of activity in the nonflight period.

lations. Where data were not available (e.g., T_{roost}), we made estimates derived from climatic data for the sites in question and the periods when measurements had been made. To estimate flight costs, we multiplied values derived from equation (5) by the time spent in flight. To estimate resting MR, we made several assumptions, thereby generating four alternative estimates and thus four estimates of total expenditures when these were combined with flight costs. In the first model, we assumed that the bats would find microhabitats that allowed metabolism to run at BMR when they were at rest. We thus multiplied the time spent at rest by the predicted BMR based on M_b using equation (1). Bats, however, do not spend their roosting times completely inactive. For *M. lucifugus*, 79% of the roost period is spent resting, with 21% of the time spent grooming and in movement (Burnett and August 1981), and some bat species may spend up to 40% of their time grooming (McLean and Speakman 1997), yet we have no data on the costs of these activities. For model 2, we assumed that these activities raised the average roosting MR to 1.5 times BMR for the roosting period. For model 3, we estimated resting MR by combining data on T_a with predictions of T_{lc} and thermal conductance of the bats involved. These estimates also include estimates of roost heating due to clustering behavior where appropriate. Model 4 was based on model 3 but added the cost of roosting activities, assuming a cost 1.5 times BMR.

Log$_e$ Field Metabolic Rate
(kJ/day)

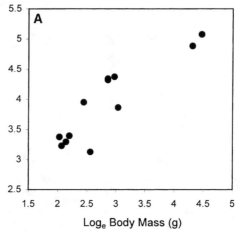

Log$_e$ Body Mass (g)

Field Metabolic Rate
(kJ/day)

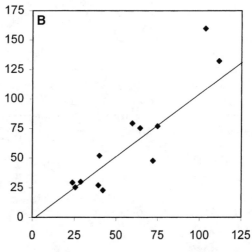

Daily Energy Expenditure
(Time/Energy budget model 4) kJ/day

Figure 10.12. *A,* Daily energy demands of bats in the field measured using doubly labeled water (for actual data refer to table 10.4) in relation to
body mass; *B,* comparison of the doubly labeled water estimates of daily
energy expenditure (field metabolic rate) with the time and energy budget
estimates for the same species/situation, using a time and energy budget
model that incorporates thermoregulatory demands and an estimate of the
costs of non-resting behaviors when roosting (model 4). The *solid line*
reflects equality.

We compared the estimates using these four separate models with the DLW measures of FMR for the nine species, subdividing the estimates where appropriate into homogenous groups (e.g., separating males from females [Kunz et al. 1998a]). This resulted in a total of 12 comparisons between the time-energy budget approach and the DLW method estimates (table 10.4).

The simplest model (1) underestimated the observed energy costs by a mean of 47.8% (range: −120% to +17%). Excluding the costs of activities in the roosts and thermoregulatory requirements leads to large underestimates of energy demands. Including both these costs (model 4) reduced the average discrepancy to only −5.7%, but the range of comparisons was again high (range: −54% to +45%). These discrepancies exceed by an order of magnitude the imprecision of the DLW method when it is applied to groups of individuals (averaging about 3%; reviewed in Speakman [1997]), so the deviations are unlikely to reflect inaccuracy in the DLW estimates. Including estimated costs of thermoregulation and nonflight activity increases the accuracy of TEB predictions; however, discrepancies for individual species can be large (fig. 10.12B). This has also generally proved to be the case in other mammalian groups (e.g., Corp et al. 1999) and does not appear to be unique to bats. At present, the underlying cause of these discrepancies is obscure. The data reviewed here and the allometric equations derived therefrom provide a method for generating estimates of daily energy expenditures of bats that differ on average from the measured demands by DLW by 24.5% (range = −54% to +46%). These figures are remarkably similar to the errors that occur simply using the scaling relationship based on DLW estimates of FMR to predict daily energy demands. At present, there would appear to be no benefit gained by applying a detailed TEB approach to estimating the energy demands of free-living bats. However, because the TEB approach allows us to partition costs between different thermoregulatory and physical activities, it should provide insight into just how bats regulate expenditures to achieve energy balance as environmental conditions change.

Acknowledgments

We are grateful to Tom Kunz and Brock Fenton for the invitation to write this chapter. Ariovaldo Cruz-Neto kindly provided some of his unpublished data on BMR of phyllostomid bats. Catherine Hambly, Mickael Henry, Murray Humphries, Colin Selman, and three anonymous referees made constructive comments on various versions of this chapter. We gratefully acknowledge support from Natural Environmental Research Council (United Kingdom), Biological Biotechnological Sciences Research Council (United Kingdom), Les Fonds FCAR (Québec), and Natural Sciences and Environmental Research Council (Canada).

Literature Cited

Adkins, R. M., and R. L. Honeycutt. 1991. Molecular phylogeny of the superorder Archonta. Proceedings of the National Academy of Sciences of the USA, 88:10317–10321.

Adolph, E. F. 1947. Tolerance of heat and dehydration in several species of small mammals. American Journal of Physiology, 151:564–575.

Allard, M. W., B. E. McNiff, and M. M. Miyamoto. 1992. Support for interordinal eutheria relationships with an emphasis on primates and their Archontan relatives. Molecular and Phylogenetic Evolution, 5:78–88.

Ammerman, L. K., and D. M. Hillis. 1992. A molecular test of bat relationships: monophyly or diphyly? Systematic Biology, 41:222–232.

Aschoff, J. 1981. Thermal conductance in mammals and birds: its dependence on body size and circadian phase. Comparative Biochemistry and Physiology A, 69:611–619.

Audet, D., and M. B. Fenton. 1988. Heterothermy and the use of torpor by the bat, *Eptesicus fuscus* (Chiroptera: Vespertilionidae): a field study. Physiological Zoology, 61:197–204.

Audet, D., and D. W. Thomas. 1997. Facultive hypothermia as a thermoregulatory strategy in the phyllostomid bats *Carollia perspicillata* and *Sturnira lilium*. Journal of Comparative Physiology B, 167:146–152.

Avery, M. I. 1985. Winter activity of pipistrelle bats. Journal of Animal Ecology, 54: 721–728.

Baker, R. J., M. J. Novacek, and N. B. Simmons. 1991. On the monophyly of bats. Systematic Zoology, 40:216–231.

Baker, W. W., S. G. Marshall, and V. B. Baker. 1968. Autumn fat deposition in the evening bat (*Nycticeus humeralis*). Journal of Mammalogy, 49:314–317.

Barclay, R. M. R., M. C. Kalcounis, L. H. Crampton, C. Stefan, M. J. Vonhof, L. Wilkinson, and R. M. Brigham. 1996. Can external radio transmitters be used to assess body temperature and torpor in bats. Journal of Mammalogy, 77:1102–1106.

Bartels, W., B. S. Law, and F. Geiser. 1998. Daily torpor and energetics in a tropical mammal, the northern blossom-bat *Macroglossus minimus* (Megachiroptera). Journal of Comparative Physiology B, 168:233–239.

Bartholomew, G. A., W. R. Dawson, and R. C. Lasiewski. 1970. Thermoregulation and heterothermy in some of the smaller flying foxes (Megachiroptera) of New Guinea. Zeitschrift für Vergleichende Physiologie, 70:196–209.

Bartholomew, G. A., J. W. Hudson, and T. R. Howell. 1962. Body temperature, oxygen consumption, evaporative water loss, and heart rate in the poor-will. Condor, 64:141–146.

Bartholomew, G. A., P. Leitner, and J. E. Nelson. 1964. Body temperature, oxygen consumption and heart rate in three species of Australian flying foxes. Physiological Zoology, 37:179–198.

Bell, G. P., G. A. Bartholomew, and K. A. Nagy. 1986. The roles of energetics, water economy, foraging behavior and geothermal refugia in the distribution of the bat, *Macrotus californicus*. Journal of Comparative Physiology, B, 156:441–450.

Bennett, P. M., and P. H. Harvey. 1985. Brain size development and metabolism of birds and mammals. Journal of Zoology (London), 207:491–509.

Bishop, C. M. 1997. Heart mass and the maximum cardiac output of birds and mammals: implications for estimating the maximum aerobic power input of flying animals. Philosophical Transactions of the Royal Society of London B, 352:447–456.

Blaxter, K. 1989. Energy Metabolism in Animals and Man. Cambridge University Press, Cambridge.

Bonaccorso, F. J., A. Arends, M. Genoud, D. Cantoni, and T. Morton. 1992. Thermal ecology of moustached and ghost-faced bats (Mormoopidae) in Venezuala. Journal of Mammalogy, 73:365–378.

Bonaccorso, F. J., and B. K. McNab. 1997. Plasticity of energetics in blossom bats (Pteropodidae): impact on distribution. Journal of Mammalogy, 78:1073–1088.

Bozinovic, F. 1992. Rate of basal metabolism of grazing rodents from different habitats. Journal of Mammalogy, 73:379–384.

Brack, V., and J. W. Twente. 1985. The duration of the period of hibernation of three species of vespertilionid bats. I. Field studies. Canadian Journal of Zoology, 63:2952–2954.

Brigham, R. M. 1987. The significance of winter activity by big brown bats (Eptesicus fuscus): the influence of energy reserves. Canadian Journal of Zoology, 65:1240–1242.

Bronner, G. N., S. K. Maloney, and R. Buffenstein. 1999. Survival tactics within thermally-challenging roosts: heat tolerance and cold sensitivity in the Angolan free-tailed bat, Mops condylurus. South African Journal of Zoology, 34:1–10.

Burnett, C. D., and P. V. August. 1981. Time and energy budgets for day-roosting in a maternity colony of Myotis lucifugus. Journal of Mammalogy, 62:758–766.

Canals, M., M. Rosenmann, and F. Bozinovic. 1989. Geometric aspects of the energetic effectiveness of huddling in small mammals. Acta Theriologica, 42:321–328.

Carpenter, R. E. 1968. Salt and water metabolism in the marine fishing bat, Pizonix vivesi. Comparative Biochemistry and Physiology, 24:951–964.

Carpenter, R. E. 1985. Flight physiology of flying foxes. Journal of Experimental Biology, 114:619–647.

Carpenter, R. E. 1986. Flight physiology of intermediate sized fruit bats (Pteropodidae). Journal of Experimental Biology, 120:79–103.

Carpenter, R. E., and J. B. Graham. 1967. Physiological responses to temperature in the long-nosed bat, Leptonycteris sanborni. Comparative Biochemistry and Physiology A, 22:709–722.

Casey, T. M., P. C. Withers, and K. K. Casey. 1979. Metabolic and respiratory responses of Arctic animals to ambient temperature during the summer. Comparative Biochemistry and Physiology A, 64:331–341.

Chaffee, R. R., and J. C. Roberts. 1971. Temperature acclimation in birds and mammals. Annual Review of Physiology, 33:155–202.

Chappell, M. A., and R. C. Roverud. 1990. Temperature effects on metabolism ventilation and oxygen extraction in a Neotropical bat. Respiratory Physiology, 81:401–412.

Christensen, E. 1947. Migration and hibernation of Tadarida mexicana. Journal of Mammalogy, 28:59–60.

Coburn, D. K., and F. Geiser. 1998. Seasonal changes in energetics and torpor patterns in the subtropical blossom-bat Syconycteris australis (Megachiroptera). Oecologia, 113:467–473.

Contreras, L. C. 1984. Bioenergetics of huddling: test of the psycho-social hypothesis. Journal of Mammalogy, 65:256–262.

Corp, N., M. L. Gorman, and J. R. Speakman. 1999. Daily energy expenditure of free-living male wood mice in different habitats and seasons. Functional Ecology, 13: 585–593.

Cowles, R. B. 1947. Vascular changes in the wings of bats. Science, 105:362–363.

Cripps, A. W., and V. J. Williams. 1975. The effect of pregnancy and lactation on food intake, gastrointestinal anatomy and the absorptive capacity of the small intestine in the albino rat. British Journal of Nutrition, 33:17–32.

Daan, S., B. M. Barnes, and A. M. Strijkstra. 1991. Warming up for sleep?–ground squirrels sleep during arousals from hibernation. Neuroscience Letters, 128:265–268.

Davis, W. H. 1970. Hibernation: ecology and physiological ecology. Pp. 265–331 in: Biology of Bats (W. A. Wimsatt, ed.). Vol. 3. Academic Press, New York.

Davis, W. H., and H. B. Hitchcock. 1965. Biology and migration of the bat Myotis lucifugus in New England. Journal of Mammalogy, 46:296–313.

Davis, W. H., and O. B. Reite. 1967. Responses of bats from temperate regions to changes in ambient temperature. Biological Bulletin, 132:320–328.

Dawson, T. J., and J. Hulbert. 1970. Standard metabolism, body temperature and surface areas of Australian marsupials. American Journal of Physiology, 218:1233–1238.

Dwyer, P. D. 1962. Studies on the two New Zealand bats. Zoological Publication of Victoria University, Wellington, 28:1–28.

Eisenberg, J. F. 1981. The Mammalian Radiations. Athlone Press, London.

Eisentraut, M. 1934. Der Winterschlaf der Fledermause mit besonderer Berucksichigung der Warmregulation. Zeitschrift für Morphologie Okologie Tiere, 29: 231–267.

Eisentraut, M. 1960. Heat regulation in primitive mammals and in tropical species. Bulletin of the Museum of Comparative Zoology, Harvard University, 124:31–43.

Elgar, M. A., and P. H. Harvey. 1987. Basal metabolic rates in mammals: allometry, phylogeny and ecology. Functional Ecology. 1:25–36.

Ellis, H. I. 1984. Energetics of free-ranging seabirds. Pp. 203–233 in: Seabird Energetics (G. C. Whittow and H. Rahn, eds.). Plenum Press, New York.

Entwistle, A., P. A. Racey, and J. R. Speakman. 1996. Habitat exploitation of a gleaning bat, Plecotus auritus. Philosophical Transactions of the Royal Society of London B, 351:921–931.

Ewing, W. G., E. H. Studier, and M. J. O'Farrell. 1970. Autumn fat deposition and gross body composition in three species of Myotis. Comparative Biochemistry and Physiology, 36:119–129.

Fell, B. F., K. A. Smith, and R. M. Campbell. 1963. Hypertrophic and hyperplastic changes in the alimentary canal of the lactating rat. Journal of Pathology and Bacteriology, 85:179–188.

Felsenstein, J. 1985. Phylogenies and the comparative method. American Naturalist, 125:1–15.

Fenton, M. B. 1992. Bats. Facts on File, New York.

Fenton, M. B., and D. W. Thomas. 1985. Migrations and dispersal of bats (Chiroptera). Contributions in Marine Sciences, 68:409–424.

Findley, J. S. 1992. Bats: A Community Perspective. Cambridge University Press, Cambridge.

Fisher, K. C., and J. F. Manery. 1967. Water and electrolyte metabolism in heterotherms. Pp. 235–279 in: Mammalian Hibernation (K. C. Fisher, A. R. Dawe, C. P. Lyman, and E. Schonbaumn, eds.). Vol. 3. Oliver & Boyd, Edinburgh.

Garland, R. J., and W. K. Milsom. 1994. End-tidal gas composition is not correlated with episodic breathing in hibernating ground squirrels. Canadian Journal of Zoology, 72:1141–1148.

Geiser, F. 1988. Reduction of metabolism during hibernation and daily torpor in mammals and birds: temperature effect or physiological inhibition? Journal of Comparative Physiology B, 158:25–38.

Geiser, F. 1994. Hibernation and daily torpor in marsupials: a review. Australian Journal of Zoology, 42:1–16.

Geiser, F., M. L. Augee, and J. K. Raison. 1984. Thermal response of liver mitochondrial membranes of the heterothermic bat (Miniopterus schreibersii) in summer and winter. Journal of Thermal Biology, 9:183–188.

Geiser, F., K. Cockburn, G. Körtner, and B. S. Law. 1996. Thermoregulation, energy metabolism and torpor in blossom bats, Syconycteris australis (Megachiroptera). Journal of Zoology (London), 239:583–590.

Geiser, F., and E. J. McMurchie. 1984. Differences in thermotropic behavior of mitochondrial membrane respiratory enzymes from homeothermic and heterothermic endotherms. Journal of Comparative Physiology B, 155:125–133.

Geiser, F., and T. Ruf. 1995. Hibernation versus daily torpor in mammals and birds: physiological variation and classification of torpor patterns. Physiological Zoology, 68:935–966.

Genoud, M., and F. J. Bonaccorso. 1986. Temperature regulation, rate of metabolism and roost temperature in the greater white-lined bat Saccopteryx bilineata (Emballonuridae). Physiological Zoology, 59:49–54.

Genoud, M., F. J. Bonaccorso, and A. Arends. 1990. Rate of metabolism and temperature regulation in two small tropical insectivorous bats (Peropteryx macrotis and Natalus tumidirostris). Comparative Biochemistry and Physiology A, 97:229–234.

Grinevitch, L., S. L. Holroyd, and R. M. R. Barclay. 1995. Sex-differences in the use of daily torpor and foraging time by big brown bats (Eptesicus fuscus) during the reproductive season Journal of Zoology (London), 235:301–309.

Hails, C. J. 1979. A comparison of flight energetics of hirundines and other birds. Comparative Biochemistry and Physiology A, 63:581–585.

Haim, A., and I. Izhaki. 1993. The ecological significance of resting metabolic rate and non-shivering thermogenesis in rodents. Journal of Thermal Biology, 18.71–82.

Hall, M. 1832. XVI. On hybernation. Philosophical Transactions of the Royal Society of London, 1:335–360.

Hamilton, J. L., and R. M. R. Barclay. 1995. Patterns of daily torpor and day roost selection by male and female big brown bats (Eptesicus fuscus). Canadian Journal of Zoology, 72:744–749.

Hanus, K. 1959. Body temperatures and metabolism of bats at different environmental temperatures. Physiologia Bohemoslovaca, 8:250–259.

Hart, J. S. 1953a. The relation between thermal history and cold resistance in certain species of rodents. Canadian Journal of Zoology, 31:80–97.

Hart, J. S. 1953b. Energy metabolism of the white footed mouse Peromyscus leucopus noveboracensis after acclimation at various environmental temperatures. Canadian Journal of Zoology, 31:99–105.

Hart, J. S. 1971. Rodents. Pp. 1–149 in: Comparative Physiology of Thermoregulation (G. C. Whittow, ed.). Vol. 1. Academic Press New York.

Harvey, P. H., and M. A. Elgar. 1987. In defence of the comparative method. Functional Ecology, 1:160–161.

Harvey, P. H., and M. D. Pagel. 1991. The Comparative Method in Evolutionary Biology. Oxford University Press, Oxford.

Hays, G. C., P. I. Webb, and J. R. Speakman. 1991. Arrhythmic breathing in torpid pipistrelle bats, *Pipistrellus pipistrellus*. Respiration Physiology, 83:185–192.

Heldmaier, G. 1969. Die Thermogenese der Mausohrfledermaus (*Myotis myotis* Borkh.) beim Erwachen aus dem Winterschlaf. Zeitschrift für Vergleichende Physiologie, 63:59–84.

Heldmaier, G., and T. Ruf. 1992. Body temperature and metabolic rate during natural hypothermia in endotherms. Journal of Comparative Physiology, B, 162:696–706.

Heldmaier, G., R. Steiger, and T. Ruf. 1993. Suppression of metabolic rate in hibernation. Pp. 545–548 *in:* Life in the Cold: Ecological, Physiological and Molecular Mechanisms (C. Carey, G. L. Florant, B. A. Wunder, and B. Horwitz, eds.). Westview Press, Boulder, Colo.

Helversen, O. von, and H.-U. Reyer. 1984. Nectar intake and energy expenditure in a flower visiting bat. Oecologia, 63:178–184.

Herreid, C. F. 1963. Metabolism of the Mexican free-tailed bat. Journal of Cellular and Comparative Physiology, 61:201–207.

Herreid, C. F., and B. Kessel. 1967. Thermal conductance in birds and mammals. Comparative Biochemistry and Physiology, 21:404–414.

Herreid, C. F., and E. H. Schlenker. 1980. Energetics of mice in stable and unstable social conditions: evidence of an air-borne factor affecting metabolism. Animal Behaviour, 28:20–28.

Herreid, C. F., and K. Schmidt-Nielsen. 1966. Oxygen consumption, temperature, and water loss in bats from different envionments. American Journal of Physiology, 211:1108–1112.

Hitchcock, H. B., R. Keen, and A. Kurta. 1984. Survival rates of *Myotis leibi* and *Eptesicus fuscus* in southeastern Ontario. Journal of Mammalogy, 65:126–130.

Hock, R. J. 1951. The metabolic rates and body temperatures of bats. Biological Bulletin, 101:289–299.

Hosken, D. J. 1997. Thermal biology and metabolism of the greater long-eared bat, *Nyctophilus major* (Chiroptera: Vespertilionidae). Australian Journal of Zoology, 45: 145–156.

Hosken, D. J., and P. C. Withers. 1997. Temperature regulation and metabolism of an Australian bat *Chalinolobus gouldii* (Chiroptera: Vespertilionidae) when euthermic and torpid. Journal of Comparative Physiology B, 167:71–80.

Hosken, D. J., and P. C. Withers. 1999. Metabolic physiology of euthermic and torpid lesser long-eared bats *Nyctophilus geoffroyi* (Chiroptera: vespertilionidae). Journal of Mammalogy, 80:42–52.

Kayser, C. 1961. The Physiology of Natural Hibernation. Pergamon Press, New York.

Kayser, C. 1965. Hibernation. Pp. 235–279 *in:* Physiological Mammalogy (W. Mayer and R. van Gelder, eds.). Academic Press, New York.

Keen, R., and H. B. Hitchcock. 1980. Survival and longevity of the little brown bat (*Myotis lucifugus*) in southeastern Ontario. Journal of Mammalogy, 61:1–7.

Kennedy, G. C., W. M. Pearce, and D. M. V. Parrott. 1958. Liver growth in the lactating rat. Journal of Endocrinology, 17:158–160.

Kleiber, M. 1961. The Fire of Life: An Introduction to Animal Energetics. John Wiley & Sons, New York.

Kluger, M. J., and J.'E. Heath. 1970. Vasomotion in the bat wing: a thermoregulatory response to internal heating. Comparative Biochemistry and Physiology, 32:219–226.

Krulin, G. S., and J. A. Sealander. 1972. Annual lipid cycle of the gray bat *Myotis grisescens*. Comparative Biochemistry and Physiology A, 42:537–549.

Kulzer, E., J. E. Nelson, J. L. McKean, and F. P. Möhres. 1970. Untersuchungen über die Temperaturregulation Australischer Fledermäuse (Microchiroptera). Zeitschrift für Vergleichende Physiologie, 69:426–454.

Kunz, T. H. 1980. Energy budgets of free-living bats. Pp, 369–392 *in:* Proceedings of the Fifth International Bat Research Conference (D. E. Wilson and A. L. Gardner, eds.). Texas Tech Press, Lubbock.

Kunz, T. H. 1982. Roosting ecology of bats. Pp. 1–56 *in:* Ecology of Bats (T. H. Kunz, eds.). Plenum Press, New York.

Kunz, T. H., S. K. Robson, and K. A. Nagy. 1998a. Economy of harem maintenance in the greater spear-nosed bat, *Phyllostomus hastatus*. Journal of Mammalogy, 79:631–642.

Kunz, T. H., J. A. Wrazen, and C. D. Burnett. 1998b. Changes in body mass and fat reserves in pre-hibernating little brown bats (*Myotis lucifugus*). Ecoscience, 5:8–17.

Kurta, A. 1991. Torpor patterns in food deprived *Myotis lucifugus* (Chiroptera: Vespertilionidae) under simulated roost conditions. Canadian Journal of Zoology, 69: 255–257.

Kurta, A., G. P. Bell, K. A. Nagy, and T. H. Kunz. 1989. Energetics of pregnancy and lactation in free-ranging little brown bats (*Myotis lucifugus*). Physiological Zoology, 62:804–818.

Kurta, A., K. A. Johnson, and T. H. Kunz. 1987. Oxygen consumption and body temperature of female little brown bats (*Myotis lucifugus*) under simulated roost conditions. Physiological Zoology, 60:386–397.

Kurta, A., and T. H. Kunz. 1987. Size of bats at birth and maternal investment during pregnancy. Symposia of the Zoological Society of London, no. 57:79–106.

Kurta, A., and T. H. Kunz. 1988. Roosting metabolic rate and body temperature of male little brown bats, *Myotis lucifugus*, in summer. Journal of Mammalogy, 69:645–651.

Kurta, A., T. H. Kunz, and K. A. Nagy. 1990. Energetics and water flux of free-ranging big brown bats (*Eptesicus fuscus*) during pregnancy and lactation. Journal of Mammalogy, 71:59–75.

Lancaster, W. C., S. C. Thomson, and J. R. Speakman. 1997. Wing temperature in flying bats measured by infrared thermography. Journal of Thermal Biology, 22:106–116.

Lasiewiski, R. C., and W. R. Dawson. 1967. A re-examination of the relation between standard metabolic rate and body weight in birds. Condor, 69:13–23.

Law, B. S. 1994. Climatic limitation of the southern distribution of the common blossom bat, *Syconycteris australis* in New South Wales. Australian Journal of Ecology, 19:366–374.

Leitner. P. 1966. Body temperature, oxygen consumption, heart rate and shivering in the California mastiff bat, *Eumops perotis*. Comparative Biochemistry and Physiology, 19:431–443.

Leitner, P., and J. E. Nelson. 1967. Body temperature, oxygen consumption and heart rate in the Australian false vampire bat, *Macroderma gigas*. Comparative Biochemistry and Physiology, 21:65–74.

Licht, P., and P. Leitner. 1967a. Behavioral responses to high temperatures in three species of Californian bats. Journal of Mammalogy,48:52–61.

Licht. P., and P. Leitner. 1967b. Physiological responses to high environmental temperatures in three species of microchiropteran bats. Comparative Biochemistry and Physiology, 22:371–387.

Lifson, N., and R. McClintock. 1966. Theory and use of the turnover rates of body water for measuring energy and material balance. Journal of Theoretical Biology, 12:46–74.

Lyman, C. P., R. C. Obrien, G. C. Greene, and E. P. Papfrangos. 1981. Hibernation and longevity in the Turkish hamster Mesocriecetus brandti. Science, 212:668–670.

Malan, A. 1980. Enzyme regulation, metabolic rate and acid-base state in hibernation. Pp. 487–501 in: Animals and Environmental Fitness (R. Gilles, ed.). Pergamon Press, Oxford.

Malan, A. 1986. pH as a control factor in hibernation. Pp. 61–70 in: Living in the Cold: Biochemical and Physiological Adaptations (H. C. Heller, X. J. Musacchia, and L. H. Wang, eds.), Elsevier, New York.

Malan, A. 1988. pH and hypometabolism in mammalian hibernators. Canadian Journal of Zoology, 66:95–98.

Masman, D., and M. Klaasen. 1987. Energy expenditure during free-flight in trained and free-living Eurasian kestrels (Falco tinnunculus). Auk, 104:603–616.

McLean, J. A., and J. R. Speakman. 1997. Non-nutritional maternal support in the brown long-eared bat. Animal Behaviour, 54:1193–1204.

McLean, J. A., and J. R. Speakman. 1999. Daily energy demands of brown long-eared bats suggest that females use metabolic compensation during lactation. Functional Ecology, 13:360–372.

McLean, J. A., and J. R. Speakman. 2000. Effects of body mass and reproduction on the basal metabolic rate of brown long-eared bats (Plecotus auritus). Physiological and Biochemical Zoology, 73:112–121.

McMillan, R. E., and D. S. Hinds. 1992. Standard, cold-induced metabolism of rodents and exercise induced. Pp. 16–33 in: Mammalian Energetics: Interdisciplinary Views of Metabolism and Reproduction. (T. E. Tomasi and T. H. Horton, eds.). Comstock, Ithaca, N.Y.

McNab, B. K. 1966. The metabolism of fossorial rodents: a study of convergence. Ecology, 47:712–733.

McNab, B. K. 1969. The economics of temperature regulation in Neotropical bats. Comparative Biochemistry and Physiology, 31:227–268.

McNab, B. K. 1974. The behavior of temperate cave bats in a sub-tropical environment. Ecology, 55:943–958.

McNab, B. K. 1980a. Food habits, energetics and the population biology of mammals. American Naturalist 116:106–124.

McNab, B. K. 1980b. On estimating thermal conductance in endotherms. Physiological Zoology, 53:145–156.

McNab, B. K. 1982. Evolutionary alternatives in the physiological ecology of bats. Pp. 151–200 in: Ecology of Bats (T. H. Kunz, ed.). Plenum Press, New York.

McNab, B. K. 1986. The influence of food habits on the energetics of eutherian mammals. Ecological Monographs, 56:1–19.

McNab, B. K. 1987. Basal metabolic rate and phylogeny. Functional Ecology, 1:159–167.

McNab, B. K. 1988. Food habits and the basal rate of metabolism in birds. Oecologia, 77:343–349.

McNab, B. K. 1989. Temperature regulation and rate of metabolism in three Bornean bats. Journal of Mammalogy, 70:153–161.

McNab, B. K. 1992. A statistical analysis of mammalian rates of metabolism. Functional Ecology, 6:672–679.

McNab, B. K., and F. J. Bonaccorso. 1995. The energetics of pteropodid bats. Symposia of the Zoological Society of London, no. 67:111–122.

McNeil Alexander, R. M. 1999. Energy for Animal Life. Oxford University Press, Cambridge.

Menaker, M. 1964. Frequency of spontaneous arousal from hibernation in bats. Nature, 203:540–541.

Milsom, W. K., S. Osborne, P. F. Chan, J. D. Huner, and J. Z. McLeod, 1993. Sleep, hypothermia, and hibernation: metabolic rate and control of breathing pattern in golden mantled ground squirrels. Pp. 233–240 *in:* Life in the Cold: Ecological, Physiological and Molecular Mechanisms (C. Carey, G. L. Florant, B. A. Wunder, and B. Horwitz, eds.). Westview Press, Boulder, Colo.

Morrison, P. 1959. Body temperatures in some Australian mammals. I. Chiroptera. Biological Bulletin, 116:484–497.

Morrison, P., and F. A. Ryser. 1951. Temperature and metabolism in some Wisconsin mammals. Federation Proceedings, 10:93–97.

Nagy, K. A. 1980. CO_2 production in animals: analysis of potential errors in the doubly labeled water method. American Journal of Physiology, 238:R466–R473.

Nagy, K. A., I. A. Girard, and T. K. Brown. 1999. Energetics of free-ranging mammals, reptiles and birds. Annual Review of Nutrition, 19:247–277.

Noll, U. G. 1979. Body temperature, oxygen consumption, noradrenaline response and cardiovascular adaptations in the flying fox, *Rousettus aegyptiacus.* Comparative Biochemistry and Physiology A, 63:79–88.

Norberg, U. M. 1990. Vertebrate Flight. Springer-Verlag, Berlin

Norberg, U. M., T. H. Kunz, J. Steffensen, Y. Winter, and O. von Helversen. 1993. The cost of hovering and forward flight in a nectar-feeding bat, *Glossophaga soricina,* estimated from aerodynamic theory. Journal of Experimental Biology, 182.207–227.

Norberg, U. M., and J. M. V. Rayner. 1987. Ecological morphology and flight in bats (Mammalia: Chiroptera): wing adaptations, flight performance, foraging strategy and echolocation. Philosophical Transactions of the Royal Society of London B, 316: 335–427.

Ochoa, H., and T. H. Kunz. 1999. Behavioral thermoregulation in the island flying fox *Pteropus hypomelanus.* Journal of Thermal Biology, 24:15–20.

O'Farrell, M. J., and E. H. Studier. 1970. Fall metabolism in relation to ambient temperature in three species of *Myotis.* Comparative Biochemistry and Physiololgy, 35: 697–703.

Oliver, J. E., and R. W. Fairchild. 1984. The Encyclopedia of Climatology. Van Nostrand Reinhold, New York.

Pagel, M. D., and P. H. Harvey. 1988. Recent developments in the analysis of comparative data. Quarterly Review of Biology, 63:413–440.

Pagels, J. F. 1975. Temperature regulation, body weight and changes in total body fat of the free-tailed bat, *Tadarida brasiliensis cynocephala* (Le Conte). Comparative Biochemistry and Physiology A, 50:237–246

Park, K. J., G. Jones, and R. D. Ransome. 1999. Winter activity of a population of greater horseshoe bats (*Rhinolophus ferrumequinum*). Journal of Zoology (London), 248:419–427.

Pennycuick, C. J. 1969. The mechanics of bird migration. Ibis, 111:525–556.

Pennycuick, C. J. 1989. Bird Flight Performance. Oxford University Press, Oxford.

Peters, R. H. 1983. The Ecological Implications of Body Size. Cambridge University Press, Cambridge.

Pettigrew, J. D. 1986. Flying primates? megabats have the advanced pathway from eye to midbrain. Science, 231:1304–1306.

Pettigrew, J. D. 1991. Wings or brain? convergent evolution in the origins of bats. Systematic Zoology, 40:199–216.

Pettigrew, J. D., B. G. M. Jamieson, S. K. Robson, L. S. Hall, K. I. McAnally, and H. M. Cooper. 1989. Phylogenetic relations between microbats, megabats and primates (Mammalia: Chiroptera and Primates). Philosophical Transactions of the Royal Society of London B, 325:489–559.

Pierson, E. D. 1986. Molecular systematics of the Microchiroptera: higher taxon relationships and biogeography. Ph.D. Dissertation. University of California, Berkeley.

Polskey, G. R., and J. R. Sealander. 1979. Lipid deposition and withdrawal during hibernation in *Pipistrellus subflavus* (Chiroptera: Vespertilionidae). Southwestern Naturalist, 24:71–78.

Racey, P.A. 1973. Environmental factors affecting the length of gestation in heterothermic bats. Journal of Reproduction and Fertility (Suppl.), 19:175–189.

Racey, P. A. 1982. Ecology of bat reproduction. Pp. 57–104 *in:* Ecology of Bats (T. H. Kunz, eds.). Plenum Press, New York.

Racey, P. A., and J. R. Speakman. 1987. The energetics of pregnancy and lactation in heterothermic bats. Symposia of the Zoological Society of London, 54:101–129.

Racey, P. A., and S. M. Swift. 1981. Variations in gestation length in a colony of pipistrelle bats (*Pipistrellus pipistrellus*) from year to year. Journal of Reproduction and Fertility, 61:123–129.

Ransome, R. D. 1990. The Natural History of Hibernating Bats. Helm Bromly, Kent.

Rayner, J. M. V. 1979. A new approach to animal flight mechanics. Journal of Experimental Biology, 80:17–54.

Rayner, J. M. V. 1990. The mechanics of flight and bird migration performance. Pp. 283–299 *in:* Bird Migration: Physiology and Ecophysiology (E. Gwinner, ed.). Springer-Verlag, Berlin.

Rayner, J. M. V., G. Jones, and P. M. Hughes. 1989. Load carrying by flying bats. Pp. 235–246 *in:* European Bat Research, 1987 (V. Hának, I. Horáček, and J. Gaisler, eds.). Charles University Press, Prague.

Rayner, J. M. V., and A. L. R. Thomas. 1991. On the vortex wake of an animal flying in a confined volume. Philosophical Transactions of the Royal Society of London B, 334:107–117.

Reynolds, P. S., and R. M. Lee. 1996. Phylogenetic analysis of avian energetics: passerines and non-passerines do not differ. American Naturalist, 147:735–759.

Reeder, W. G., and R. B. Cowles. 1951. Aspects of thermoregulation in bats. Journal of Mammalogy, 32:389–403.

Rodríguez-Durán, A. 1994. Metabolic rates and thermal conductance in four species of Neotropical bats roosting in hot caves. Comparative Biochemistry and Physiology A, 110:347–357.

Rothe, H. J., W. Biesel, and W. Nachtigall. 1987. Pigeon flight in a wind tunnel. II. Gas exchange and power requirements. Journal of Comparative Physiology B, 157: 99–109.

Roverud, R. C., and M. A. Chappell. 1991. Energetic and thermoregulatory aspects of clustering behavior in the Neotropical bat *Noctilio albiventris*. Physiological Zoology, 64:1527–1541.

Schmidt-Nielsen, K. 1990. Animal Physiology: Adaptation and Environment, 5th ed. Cambridge University Press, Cambridge.

Scholander, P. F., R. Hock, V. Walter, and L. Irving. 1950. Adaptation to cold in arctic and tropical mammals and birds in relation to body temperature, insulation and basal metabolic rate. Biological Bulletin, 99:259–271.

Shimoizumi, J. 1959. Studies of the hibernation of bats. Scientific Reports, Tokyo Kyoiku Diagaku, Section B, 9:1–35

Shump, K. A., and A. U. Shump. 1980. Comparative insulation in vespertilionid bats. Comparative Biochemistry and Physiology A, 66:351–354.

Simmons, N. B. 1998. A reappraisal of interfamilial relationships of bats. Pp. 3–26 *in:* Bat Biology and Conservation (T. H. Kunz and P. A. Racey, eds.). Smithsonian Institution Press, Washington, D.C.

Simmons, N. B., and J. H. Geisler. 1998. Phylogenetic relationships of *Icaronycteris, Archaeonycteris, Hassianycteris,* and *Paleochiropteryx* to extant bat lineages, with comments on the evolution of echolocation and foraging strategies in microchiroptera. Bulletin of the American Museum of Natural History, 235:1–182.

Snyder, G. K., and J. R. Nestler. 1990. Relationships between body temperature, thermal conductance and energy metabolism during daily torpor and hibernation in rodents. Journal of Comparative Physiology A, 66:351–354.

Speakman, J. R. 1993. The evolution of echolocation for predation. Symposia of the Zoological Society of London, no. 65:39–63.

Speakman, J. R. 1995. Energetics and the evolution of body size in small terrestrial mammals. Symposia of the Zoological Society of London, no. 69:63–81.

Speakman, J. R. 1997. Doubly-labelled Water: Theory and Practice. Chapman & Hall, London.

Speakman, J. R. 1998. Doubly-labeled water: history and theory of the technique. American Journal of Clinical Nutrition, 68:S932–938.

Speakman, J. R. 2000. The cost of living: factors influencing the daily energy demands of small mammals. Advances in Ecological Research, 30:177–297.

Speakman, J. R., G. C. Hays, and P. I. Webb. 1994. Is hyperthermia a constraint on the diurnal activity of bats. Journal of Theoretical Biology, 171:325–341.

Speakman, J. R., W. C. Lancaster, W. Ward, G. Jones, and K. C. Cole. 2003. Energy costs of echolocation in stationary bats. *In:* Echolocatioon in Bats and Dolphins (J. Thomas, C. F. Moss, and M. Vater, eds.). University of Chicago Press, Chicago (in press).

Speakman, J. R., and J. McQueenie. 1996. Limits to sustained metabolic rate: the link between food intake, BMR, and morphology in reproducing mice (*Mus musculus*). Physiological Zoology, 69:746–769.

Speakman, J. R., and P. A. Racey. 1987. The energetics of pregnancy and lactation in the brown long-eared bat, *Plecotus auritus*. Pp. 368–393 *in:* Recent Advances in the Study of Bats (M. B. Fenton, P. Racey, and J. M. V. Rayner, eds.). Cambridge University Press, Cambridge.

Speakman, J. R., and P. A. Racey. 1989. Hibernal ecology of the pipistrelle bat: energy expenditure, water requirements and mass loss, implications for survival and the function of winter emergence flights. Journal of Animal Ecology, 58:797–814.

Speakman, J. R., and P. A. Racey. 1991. No cost of echolocation for bats in flight. Nature, 350:421–423.

Speakman, J. R., and F. Rossi. 1999. No support for socio-physiological effect on the metabolism of white mice. Functional Ecology 13:373–382.

Speakman, J. R., P. I. Webb, and P. A. Racey. 1991. Effects of disturbance on the energy expenditure of hibernating bats. Journal of Applied Ecology, 28:1087–1104.

Steffen, J. M., and M. L. Riedesel. 1982. Pulmonary ventilation and cardiac activity in hibernating and arousing golden-mantled ground squirrels (*Spermophilus lateralis*). Cryobiology, 19:83–91.

Stones, R. C., and J. E. Wiebers. 1967. A review of temperature regulation in bats (Chiroptera). American Midland Naturalist, 74:155–167.

Strelkov, P. P. 1969. Migratory and stationary bats (Chiroptera) of the European part of the Soviet Union. Acta Biologica Cracoviensis (Series Zoologia), 14:393–439.

Strijkstra, A., and S. Daan. 1997. Ambient temperature during torpor affects NREM sleep EEG during arousal episodes in hibernating European ground squirrels. Neuroscience Letters, 221:177–180.

Strijkstra, A., and S. Daan. 1998. Dissimilarity of slow-wave activity enhancement by torpor and sleep deprivation in a hibernator. American Journal of Physiology, 275:R1110–1117.

Studier, E. H. 1981. Energetic advantages of slight drops in body temperature in little brown bats *Myotis lucifugus*. Comparative Biochemistry and Physiology A, 70: 537–540.

Studier, E. H., and M. J. O'Farrell. 1972. Biology of *Myotis thysanodes* and *M. lucifugus* (Chiroptera: Vespertilionidae). I. Thermoregulation. Comparative Biochemistry and Physiology A, 41:567–596.

Studier, E. H., and M. J. O'Farrell. 1976. Biology of *Myotis thysanodes* and *M. lucifugus* (Chiroptera: Vespertilionidae). III. Metabolism, heart rate, breathing rate, evaporative water loss and general energetics. Comparative Biochemistry and Physiology A, 54:423–432.

Studier, E. H., and M. J. O'Farrell. 1980. Physiological ecology of *Myotis*. Pp. 415–424 *in:* Proceedings of the Fifth International Bat Research Conference (D.E. Wilson and A. L. Gardner, eds.). Texas Tech Press, Lubbock.

Studier, E. H., and D. E. Wilson. 1970. Thermoregulation in some Neotropical bats. Comparative Biochemistry and Physiology, 34:251–262.

Studier, E. H., and D. E. Wilson. 1979. Effects of captivity on thermoregulation and metabolism in *Artibeus jamaicensis* (Chiroptera: Phylostomatidae). Comparative Biochemistry and Physiology A, 62:347–350.

Swift, S. M. 1980. Activity patterns of pipistrelle bats (*Pipistrellus pipistrellus*) in northeast Scotland. Journal of Zoology (London), 190:285–295.

Symonds, M. R. E. 1999. Life histories of Insectivora: the role of phylogeny, metabolism, and sex differences. Journal of Zoology (London), 249:315–337.

Szewczak, J. M., and D. C. Jackson. 1992. Ventilatory response to hypoxia and hypercapnia in the torpid bat, *Eptesicus fuscus*. Respiration Physiology, 88:217–232.

Tähti, H., and A. Soivio. 1975. Blood gas concentration, acid-base balance and blood pressure in hedgehogs in the active state and in hibernation with periodic respiration. Annales Zoologici Fennici, 12:188–192.

Thomas, D. W. 1984. Fruit intake and energy budgets of frugivorous bats. Physiological Zoology, 57:457–467.

Thomas, D. W. 1988. Data attributed to Thomas presented in T. H. Fleming. The Short-tailed Fruit Bat: A Study in Plant Animal Interactions. University of Chicago Press, Chicago.

Thomas, D. W. 1993. Lack of evidence for a biological "alarm clock" in bats (*Myotis* spp.) hibernating under natural conditions. Canadian Journal of Zoology, 71:1–3.

Thomas, D. W. 1995a. Hibernating bats are sensitive to non-tactile human disturbance. Journal of Mammalogy, 76:940–946.

Thomas, D. W. 1995b. The physiological ecology of hibernation in vespertilionid bats. Symposia of the Zoological Society of London, no. 67:233–244.

Thomas, D. W., and G. P. Bell. 1986. Thermoregulation strategies and the distribution of bats along climatic gradients. Bat Research News, 27:39.

Thomas, D. W., and D. Cloutier. 1992. Evaporative water loss by hibernating little brown bats, *Myotis lucifugus*. Physiological Zoology, 65:433–456.

Thomas, D. W., D. Cloutier, and D. Gagne. 1990a. Arrythmic breathing, apnea and non-steady state oxygen uptake in hibernating little brown bats, *Myotis lucifugus*. Journal of Experimental Biology, 149:395–406.

Thomas, D. W., M. Dorais, and J. M. Bergeron. 1990b. Winter energy budgets and the cost of arousals for hibernating little brown bats, *Myotis lucifugus*. Journal of Mammalogy, 71:475–479.

Thomas, D. W., and F. Geiser. 1997. Periodic arousals in hibernating mammals: is evaporative water loss involved? Functional Ecology, 11:585–591.

Thomas, S. P. 1975. Metabolism during flight of two species of bats, *Phyllostomus hastatus* and *Pteropus gouldii*. Journal of Experimental Biology, 63:273–292.

Thomas, S. P. 1981. Ventilation and oxygen extraction in the bat *Pteropus gouldii* during rest and steady flight. Journal of Experimental Biology, 94:231–250.

Thomas, S. P. 1987. The physiology of bat flight. Pp. 75–99 *in:* Recent Advances in the Study of Bats (M. B. Fenton, P. Racey, and J. M. V. Rayner, eds.). Cambridge University Press, Cambridge.

Trachsel, L., D. M. Edgar, and H. C. Heller. 1991. Are ground squirrels sleep deprived during hibernation? American Journal of Physiology, 260:R1123–R1129.

Trayhurn, P., J. S. Keith, P. A. Racey, and A. M. Burnett. 1991. Immunological identification of uncoupling protein in interscapular "brown" adipose tissue of suckling and adult pipistrelle bats (*Pipistrellus pipistrellus*). Comparative Biochemistry and Physiology B, 99:317–320.

Trune, D. R., and C. N. Slobodchikoff. 1976. Social effects of roosting on metabolism of the pallid bat (*Antrozous pallidus*). Journal of Mammalogy, 57:656–663.

Tucker, V. A. 1966. Oxygen consumption of a flying bird. Science, 154:150–151.

Tucker, V. A. 1972. Metabolism during flight in the laughing gull, *Larus attricilla*. American Journal of Physiology, 222:237–245.

Twente, J. W., and J. Twente. 1987. Biological alarm clock arouses hibernating big brown bats, *Eptesicus fuscus*. Canadian Journal of Zoology, 65:1668–1674.

Twente, J. W., J. Twente, and V. Brack. 1985. The duration of the period of hibernation of three species of vespertilionid bats. II. Laboratory studies. Canadian Journal of Zoology, 63:2955–2961.

van Valen, L. 1979. The evolution of bats. Evolutionary Theory, 4:104–121.

Voigt, C. C. 2000. Intraspecific scaling of flight power in the bat *Glossophaga soricina* (Phyllostomidae) and the upper limit of body size in nectar-feeding bats. Journal of Comparative Physiology B, 170:403–411.

Voigt, C. C., and Y. Winter. 1999. The energetics of hovering flight in glossophagine bats (Phyllostomidae: glossophaginae) and its scaling in bats, hummingbirds and moths. Journal of Comparative Physiology B, 169:38–48.

Wang, L. H. C. 1978. Factors limiting maximum cold induced heat production. Life Sciences, 23:2089–2098.

Ward, S. M., U. Moller, D. M. Jackson, J. M. V. Rayner, W. Nachtigall, and J. R. Speakman. 1999. Power requirements for starling flight in a wind tunnel. Biologia e Conservazione della Fauna, 102:335–339.

Webb, P. I., G. C. Hays, J. R. Speakman, and P. A. Racey. 1992. The functional significance of ventilation frequency, and its relationship to oxygen demand in the resting brown long-eared bat (*Plecotus auritus*). Journal of Comparative Physiology B, 162:144–147.

Webb, P. I., J. R. Speakman, and P. A. Racey. 1993. The implication of small reductions in body temperature for radiant and convective heat loss in resting endothermic brown long-eared bat, *Plecotus auritus*. Journal of Thermal Biology, 18:131–135.

Webb, P. I., J. R. Speakman, and P. A. Racey. 1996. How hot is a hibernaculum? a review of temperatures selected by bats during winter hibernation. Canadian Journal of Zoology, 74:761–765.

Whitaker, J. O., and L. J. Rissler. 1993. Do bats feed in winter? American Midland Naturalist, 129:200–203.

White, G. 1786. The Natural History and Antiquities of Selbourne in the County of Southampton. Swan, Sonnenschein & Lowrey & Co., London.

Wilde, C. J., M. A. Kerr, C. H. Knight, and P. A. Racey. 1995. Lactation in vespertilionid bats. Symposia of the Zoological Society of London, no. 67:139–149.

Wilde, C. J., C. H. Knight, and P. A. Racey. 1999. Influence of torpor on milk protein composition and secretion in lactating bats. Journal of Experimental Zoology, 284:35–41.

Winter, Y. 1998. Energetic cost of hovering flight in a nectar feeding bat measured with fast response respirometry. Journal of Comparative Physiology B, 168:434–444.

Winter, Y., and O. von Helversen. 1998. The energy cost of flight: do small bats fly more cheaply than birds? Journal of Comparative Physiology B, 168:105–111.

PART III

MACROECOLOGY

CHAPTER 11

Evolution of Ecological Diversity in Bats

Nancy B. Simmons and Tenley M. Conway

Introduction

Ask a group of biologists what the term "diversity" means, and you are likely to get several different answers. Some would focus on species richness, while others might highlight morphological or ecological variation or the complexity of biotic interactions or discuss the importance of patterns of species abundance. Regardless, few would disagree with the assessment that Chiroptera is a diverse group of mammals. With approximately 1,100 extant species recognized worldwide, bats are second only to rodents in terms of total numbers of species (Wilson and Reeder 2003). Single localities in the Neotropics may support as many as 110 sympatric bat species, a number far exceeding that known for any other mammalian order anywhere in the world (Voss and Emmons 1996). As is abundantly demonstrated by the chapters in this book, bat species may vary significantly in body size, morphology, diet, foraging and flight behavior, sensory ecology, roosting habits, hibernation and migration patterns, social behavior, life-history strategies, and reproductive biology. Indeed, it is the great numerical and ecological diversity of bats that has attracted many researchers to this group.

Modern classifications of bats differ in many details but generally agree in the subdivision of extant Chiroptera into two suborders: Megachiroptera, containing one family and at least 188 species, and Microchiroptera, including 18 families and at least 917 species (e.g., McKenna and Bell 1997; Simmons 1998, 2003; Simmons and Geisler 1998). The two suborders are characterized by different sensory and feeding ecologies. Megachiroptera (Old World fruit bats) is a clade consisting of fruit- and nectar-feeding bats that use vision and olfaction to locate their food (Dumont, this volume; Helversen and Winter, this volume). In contrast, all Microchiroptera (echolocating bats) are capable of sophisticated echolocation, and many use echolocation to detect, track, and evaluate flying insect prey (Jones and Rydell, this volume).[1] These basic distinctions are common knowledge, as is the perception

---

1. We follow Simmons and Geisler (1998) in using the term "sophisticated echolocation" to describe systems that can be used for orientation in cluttered environments (e.g., within vegetation or under the

that megachiropterans (megabats) are large-bodied inhabitants of the Old World tropics, while microchiropterans (microbats) are smaller and more cosmopolitan. However, these stereotypes are somewhat misleading because they overlook many of the more interesting patterns of ecological diversity, including the highly mosaic nature of ecological and biogeographical variation in bats.

Most bat families and subfamilies are recognized based on shared morphological and ecological traits that are found in all or most members of the group (Corbet and Hill 1992; Griffiths 1994; Griffiths and Smith 1991; Griffiths et al. 1992; Hill 1974; Koopman 1984, 1994; Miller 1897, 1907; Van Valen 1979; Simmons 1998; Wetterer et al. 2000). As judged purely by comparisons of species richness (fig. 11.1), some groups appear to have been more successful than others. Several families are monotypic or nearly so (e.g., Craseonycteridae, Myzopodidae, Mystacinidae, Noctilionidae, Furipteridae), while others include more than 100 species (e.g., Pteropodidae, Rhinolophidae, Phyllostomidae, Vespertilionidae). However, species richness is only one measure of the evolutionary or ecological "success" of a group. Other measures might include population size, breadth of geographic range, longevity (of individuals, populations, species, or higher-level taxa), reproductive success rates, energetic efficiency, or any number of other features that may affect fitness or success in ecological or evolutionary terms. Teasing apart the various factors involved and understanding how past events have shaped the ecologies of bat species are among the greatest challenges in chiropteran biology.

It has become increasingly clear in recent years that phylogenetic hypotheses provide a critical framework for interpreting patterns of evolution in biological systems. This is not just true of efforts to understand minutiae of dental evolution or changes in genes but also applies to studies of ecology and behavior (e.g., Brooks and McLennan 1991; Swofford and Maddison 1992; Wenzel 1992). Mapping the taxonomic distribution of character states (which may include ecological traits or behavioral features) on phylogenetic trees provides a means of reconstructing evolutionary patterns of transformation,

---

forest canopy) and for detection, tracking, and evaluation of moving objects including prey. Calls associated with sophisticated echolocation are produced in the larynx and are often (but not always) highly structured in terms of call length, frequency, and type of modulation. Call structure, duration, and pulse interval may be intentionally varied by the animal depending on the circumstances (e.g., searching for prey, approaching prey, attacking prey), and call structure is often species specific. There is a strong correlation between call type and flight/foraging strategies, in part because the information content of returning echoes varies depending on call structure and pulse interval. However, all sophisticated echolocators apparently record and process information of several types, including relative timing of pulse and echo (including different arrival times at the right and left ears), differences in intensity between pulse and echo, and (in some cases) frequency shifts between pulse and echo. Sophisticated echolocation apparently provides the animal with a detailed acoustic map of its environment, a map that changes moment by moment depending on movement of both the echolocator and the objects in its environment. This form of echolocation occurs in all extant microchiropterans but not in megachiropterans.

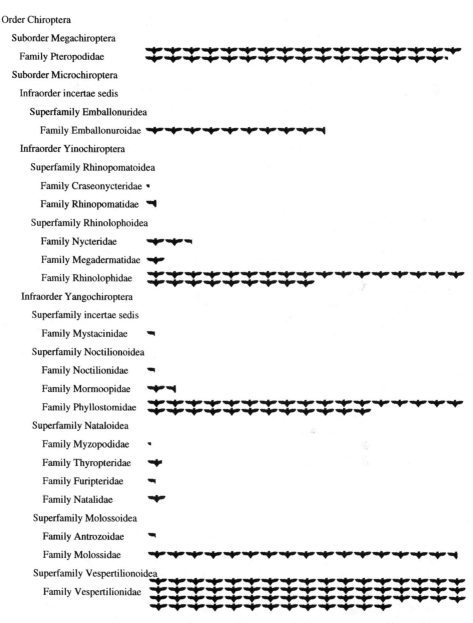

Figure 11.1. Comparison of species richness among extant families of bats. Each bat icon represents five extant species. There is no typical level of diversity for a bat family; note that family-level species richness varies from one species (Craseonycteridae, Myzopodidae) to more than 300 species (Vespertilionidae). The classification employed here follows that of Simmons (1998) and Simmons and Geisler (1998).

which may in turn lead to hypotheses of correlation, process, and even causation of change (Brooks and McLennan 1991; Swofford and Maddison 1992; Wenzel 1992). Covariation among characters can be investigated using computer programs like MacClade (Maddison and Maddison 1992), and comparative-method techniques such as phylogenetically independent contrasts can be used to test hypotheses of concerted evolution among traits (Felsenstein 1985; Garland et al. 1992, 1993, 1999; Kirsch and Lapointe 1997; Martins and Garland 1991). Computer simulations can be used to obtain null distributions of test statistics (Garland et al. 1993, 1999), and maximum-likelihood methods can be used to estimate and evaluate accuracy of ancestral state reconstructions (Schluter 1997).

Only a few studies of bats have explicitly used phylogenetic methods to examine patterns of evolution in ecology and behavior. These include studies of the evolution of feeding habits (e.g., Ferrarezzi and Gimenez 1996; Kirsch and Lapointe 1997; Simmons and Wetterer 2000; Wetterer et al. 2000), physiology and energetics (McNab and Bonaccorso 1995), echolocation and foraging strategies (e.g., Simmons and Geisler 1998), and analyses of biogeographic history (e.g., Baker et al. 1994; Hand and Kirsch 1998; Kirsch et al. 1998; Lim and Engstrom 1998; Simmons 1998). In the context of these studies, the goal of the present chapter is to explore patterns of variation in bat ecology from a phylogenetic perspective, with an emphasis on the mosaic pattern of distribution of different ecological and behavioral traits in different clades. The phylogenetic, geographic, and temporal origins of major ecological complexes will be discussed in this evolutionary context. We will examine the role that phylogenetic history may have played in the evolution of ecological diversity in bats and illustrate the potential value of a phylogenetic approach for understanding the ecologies of living taxa. Because modern bat species and faunas arose as the result of evolutionary processes, it is important to consider evolutionary history when seeking explanations for biological phenomena at all levels.

Early Evolution of Bats

The Fossil Record and "Ghost" Lineages

The evolutionary origins of bat diversity remain poorly understood, but the roots of the chiropteran radiation clearly go back to at least the early Tertiary. Twenty-four genera of Eocene bats are currently recognized, and at least nine of these were present in the early Eocene (McKenna and Bell 1997; Simmons and Geisler 1998). The oldest known bat is *Icaronycteris index* from late Wasatchian deposits in the Green River Formation of North America, thought to be at least 53 million years old (Jepsen 1966, 1970; Simmons and Geisler 1998). No unambiguous fossils of bats have been found in Paleocene deposits, but occurrence of bats in the Paleocene would not be surprising. Early Eocene bats are known from four continents (North America, Europe, Africa, and Australia [Beard et al 1992; Hand et al. 1994; Simmons and Geisler 1998]), suggesting that Chiroptera was well established throughout much of the world by that time.

Four genera of Eocene bats (*Icaronycteris, Archaeonycteris, Hassianycteris,* and *Palaeochiropterx*) are known from complete or nearly complete skeletons. In a recent phylogenetic analysis, Simmons and Geisler (1998) found that these taxa form a series of consecutive sister taxa to a clade containing all extant microchiropterans (fig. 11.2). Cranial characters and stomach contents suggest that all these fossil taxa could echolocate, indicating that this "key innovation" of microchiropterans was present by the early Eocene (Habersetzer and Storch 1992; Habersetzer et al. 1994; Novacek 1985, 1987; Simmons and Geisler 1998). Assuming subordinal monophyly, the divergence between the lineages leading to extant Megachiroptera and Microchiroptera logically must predate the age of the earliest fossils referable to either group, so this divergence must have occurred prior to the origin of the oldest currently known Eocene bat fossil—more than 53 million years ago.

The earliest appearance of a lineage in the fossil record provides only a minimum age for that group, not an indication of its actual time of origin. Each lineage in a pair of sister taxa must be the same age since they originated in a single speciation event, but it is rare that sister taxa appear simultaneously in the fossil record (Benton and Hitchin 1997; Hulsenbeck 1994; Norell 1992; Padian et al. 1994). Typically, one taxon of a pair of sister taxa has a longer record than the other due to the stochastic nature of fossil preservation and discovery or because differences in distribution or ecology affect the chances of preservation and recovery (Norell 1992). Information from phylogenies can be used to reconstruct "ghost lineages," those parts of evolutionary lineages that must have existed but for which we currently have no fossils (Norell 1992; Novacek 1996; Padian et al. 1994). Application of this method to bats has yielded somewhat surprising results. Fossils referable to six major family-level lineages are known from middle–late Eocene deposits, and reconstruction of ghost lineages using the phylogeny from Simmons and Geisler (1998) leads to the conclusion that at least six more extant lineages were present by the end of the Eocene (fig. 11.2). All of the microchiropteran superfamilies recognized by Simmons and Geisler (1998) must have been distinct by this time (i.e., Emballonuroidea, Rhinopomatoidea, Rhinolophoidea, Noctilionoidea, Nataloidea, Molossoidea, Vespertilionoidea), as were at least five extant families (i.e., Emballonuridae, Megadermatidae, Rhinolophidae, Molossidae, Vespertilionidae). As many of these groups are characterized by distinct ecological habits, these observations suggest that many suites of ecological and behavioral traits present in different bat taxa have had a long evolutionary history.

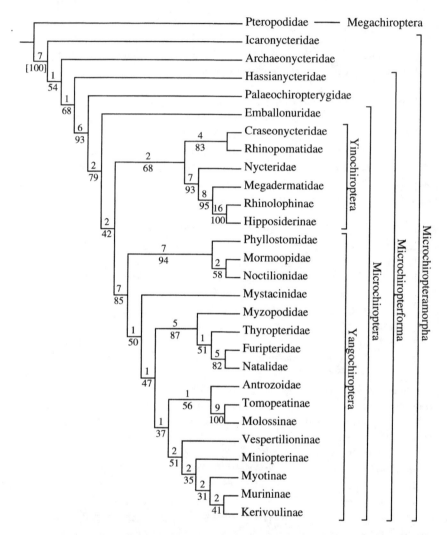

Figure 11.2. Phylogenetic tree of bats from Simmons and Geisler (1998, fig. 36) showing relationships among extant family-level taxa and selected fossil taxa. This was the single most parsimonious tree that resulted from a parsimony analysis of 195 morphological characters, 12 rDNA restriction site characters, and one character based on the number of R-1 tandem repeats in the mtDNA d-loop region. The numbers above internal branches are decay values; numbers below the branches are bootstrap values that resulted from an analysis in which Chiroptera was constrained to be monophyletic. Group names shown on the right are those proposed by Simmons and Geisler (1998) in their higher-level classification. All of the taxa shown here are believed to be monophyletic with the possible exception of Vespertilioninae (see Simmons 1998).

Evolution of Flight, Echolocation, and Foraging Habits in Eocene Bats

Simmons and Geisler (1998; fig. 11.2) used their proposed phylogeny as a basis for evaluating previous hypotheses on the evolution of flight, echolocation, and foraging strategies. Based on character distributions on the phylogenetic tree, they proposed that flight evolved before echolocation and that the first bats probably used vision for orientation in their arboreal/aerial environment. The evolution of flight was apparently followed by the origin of low duty-cycle laryngeal echolocation in early members of the microchiropteran lineage. Simmons and Geisler (1998) suggested that this system was probably simple at first, permitting orientation and obstacle detection but not detection or tracking of airborne insect prey. The energy costs of echolocation to flying bats are relatively low as a result of the mechanical coupling of ventilation and flight (Speakman 1993; Speakman and Racey 1991; Speakman and Thomas, this volume). The benefits of aerial insectivory may be substantial, and a more sophisticated low duty-cycle echolocation system capable of detecting, tracking, and assessing airborne prey probably evolved quite rapidly after the initial "discovery" of echolocation by bats. The need for an increasingly derived auditory system, together with limits on body size imposed by the mechanics of flight, echolocation, and prey capture, may have resulted in reduction and simplification of the visual system in microchiropteran bats as echolocation became increasingly important (Simmons and Geisler 1998).

The scenario described above, which is a modified version of Norberg's (1989) "flight first" theory, is consistent with bat monophyly and Simmons and Geisler's (1998) phylogeny. Two competing hypotheses, the "echolocation first" theory (Fenton et al. 1995; Hill and Smith 1984) and the "tandem evolution" theory (Speakman 1993), appear less likely when viewed in light of bat monophyly. Both of the latter theories were developed in the context of bat diphyly (e.g., Pettigrew 1995; Pettigrew et al. 1989), a phylogenetic hypothesis refuted in recent years (see Simmons [1994, 1998], Simmons and Geisler [1998], and references cited therein). Assuming that bats (and both suborders) are monophyletic, the echolocation first and tandem evolution theories require either independent gain of flight in the two bat suborders (i.e., absence of flight in the common ancestor of Megachiroptera and Microchiroptera) or loss of echolocation in Megachiroptera. Like Norberg (1989) and Simmons and Geisler (1998), we prefer the less complex, more parsimonious explanation that flight evolved before echolocation and that each of these specializations was retained in all subsequent descendants including various Eocene bats.

Drawing on data from several previous studies (e.g., Habersetzer and Storch 1987, 1989, 1992; Habersetzer et al. 1992, 1994; Norberg 1989), Simmons and Geisler (1998) reconstructed foraging strategies of *Icaronycteris, Archaeonycteris, Hassianycteris,* and *Palaeochiropteryx* based on postcranial osteology and wing form, cochlear size, and stomach contents (table 11.1). As noted above, their analyses confirmed previous suggestions that these taxa used echolocation. In the context of the phylogeny, Simmons and Geisler (1998) proposed that foraging behavior in the microchiropteran lineage evolved in a series of steps. In the earliest stage (stage 1), bats might have gleaned food objects during short flights from a perch using vision for orientation and obstacle detection. Prey detection would have been by passive means, including vision, olfaction, and/or listening for prey-generated sounds. This stage is not represented by any known fossils but probably represents the lifestyle of the earliest bats (e.g., those ancestral to both megabats and microbats). Although not explicitly discussed by Simmons and Geisler (1998), it is possible—indeed likely—that these early bats were opportunistic omnivores, feeding on insects, fruit, and flower products as available in the local environment. This sort of eclectic diet and flexible foraging behavior occurs today in mystacinids, antrozoids, and many phyllostomids (e.g., Daniel 1976, 1990; Gardner 1977; Hermanson and O'Shea 1983; Johnson 1997; Simmons and Wetterer 2000).

Until recently, most bat biologists assumed that the earliest bats were entirely or predominantly insectivorous (e.g., Gillette 1975; Hill and Smith 1984; Legendre and Lapointe 1995; Smith 1976). In contrast to this view, Ferrarezzi and Gimenez (1996, 79) argued that the diet of the earliest bats was herbivory (phytophagy), which they defined as "the habit of feeding exclusively or predominantly on vegetal material, including fruits, flowers, and floral products, buds, and leaves." They reached this conclusion by mapping dietary habits in a nonadditive fashion on a cladogram that assumed monophyly of both Chiroptera and Volitantia (Chiroptera+Dermoptera). However, Ferrarezzi and Gimenez (1996) also discussed another possibility—that the earliest bats were omnivorous—and noted that this hypothesis was equally parsimonious if omnivory was treated as an alternative state intermediate between insectivory and herbivory.

We consider it unlikely that the earliest bats were exclusively herbivorous for several reasons. First, there is no evidence that dental specializations for herbivory (including frugivory and/or nectarivory) were present in any bat lineage prior to the Miocene. Many small-bodied extant mammals thought to be potential close relatives of bats (e.g., tree shrews, small primates, some lipotyphlans, some small carnivorans, many rodents) are facultative omnivores despite the fact that they belong to groups often labeled as "insectivorous," "herbivorous," or "carnivorous" (Nowak 1991 and references cited therein). Indeed, opportunistic omnivory may be at least as common (perhaps more common) than strict insectivory, strict carnivory, or strict herbivory in small-bodied mammals (e.g., those with adult body mass less than 200 g [Fleagle 1978; Kay 1973; Kay and Hylander 1978; Landry 1970; Nowak 1991]). Reasons for this pattern are unclear but may be linked to metabolic rates and the need to obtain adequate quantities of protein, calcium, and trace minerals (Fleagle

Table 11.1. Foraging ecologies of Eocene bats: observations and hypotheses

	Icaronycteris	Archaeonycteris	Hassianycteris	Palaeochiropteryx
Wing morphology	Low aspect ratio; low wing tip indices; moderate wing loading	Low aspect ratio; low wing tip indices; moderate wing loading	Low aspect ratio; low wing tip indices; high wing loading	Low aspect ratio; low wing tip indices; moderate wing loading
Calcar	Absent	Absent	Present	Present
Cochlear size	Moderate	Moderate	Large	Large
Stomach contents	Unknown	Coleoptera	Macrolepidoptera; Coleoptera; Blattoidea	Microlepidoptera; Tricoptera; scale-bearing Diptera
Proposed mode of orientation and obstacle detection	Echolocation and vision	Echolocation and vision	Echolocation and vision	Echolocation and vision
Proposed foraging habitat	Near ground and vegetation; along edges of lakes	Near ground and vegetation; along edges of lakes	Well above ground in forest clearings, above forest canopy, and over lakes	Near ground and vegetation; along edges of lakes
Proposed principal foraging style(s)	Gleaning prey from surfaces during short flights from a perch	Gleaning prey from surfaces during short flights from a perch	Fast aerial hawking; prey capture using calcar-supported uropatagium	Perch hunting for both flying and nonflying prey: slow aerial hawking; prey capture using calcar-supported uropatagium
Proposed mode of detecting and tracking prey	Vision and/or listening for prey-generated sounds	Vision and/or listening for prey-generated sounds	Echolocation	Echolocation (flying prey); vision and/or listening for prey-generated sounds (nonflying prey)
Proposed echolocation duty cycle	Low duty cycle	Low duty cycle	Low duty cycle	Low duty cycle
Proposed type of echolocation call	Short (≤2ms) multiharmonic calls or broad band FM calls in 30–90 kHz range	Short (≤2ms) multiharmonic calls or broad band FM calls in 30–90 kHz range	Narrowband FM calls or FM calls with both steep and shallow sweeps; calls under 30kHz	Broad band FM calls in 30–90 kHz range

Source. Reproduced from Simmons and Geisler (1998, table 9); see that publication for data and chain of inferences used to develop these hypotheses.

1978; Kay 1973; Kay and Hylander 1978; Landry 1970). Although extant Dermoptera are folivores (Wischusen and Richmond 1998), early Tertiary members of this clade (e.g., paromomyids, plesiadapiforms) were probably opportunistic omnivores that fed on a variety of items including arthropods, nuts, seeds, and even tree exudates (Beard 1990, 1993; Biknevicius 1986; Kay et al. 1990). We expect that the earliest bats, like many other small-bodied mammals, met their dietary needs with a combination of food items including both insects and plant products. In this respect, we agree with McNab (1982) and Kirsch and Lapointe (1997, fig. 9.7) who suggested that the "*ur*-bat" (the common ancestor of Megachiroptera and Microchiroptera) was a multiple-resource user who ate both insects and plant products.

At some point after the evolution of powered flight, the ancestral bat lineage apparently split into two lineages, one of which ultimately gave rise to extant Megachiroptera, the other to extant Microchiroptera. Simmons and Geisler (1998) named the former clade Megachiropteramorpha, and the latter Microchiropteramorpha (fig. 11.2). Basal members of both lineages presumably retained ancestral foraging habits (e.g., stage 1). Within Megachiropteramorpha, subsequent evolution led to specializations for frugivory and nectarivory, which were almost certainly present in the earliest members of the crown clade, Megachiroptera (the smallest clade containing all extant Pteropodidae). Like their ancestors, early megachiropteramorphs probably used vision for orientation and obstacle detection, and detected food objects by passive means including vision and olfaction. These habits have been retained in all extant Megachiroptera.

Evolution in the other major bat lineage, Microchiropteramorpha, proceeded along different lines. Simmons and Geisler (1998) suggested that the next stage of the evolution of foraging habits in this group involved the use of a new system—echolocation—for orientation and obstacle detection. Bats at this stage (stage 2) would have continued to forage as their ancestors had done, gleaning stationary prey objects during short flights from a perch. They may have captured flying prey at times, but prey detection would have been by passive means, not echolocation. *Icaronycteris* and *Archaeonycteris* probably represent examples of this stage of evolution (Simmons and Geisler 1998).

With the passage of time, subsequent evolutionary refinements of the echolocation system apparently began to make detection, tracking, and evaluation of flying prey possible, thus leading to yet another step in the evolution of foraging behaviors. At this stage (stage 3), bats probably continued to hunt from perches for stationary prey but also used echolocation to detect and track flying prey (Simmons and Geisler 1998). To exploit this method fully, these bats would have had to increase signal strength to maximize range and provide the necessary time to track and evaluate targets (Fenton et al. 1995). Environmental clutter (which produces many distracting echoes) would have presented difficulties at this stage, so Simmons and Geisler (1998) suggested that flycatching using echolocation to detect and track prey probably evolved in bats that frequented forest gaps and the edges of forests along lakes and rivers, places where vegetation (with potential perches) lies adjacent to relatively clutter-free open spaces. Alternatively, flycatching might have first evolved among bats that favored canopy-covered woodland habitats with little understory vegetation (see Swartz et al., this volume), another habitat where multiple potential perches lie adjacent to areas of uncluttered open space (M. B. Fenton, personal communication). No examples of bats clearly representing stage 3 are known from the fossil record, although *Icaronycteris* and/or *Archaeonycteris* may have foraged this way at times (Simmons and Geisler 1998).

According to Simmons and Geisler's (1998) hypothesis, the next evolutionary stage (stage 4) brought continued refinement of the echolocation and flight systems as well as evolution of a calcar-supported uropatagium useful for capturing prey on the wing. Microchiropteraform bats at this stage probably used a combination of perch hunting and continuous aerial hawking. Stationary prey would have been detected by passive means, and aerial prey by echolocation. This combination of foraging strategies was likely used by the common ancestor of *Hassianycteris*, *Palaeochiropteryx*, and extant Microchiroptera and was subsequently retained in *Palaeochiropteryx* (Simmons and Geisler 1998).

The final evolutionary stage (stage 5) proposed by Simmons and Geisler (1998) brought exclusive reliance on continuous aerial hawking using echolocation for prey detection, tracking, and evaluation. This foraging strategy was apparently used by *Hassianycteris* and the common ancestor of extant microchiropterans. In this context, extant bats that rely on continuous aerial hawking have simply retained the condition primitive for Microchiroptera (note that Microchiroptera is defined as the clade stemming from the most recent common ancestor of Emballonuridae + Yangochiroptera and, thus, represents the smallest clade including all extant non-pteropodid bats; see fig. 11.2).

In evaluating Simmons and Geisler's (1998) hypotheses about the evolution of echolocation and foraging habits, it is important to remember that their scenario is highly dependent on structure of the phylogenetic tree (fig. 11.2), which is itself a hypothesis. Although relationships of the fossil taxa are fully resolved and some groups are quite well supported, inclusion of additional taxa or character data will doubtless result in modifications to the tree. Major topological rearrangements may occur, especially when morphological and molecular sequence data are combined. Several molecular studies have suggested that rhinolophoids may be more closely related to Megachiroptera than to other echolocating bats and that Microchiroptera thus may not be monophyletic (Baker et al. 1997; Hutcheon et al. 1998; Kirsch 1996; Kirsch and Hutcheon 1997; Kirsch et al. 1998; Pettigrew and Kirsch, 1995; Porter et al. 1996). While this appears relatively unlikely (Simmons 1998, 2000), it nevertheless illustrates the uncertain nature of our understanding of higher-level bat phylogeny. Changes in Simmons and Geisler's (1998) hypotheses of inter-familial relationships may alter our perceptions of the sequence or direction of changes in both morphology and ecology of early bats. The same is true for the ecological reconstructions for the fossil bats (table 11.1), which may require modifications as new data are gathered in the future. In this context, the Simmons and Geisler (1998) hypothesis for the evolution of echolocation and foraging should be treated as a sort of "best guess" working hypothesis about the pattern and sequence of changes that occurred in the early stages of bat evolution. Additional testing using more detailed and comprehensive phylogenies and paleoecological reconstructions is a high priority for future research in this area.

The timing of the evolutionary transformations in foraging strategies remains unclear, but available evidence suggests that evolution occurred rapidly after the origin of echolocation. Simmons and Geisler (1998) noted that representatives of three different stages are present at Messel, an early/middle Eocene fossil locality in Germany. This fauna included at least seven bats, including two species of *Archaeonycteris* (stage 2), two species of *Palaeochiropterx* (stage 4), and three species of *Hassianycteris* (stage 5). Co-occurrence of these forms in a single fauna—especially so soon after the apparent origin of echolocation—suggests that many critical transformations probably took place in a burst of evolutionary change in the early Eocene. This adaptive radiation, which led to the origination of many modern groups by the end of the Eocene (see "The Fossil Record and 'Ghost' Lineages" above), subsequently gave rise to almost one-fourth of extant mammal species, an unparalleled contribution to mammalian biodiversity.

Evolution of Foraging Habits in Extant Families
Gleaning, Passive Prey Detection, and Perch Hunting

If we accept that the primitive foraging strategy for Microchiroptera consisted of continuous aerial hawking for insects, this leads to interesting hypotheses concerning evolution of foraging within several extant families. Specifically, Simmons and Geisler's (1998) analysis suggests that gleaning, passive prey detection, and perch hunting among extant microchiropterans represent secondarily derived specializations rather than retentions of primitive habits. Each of these behaviors apparently evolved independently several times in Microchiroptera, occurring in different combinations in different lineages.

Mapping foraging habits on Simmons and Geisler's tree indicates that passive prey detection secondarily evolved at least five or six times—in phyllostomids, mystacinids, antrozoids, vespertilionids, and either independently in nycterids and megadermatids or in the common ancestor of Rhinolophoidea. Similarly, gleaning must have evolved independently at least four times within Microchiroptera—in phyllostomids, antrozoids, vespertilionids, and rhinolophoids. Perch hunting apparently evolved at least three times—in phyllostomids, vespertilionids, and rhinolophoids. In this case, use of an alternative phylogenetic tree does not significantly alter the evolutionary conclusions. Mapping the same data on the DNA hybridization tree of Kirsch et al. (1998) indicates that passive prey detection must have evolved independently at least four times, gleaning at least three times, and perch hunting at least three times. Although passive prey detection, gleaning, and perch hunting are linked in some taxa (e.g., most megadermatids and nycterids), they are decoupled in other forms (Simmons and Geisler 1998). For example, *Mystacina* apparently uses passive prey detection but does not glean or hunt from perches; instead, it approaches its prey "on foot" (B. Lloyd, personal communication). *Antrozous pallidus* exhibits similar habits but, in addition, gleans prey from surfaces (Hermanson and O'Shea 1983; Johnston 1997). Rhinolophids sometimes hunt from perches and glean, but apparently do not use passive cues to detect their prey (Norberg and Rayner 1987). *Lavia frons* (a megadermatid) hunts from perches but apparently does not glean or use passive prey detection (Vaughan and Vaughan 1986). These observations suggest that passive prey detection, gleaning, and perch hunting evolved in different ways in different microchiropteran lineages.

The mosaic nature of the distribution of the behavioral traits discussed above illustrates the importance of understanding phylogenetic history when attempting to interpret ecologies of extant species. It may be tempting to assume that ecological or behavioral generalizations based on one taxon may apply equally to another, but this is not always the case, particularly if the taxa in question are not close phylogenetic relatives. If a behavioral trait or complex evolved independently in two lineages (something that cannot be ascertained without a phylogenetic analysis), then those behaviors are not homologous. Accordingly, generalizations may not be appropriate unless care is taken to evaluate the underlying assumptions. On the positive side, however, multiple cases of convergence offer additional degrees of freedom in testing hypotheses about ecological and behavioral evolution. From such cases we may learn considerably more about how evolutionary change occurs than can be learned from studying unique events (i.e., ecologies or behaviors that evolved only once in bats). Unfortunately, the quality and quantity of data presently available for bats make it difficult to test effectively for correlated trait evolution using statistical methods such as the concentrated-changes test. These methods require fully resolved trees (preferably at the species level) and complete or near-complete data for the traits under consideration. In the present case, we know that there is significant within-family variation in foraging habits, so species-level phylogenies (which do not currently exist) would be required to test effectively for correlated evolution in passive prey detection, gleaning, and perch hunting. A further problem is that comparable ecological data are not available for all relevant species. It is hoped that these difficulties will be overcome in the future with correlated interdisciplinary field and systematic studies.

High Duty Cycle of Echolocation and Flutter Detection

The majority of extant microchiropterans use what is termed "low duty-cycle" echolocation, a system in which emitted pulses and returning echoes are separated in time (Fenton 1994, 1995; Fenton et al. 1995; Jones and Rydell, this volume). Low duty-cycle bats use a wide variety of echolocation call types, most of which include FM (frequency modulated) components; search phase call sequences are characterized by short signal pulses with relatively long gaps between them (Fenton 1994, 1995; Fenton et al. 1995; Schnitzler and Kalko 1998). In contrast, rhinolophines, hipposiderines, and one mormoopid (*Pteronotus parnellii*) use "high duty-cycle" echolocation, in which the pulse and echo are separated in frequency rather than time (Fenton 1994, 1995; Fenton et al. 1995). These bats produce long constant-frequency (CF) echolocation signals that overlap with returning echoes, and they utilize the Doppler effect, which shifts the frequency of returning echoes to a lower frequency than that of the original pulses (Fenton 1994, 1995; Fenton et al. 1995; Neuweiler 1989, 1990; Schnitzler 1987; Schnitzler and Kalko 1998). Self-deafening is reduced because the emitted pulse is dominated by frequencies outside the acoustic fovea (zone of maximum hearing sensitivity), while both the external and inner ears are sharply tuned to the frequencies of the returning echoes (Fenton 1994; Fenton et al. 1995; Neuweiler 1990; Neuweiler et al. 1980; Obrist et al. 1993; Schnitzler 1987; Schuller and Pollack 1976; Vater et al. 1985). High duty-cycle echolocation appears to be uniquely well suited for detection of fluttering insects in clutter and classification of insects according to their specific flutter patterns (Schnitzler 1987; Schnitzler and Kalko 1998). Bats that use this system ("narrow-space CF bats" in Schnitzler and Kalko's terminology) can forage for fluttering insects in dense forest close to vegetation or the ground, something that is beyond the limits of most echolocating bats (Fenton 1994, 1995; Neuweiler 1989, 1990; Schnitzler 1987; Schnitzler and Kalko 1998).

There is a general consensus that high duty-cycle echolocation evolved twice in Microchiroptera, once in the lineage leading to Rhinolophidae (including Hipposiderinae) and once within the genus *Pteronotus*, in both cases apparently originating from a low duty-cycle system (Fenton et al. 1995; Pye 1980; Simmons 1980; Simmons and Geisler 1998; Simmons and Stein 1980). Extant bats that are high duty-cycle echolocators are characterized by a cochlea with an exceptionally large basal turn (Habersetzer and Storch 1992). None of the basal Eocene bats discussed above (*Icaronycteris, Archaeonycteris, Palaeochiropteryx, Hassianycteris*) show such cochlear enlargement, suggesting that the most primitive echolocating bats used a low duty-cycle system as discussed earlier (Habersetzer and Storch 1992; Simmons and Geisler 1998).

The time of origin of high duty-cycle echolocation in rhinolophids and within *Pteronotus* is not known, but phylogenetic information and fossil data are relevant to this question. All known extant Rhinolophinae and Hipposiderinae are high duty-cycle echolocators, and these two lineages are undoubtedly sister taxa (Simmons 1998; Simmons and Geisler 1998). The most parsimonious explanation for this pattern is that the most recent common ancestor of Rhinolophinae and Hipposiderinae was a high duty-cycle echolocator. The earliest fossils of both lineages occur in middle Eocene deposits (Hand and Kirsch 1998; McKenna and Bell 1997; Simmons and Geisler 1998), suggesting that high duty-cycle echolocation evolved in the rhinolophid lineage sometime in the early Eocene.

A recent phylogenetic analysis of fossil and living Hipposiderinae found that the late Eocene fossil *Palaeophyllophora querci* nests well up within Hipposiderinae (Hand and Kirsch 1998). This result suggests that much of the genus- and species-level differentiation within the subfamily must have taken place in the middle Eocene. If so, the adaptive radiation of this group (which now contains nine genera and more than 65 species) probably occurred very soon after the origin of the high duty-cycle system. The same may also be true of Rhinolophinae, which has a similarly extensive fossil record and currently contains more than 60 species. Use of high duty-cycle echolocation may have effectively opened a previously unexploited set of ecological niches (for nocturnal aerial predators foraging for fluttering insects in dense forest close to vegetation or the ground) when this system originated in the Eocene ancestors of Hipposiderinae and Rhinolophinae. As such, the evolution of high duty-cycle CF echolocation may have been the "key innovation" (sensu Liem 1973) responsible for the diversification and continued success of Rhinolophidae in the Tertiary. Although not discussed by Simmons and Geisler (1998), the evolution of high duty-cycle echolocation effectively constitutes a "stage 6" in this history of the evolution of foraging strategies in Eocene members of the microchiropteran lineage, and it apparently resulted in changes nearly as significant (in terms of ultimate impact on the world's bat faunas) as many of the previous stages.

This raises an interesting question: If Old World rhinolophid diversity evolved in part as a result of acquisition of high duty-cycle echolocation, why was there not a similar burst of diversification in New World mormoopids following the evolution of high duty-cycle echolocation in that group? Rhinolophids make up approximately 17% of extant microchiropteran species; in contrast, high duty-cycle mormoopids (one species) make up only 0.1% of Microchiroptera (Koopman 1993). The explanation for this pattern may lie in the different opportunities open to each lineage (Rhinolophidae and *Pteronotus*) subsequent to the "discovery" of high duty-cycle echolocation. In the case of Rhinolophidae, there may have been few potential competitors capable of similarly effective nocturnal aerial foraging for fluttering insects in dense vegetation when high duty-cycle echolocation evolved in this lineage (presumably

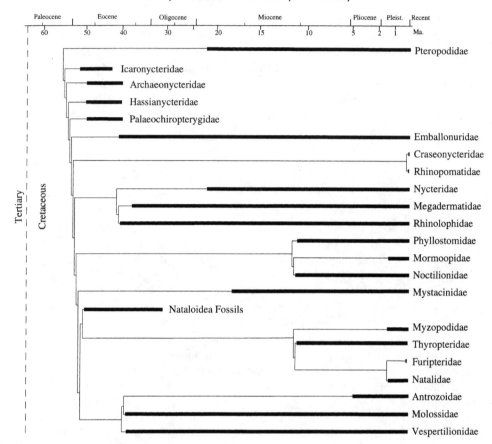

Figure 11.3. Reconstruction of the temporal pattern of diversification of the major lineages of bats based on Simmons and Geisler's (1998) phylogenetic tree (see fig. 11.1). *Thick lines* indicate time ranges documented by fossils; *thin lines* represent "ghost lineages" that must have existed given presumed phylogenetic relationships (see text for discussion). The time scale follows that of McKenna and Bell (1997, fig. 1). Data on fossil ranges are based on Beard et al. (1992), Czaplewski (1997), McKenna and Bell (1997), Sigé (1991), Simmons and Geisler (1998), and references cited therein.

the middle Eocene). The Old World tropics of the Eocene would have offered these bats enormous untapped resources on both the local and broad geographic scale. Subsequent evolution in the following 40 million years resulted in the rhinolophid diversity that we see today. In contrast, evolution of high duty-cycle echolocation in *Pteronotus* was probably a much more recent evolutionary event. Although Noctilionoidea must have been distinct since the middle Eocene and Mormoopidae since the late Miocene (fig. 11.3), the fossil record of *Pteronotus* extends only to the Pleistocene (McKenna and Bell 1997). This suggests that high duty-cycle echolocation evolved quite recently in *Pte-*

ronotus, perhaps within the past few million years. The absence of a subsequent adaptive radiation in this lineage may be due, in part, to the timescale—perhaps there simply has not yet been enough evolutionary time. However, another factor may have also played a role this case: presence of preexisting ecological competitors.

The Neotropics was probably an ecologically crowded place (from the bat perspective) by the time high duty-cycle echolocation evolved in *Pteronotus.* "Narrow-space FM bats" are those that use a form of low duty-cycle echolocation (short duration, broadband, uni- or multiharmonic, overlap-sensitive FM signals) to operate in highly cluttered forest environments (Schnitzler and Kalko 1998). All are gleaners, and most or all probably rely on passive cues to detect their prey (Schnitzler and Kalko 1998). Phyllostomids are narrow-space FM bats, and many (particularly phyllostomines) are adept at locating prey by the sounds that they make (e.g., fluttering or crashing into vegetation [Schnitzler and Kalko 1998]). The earliest known phyllostomids include at least three taxa from the La Venta fauna of Columbia, which is thought to be 11.0–13.8 Ma in age (Czaplewski 1997). *Notonycteris magdalenensis* is a phyllostomine thought to be closely related to *Vampyrum* and *Chrotopterus* (Savage 1951); also known from the fauna are specimens referred to *Tonatia,* Phyllostominae *incertae sedis,* and Glossophaginae *incertae sedis* (Czaplewski 1997). If these assessments are correct, the radiation of gleaning phyllostomids must have been well underway by the middle Miocene. The forest world in which high duty-cycle echolocation evolved in *Pteronotus* may thus have been very different from that in which Eocene rhinolophids evolved. In the face of considerable competition from narrow-space FM phyllostomids good at foraging in dense vegetation by listening for prey-generated sounds, high duty-cycle echolocation may not have accrued as large a selective advantage in the *Pteronotus parnellii* lineage as it apparently did in rhinolophids.

The different evolutionary histories of endemic bat lineages may therefore help to explain why dense forest understory habitats in the Old World and Neotropics support assemblages of insectivorous bats that differ significantly in foraging habits. In Old World tropical forests, understory faunas are dominated by high duty-cycle bats (rhinolophids) that feed principally on flying prey, and there are relatively few gleaners (nycterids and some megadermatids and vespertilionids). In contrast, understory insectivorous bat faunas in the Neotropics are dominated by gleaners (phyllostomids, some vespertilionids), and there is only one high duty-cycle species (*Pteronotus parnellii*). As suggested above, these differences in proportions may reflect the timing of origin of high duty-cycle echolocation in endemic bat lineages (relatively early origin in the Old World, late origin in the Neotropics). However, this explanation does not adequately explain why the Old World gleaners—particularly nycterids and megadermatids, which are ecologically similar to phyllostomids—failed to become as diverse as their Neotropical counterparts. The fossil record of megadermatids extends back to the late Eocene, and reconstruction of ghost lineages indicates that both Nycteridae and Megadermatidae must have been distinct by the middle Eocene (fig. 11.3). Complex competitive interactions (probably involving rhinolophids), variations in reproductive success, differences in the pool of potential prey species, and many other factors may have contributed to limiting diversification of nycterids and megadermatids. Regardless, understanding the differences between Old World and Neotropical bat faunas clearly requires an evolutionary perspective.

Evolution of Dietary Habits in Phyllostomidae

Few topics in evolutionary biology of bats have received as much attention as the origins of major dietary ensembles in Phyllostomidae. Phyllostomids as a group encompass more dietary diversity than any other chiropteran family, and researchers have long sought to understand the patterns and the processes that produced this remarkable radiation (for a historical review, see Ferrarezzi and Gimenez [1996] and Wetterer et al. [2000]). Recent contributions in this area have used phylogenies of the family to reconstruct the evolutionary history of diet change in Phyllostomidae. Ferrarezzi and Gimenez (1996) mapped seven feeding habits (strict insectivory, predominant insectivory, predominant carnivory, strict frugivory, predominant frugivory, predominant nectarivory, and sanguinivory) on a summary cladogram derived from phylogenetic analyses by Baker et al. (1989), Gimenez et al. (1996), Honeycutt and Sarich (1987), Lim (1993), and Owen (1987). Results of Ferrarezzi and Gimenez's (1996) study suggested that predominant insectivory is primitive for phyllostomids, and that most of the other feeding habits evolved only once within the group.

Wetterer et al. (2000) mapped dietary habits on a new phylogenetic tree derived from a comprehensive analysis of morphological, restriction site, and sex-chromosome characters of phyllostomids and reached broadly similar conclusions about the evolution of diet in the family. Several of the major dietary ensembles are limited to single clades, suggesting that each evolved only once within Phyllostomidae. However, tree topology within subfamilies led Wetterer et al. (2000) to conclude that patterns of dietary evolution within some clades may have been somewhat more complex than suggested by Ferrarezzi and Gimenez (1996). For example, predominant carnivory may have evolved twice (or there may have been a reversal) among phyllostomines, and degrees of reliance on fruit (as opposed to a mixed diet including insects and/or nectar and pollen) apparently shifted several times within various branches of Stenodermatinae. In the case of carnivory, body size appears to be a critical factor. Among lineages of gleaning bats, large body size and inclusion of vertebrates in the diet are highly correlated (Norberg and Fenton 1988). Large gleaning bats of several families depend heavily on small vertebrate prey and supplement this diet with insects (Norberg and Fenton 1988). In contrast, the diet of medium-sized gleaners consists mostly of insects, occasionally supplemented with vertebrates if the prey is small enough (Norberg and Fenton 1988). Thus, there seems to be no clear division between "insectivory" and "carnivory" among gleaning bats (Norberg and Fenton 1988). In this context, the evolution of carnivory (inclusion of vertebrates in the diet) in phyllostomids may reflect evolution of larger body size rather than a major dietary shift.

Additional studies of evolution of dietary habits have focused on particular functional complexes. Ferrarezzi and Gimenez (1996) investigated the evolution of blood-feeding habits in desmodontines by mapping favored prey (avian or mammalian), foraging site (arboreal or terrestrial), and specificity of plasminogen activation of saliva (activates both avian and mammalian plasminogens or just mammalian plasminogens) on a phylogeny. The results suggest that feeding on avian prey approached in an arboreal milieu is probably primitive for vampires, and the habits of *Desmodus* (which preys principally or entirely on mammals that they approach on the ground) are probably derived. Schutt (1998) independently reached a similar conclusion based on comparative morphology and expanded this idea into a detailed "arboreal-feeding hypothesis" by integrating information from morphology, feeding behavior, and the fossil history of phyllostomids. Schutt's (1998) hypothesis, which was also framed in a phylogenetic context, suggests that blood-feeding desmodontines evolved from arboreal, omnivorous phyllostomids. In another study of unusual feeding specializations in phyllostomids, Simmons and Wetterer (2002) concluded from phylogenetic mapping that cactophily (feeding on cactus products including nectar, pollen, and fruit) is a derived behavior in glossophagines that apparently evolved multiple times within that monophyletic subfamily. In each case, consideration of phylogenetic relationships of relevant taxa has contributed to our understanding of the evolution of their ecology and behavior.

Studies of dietary evolution in phyllostomids have been limited by two critical factors: lack of species-level phylogenies for many groups and absence of detailed information on diet for many species. Ongoing studies are rapidly addressing our lack of phylogenetic information, but the latter problem is less easily solved. As we have gained increasing insight into bat ecology and evolution, it has become increasingly clear that these animals often exhibit considerable flexibility in their habits and requirements. For example, the diet of *Glossophaga soricina* consists of fruit, nectar, pollen, flower parts, and insects, with relative proportions varying with season and locality (Alvarez et al. 1991; Gardner 1977; Heithaus et al. 1975; Helversen and Winter, this volume). Many other phyllostomids apparently have equally eclectic diets, but details of resource use are known for only a few species. Additional basic fieldwork

in bat ecology will be necessary before we can hope to achieve a more sophisticated understanding of the evolution of diet and foraging strategies in
phyllostomids.

Evolution of Nectar Feeding in Pteropodidae

The only other bat group known for extensive use of fruits and flower products is Pteropodidae, the "Old World fruit bats." It has long been assumed that
the primitive dietary habit for Pteropodidae is frugivory and that nectarivory
is a specialization that subsequently evolved within one or more lineages in
that clade (Andersen 1912; Ferrarezzi and Gimenez 1996). While some phylogenetic analyses of Pteropodidae have provided support for this idea by indicating that nectarivorous taxa nest within otherwise frugivorous clades (e.g.,
Hood 1989; Kirsch et al. 1995), others studies have not confirmed this conclusion. Based on the structure of a phylogenetic tree derived from restriction
fragment-length polymorphism data, Colgan and Flannery (1995) suggested
that frugivorous pteropodids might actually have evolved from a nectarivorous ancestor. Freeman (1995) concluded that either scenario (frugivory →
nectarivory or nectarivory → frugivory) seems equally likely from the perspective of craniodental morphometrics. Elaborating further, Freeman (1998,
152) stated: "I believe all megachiropterans evolved from a long-tongued ancestor, which could easily have been a nectarivore. Both nectarivores and frugivores have narrow elongated palates with space between relatively simple,
noncomplex teeth that do not register, fused mandibles, shallower, less distinct jaws joints, and thegosed upper canines. This suite of characters argues
for a tongue-feeding ancestor that may have been either a nectarivore or a
tongue-feeding frugivore or both."

Kirsch and Lapointe (1997) subsequently took a novel approach to the
problem by considering intermediate conditions between strict frugivory and
strict nectarivory, mapping the percentage of nectar in the diet (0%, 20%, 50%,
100%) on alternative phylogenetic trees derived from the work of Anderson
(1912), Kirsch et al. (1995), and Springer et al. (1995). They found high degrees
of homoplasy no matter what tree or assumption set (ordered or unordered
transformations) was used, and concluded that varying degrees of nectarivory
had evolved multiple times in Pteropodidae. Perhaps even more interesting is
their conclusion that "mixed feeding"—including perhaps 50% nectar in the
diet—might have characterized the earliest pteropodids. This hypothesis
agrees well with Freeman's (1995, 1998) conclusions and is also consistent with
our hypothesis (see above) that the earliest bats were opportunistic feeders
that included many foodstuffs in their diet. Early megachiropteramorphs
(which were probably opportunistic omnivores) and pteropodids (herbivores
that may have fed on both fruit and nectar) each probably had more generalized feeding habits than most of their descendants. In this context, strict fru-

givory and strict nectarivory both represent conditions derived within Pteropodidae; neither is primitive for the family.

Kirsch and LaPointe (1997) also used phylogenetically based statistical methods to investigate the relationship among features of molar morphology (length and width of M1), mandibular morphology (length of the mandible and height of the coronoid process), and diet (percentage of nectar in the diet) within Pteropodidae. Independent-contrasts analyses detected no significant correlations among these features when analyzed in the context of the morphology-based phylogenies of Anderson (1912) and Springer et al. (1995). In contrast, Kirsch and LaPointe (1997) found significant correlations for all comparisons when analyzed in the context of various DNA hybridization trees. They interpreted these results as suggesting that nectarivory is functionally related to morphology and that it is not merely present in a variety of fruit bats as an accidental consequence of shared ancestry (Kirsch and La-Pointe 1997). As a further test, Kirsch and LaPointe (1997) used a series of randomized trees to investigate the possibility that significant correlations might be detected on any tree, so long as it was not generated from craniodental or lingual morphology. They found that none of the significant correlations detected in the context of the DNA hybridization trees were more extreme that those obtained from random trees. From this they concluded that anatomy and diet, while they may be correlated with each other, cannot be demonstrated to be so by their molecular trees (Kirsch and LaPointe 1997).

As with phyllostomids, studies of evolution of pteropodids continue to be limited by lack of species-level phylogenies and absence of adequate ecological data for many taxa. Continuing phylogenetic studies are addressing the lack of phylogenetic information, but much work remains to be done, particularly on more speciose genera such as *Pteropus*. Obtaining ecological data on pteropodid species—while they remain extant—is of critical importance. The diet of many species is unknown, and some of what is assumed may be incorrect. Pollen, leaves, and insects may play a larger role in the diets of pteropodids than previously thought (Courts 1998, and references cited therein). It is clear that more field studies are necessary to fill the many gaps in our knowledge of these bats.

Chiropteran Body Size

The evolution of body size in bats has received increased attention in recent years (Barclay and Harder, this volume; Jones and MacLarnon 2000; Swartz et al., this volume; Willig et al., this volume). In a comparative study of flying vertebrates, Rayner (1981, 166) observed that bats may be considered unusual since they seem limited to smaller body sizes than are other groups: "No bat weighs more than about 1.5 kg, while birds weigh up to 10 or 12 kg (although sustained flapping flight is rare in birds more than about 4 kg); the smallest passerine bird is the goldcrest (4 g), while the smallest bat is about 1.5 g (probably also the smallest of all mammals)." Rayner (1981) suggested that the upper bound on the size limits of bats might be a result of the mechanics and metabolic requirements for sustained flight. Virtually all large birds use gliding and soaring (relatively inexpensive habits in terms of metabolic requirements) for much of their activity, and many smaller birds soar when conditions are suitable (Rayner 1981). Bats cannot use thermal soaring because they are active at night when convective air currents are absent (Rayner 1981). In the absence of very long wings (perhaps impossible due to design constraints) or wings with slits for drag reduction (such as provided by the separated primary feathers of birds), the glide performance of a bat of large mass may be too poor for gliding to be a useful means of sustaining flight (Rayner 1981). It is also possible that the mammalian metabolic system may not be able to provide sufficient power for large bats to fly (Rayner 1981; Speakman and Thomas, this volume).

In the absence of agreement concerning relationships of bats to other mammalian orders, it is not clear whether the chiropteran lineage underwent a size reduction early in its history (e.g., in the late Paleocene or early Eocene), or simply never evolved larger body sizes. If we accept that bats probably evolved from gliding ancestors (see discussion in Simmons 1995), it seems likely that no significant reduction in body size was required by the origin of flapping flight in the earliest bats. Living mammalian gliders (which represent three orders and at least six independent evolutionary lineages) all have mean body masses between 10 g and 1.5 kg, broadly similar to the range seen in bats (Rayner 1981). It therefore seems likely that the earliest bats weighed less than 1.5 kg. But were they 10 g, 50 g, 100 g, or 1 kg? This question cannot be adequately answered without additional phylogenetic information (e.g., concerning the sister group of bats) or fossil evidence (e.g., fossils of bats that predate the split between the megachiropteran and microchiropteran lineages).

Although there is overlap between the two bat suborders, body mass tends to be much greater in megachiropterans than in microchiropterans (fig. 11.4). Jones (1996) calculated that mean body mass for pteropodids (at least those in his sample of 55 species) is more than 200 g, while the largest microchiropteran species (*Cheiromeles torquatus*) has a mean mass of only approximately 160 g. The smallest megachiropterans (e.g., *Macroglossus*, *Syconycteris*) have a mean body mass of around 15 g (Norberg and Rayner 1987). In contrast, more than 70% of microchiropteran species have a mean body mass of less than 20 g, and almost 40% weigh less than 10 g (Jones 1996, Jones and MacLarnon 2000). The smallest known microchiropteran, *Craseonycteris thonglongyai*, has an adult body mass of only about 2 g (Jones 1996). These differences in body size constitute one of the most widely cited differences between Megachiroptera and Microchiroptera.

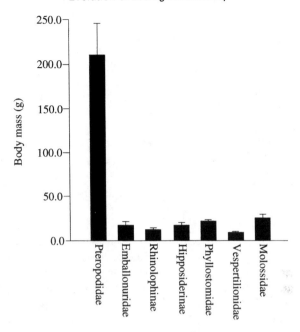

Figure 11.4. Body masses (mean ± SE) for selected families and subfamilies of bats where data are available for more than 20 species (redrawn from Jones 1996, fig. 2). Sample sizes are as follows: Pteropodidae, n = 55; Emballonuridae, n = 31; Rhinolophinae, n = 22; Hipposiderinae, n = 23; Phyllostomidae, n = 106; Vespertilionidae, n = 150; and Molossidae, n = 44. Statistical analyses of these data by Jones (1996) indicated that pteropodids are significantly heavier than any other group, molossids are significantly heavier than rhinolophids and vespertilionids, and phyllostomids are significantly heavier than vespertilionids.

Evolution of Body Size in Megachiroptera

As noted above, body size in extant megachiropterans ranges from small (e.g., *Macroglossus minimus*, 12–18 g., length of forearm = 39–45 mm) to very large (e.g., *Pteropus vampyrus*, mean mass = 1.2 kg; length of forearm = 179–220 mm [Koopman 1994; Lekagul and McNeely 1977; Norberg and Rayner 1987]). Pettigrew et al. (1989) suggested that the ancestral megachiropteran was moderately large, with a forearm length in the 70–150 mm range. Living pteropodids in this size class have body masses ranging from about 60 to 500 g (Bergmans 1997; Bergmans and Rozendaal 1988; Flannery 1995; Ingle and Heaney 1992; Kingdon 1974; Norberg and Rayner 1987; Nowak 1991; Smith and Hood 1983; Thomas and Marshall 1984). One observation cited by Pettigrew et al. (1989) in support of the "moderate size" hypothesis was the size of the Oligocene fossil bat *Archaeopteropus,* then considered to represent an early megachiropteran. However, Schutt and Simmons (1998) recently showed that *Archaeopteropus* is probably a microchiropteramorph, so size of this taxon is not relevant to understanding the evolution of body size in Megachiroptera.

Assuming that the earliest megachiropteran had a moderate body size, Pettigrew et al. (1989, 533) went on to suggest that both small body size and large body size are derived features within the suborder. These hypotheses can be tested by considering body size data in the context of recent phylogenies of Pteropodidae (e.g., those of Springer et al. [1995] and Kirsch and Lapointe [1997]). Body mass data are available for only about one-third of extant species (Barclay and Harder, this volume; Jones 1996; Jones and MacLarnon 2000), thus in the present analyses we used length of forearm as an alternative measure of body size. Our data source was Koopman (1994), who reported maximum and minimum lengths of forearm for all bat species. Because sample size, mean, and standard deviation of forearm measurements were not provided by Koopman (1994), we used the midpoint value (that value which lies halfway between the minimum and maximum reported values) for each species as a measure of length of forearm for that taxon. Although far from ideal, this approach permitted us to score every species in the genera included in the Kirsch and Lapointe (1997) and Springer et al. (1995) phylogenetic trees. Based on the range of values obtained, we defined a series of character states describing progressive size classes (e.g., midpoint forearm length $\leq$ 50 mm, 50.1–60.0 mm, 60.1–80.0 mm, etc.). Each genus was then scored based on the range of sizes found among its constituent species; taxonomic polymorphism was accounted for by attributing more than one character state to some genera (e.g., *Pteropus,* which contains species that fall into three different size classes). These data were then mapped in an additive fashion on phylogenies based on morphology (from Springer et al. 1995) and DNA hybridization data (Kirsch and LaPointe 1997). Although the phylogenies differ substantially, the outcome was the same: both trees indicate that a moderately small length of forearm (midpoint value = 60.1–80.0 mm) is primitive for Pteropodidae (fig. 11.5). Among modern pteropodids, bats in this forearm size class have adult body masses that range from 25 to 120 g (Bergmans 1997; Bergmans and Rozendaal 1988; Flannery 1995; Ingle and Heaney 1992; Kingdon 1974; Norberg and Rayner 1987; Nowak 1991; Smith and Hood 1983; Thomas and Marshall 1984).

An alternative approach to scoring measurement data as a series of discrete character states (such as we did in the example above) is to deal directly with continuous character data. Several methods exist for analyzing continuous character data in a phylogenetic context, including "linear" and "squared-change" parsimony (Maddison 1991; Maddison and Maddison 1992) and maximum-likelihood estimation and evaluation of ancestral character states (Schluter et al. 1997). As a test of the results of the discrete-state character mapping exercise described above, we scored all of the taxa in the Kirsch and

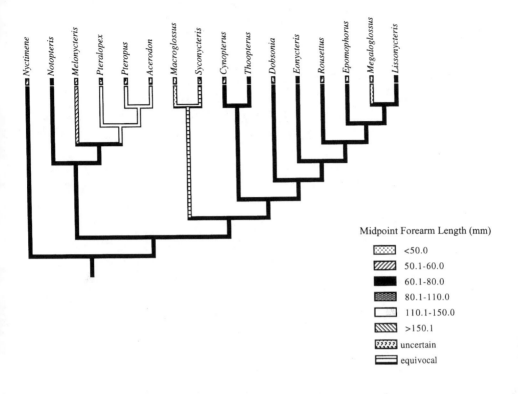

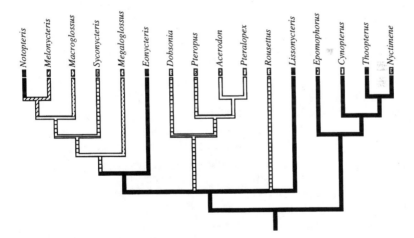

Figure 11.5. Evolution of forearm length in pteropodid bats, as inferred from alternative phylogenetic trees. The tree on the *bottom* was derived from an analysis of morphological data (Springer et al. 1995); that on the *top* was based on DNA hybridization data (Kirsch and Lapoint 1997). Values for forearm length are midpoint values; see text for explanation. Despite major differences in tree topology, both reconstructions indicate that a forearm length between 60.1 and 80.0 mm is probably primitive for Pteropodidae. Extant pteropodids in this size class have adult body masses that range from 25 to 120 g (Bergmans 1997; Bergmans and Rozendaal 1988; Flannery 1995; Ingle and Heaney 1992; Kingdon 1974; Norberg and Rayner 1987; Nowak 1991; Smith and Hood 1983; Thomas and Marshall 1984).

Lapointe (1997) and Springer et al. (1995) phylogenetic trees for a continuous character using the complete range of forearm variation known for the genus (e.g., 86–220 mm for *Pteropus;* data from Koopman [1994]). We analyzed these data using both linear and squared-change parsimony as implemented in MacClade (Maddison and Maddison 1992). For the Kirsch and LaPointe (1997) tree, reconstructed ancestral values for length of forearm ranged from 69 to 74 mm. Various versions of the Springer et al. (1995) tree (with different resolution of the polytomy) yielded reconstructed ancestral forearm length values of 59–79 mm. These values are consistent with those derived from discrete-state midpoint value mapping (fig. 11.5). As Pettigrew et al. (1989) suggested, both larger and smaller body sizes appear to be derived within the family. Regardless of which phylogeny or method one chooses, the pattern of body size evolution appears to have been complex both within and between genera. Several genera (e.g., *Pteropus, Acerodon, Pteralopex, Dobsonia, Epomophorus, Nyctimene*) include species of many different sizes. Understanding patterns of evolution at these lower taxonomic levels will require more detailed species-level phylogenies than are presently available.

Evolution of Body Size in the Microchiroptera

Most workers would agree with Jones's (1996, 114) statement that "microchiropterans really are miniature vertebrates." This is clearly true of the aerial insectivores, which are typically very small (<30 g, adult body mass). Only four species in this foraging category regularly achieve adult body masses of more than 100 g (i.e., *Taphozous peli, Hipposideros commersoni, Scotophilus nigrita,* and *Cheiromeles torquatus* [Arita and Fenton 1997; Barclay and Brigham 1991; Black 1974; Fenton and Fleming 1976; Jones 1996, Krazanowski 1977; McNab 1969; Simmons and Geisler 1998]). Maximum body size may be limited in microchiropteran aerial insectivores for several reasons, including (1) the small size of available prey (McNab 1969; Black 1974); (2) the mechanics of flight and aerial prey capture (Jones 1996; Norberg 1986, 1994; Norberg and Rayner 1987); (3) the effective range of echolocation calls (relatively short) and the concomitant need to be maneuverable and agile enough to catch small prey at short range (Barclay and Brigham 1991); and (4) the coupling of flight and echolocation, which may limit the ability of large bats (which have lower wingbeat frequencies and thus lower call-repetition rates) to detect small flying prey (Jones 1994, 1996; although see Heller 1995).

Barclay and Brigham (1991) argued that aerial insectivores are smaller than bats that forage by gleaning prey from surfaces. While it is true that all extant bats with mean body mass less than 5 g are aerial insectivores, Jones (1996) noted that Barclay and Brigham's (1991) failure to control for phylogenetic effects may have biased their results. Jones (1996) found that there is no significant difference between mean body masses of gleaners and aerial insectivores within Vespertilionidae, which is the only family that contains a sufficient number of insectivorous species to make meaningful comparisons possible.

Regardless of the evolutionary forces that have shaped phylogenetic changes in body size, it is clear that body size significantly affects and may constrain the ecology of bat species (Barclay and Brigham 1991; Barclay and Harder, this volume; Bogdanowicz et al. 1999; Jones 1994, 1996; Jones and MacLarnon 2000; Norberg and Fenton 1988; Norberg and Rayner 1987; Rayner 1981; Simmons and Wetterer, in press; Swartz et al., this volume; Willig et al., this volume). Most aspects of flight mechanics are significantly affected by changes in total mass and mass distribution, and relatively large body size may be an important factor in facilitating long-distance flight and migration in some species (Fleming and Eby, this volume; Norberg and Rayner 1987; Simmons and Wetterer 2000). Echolocation call structure is often influenced by body size, with larger bat species tending to use lower frequency calls than smaller congeners (Bogdanowicz et al. 1999; Jones 1996; Jones and Rydell, this volume). Pulse repetition rates are also influenced by size via wingbeat frequency, which is typically lower in larger bats (Jones 1994, 1996).

There is some disagreement concerning the effects of these factors on dietary niche breadth. Barclay and Brigham (1991) suggested that larger aerial insectivores seem to be limited to relatively large insect prey (e.g., moths and large beetles), while smaller bats can exploit either large or small prey. However, Aldridge and Rautenbach (1987) and Jones (1996) found that the opposite is true in African and European bat faunas: large aerial insectivorous species appear to exploit both large and small prey, while smaller-sized species are limited to only small prey. As noted earlier, body size is also an important predictor of the degree to which predatory bats take prey other than arthropods; bats that include small vertebrates in their diet tend to be larger than those that are strictly insectivorous (Norberg and Fenton 1988). Size may also constrain feeding methods in frugivorous and nectarivorous taxa, influencing such factors as whether a bat perches or hovers when feeding at flowers and whether fruits are carried to feeding roosts or consumed in situ. Choice of roosts may also be influenced by body size (e.g., large bats may not be able to fit in small cracks or roost in modified foliage [Kunz 1982; Kunz and Lumsden, this volume]).

Body sizes of the earliest members of the lineage leading to Microchiroptera vary significantly, but all have lengths of forearm between 38.0 and 85.0 mm and probably weighed between 6 and 100 g (table 11.2). These data suggest that evolution of extremely small body size (forearm length [FA] $\leq$ 35 mm, mean adult body mass $\leq$ 5 g) and large size (FA $\geq$ 85 mm, mean adult body weight $\geq$ 100 g) are derived traits. A very limited number of species fall into these size categories, the combined total of which is probably less than 10% of extant microchiropteran species (Jones 1996; Jones and MacLarnon 2000). Bats in the "very small" category defined above occur in Emballonuridae,

Table 11.2. Body sizes of Eocene microchiropteramorph bats

Taxon	Length of Forearm (mm)	Estimated Body Mass (g)
Icaronycteris index	46.3–48.0	8.9–33.2 (15.4)
Archaeonycteris trigonodon	52.5–58.2	16.3–57.5 (23.2–32.5)
Hassianycteris messelensis	64.5–69.5	15.5–45.5 (45.5)
Hassianycteris magna	81.5	41.2–97.8 (97.8)
Palaeochiropteryx tupaiodon	38.6–46.0	6.9–15.1 (8.5–15.1)
Palaeochiropteryx spiegeli	43.5–48.7	9.6–18.2 (12.6–18.2)

Sources. Forearm length data are from Habersetzer and Storch (1987). Body mass estimates are from Norberg (1989).

Note. The values listed for estimated body mass indicate the range of estimates derived using five different regression equations based on wingspan (two reconstructions), length of radius, length of tibia, and length of head + body in extant microchiropterans. Norberg (1989) argued that the regression equation based on radius length probably yields the most realistic estimate of body mass; these values are given in parentheses.

Craseonycteridae, Rhinolophidae, Thyropteridae, Natalidae, Furipteridae, and Vespertilionidae; "large" bats occur in Megadermatidae, Rhinolophidae, Phyllostomidae, and Molossidae (Barclay and Brigham 1991; Koopman 1994; Norberg and Rayner 1987; Simmons and Voss 1998).

Little can be concluded in the absence of a thorough species-level analysis, but one pattern stands out when available body size data are considered in a broad phylogenetic context: both very small and large body sizes (as defined above) seem to have evolved independently multiple times in Microchiroptera. Very small body size appears correlated with the origin of one small radiation, that of the Neotropical nataloids (all of which are tiny bats; adult body mass ≤ 7 g). However, in all other instances, exceptionally small or large body size appears in only a few members of lineages that are otherwise characterized by small to moderate body sizes. For example, very small body size characterizes some Neotropical emballonurids (e.g., *Rhynchonycteris, Centronycteris,* some *Saccopteryx*), but these are highly derived forms that nest well up within the emballonurid clade in phylogenetic analyses (e.g., Barghoorn 1977; Griffiths and Smith 1991; Robbins and Sarich 1988). In the context of recent hypotheses of interfamilial and subfamilial relationships within Microchiroptera (e.g., Simmons 1998; Simmons and Geisler 1998; Volleth and Heller 1994), it appears that very small body size (FA ≤ 35 mm, mean adult body mass ≤ 5 g) probably evolved independently in Craseonycteridae, Rhinolophidae, Neotropical Nataloidea, and probably several times in Vespertilionidae. Similarly, phylogenies of Megadermatidae, Rhinolophidae, Phyllostomidae, and Molossidae (e.g., Bogdanowicz and Owen 1998; Freeman 1981; Griffiths et al. 1992; Hand 1985; Hand and Kirsch 1998; Legendre 1985; Wetterer et al. 2000) indicate that large-bodied species (FA ≥ 85 mm, mean adult body mass ≥ 100 g) were independently derived from within smaller-bodied clades in each of these families.

Analyses of body size at the family level are hampered by differences in species richness among families. Jones (1996) analyzed mean body masses in the most speciose microchiropteran families (fig. 11.4) and found that molossids are significantly heavier than rhinolophids and vespertilionids and that phyllostomids are significantly heavier than vespertilionids. The biological significance of these patterns is not clear, however, since only mean values were compared and polarity of change within families was not considered.

How has body size influenced subsequent evolutionary history of microchiropteran clades? Addressing this question adequately will require detailed ecological, phylogenetic, and morphometric data that are not yet available and sophisticated phylogenetic analyses far beyond those yet completed. Several evolutionary models suggest that smaller-bodied taxa should be more speciose than larger-bodied forms, and speciation rates may be higher in smaller forms (e.g., Dial and Marzluff 1988; Hutchinson and MacArthur 1959; MacArthur and Levins 1964; May 1978; Van Valen 1973). To a large extent, this may have been the case in microchiropteran bats, although Jones (1996) noted that the small number of species in the very smallest size class (<5 g) argues against a simple interpretation of the bat data. However, Jones (1996) noted that the pattern in Microchiroptera might be explained by the interplay of speciation and extinction rates since extinction rates are expected to be highest for extreme-sized species (Dial and Marzluff 1988). All of these hypotheses clearly require additional exploration and testing.

Evolution of Biogeographic Patterns

Several recent studies have integrated phylogenies of bats with analyses of biogeographic history (e.g., Baker et al. 1994; Hand and Kirsch 1998; Kirsch et al. 1998; Lim and Engstrom 1998; Simmons 1998). The goals of these analyses have been to reconstruct patterns of origination, diversification, dispersal, and vicariance through space and time. Such studies are typically most informative when conducted at lower taxonomic levels because higher-level taxa (e.g., genera, subfamilies) often have very large, overlapping geographic ranges.

A nice example of how a phylogeny may be used to better understand the biogeographic history of a group of closely related species is Baker et al.'s (1994) analysis of *Chiroderma*. This Neotropical genus includes five species with very different geographic ranges. Two species (*C. doriae* from southern Brazil and *C. improvisum* from the Lesser Antilles) have restricted ranges that do not overlap with those of the more widespread species *C. salvini, C. trinitatum,* and *C. villosum* (although *C. doriae* and *C. villosum* are sympatric in some areas). Similarities in size and form of the anterior dentition have suggested that *C. improvisum* and *C. doriae* might be sister taxa (Baker and Genoways 1976), leaving the biogeographic history of this group—and mechanisms of dispersal and divergence—in question. However, a more recent study of *Chiroderma* based on gene sequence data resulted in a different hypothesis that has implications for understanding speciation in the genus. Using cytochrome *b* sequence data, Baker et al. (1994) constructed a well-supported phylogeny, which they used to evaluate previous hypotheses of geographic history of the group (fig. 11.6). They found that *C. doriae* and *C. improvisum* are not sister taxa but are, instead, more closely related to species that occur in regions geographically adjacent to their ranges (*C. trinitatum* and *C. villosum*, respectively). *Chiroderma* thus contains two pairs of sister taxa that occupy adjacent but disjunct ranges (fig. 11.6), suggesting that they may have originated through allopatric speciation (Baker et al. 1994). It seems likely that the common ancestor of the five extant species had a range that included northern South America, but details of the primitive geographic range (and the speciation mechanisms involved) remain unclear in the absence of information about the sister group of *Chiroderma*. Baker et al. (1994) used *Uroderma* and *Platyrrhinus* to root their tree, but Wetterer et al. (2000) found that a clade comprising *Vampyressa* and *Ectophylla* may be the sister group of *Chiroderma*. Regardless, species-level phylogenies of the taxa in question must be completed before we begin to speculate about the biogeographic origins of the genus *Chiroderma*.

As implied above, phylogeny-based reconstructions of biogeographic history may have implications that reach beyond pattern to process—processes of speciation, vicariance, and responses to habitat change, just to name a few. However, these areas of research have not yet been investigated. For example, it would be interesting to investigate possible barrier mechanisms that might explain the apparently allopatric pattern of speciation in *Chiroderma*. Data on climate change and shifting habitat distributions might be combined with behavioral data, information on habitat requirements, and estimates of divergence (e.g., percent sequence divergence within and among populations and species) to develop a more comprehensive hypothesis of the evolutionary history of the genus.

A new focus of study, "phylogeography," has recently emerged at the interface of systematics and population genetics. Using molecular markers and DNA sequence data, these studies strive to understand the genetic and geographic history of populations in an evolutionary context. Only a few studies on bats have been completed (e.g., Hisheh et al. 1998; Phillips et al. 1991; Schmitt et al. 1995), but these have proven promising. In the future, insights derived from phylogeographic studies may be combined with ecological information to provide an increasingly sophisticated view of how evolution has operated to shape bat diversity.

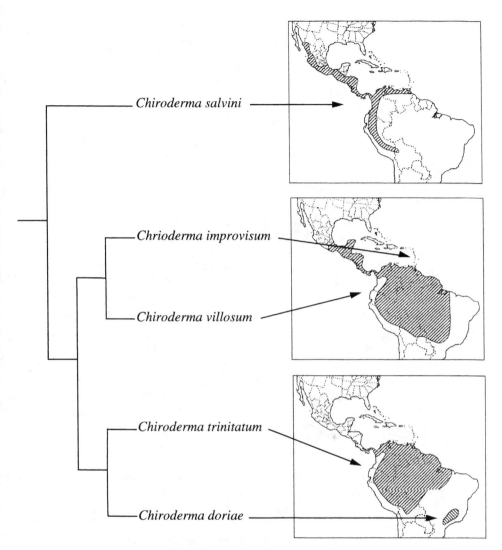

Figure 11.6. Phylogeny and biogeography of *Chiroderma* (after Baker et al. 1994). The phylogeny shown on the *left* was derived from a parsimony analysis of cytochrome *b* gene sequence data; the range maps (redrawn from Baker et al. 1994) summarize geographic data from Baker and Genoways (1976), Hall (1981), Koopman (1982) and Taddei (1979, 1980). This analysis indicates that *Chiroderma* contains two pairs of sister taxa (*C. trinitatum* and *C. doriae, C. villosum* and *C. improvisum*) that occupy adjacent but disjunct ranges. This pattern suggests that these species may have originated as the result of allopatric speciation (Baker et al. 1994).

Discussion and Conclusions

The diversity of extant and fossil bats is both a blessing and a curse to those who strive to understand their ecology and evolution. When considered in a historical context, bats present an almost textbook case of mosaic evolution. Many clades are very old, yet much of the extant diversity may be traced to radiations that could have occurred very rapidly in geological time. Many traits have evolved more than once, reversals are common, and our understanding of geographical patterns of diversification are sketchy at best. The potential power of phylogenetic analyses for better understanding the ecologies of living taxa is clear, but quality of results depends entirely on the quality of the ecological data input and the resolution and strength of the phylogenies used. However, these limitations should not discourage interested scientists from pursuing studies of bat evolutionary ecology. Indeed, both systematists and ecologists benefit in the course of such studies. The phrase "reciprocal illumination" has been used to describe how information from fossils can contribute positively to studies of extant organisms and vice versa (Gould 1995). The same is true of ecological and phylogenetic studies. We focused in this chapter on how phylogenies have been used to develop and evaluate hypotheses about ecology and behavior of bats, but the reverse can also be enlightening— ecological data can play a role in developing and testing phylogenetic hypotheses. For example, features of echolocation call structure appear to hold promise as phylogenetic characters at several different taxonomic levels (Simmons and Kalko 1998). Behavioral data are playing an increasingly important role in systematics (e.g., Carpenter 1989; Codington 1986; Paterson et al. 1995; Wenzel 1992), and Chiroptera provides fertile ground for work in this area.

One of the principal difficulties facing researchers interested in evolutionary ecology of bats is lack of adequately detailed consensus phylogenies. Simmons and Geisler's (1998) family-level phylogeny was used as a basis for many hypotheses discussed above, but this tree has many shortcomings and it is not accepted by all researchers (e.g., see Kirsch et al. 1998). The same is true of most of the other phylogenetic hypotheses discussed above. As is evident from a glance at any issue of *Bat Research News,* phylogenetic work on bats is continuing at a rapid pace, and each year brings new controversies as well as additional refinements in our understanding of bat relationships. All groups of bats would benefit from additional phylogenetic study. In our opinion, the most productive studies are likely to be those that consider multiple data sets (including both morphological and molecular data) sampled at the species level. Such comprehensive studies are rare, so this field offers almost limitless opportunities for future research.

Another area that offers great potential for future contributions is the phylogenetically based statistical analysis of correlations among various combinations of morphological, ecological, and behavioral traits. Methods including phylogenetically independent contrasts, Monte Carlo computer simulations, and phylogenetic autocorrelation offer means of rigorously addressing a variety of topics including form of allometric relationships, correlations between physiological phenotypes and behavioral/ecological characteristics, analyses of where and when particular traits may have evolved, and tests of whether rates of morphological or physiological evolution have differed among lineages (see summary in Garland et al. 1999). Several studies of bats using phylogenetically based statistical methods have been initiated recently (e.g., Cruz-Neto and Abe 1999; Hutcheon et al. 1999; Jones and MacLarnon 2000). It seems likely that such studies will significantly influence future work in many areas, including studies of foraging strategies, echolocation call structure, body size, physiology, and life-history strategies. Opportunities for productive integration of phylogenetic information into ecological and behavioral research programs are numerous and are presently limited more by lack of data than absence of appropriate analytical methods.

To make interdisciplinary studies most fruitful, it is important that systematists consider ecological and behavioral patterns (and interesting scientific questions in these fields) when planning phylogenetic analyses. Similarly, ecologists should consider phylogeny when planning ecological studies. Even very simple steps may be important, such as including ecologically well-known or divergent taxa in a systematic study (rather than just choosing taxonomic exemplars at random) or including ecological comparisons with sister taxa in a behavioral study (rather than making comparisons only with sympatric taxa or presumed ecological equivalents). In this way, the potential contributions of data and interpretations can be maximized, and relatively small steps forward (e.g., a new phylogeny or new ecological data on a species) can lead to significant advances in our understanding of bat history and diversity.

Acknowledgments

The discussions presented here would not have been possible without the contributions of many individuals who have worked on evolution and ecology of bats. We especially thank our colleagues R. Baker, L. Davalos, J. Dunlop, B. Fenton, T. Fleming, J. Geisler, T. Griffiths, E. Kalko, J. Kirsch, T. Kunz, M. Novacek, M. Rockman, W. Schutt, S. Swartz, R. Van Den Bussche, A. Wetterer, and J. Wible for many hours of discussion on these topics. Our thanks also to T. Kunz and B. Fenton for inviting us to contribute this chapter. B. Fenton, W. Schutt, and two anonymous reviewers read earlier versions of this manuscript. We thank them for their comments, which significantly improved this contribution. This study was supported in part by National Science Foundation Research Grants DEB-9106868 and DEB-9873663 to NBS.

Literature Cited

Alvarez, J., M. R. Willig, J. K. Jones, Jr., and W. D. Webster. 1991. Glossophaga soricina. Mammalian Species, 379:1–7.

Andersen, K. 1912. Catalogue of the Chiroptera in the Collection of the British Museum. Vol. 1. Megachiroptera. 2d ed. British Museum, London.

Arita, H. T., and M. B. Fenton. 1997. Flight and echolocation in the ecology and evolution of bats. Trends in Ecology and Evolution, 12:53–58.

Baker, R. J., and H. H. Genoways. 1976. A new species of Chiroderma from Guadelupe, West Indies (Chiroptera: Phyllostomatidae). Occasional Papers, the Museum, Texas Tech University, 39:1–9.

Baker, R. J., C. S. Hood, and R. L. Honeycutt. 1989. Phylogenetic relationships and classification of the higher categories of the New World bat family Phyllostomidae. Systematic Zoology, 3:228–238.

Baker, R. J., J. L. Longmire, M. Maltbie, M. J. Hamilton, and R. A. Van Den Bussche. 1997. DNA synapomorphies for a variety of taxonomic levels from a cosmid library from the New World bat Macrotus waterhousii. Systematic Biology, 46:579–589.

Baker, R. J., V. A. Taddei, J. L. Hudgeons, and R. A. Van Den Bussche. 1994. Systematic relationships within Chiroderma (Chiroptera: Phyllostomidae) based on cytochrome b sequence variation. Journal of Mammalogy 75:321–327.

Barclay, R. M. R., and R. M. Brigham. 1991. Prey detection, niche breadth, and body size in bats: why are aerial insectivorous bats so small? American Naturalist, 137:693–703.

Barghoorn, S. F. 1977. New material of Vespertiliavus Schlosser (Mammalia, Chiroptera) and suggested relationships of emballonurid bats based on cranial morphology. American Museum Novitates, 2618:1–29.

Beard, K. C. 1990. Gliding behaviour and palaeoecology of the alleged primate family Paromomyidae (Mammalia, Dermoptera). Nature, 345:340–341.

Beard, K. C. 1993. Origin and evolution of gliding in early Cenozoic Dermoptera (Mammalia, Primatomorpha. Pp. 63–90 in: Primates and Their Relatives in Phylogenetic Perspective (R. D. E. MacPhee, ed.). Plenum Press, New York.

Beard, K. C., B. Sigé, and L. Krishtalka. 1992. A primitive vespertilionoid bat from the early Eocene of central Wyoming. Comptes Rendus de l'Académie des Sciences Paris, 314:735–741.

Benton, M. J., and R. Hitchin. 1997. Congruence between phylogenetic and stratigraphic data on the history of life. Proceedings of the Royal Society of London B., 264: 885–890.

Bergmans, W. 1997. Taxonomy, biogeography of African Fruit Bats (Mammalia, Megachiroptera). 5. The genera Lissonycteris Andersen, 1912, Myonycteris Matschie, 1899 and Megaloglossus Pagenstecher, 1985; general remarks and conclusions; annex: key to all species. Beaufortia, 47:11–90.

Bergmans, W., and F. G. Rozendaal. 1988. Notes on collection of fruit bats from Sulawesi and some off-lying islands (Mammalia, Megachrioptera). Zoologische Verhandelingen (Leiden), 248:1–74.

Biknevicius, A. R. 1986. Dental function and diet in the Carpolestidae (Primates, Plesiadapiformes). American Journal of Physical Anthropology, 71:157–171.

Black, H. L. 1974. A north temperate bat community: structure and prey populations. Journal of Mammalogy, 55:138–157.

Bogdanowicz, W., M. B. Fenton, and K. Daleszczyk. 1999. The relationship between echolocation calls, morphology, and diet in insectivorous bats. Journal Zoology (London), 247:381–393.

Bogdanowicz, W., and R. D. Owen. 1998. In the Minotaur's labyrinth: phylogeny of the bat family Hipposideridae. Pp. 27–42 *in:* Bat Biology and Conservation (T. H. Kunz and P. A. Racey, eds.). Smithsonian Institution Press, Washington, D.C.

Brooks, D. R., and D. A. McLennan. 1991. Phylogeny, Ecology, and Behavior: A Research Program in Comparative Biology. University of Chicago Press, Chicago.

Carpenter, J. M. 1989. Testing scenarios: wasp social behavior. Cladistics, 10:295–304.

Coddington, J. A. 1986. The monophyletic origin of the orb web. Pp. 319–363 *in:* Spiders, Webs, Behavior, and Evolution (W. A. Sherar, ed.). Stanford University Press, Stanford, Calif.

Colgan, D. J., and T. F. Flannery. 1995. A phylogeny of Indo-west Pacific Megachiroptera based on ribosomal DNA. Systematic Biology, 44:209–220.

Corbet, G. B., and J. E. Hill. 1992. The Mammals of the Indomalayan Region: A Systematic Review. Oxford University Press, Oxford.

Courts, S. E. 1998. Dietary strategies of Old World fruit bats (Megachiroptera, Pteropodidae): how do they obtain sufficient protein? Mammal Review, 28:185–194.

Cruz-Neto, A. P., and A. S. Abe. 1999. The relationship between body mass, phylogeny, diet, and basal metabolic rate in phyllostomid bats: a computer-simulation approach. Bat Research News, 40:167. (Abstract.)

Czaplewski, N. J. 1997. Chiroptera. Pp. 410–431 *in:* Vertebrate Paleontology in the Neotropics: The Miocene Fauna of La Vent, Columbia (R. F. Kay, R. H. Madden, R. L. Cifelli, and J. J. Flynn, eds.). Smithsonian Institution Press, Washington, D.C.

Daniel, M. J. 1976. Feeding by the short-railed bat (*Mystacina tuberculata*) on fruit and possibly nectar. New Zealand Journal of Zoology, 3:391–398.

Daniel, M. J. 1990. Family Mystacinidae. Pp. 122–136 *in:* The Handbook of New Zealand Mammals (C. M. King, ed.). Oxford University Press, Aukland.

Dial, K. P., and J. M. Marzluff. 1988. Are the smallest organisms the most diverse? Ecology, 69:1620–1624.

Felsenstein, J. 1985. Phylogenies and the comparative method. American Naturalist, 125:1–15.

Fenton, M. B. 1994. Echolocation: its impact on the behavior and ecology of bats. Ecoscience, 1:21–30.

Fenton, M. B. 1995. Natural history and biosonar signals. Pp. 37–86 *in:* Hearing in Bats (A. N. Popper and R. R. Fay, eds.). Springer-Verlag, New York.

Fenton, M. B., D. Audet, M. K. Obrist, and J. Rydell. 1995. Signal strength, timing, and self-deafening: the evolution of echolocation in bats. Paleobiology, 21:229–242.

Fenton, M. B., and T. H. Fleming. 1976. Ecological interactions between bats and nocturnal birds. Biotropica, 8:104–110.

Ferrarezzi, H., and E. A. Gimenez. 1996. Systematic patterns and the evolution of feeding habits in Chiroptera (Archonta: Mammalia). Journal of Comparative Biology, 1:75–94.

Flannery, T. F. 1995. Mammals of the South-West Pacific and Moluccan Islands. Cornell University Press, Ithaca, N.Y.

Fleagle, J. G. 1978. Size distribution of living and fossil primate faunas. Paleobiology, 4:67–76.

Freeman, P. W. 1981. A multivariate study of the family Molossidae (Mammalia: Chiroptera): morphology, ecology, evolution. Fieldiana, 7:1–173.

Freeman, P. W. 1995. Nectarivorous feeding mechanisms in bats. Biological Journal of the Linnean Society, 56:439–463.

Freeman, P. W. 1998. Form, function, and evolution in skulls and teeth of bats. Pp. 140–156 in: Bat Biology and Conservation (T. H. Kunz and P. A. Racey, eds.). Smithsonian Institution Press, Washington, D.C.

Gardner, A. L. 1977. Feeding habits. Pp. 293–350 in: Biology of Bats of the New World Family Phyllostomatidae (R. J. Baker, J. K. Jones, Jr., and D. C. Carter, eds.). Pt. 2. Special Publications, the Museum, Texas Tech University, 13. Texas Tech University Press, Lubbock.

Garland, T., A. W. Dickerman, C. M. Janis, and J. A. Jones. 1993. Phylogenetic analysis of covariance by computer simulation. Systematic Biology, 42:265–292.

Garland, T., P. H. Harvey, and A. R. Ives. 1992. Procedures for the analysis of comparative data using phylogenetically independent contrasts. Systematic Biology, 41:18–32.

Garland, T., P. E. Midford, and A. R. Ives. 1999. An introduction to phylogenetically based statistical methods, with a new method for confidence intervals on ancestral values. American Zoologist, 39:374–388.

Gillette, D. D. 1975. Evolution of feeding strategies in bats. Tebiwa, 18:39–48.

Gimenez, E. A., H. Ferrarezzi, and V. A. Taddei. 1996. Lingual morphology and cladistic analysis of the New World nectar-feeding bats (Chiroptera: Phyllostomidae). Journal of Comparative Biology, 1:41–64.

Gould, G. C. 1995. Hedgehog phylogeny (Mammalia, Erinaceidae)—the reciprocal illumination of the quick and the dead. American Museum Novitates, 3131:1–45.

Griffiths, T. A. 1994. Phylogenetic systematics of slit-faced bats (Chiroptera, Nycteridae), based on hyoid and other morphology. American Museum Novitates, 3090:1–17.

Griffiths, T. A., and A. L. Smith. 1991. Systematics of emballonuroid bats (Chiroptera: Emballonuridae and Rhinopomatidae), based on hyoid morphology. Bulletin of the American Museum of Natural History, 206:62–83.

Griffiths, T. A., A. Truckenbrod, and P. J. Sponholtz. 1992. Systematics of megadermatid bats (Chiroptera, Megadermatidae), based on hyoid morphology. American Museum Novitates, 3041:1–21.

Habersetzer, J., G. Ricter, and G. Storch. 1992. Reprint. Bats: already highly specialized insect predators. Pp. 181–191 in: Messel: An Insight into the History of Life and of the Earth (S. Schall and W. Ziegle, eds.). Translated by M. Shaffer-Fehre. Clarendon Press, Oxford. Original edition, Waldemar Kramer, Frankfurt; 1988.

Habersetzer, J., G. Ricter, and G. Storch. 1994. Paleoecology of early middle Eocene bats from Messel, FRG: aspects of flight, feeding and echolocation. Historical Biology, 8:235–260.

Habersetzer, J., and G. Storch. 1987. Klassifikation und funktionelle Flügelmorphologie paläogener Fledermaüse (Mammalia, Chiroptera). Courier Forschinstitute Senkenberg, 91:11–150.

Habersetzer, J., and G. Storch. 1989. Ecology and echolocation of the Eocene Messel bats. Pp. 213–233 in: European Bat Research, 1987 (V. Hanák, T. Horácek, and J. Gaisler, eds.). Charles University Press, Prague.

Habersetzer, J., and G. Storch. 1992. Cochlea size in extant Chiroptera and middle Eocene Microchiroptera from Messel. Naturwissenschaften, 79:462–466.

Hall, E. R. 1981. The Mammals of North America. Vol. 1. John Wiley & Sons, New York.

Hand, S. J. 1985. New Miocene megadermatids (Chiroptera: Megadermatidae) from Australia with comments on megadermatid phylogenetics. Australian Mammalogy, 8:5–43.

Hand, S. J., and J. A. W. Kirsch. 1998. A southern origin for the Hipposideridae (Microchiroptera)? evidence from the Australian fossil record. Pp. 72–90 in: Bat Biology and Conservation (T. H. Kunz and P. A. Racey, eds.). Smithsonian Institution Press, Washington, D.C.

Hand, S. J., M. Novacek, H. Godthelp, and M. Archer. 1994. First Eocene bat from Australia. Journal of Vertebrate Paleontology, 14:375–381.

Heithaus, E. R., T. H. Fleming, and P. A. Opler. 1975. Foraging patterns and resource utilization in seven species of bats in a seasonal tropical rainforest. Ecology, 56:841–854.

Heller, K.-G. 1995. Echolocation and body size in insectivorous bats: the case of the giant naked bat Cheiromeles torquatus (Molossidae). Le Rhinolophe, 11:27–38.

Hermanson, J. W., and T. J. O'Shea. 1983. Antrozous pallidus. Mammalian Species, 213:1–8.

Hill, J. E. 1974. A new family, genus, and species of bat (Mammalia: Chiroptera) from Thailand. Bulletin of the British Museum (Natural History) Zoology, 27:301–336.

Hill, J. E., and J. D. Smith. 1984. Bats: A Natural History. University of Texas Press, Austin.

Hisheh, S., M. Westerman, and L. H. Schmitt. 1998. Biogeography of the Indonesian archipelago: mitochondrial DNA variation in the fruit bat, Eonycteris spelaea. Biological Journal of the Linnean Society, 65:329–345.

Honeycutt, R. L., and V. M. Sarich. 1987. Albumin evolution and subfamilial relationships among New World leaf-nosed bats (Family Phyllostomidae). Journal of Mammalogy, 68:508–517.

Hood, C. S. 1989. Comparative morphology and evolution of the female reproductive tract in macroglossine bats (Mammalia, Chiroptera). Journal of Morphology, 199:207–221.

Huelsenbeck, J. P. 1994. Comparing the stratigraphic record to estimates of phylogeny. Paleobiology, 20:470–483.

Hutcheon, J. M., T. Garland, and J. A. W. Kirsch. 1999. The phylogeny of echolocation. Bat Research News, 40:174–175. (Abstract.)

Hutcheon, J. M., J. A. W. Kirsch, and J. D. Pettigrew. 1998. Base-compositional biases and the bat problem. III. The question of microchiropteran monophyly. Philosophical Transactions of the Royal Society of London B, 353:607–617.

Hutchinson, G. E., and R. H. MacArthur. 1959. A theoretical ecological model of size distributions among species of animals. American Naturalist, 93:117–125.

Ingle, N. R., and L. R. Heaney. 1992. A key to the bats of the Philippine Islands. Fieldiana (Zoology), n.s., 69:1–44.

Jepsen, G. L. 1966. Early Eocene bat from Wyoming. Science, 154:1333–1339.

Jepsen, G. L. 1970. Bat origin and evolution. Pp.1–65 in: Biology of Bats (W. A. Wimsatt, ed.). Vol. 2. Academic Press, New York.

530 Nancy B. Simmons and Tenley M. Conway

Johnston, D. S. 1997. Foraging flexibility in the pallid bat (*Antrozous pallidus*). Ph.D. Dissertation. York University, North York.

Jones, G. 1994. Scaling of wingbeat and echolocation pulse emission rates in bats: why are aerial insectivorous bats so small? Functional Ecology, 8:450–457.

Jones, G. 1996. Does echolocation constrain the evolution of body size in bats? Symposia of the Zoological Society of London, no. 69:111–128.

Jones, K. E., and A. MacLarnon. 2001. Bat life-histories: testing models of life-history evolution in mammals using a comparative phylogenetic approach. Evolutionary Ecology Research, 3:465–476.

Kay, R. F. 1973. Mastication, molar tooth structure, and diet in primates. Ph.D. Dissertation. Yale University, New Haven, Conn.

Kay, R. F., and W. Hylander. 1978. The dental structure of arboreal folivores with special reference to Primates and Phalangeroidea (Marsupialia). Pp. 171–191 *in:* The Biology of Arboreal Folivores (G. G. Montgomery, ed.). Smithsonian Institution Press, Washington D.C.

Kay, R. F., R. W. Thorington, and P. Houde. 1990. Eocene plesiadapiform shows affinities with flying lemurs not primates. Nature, 345:342–344.

Kingdon, J. 1974. East African Mammals. Vol. 2, Pt. A. University of Chicago Press, Chicago.

Kirsch, J. A. W. 1996. Bats are monophyletic; megabats are monophyletic; but are microbats also? Bat Research News, 36:78.

Kirsch, J. A. W., T. F. Flannery, M. S. Springer, and F. J. Lapointe. 1995. Phylogeny of the Pteropodidae (Mammalia: Chiroptera) based on DNA hybridization, with evidence for bat monophyly. Australian Journal of Zoology, 43:395–428.

Kirsch, J. A. W., and J. M. Hutcheon. 1997. Further on the possibility that microchiropterans are paraphyletic. Bat Research News, 37:138. (Abstract.)

Kirsch, J. A. W., J. M. Hutcheon, D. G. P. Byrnes, and B. D. Lloyd. 1998. Affinities and historical zoogeography of the New Zealand short-tailed bat, *Mystacina tuberculata* Gray 1843, inferred from DNA hybridization comparisons. Journal of Mammalian Evolution, 5:33–64.

Kirsch, J. A. W., and F.-J. Lapointe. 1997. You aren't (always) what you eat: evolution of nectar-feeding among Old World fruit bats. Pp. 313–330 *in:* Molecular Evolution and Adaptive Radiation (T. J. Givnish and K. J. Sytsma, eds.). Cambridge University Press, Cambridge.

Koopman, K. F. 1982. Biogeography of the bats of South America. Pp. 273–302 *in:* Mammalian Biology in South America (M. A. Mares and H. H. Genoways, eds.). Special Publication Series, Pymatuning Laboratory of Ecology, University of Pittsburgh, 6:1–539.

Koopman, K. F. 1984. Bats. Pp. 145–186 *in:* Orders and Families of Recent Mammals (S. Anderson and J. K. Jones, Jr., eds.). Wiley, New York.

Koopman, K. F. 1993. Order Chiroptera. Pp. 137–242 *in:* Mammal Species of the World: A Taxonomic and Geographic Reference (D. E. Wilson and D. M. Reeder, eds.). Smithsonian Institution Press, Washington, D.C.

Koopman, K. F. 1994. Chiroptera: Systematics. Pt. 60 of Mammalia, vol. 8 of Handbook of Zoology. Walter de Gruyter, New York.

Krazanowski, A. 1977. Weight classes of Palearctic bats. Acta Theriologica, 22:365–370.

Kunz, T. H. 1982. Roosting ecology of bats. Pp. 1–55 *in:* Ecology of Bats (T. H. Kunz, ed.). Plenum Press, New York.

Landry, S. O. 1970. The Rodentia as omnivores. Quarterly Review of Biology, 45: 351–372.

Legendre, S. 1985. Molossidés (Mammalia, Chiroptera) cénezoiques de l'Ancien et du Nouveau Monde; statut systématique; intégration phylogénique des données. Neues Jahrbuch für Geologie und Paläontologie Abhandlungen. 170:205–227.

Legendre, P., and F.-J. Lapointe. 1995. Matching behavioral evolution to brain morphology. Brain, Behavior and Evolution, 45:110–121.

Lekagul, B., and J. A. McNeely. 1977. Mammals of Thailand. Association for the Conservation of Wildlife, Bangkok.

Liem, K. 1973. Evolutionary strategies and morphological innovations: cichlid pharyngeal jaws. Systematic Zoology, 22:425–441.

Lim, B. 1993. Cladistic reappraisal of Neotropical stenodermatine bat phylogeny. Cladistics, 9:147–165.

Lim, B., and M. D. Engstrom. 1998. Phylogeny of Neotropical short-tailed fruit bats, *Carollia* spp.: phylogenetic analysis of restriction site variation in mtDNA. Pp. 43–58 *in:* Bat Biology and Conservation (T. H. Kunz and P. A. Racey, eds.). Smithsonian Institution Press, Washington, D.C.

MacArthur, R. H., and R. Levins. 1964. Competition, habitat selection, and character displacement in a patchy environment. Proceedings of the National Academy of Sciences of the USA, 51:1207–1210.

Maddison, W. P. 1991. Squared-change parsimony reconstructions of ancestral states for continuous-valued characters on a phylogenetic tree. Systematic Zoology, 40:304–314.

Maddison, W. P., and D. R. Maddison. 1992. MacClade, version 3.0. Sinauer Associates, Sunderland, Mass.

Martins, E. P., and T. Garland. 1991. Phylogenetic analyses of the correlated evolution of continuous characters: a simulation study. Evolution, 45:543–557.

May, R. M. 1978. The dynamics and diversity of insect faunas. Pp. 188–204 *in:* Diversity of Insect Faunas (L. A. Mound and N. Waloff, eds.). Blackwell Scientific Publications, Oxford.

McKenna, M. C., and S. K. Bell. 1997. Classification of Mammals above the Species Level. Columbia University Press, New York.

McNab, B. K. 1969. The economics of temperature regulation in Neotropical bats. Comparative Biochemistry and Physiology, 31:227–268

McNab, B. K. 1982. Evolutionary alternatives in the physiological ecology of bats. Pp. 151–200 *in:* Ecology of Bats (T. H. Kunz, ed.). Plenum Press, New York.

McNab, B. K., and F. J. Bonaccorso. 1995. The energetics of pteropodid bats. Symposia of the Zoological Society of London, no. 67:111–122.

Miller, G. S. 1897. Revision of the North American bats of the family Vespertilionidae. North American Fauna, 13:5–135.

Miller, G. S. 1907. The families and genera of bats. U.S. National Museum Bulletin, 57:1–282.

Neuweiler, G. 1989. Foraging ecology and audition in echolocating bats. Trends in Ecology and Evolution, 4:160–166.

Neuweiler, G. 1990. Auditory adaptations for prey capture in echolocating bats. Physiological Review, 70:615–641.

Neuweiler, G., V. Bruns, and G. Schuller. 1980. Ears adapted for the detection of motion, or how echolocating bats have exploited the capabilities of the mammalian auditory system. Journal of the Acoustic Society of America, 68:741–753.

Norberg, U. M. 1986. On the evolution of flight and wing form in bats. Pp. 13–26 *in:* Bat Flight—Fledermausflug (W. Nachtigall, ed.). Bionia Report 5. Gustav Fischer, Stuttgart.

Norberg, U. M. 1989. Ecological determinates of bat wing shape and echolocation call structure with implications for some fossil bats. Pp. 197–211 *in:* European Bat Research, 1987 (V. Hanák, T. Horácek, and J. Gaisler, eds.). Charles University Press, Prague.

Norberg, U. M. 1994. Wing design, flight performance, and habitat use in bats. Pp. 205–239 *in:* Ecological Morphology: Integrative Organismal Biology (P. C. Wainwright and S. M. Reilly, eds.). University of Chicago Press, Chicago.

Norberg, U. M., and M. B. Fenton. 1988. Carnivorous bats? Biological Journal of the Linnean Society, 33:383–394.

Norberg, U. M., and J. M. V. Rayner. 1987. Ecological morphology and flight in bats (Mammalia: Chiroptera): wing adaptations, flight performance, foraging strategy and echolocation. Philosophical Transactions of the Royal Society of London B, 316:335–427.

Norell, M. A. 1992. Taxic origin and temporal diversity: the effect of phylogeny. Pp. 89–118 *in:* Extinction and Phylogeny (M. J. Novacek and Q. D. Wheeler, eds.). Columbia University Press, New York.

Novacek, M. J. 1985. Evidence for echolocation in the oldest known bats. Nature, 315:140–141.

Novacek, M. J. 1987. Auditory features and affinities of the Eocene bats *Icaronycteris* and *Palaeochiropteryx* (Microchiroptera, *incertae sedis*). American Museum Novitates, 2877:1–18.

Novacek, M. J. 1996. Paleontological data and the study of adaptation. Pp. 311–359 *in:* Adaptation (M. D. Rose and G. V. Lauder, eds.). Academic Press, New York.

Nowak, R. M. 1991. Walker's Mammals of the World. Vol. 1. 5th ed. Johns Hopkins University Press, Baltimore

Obrist, M. K., M. B. Fenton, J. L. Egar, and P. A. Schlegel. 1993. What ears do for bats: a comparative study of pinna sound pressure transformation in Chiroptera. Journal of Experimental Biology, 180:119–152.

Owen, R. D. 1987. Phylogenetic analyses of the bat subfamily Stenodermatinae (Mammalia: Chiroptera). Special Publications, the Museum, Texas Tech University, 26:1–65.

Padian, K., D. R. Lindberg, and P. D. Polly. 1994. Cladistics and the fossil record: the uses of history. Annual Review of Earth and Planetary Sciences, 22:63–91.

Paterson, A. M., G. P. Wallis, and R. D. Gray. 1995. Penguins, petrels, and parsimony—does cladistic analysis of behavior reflect seabird phylogeny? Evolution, 49:974–989.

Pettigrew, J. D. 1995. Flying primates: crashed, or crashed through? Symposia of the Zoological Society of London, no. 67:3–26.

Pettigrew, J. D., B. G. M. Jamieson, S. K, Robson, L. S. Hall, K. I. McAnally, and H. M.

Cooper. 1989. Phylogenetic relations between microbats, megabats and primates (Mammalia: Chiroptera and Primates). Philosophical Transactions of the Royal Society of London B, 325:489–559.

Pettigrew, J. D., and J. A. W. Kirsch. 1995. Flying primates revisited: DNA hybridization with fractionated, GC-enriched DNA. South African Journal of Science, 91:477–482.

Pettigrew, J. D., and J. A. W. Kirsch. 1998. Base-compositional biases and the bat problem. I. DNA-hybridization melting curves based on AT- and GC-enriched tracers. Philosophical Transactions of the Royal Society of London B, 353:369–79.

Phillips, C. J., D. E. Pumo, H. H. Genoways, P. E. Ray, and C. A. Briskey. 1991. Mitochondrial DNA evolution and phylogeography in two Neotropical fruit bats, *Artibeus jamaicensis* and *Artibeus lituratus*. Pp. 97–123 *in:* Latin American Mammalogy (M. A. Mares, and D. J. Schmidly, eds.). University of Oklahoma Press, Norman.

Porter, C. A., M. Goodman, and M. J. Stanhope. 1996. Evidence on mammalian phylogeny from sequences of exon 28 of the von Willebrand Factor gene. Molecular Phylogeny and Evolution, 5:89–101.

Pye, D. 1980. Adaptiveness of echolocation signals in bats: flexibility in behavior and in evolution. Trends in Neuroscience, 3:232–235.

Rayner, J. M. V. 1981. Flight adaptations in vertebrates. Symposia of the Zoological Society of London, no. 48:137–172.

Robbins, L. W., and V. M. Sarich. 1988. Evolutionary relationships in the family Emballonuridae (Chiroptera). Journal of Mammalogy, 69:1–13.

Savage, D. E. 1951. A Miocene phyllostomatid bat from Colombia, South America. University of California Publications, Bulletin of the Department of Geological Sciences, 28:357–366.

Schluter, D., T. Price, A. Mooers, and D. Ludwig. 1997. Likelihood of ancestral states in adaptive radiation. Evolution, 51:1699–1711.

Schmitt, L. H., D. J. Kitchener, and R. A. How. 1995. A genetical perspective of mammalian variation and evolution in the Indonesian archipelago: biogeographic correlates in the fruit bat genus *Cynopterus*. Evolution, 49:399–412.

Schnitzler, H.-U. 1987. Echoes of flutter insects: information for echolocating bats. Pp. 226–243 *in:* Recent Advances in the Study of Bats (M. B. Fenton, P. Racey, and J. M. V. Rayner, eds.). Cambridge University Press, Cambridge.

Schnitzler, H.-U., and E. K. V. Kalko. 1998. How echolocating bats search and find food. Pp. 183–196 *in:* Bat Biology and Conservation (T. H. Kunz and P. A. Racey, eds.). Smithsonian Institution Press, Washington, D.C.

Schuller, G., and G. Pollack. 1976. Disproportionate frequency representation in the inferior colliculus of Doppler-compensating greater horseshoe bats: evidence of an acoustic fovea. Journal of Comparative Physiology A, 132:47–54.

Schutt, W. A. 1998. Chiropteran hindlimb morphology and the origin of blood feeding in bats. Pp. 57–168 *in:* Bat Biology and Conservation (T. H. Kunz and P. A. Racey, eds.). Smithsonian Institution Press, Washington, D.C.

Schutt, W. A., and N. B. Simmons. 1998. Morphology and homology of the chiropteran calcar, with comments on the phylogenetic relationships of *Archaeopteropus*. Journal of Mammalian Evolution, 5:1–32.

Sigé, B. 1991. Morphologie dentaire lactéale d'un chiroptère de l'Éocène inférieur-moyen d'Europe. Geobios, 13:231–236.

Simmons, J. A. 1980. Phylogenetic adaptations and the evolution of echolocation in bats.

Pp. 267–278 *in:* Proceedings Fifth International Bat Research Conference (D. E. Wilson and A. L. Gardner, eds.). Texas Tech Press, Lubbock.

Simmons, J. A., and R. A. Stein. 1980. Acoustic imaging in bat sonar: echolocation signals and the evolution of echolocation. Journal of Comparative Physiology A, 135:61–84.

Simmons, N. B. 1994. The case for chiropteran monophyly. American Museum Novitates, 3103:1–54.

Simmons, N. B. 1995. Bat relationships and the origin of flight. Symposia of the Zoological Society of London, no. 67:27–43.

Simmons, N. B. 1998. A reappraisal of interfamilial relationships of bats. Pp. 3–26 *in:* Bat Biology and Conservation (T. H. Kunz and P. A. Racey, eds.). Smithsonian Institution Press, Washington, D.C.

Simmons, N. B. 2003. Order Chiroptera. *In:* Mammal Species of the World: A Taxonomic and Geographic Reference. 2d ed. (D. E. Wilson and D. M. Reeder, eds.). Smithsonian Institution Press, Washington, D.C. (in press).

Simmons, N. B., and J. H. Geisler. 1998. Phylogenetic relationships of *Icaronycteris, Archaeonycteris, Hassianycteris,* and *Palaeochiropteryx* to extant bat lineages, with comments on the evolution of echolocation and foraging strategies in Microchiroptera. Bulletin of the American Museum of Natural History, 235:1–182.

Simmons, N. B., and E. K. V. Kalko. 1998. Is there phylogenetic signal in bat echolocation calls? Bat Research News, 39:102.

Simmons, N. B., and R. S. Voss. 1998. The mammals of Paracou, French Guiana: a Neotropical lowland rainforest fauna. 1. Bats. Bulletin of the American Museum of Natural History, 237:1–219.

Simmons, N. B., and A. L. Wetterer. 2002. Phylogeny and convergence in cactophilic bats. Pp. 87–121 *in:* Evolution, Ecology, and Conservation Biology of Columnar Cacti and Their Mutualists (T. H. Fleming and A. Valiente-Banuet, eds.). University of Arizona Press, Tucson.

Smith, J. D. 1976. Chiropteran evolution. Pp. 46–49 *in:* Biology of Bats of the New World Family Phyllostomatidae. (R. J. Baker, J. K. Jones, Jr., and D.C. Carter, eds.). Pt. 1. Special Publications, the Museum, Texas Tech University, 13. Texas Tech University Press, Lubbock.

Smith, J. D., and C. S. Hood. 1983. A new species of tube-nosed fruit bat (Nyctimene) from the Bismark Archipelago, Papua New Guinea. Occasional Papers, the Museum, Texas Tech University, 81:1–14.

Speakman, J. R. 1993. The evolution of echolocation for predation. Symposia of the Zoological Society of London, no. 65:39–63.

Speakman, J. R., and P. A. Racey. 1991. No cost of echolocation for bats in flight. Nature, 350:421–423.

Springer, M. S., L. J. Hollar, and J. A. W. Kirsch. 1995. Phylogeny, molecules versus morphology, and rates of character evolution among fruit bats (Chiroptera: Megachiroptera). Australian Journal of Zoology, 43:557–582.

Swofford, D. L., and W. P. Maddison. 1992. Phylogeny, character-state reconstructions, and evolutionary inference. Pp. 186–223 *in:* Systematics, Historical Ecology, and North American Freshwater Fishes (R. L. Maden, ed.). Stanford University Press, Palo Alto, Calif.

Taddei, V. A. 1979. Phyllostomidae (Chiroptera) do Norte-Ocidental do Estado de São Paulo. III. Stenodermatinae. Ciêncie e Cultura, 31:900–914.

Taddei, V. A. 1980. Aspectos da biologia de *Chiroderma doriae,* Thomas, 1891 (Chiroptera, Phyllostomidae). Anais da Academia Brasileira de Ciêncieas, 52:643–644.

Thomas, D. W., and A. G. Marshall. 1984. Reproduction and growth in three species of west African fruit bats. Journal of Zoology (London), 202:265–281.

Van Valen, L. 1973. Body size and numbers of plants and animals. Evolution, 27:27–35.

Van Valen, L. 1979. The evolution of bats. Evolutionary Theory, 4:104–121.

Vater, M., A. S. Feng, and M. Betz. 1985. An HPR-study of the frequency-place map of the horseshoe bat cochlea: morphological correlates of the sharp tuning to a narrow frequency band. Journal of Comparative Physiology A, 157:671–686.

Vaughan, T. A., and R. P. Vaughan. 1986. Seasonality and the behavior of the African yellow-winged bat. Journal of Mammalogy, 67:91–102.

Volleth, M., and K.-G. Heller. 1994. Phylogenetic relationships of vespertilionid genera (Mammalia: Chiroptera) as revealed by karyological analysis. Zeitschrift für Zoologische Systematik und Evolutionsforschung, 32:11–34.

Voss, R. S., and L. H. Emmons. 1996. Mammalian diversity in Neotropical lowland rainforests: a preliminary assessment. Bulletin of the American Museum of Natural History, 230:1–115.

Wenzel, J. W. 1992. Behavioral homology and phylogeny. Annual Review of Ecology and Systematics, 23:361–381.

Wetterer, A. L., M. V. Rockman, and N. B. Simmons. 2000. Phylogeny of phyllostomid bats: data from diverse morphological systems, sex chromosomes, and restriction sites. Bulletin of the American Museum of Natural History, 258:1–200.

Wilson, D. E., and D. M. Reeder, eds. 2003. Mammal Species of the World: A Taxonomic and Geographic Reference. 2d ed. Smithsonian Institution Press, Washington, D.C. (in press).

Wischusen, E. W., and M. E. Richmond. 1998. Foraging ecology of the Philippine flying lemur (*Cynocephalus volans*). Journal of Mammalogy, 79:1288–1295.

CHAPTER 12

Trophic Strategies, Niche Partitioning, and Patterns of Ecological Organization

Bruce D. Patterson, Michael R. Willig, and Richard D. Stevens

Introduction

The tremendous variety of foods that bats exploit, coupled with the various foraging techniques and roosting structures they utilize, have led to remarkable levels of abundance and diversity. Here, we examine elements underlying ecological organization of bat. First, we review an enormous literature that describes and analyzes the principal trophic strategies of bats and the salient features of each type of resource. In doing so, we attempt to identify prevailing patterns of niche partitioning and to determine the special opportunities or constraints that appear to be associated with each feeding strategy. Bats respond to these opportunities and constraints at multiple levels of organization: individual variation, abundance and range limits of species, and the amalgamation of local assemblages into faunas. Where possible, we identify this kind of variation and its corollaries, but the mechanisms that underlie many ecological patterns involving bats mostly remain elusive.

The geographical context of assemblages also helps to shape their variation at broader spatial scales. We follow the discussion of resource partitioning and coexistence at local scales with analyses of latitudinal patterns. These analyses and accompanying discussion focus on gradients in local diversity and the role of character displacement in structuring them.

In the concluding remarks, we relate our findings to bat ecology and conservation. We also call attention to two research areas that appear particularly promising for further ecological research.

Conceptual Underpinnings and Terminology

Controversy surrounds the taxonomic, spatial, and temporal delineation of ecological communities. Most concepts of community involve a set of populations or species that is likely to interact through competitive, trophic, or mutualistic associations. These interactions occur among species that share temporal and spatial domains (Ricklefs 1979). In this expansive sense, a "community" is the suite of all taxa that exist contemporaneously at a given place, or the biotic portion of an ecosystem. A community cannot be defined by taxonomic criteria (i.e., "bat communities") because communities must include both autotrophs and heterotrophs. So defined, few communities have been enumerated or studied because it is too costly, time consuming, and taxonomically challenging to do so in any but the simplest systems.

Ecologists commonly study subsets of ecological communities, often because their interests and expertise are circumscribed either by taxa (e.g., bats, birds, bacteria) or by processes (e.g., frugivory, carnivory, detritivory). Fauth et al. (1996) proposed a nomenclature for these subsets that is adopted here (fig. 12.1). An "assemblage" is a subset of the community defined only by taxonomic constraints. Thus, all mammalian (or avian or fungal) components of a community constitute assemblages. In contrast, a "guild" is a subset of the community defined by some type of functional characteristics (e.g., habitat, foraging mode, and diet [Root 1967, 1973]). To study all frugivores or all gleaning insectivores within a community is to target guilds. However, even these more limited groupings are difficult to study in nature. Few community ecologists study the structure or diverse interactions among all the mammals in an ecosystem or their composition and interactions among all insectivorous species (including, e.g., arthropods, amphibians, reptiles, birds, and mammals). Rather, they study components of the community stemming from cross-classifications based on both taxonomy and function (e.g., detritivorous insects, nectarivorous birds, granivorous mammals). These cross-classified groupings constitute "ensembles" (Fauth et al. 1996; fig. 12.1).

By this terminology, bat ecology is the ecological study of assemblages of chiropterans. Studies of frugivorous bats or insectivorous bats focus on ensembles. These distinctions are necessary because rejection or acceptance of hypotheses concerning community organization may depend on whether the research focused on ensembles or assemblages. For example, patterns of body size in frugivorous bats may be difficult to comprehend if interactions with other frugivorous animals such as birds are important but not addressed in a study's design. Indeed, emphasizing that frugivorous bats constitute an ensemble rather than a guild is a terminological reminder that other frugivores may exert significant impacts on the composition and structure of these interacting groups of bats, either through joint exploitation of resources (e.g., Palmeirim et al. 1989) or through territoriality (e.g., Miller 1962).

Logistics commonly force ecologists to study incomplete or heterogeneous collections of bat populations. Often, biologists study the composition and structure of bat faunas occupying geopolitical units (e.g., states, provinces, departments) or large islands (e.g., Puerto Rico). Such studies are unlikely to resolve hypotheses concerning community-level phenomena (Willig 1986). Included populations may never interact with one another, or interactions may

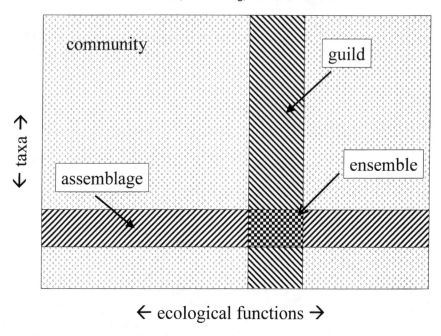

Figure 12.1. Conceptual partitions of an ecological community (after Fauth et al. 1996).

take place with excluded populations. The scale at which patterns are sought in the field should correspond to the spatial and temporal scales of mechanisms thought to produce them.

In some localities, bat diversity may be so great that sampling all elements becomes impractical. Myers and Wetzel (1983, fig. 3) documented the slow accumulation of bat species in the Chaco Boreal of Paraguay, which eventually reached an asymptote of 20 species after 60 nights of collecting. However, 28 species are thought to occur in the area. This problem is particularly important in studies of tropical assemblages, where whole segments of faunas often escape detection and monitoring. Besides necessitating certain types of sampling methodologies, such features may also dictate the choice of sampling units and scales of analysis (e.g., Patterson et al. 1998).

If mechanisms that transpire within communities (e.g., competition, predation, mutualism) are of interest, then communities or their subsets should be the focus of study. Although studies of broadly defined geographic units provide important information concerning biogeography, and may inform studies of community composition (e.g., Arita 1997), it is unclear how patterns and processes operating at a geographic level affect mechanisms of coexistence at local levels (Allen and Starr 1982). No single spatial or temporal scale is proper for studying communities and their component ensembles or assemblages. Rather, the guiding principle should be that the area is not so large, or the time so long, as to contain species with no likelihood of interacting. Similarly, the area should be sufficiently large and the time adequately expansive to contain all of the species whose interactions affect population dynamics.

Finally, it should be obvious from the foregoing that ecologists interested in studying bat assemblages at appropriate spatial and temporal scales are largely limited to observation and comparison, not experiment and manipulation. Experimental community ecology calls for replicated grids, controlled resource levels, and reciprocal-removal experiments that are practically impossible to use in studying bats. For example, each night while foraging, a nectarivorous bat (*Leptonycteris curasoae*) may travel dozens of kilometers, whereas a granivorous pocket mouse (*Chaetodipus baileyi*) foraging below it may move only dozens of meters. The wide-ranging behavior of bats often precludes elegant experimentation and decisive refutation of hypotheses. However, the diversity and importance of ecological roles played by bats in natural communities ensures that elucidating their ecological roles will constitute an important, even vital, step in ecological understanding and environmental management.

Bat Ensembles and Their Characteristics

There are many possible ways to classify coexisting bats that are most likely to interact with one another. Major food types, principal roosting habits, echolocation type, and flight strata have all been used singly or in combination to identify interacting bat ensembles (Findley 1993; Norberg and Rayner 1987; Patterson et al. 1996; Willig 1986). The suitability of these definitions differs widely with the assemblage and research goals under study. For example, those investigating aerial insectivores or the role of echolocation in foraging will prefer the trivariate "guild" classification of Kalko and coworkers (Kalko 1998), which is based on food, feeding method, and habitat structure. Because habitat structure represents an important avenue of resource partitioning among aerial insectivores, it is essential in any ecological classification of these bats (see below; Jones and Rydell, this volume; Norberg and Rayner 1987). However, a univariate classification based on food type alone may sometimes be adequate because bat carnivores, piscivores, sanguinivores, frugivores, nectarivores, and omnivores (as well as a few insectivores) all forage via gleaning in relatively confined spaces.

We use a broad classification of bat diets (Ferrarezzi and Amaral Gimenez 1996) to organize our discussion of bat ensembles. Herbivory and animalivory are treated as two main classes of trophic adaptation and behavior, with various subdivisions recognized within each. Subclasses of herbivory exploited by bats include frugivory, feeding predominantly or solely on fruits; nectarivory, feeding on nectar as well as on other floral products and parts, such

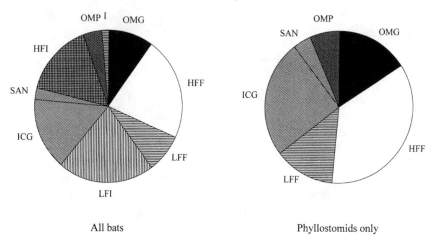

All bats   Phyllostomids only

Figure 12.2. Trophic diversity of bat assemblages in Andean foothills of Peru. Graphs represent 123 species of bats, and 76 species of phyllostomids, that inhabit forests at or below 1,000 m in the Manu National Park and Biosphere Reserve (data from tables I and II of Patterson et al. 1996). Trophic categories are *HFF*, high-flying frugivore; *HFI*, high-flying insectivore; *ICG*, insecti-carnivore; *I*, insecti-piscivore; *LFF*, low-flying frugivore; *LFI*, low-flying insectivore; *OMG*, omnivorous nectarivore; *OMP*, omnivorous predator; and *SAN*, sanguinivore. The phyllostomid radiation exploits nearly the full range of trophic procurement strategies utilized by Neotropical bats.

as pollen and petals; and folivory, consumption of leaves, including buds and other green parts of plants. Animalivory includes generalized insectivory, carnivory, piscivory, and sanguinivory (see also Francis 1990; Muñoz-Saba et al. 1995, 1997). Simmons and Conway (this volume) discuss the evolution of various feeding adaptations within Chiroptera. Figure 12.2 contains a classification of food habits among bats in exceptionally rich assemblages in southeastern Peru.

The utility of any dietary classification is limited by behavioral plasticity and individual variation. We are only now learning about components of variability, which may be highly deterministic. For example, species adapted to operate in closed habitats might exploit edge and open habitats, but those adapted to open habitats might have limited access to closed or edge habitats because of sensory prerequisites for navigating in "clutter" (Fenton 1990). Such attributes seem likely to structure interactions of bats in ways scarcely appreciated at present.

Although we have chosen food as a framework for considering ecological interactions among bats, a variety of additional ecological interactions help to shape bat behavior, ecology, and ultimately their integration into ecological assemblages. Predation surely has pervasive effects on bat ecology, possibly limiting the extent of the chiropteran radiations (Rydell and Speakman 1995). Modern bats are almost exclusively nocturnal—the ecological diversity of species with this habit would seem to invalidate alternative explanations based on resource limitation or hyperthermia during daylight flight. Competition with birds is an unlikely explanation for nocturnal habits because most groups of aerial-feeding birds evolved after bats had adapted to night hunting. Because bats do not become diurnal on small islands lacking predators (e.g., *Nyctalus* on the Azores), other factors may also be operational (Speakman 1995).

Predation serves to limit bat foraging ranges, activity periods, and courtship displays (Fenton 1995). It affects social behavior and interactions, including the adoption of feeding roosts, alternation of day roosts, and emergence behavior (Kunz and Lumsden, this volume). In South Africa, small colonies of *Epomophorus wahlbergi* (Pteropodidae) and *Scotophilus viridis* (Vespertilionidae) utilize roost switching and unpredictable "burst" emergence to foil predators (Fenton et al. 1985; see also Acharya 1992; Morrison 1980). Roosting in large colonies decreases the risk of predator attack on any one individual (Fenton et al. 1994). Bats emerging from roosts show the evasive behavior patterns predicted by the selfish-herd principle (Fenton 1995). Morrison (1980) noted that *Artibeus lituratus* and *Vampyrodes caraccioli* change roost sites on almost a daily basis, but roosts are not shifted to minimize distance to fruiting trees. Sit-and-wait predators such as owls seem to be attracted to fruiting trees, and use of remote feeding roosts may reflect the importance of predation (Morrison 1980). Bats often suspend or greatly reduce feeding passes to fruiting trees during periods of bright moonlight. Some phyllostomids, including *Artibeus lituratus* and *Carollia perspicillata*, quickly investigate alarm calls issued by other bats, including members of other species (Ryan et al. 1985).

Although roosting behavior can evolve in response to predation, other pressures may also affect it. Many bats are found in characteristic roosting associations with other species (Kunz 1982; Kunz and Lumsden, this volume), which may influence the survival or persistence of codependent species (Campanhã and Fowler 1993). Some roosting associations comprise species belonging to two or more trophic ensembles (Wohlgenant 1994). In addition, bats that depend on plants for roosts may indirectly depend on other bat species that pollinate their flowers or disperse their seeds. Such interactions seem particularly obvious in the sequence of species recolonizing Krakatau (Whitaker and Jones 1994), with *Cynopterus sphinx* as an important disperser of pioneer plants used by later colonists (Kunz 1996). Obviously, none of these ecological interactions are anticipated in our trophic-based classification. We urge biologists to ensure that the classification they employ in analyzing their own data is meaningfully related to known patterns of resource exploitation.

Animalivory
Arthropods

Most bats in temperate zones exploit arthropods, but tropical bats have strikingly diversified trophic habits. Therefore, it is surprising to many that arthropod consumers also can dominate tropical communities. Francis (1990) determined that insectivorous species made up 75%–87% of the bat fauna and 88%–93% of the individuals inhabiting the understory of lowland dipterocarp rainforest in Malaysia, even though his samples were strongly biased toward capturing frugivores. Almost two-thirds of the 78 bat species at Paracou, French Guiana, depend on insects ("aerial insectivores"+"gleaning animalivores" = 64%; Simmons and Voss 1998), and aerial insectivores may constitute 30%–50% of the species (but less of the biomass) in local assemblages in the Neotropics (Kalko 1998). Heller and Volleth (1995) concluded that the richest ensembles of insectivorous bats are roughly equivalent in number in the New and Old Worlds.

Insectivorous bats are usually subdivided by foraging method into aerial insectivores (those capturing insects on the wing during flight) and foliage gleaners (those flying slowly over substrates and taking prey from them). Foraging in different microhabitats often exposes organisms to novel prey distributions (e.g., taxonomic composition and size spectra) and entails different demands on sensory and locomotory systems. Findley (1993) classified insectivorous bats based on foraging station: sallying and gleaning, forest and clearing, water surface, and open air (see also Decher 1997). Kalko et al. (1996) divided aerial insectivores into three groups (calling these "guilds") based on whether the foraging space is uncluttered, background cluttered, or highly cluttered. Simmons and Geisler (1998) inferred that insectivorous foraging strategies of bats evolved sequentially, from gleaning stationary prey from a perch, to using perches to hunt stationary and flying prey, and finally to aerial hawking.

The ecological relationships and interactions of arthropod-feeding bats are still poorly understood, in part owing to the methodological challenges that must be overcome to study them. Many insectivorous species fly high above the ground and are difficult to capture in mist nets, so that they are apt to be underrepresented in inventory or ecological studies (Patterson et al. 1996; Kalko 1998). The echolocation calls of others are so quiet that they can only be monitored by remote sensing equipment at a few meters distance (Fenton et al. 1992). In addition, many insectivorous bats are known to exhibit high degrees of ecological specialization and niche subdivision (see below). Despite this, much progress has been made since 1982, when an important review concluded: "A consideration of the available literature [on insectivorous bats] provides no convincing evidence that bats specialize by the timing of activity, diet, use of habitat, foraging strategy or morphology. . . . There is still no clear picture of how sympatric insectivorous bats partition food resources, or if, indeed, they do" (Fenton 1982, 280).

The biophysics of echolocation appears to constrain body size in bats (Jones 1996). In turn, body size can influence call frequency and respiration rates (Jones 1999). To exploit the resources represented by nocturnal flying insects, bats must use high-frequency sounds to give strong echoes from small targets, and they must be small in order to produce ultrasound. In addition, wingbeat frequency decreases as body size increases, and aerial insectivores typically produce no more than one pulse per wingbeat. Larger insectivorous species might not produce echolocation calls at a rate sufficient to capture enough insects. These biophysical relationships explain why ecomorphological classifications of bats also separate them into groupings based on echolocation calls (e.g., Habersetzer 1986). In addition, Bogdanowicz et al. (1999) showed that the dominant frequencies of bat echolocation calls can be correlated with the incidence of different moths in their diets, but that this relationship holds only for low duty-cycle echolocators (Vespertilionidae and Molossidae), not for species of high duty-cycle bats (Rhinolophidae and Hipposideridae).

Norberg and Rayner (1987) showed that insectivorous bats exhibit highly diversified morphologies and correspondingly distinctive flight and foraging behaviors. In general, bats that hawk high-flying insects have small pointed wings, making them agile, fast, and cost-efficient flyers. Gleaners and bats hunting for insects in vegetation have very short and rounded wing tips on short, broad wings, making them highly maneuverable during slow flight. Insectivorous species that forage in open areas often have longer wings, and hence lower cost of transport, than those flying in cluttered environments. Adaptive variation in wing morphology was also correlated with corresponding echolocation call structure. Insectivorous bats that differ in body size and wing shape often choose different foraging habitats and harvest different subsets of food resources (e.g., Saunders and Barclay 1992). Analyzing Indian bat assemblages, Habersetzer (1986) determined that differences in echolocation call structure were correlated with differences in wing morphology but provided clearer separation of coexisting species. Heller and Helversen (1989) and Kingston et al. (2000) documented partitioning of frequency calls by rhinolopid bats.

Bats that take prey from surfaces often take different taxa than those hunting airborne targets, and such foraging differences may determine both the taxonomic identity of prey species and their energetic values (Fenton 1995). Gleaners frequently use prey-generated sound to locate prey (Belwood and Morris 1987; Fiedler 1979; LaVal and LaVal 1980), whereas hunters of flying insects often rely on echolocation. Gleaning insectivores generally have more restricted geographic ranges than do aerial foragers (Arita et al. 1997). Humphrey et al. (1983) studied surface-gleaning bats in Panama, and found that gleaning omnivores exploited resources differentiated by combinations of food type, food size, and vertical and horizontal foraging habitat. Five types of food constituted important niche differences for guild members, but beetles formed an important part of the diet for all insect gleaners (see also Kalko et al. 1999).

There is now compelling evidence to indicate that insectivorous bats are especially sensitive to deforestation and other forms of habitat disturbance in both the Neotropics (Brosset et al. 1996; dos Reis and Muller 1995; Fenton et al. 1992) and the Paleotropics (Zubaid 1993). For example, Wilson et al. (1996) determined that taxa such as Emballonuridae, insectivorous Phyllostominae, Thyropteridae, Furipteridae, and Vespertilionidae are all more abundant in undisturbed than in disturbed forests in Peru. Similar patterns, perhaps not quite so sharply developed, also characterize temperate regions (Eichstadt 1997; Law et al. 1999). This special vulnerability presumably derives from altered resource distributions in modified or converted forests (Fenton et al. 1992). In contrast, light or moderate levels of disturbance may actually increase the number of pteropodid species living in Philippine rainforests (Utzurrum 1998) or the abundance of carolline phyllostomids living in Neotropical forests (Wilson et al. 1996), provided that forest structure and composition remain essentially intact.

Differential habitat use may be a principal avenue of resource partitioning by insectivorous bats. McKenzie and colleagues (McKenzie and Rolfe 1986, McKenzie and Start 1989) characterized the structure of bat ensembles in mangrove communities in Australia using aerodynamic attributes. They found almost no overlap among flight morphologies of species that foraged in a given stand of mangroves. Morphological differences could be tied to differences in both vertical stratum and foraging microhabitats (also true of *Myotis myotis* and *M. blythii* where they occur in sympatry; Arlettaz and Perrin 1995). Differential abundance of arthropods in different habitat patches and strata may ultimately account for the morphological differentiation of bats, such as beetle-feeding and moth-feeding molossids (Findley and Black 1983, Freeman 1979).

In upland rainforest in Queensland, use of space by insectivorous bats was predictable from their ecomorphologies (Crome and Richards 1988). Bats were divisible into closed-canopy specialists (low aspect ratio and wing loading), gap specialists (high aspect ratios and wing loadings), and gap incorporators (intermediate). No patterns in assemblage structure were correlated with gap age, suggesting that the exclusion of gap species in the course of ecological successions is a slow process.

Kalko (1995) examined the echolocation calls produced by Neotropical Emballonuridae, finding them highly structured with respect to habitat. *Diclidurus* and *Peropteryx* spp. exploited open, uncluttered space with low frequency, shallow modulated echolocation calls. *Saccopteryx bilineata, S. leptura,* and *Cormura brevirostris* used medium-frequency echolocation signals in "cluttered habitats" filled by vegetation. Foraging within a meter or two of small forest streams, *Rhynchonycteris naso* used high-frequency echolocation signals similar in structure to those of noctilionid bats (that also forage in similar situations). Parallel structure may exist for vespertilionids and molossids but not phyllostomids, which represented the largest and most diverse group in terms of diet, yet emitted rather uniform signals.

Horizontal segregation may characterize north temperate bat ensembles, which are often dominated by maneuverable, relatively slow-flying species (Arlettaz 1999). Although some species hovered in complex, structurally diverse habitats where insect density and diversity appeared high, most were open-air foragers incapable of very slow or hovering flight. The latter species foraged on the wing over large areas. Terrestrial foragers (i.e., *Antrozous pallidus*) and over-water foragers (i.e., *Myotis yumanensis*) constituted separate trophic classes (see also Black 1974). Habitat generalists, such as *Lasiurus cinereus semotus* in Hawaii, may use different flight speeds and echolocation frequencies to forage in open versus cluttered habitats (Jacobs 1999), enabling the bats to capture a different spectrum of prey body sizes in each habitat.

In Neotropical communities, vertical stratification may be more important in organizing interactions. Handley (1967) was among the first to substantiate this hypothesis, comparing the relative capture success of nets placed on the ground and in the canopy of Amazonian forest near Belém, Brazil. He encountered generally higher capture rates in the canopy nets. The species that were taken solely in the canopy nets were, without exception, poorly known, supposedly uncommon, and irregularly distributed. In this instance, vertical stratification of flight activity has faunistic as well as ecological implications.

In Panama, food is limiting at some periods of the year, and flight activity of insectivorous species is vertically stratified (Bonaccorso 1979). *Mimon crenulatum, Trachops cirrhosus,* and *Tonatia sylvicola* were netted significantly more often near the ground than at subcanopy level, whereas *Tonatia saurophila* generally was captured at subcanopy level. Vertical segregation of potential competitors is accompanied by some horizontal or macrohabitat segregation, with the partial displacement of *Trachops* to creek forest and *Tonatia sylvicola* to forest patches between creeks. Finally, some evidence suggested that species with similar feeding habits are active at different times (Bonaccorso 1979; see also Ascorra et al. 1996).

Eichstadt (1997) studied resource partitioning by eight sympatric species in Europe, all roosting together within a 30-ha forest. Three groups were evident, based on diet choice: several opportunists taking small flying insects, a flightless beetle specialist (*Myotis myotis*), and a moth specialist (*Plecotus auritus*). Differences existed between opportunist and specialist species in habitat use. Opportunistic species foraged together on superabundant resources over wetlands and lakes, whereas specialists foraged on defensible resources in territories close to their roosts. Because opportunists feed on abundant prey that is not depleted greatly by feeding activities, defense is impractical or too costly.

Eichstadt (1997) found species differences in temporal activity periods, with long-distance commuters leaving day-roost areas first. However, it is questionable whether such differences enabled the bats to exploit different resources. Extended studies of foraging by *Lasiurus cinereus* and *L. borealis* at streetlights revealed no consistent evidence of temporal partitioning (Hickey et al. 1996). Over 3 yr, the food niche of larger *L. cinereus* (25–35 g) was consistently broader than for *L. borealis* (7–13 g). Niche overlap varied among years and was highest when resources were unusually low.

Behavioral opportunism is responsible for some of these shifts. Husar (1976) studied the diets of *Myotis evotis* and *M. auriculus* in New Mexico and documented that the diets of bats are highly labile. Although exploiting highly similar diets in allopatry, these two species are more specialized where their ranges overlap, *M. evotis* relying more heavily on beetles and *M. auriculus* on moths. Furthermore, each species exhibits sexual differences in foods taken in allopatry but not sympatry.

Other Animal Foods

Animalivorous (or carnivorous) bats include members of three families: Nycteridae, Megadermatidae, and Phyllostomidae. They exploit a variety of other organisms as food, but arthropods remain an important component (Norberg and Fenton 1988). Because they sometimes carry heavy loads, they tend to be heavy, with low aspect ratios and low relative wing loadings (Norberg and Fenton 1988). Nevertheless, they otherwise resemble other arthropod-feeding species that forage by perching, hunting, and gleaning (Norberg and Rayner 1987) and sometimes supplement their diets with fruit (Ferrarezzi and Amaral Gimenez 1996). Fenton et al. (1990) found that *Nycteris grandis* fed primarily on frogs, bats, and arthropods and occasionally took birds and fish. Lack of carnassial teeth and long handling times for larger prey (Fenton et al. 1983) may limit exploitation of carnivory by bats.

Although fishing bats also consume small vertebrates, fishing bats forage differently and do not fly in clutter (Norberg and Fenton 1988). All piscivorous bats supplement their diets with arthropods, either aquatic or nonaquatic insects caught on the water surface or else crustaceans (Ferrarezzi and Amaral Gimenez 1996). Piscivores forage over long stretches of open water, selecting for very long wings (low flight power and cost of transport) with long rounded tips for control and stability (Norberg and Rayner 1987).

Finally, blood feeding or sanguinivory is the most specialized diet exploited by bats, necessitating all sorts of correlated adaptations. However, even vampire bats supplement their diets with insects (Ferrarezzi and Amaral Gimenez 1996). The ecological mechanisms facilitating coexistence of the three living species are known, but the origins of sanguinivory itself remain speculative (reviewed in Schutt 1998). The fossil record, representing only members of the genus *Desmodus,* sheds little light on transformation series (Freeman 2000). Preliminary data suggest that *Desmodus rotundus* prefers mammalian prey and the other two vampire species prefer birds (see Gardner 1977; Greenhall and Schmidt 1988; Greenhall and Schutt 1996). In addition, the two bird feeders may be distinguished by roosting sites: *Diaemus youngi* prefers tree cavities, whereas *Diphylla ecaudata* roosts in caves, as does *Desmodus* (Dalquest 1955; Uieda 1993).

Herbivory

Over half of the world's living bat species use plants exclusively or opportunistically as resources for food and/or shelter (Kunz 1996). Roughly a third visit plants for nectar, pollen, or fruit (Fleming 1993). We recognize the following subdivisions: frugivory, feeding predominantly or solely on fruits; nectarivory, feeding on nectar as well as other floral products and parts, such as pollen and petals; and folivory, consumption of leaves, buds, and other green parts of plants (Ferrarezzi and Amaral Gimenez 1996). Omnivorous species also exploit diets partly comprising vegetal material (usually fruits and/or flowers).

Herbivory by bats has involved a host of morphological adaptations (Kunz and Ingalls 1994), developed over more than 50 million years of evolutionary association (Marshall 1983). These have included novel dental features (for microchiropterans, cf. Ferrarezzi and Amaral Gimenez 1996) to crush fruit, not triturate chitinous exoskeletons (Andersen 1908). Compared to insectivorous relatives, herbivorous species possess structurally complex stomachs (Forman 1990) that are more specialized and compartmentalized (Perrin and Hughes 1992). Herbivorous species are also distinguished by relatively large brains, apparently an aid in locating and exploiting rich but highly dispersed food resources (Eisenberg and Wilson 1978). "Brainy" herbivorous bats are sharply distinguished from insectivorous species, which have smaller-than-expected brains because of constraints on body mass imposed by requirements for highly maneuverable flight. Neurobiology of bats is treated in far greater detail in Baron et al. (1996), with ecological patterns of brain morphology in bats detailed in the third volume.

Most herbivorous bats are members of the Pteropodidae and Phyllostomidae, whose evolutionary radiations have taken place independently and in geographic isolation. The Pteropodidae are exclusively Old World in distribution, whereas the Phyllostomidae are practically confined to the Neotropics. Although the Pteropodidae are predominantly herbivorous, with major radiations of nectarivorous and frugivorous species, the phyllostomids employ both of these strategies (Gardner 1977; Wilson 1973), as well as a number of others (see Kalko et al. 1996; Patterson et al. 1996).

The differential success and radiations of the Pteropodidae and Phyllostomidae may be partly attributed to differences in their sensory systems (Heller and Volleth 1995). Although phylogenetic relationships among "yangochiropteran" families remain poorly understood, there is strong support for a group containing the Neotropical endemic families Noctilionidae, Mormoopidae, and Phyllostomidae (Simmons 1998) and the New Zealand endemic

Mystacina (Kirsch et al. 1998). All these taxa depend to varying degrees on echolocation, as do the potential sister clades of this group, whereas this ability is limited in Pteropodidae to some cave-roosting *Rousettus*. Perhaps in consequence, the number of coexisting phyllostomids exceeds the number of syntopic pteropodids in most tropical lowland forest sites. Differences between groups are especially marked in the understory ensemble, where a premium is placed on the ability to fly amid structural complexity (Francis 1994; Heller and Volleth 1995; Kingston et al. 2000; Zubaid 1994).

The diversity of herbivorous bats is highest in the tropics. Although few studies have documented annual cycles of tropical bat ensembles, the importance of annual variation can be gleaned from studies of vegetation. Most tropical shrubs and trees produce leaves and flowers in bursts, with seasonal variation in production of leaves, flowers, and fruits as adaptations to biotic or abiotic factors (van Schaik et al. 1993). Whereas abiotic factors promote synchronous responses by a flora, biotic factors can select for either clumping or staggering of events. Competition for food resources typically leads to staggered exploitation patterns, whereas predation often leads to synchronized activities, as individuals and groups respond to external predation threats (in this way resembling many abiotic factors, such as seasonal rains or cold periods that produce flora-wide responses). In seasonally dry forests, many plants concentrate leafing and flowering at the onset of the rainy season with synchronous fruiting (an adaptation that reduces seedling mortality at the start of the following dry season).

At the level of the forest community, phenological variation leads to dietary switching by primary consumers, as well as to seasonal breeding, range changes, migration, or aestivation (van Schaik et al. 1993). Certain plant products—keystone resources—act as mainstays of the primary consumer community during periods of resource scarcity. "Nomadism [as a response of vertebrates to seasonal or local resource scarcity] may be more prevalent in bats than is currently realized" (van Schaik et al. 1993, 370). Patchiness of tropical trees and the enhanced vagility of flight may explain the numerical predominance of very rare bat species in many local ensembles (cf. Handley et al. 1991). Apparent segregation of male and female *Otopteropus cartilagonodus* along elevational gradients in the Philippines (Ruedas et al. 1994) represents a solution to intraspecific competition that is available only to highly vagile species.

In turn, bats are particularly important in maintenance of tropical forests because (1) conspecific trees are often spatially isolated and dioecious; (2) the longevity of tree species, together with intense pressure from competitors, predators, and pathogens, places a premium on genetic recombination; (3) both tree species and their pollinators are highly diverse; and (4) biotic pollinators are more important than they are in temperate regions (Bawa 1990).

Nectar and Pollen

Radiations of bats to exploit the nectar and pollen produced by flowering plants have taken place in both the Old World (Pteropodidae: "macroglossines") and New World (Phyllostomidae: Glossophaginae, Lonchophyllinae, and Brachyphyllinae). Nectarivorous species are distinguished by their greatly elongated heads and tongues, representing extremes in both categories for the chiropteran radiations. Kitchener et al. (1990) showed correlations between tongue morphology (including details of tip morphology as well as overall tongue length) and dietary breadth for the six common species of nectarivorous pteropodids on Lombok, Indonesia.

The number of plant species pollinated by bats and the number of bat species involved as pollen vectors are not known for any tropical wet forest. Bats encompassed nearly 4% of 52 animal species pollinating woody canopy plants at La Selva (Costa Rica), and the same fraction of 220 pollinators of subcanopy and understory plants there (Bawa 1990). Pollination by vertebrates is more important in the tropics than in the temperate zone. However, along elevational gradients within the tropics, pollination by hummingbirds actually increases from lowland tropical forest into montane tropical forests (Bawa 1990), and nectarivores make up proportionally larger fractions of bat faunas ascending elevational gradients in Peru (Patterson et al. 1996) and Venezuela (Soriano et al. 1999). Although flowers may receive a wide range of visitors, only one or two—often members of the same order—act as effective vectors (Bawa 1990; Helversen and Winter, this volume). For plants, bats may be more expensive energetically than other pollinators, but these energetic costs are offset by the wide distribution given pollen by widely foraging bats (Start and Marshall 1976). Bat pollination is found in many plant families but is particularly common or well studied among the Bombacaceae (Bawa 1990).

Start and Marshall (1976) described a West Malaysian ensemble containing three species. *Eonycteris spelaea* roosts in caves, travels considerable distances to feed, forages in flocks, and obtains food from diverse, scattered, seasonally flowering sources. In contrast, *Macroglossus minimus* and *M. sobrinus* roost singly or in well-spaced groups close to food sources (*Sonneratia* sp. and *Musa* sp., respectively), and forage individually on sources that flower aseasonally. Heithaus et al. (1975) studied plant-visiting bats in Guanacaste, Costa Rica, where floral resources varied seasonally. Although most species were nectarivorous periodically, only *Glossophaga soricina* maintained this diet during periods of nectar scarcity (see Helversen and Winter, this volume).

Fruit resources were more uniformly abundant throughout the year and consequently were more finely divided by bat species. If competition for food has structured this bat assemblage, fruit has been more important than nectar and pollen in controlling diversity. However, flowering times of plant species were staggered, suggesting competition among bat-pollinated species

(Heithaus et al. 1975). Asynchronous peaks of flowering for different species of plants in the diet of *Glossophaga longirostris* allow plant species to use the same seed disperser and pollinator without suffering interspecific competition for such "services" (Soriano et al. 1991).

At a finer scale, *Artibeus jamaicensis, Sturnira lilium, Phyllostomus discolor,* and *G. soricina* all pollinated *Bauhinia* flowers at Guanacaste (Heithaus et al. 1974). Spatial partitioning of *Bauhinia* flowers is considerable and perhaps is the dominant form of resource partitioning (Fischer 1992; Heithaus et al. 1974). Malaysian nectarivorous bats also partition resources spatially, maintaining the effectiveness of this strategy by trapline foraging and territorial defense. Patterns of visitation at *Musa* and *Oroxylum* are typical of trapline foragers (Gould 1978). The phyllostomid *Lonchophylla bokermanni* is also a trapliner, habitually foraging in open habitats for *Encholirium glaziovii*, a ground-dwelling bromeliad (Sazima et al. 1989). Although other flower-visiting bats occur in southeastern Brazil and are not mechanically precluded from exploiting this bromeliad, none seems to compete with *L. bokermanni*. However, this strategy is not universally used by nectarivorous species—visits to *Parkia* flowers by Old World pteropodids are more frequent and opportunistic than in these other plants and sometimes involve agonistic interactions among would-be pollinators (Gould 1978).

Temporal partitioning of nectar and pollen resources appears less common than spatial partitioning. Heithaus et al. (1974) suggested that vertical and temporal differentiation may enhance spatial partitioning. *Phyllostomus discolor* began foraging earlier in the evening and confined its visits to flowers >2.5 m aboveground. This could leave more nectar later in the evening for *G. soricina* and *Anoura caudifer* in flowers below this level because an individual bat does not exhaust the nectar supply of a flower on a given visit. In some situations, however, little advantage may be gained by such refined resource partitioning: on the island of Curaçao, *Leptonycteris curasoae* and *G. longirostris* are the principal pollinators of at least two of three columnar cacti that grow on the island (Petit 1997). Fully 85%–91% of dietary samples contained seeds and pollen of cacti. Parturition and lactation in both species are timed to correspond to peaks in the seasonal phenology of the two abundant species of cactus.

Nectarivorous bats sometimes forage in cohesive but unorganized flocks to search for resources more efficiently and to exploit them more thoroughly (Howell 1979). Communal feeding in *Leptonycteris* permits intermittent communal roosting, in turn leading to more efficient digestion (via shared thermoregulatory costs on cold desert nights). By following "leader" bats from a depleted resource patch without individually confirming its depletion, all bats save energy, and these savings are enhanced because only lead bats appear to use echolocation.

Anthropogenic extinctions of birds have increased the importance of bats as dispersers of large-seeded fruits and reduced the guild of pollinators and dispersers of small-seeded fruits (Rainey et al. 1995). Resource defense, more common among megachiropteran frugivores than among microchiropterans, may structure patterns of pollination and seed dispersal as bat density and movements are determined by local resource abundance and behavioral responses to them (Richards 1995). Where plants depend on a single species of bats for pollination, such as insular populations of *Ceiba pentandra* pollinated by *Pteropus,* behavioral responses such as establishment of feeding territories, can structure patterns of plant reproduction (Elmqvist et al. 1992).

The ecological limits of nectarivory are difficult to specify. For example, Muñoz-Saba et al. (1997) determined that *Artibeus jamaicensis, A. lituratus, A. obscurus, G. soricina, Carollia perspicillata, Sturnira lilium, Uroderma bilobatum,* and *U. magnirostrum* all forage on nectar and pollen in the Serrania La Macarena. Collectively, their diet is 83% fruit, 8% insects, and 9% pollen and nectar. In addition to pollen from plants obviously adapted for chiropteran pollination, these bat species also carried pollen grains of Compositae, Myrtaceae, Gramineae, and Euphorbiaceae. Valiente-Banuet et al. (1996) documented some pollen from an abundant cactus, *Neobuxbaumia,* on *A. jamaicensis* but showed that *Leptonycteris* and *Choeronycteris* were the plant's only pollinators. An even more diverse array of visitors (including various birds and all three bats) disperses its seeds.

In certain areas, terrestrial mammals rival or exceed the importance of bats as pollinators. Despite a diversity of potential bat pollinators, several Amazonian tree species are evidently pollinated by nonvolant mammals, including *Ceiba, Ochroma,* and *Quararibea* (all Bombacaceae [Janson et al. 1981]). Elsewhere, *Banksia* flowers are visited by two Australian marsupials, *Petaurus breviceps* and *Antechinus stuartii* (Goldingay et al. 1991). Both species of marsupial depart with pollen loads comparable to those of nectar-feeding birds.

Fruits

Terborgh (1986) estimated that frugivorous species represent 80% of the mammalian and avian biomass at Cocha Cashu, Peru, and frugivorous bats tend to dominate assemblages in lowland Neotropical forests (Kalko 1998 and references therein). Depending on habitat, 50% to >90% of the species of tropical shrubs and trees rely on fruit-eating vertebrates to disperse their seeds (Howe and Smallwood 1982). Seeds of Araceae, Guttoferae, Hypericaceae, Moraceae, Piperaceae, and Solanaceae were recovered from the feces of 15 species of frugivorous phyllostomids from Loreto, Peru (Ascorra and Wilson 1992), and Ascorra et al. (1996) demonstrated comparable diversity of seeds dispersed by bats in Madre de Dios, Peru. The dispersion of tropical plants is every bit as much a consequence of the foraging, dietary, and home-range patterns of frugivores as the latter are a product of the former.

At the same time, despite comparable diversity of insectivorous species, the understory of dipterocarp forests in Southeast Asia supports relatively few species of obligate frugivores, at least in nonmasting years (Francis 1990). Some Asian frugivores are astonishingly catholic in their habits. Tan et al. (1998) documented fruits of 54 plant species, leaves of 14 species, and flower parts of four species in the diets of *Cynopterus brachyotis* in secondary habitats in Malaysia. Opportunities for partitioning resources in the company of such trophic generalists appear limited.

Marshall (1985) reviewed feeding habits of pteropodid bats throughout the Old World and concluded that their food tends to be conspicuous, clumped spatially and temporally, and abundant and easily harvested within these clumps. Consequently, there is no need to specialize on food resources as insectivorous species must do, and there is time for complex mating behaviors such as lekking, an activity unknown in insectivorous species. In addition, flock foraging is possible, even advantageous, in harvesting clumped resources.

In addition, species of pteropodid frugivores tend to be wider-ranging than their phyllostomid counterparts, as shown by comparisons of mean ranges and maximum movements of 14 species of Asian and African pteropodids and six species of phyllostomids (Heideman and Heaney 1989). Whether these are real taxonomic differences or are instead related to the body sizes, group sizes, and/or resource distributions of each situation remain unknown.

Fleming and Williams (1990) showed that at least 28 species of vertebrates (15 diurnal, 13 nocturnal) eat *Cecropia peltata* fruit in a Costa Rican tropical dry forest. An equal amount of fruit is removed day and night. Owing to gentler treatment of ingested seeds, bats and birds were more effective dispersers of this species than were monkeys. Digestion by *Carollia* has either no effect (e.g., *Piper, Chlorophora,* and *Muntingia*) or a positive effect (*Cecropia* and *Solanum*) on seed germination (Fleming 1988; see also Iudica and Bonaccorso [1997] for *Sturnira lilium*).

Bats tend to be wasteful seed dispersers, often defecating in dark areas with little opportunity for subsequent germination; they are probably more efficient at dispersing pollen than seeds (Fleming and Sosa 1994). However, given the large numbers of seeds ingested by individual bats, their density in the forest, and the length of the fruiting season, bats are highly successful at dispersing seeds (Fleming 1988). In the Old World, pteropodids ingest small seeds and transport them substantial distances; larger seeds are frequently carried with fruits to feeding roosts, farther than is practicable for avian frugivores. Only the large pteropodids are capable of dispersing the largest seeds (Kunz 1996).

Fruit resources in some tropical forests are present and more or less uniformly abundant throughout the year, so that fruits may be more finely divided by bat species than are floral resources. Mean size of fruit and its variance can be positively correlated with body size of bats (Heithaus et al. 1975), and species may specialize on the types of fruits eaten, provided that their phenology makes them available throughout the year. Marinho-Filho (1991) documented the seasonal abundance and phenology of *Piper* and *Solanum*, which are principal dietary components of *Carollia* and *Sturnira*, respectively, and are implicated in the coexistence of these species. Kalko et al. (1996) described the characteristics of figs eaten by frugivorous bats in the New and Old World tropics. Experiments by Rieger and Jakob (1988) in Costa Rica suggest that olfactory cues are also used in foraging decisions by phyllostomid frugivores.

Fruit and seed characteristics may serve as the basis of fruit choice (and hence resource partitioning), as shown by Gautier-Hion et al. (1985; also see Dumont, this volume; Muller and dos Reis 1992). Size-dependent partitioning of fruit resources by bats may affect the spatial distribution of the trees on which they feed. Utzurrum (1995) documented differences in fruit handling by Philippine bats. Large bats (>150 g) consume resources on the spot, producing splats and ejecta directly beneath the source tree. Smaller bats use feeding roosts in the immediate vicinity of, but not directly beneath, the focal feeding tree. Clumps of seeds found beneath these roost trees were mixed, producing a more heterogeneous pattern of seed scattering than would be expected if all bats behaved the same. Their importance as seed dispersers is confirmed by seed germination tests, in which gut-passed seeds (splats) of *Ficus chrysolepis* show higher percentages of germination than do seeds from fruits or ejecta (Utzurrum 1995).

In Amazonian Brazil, Handley (1967) noted reciprocal abundance patterns of two common species of *Artibeus* in the canopy and at ground level. *Artibeus lituratus* is usually captured far less frequently on the ground than its congener *A. jamaicensis*. However, *A. lituratus* dominated canopy samples. Ascorra et al. (1996) devoted substantial effort to aerial sampling, presenting the number of captures—recorded on the ground as well as in elevated nets of various heights—for 50 bat species in Manu, Peru (see also Simmons and Voss 1998). Among primarily frugivorous species, only *Chiroderma trinitatum* and *C. villosum* were captured solely in aerial nets (Ascorra et al. 1996), and both are taken in nets on the ground in that region (B. Patterson and S. Solari, personal observation). However, several frugivores and omnivores (e.g., *A. lituratus* and *Phyllostomus hastatus*) were more commonly taken in elevated nets, indicating vertically stratified activity patterns (Ascorra et al. 1996; Simmons and Voss 1998).

Vertical stratification of forest-inhabiting bats in lowland rainforest of Luzon, Philippines, was documented by Ingle (1993). She found that pteropodids made up 95% of captures in subcanopy nets (placed 3–16 m aboveground in natural forest gaps) but only 25% of bat captures in nets placed 0–3 m aboveground. Megabats appeared to be substantially more active in the subcanopy, and this may coincide with the vertical distribution of their food. In dipterocarp rainforest in Malaysia, Francis (1994; see also Zubaid 1994) also documented differential use of vertical space by bats. Nets on the ground in primary forests captured only two individual bats, whereas subcanopy nets captured bats at rates up to 100 times those on the ground. Diversity indices calculated from samples also showed a positive correlation with height above the ground. None of the species captured on the ground was restricted to that level, practically the inverse of patterns known from the Neotropics. Both of the common ground species were also found in the subcanopy.

Weak vertical stratification of Paleotropical pteropodids relative to Neotropical phyllostomids may reflect more fruit resources in the canopy of dipterocarp forests (e.g., Fleming 1988) but might also reflect the more limited ability of pteropodids to navigate in cluttered understory habitats (Francis 1994; Zubaid 1994). Certainly, Neotropical frugivores that feed on canopy plants often fly near the ground (Ascorra et al. 1996), perhaps to avoid predators or to reach roosting sites. Vertical stratification of flight activity undoubtedly contributes to our conceptions of relative abundance because the latter are fashioned from capture or ultrasonic recordings. Given the vulnerability of bats to human-induced changes in habitat quality and structure, either directly through clear-cutting or indirectly through pollution, far more work is needed.

Folivory

Bats chew leaves of at least 44 species of plants in 23 different families (Kunz and Diaz 1995). At least 17 species of Old World pteropodids and four plant-visiting bats in the Neotropics are known to feed on leaves, although none depends exclusively on them (Kunz and Ingalls 1994). Feeding on leaves involves mastication, swallowing liquids, and expelling fibrous residue (spats) (Lowry 1989). The digestive tracts of Old World fruit bats are designed for a largely liquid diet and rapid gut passage, and so require little if any modification for a leaf diet (Kunz and Ingalls 1994). In addition, plant-visiting bats are preadapted in having longer stomachs and intestines than do their insectivorous counterparts. Longer guts of plant-visiting bats offer added surface area for absorbing protein from the ingested fruits and leaves (Tedman and Hall 1985). The principal contribution of folivory to coexistence of bat species may be in supplementing the nutrient value of herbivorous diets, which are otherwise inadequate for bats in terms of mineral nutrients or proteins (Rajamani et al. 1999; Ruby et al. 2000).

Clearly, evolutionary diversification has resulted in bat species occupying a variety of ecological niches. As a consequence, members of the Chiroptera provide a diversity of ecological services (e.g., seed dispersal, flower pollination) that may be important to the overall composition, structure, and function of ecosystems. Nonetheless, natural selection, operating within the context of local environments and historical constraints, has produced considerable taxonomic and functional heterogeneity at the level of local bat assemblages. Indeed, much of the theoretical and empirical research pertinent to contemporary community ecology has focused on identifying recurrent patterns in local assemblages and identifying their underlying mechanistic bases. In the past, patterns of bat species composition within assemblages often were explained in light of such prevailing ecological theory, but research on bats lagged behind other taxonomic subdisciplines of ecology, such as those focusing on birds, rodents, or plants. In the past few decades, bat ecologists have begun to make rapid strides in documenting patterns and exploring their mechanistic bases. In the sections that follow, we rely heavily on the work of empirical bat ecologists to refine our understanding of both pattern and process as it relates to chiropteran components of local assemblages.

Broad-Scale Patterns in the Organization of Local Bat Assemblages and Ensembles

For 25 yr, ecologists have argued whether ecological assemblages are equilibrial and structured by deterministic processes or nonequilibrial and assembled by stochastic processes (Cody and Diamond 1975; Diamond and Case 1986; Gee and Giller 1987; Kikkawa and Anderson 1986; Polis 1991; Ricklefs and Schluter 1993; Strong et al. 1984). Although competition, predation, and mutualism have all been championed as mechanisms determining the formation and structure of local assemblages, evidence for the hegemony of one mechanism, or even for equilibrial conditions, is controversial. In part, the failure to find a dominant factor may be a consequence of spatial variation in environmental characteristics (e.g., temperature, precipitation, or insolation), which prevents any particular mechanism from operating with equal efficacy in all locations. Indeed, as the discipline of ecology has matured as a science, the search for a dominant mechanism structuring communities has been replaced with the deeper goal of understanding the environmental or evolutionary contexts in which equilibrial communities are most likely (Pickett et al. 1994).

While the controversy concerning the mechanistic basis of community organization has waxed and waned, bat ecologists intensively sampled the composition of bats in many local assemblages throughout the New World (fig. 12.3; table 12.1). We examined spatial variation in the biodiversity of assemblages and ensembles of bats at numerous sites ranging from 42° N latitude to 24° S latitude (Stevens and Willig, unpublished data), selected for analysis those that met certain spatial and temporal criteria: (1) each site must represent a single biome, with reasonable evidence to suggest sampling has been adequate to uncover potentially interacting bat species (i.e., sampling based on multiple sites within an area <1,000 km²); (2) sampling must have been undertaken on a regular basis in all seasons during which bats are active;

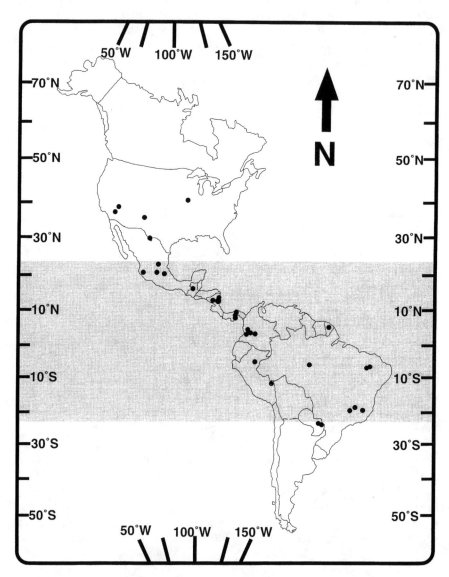

Figure 12.3. Map indicating the location of 32 bat assemblages (table 12.1) for which comparable and reliable data on species richness are available in the New World. The shaded region encompasses the tropics. Modified from Stevens and Willig (unpublished).

Table 12.1. Environmental characteristics of each of 32 bat assemblages used to evaluate patterns of diversity

Community	Latitude	Longitude	Habitat	References
Iowa, USA*	42.25°N	93.00°W	Temperate riparian	Kunz 1973
California, USA*	36.50°N	117.25°W	Temperate desert	Supernant 1977
Nevada, USA	36.20°N	115.20°W	Temperate desert	O'Farrell and Bradley 1970
New Mexico, USA	34.87°N	107.35°W	Temperate desert	Black 1974
Big Bend Ranch, USA	29.75°N	103.75°W	Temperate desert	Yancey 1996
Queretaro, Mexico	21.12°N	99.33°W	Montane tropical forest	Navarro and L.-Paniagua 1995
Manantlan, Mexico	19.33°N	104.00°W	Montane tropical forest	Iniguez Davalos 1993
Ixtapan del Oro, Mexico	19.25°N	100.16°W	Montane tropical forest	Alvarez and Alvarez-Castaneda 1996
Los Tuxtlas, Mexico	18.42°N	95.00°W	Wet tropical forest	Estrada et al. 1993
Chiapas, Mexico*	16.10°N	95.00°W	Wet tropical forest	Medellin 1993
Guanacaste-1, Costa Rica*	9.47°N	85.15°W	Wet tropical forest	LaVal and Fitch 1977
Guanacaste-2, Costa Rica*	9.47°N	85.15°W	Wet tropical forest	Fleming et al. 1972
Puntarenas, Costa Rica*	10.00°N	84.75°W	Montane tropical forest	LaVal and Fitch 1977
Heredia, Costa Rica*	10.5°N	83.75°W	Wet tropical forest	LaVal and Fitch 1977
Sherman, Panama*	9.33°N	79.95°W	Wet tropical forest	Fleming et al. 1972
Rodman, Panama*	8.95°N	79.62°W	Dry tropical forest	Fleming et al. 1972
BCI, Panama*	9.17°N	79.80°W	Wet tropical	Handley et al. 1991
Paroucou, French Guiana	5.27°N	52.92°W	Wet tropical forest	Simmons and Voss 1998
Hormiguero, Colombia	4.00°N	76.00°W	Wet tropical forest	Thomas 1972
Zabelitas, Colombia*	4.00°N	76.50°W	Wet tropical forest	Thomas 1972
Marcarena, Colombia	3.25°N	73.92°W	Wet tropical forest	Sanchez-Palomino et al. 1993
Manaus, Brazil	3.00°N	60.00°W	Wet tropical forest	dos Reis 1984
Pance, Colombia*	3.00°N	76.00°W	Montane tropical forest	Thomas 1972
Jenero Herrera, Peru*	4.92°S	73.75°W	Wet tropical forest	Ascorra et al., personal communication
Edaphic Cerrado, Brazil*	7.23°S	39.38°W	Tropical woodland-savannah	Willig 1982
Caatinga, Brazil*	7.58°S	39.66°W	Dry tropical forest	Willig 1982
Manu, Peru	11.93°S	71.28°W	Wet tropical forest	Ascorra et al. 1996
Linhares, Brazil	19.01°S	40.30°W	Wet semitropical forest	Peracchi and Albuquerque 1993
Panga, Brazil	19.25°S	48.38°W	Wet semitropical forest	Pedro and Taddei 1997
Minas Gerais, Brazil	19.83°S	41.83°W	Wet semitropical forest	Moura de Souza Aguiar 1994
Rio Verde, Paraguay	23.00°S	56.12°W	Dry semitropical forest	Stevens and Willig 2002
Mbaracayu, Paraguay	24.00°S	55.50°W	Wet semitropical forest	Stevens and Willig 2002

*Denotes the 15 communities analyzed by Stevens and Willig (1999, 2000a, 2000b).

(3) data must be gathered within a 1–5-yr period, raising the likelihood of including rare species and diminishing the likelihood of sampling species that do not interact because of temporal turnover; and (4) finally, a site was included in analyses only if empirical collector's curves (numbers of species as a function of number of individuals) for that site revealed that the observed species richness had reached an asymptote. Although each criterion is somewhat

subjective, collectively they increase the likelihood that the species present at the 32 sites potentially interact.

Gradients of Biodiversity

Four indices were used to characterize the nature of biodiversity at each site (Magurran 1988). Species richness (S) is simply the number of different species. Species dominance, as estimated by the Berger-Parker index (BP), is the proportional abundance of the species with the highest number of captures. Species diversity as estimated by the Shannon-Weiner index (H'), is sensitive to both S and the equitability with which individuals are apportioned among species, and is given by

$$H' = - \quad p_i (\log p_i),$$

where p_i is the proportional abundance of species i, and the summation is over all S species in the assemblage. Species evenness (E), as estimated by Pielou's ratio (Pielou 1969), is given by

$$E = \frac{H'}{H_{max}} = \frac{H'}{\log S}.$$

This equation reflects the degree to which the diversity of a community is at its maximum, given the number of species that it contains. Assemblages with high evenness have an equitable distribution of individuals among species, whereas assemblages with low evenness are characterized by heterogeneity in the proportional abundances of species.

Bat assemblages ranged in richness by almost an order of magnitude, from <10 species in more temperate latitudes to almost 80 species in tropical latitudes. As expected, based on biogeographic studies of latitudinal gradients of diversity (Willig and Sandlin 1991; Willig and Selcer 1989), species richness of an assemblage increased significantly ($P < 0.001$) toward the tropics (fig. 12.4A). In contrast, neither species dominance (fig. 12.4B) nor species evenness (fig. 12.4C), exhibited a significant latitudinal gradient ($P > 0.118$). Mostly as a consequence of its richness component, species diversity significantly increased ($P = 0.003$) toward the tropics (fig. 12.4D).

In part, species richness of a local assemblage of bats is related to the regional richness of bats (fig. 12.5A). Indeed, species richness of an assemblage is associated positively with the number of species whose geographic ranges overlap the location of a site ($P < 0.001$). Moreover, as the regional richness of the bat fauna increases, so does the variation in the number of species that coexist in a local assemblage. As a consequence, only a third of the variation in the richness of a local assemblage is accounted for by variation in its regional richness ($r^2 = 0.376$).

Although primary gradients of temperature, precipitation, and insolation

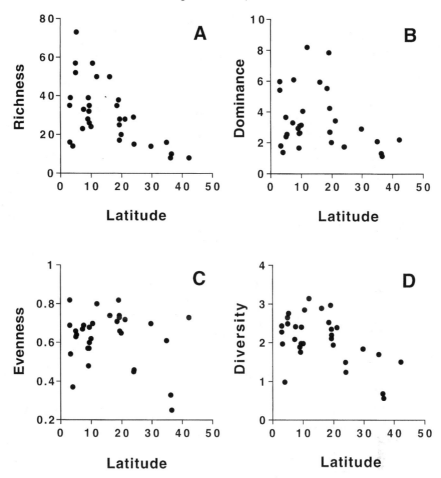

Figure 12.4. Latitudinal gradients in species richness (A), dominance (B), evenness (C), and diversity (D) based on 32 assemblages in the New World (see fig. 12.1 and table 12.1). The relationship between richness and latitude, as well as between diversity and latitude is significant, with tropical regions supporting higher biodiversity. Modified from Stevens and Willig (unpublished).

are related to gradients of latitude, other biogeographic factors (e.g., mountain ranges, ocean currents, edaphic features, distance) create spatial heterogeneity in environmental characteristics that are independent of latitude but affect species richness as well. Neighboring but strongly differentiated Amazonia and the caatinga in central Brazil are a case in point. Clearly, not all tropical assemblages should be expected to be equally species rich as a consequence. In fact, tropical regions might be expected to be most variable in bat species richness at the local level (cf. Janzen 1967). In general, tropical latitudes support a

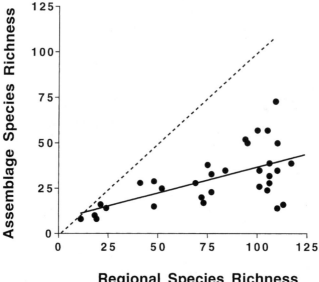

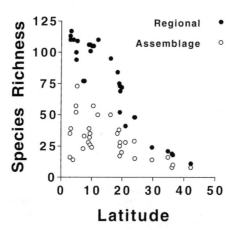

Figure 12.5. *A,* Graphic representation of the association between species richness of local assemblages and the species richness of regional faunas (defined by the number of species ranges that intercept the geographic location of the assemblage). The *solid line* represents the best-fit line from regression analysis, whereas the *dashed line* represents the upper boundary of assemblage richness based on the geographic ranges of species. As regional richness increases, local assemblages contain proportionally fewer of the possible species, suggesting a gradient in ecological filters. *B,* Latitudinal gradients in local (assemblage) and regional richness. The regional gradient is stronger (steeper slope and higher r^2) than the local gradient, mostly as a consequence of greater variation among assemblages at tropical latitudes. Modified from Stevens and Willig (unpublished).

greater number of life zones or plant formations than do cool temperate or boreal regions (Holdridge 1967). The primary environmental mechanisms that drive the latitudinal increase in species richness toward the tropics (i.e., solar irradiance, temperature, stress, or variability), may similarly be the ones that affect the greater variability in the tropics (i.e., the mean and variation in critical environmental characters are positively correlated).

Ecomorphological Structure

Morphologies of species evolve as a consequence of natural selection, so differences in size and shape should be adaptive with respect to ecological function (Peters 1983; Swartz et al., this volume). Morphology can be quantified more easily than direct ecological measures, such as reproductive power (Brown et al. 1996), reproductive stress (Clarke 1995; Palmer and Strobeck 1986), or ecological overlap (MacArthur and Levins 1967; Schoener 1974). It is also possible to use powerful statistical analyses to analyze morphological characters because, after log transformation, they frequently have normal distributions and homogeneous variances (Ricklefs and Travis 1980).

Morphological variation often provides considerable insight into ecological variation within and among species for animals in general (Blackburn et al. 1993; Dayan and Simberloff 1994; Fa and Purvis 1997; Hespenheide 1973) and for bats in particular (Findley and Black 1983; Freeman 1981, 1984, 1988, 1992; Swartz et al., this volume). Consequently, morphological patterns have been used to approximate the underlying ecological patterns in many bat faunas (e.g., Aldridge and Rautenbach 1987; Fenton 1972; Findley 1973 1976; Findley and Black 1983; McNab 1971; Schum 1984; Tamsitt 1967; Willig 1982 1986; Willig and Moulton 1989), despite the many simplifying assumptions that this approach entails.

In general, assemblages of closely related bat species typically comprise a relatively large number of specialists with invariant attributes clustered tightly around a centroid, whereas a smaller number of distinctive, variable species occupy more distant, peripheral niches (Findley 1976; Findley and Black 1983; Fleming 1986). Increases in species richness are facilitated by increasing the size of the total morphological hypervolume occupied by the assemblage, as well as by increases in the degree of species packing within constituent ensembles. Findley (1976), for example, compared three tropical bat faunas and found them to be essentially alike in packing, dispersion, and average degree of morphological overlap. In a study of ensembles of insectivorous-carnivorous bats from the New World, Schum (1984) found that species packing (i.e., distances between nearest neighbors in ecomorphological space) increased with increasing species richness. Moreover, Findley (1993) noted that a negative exponential relationship best described the decrease in mean nearest-neighbor distances with increasing species richness in those ensembles, suggesting limits on the degree of packing or similarity that can be accomplished within an ensemble.

Alternatively, novel ensembles may be accumulating within assemblages along species richness and latitudinal gradients. Stevens and Willig (2000b) presented evidence for this based on 15 assemblages from the New World (table 12.1). The number of ensembles per assemblage ranged from one in Iowa, to seven in Chiapas, Sherman, Barro Colorado Island, edaphic Cerrado, and Caatinga communities. Moreover, as the species richness of an assemblage increased, the number of constituent ensembles increased as well (Spearman's $r_s = 0.52$, $P = 0.023$). The ecomorphological characteristics of species in the different ensembles are quite distinctive (Stevens and Willig 1999, fig. 1). The addition of ensembles explains why increases in species richness consistently enlarge the ecomorphological hypervolume of the resultant assemblage.

Size Assortment and Character Displacement

The degree to which morphological patterns within assmblages are the product of biotic interactions or any other mechanism is unclear. Indeed, the similarity in structure between real ensembles or assemblages and those generated by random selection of species from regional species pools has been noted for over a quarter century (Ricklefs and Travis 1980; Van Valen 1973). This observation applies equally well to bat faunas (Fleming 1986; Schum 1984; Willig and Moulton 1989). Nonetheless, comprehensive or broad-scale evaluation of the degree to which ecomorphological patterns are a consequence of deterministic patterns have only recently been realized for bat assemblages and ensembles.

Morphological Structure and Competition

In a recent study, Stevens and Willig (1999) extended powerful simulation approaches (Moulton and Pimm 1987; Willig and Moulton 1989) to assess the ubiquity of deterministic structure for 15 sites throughout North, Central, and South America (table 12.1). They evaluated the degree to which morphological overdispersion—an indication of competitively induced structure (Hutchinson 1959; MacArthur and Levins 1967)—characterized each of five ensembles of bats (aerial insectivores, frugivores, gleaning animalivores, molossid insectivores, and nectarivores). If species that are morphologically alike jointly consume limiting resources, they should experience interspecific competition. If sufficiently intense, these interactions should favor divergence between competitors, resulting in character displacement. Alternatively, local extinction might characterize species not sufficiently distant from nearest neighbors in ecomorphological space. In either scenario, ecomorphological distances between species should be larger than expected by chance. Similarly, the distances between species would be more homogenous than expected by chance alone as a result of the elimination or displacement of nearest neighbors that exceed some degree of limiting similarity. In summary, ecological equilibria should occur within an ensemble when competitive interactions are minimized by hyperdispersed morphological patterns, leading to similar ecomorphological distances between nearest neighbors.

Overdispersion in at least one bat ensemble was detected at 10 of 15 sites in the New World. Similarly, overdispersion for at least one site was detected for four of five ensembles. Although overdispersed ecomorphological patterns exist in bat ensembles, they are not a common characteristic of the structure. Indeed, no conspicuous pattern existed as to which sites or ensembles exhibited overdispersion. Consequently, Stevens and Willig (1999) conducted a meta-analysis, combining results for all ensembles within an assemblage or combining results for all sites at which an ensemble occurs in the New World. Three of 15 assemblages (California, Guanacaste, and Heredia) exhibited larger distances between nearest neighbors than expected by chance alone. Similarly, three of 15 assemblages exhibited less variable distances between nearest neighbors than expected by chance alone (i.e., Sherman, Barro Colorado Island, and Caatinga).

Finally, the aerial insectivore and frugivore ensembles exhibited structure consistent with a competitive mechanism, based on larger and less variable distances, respectively, between nearest neighbors. Environmental factors such as spatial heterogeneity and climatic variability, or taxon-specific factors such as high mobility (enhancing the likelihood of the rescue effect, sensu Brown and Kodric-Brown 1977), may countermand mechanisms associated with deterministic processes, such that overdispersion is at best a transitory outcome of the effect of competitive interactions on ecomorphological structure.

Density Compensation and Competition

Competitive interactions can act as agents of natural selection in one of two general ways. Competition can cause the morphological evolution of populations via character displacement. Alternatively, competition can contribute to the local extinction of populations of species that exceed the limits of similarity that a local environment can support. However, competition may not be sufficiently intense to cause the local extinction of species but may alter the abundances of local populations of interacting species.

Indeed, it is a general principle in population biology that species that experience more competitive pressure should exhibit lower abundances (Lotka 1932; Volterra 1931) and that the consequence of competitive interactions at equilibrium need not be exclusion or local extinction. If morphological similarity is a surrogate for ecological similarity, then those species that are most dissimilar from other species in an ensemble should experience the least competitive pressure and enjoy the highest abundance, all other things being equal. This phenomenon is known as density compensation (Crowell 1962; Hawkins and MacMahon 1989; Root 1973). As a result, the abundance of a species and its morphological separation from competitors should have a positive association, and the magnitude of the association should be greater than produced by stochastic processes.

A simulation model (Stevens and Willig 2000a) was used to assess whether density compensation affects ensemble structure. The model can also be used to determine whether the competitive effects producing density compensation are mediated by interactions between nearest neighbors or are more diffuse in nature. The model was applied to data for five ensembles in 15 bat assemblages occurring throughout the New World (fig. 12.3; table 12.1). Nonrandom associations between abundance and morphological distance were detected in five assemblages (Guanacaste, Sherman, Barro Colorado Island, Peru, and edaphic Cerrado) and in three feeding guilds (aerial insectivores, gleaning animalivores, and frugivores). Nonetheless, no assemblage was strongly structured in all its constituent ensembles, and no ensemble was deterministically structured across most locations. There was no pattern as to which sites or which ensembles consistently evinced signs of density compensation.

A meta-analysis of the data suggests that gleaning animalivores are most affected by interactions between nearest neighbors, and the assemblage at Sherman exhibits a strong pattern of density compensation in all constituent ensembles from both diffuse and near-neighbor perspectives. Moreover, the likelihood of significant deterministic structure was unrelated to the number of species in an ensemble or to the number of species in the assemblage from which the ensemble was obtained (Stevens and Willig 2000b).

Methodological Constraints

Understanding pattern and process at the level of local assemblages is predicated on accurate estimation of attributes such as species composition and density. As a result, all assemblage-level research suffers from a number of constraints and biases related to sampling methodologies and effort, and this is particularly true for studies of bats. Characteristics related to temporal (nocturnal activity), spatial (vertical stratification), and sensory (ultrasonic communication) activities of bats make it difficult to detect their presence and estimate their abundance, compared to the situation for many other groups of animals or plants. As a consequence, we have been careful to utilize rigorous criteria in selecting bat assemblages for analysis. Nonetheless, we have likely underestimated species richness and excluded rare species from our representations of local assemblages.

We cannot evaluate, per se, the accuracy of estimates used in the previous broad-scale analyses of bat assemblages. To do so would require knowledge of the actual taxonomic composition and species densities at local sites. If we had such "parametric information," we (as well as the original authors) could have used it in analyses. Thus, caution should be applied in accepting patterns and process as real in ecological studies at the assemblage level. Nonetheless, resampling methodologies or jackknife procedures can be used to assess the degree to which detected patterns are predicated on sampling effort alone. Initial sensitivity analyses that focus on attributes of diversity (species richness, species evenness, species dominance, and species diversity) with sampling effort suggest such patterns may be quite robust for bat assemblages.

The essence of resampling methodologies or jackknife procedures is that it is easier and more accurate to predict the effects of sampling at lower levels of effort than it is to extrapolate to more intense levels of effort. For each of 32 bat assemblages (table 12.1), Stevens and Willig (2001) evaluated the extent to which sampling intensity affected the detection of latitudinal gradients in 15 different indexes of species richness, evenness, dominance, or diversity. Regardless of index, the statistical association between latitude and an index of biodiversity was insensitive to sampling effort to about 50% of the original effort. For example, based on all data at each site, the correlation between species richness and latitude was approximately -0.59 and significant ($P < 0.05$). Incremental reductions in sampling (5% steps) to 50% of the original effort gradually decreased the average correlation to -0.58, but all simulated correlations remained significant.

In a similar fashion, the correlation between Hill's index of evenness and latitude based on all the data at each site was approximately 0.11 and nonsignificant at $\Delta = 0.05$. The correlation gradually decreased with reductions in sampling effort, but at 50% of the original sampling intensity, the correlation remained 0.11 with all simulated correlations not significant. Although variation in the magnitude of the simulated correlations generally increased with decreasing sampling intensity, decisions of significance or nonsignificance were unaffected for the most part. If simulations to lower levels of effort provide insight to the patterns that would be detected with greater effort, then the gradients in diversity, which we detected, are reasonably robust and unlikely to be altered appreciably by increased effort at the local level.

Additional application of these numerical techniques to other measures of pattern at the assemblage level represents an exciting area of future research. Equally important, as new technologies become refined (e.g., bat detectors) or the application of canopy netting becomes prevalent, more accurate estimations of assemblage composition and abundance may be forthcoming. Thereafter, comparison of patterns based on ground netting to those based on combined approaches will provide deeper insight to the spatial and temporal dynamics of bat assemblages.

Overview and Prospectus

Bat species achieve remarkable degrees of local diversity and regional abundance that are enabled by their ability to exploit extensive home ranges (e.g.,

Handley et al. 1991). In some regions, individuals of up to 100 species of bats may have at least partially overlapping home ranges (e.g., Patterson et al. 1996; Simmons and Voss 1998). No other group of mammals in the 230-million-year history of the group has approached this degree of alpha diversity. Flight (or more precisely the enhanced vagility it provides) is fundamental to this eco-logical overlap. Lacking flight and thus being more confined to given locales, terrestrial mammals must satisfy their ecological requirements within a smaller spatial "footprint," leading to finer-grained (sensu Levins 1968), less specialized patterns of resource utilization. As shown in the preceding survey, most trophic groupings of bats rely to some degree on exploitation of diffuse resources. In areas where diversity is greatest—the tropics—partitioning of resources may involve a combination of spatial, temporal, and food-specific factors. The ability of bats to move quickly and efficiently over considerable distances in search of food, roosts, and mates has permitted such extensive and refined partitioning of resources.

Ecological resource partitioning sustains bat diversity at high levels but also leaves them vulnerable to rapid endangerment and extinction in a hu-man-dominated world. As humans convert an ever-greater fraction of natural habitats, all kinds of organisms are affected (Willig and Walker 1999), but three groupings are especially vulnerable to landscape changes. First, those with small geographic ranges may find their entire ranges denuded and may be left without appropriate habitats. Second, species that are highly specialized in terms of food preferences or ecological requirements may be vulnerable to the disappearance of these resources or conditions, even if the habitat in which they live remains largely unaltered by human activity. Finally, even species with large geographic ranges can be vulnerable to anthropogenic modification of the landscape if they exploit decidedly diffuse resources.

It has been argued that the great diversity of tropical forests is built on the highly dispersed distributions of many plant species (Condit et al. 2000). To exploit these dispersed species requires adaptations to and reliance on diffuse resources. Bats, more than any other group of terrestrial mammals, excel in the efficient exploitation of diffuse resources. However, as tropical forests become fragmented, the commuting costs while foraging necessarily increase as bats traverse increasingly large, unsuitable matrix habitats.

Even though the high vagility associated with flight allows bats to exploit fragments, seemingly protecting them from the effects of deforestation, their reliance on already-diffuse resources leaves bats highly vulnerable, as in-creasingly distant resources become too expensive to harvest. Sadly, the same phenomenon applies to some bats in the temperate zone, where bat species constitute disproportionate fractions of the endangered species lists for Eu-rope and North America. Temperate-zone bats are often doomed by their spe-cialized roosting habits, which make disruptions of breeding or hibernation cycles all-too-common—often lethal—occurrences. Tropical bats, in contrast,

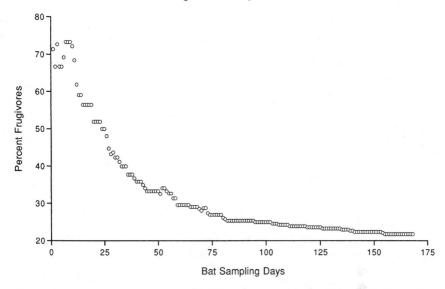

Figure 12.6. Differential capture success for different bat ensembles is well known. Differential accumulation of bats at Paracou, French Guiana, produces remarkably different estimations of the diversity and community importance of frugivores, depending on the sampling intensity employed (Simmons and Voss 1998, fig. 77). Greater sampling efforts over realistic spatial and temporal scales will be necessary to understand all tropical communities. However, bat ecologists must be especially committed to this effort, given the vagility, elusiveness, and ecological diversity of bats.

particularly insectivorous species, respond to habitat conversion with reductions in both species richness and abundance (Brosset et al. 1996; Estrada et al. 1993; Vaughan and Hill 1996; Wilson et al. 1996). Some, like many phyllostomines, appear well suited to serve as "indicator species," indexing by their presence and abundance the quality of habitat remaining in a region (e.g., Fenton et al. 1992).

Because bats are far ranging, nocturnal, and operate in a perceptual realm organized mostly by sound rather than sight (except in pteropodids), they are often difficult to sample. Most ecological studies of bats have targeted only part of an assemblage—few have targeted guilds or ensembles. Assemblage-wide inferences from such incomplete samples can lead to serious misconceptions. Figure 12.6 (from Simmons and Voss 1998, fig. 77) illustrates that the proportion of species that are frugivores in a Neotropical bat fauna decreases monotonically with sampling, from a predominating 70% after a week of sampling to an asymptotic value of 22%. Although it is customary to ascribe patterns such as these to the different perceptual abilities of bats or to the flight strata that they utilize, differential apparency to sampling is also tied to the diffuse space-use patterns under discussion.

Finally, Kalko (1998) observed that future studies of bat "communities" (assemblages by our definition) should combine macroecological and mechanistic studies. Our review concurs that far more progress has been made in assembling macroecological patterns than in testing the mechanisms thought to produce them. Certainly, the large spatial scale needed for such tests offers a powerful explanation for these trends. Kalko's discussion emphasized the methodological challenges to be overcome in order to understand the ecological interactions of tropical bat diversity. Although all are important, we would emphasize one: ecologists must be prepared to compare, contrast, and ultimately integrate ecological patterns of bats with those shown by other taxonomic groups. Comparisons of gross macroecological patterns have been revealing (Graham 1990; Patterson et al. 1998), but we need to go further.

Comparisons must extend to the level of resource partitioning and species-abundance patterns if we are to understand their mechanistic basis. Only a few comparative studies exist (e.g., Gorchov et al. 1995; Palmeirim et al. 1989), but efforts underway in Panama by Kalko, Robinson, and von Staden (Kalko 1998), in Peru by Patterson, Stotz, Solari, and Hodgkison, and in Malaysia by Kingston, Jones, Zubaid, and Kunz (Kingston et al. 2000) promise to extend them. Only when bat ensembles can be placed into a community-wide context will bat studies take their rightful place in ecology's pantheon.

Acknowledgments

We are grateful to two anonymous reviewers as well as to Tom Kunz, Jodi Sedlock, and Sergio Solari for their help and insights in developing more balanced and complete references to the enormous literature on ecological interactions of bats. We thank Tom Kunz and Brock Fenton for their invitation to participate in this important undertaking and for their comments on the text.

Literature Cited

Acharya, L. 1992. Epomophorus wahlbergi. Mammalian Species, 394:1–4.
Aldridge, H. D. J. N., and I. L. Rautenbach. 1987. Morphology, echolocation and resource partitioning in insectivorous bats. Journal of Animal Ecology, 56:763–778.
Allen, T. F. H., and T. B. Starr. 1982. Hierarchy: Perspectives for Ecological Complexity. University of Chicago Press, Chicago.
Alvarez, T., and S. T. Alvarez-Castaneda. 1996. Aspectos biologicos y ecologicos de los murcielagos de Ixtapan Del Oro, Estado de Mexico, Mexico. Pp. 169–182 in: Contributions in Mammalogy: A Memorial Volume Honoring J. Knox Jones, Jr. (H. H. Genoways and R. J. Baker, eds.). Museum of Texas Tech University, Lubbock.
Andersen, K. 1908. A monograph of the chiropteran genera Uroderma, Enchisthenes, and Artibeus. Proceedings of the Zoological Society of London, 21:204–319.
Arita, H. T. 1997. Species composition and morphological structure of the bat fauna of Yucatan, Mexico. Journal of Animal Ecology, 66:83–97.
Arita, H. T., F. Figueroa, A. Frisch, P. Rodríguez, and K. Santos-del-Prado. 1997. Geographical range size and the conservation of Mexican mammals. Conservation Biology, 11:92–100.

Arlettaz, R. 1999. Habitat selection as a major resource partitioning mechanism between the two sympatric sibling bat species *Myotis myotis* and *Myotis blythii*. Journal of Animal Ecology, 68:460–471.

Arlettaz, R., and N. Perrin. 1995. The trophic niches of sympatric sibling *Myotis myotis* and *M. blythii:* do mouse-eared bats select prey? Symposia of the Zoological Society of London, no. 67: 361–376.

Ascorra, C. F., S. Solari, and D. E. Wilson. 1996. Diversidad y ecologia de los quiropteros en Pakitza. Pp. 593–612 *in:* Manu: The Biodiversity of Southwestern Peru (D. E. Wilson and A. Sandoval, eds.). Smithsonian Institution Press, Washington, D.C.

Ascorra, C., and D. E. Wilson. 1992. Bat frugivory and seed dispersal in the Amazon, Loreto, Peru. Publicaciones del Museo de Historia Natural, Universidad Nacional Mayor de San Marcos, Serie A Zoología, 43:1–6.

Baron, G., H. Stephan, and H. D. Frahm. 1996. Comparative Neurobiology in Chiroptera. 3 vols. Birkhäuser Verlag, Basel.

Bawa, K. S. 1990. Plant-pollinator interactions in tropical rain forests. Annual Review of Ecology and Systematics, 21:399–422.

Belwood, J. J., and G. K. Morris. 1987. Bat predation and its influence on calling behavior in Neotropical katydids. Science, 238:64–67.

Black, H. L. 1974. A north temperate bat community: structure and prey populations. Journal of Mammalogy, 55:138–157.

Blackburn, T. M., V. K. Brown, B. M. Doube, J. D. D. Greenwood, J. H. Lawton, and N. E. Stork. 1993. The relationship between abundance and body size in natural animal assemblages. Journal of Animal Ecology, 62:519–528.

Bogdanowicz, W., M. B. Fenton, and K. Daleszczyk. 1999. The relationship between echolocation calls, morphology and diet in insectivorous bats. Journal of Zoology (London), 247:381–393.

Bonaccorso, F. J. 1979. Foraging and reproductive ecology in a Panamanian bat community. Bulletin of the Florida State Museum, Biological Sciences, 24:359–408.

Brosset, A., P. Charles-Dominique, A. Cockle, J.-F. Cosson, and D. Masson. 1996. Bat communities and deforestation in French Guiana. Canadian Journal of Zoology, 74:1974–1982.

Brown, J. H., and A. Kodric-Brown. 1977. Turnover rates in insular biogeography: effects of immigration on extinction. Ecology, 63:391–400.

Brown, J. H., M. L. Taper, and P. A. Marquet. 1996. Darwinian fitness and reproductive power: reply to Kozlowski. American Naturalist, 147:1092–1097.

Campanhã, R. A. da C., and H. G. Fowler. 1993. Roosting assemblages of bats in arenitic caves in remnant fragments of Atlantic forest in southeastern Brazil. Biotropica, 25:362–365.

Clarke, G. M. 1995. Relationships between developmental stability and fitness: application for conservation biology. Conservation Biology, 9:18–24.

Cody, M. L., and J. M. Diamond, eds. 1975. Ecology and Evolution of Communities. Belknap Press, Cambridge, Mass.

Condit, R., P. Ashton, P. Baker, S. Bunyavejchewin, S. Gunatilleke, N. Gunnatilleke, S. Hubbell, et al. 2000. Spatial patterns in the distribution of tropical tree species. Science, 288:1414–1418.

Crome, F. H. J., and G. C. Richards. 1988. Bats and gaps: microchiropteran community structure in a Queensland rain forest. Ecology, 69:1960–1969.

Crowell, K. L. 1962. Reduced interspecific competition among birds of Bermuda. Ecology, 43:75–88.

Dalquest, W. W. 1955. Natural history of the vampire bats of eastern Mexico. American Midland Naturalist, 53:79–87.

Dayan, T., and D. Simberloff. 1994. Morphological relationships among coexisting heteromyids: an incisive dental character. American Naturalist, 143:462–477.

Decher, J. 1997. Bat community patterns on the Accra Plains of Ghana, West Africa. Zeitschrift für Säugetierkunde, 62:129–142.

Diamond, J. M., and T. Case, eds. 1986. Community Ecology. Harper & Row, New York.

dos Reis, N. R. 1984. Estructura de comunidade de morçegos na região de Manaus, Amazonas. Revista Brasileira de Biologia, 44:247–254.

dos Reis, N. R., and M. F. Muller. 1995. Bat diversity of forests and open areas in a subtropical region of south Brazil. Ecología Austral, 5:31–36.

Eichstadt, H. 1997. Ressourcennutzung und Nischengestaltung einer Fledermausgemeinschaft im Nordosten Brandenburgs. Säugetierkundliche Mitteilungen, 40: 3–171.

Eisenberg, J. F., and D. E. Wilson. 1978. Relative brain size and feeding strategies in the Chiroptera. Evolution, 32:740–751.

Elmqvist, T., P. A. Cox, W. E. Rainey, and E. D. Pierson. 1992. Restricted pollination on oceanic islands: pollination of *Ceiba pentandra* by flying foxes in Samoa. Biotropica, 24:15–23.

Estrada, A., R. Coates-Estrada, and D. Meritt, Jr. 1993. Bat species richness and abundance in tropical rain forest fragments and in agricultural habitats at Los Tuxtlas, Mexico. Ecography, 16:309–318.

Fa, J. E., and A. Purvis. 1997. Body size, diet, and population density in Afrotropical forest mammals: a comparison with Neotropical species. Journal of Animal Ecology, 66:98–112.

Fauth, J. E., J. Bernardo, M. Camara, W. J. Resetarits, J. Van Buskirk, and S. A. McCollum. 1996. Simplifying the jargon of community ecology: a conceptual approach. American Naturalist, 147:282–286.

Fenton, M. B. 1972. The structure of aerial-feeding bat faunas as indicated by ear and wing elements. Canadian Journal of Zoology, 50:287–296.

Fenton, M. B. 1982. Echolocation, insect hearing, and feeding ecology of insectivorous bats. Pp. 261–285 *in:* Ecology of Bats (T. H. Kunz, ed.). Plenum Press, New York.

Fenton, M. B. 1990. The foraging behaviour and ecology of animal-eating bats. Canadian Journal of Zoology, 68:411–422.

Fenton, M. B. 1995. Constraint and flexibility—bats as predators, bats as prey. Symposia of the Zoological Society of London, no. 67:277–289.

Fenton, M. B., L. Acharya, D. Audet, M. B. C. Hickey, C. Merriman, M. K. Obrist, D. M. Syme, and B. Adkins. 1992. Phyllostomid bats (Chiroptera: Phyllostomidae) as indicators of habitat disruption in the Neotropics. Biotropica, 24: 440–446.

Fenton, M. B., R. M. Brigham, A. M. Mills, and I. L. Rautenbach. 1985. The roosting and foraging areas of *Epomophorus wahlbergi* (Pteropodidae) and *Scotophilus viridis* (Vespertilionidae) in Kruger National Park, South Africa. Journal of Mammalogy, 66: 461–468.

Fenton, M. B., C. L. Gaudet, and M. L. Leonard. 1983. Feeding behaviour of *Nycteris grandis* and *Nycteris thebaica* (Nycteridae) in captivity. Journal of Zoology (London), 200:347–354.

Fenton, M. B., I. L. Rautenbach, S. E. Smith, C. M. Swanepoel, J. Grosell, and J. van Jaarsveld. 1994. Raptors and bats: threats and opportunities. Animal Behaviour, 48:9–18.

Fenton, M. B., C. M. Swanepoel, R. M. Brigham, J. Cebek, and M. B. C. Hickey. 1990. Foraging behavior and prey selection by large slit-faced bats (*Nycteris grandis*; Chiroptera: Nycteridae). Biotropica, 22:2–8.

Ferrarezzi, H., and E. do Amaral Gimenez. 1996. Systematic patterns and the evolution of feeding habits in Chiroptera (Archonta: Mammalia). Journal of Comparative Biology, 1:75–94.

Fiedler, J. 1979. Prey catching with and without echolocation in the Indian false vampire bat (*Megaderma lyra*). Behavioral Ecology and Sociobiology, 6:155–160.

Findley, J. S. 1973. Phenetic packing as a measure of faunal diversity. American Naturalist, 107:580–584.

Findley, J. S. 1976. The structure of bat communities. American Naturalist, 110:129–139.

Findley, J. S. 1993. Bats: A Community Perspective. Cambridge Studies in Ecology, Cambridge University Press, Cambridge.

Findley, J. S., and H. L. Black. 1983. Morphological and dietary structuring of a Zambian insectivorous bat community. Ecology, 64:625–630.

Fischer, E. A. 1992. Foraging of nectarivorous bats on *Bauhinia ungulata*. Biotropica, 4:1–4.

Fleming, T. H. 1986. The structure of Neotropical bat communities: a preliminary analysis. Revista Chilena de Historia Natural, 59:135–150.

Fleming, T. H. 1988. The Short-Tailed Fruit Bat: A Study in Plant-Animal Interactions. University of Chicago Press, Chicago.

Fleming, T. H. 1993. Plant-visiting bats. American Scientist, 81:460–467.

Fleming, T. H., E. T. Hooper, and D. E. Wilson. 1972. Three Central American bat communities: structure, reproductive cycles, and movement patterns. Ecology, 53:555–569.

Fleming, T. H., and V. J. Sosa. 1994. Effects of nectarivorous and frugivorous mammals on reproductive success of plants. Journal of Mammalogy, 75:845–851.

Fleming, T. H., and C. F. Williams. 1990. Phenology, seed dispersal, and recruitment in *Cecropia peltata* (Moraceae) in Costa Rican tropical dry forest. Journal of Tropical Ecology, 6:163–178.

Forman, G. L. 1990. Comparative macro- and micro-anatomy of stomachs of macroglossine bats (Megachiroptera: Pteropodidae). Journal of Mammalogy, 71:555–565.

Francis, C. M. 1990. Trophic structure of bat communities in the understorey of lowland dipterocarp rain forest in Malaysia. Journal of Tropical Ecology, 6:421–431.

Francis, C. M. 1994. Vertical stratification of fruit bats (Pteropodidae) in lowland dipterocarp rainforest in Malaysia. Journal of Tropical Ecology, 10:523–530.

Freeman, P. W. 1979. Specialized insectivory: beetle-eating and moth-eating molossid bats. Journal of Mammalogy, 60:467–479.

Freeman, P. W. 1981. Correspondence of food habits and morphology in insectivorous bats. Journal of Mammalogy, 62:166–173.

Freeman, P. W. 1984. Functional cranial analysis of large animalivorous bats (Microchiroptera). Biological Journal of the Linnean Society, 21:387–408.

Freeman, P. W. 1988. Frugivorous and animalivorous bats (Microchiroptera): dental and cranial adaptations. Biological Journal of the Linnean Society, 33:249–272.

Freeman, P. W. 1992. Canine teeth of bats (Microchiroptera): size, shape, and role of crack propagation. Biological Journal of the Linnean Society, 45:97–115.

Freeman, P. W. 2000. Macroevolution in Microchiroptera: recoupling morphology and ecology with phylogeny. Evolutionary Ecology Research, 2:317–335.

Gardner, A. L. 1977. Feeding habits. Pp. 293–350 in: Biology of Bats of the New World Family Phyllostomatidae, pt. II (R. J. Baker, J. K. Jones, and D. C. Carter, eds.). Special Publications, the Museum, no. 13. Texas Tech Press, Lubbock.

Gautier-Hion, A., J.-M. Duplantier, R. Quris, F. Feer, C. Sourd, J.-P. Decoux, G. Dubost, et al. 1985. Fruit characters as the basis of fruit choice and seed dispersal in a tropical forest vertebrate community. Oecologia (Berlin), 65:324–337.

Gee, J. H. R., and P. S. Giller, eds. 1987. Organization of Communities: Past and Present. Blackwell Scientific Publications, Oxford.

Goldingay, R. L., S. M. Carthew, and R. J. Whelan. 1991. The importance of non-flying mammals in pollination. Oikos, 61:79–87.

Gorchov, D. L., F. Cornejo, C. F. Ascorra, and M. Jaramillo. 1995. Dietary overlap between frugivorous birds and bats in the Peruvian Amazon. Oikos, 74:235–250.

Gould, E. 1978. Foraging behavior of Malayan nectar-feeding bats. Biotropica, 10:184–193.

Graham, G. L. 1990. Bats versus birds: comparisons among Peruvian vertebrate faunas along an elevational gradient. Journal of Biogeography, 17:657–668.

Greenhall, A. M., and U. Schmidt, eds. 1988. Natural History of Vampire Bats. CRC Press, Boca Raton, Fla.

Greenhall, A. M., and W. A. Schutt, Jr. 1996. Diaemus youngi. Mammalian Species, 533:107.

Habersetzer, J. 1986. Vergleichende flügelmorphologische Untersuchungen an einer Fledermausgesellschaft in Maurai. Pp. 75–104 in: Bat Flight—Fledermausflug (W. Nachtigall, ed.). Bionia Report no. 5. G. Fischer, Stuttgart.

Handley, C. O., Jr. 1967. Bats of the canopy of an Amazonian forest. Atas Simposio, Biota Amazonica 5 (Zoologia), 211–215.

Handley, C. O., Jr., D. E. Wilson, and A. L. Gardner, eds. 1991. Demography and Natural History of the Common Fruit Bat Artibeus jamaicensis on Barro Colorado Island, Panama. Smithsonian Institution Press, Washington, D.C.

Hawkins, C. P., and J. A. MacMahon. 1989. Guilds: the multiple meanings of a concept. Annual Review of Ecology and Systematics, 34:423–451.

Heideman, P. D., and L. R. Heaney. 1989. Population biology and estimates of abundance of fruit bats (Pteropodidae) in Philippine submontane rainforest. Journal of Zoology (London), 218:565–586.

Heithaus, E. R., T. H. Fleming, and P. A. Opler. 1975. Foraging patterns and resource utilization in seven species of bats in a seasonal tropical forest. Ecology, 56:841–854.

Heithaus, E. R., P. A. Opler, and H. G. Baker. 1974. Bat activity and pollination of Bauhinia pauletia: plant-pollinator coevolution. Ecology, 55:412–419.

Heller, K.-G., and O. von Helversen. 1989. Resource partitioning of sonar frequency bands in rhinolophid bats. Oecologia, 80:178–186.

Heller, K.-G., and M. Volleth. 1995. Community structure and evolution of insectivorous bats in the Palaeotropics and Neotropics. Journal of Tropical Ecology, 11:429–442.

Hespenheide, H. A. 1973. Ecological inferences from morphological data. Annual Review of Ecology and Systematics, 3:213–229.

Hickey, M. B. C., L. Acharya, and S. Pennington. 1996. Resource partitioning by two species of vespertilionid bats (*Lasiurus cinereus* and *Lasiurus borealis*) feeding around street lights. Journal of Mammalogy, 77:325–334.

Holdridge, L. R. 1967. Life Zone Ecology. Tropical Science Center, San Jose, Costa Rica.

Howe, H. F., and J. Smallwood. 1982. Ecology of seed dispersal. Annual Review of Ecology and Systematics, 13:201–228.

Howell, D. J. 1979. Flock foraging in nectar-feeding bats: advantages to the bats and to the host plants. American Naturalist, 114:23–49.

Humphrey, S. R., F. J. Bonaccorso, and T. L. Zinn. 1983. Guild structure of surface-gleaning bats in Panama. Ecology, 64:284–294.

Husar, S. L. 1976. Behavioral character displacement: evidence of food partitioning in insectivorous bats. Journal of Mammalogy, 57:331–338.

Hutchinson, G. E. 1959. Homage to Santa Rosalia, or why are there so many kinds of animals? American Naturalist, 93:145–159.

Ingle, N. 1993. Vertical stratification of bats in a Philippine rainforest. Asia Life Sciences, 2:215–222.

Iniguez Davalos, L. I. 1993. Patrones ecoogicos en la comunidad de murcielagos de la Sierra de Manantlan. Pp. 355–370 *in:* Avances en el Estudio de los Mamiferos de Mexico (R. A. Medellin and G. Ceballos, eds.). Publicaciones Especiales, Asociación Mexicana de Mastozoologia, vol. 1. Asociación Mexicana de Mastozoologia, Mexico City.

Iudica, C. A., and F. J. Bonaccorso. 1997. Feeding of the bat, *Sturnira lilium,* on fruits of *Solanum riparium* influences dispersal of this pioneer tree in forests of northwestern Argentina. Studies on Neotropical Fauna and Environment, 32:4–6.

Jacobs, D. S. 1999. The diet of the insectivorous Hawaiian hoary bat (*Lasiurus cinereus semotus*) in an open and a cluttered habitat. Canadian Journal of Zoology, 77:1603–1608.

Janson, C. H., J. Terborgh, and L. H. Emmons. 1981. Non-flying mammals as pollinating agents in the Amazonian forest. Biotropica, 13:1–6.

Janzen, D. H. 1967. Why mountain passes are higher in the tropics. American Naturalist, 101.233–249.

Jones, G. 1996. Does echolocation constrain the evolution of body size in bats? Symposia of the Zoological Society of London, no. 69:111–128.

Jones, G. 1999. Scaling of echolocation call parameters in bats. Journal of Experimental Biology, 202:3359–3367.

Kalko, E. K. V. 1995. Echolocation signal design, foraging habitats and guild structure in six Neotropical sheath-tailed bats (Emballonuridae). Symposia of the Zoological Society of London, no. 67:259–273.

Kalko, E. K. V. 1998. Organisation and diversity of tropical bat communities through space and time. Zoology, 101:281–297.

Kalko, E. K. V., D. Friemel, C. O. Handley, Jr., and H.-U. Schnitzler. 1999. Roosting and foraging behavior of two Neotropical gleaning bats, *Tonatia silvicola* [*sic*] and *Trachops cirrhosus.* Biotropica, 31:344–353.

Kalko, E. K. V., C. O. Handley, Jr., and D. Handley. 1996. Organization, diversity, and long-term dynamics of a Neotropical bat community. Pp. 503–553 *in:* Long-Term

Studies of Vertebrate Communities (M. L. Cody and J. A. Smallwood, eds.). Academic Press, San Diego, Calif.

Kikkawa, J., and D. J. Anderson. 1986. Community Ecology: Pattern and Process. Blackwell Scientific Publications, Oxford.

Kingston, T., G. Jones, A. Zubaid, and T. H. Kunz. 2000. Resource partitioning in rhinolophid bats revisited. Oecologia, 124:332–342.

Kirsch, J. A. W., J. M. Hutcheon, D. G. P. Byrnes, and B. D. Lloyd. 1998. Affinities and historical zoogeography of the New Zealand short-tailed bat, *Mystacina tuberculata* Gray 1843, inferred from DNA-hybridization comparisons. Journal of Mammalian Evolution, 5:33–64.

Kitchener, D. J., A. Gunnell, and Maharadatunkamsi. 1990. Aspects of the feeding biology of fruit bats (Pteropodidae) on Lombok Island, Nusa Tenggara, Indonesia. Mammalia, 54:561–578.

Kunz, T. H. 1973. Resource utilization: temporal and spatial components of bat activity in central Iowa. Journal of Mammalogy, 54:14–32.

Kunz, T. H. 1982. Roosting ecology of bats. Pp. 1–55 *in:* Ecology of Bats (T. H. Kunz, ed.). Plenum Press, New York.

Kunz, T. H. 1996. Obligate and opportunistic interactions of Old-World tropical bats and plants. Pp. 37–65 *in:* Conservation and Faunal Biodiversity in Malaysia (Z. A. A. Hasan and Z. Akbar, eds.). Penerbit Universiti Kebangsaan Malaysia, Bangi.

Kunz, T. H., and C. A. Diaz. 1995. Folivory in fruit-eating bats, with new evidence from *Artibeus jamaicensis* (Chiroptera: Phyllostomidae). Biotropica, 27:106–120.

Kunz, T. H., and K. A. Ingalls. 1994. Folivory in bats: an adaptation derived from frugivory. Functional Ecology, 8:665–668.

LaVal, R. K., and H. S. Fitch. 1977. Structure, movements, and reproduction in three Costa Rican bat communities. Occasional Papers, Museum of Natural History, University of Kansas, 69:1–28.

LaVal, R. K., and M. L. LaVal. 1980. Prey selection by a Neotropical foliage-gleaning bat, *Micronycteris megalotis*. Journal of Mammalogy, 61:327–330.

Law, B. S., J. Anderson, and M. Chidel. 1999. Bat communities in a fragmented forest landscape on the south-west slopes of New South Wales, Australia. Biological Conservation, 88:333–345.

Levins, R. 1968. Evolution in Changing Environments. Monographs in Population Biology, no. 2. Princeton University Press, Princeton, N.J.

Lotka, A. J. 1932. The growth of mixed populations: two species competing for a common food supply. Journal of the Washington Academy of Sciences, 22:461–469.

Lowry, J. B. 1989. Green-leaf fractionation by fruit bats: is this feeding behavior a unique nutritional strategy for herbivores? Australian Wildlife Research, 16:203–206.

MacArthur, R. H., and R. Levins. 1967. The limiting similarity, convergence, and divergence of coexisting species. American Naturalist, 101:377–385.

Magurran, A. E. 1988. Ecological Diversity and its Measurement. Princeton University Press, Princeton, N.J.

Marinho-Filho, J. S. 1991. The coexistence of two frugivorous bat species and the phenology of their food plants in Brazil. Journal of Tropical Ecology, 7:59–67.

Marshall, A. G. 1983. Bats, flowers, and fruit: evolutionary relationships in the Old World. Biological Journal of the Linnean Society, 20:112–135.

Marshall, A. G. 1985. Old World phytophagous bats (Megachiroptera) and their food plants: a survey. Zoological Journal of the Linnean Society, 83:351–369.

McKenzie, N. L., and J. K. Rolfe. 1986. Structure of bat guilds in the Kimberley mangroves, Australia. Journal of Animal Ecology, 55:401–420.

McKenzie, N. L., and A. N. Start. 1989. Structure of bat guilds in mangroves: environmental disturbances and determinism. Pp. 167–178 in: Patterns in the Structure of Mammalian Communities (D. W. Morris, Z. Abramsky, B. J. Fox, and M. R. Willig, eds.). Special Publications, the Museum, no. 28. Texas Tech Press, Lubbock.

McNab, B. K. 1971. The structure of tropical bat faunas. Ecology, 52:352–358.

Medellin, R. A. 1993. Estructura y diversidad de una comunidad de murcielagos en el tropical humedo Mexicano. Pp. 333–354 in: Avances en el Estudio de los Mamiferos de Mexico (R. A. Medellin and G. Ceballos, eds.). Publicaciones Especiales, Asociación Mexicana de Mastozoologia, vol. 1. Asociación Mexicana de Mastozoologia, Mexico City.

Miller, D. 1962. Daytime attack on a bat by blackbirds. Journal of Mammalogy, 43:546.

Morrison, D. W. 1980. Foraging and day-roosting dynamics of canopy fruit bats in Panama. Journal of Mammalogy, 61:20–29.

Moulton, M. P., and S. L. Pimm. 1987. Morphological assortment in introduced Hawaiian passerines. Evolutionary Ecology, 1:41–62.

Moura de Souza Aguiar, L. 1994. Comunidades de Chiroptera em tres areas de Mata Atlantica em diferentes estadios de sucessao-Estacao Biologica de Caratinga, Minas Gerais. Ph.D. Dissertation. Universidade Federal de Minas Gerais, Minas Gerais, Brazil.

Muller, F. M., and N. R. dos Reis. 1992. Partição de recursos alimentares entre quatro espécies de morcegos frugívoros (Chiroptera, Phyllostomidae). Revista Brasiliera Zoologia, 9:345–355.

Muñoz-Saba, Y., A. Cadena, and J. O. Rangel-Ch. 1995. Gremios de murcielagos forrajeadores de nectar-pollen en un bosque de galeria de la Serrania La Macarena-Colombia. Caldasia, 17:459–461.

Muñoz-Saba, Y., A. Cadena, and J. O. Rangel-Ch. 1997. Ecología de los murcielagos antofilos del sector la Curia, Serrania La Macarena (Colombia). Revista de la Academia Colombiana de Ciencias Exactas, Fisicas y Naturales, 21:473–486.

Myers, P., and R. M. Wetzel. 1983. Systematics and zoogeography of the Chaco Boreal. Miscellaneous Publications, Museum of Zoology, University of Michigan, 165:1–59.

Navarro, D., and L. Leon-Paniagua. 1995. Community structure of bats along an altitudinal gradient in tropical eastern Mexico. Revista Mexicana de Mastozoología, 1:9–21.

Norberg, U. M., and M. B. Fenton. 1988. Carnivorous bats? Biological Journal of the Linnean Society, 33: 383–394.

Norberg, U., and J. M. V. Rayner. 1987. Ecological morphology and flight in bats (Mammalia: Chiroptera): wing adaptations, flight performance, foraging strategy and echolocation. Philosophical Transactions of the Royal Society of London B, 316: 335–427.

O'Farrell, M. J., and W. G. Bradley. 1970. Activity patterns of bats over a desert spring. Journal of Mammalogy, 51:18–26.

Palmer, A. R., and C. Strobeck. 1986. Fluctuating asymmetry: measurement, analysis, and patterns. Annual Review of Ecology and Systematics, 17:391–421.

Palmeirim, J. M., D. L. Gorchov, and S. Stoleson. 1989. Trophic structure of a neotropical frugivore community: is there competition between birds and bats? Oecologia, 79:403–411.

Patterson, B. D., V. Pacheco, and S. Solari. 1996. Distributions of bats along an elevational gradient in the Andes of south-eastern Peru. Journal of Zoology (London), 240:637–658.

Patterson, B. D., D. Stotz, S. Solari, J. Fitzpatrick, and V. Pacheco. 1998. Contrasting patterns of elevational zonation for birds and mammals in the Andes of southeastern Peru. Journal of Biogeography, 25:593–607.

Pedro, W. A., and V. A. Taddei. 1997. Taxonomic assemblage of bats from Panga Reserve, southeastern Brazil: abundance patterns and trophic relations in the Phyllostomidae (Chiroptera). Boletin do Museo Biologico Mello Leitao, 6:3–21.

Peracchi, A. L., and S. T. de Albuquerque. 1993. Quiropteros do Municipio de Linhares, Estado do Espírito Santo, Brasil (Mammalia, Chiroptera). Revista Brasileira de Biologia, 53:575–581.

Perrin, M. R., and J. J. Hughes. 1992. Preliminary observations on the comparative gastric morphology of selected Old World and New World bats. Zeitschrift für Säugetierkunde, 57:257–268.

Peters, R. H. 1983. The Ecological Implications of Body Size. Cambridge University Press, Cambridge.

Petit, S. 1997. The diet and reproductive schedules of Leptonycteris curasoae curasoae and Glossophaga longirostris elongata (Chiroptera: Glossophaginae) on Curaçao. Biotropica, 29:214–223.

Pickett, S. T. A., J. Kolasa, and C. G. Jones. 1994. Ecological understanding: the nature of theory and the theory of nature. Academic Press, San Diego, Calif.

Pielou, E. C. 1969. An Introduction to Mathematical Ecology. John Wiley & Sons, New York.

Polis, G. A. 1991. Desert communities: an overview of patterns and processes. Pp. 1–26 in: The Ecology of Desert Communities (G. A. Polis, ed.). University of Arizona Press, Tucson.

Rainey, W. E., E. D. Pierson, T. Elmqvist, and P. A. Cox. 1995. The role of flying foxes (Pteropodidae) in oceanic island ecosystems of the Pacific. Symposia of the Zoological Society of London, no. 67:47–62.

Rajamani, L., A. Aminah, A. Zubaid, K. H. Tan, and T. H. Kunz. 1999. Chemical composition of leaves consumed by the lesser dog-faced fruit bat, Cynopterus brachyotis, in peninsular Malaysia. Acta Chiropterologica, 1:209–214.

Richards, G. C. 1995. A review of ecological interactions of fruit bats in Australian ecosystems. Symposia of the Zoological Society of London, no. 68:70–96.

Ricklefs, R. E. 1979. Ecology. Chiron Press, Newton, Mass.

Ricklefs, R. E., and D. Schluter, eds. 1993. Species Diversity in Ecological Communities. University of Chicago Press, Chicago.

Ricklefs, R. E., and J. Travis. 1980. A morphological approach to the study of avian community organization. Auk, 97:321–338.

Rieger, J. F., and E. M. Jakob. 1988. The use of olfaction in food location by frugivorous bats. Biotropica, 20:161–164.

Root, R. B. 1967. The niche exploitation pattern of the blue-gray gnatcatcher. Ecological Monographs, 37:317–350.

Root, R. B. 1973. Organization of the plant-arthropod association in simple and diverse habitats: the fauna of collards (Brassica oeracea). Ecological Monographs, 43:95–124.

Ruby, J., P. T. Nathan, J. Balasingh, and T. H. Kunz. 2000. Chemical composition of fruits and leaves eaten by the short-nosed fruit bat, *Cynopterus sphinx*. Journal of Chemical Ecology, 26:2825–2841.

Ruedas, L. A., J. R. Dembowski, and R. V. Sison. 1994. Morphological and ecological variation in *Otopteropus cartilagonodus* Kock, 1969 (Mammalia: Chiroptera: Pteropodidae) from Luzon, Philippines. Proceedings of the Biological Society of Washington, 107:1–16.

Ryan, J. M., D. B. Clark, and J. A. Lackey. 1985. Response of *Artibeus lituratus* (Chiroptera: Phyllostomidae) to distress calls of conspecifics. Journal of Mammalogy, 66:179–181.

Rydell, J., and J. R. Speakman. 1995. Evolution of nocturnality in bats: potential competitors and predators during their early history. Biological Journal of the Linnean Society, 54:183–191.

Sanchez-Palomino, P., P. Rivas-Pava, and A. Cadena. 1993. Composición, abundancia y riqueza de especies de la comunidad de murcielagos en bosques de galeria en la Serrania de la Marcarena. Caldasia, 17:301–312.

Saunders, M. B., and R. M. R. Barclay. 1992. Ecomorphology of insectivorous bats: a test of predictions using two morphologically similar species. Ecology, 73:1335–1345.

Sazima, I., S. Vogel, and M. Sazima. 1989. Bat pollination of *Encholirium glaziovii*, a terrestrial bromeliad. Plant Systematics and Evolution, 168:167–179.

Schoener, T. W. 1974. Some methods for calculating competition coefficients from resource-utilization spectra. American Naturalist, 108:332–340.

Schum, M. 1984. Phenetic structure and species richness in North and Central American bat faunas. Ecology, 65:1315–1324.

Schutt, W. A., Jr. 1998. Chiropteran hindlimb morphology and the origin of blood feeding in bats. Pp. 157–168 *in:* Bat Biology and Conservation (T. H. Kunz and P. A. Racey, eds.). Smithsonian Institution Press, Washington, D.C.

Simmons, N. B. 1998. A reappraisal of interfamilial relationships of bats. Pp. 3–26 *in:* Bat Biology and Conservation (T. H. Kunz and P. A. Racey, eds.). Smithsonian Institution Press, Washington, D.C.

Simmons, N. B., and J. H. Geisler. 1998. Phylogenetic relationships of *Icaronycteris, Hassianycteris,* and *Palaeochiropteryx* to extant bat lineages, with comments on the evolution of echolocation and foraging strategies in Microchiroptera. Bulletin of the American Museum of Natural History, 235:1–182.

Simmons, N. B., and R. S. Voss. 1998. The Mammals of Paracou, French Guiana: a Neotropical lowland rainforest fauna. 1. Bats. Bulletin of the American Museum of Natural History, 237:1–219.

Soriano, P. J., A. D. de Pascual, J. Ochoa-G., and M. Aguilera. 1999. Biogeographic analysis of the mammal communities in the Venezuelan Andes. Interciencia, 24:17–25.

Soriano, P. J., M. Sosa, and O. Rossell. 1991. Hábitos alimentarios de *Glossophaga longirostris* Miller (Chiroptera: Phyllostomidae) en una zona árida de los Andes venezolanos. Revista de Biología Tropical, 39:263–268.

Speakman, J. R. 1995. Chiropteran nocturnality. Symposia of the Zoological Society of London, no. 67:187–201.

Start, A. N., and A. G. Marshall. 1976. Nectarivorous bats as pollinators of trees in western Malaysia. Pp. 141–150 *in:* Tropical Trees: Variation, Breeding and Conservation (J. Burley and B. T. Styles, eds.). Linnean Society Symposium Series 2. Linnean Society of London and Academic Press, London.

Stevens, R. D., and M. R. Willig. 1999. Size assortment in New World bat communities. Journal of Mammalogy, 80:644–658.

Stevens, R. D., and M. R. Willig. 2000a. Community structure, abundance, and morphology. Oikos, 88:48–56.

Stevens, R. D., and M. R. Willig. 2000b. Density compensation in bat communities. Oikos, 89:367–377.

Stevens, R. D., and M. R. Willig. 2002. Geographical ecology at the community level: perspectives on the diversity of New World bats. Ecology, 83:545–560.

Strong, D. R., D. Simberloff, L. G. Abele, and A. B. Thistle, eds. 1984. Ecological Communities: Conceptual Issues and the Evidence. Princeton University Press, Princeton, N.J.

Supernant, H. R. 1977. Nocturnal activity patterns in a bat fauna of southern California with comments on the physiological ecology of *Pipistrellus hesperus*. M.S. Thesis, California State University, Fullerton.

Tamsitt, J. R. 1967. Niche and species diversity in Neotropical bats. Nature, 213: 784–786.

Tan, K. H., A. Zubaid, and T. H. Kunz. 1998. Food habits of *Cynopterus brachyotis* (Muller) (Chiroptera: Pteropodidae) in Peninsular Malaysia. Journal of Tropical Ecology, 14:299–307.

Tedman, R. A., and L. S. Hall. 1985. The morphology of the gastrointestinal tract and food transit time in the fruit bats *Pteropus alecto* and *P. poliocephalus* (Megachiroptera). Australian Journal of Zoology, 33:625–640.

Terborgh, J. W. 1986. Community aspects of frugivory in tropical forests. Pp. 371–384 *in:* Frugivores and Seed Dispersal (A. Estrada and T. H. Fleming, eds.). Kuk, Dordrecht,.

Thomas, M. E. 1972. Preliminary study of the annual breeding patterns and population fluctuations of bats in three ecologically distinct habitats in southwestern Colombia. Ph.D. Dissertation. Tulane University, New Orleans.

Uieda, W. 1993. Comportamento alimentar do morcego hematófago *Diaemus youngi*, em aves domésticas. Revista Brasileira de Biologia, 53:529–538.

Utzurrum, R. C. B. 1995. Feeding ecology of Philippine fruit bats: patterns of resource use and seed dispersal. Symposia of the Zoological Society of London, no. 67:63–78.

Utzurrum, R. C. B. 1998. Geographical patterns, ecological gradients, and the maintenance of tropical fruit bat diversity: the Philippine model. Pp. 342–353 *in:* Bat Biology and Conservation (T. H. Kunz and P. A. Racey, eds.). Smithsonian Institution Press, Washington, D.C.

Valiente-Banuet, A., M. del Coro Arizmendi, A. Rojas-Martinez, and L. Dominguez-Canseco. 1996. Ecological relationships between columnar cacti and nectar-feeding bats in Mexico. Journal of Tropical Ecology, 12:103–119.

van Schaik, C. P., J. W. Terborgh, and S. J. Wright. 1993. The phenology of tropical forests: adaptive significance and consequences for primary consumers. Annual Review of Ecology and Systematics, 24:353–377.

Van Valen, L. 1973. Body size and numbers of plants and animals. Evolution, 27:27–35.

Vaughan, N., and J. E. Hill. 1996. Bat (Chiroptera) diversity and abundance in banana plantations and rain forest, and three new records for St. Vincent, Lesser Antilles. Mammalia, 50:441–447.

Volterra, V. 1931. Variations and fluctuations of the numbers of individuals in animal species living together. Pp. 409–448 *in:* Animal Ecology (R.N. Chapman, ed.). McGraw Hill, New York.

Whitaker, R. J., and S. J. Jones. 1994. The role of frugivorous bats and birds in the rebuilding of a tropical rainforest ecosystem, Krakatau, Indonesia. Journal of Biogeography, 21:245–258.

Willig, M. R. 1982. A comparative ecological study of Caatingas and Cerrado chiropteran communities: composition, structure, morphometrics, and reproduction. Ph.D. Dissertation. University of Pittsburgh, Pittsburgh.

Willig, M. R. 1986. Bat community structure in South America: a tenacious chimera. Revista Chilena de Historia Natural, 59:151–168.

Willig, M. R. and M. A. Moulton. 1989. The role of stochastic and deterministic processes in the structure of Neotropical bat communities. Journal of Mammalogy, 70:323—329.

Willig, M. R., and E. A. Sandlin. 1991. Gradients of species density and turnover in New World bats: a comparison of quadrate and band methodologies. Pp. 81–96 *in:* Latin American Mammalogy: History, Biodiversity, and Conservation (M. A. Mares and D. J. Schmidly, eds.). University of Oklahoma Press, Norman.

Willig, M. R., and K. W. Selcer. 1989. Bat species density gradients in the New World: a statistical assessment. Journal of Biogeography, 16:189–195.

Willig, M. R., and L. R. Walker. 1999. Disturbance in terrestrial ecosystems: salient themes, synthesis, and future directions. Pp. 747–767 *in:* Ecosystems of Disturbed Ground (L. R. Walker, ed.). Elsevier Press, Amsterdam.

Wilson, D. E. 1973. Bat faunas: a trophic comparison. Systematic Zoology, 22: 14–29.

Wilson, D. E., C. F. Ascorra, and S. Solari. 1996. Bats as indicators of habitat disturbance. Pp. 613–626 *in:* Manu: The Biodiversity of Southeastern Peru (D. E. Wilson and A. Sandoval, eds.). Smithsonian Institution Press, Washington, D.C.

Wohlgenant, T. J. 1994. Roost interactions between the common vampire bat (*Desmodus rotundus*) and two frugivorous bats (*Phyllostomus discolor* and *Sturnira lilium*) in Guanacaste, Costa Rica. Biotropica, 26:344–348.

Yancey, F. D. 1996. The mammals of Big Bend Ranch State Park. Ph.D. Dissertation. Texas Tech University, Lubbock.

Zubaid, A. 1993. A comparison of the bat fauna between a primary and fragmented secondary forest in peninsular Malaysia. Mammalia, 57:202–206.

Zubaid, A. 1994. Vertical stratification of pteropodid bats in a Malaysia lowland rainforest. Mammalia, 58:309–311.

CHAPTER 13

Patterns of Range Size, Richness, and Body Size in the Chiroptera

Michael R. Willig, Bruce D. Patterson, and Richard D. Stevens

Introduction

Macroecology seeks to quantify broad-scale patterns of species richness, species range size, body size, and abundance and to understand them from ecological and historical perspectives (Brown 1995). It primarily examines patterns that occur at regional to global spatial scales—patterns that have been produced as a consequence of mechanisms operating over decades to millennia. Consequently, macroecology is concerned with the statistical distribution of variables and is nonexperimental by nature; nevertheless, it is subject to rigorous assessment by comparative studies.

Studies of bats provide important insights into macroecological patterns and the mechanisms that give rise to them. This derives from the high taxonomic and ecological diversity of the order Chiroptera (Simmons 2003), its broad distribution on all continents except Antarctica (Koopman 1984), and the tremendous spatial variation that exists in the abundances of species and their assembly into local communities (Findley 1993; Stevens and Willig 1999, 2000, 2002). Moreover, as the only volant mammals, bats are excellent candidates for comparative macroecological studies with nonvolant mammals, such as rodents, or with other volant homeotherms (i.e., birds).

Bats constitute approximately one-fifth of the living species of mammals recognized by Wilson and Reeder (2003). Only the Rodentia is more species-rich, with more than 40% of all mammalian species. Although the Chiroptera now includes 18 families and 202 genera, almost 80% of the species are exclusively tropical, with the family Vespertilionidae making the largest contribution to temperate-zone faunas. Paralleling the taxonomic richness of the order, ecological diversity is equally impressive, with bats occupying frugivorous, nectarivorous, animalivorous, insectivorous, piscivorous, and sanguinivorous ensembles (guilds).

Studies of Chiroptera are only beginning to contribute to macroecological understanding. In part, this may be a consequence of the difficulty of quantifying the abundances of local populations of small, vagile, nocturnal species, especially those in highly diverse and vertically stratified tropical ecosystems.

Moreover, the spatial distribution of species at the continental scale is only available for the New World (see references in Willig and Lyons [1998] or Willig and Gannon [1997]) and Australia (Strahan 1995) in an easily accessible format with reasonably accurate detail. Nonetheless, the foundational work of Corbet and Hill (1992) and Corbet (1978) for Eurasia, Kingdon (1997) for Africa, and Strahan (1983, 1995) for Australia represent productive points of departure for macroecological analyses of Old World mammals. Thus, by necessity, most of the discussion that follows emphasizes patterns in North and South America. Analyses from Africa, Australia, and the Philippines are included for illustrative purposes.

Macroecological conclusions based on bats in the New World may be widely applicable to bats in general. Nonetheless, the considerable differences in the geography of landmasses or the systematics and evolutionary history of bats in Old and New Worlds offer significant challenges. For example, the shapes, sizes, and connectivity of Old World continents are quite distinct from those in the New World. The New World comprises two large interconnected continents, a conterminous tropic region, and few large associated islands. The Old World comprises three disconnected continents (Africa, Eurasia, and Australia), three distinct tropical areas, and a multitude of associated large islands. Similarly, all New World bats are members of the Microchiroptera, whereas Old World bats include members in both the Micro- and Macrochiroptera. The world is not a simple place. Profitable research agendas of the future would assemble data for all continents, quantify patterns, and compare that information with data for other taxa, resulting in a more synthetic and comprehensive understanding that distinguishes broad-scale biogeographic and macroecological patterns from patterns that are regional or idiosyncratic in nature.

Patterns of Species Range Size

The emergence of macroecology (Brown 1995) as a dynamic research focus in contemporary ecology has provided incentive to document patterns in the distribution of species range sizes, determine the mechanisms that effect such patterns, and understand their consequences to local species assemblages. Within a particular taxonomic group, species range sizes can be quite variable. However, complications associated with the absence of a universal metric of range size (e.g., latitudinal extent vs. geographic range area), biases associated with underestimating the range sizes of species with low local densities, the dynamic nature of species ranges over time, and uneven sampling intensities for species in a taxon reduce the number of comparative studies that are possible. Nonetheless, a growing body of evidence suggests that species range-size distributions are right-skewed or "hollow curves," which appear to be normally distributed when frequency is evaluated for log-transformed range sizes (Gaston 1994; Pagel et al. 1991; Rapoport 1994). Lognormal distributions could be the consequence of the interaction of a multitude of mechanisms acting independently and in concert or could be a consequence of a few processes related to the evolutionary development and age of a higher taxon (Gaston 1994; Lawton 1993), the manner in which the likelihood of speciation or extinction is associated with range size (Chown 1997; Tokeshi 1993), or any of a broad suite of ecological processes (Gaston and Kunin 1997; Kunin and Gaston 1993). For example, some of the variation in range sizes for a higher taxon may be related to latitude and its correlates. The propensity for range sizes of species to be smaller in the tropics than toward the poles has been designated as "Rapoport's rule" and has been used to account for the latitudinal gradient in species density (Brown 1995; Stevens 1989). However, a growing body of empirical and theoretical research suggests that the empirical relationship is equivocal (Gaston 1999; Gaston et al. 1998; Rohde et al. 1993; Roy et al. 1994) and may be spurious (Colwell and Hurtt 1994; Colwell and Lees 2000; Lyons and Willig 1997).

Variation in Range Size

Lyons and Willig (1997) determined the latitudinal extent (in degrees) of the distributional ranges of 255 species of bats (fig. 13.1) in the continental New World. Unfortunately, similar distributional ranges have not been compiled for bats on other continents. Consequently, comparisons of Old World versus New World faunas, or Megachiroptera versus Microchiroptera, await resolution of taxonomic and distributional status.

In the New World, species differ in latitudinal range size by over two orders of magnitude (fig. 13.1), from 1° of latitude for *Myotis cobanensis* (known from a single collection locality) and *Rhogeessa gracilis*, to over 100° of latitude for *Lasiurus cinereus* (mean = 28.9; SD = 19.1; SEM = 1.2). Similarly, central tendencies and dispersions of latitudinal range sizes differ among New World bat families (table 13.1), with the species of Noctilionidae and Natalidae having the largest and smallest mean latitudinal range sizes, respectively, and the species of Molossidae and Noctilionidae having the most and least variable latitudinal range sizes, respectively (fig. 13.2*A–E*).

Like the situation for most terrestrial taxa, the distribution of latitudinal range sizes for all New World bats differs from a lognormal distribution and is left-skewed, with a greater-than-expected number of small-sized ranges (skewness, $g_1 = -0.9221$, $P < 0.001$; kurtosis, $g_2 = -0.480$, $0.2 > P > 0.1$). Moreover, the latitudinal ranges of three of the four chiropteran families with more than 15 species in the New World exhibited significant deviations from a lognormal distribution and were left skewed (Phyllostomidae, $g_1 = -0.855$, $0.001 < P < 0.002$ and $g_2 = -0.786$, $P > 0.2$; Vespertilionidae, $g_1 = -0.868$, $0.005 < P < 0.01$ and $g_2 = 0.218$, $P > 0.2$; Molossidae, $g_1 = -0.953$, $0.05 < P$

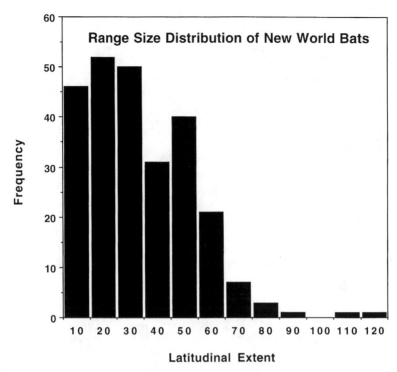

Figure 13.1. Frequency distribution of latitudinal extents (in degrees) for the distributional ranges of 255 species of New World bats.

Table 13.1. Descriptive statistics for latitudinal extent (in degrees) of range sizes for continental New World bat species by family

Family	Number of species	Mean	SD	SE
Emballonuridae	18	26.5	15.4	3.6
Furipteridae	2	31.0	14.1	10.0
Molossidae	29	37.7	23.5	4.4
Mormoopidae	5	40.6	5.9	2.7
Natalidae	2	17.5	9.2	6.5
Noctilionidae	2	52.0	8.5	6.0
Phyllostomidae	128	27.0	16.1	1.4
Thyropteridae	2	37.0	11.3	8.0
Vespertilionidae	67	28.0	23.0	2.8

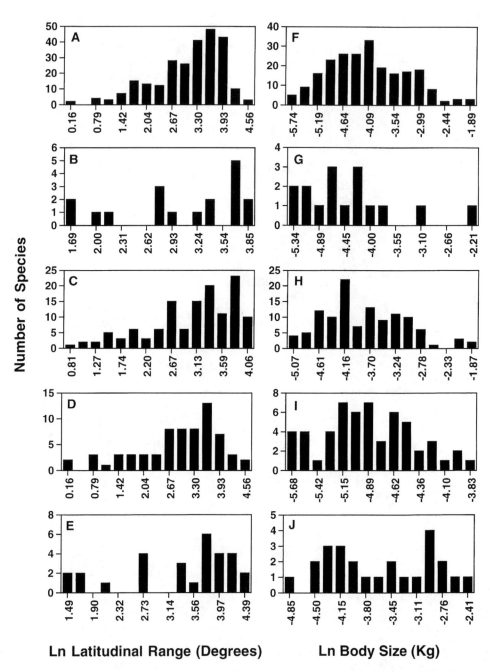

Number of Species

Ln Latitudinal Range (Degrees)

Ln Body Size (Kg)

Figure 13.2. Comparison of frequency distributions of the natural logarithm (log) of range size in degrees of latitude (*left col.*) and natural log of body size in kilograms (*right col.*) for all *bats* (A, F), emballonurids (B, G), phyllostomids (C, H), vespertilionids (D, I), and molossids (E, J) from the New World.

< 0.02 and $g_2 = -0.301$, $P > 0.20$). In contrast, the distribution of latitudinal range sizes for the Emballonuridae did not differ from a log normal ($g_1 = -0.732$, $0.2 > P > 0.1$; $g_2 = -0.786$, $0.2 > P > 0.1$), but the fewer species in the family results in lower statistical power to detect deviations if they are real (note that g_2 is negative in this situation also). Thus, New World bats in general, as well as constituent larger families, tend to be more constrained with respect to their maximum range size than with respect to their minimum range size. Genetic differentiation by distance or geographic barriers to gene flow may enhance the likelihood of speciation by broad-ranging taxa, resulting in less variation than expected based on a lognormal distribution.

Much of the variation in the latitudinal extent of species ranges may be a consequence of differences among species in body mass. Indeed, several researchers have postulated or observed that larger-bodied species have larger geographic ranges than do small-bodied species (Arita et al. 1990; Brown and Maurer 1987, 1989; Lawton et al. 1994; Rapoport 1975, 1982; Taylor and Gotelli 1994). Two general mechanisms have been hypothesized to account for such patterns. The first, espoused by Brown and Maurer (1987), is that the upper limits to range size are determined by the available continental land area, whereas the lower limit is defined by the minimum viable population size of the taxon. Because home-range size scales positively and population density scales negatively with body mass, the minimum range size of species would tend to increase with increasing body size. Because of small total population sizes, large-bodied species with small geographic ranges would have high probabilities of extinction, thus predisposing a positive correlation between body mass and range size.

A second mechanism, suggested by Gaston (1990), regards small-bodied species as more sensitive to density-independent perturbations, diminishing the likelihood of their persistence over broad geographic areas and, thereby, affecting a positive association between body mass and range size. Conversely, Gaston (1988) and Gaston and Lawton (1988a, 1988b) suggested a mechanism whereby small body size would lead to larger geographic distributions. This mechanism is based on the observation that the intrinsic rate of population increase decreases with increasing body size. Thus, small-bodied species that colonize a site should be able to increase more rapidly to population densities that are sufficiently large to avoid stochastic extinction. At the same time, the smaller equilibrium densities of large species would make them more susceptible to stochastic extinction, regardless of their rates of increase. In general, the salient differences between the two sets of ideas are predicated on whether species persist because of the size of their total populations throughout their distribution (positive correlations between body mass and range size) or persist as a consequence of local abundance and demographic attributes (negative correlations between body mass and range size).

Despite theory calling for positive or negative relationships on a log-log scale (fig. 13.3), the latitudinal extent of a species range is not related significantly to body mass for all New World bats ($r = 0.101, P > 0.05$). Moreover, the lack of a significant relationship is a not consequence of the different families evincing strong but opposing patterns. Indeed, the relationship between log of latitudinal extent and body mass for the Emballonuridae (fig. 13.3B), Phyllostomidae (fig. 13.3C), and Molossidae (fig. 13.3E) are all non-significant. Only the Vespertilionidae (fig. 13.3D) shows a significant and positive correlation ($r = 0.331, 0.01 > P > 0.001$). The same statistical inferences follow if all analyses are conducted in arithmetic space. A number of conclusions are possible based on these results. For bats in general, and for the Emballonuridae, Phyllostomidae, and Molossidae in particular, neither mechanism may be in operation, or the mechanisms may counteract each other, resulting in the absence of a clear statistical pattern. Because critical life-history traits of bats (e.g., litter size or number of litters per year) do not vary with body size in the same manner as in other taxa (e.g., Barclay and Herder, this volume) or are unknown (e.g., association between body size and either home-range size or population density), the demographic linkages required to produce an association between body size and range size may be inoperative or weakened in the Chiroptera in general. Alternatively, latitudinal extent may not be an appropriate metric for assessing range size, and analyses with actual range areas may yield different insights.

In a broad analysis of 150 species of New World bats, Arita (1993) documented that the distribution of species was not mutually independent among categories (three-dimensional contingency test, $X^2 = 20.78, P < 0.05$) defined by species range size (above and below the median range area), abundance (above or below the median abundance), and taxonomy (Emballonuridae, Phyllostomidae and associated taxa, Vespertilionidae and associated taxa, and Molossidae). Decomposition of the three-way interaction into a component assessing the partial independence of taxonomy from distribution and abundance was rejected as well ($X^2 = 17.03, P < 0.05$), suggesting that there are taxonomic differences in the allocation of species to conditions of "rarity" as defined by a classification based on abundance and distribution (i.e., locally abundant but restricted distribution, locally abundant and widespread distribution, locally scarce and restricted distribution, locally scarce but widespread distribution). Species in the Phyllostomidae generally are equally common in all categories of rarity, except that they are infrequent in the locally scarce and restricted category. In contrast, each of the other three groups disproportionately occurred in two categories of rarity: locally rare but widespread distributions or locally rare and restricted distributions. Using an index of rarity based on both local abundance and distributional area, Arita (1993) identified a list of rare species that would be of special conservation

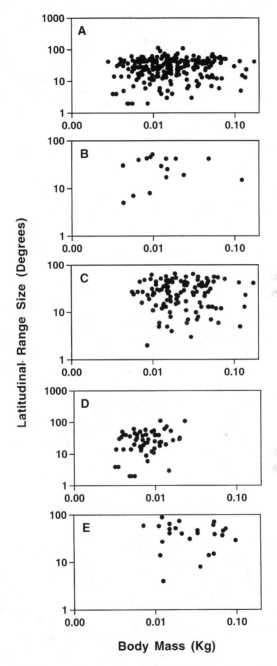

Figure 13.3. Correlation between log-transformed latitudinal extent (degrees) and body size (kilograms) for all bats (A; r = 0.101, n = 255, P > 0.05), Emballonuridae (B; r = 0.169, n = 16, P > 0.05), Phyllostomidae (C; r = −0.004, n = 115, P > 0.05), Vespertilionidae (D; r = 0.311, n = 56, P < 0.01), and Molossidae (E; r = −0.022, n = 25, P > 0.05) in the New World.

Table 13.2. List of Neotropical bats species of special conservation concern based on considerations of patterns of local abundance and geographic range area (after Arita, 1993)

Family	Species
Emballonuridae	*Balantiopteryx plicata*
	Diclidurus ingens
	Peropteryx kappleri
Mormoopidae	*Mormoops megalophylla*
Phyllostomidae	*Glyphonycteris daviesi*
	Lampronycteris brachyotis
	Lonchorhina orinocensis
	Micronycteris schmidtorum
	Phyllostomus latifolius
	Tonatia carrikeri
	T. evotis
	T. schulzi
	Anoura cultrata
	Choeroniscus minor
	Lichonycteris obscura
	Scleronycteris ega
	Centurio senex
	Sturnira bidens
	S. bogotensis
	S. erythromos
	Vampyressa brocki
	V. nymphaea
	Platyrrhinus brachycephalus
	P. infuscus
Natalidae	*Natalus tumidirostris*
Vespertilionidae	*Bauerus dubiaquercus*
	Eptesicus diminutus
	E. fuscus
	Lasiurus intermedius
	Myotis nesopolus
	M. oxyotus
Molossidae	*Eumops dabbenei*
	E. hansae
	Molossops greenhalli
	Promops centralis
	P. nasutus

Note. Taxonomic designations are updated based on Simmons (1996) and Wetterer et al. (2000).

concern, targeting them as conservation priorities and as foci for studies of autecology and natural history (table 13.2).

In a study restricted to species whose distributions were wholly North American, Pagel et al. (1991) documented a hollow curve for bats based on an arithmetic plot of the geographic areas of their distributions. Almost 70% of the 91 North American species occupied less than 7% of the continental area, and less than 5% of the species had ranges that included more than half of the continental area (median, 1.9%; average, 6.5%). In a more regional analysis, Arita et al. (1997) documented a right-skewed distribution of range areas for bats as well as for nonvolant mammals based on arithmetic plots. Indeed, Anderson (1977) noted a hollow curve for all mammalian orders, including bats, and subsequently developed a quantitative model to account for variation in range size distributions that proved equally applicable to birds, reptiles, amphibians, and fish (Anderson 1985). Nonetheless, caution must be exercised in considering and interpreting analyses that only consider a portion of a particular landmass or a portion of the geographic ranges of species. Moreover, exclusion of bats with ranges that extend into South America likely biased detection of pattern in an appreciable manner, especially given the increase in species richness toward tropical areas.

Smith et al. (1994) explored macroecological patterns of mammalian taxa prior to the arrival of Europeans in mainland Australia based on distributional data in Strahan (1983). For 37 species of bats whose distributions are limited to Australia or Tasmania (thereby excluding 20 species with broader distributions in Asia and the Pacific), they examined range size characteristics within mainland Australia. Compared to most other mammalian orders (Dasyuromorphia, Diprotodontia, and Rodentia but not Peramelemorphia), bats generally have quite large ranges, almost twice the size of those in other orders (table 13.3; see also Rapoport 1982, table 2.1). In addition, they have the least right-skewed and most platykurtic distribution of range sizes of all Australian orders. Only 13 of 214 mammal species (6%) have range distributions that are large enough to make up at least half of the mainland; eight of them (62%) are bats. In contrast, the great majority of Australian mammals (87%) occupy less than a third of the continent, and only 14% of these are bats. The

Table 13.3. Comparison of the range-size characteristics of bats to that of the other main orders of Australian mammals whose ranges are restricted to Australia or Tasmania (after Smith et al. 1994)

Order	Number of Species	Mean	Median	Number of Ranges	
				Large	Small
Chiroptera	37	24.9	10.6	8	26
Dasyuromorphia	39	13.2	6.7	1	35
Peramelemorphia	10	20.0	14.8	1	8
Diprotodontia	74	10.5	4.7	3	67
Rodentia	52	8.4	3.4	0	48
All species*	214	13.4	6.2	13	186

Note. Descriptive statistics are for percent of area of mainland Australia. Large and small ranges are arbitrarily defined as those occupying at least half or no more than 30% of the continental area, respectively.

*Statistics for all species include data on platypus and marsupial mole, which are not included in table otherwise.

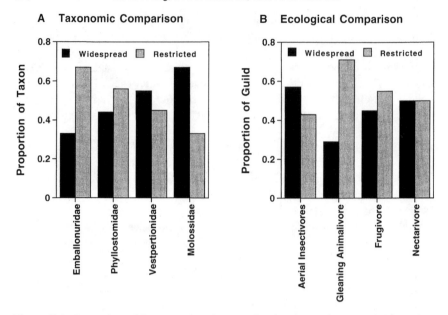

Figure 13.4. Comparison of the proportion of restricted and widespread species of Mexican bats within taxonomic (*A*) or ecological (*B*) groups (from Arita et al. 1997). See text for details concerning criteria for restricted and widespread classifications. Group richness on which the proportion of species is based is nine for Emballonuridae, 62 for Phyllostomidae (including species in the Noctilionidae and Mormoopidae; hence the Noctilionoidea), 47 for Vespertilionidae (including Natalidae and Thyropteridae), 18 for Molossidae, 76 for aerial insectivores, 21 for gleaning animalivores, 22 for frugivores, and 12 for nectarivores (the two noctilionids and three desmodontines were not included in analyses because of low species richness).

observation that Australian mammals on average occupy a larger proportion of their continental landmass than is the case for North American and Palearctic mammals may be a consequence of the reduced topographic heterogeneity there (Smith et al. 1994) but may reflect the disproportionate contribution of bats to the right-hand tail of the range-size distribution in Australia, compared to the situations in Nearctic or Palearctic regions.

The extent to which variation in the area of species ranges is associated with taxonomic affiliation or ecological function has been assessed for Mexican bats by Arita et al. (1997). They classified bats into two broad categories: widespread and restricted. Widespread bats are those species with range areas that are greater than the median range of all bats in Mexico, whereas restricted species are those ranging over areas smaller than the median (Arita 1993). No differences in the proportion of widespread versus restricted species existed among four taxa (fig. 13.4*A*): Emballonuridae, Phyllostomidae (plus Mormoopidae and Noctilionidae), Vespertilionidae (plus Natalidae and Thyropteridae), and Molossidae ($X^2 = 4.56$, df = 3, $P > 0.05$). In contrast, highly significant differences ($X^2 = 14.39$, df = 3, $P < 0.01$) occurred among feeding categories (i.e, aerial insectivores, gleaning animalivores, frugivores, and nectarivores; fig. 13.4B). Thus, the ecological characteristics of bats are associated more intimately with the size of their geographic distributions than are the phylogenetic affinities of the species, per se.

In a comparative study of South American mammals, Ruggiero (1994) documented considerable variability in the area of the geographic ranges of bats. Along with hystricognath rodents and carnivores, the frequency distribution of range size for bats differed significantly from a lognormal distribution. More specifically, widespread taxa were more common in bats than predicted by a lognormal distribution. Indeed, a quarter of the species had ranges less than or equal to 900,000 km^2, half of the species had ranges between 900,000 and 9,200,00 km^2, and a quarter of the species had ranges greater than 9,200,000 km^2. By convention (Rapoport 1975, 1982), these groups of species, as classified by their quartile representation in the Chiroptera with regard to range size, are microareal (i.e., lower quartile), mesoareal (middle two quartiles), and macroareal (upper quartile) species. Microareal species are not randomly distributed throughout South America (Ruggiero 1994). Rather, they are clustered at the coasts of the continent and commonly are associated with the tropical Andes. This has important ramifications for conservation because areas of highest richness (e.g., Amazonian lowlands) often harbor species with the highest vulnerability to extinction (i.e., microareal species).

Latitudinal Gradients

Early work concerning the propensity for the geographic ranges of species to become progressively smaller toward the equator resulted in the formulation of Rapoport's rule (Stevens 1989). Subsequent work identified significant complications for assessing the veracity of the rule. For example, detection of the pattern may be predicated on the measure of central tendency used for its quantification (Roy et al. 1994). Moreover, violations of spatial independence and phylogenetic independence among latitudinal measures compromise statistical assessments (Lecher and Harvey 1994; Rohde et al. 1993; Taylor and Gotelli 1994). Finally, geometric constraints may predispose the distribution of species ranges to evince a pattern coincident with the predictions of Rapoport's rule (Colwell and Hurtt 1994; Colwell and Lees 2000; Lyons and Willig 1997).

For bats, Ruggiero (1994) was the first to document the empirical pattern of decreasing mean latitudinal range size toward the tropics using Stevens's method of plotting the mean latitudinal range of all species that occur within a band versus the latitude of the band (Stevens 1989). In Ruggierio's analysis, species whose ranges extended into Central America were truncated and estimated by only their South American areas, which would underestimate the distribution of tropical species. Ruggierio also documented the increase in variability in latitudinal extent away from the tropics. A similar empirical pattern was documented by Lyons and Willig (1997) using Stevens's method for

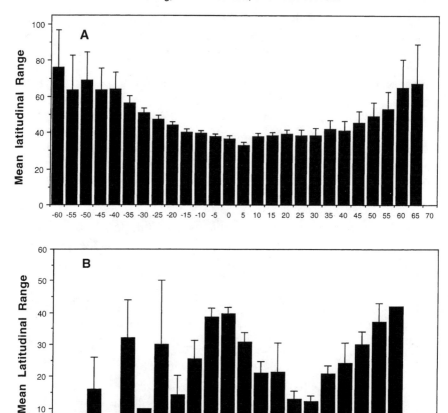

Figure 13.5. Latitudinal gradients of mean range size of New World bats based on (*A*) Stevens's method and (*B*) the midpoint method (see text for descriptions of methods) from Lyons and Willig (1997). Negative values for latitude represent the southern hemisphere, whereas positive values represent the northern hemisphere. *Vertical bars* represent 1 SEM; bars without standard errors represent latitudinal bands occupied by single species.

the entire New World (fig. 13.5*A*). Nonetheless, the spatial nonindependence of species distributions tends to smooth the pattern and may affect statistical significance, even if it cannot create a pattern per se (Pagel et al. 1991). The midpoint method of Rohde et al. (1993) circumvents this problem by calculating the midlatitudinal point in the distribution of each species in a fauna, and evaluating the degree to which the extent of latitudinal ranges are a function of their midlatitudes. An evaluation of Rapoport's rule based on the midpoint method for New World bats (Lyons and Willig 1997) confirmed that range size decreases with decreasing latitude in North America to about 20° N latitude; however, the pattern in South America did not conform to the predictions of Rapoport's rule (fig. 13.5B). Thus, both Stevens's method and the midpoint method failed to consider constraints imposed by the geometry of the continental landmasses on which a fauna occurs (Colwell and Hurtt 1994).

Based on a series of simulation analyses, it appears that the variation in range sizes of New World bats differs from that produced by stochastic processes, and the rate at which range size decreases with increasing latitude was not as precipitous as predicted by a null model (Lyons and Willig 1997). Because of geometric constraints, tropical species are predisposed to have large ranges and extratropical species are predisposed to have small ranges (fig. 13.6), and they do (i.e., the correlation between latitudinal range size and midlatitude is −0.233). However, the spirit of Rapoport's rule is met because the correlation is much less negative than produced by chance (fig. 13.7). Tropical species have smaller ranges and extratropical species have larger ranges than those produced by stochastic processes alone.

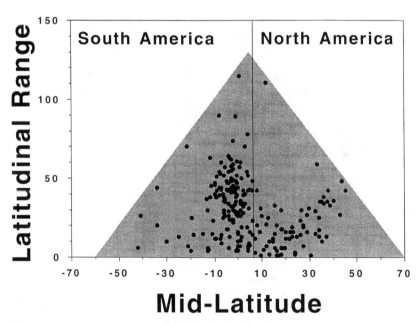

Figure 13.6. Graphical representation of the relationship between latitudinal range size and midlatitude for New World bats (Lyons and Willig 1997). The *vertical line* represents the Isthmus of Panama. Negative values for latitude represent the southern hemisphere, whereas positive values represent the northern hemisphere. The *triangular shaded area* represents the values of range size that are possible given geometric constraints, whereas the *dots* represent the actual latitudinal ranges of bat species.

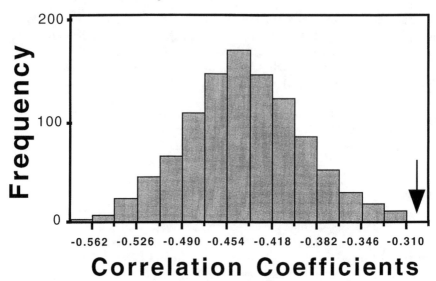

Correlation Coefficients

Figure 13.7. The significance of the correlation between latitudinal range size and midlatitude for New World bats ($n = 244$) was evaluated using a simulation approach because of biases related to geometric constraints on range size (Lyons and Willig 1997). In the simulation model, the northern and southern termini of 244 ranges were produced at random, and the correlation between latitudinal range size and latitudinal midpoint were calculated. These random correlation coefficients were generated 1,000 times, producing a probability density function (histogram). The actual correlation coefficient (*arrow*) was among the rarest of outcomes produced by chance ($r = -0.233, P < 0.001$), suggesting that although the empirical pattern is for range size to increase with decreasing latitude, the ranges of tropical species are smaller and the ranges of extratropical species are larger than geography of the New World would impose because of geometric constraints.

Elevational Gradients

Prior to the 1990s, elevational gradients in species range size received little attention in the literature; the focus was on the description of pattern rather than on an assessment of causal mechanisms. The extension of Rapoport's rule to elevational gradients by Stevens (1992) provided renewed interest in such patterns by suggesting alternative mechanisms to those historically thought to produce the elevational gradient in richness (see MacArthur 1972) and by linking latitudinal and elevational mechanisms thought to effect patterns of richness and range size. A number of climatic characteristics are known to affect range limits of species; these include humidity, precipitation, evapotranspiration, and temperature (Stevens and Fox 1991). Because the breadth of climatic conditions experienced by mountain residents in an individual's lifetime increases with increasing elevation (Adams et al. 1920; Merriam 1894), high-elevation species must be adapted to a broader range of climatic conditions than are their lowland counterparts, and this, in turn, facilitates broad range size in high-elevation species. Indeed, Stevens's (1992) claim of parallel gradients of species richness (decreasing with increasing elevation) and species range size (increasing with increasing elevation) for a wide range of montane organisms (i.e., trees, mammals, birds, reptiles, insects, and amphibians) catalyzed research on elevational gradients in range size from a conceptual perspective. Nonetheless, considerable controversy surrounds these issues (Rahbek 1997).

Elevational gradients in range size for bats are well documented at two sites: Manu National Park and Biosphere Reserve in the Andes of southeastern Peru (Patterson et al. 1996, 1998) and the Sierra de Manantlan Biosphere Reserve in the Sierra Madre del Sur of west-central Mexico (Iniguez Davalos 1993). Both sites span broad elevational profiles of more than 2,000 m and include a number of botanical life zones; nonetheless, the way in which range size varies with elevation differs between sites.

The elevational gradient in Manu traversed approximately 3,000 m and included five major life zones: lowland rainforest, montane rainforest, cloud forest, elfin forest, and puna. Species richness declined monotonically with increasing elevation (fig. 13.8A). Moreover, the elevational ranges of the 129 bat species showed a striking pattern (fig. 13.8B) of increasing elevational amplitude with increasing elevation, with no indication of a mid-elevational peak in range amplitude. Indeed, after classifying species into pentiles based on elevational midpoints (fig. 13.8C), pentile rank accounted for significant variation in mean elevational amplitude ($P < 0.003$, adjusted $R^2 = 0.95$). Most species (111 of 129) extend into the lowlands (< 500 m) and only a few species have elevational maxima above 1,000 m. In fact, only five species ($<4\%$ of the fauna)—*Sturnira bidens* (1,990–2,865 m), *S. nana* (1,600–1,700 m), *Anoura* sp. nov. (3350–3450 m), *Carollia* sp. nov. (1,700–2,250 m), and *Mormopterus phrudus* (1,850 m)—have their entire distributions above 1,500 m. The elevational amplitude of species occupying the highest elevations (top pentile) is approximately two-thirds of the length of the entire gradient (mean amplitude, 2,196 m). Taken together, these data provide strong support for "Stevens's rule" of range size and elevational gradients. Indeed, because bats have no center of endemism on the altiplano, and because highland species do not replace lowland species along the gradient, agreement with Stevens's rule is almost guaranteed. Birds and murid rodents, which exhibit distinctly endemic faunas in the altiplano, fail to show such correlations between elevation and range amplitude.

The elevational gradient in the Sierra de Manantlan of Mexico traversed over 2,500 m (400–2,980 m) and included savanna, tropical deciduous forest, tropical semideciduous forest, gallery forest, oak forest, pine-oak forest, pine forest, and montane mesophytic forest. Although 55 species of bats occurred in the general region, Iniguez Davalos (1993) analyzed the distributional amplitudes of only the 27 species actually collected during his study. Species richness in the Sierra de Manantlan varies with elevation in an idiosyncratic manner (fig. 13.9A): low elevations harbor intermediate numbers of species;

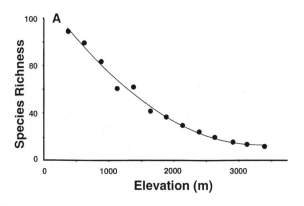

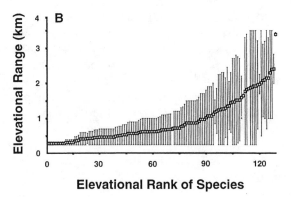

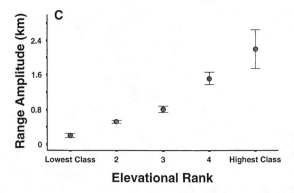

Figure 13.8. Elevational gradients of bats in the Manu National Park and Biosphere Reserve of southeastern Peru (from Patterson et al. 1998). *A,* Species richness (*S*) decreases with elevation (*E*) along a 3,000-m transect ($S = 138.7005 - 0.07835E + 0.000011E^2$; $P < 0.001$ for all regression coefficients; adjusted $R^2 = 0.99$). *B,* Elevational range profiles (*vertical lines*) of each species arranged according to midelevation (*squares*). *C,* Range amplitude (mean, *filled circle*; standard error, *vertical bar*) as a function of elevation for each of five classes corresponding to pentiles of species ranks by elevational midpoint.

Sierra de Manantlan

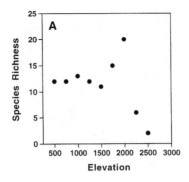

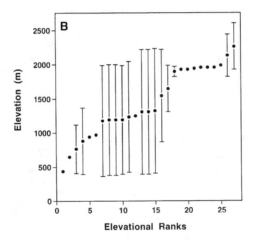

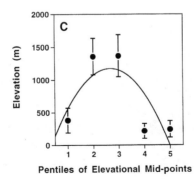

Figure 13.9. Elevational amplitudes of bats from the Sierra de Manantlan Biosphere Reserve in the Sierra Madre del Sur of west central Mexico (derived from Iniguez Davalos 1993). *A*, Species richness is unrelated to elevation. *B*, Elevational range profiles (*vertical lines*) of each species arranged according to mid-elevation (*filled circles*). *C*, Range amplitude (mean, *filled circle*; standard error, *vertical bar*) as a function of elevation for each of five classes corresponding to pentiles of species ranks by elevational midpoint.

high elevations harbor few species; and mid-elevations support the highest number of species. In contrast to the pattern showed by Andean bats, the elevational amplitudes of species shows a modal distribution (fig. 13.9B), with largest elevational amplitudes exhibited by species occupying the second and third pentiles (fig. 13.9C). Species with the smallest elevational amplitudes had midpoints in their elevational distribution located either below 1,250 m (five species) or above 1,750 m (eight species). In contrast, species with the broadest elevational amplitudes had elevational midpoints at intermediate elevations. Indeed, pentile rank accounted for appreciable variation in mean elevational amplitude as expressed by a second-degree polynomial (adjusted $R^2 = 0.61$). Less than half of the species extended into the lowlands (<500 m), although over half of the species have elevational minima below 1,000 m. Over a third of the fauna have elevational minima above 1,750 m. Clearly bat distributions in the Sierra Manantlan do not conform to Stevens's rule. These data are consistent with the idea that discrete highland and lowland faunas meet on the slopes of the Sierra Manantlan. Mid-elevation species, by virtue of being eurytopic, exhibit the greatest elevational amplitude, regardless of their faunal (or historical) affinities.

Summary and Prospectus

A number of generalities are possible regarding the geographic range sizes of bats. First, range sizes are relatively large compared to those of other mammalian taxa. Second, range size varies considerably but independently of body size. Third, range size is associated positively with species abundances. Fourth, range size is related to elevation and latitude, but patterns are variable (monotonic vs. modal) and are context dependent. Fifth, the propensity to be wide ranging is constrained more by ecological than by phylogenetic considerations, although history clearly defines some elevational patterns. Sixth, the geographic range size is more constrained by considerations of being large than of being small.

These descriptive patterns provide a springboard for future research investigations. Methodological issues regarding the consequences of employing different metrics of range size need to be evaluated, with mechanistic explanations for patterns linked more explicitly with particular metrics. Dissecting latitudinal patterns of range size into components related to consequences of empirical variability in range size and actual geographic placement of ranges should help illuminate underlying mechanistic processes. At least as important, increased attention needs to be devoted to assessing the effects of phylogeny on patterns of range size and, then, to decoupling those effects prior to assessments of biogeographic or ecological mechanisms. Independent contrasts and other explicit statistical tools that account for phylogenetic pattern clearly can contribute to this. Scale dependence affects most macroecological patterns, and those of range size likely are not immune. Although most studies of range size focus on the level of regional faunas, studies that focus on the local level will likely offer new perspectives concerning mechanisms guiding the local assembly of species into communities. More specifically, the utility of linking considerations of range size with abundance as a way to provide insight into conservation status and likelihood of extirpation is an exciting area for future research. Furthermore, little is known about the internal configuration of a species range, and additional focus on the way in which local abundance differs throughout the ranges of widespread versus more narrow-ranging species has both conservation relevance and utility in building foundational theory in macroecology.

Patterns of Species Richness

Broad-scale patterns of species richness have received considerable attention during the past quarter century. This attention is a consequence of revitalized interest in biodiversity and the mechanisms giving rise to it, as well as of increasing concern about the unprecedented rate of species loss at the global scale (e.g., Myers and Giller 1988; Rosenzweig 1995). Areographic patterns of species richness for bats are well documented, at least with respect to latitude in North and South America, with equally promising research forthcoming for Africa. Elevational patterns have been described as well but at only a small number of sites, so the generality of patterns is less clear than those for latitude.

Latitudinal Gradients

The study of mammals has contributed substantively to quantification of the gradient of increasing species richness from high to low latitudes. Early research focused on the latitudinal gradient in North America (e.g., Hagmeier 1966; Hagmeier and Stults 1964; McCoy and Connor 1980; Pagel et al. 1991; Simpson 1964; Wilson 1974). Although patterns were striking, it was unclear whether the gradient in richness was a product of latitude and its environmental correlates (i.e., temperature, insolation, and seasonality) or a consequence of the unique shape or history of North America. Fortunately, recent efforts have considered South America (e.g., Mares and Ojeda 1982; Ruggiero 1994) or have quantified patterns throughout the New World (Kaufman 1995; Kaufman and Willig 1998; Willig and Gannon 1997; Willig and Sandlin 1991; Willig and Selcer 1989), conclusively documenting that species richness significantly increases toward the tropics in both North and South America.

Findley (1993, fig. 6.3), created a global map of bat species richness based on the number of species in 500×500-km^2 quadrats. Even at this broad scale, it is clear that latitudinal gradients are not similar on all continents. The Neotropics near northwestern Brazil and the Guianas (120 species) and the Paleotropics near the Indomalayan region (100 species) harbor the greatest richnesses of bats, with reasonably symmetrical reductions in diversity toward the poles. More recent works (Patterson et al. 1996, 1998) have described faunas that are even more species rich from smaller areas in Peru. In contrast to the general pattern in the New World, a latitudinal transect through Africa and Europe shows no regular pattern. Richness is low in arid regions of the Arabian Peninsula and Saharan Africa; increases in the savannas and grasslands to the south (ca. 60 species); and actually declines in the equatorial rainforests of Central and West Africa!

The Chiroptera make a strong contribution to the overall mammalian pattern in the New World (Kaufman 1995; McCoy and Connor 1980; Wilson 1974). The number of species in a 1° band ranges from 80 to 150 species between 0° and 20° latitude, from 10 to 80 species between 20° and 40° latitude, and from zero to 20 species between 40° and 60° latitude (fig. 13.10). Although the details of the form of the increase in richness toward the tropics depend on the size and shape of the sampling units (e.g., cf. Kaufman and Willig 1998; Lyons and Willig 1999; Willig and Lyons 1998; Willig and Sandlin 1991; Willig and Selcer 1989), a rapid increase in species at midlatitudes is characteristic of the order as a whole.

Analyzing patterns of species density for African mammals, Castro-Arellano (1997) documented considerable variability in the latitudinal gradient among orders. Of the 198 species of bats in mainland Africa excluding Madagascar, 156 were endemic to the continent. His quantitative analyses, based on 165 species (124 endemics) distributed over 167 square quadrats (445 km on a side), revealed that over half of the variation in species density was related to latitude, with an additional 5.5% related to longitude. Given that latitude had such a dominant effect on species density, he more precisely evaluated the form of the relationship. He showed that the rate of increase in species toward the tropics was not constant. Rather, it is characterized by a second-degree polynomial in the northern hemisphere (accounting for almost three-quarters of the variation in species density) and a third-degree polynomial in the southern hemisphere (accounting for about 65% of the variation in species density). Thus, like North and South American bats, those in Africa show a strong latitudinal gradient. But unlike the symmetrical pattern north and south of the equator in the New World (Kaufman and Willig 1998), strong asymmetry characterizes the gradient north and south of the equator in Africa.

Species richness of a large area, such as a latitudinal band (gamma diversity), can be a consequence of two factors: alpha diversity and beta diversity (Whittaker 1972). Alpha diversity is the species richness of a standardized sampling unit (e.g., quadrat), whereas beta diversity reflects turnover in species composition between adjacent sampling units (e.g., quadrats within bands). Some have suggested that tropical regions contain more species because of high beta diversity in tropical areas compared to temperate counterparts

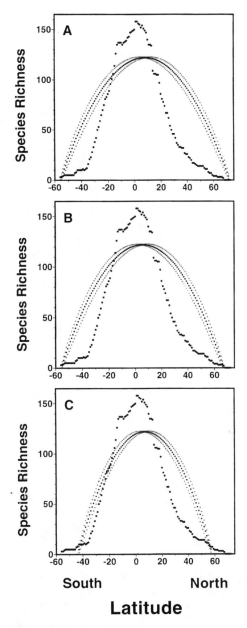

South   **North**

Latitude

Figure 13.10. Comparisons of actual species richness gradients of bats (*filled circles* represent counts of species whose ranges overlap a 1-degree latitudinal band) and those predicted by a stochastic model (inner curves represents actual predictions, outer lines define the 95% confidence bands) for three different latitudinal domains: *A*, the continental New World; *B*, the distributional limits for all bats; and *C*, the smallest latitudinal distribution within which 95% of all species occurs (from Willig and Lyons 1998).

(Rohde 1992; Rosenzweig 1995). For New World bats (Willig and Sandlin 1991), this hypothesis is unsupported, at least at large spatial scales; species turnover does not increase toward tropical latitudes for any of the larger families in the New World (i.e., Phyllostomidae, Emballonuridae, Mormoopidae, Vespertilionidae, or Molossidae). In fact, tropical regions have higher richness despite an empirical latitudinal gradient in which beta diversity decreases toward the tropics, primarily as a consequence of patterns in families other than the Phyllostomidae, particularly in the Vespertilionidae.

The form, strength, and peak of the latitudinal gradient in bat species richness is taxon dependent for analyses based on quadrats of equal size (250,000 km^2) in the New World (Willig and Selcer 1989). However, at the spatial scale of that analysis, variation in bat species richness was little affected by variation in longitude, biome richness, or land area. The Phyllostomidae and Molossidae each exhibited a linear increase in richness with decreasing latitude, whereas the Vespertilionidae exhibited a unimodal pattern. Phyllostomids increased most rapidly toward the tropics, with about 73% of the variation in richness related to latitude. Molossids increased most slowly toward the tropics, with about 72% of the variation in richness related to latitude. Moreover, the unimodal pattern for vespertilionids was quite platykurtic (i.e., flat); maximum richness arises at the interface of tropical and temperate regions, with only 49% of the variation related to latitude.

Latitudinal patterns of species richness may depend on the size of the sampling unit. Recently, Lyons and Willig (1999) found no evidence for scale dependence in the latitudinal gradient of species density at spatial scales between 1,000 and 25,000 km^2. Nonetheless, their approach is predicated on a priori knowledge of the form and parameterization of gradients of species richness (S) with respect to both area (A) and latitude. In a reanalysis, modeled after the approach of Pastor et al. (1996), Lyons and Willig (2002) showed that the parameters (slope, z; intercept, C) of the log form of the power function ($S = CA^z$) differ with latitude. Latitudinal variation in z represents scale dependence, whereas latitudinal variation in C is a scale-free measure of the latitudinal gradient of diversity. In general, any macroecological pattern that is affected by species richness may also exhibit scale dependence because the exponential increase in richness with area will introduce nonadditive components to the broader macroecological pattern.

Most hypotheses proposed to account for the latitudinal gradient in diversity are circular or unsubstantiated by empirical evidence (Rohde 1992), and none lead to direct quantitative predictions about the form of the relationship (Willig 2000). Nonetheless, a general null model, first advanced by Colwell and Hurtt (1994), has been applied successfully in its analytical form to empirical data for New World bats (Willig and Lyons 1998). In essence, if the distribution of the boundaries of species ranges is stochastic with respect to latitude, then species richness should increase symmetrically, from the edges of continental landmasses to their center (i.e., as a hyperbolic function). In the New World, the model predicts a tropical peak in species richness, which accounts for two-thirds or more of the latitudinal variation in bat species richness (fig. 13.10), depending on geographic domain (i.e., 67% if the New World; 71% if the latitudinal limits of the Chiroptera; 77% if the smallest latitudinal bounds that wholly contain 95% of New World bat species). Moreover, for any randomized latitudinal domain, the gradient of richness for species endemic to that domain should peak at the middle of that domain. For 20 different random latitudinal domains from within the New World, the null model accounted for significant variation in endemic bat species richness. Geometric constraints associated with the bounded nature of terrestrial landmasses predispose latitudinal gradients in the New World toward a tropical peak. Nonetheless, systematic deviations from the predictions of the null model characterize the empirical gradient. Species richness is higher in the tropics and lower in the temperate zones than predicted by the stochastic model, and the residuals are not related to the breadth of the continent at each latitude (i.e., area). Future research should explore the degree to which deviations from the predictions of such null models are related to environmental, evolutionary, or historical factors.

Elevational Gradients

The way in which species richness changes with elevation long has been of interest to biogeographers and ecologists (Brown 1995). Regional richness often is enhanced by the beta diversity that arises with changes in elevation (Brown and Lomolino 1998). Moreover, the elevational gradient may recapitulate the latitudinal gradient in richness, in part because similar mechanisms (Brown 1988) may be in operation (i.e., temperature and productivity decrease with increasing elevation and increasing latitude). Indeed, these mechanisms may be more clearly evident in cases of elevational gradients because points along an elevational gradient are closer and less likely to have undergone distinctive geological or evolutionary histories.

Nonetheless, the existence of a uniform, "general pattern" has been questioned recently, primarily based on issues related to inadequate experimental design (Rahbek 1995). Most studies of the elevational gradient fail to control for the effect of decreasing area on species richness as elevation increases. Thus, at least some of the decrease in richness with increasing elevation may be confounded by areal effects. Similarly, sites often are not stratified or evenly sampled along the elevational gradient, with lowland sites frequently excluded from analyses. Without equivalent sampling along the entire elevational landscape, it may be difficult to detect nonlinearities or modal relationships. Finally, if fewer individuals are sampled at higher elevations, "passive sampling" (Coleman et al. 1982) may obscure the actual variation of species richness by biasing richness to be higher toward the lowlands.

Mountains in Peru

The most comprehensive understanding of elevational gradients of richness for bats derives from the work of Graham (1983, 1990) and Patterson et al. (1996, 1998) in the Andes of Peru. The former author defined elevational patterns based on regional sampling of localities spanning almost 10° of latitude, whereas the latter authors assembled data from a more restricted landscape representing a 3,200-m elevational transect in the Manu Biosphere Reserve (spanning only a single degree of latitude). Despite differences in the spatial focus of their research, the general patterns were congruent in that no modality was evident in the gradient, and species richness decreased with increasing elevation (fig. 13.8A).

Building on the early work of Koopman (1978) and Tuttle (1970), Graham (1983) quantified an elevational decrease in bat species richness ($r = -0.978, P < 0.001$) and diversity ($r = -0.848, 0.001 < P < 0.01$) and an elevational increase in evenness ($r = 0.669, 0.01 < P < 0.05$) in the Andes of central Peru. Richness ranged from over 80 species at low elevations (ca. 200 m) to approximately 10 species at high elevations (ca. 3,400 m). Moreover, the rate at which bat species richness decreased with increasing elevation was dependent on the feeding ensemble associations of the bats. Frugivores (ca. eightfold) and insectivores (ca. sevenfold) showed indistinguishably fast rates of elevational decline, whereas other ensembles (i.e., piscivores, carnivores, sanguinivores, and nectarivores) declined more gradually. Some of these differences in rates among ensembles may be a consequence of the disparity in their overall species richness, especially for sanguinivores (three species maximum) and piscivores (two species maximum). Moreover, the piscivore guild is an artificial construct: evidence for a fish-eating diet in *Noctilio albiventris*, for instance, is lacking; the species pervasively consumes insects. In general, elevational decreases in temperature, foliage height diversity, and food abundance were postulated as ecological filters that exact an energetic cost on the ability of lowland species to exploit higher elevational habitats, resulting in their progressively lower richness (Graham 1983, 1990).

Despite differences in age and richness of their lineages, birds and bats have had approximately equivalent times to diversify and evolve along the eastern slopes of the Andes, making them appropriate taxa to contrast, given that they are homeotherms and the only extant vertebrate taxa that are volant. Graham (1990) documented that changes in elevation account for significant variation in the proportional richness (i.e., percent of the regional fauna) of bats (89 species, $b_1 = -0.249, r^2 = 0.94$), diurnal birds (555 species, $b_1 = -0.131\%$ per 1,000 m, $r^2 = 0.90$), and nocturnal birds (18 species, $b_1 = -0.131\%$ per 1,000 m, $r^2 = 0.62$) along the eastern slopes of the Andes. The decline in proportional richness of bats was greater than that for diurnal birds (Bonferroni adjusted $P = 0.01$) as well as for nocturnal birds (Bonferroni adjusted $P = 0.064$), but the two avian groups declined at indistinguishable rates (Bonferroni adjusted

$P > 0.99$). Two factors may contribute to these differences. First, the proportional richness of lowland bat faunas (>80%) is appreciably higher than that of either diurnal (<60%) or nocturnal (<70%) bird faunas. Second, the rate of turnover between elevational faunas is greater for birds than for bats. In fact, only a single bat species treated by Graham, *Histiotus macrotus*, is a true highland specialist (1% of the fauna), whereas approximately 10% and 31% of faunal pools are highland specialists for diurnal and nocturnal birds, respectively. Hence, bat assemblages at high elevations contain a smaller proportion of the fauna, include fewer species restricted to highlands, and consist of species with broader elevational breadths than do faunas comprising nocturnal or diurnal birds. This suggests that lowland bats have been less successful at colonizing or radiating at high elevations compared to birds.

Differences between bats and birds in their success of occupying high-elevation habitats may be related to three factors (Graham 1990; Schmidt-Nielsen 1971). First, bat and bird respiratory systems are quite different. The crosscurrent blood flow to the parabronchial lungs of birds guarantees that gases in air capillaries repeatedly meet freshly deoxygenated blood. In contrast, alveolar ducts in the lungs of bats represent a tidal ventilation system, which at best can equilibrate partial pressures of oxygen in the lungs with that in the pulmonary capillaries. Where partial pressures of oxygen are low, as at high elevations, this may represent a design constraint that limits the ability of bats to exploit high-elevation habitats. Second, rates of energy expenditure at higher elevations associated with thermoregulation are greater for bats than for birds. The generally smaller size of bats compared to birds (hence greater loss of heat because of larger surface to volume ratios) and the increased exposure of vascular tissue in the membranous wings of bats compared to the insulated, feathered wings of birds, result in a greater energetic cost to bats while exploiting cooler regions. Perhaps even more important, the obligate nocturnal habits of bats expose them to considerably lower temperatures during their peak periods of activity compared to the situation enjoyed by diurnal organisms, such as birds. Third, the elevational decline of food resources of bats may be much greater than that for birds. For example, species of trees that produce fruits and that are consumed by bats (e.g., figs and palms) become much rarer at mid- and high elevations, compared to trees that bear fruits on which birds feed (Gentry 1982). Hummingbird-pollinated plants may even increase (e.g., ericaceous plants) in richness at higher elevations, whereas few species of bat-pollinated plants are known to occur at high elevations (Fleming 1988; Gentry 1988; Heithaus 1982). Moreover, the elevational decrease in abundance and diversity of nocturnal insects is greater than that of diurnal insects (Terborgh 1977), although the available data are far from comprehensive. In summary, bat richness may decrease more rapidly than does bird richness because of a greater diminution in the diversity and quantity of resources they consume as well as because of a greater energetic cost of metabolism at higher elevations.

Patterson et al. (1996) documented that most species present in the Manu Biosphere Reserve of southeastern Peru are broadly distributed in lowland Amazonia and that highland assemblages are reduced and are proper subsets of lowland faunas. Although changes in species composition (Jaccard index) were erratic along the elevational gradient, most values were high (>0.8), with an indication of appreciable faunal turnover at the elfin forest-altiplano transition (>2,600 m). Species richness dramatically decreased with increasing elevation. Richness differed greatly among point samples and was related significantly to elevation ($b_1 = -0.060$ species per m, $R^2 = 0.851$, $0.01 < P < 0.05$), with only a single species captured at 3,200–3,300 m and 59 species captured at 300–400 m. Nonetheless, part of the elevational variation in species richness was related to differences in sampling effort (nights of netting, $r = 0.76$) and capture rates (number of individuals captured, $r = 0.72$) among elevational sites. To control for this, species presence at each elevation was inferred from elevational range data for each species based on the capture results from all sites along the transect. Again, elevation accounted for almost all of the variation among sites in species richness ($b_1 = -0.053$ species per m, $R^2 = 0.971$, $P < 0.001$). When species richness per site was further increased based on presumed or possible presence of taxa as suggested by literature records, the form of the relationship became more complex (quadratic or cubic) but in all cases the linear component remained strong and significant. In these cases, the richness of lowland sites is augmented more than is richness of highland sites, a reflection of the fact that few Amazonian species are highland specialists and lowland sites exhibit greater beta diversity. Moreover, the richness of elevational assemblages was correlated inversely to the number of endemic species. This negative correlation has important ramifications for conservation and suggests that protection of species-rich areas does not necessarily protect endemic taxa.

The manner in which richness diminishes with elevation depends on the ecological classification of the species. Nine ensembles were recognized in lowland assemblages (i.e., omnivorous predators, insectivorous piscivores, low-flying frugivores, high-flying frugivores, low-flying insectivores, high-flying insectivores, omnivorous nectarivores, sanguinivores, and insectivorous carnivores), but only six occurred above 3,000 m with the loss of high-flying insectivores, insectivorous piscivores, and omnivorous predators. High-flying insectivores composed an appreciable proportion (ca. 15%) of the fauna at low elevations, decreased rapidly around 1,000–1,500 m, and were absent above 2,000 m. Members of the insectivorous carnivore ensembles similarly represented about 15% of the fauna at low elevations and decreased rapidly at 1,000–1,500 m but persisted at low levels of richness, well into the highlands. In contrast, low-flying insectivores and high-flying frugivores were represented in the lowlands by about 20 and 25 species, respectively, and consistently decreased with elevation until 3,000–3,500 m, where they were represented by three and seven species, respectively.

Elevational patterns of bats contrasted with those of mice as well as birds (Patterson et al. 1998). By demonstrating a gradual decline in richness with increasing elevation (fig. 13.8A), it could be seen that bats were more similar to birds than to mice, whose variation in richness was unrelated to elevation. Although zonations in faunal assemblages were observed between lowland and montane faunas for all three groups, the elevational boundaries of the assemblages differed among groups (i.e., birds, 500 m; bats, 750 m; mice, 1000 m). Moreover, these faunal zones were less discrete for birds and least discrete for bats and did not correspond to vegetational ecotones. A similar lack of congruence between phytogeographic zones and bat assemblages in Venezuela was documented by Willig and Mares (1989). Differences in the elevational zonation of these three groups may reflect phylogenetic histories as well as other biological attributes and suggest limited utility of using one taxon (e.g., birds) as an indicator of general patterns in other groups or as an umbrella group on which to base conservation decisions.

Mountains in the Philippines

The numerous islands of the Philippines harbor 73 species of bats, of which about 40% are endemic to the country (Heaney 1986; Heaney et al. 1998). In the most detailed study of elevational gradients of species richness of flying foxes (Pteropodidae but excluding *Acerodon* and *Pteropus* because of their wide-ranging habits), Heaney et al. (1989) compared patterns on an oceanic (Negros) and landbridge (Leyte) island. Each island supported eight species of smaller flying fox, with seven common to both islands.

Species richness was greatest in the lowlands on both islands (seven to eight species), with substantial declines in ridgetop mossy forest (three species). Indeed, changes in species richness were associated more with changes in forest type than with changes in elevation per se. For example, the bat species richness of two mossy forest sites that differed in elevation by 550 m (i.e., the 950-m site on Leyte and the 1,500-m site on Negros) was the same (three species). The most abundant species in agricultural lands were taxa that are widespread in Southeast Asia, whereas the most abundant species in forested areas were Philippine endemics. However, elevational gradients may be difficult to quantify in an unequivocal manner from these data because of three factors. First, elevational relief is not great on either island: 950 m on Leyte and 1,500 m on Negros. Second, only four or five survey stations were established along the elevational gradient, compromising an ability to distinguish linear from nonlinear forms of the elevational relationship with richness. Finally, anthropogenic activities dominated the lowest site on Leyte (agriculture and secondary forest at 50 m) and the lower two sites on Negros (urban and agriculture at 10 m and upland farmland at 600 m). Thus, richness may be responding to a gradient of increasing anthropogenic disturbance toward the lowlands as well as to gradients of biotic and abiotic factors that are correlated with elevation.

Mountains in Mexico

The state of Oaxaca in Mexico harbors one of the most species-rich mammalian faunas in Central America and Mexico and contains many endemic species (Peterson et al. 1993; Sanchez-Cordero 1993). In a comparative study of two elevational transects in Oaxaca (640–2,600 m in Sierra Mazateca, 700–3,000 m in Sierra Mixteca), Sanchez-Cordero (2001) documented elevational changes in richness, trophic diversity, and endemism. Both elevational transects harbored similar species richness (16 species in Sierra Mazateca and 17 species in Sierra Mixteca), with 11 species shared in common.

Variation in elevation was not associated significantly with variation in bat species richness in the Sierra Mazateca ($r_s = -0.63$, $n = 5$, $P > 0.10$) but was significantly and negatively associated with elevation in the Sierra Mixteca ($r_s = -0.96$, $n = 7$, $P < 0.05$). The small number of elevational sites reduces the power of statistical analyses, especially in the Sierra Mazateca. Nonetheless, combining these results in a meta-analysis (Sokal and Rohlf 1995) suggests that a significant negative association exists between elevation and species richness (Fisher's test for combining probabilities, $X^2 = 11.98$, df $= 4$, $P < 0.05$). The similarities between transects become stronger when assessing habitat-specific patterns of richness. In the Sierra Mazateca, richness was highest in pine-oak forest and high in thorn scrub and tropical semideciduous forest and decreased abruptly in high-elevation oak and cloud forest. In the Sierra Mixteca, richness was highest in lowland tropical semideciduous forest and low in high-elevation cloud, oak, and pine-oak forests.

Paralleling changes in species richness, trophic richness was greater at lower than at higher elevations. All four bat ensembles recognized in the study (i.e., frugivores, nectarivores, insectivores, and sanguinivores) occurred at lowland elevations in thorn scrub, tropical semideciduous forest, and tropical deciduous forest. Sanguinivores and nectarivores did not extend above 1,050 m, frugivores did not occur above 2,450 m, and insectivores peaked at intermediate elevations (between 1,850 m and 2,100 m).

Summary and Prospectus

Regardless of continent or scale, latitudinal gradients of richness are qualitatively similar: species number increases within decreasing latitude. However, important quantitative differences in the form or parameterization of the latitudinal gradient exist among taxa, continents, or scales. Bat richness generally decreases with increasing elevation; however, the form and parameterization of the elevational gradient is geographically variable and sometimes modal.

The extensive documentation of gradients of richness for bats suggests more sophisticated macroecological analyses in the future. The roles of geographic constraints or stochastic processes in affecting latitudinal and elevational gradients need to be assessed more broadly, and the environmental correlates responsible for deviations from null model predictions need to be identified explicitly. Geographically explicit and scale-dependent analyses based on latitude, longitude, and elevation need to be executed on all continents using Geographic Information Systems (GIS) to identify regional hot spots of bat diversity or areas with unusually high numbers or proportions of species with threatened or endangered status. Quantitative analyses, combining information about all components of diversity at the local level (e.g., richness, evenness, dominance, diversity) and regional patterns related to the composition of faunal pools need to be conducted across a wide array of latitudes in both the New and Old Worlds to understand the interplay between biogeographic and evolutionary processes, on the one hand, and ecological processes, on the other hand. Regional, hemispheric, and global analyses of the way in which variation in productivity might contribute to latitudinal and elevational patterns of richness is ripe for future research. Similarly, the extent to which beta diversity contributes to high regional diversity ought to be explored in a latitudinally and elevationally explicit fashion. Finally, the extent to which functional diversity and species diversity are interrelated should be examined in a number of different geographic settings.

Patterns of Body Size

Body size is an incisive phenotypic character that integrates the results of evolution at the levels of morphology, physiology, behavior, and ecology (Barclay and Harder, this volume; Calder 1984; LaBarbera 1989; Peters 1983; Schmidt-Nielsen 1984; Simmons and Conway, this volume; Speakman and Thomas, this volume; Swartz et al., this volume). It can be estimated conveniently by body mass, regardless of the shape or taxonomic affiliation of species (Brown 1995). Thus, it is an ideal comparative character to examine in a macroecological context. Moreover, early work (Hutchinson and MacArthur 1959) suggested that, within a given biota, distributions of body size were right-skewed on a log-normal scale—that is, relatively small size is common, whereas extremely small or large size is rare. Subsequent empirical work generally corroborated that observation for a diversity of organisms from bacteria to mammals (e.g., Brown and Nicoletto 1991; Maurer et al. 1992).

Considerations of the Central Limit Theorem suggest that, because body-size distributions are not usually normal on a logarithmic scale, empirical patterns are not the consequence of random multiplicative factors acting in concert. The ways in which deviations from log normal differ with taxon or scale might implicate underlying mechanisms. More specifically, Brown (1995) noted three general phenomena: (1) body-size distributions are "highly modal," (2) a gradual decline in the frequency of larger body masses produces an extended tail to the right of the mode, and (3) a sharp decline in frequency occurs from the mode toward the left (i.e., small size). Moreover, as geographic scale defining the biota shrinks, the shape of the distribution changes to become more platykurtic while retaining the same range of body sizes but with gaps (i.e., some body-size classes become absent) in distributions at small spatial scales (Brown and Nicoletto 1991; Holling 1992; Maurer 1999).

Taxonomic Variation in the Distribution of Body Size

Based primarily on body-size information (body mass) from Silva and Downing (1995), New World bats, like other taxa, do not have a lognormal frequency distribution of body masses at the hemispheric scale. Rather, they exhibit significant right-skewness (g_1 = 0.328, 0.05 > P > 0.02) and platykurtosis (g_2 = −0.858, 0.05 > P > 0.02). Nonetheless, the pattern for bats (fig. 13.2F) shows a number of remarkable contrasts with patterns described for other taxa (Brown et al. 1993). First, the modal value is not as dominant as in other groups. Second, the decline in frequency to the left of the mode is gradual and approximately linear. Third, the right-hand tail is less extended and not as sharply concave.

In part, this distinctive pattern for bats may be a consequence of heterogeneity in the distribution of body sizes among the larger families (fig. 13.2G–J, representing the Emballonuridae, Phyllostomidae, Vespertilionidae, and Molossidae). For example, the distribution for phyllostomids is more like the general pattern, with a highly distinctive mode, an extensive right-hand tail, and a sharply declining left-hand tail. In contrast, the vespertilionids have a more platykurtic distribution and the molossids essentially exhibit a bimodal distribution of body masses.

Scale Dependence in Distributions of Body Size

As geographic scale changes from local, through regional, to continental or hemispheric, so do the attributes of the distribution of body size in bats (fig. 13.11). However, observed changes in the distribution of body sizes in bats are not completely coincident with the observations of Brown (1995). As predicted, the range of body masses (log of mass in grams from 0.8 to 2.0) is conserved proceeding from the distribution at the hemispheric scale (fig. 13.11A) to that of a site representing a local community (fig. 13.11F). Moreover, the body-size distribution is more highly platykurtotic (g_2 = −0.947, 0.05 > P > 0.02) in South America than at the hemispheric scale, but it is not significantly skewed to the right (g_1 = 0.157, P > 0.2). Unfortunately, species richness at smaller spatial scales is sufficiently reduced, compromising the power to detect significant kurtosis or skewness in body-size distributions (Sokal and Rohlf 1995). However, examination of the modes and moment statistics (i.e., g_1 and g_2) indicates that at the local level (e.g., interior Atlantic rainforest at Reserva Natural de M'baracayu in eastern Paraguay), distributions are least platykurtotic (g_2 = −0.437) and most right-skewed (g_1 = 0.523) compared to any other spatial scale.

Similarly, Arita and Figueroa (1999) investigated the scale dependence of

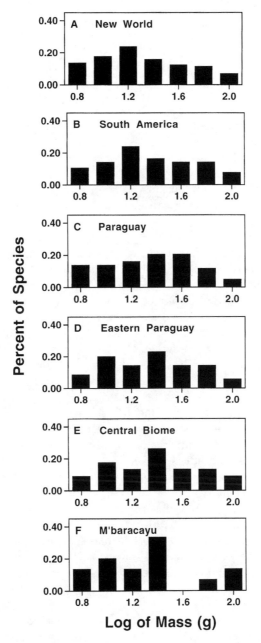

Figure 13.11. Hierarchical evaluation of the logarithmic distribution of body sizes at the level of (A) the continental New World (skewness, $g_1 = 0.328^*$; kurtosis, $g_2 = -0.858^*$), (B) South America ($g_1 = 0.157$; $g_2 = -0.947^*$), (C) Paraguay ($g_1 = 0.037$; $g_2 = -0.912$), (D) mesic eastern Paraguay ($g_1 = 0.110$; $g_2 = -0.937$), (E) the Paraguay central biome ($g_1 = 0.090$; $g_2 = -0.834$) in the mesic east, and (F) a local site (Reserva Natural de M'baracayu) in the Paraguay central biome ($g_1 = 0.523$; $g_2 = -0.437$).

patterns of body size in Mexican bats. In general, body mass ranged from 3 g in some vespertilionids, to 150 g in the carnivorous phyllostomid, *Vampyrum spectrum*. At the scale of the entire country, the body-size distribution was unimodal and right-skewed on a semilogarithmic plot. Unlike the pattern of scale dependence exhibited by nonvolant Mexican mammals in which the distribution of body sizes progressively became more even with decreasing spatial scale, for bats this was not so.

The distribution of body sizes in bats retained its right-skewness, even at the smallest spatial scale (0.5 × 0.5 degree quadrats). Moreover, the distribution of body sizes at smaller scales was indistinguishable from those based on a random selection of the same number of species from the entire Mexican bat fauna. Finally, bats differed considerably from nonvolant mammals in Mexico in that they did not exhibit the expected latitudinal increases in body size.

Correlates of Body Size

Based on a subset of 150 New World bat species representing all families and feeding assemblages, Arita (1993) documented that a high proportion of the variation among bats in body size is a consequence of phylogenetic constraints. Indeed, variation among genera within families (42.7%) and among families within the order (42.3%) accounted for most of the variation in body size (85%). Indeed, variation among species within genera accounted for no more than 15% of the total variation. This corroborates a general pattern (see Harvey and Pagel 1991) in which most of the variation in ecologically relevant traits is associated with higher levels in the taxonomic hierarchy.

After controlling for phylogenetic constraints by using means of genera in analyses (following Harvey and Pagel [1991]), Arita (1993) found no association between mean generic body mass and measures of local abundance ($r_s = 0.01$, $n = 59$, $P > 0.05$) or with area of geographic distribution ($r_s = 0.26$, $n = 59$, $P > 0.05$). This contradicts conventional wisdom (Brown 1995; Maurer 1999) and is surprising because other groups of mammals show a significant association of body mass with density (e.g., Damuth 1981; Eisenberg 1980; LaBarbera 1989; Mohr 1940; Robinson and Redford 1986) and with geographic range size (e.g., Arita et al. 1990; Brown 1981; Cristoffer 1990; Pagel et al. 1991). As expected, some large species of bat such as *Vampyrum spectrum* and *Chrotopterus auritus* are locally rare, and some small species such as *Glossophaga soricina* are relatively abundant. However, large species of *Phyllostomus* or *Artibeus* are among the most abundant, and small species such as *Centronycteris centralis* (see Simmons and Handley 1998) are extremely rare members of local assemblages. In part, the absence of an association may be a consequence of the smaller ratio of body masses in the New World Chiroptera (1 : 56; 3 g for *Furipterus horrens* and 169 g for *V. spectrum*) compared to other groups, such as nonvolant mammals (1 : 20,000) or a consequence of nonlinearities in the associations between variables (Brown and Maurer 1987). The latter explanation can be discounted in cases when nonparametric correlations fail to detect patterns.

Summary and Prospectus

Although distributions of body size have only begun to be studied in bats, with the most extensive analyses restricted to New World faunas, a number of emerging generalizations are possible. First, the distributions of body size are generally unimodal, right-skewed, and platykurtic for most families of the Chiroptera. Second, empirical patterns for bats are not congruent with those documented for terrestrial mammals, perhaps because of additional constraints imposed on bat evolution related to tradeoffs involving flight and echolocation. Third, the significant relationship between body size and abundance, at least in Mexican bats, is mostly a consequence of phylogenetic trends.

Research questions related to body mass have only begun to be explored, with many avenues of investigation worthy of pursuit. Once accurate estimates of body mass are available for most species of bat, general questions concerning taxon-specific and geographically explicit characteristics of body-size distributions should be explored, including a global assessment of phylogenetic constraints. Importantly, bats represent ideal organisms to test contentions of Brown et al. (1993) concerning the association of body-size evolution and life-history evolution. The range of masses in the Chiroptera (1,200 g *Pteropus vampyrus* to the 2 g *Craseonycteris thonglongyai*) spans the proposed optimal size (100 g) for mammals and includes many small (<20 g) taxa; moreover, phylogenies in the group are reasonably well known (see Jones and Purvis 1997). However, the preliminary assessment is that bats do not adhere to the predictions of theory based on reproductive power, and future research is required to see if alternative models (Charnov 1993; Kozlowski and Weiner 1997) are generally more applicable to the association between body mass and life-history characteristics in bats.

Overview

Although macroecological patterns for bats differ from many general mammalian patterns concerning body mass and abundance, they represent exemplars of the latitudinal gradient of richness. To understand the ubiquity of such patterns, future efforts should concentrate on the application of standardized macroecological methodologies to bat distributions in Africa, Australia, and Eurasia and on intercontinental comparisons. Equally important, research concerning bats is woefully lacking concerning elevational gradients of richness, range size, body size or abundance on all continents and, as such, represent ripe areas of exploration for future study. Indeed, the contradictory elevational patterns observed for richness and range size highlight the need for a larger number of studies at various latitudes on all continents.

For both elevational and latitudinal analyses, comparison of observed patterns with those produced by appropriate null models, and the design of quantitative analyses to tease apart various competing mechanistic hypotheses should be the emphasis of research in the future (see Hofer et al. 1999). Comparative studies with small mammals such as rodents and marsupials, or with other volant taxa, provide one avenue to explore the mechanistic bases of pattern. Alternatively, the application of analytical models incorporating phylogenetic information is apt to prove fruitful in virtually all macroecological studies of bats.

Clearly, the comparative method has revolutionized the way in which ecologists approach correlative studies like those that dominate macroecological research. Indeed, phylogenetic contingency shapes numerous phenotypic attributes important to the ecological characteristics of all organisms. However, Brown et al. (1996) make an important point: the increasingly popular approach of removing variation among species that is related to phylogenetic relationships may stymie progress in identifying recurrent patterns and understanding their mechanistic bases. Although the analysis of "phylogeny-free" residual variation removes the effect of sharing a common ancestor, it also removes the variation due to the ecology of the common ancestor as well. Ecological and phylogenetic relationships evolve together and are confounded in various ways. To the extent that this is true, phylogenetic variation cannot be removed without compromising ecological understanding. We do not advocate the suspension of comparing ecological phenomena in the context of phylogenies. Rather, we suggest that future research should explore novel ways of analyzing these two important and interrelated processes in a manner that will avoid the "either/or" interpretations of current methodologies (see Brown 1994).

Given the reasonably stable state of bat alpha taxonomy compared to many groups of plants or animals, and the growing information concerning accurate range-size distributions throughout the world, the early part of the current decade should see increasingly strong contributions by bat biologists to the study of macroecology. As ecological studies of bats at the community level increase, and as technologies for assessing bat abundance improve, broader understanding of the scale dependence of associations among range size, population density, and body mass should be forthcoming as well, from both latitudinal and elevational perspectives.

Acknowledgments

We thank M. B. Fenton and T. H. Kunz for their confidence, patience, and cooperation, especially when we were behind schedule. In addition, we thank

D. M. Kaufman, and S. K. Lyons for sharing data on body size and range distributions of New World bats; C. P. Bloch, R. D. Owen, and S. J. Presley for data on the composition of bat assemblages in Paraguay; and I. Castro-Arellano, S. K. Lyons, and V. Sanchez-Cordero for access to unpublished manuscripts. In part, this chapter was completed while Willig was a sabbatical fellow at the National Center for Ecological Analysis and Synthesis, a center funded by the National Science Foundation (grant number DEB-94-21535), the University of California, Santa Barbara, and the State of California; additional support was provided by a developmental leave from the office of the provost, Texas Tech University. The manuscript benefited by the critical comments of C. P. Bloch, S. J. Presley, and two anonymous reviewers.

Literature Cited

Adams, C. C., G. P. Burns, T. L. Hankinson, B. Moore, and N. Taylor. 1920. Plants and animals of Mount Marcy, New York. I. Ecology, 1:71–94.

Anderson, S. 1977. Geographic ranges of North American terrestrial mammals. American Museum Novitates, 2629:1–15.

Anderson, S. 1985. The theory of range-size (RS) distributions. American Museum Novitates, 2833:1–20.

Arita, H. T. 1993. Rarity in Neotropical bats: correlations with phylogeny, diet, and body mass. Ecological Applications, 3:506–517.

Arita, H. T., and F. Figueroa. 1999. Geographic patterns of body-mass diversity in Mexican mammals. Oikos, 85:310–319.

Arita, H. T., F. Figueroa, A. Frisch, P. Rodriguez, and K. Santos-del-Prado. 1997. Geographical range size and the conservation of Mexican mammals. Conservation Biology, 11:92–100.

Arita, H. T., J. G. Robinson, and K. H. Redford. 1990. Rarity in Neotropical forest mammals and its ecological correlates. Conservation Biology, 4:181–192.

Brown, J. H. 1981. Two decades of homage to Santa Rosalia: toward a general theory of diversity. American Zoologist, 21·877–888.

Brown, J. H. 1988. Species diversity. Pp. 57–89 in: Analytical Biogeography: An Integrated Approach to the Study of Animal and Plant Distribution (A. A. Myers and P. S. Giller, eds.). Chapman & Hall, New York.

Brown, J. H. 1995. Macroecology. University of Chicago Press, Chicago.

Brown, J. H., and M. V. Lomolino. 1998. Biogeography. 2d ed. Sinauer Associates, Sunderland, Mass.

Brown, J. H., P. A. Marquet, and M. L. Taper. 1993. Evolution of body size: consequences of an energetic definition of fitness. American Naturalist, 142:573–584.

Brown, J. H., and B. A. Maurer. 1987. Evolution of species assemblages: effects of energetic constraints and species dynamics on the diversification of the American avifauna. American Naturalist, 130:1–17.

Brown, J. H., and B. A. Maurer. 1989. Macroecology: the division of food and space among species on continents. Science, 243:1145–1150.

Brown, J. H., and P. F. Nicoletto. 1991. Spatial scaling of species composition: body masses of North American land mammals. American Naturalist, 138:1478–1512.

Brown, J. H., G. C. Stevens, and D. M. Kaufman. 1996. The geographic range: size, shape, boundaries, and internal structure. Annual Review of Ecology and Systematics, 27:597–623.

Brown, J. L. 1994. Review of Phylogeny, Ecology, and Behavior: A Research Program in Comparative Biology, by D. R. Brooks and D. A. McLennan. Journal of Mammalogy, 75:243–245.

Calder, W. A., III. 1984. Size, Function, and Life History. Harvard University Press, Cambridge, Mass.

Castro-Arellano, I. 1997. Patrones de distribucion geografica en los mamiferos terrestres de Africa. Bachelors Thesis, Universidad Nacional Autonoma de Mexico, Mexico City.

Charnov, E. L. 1993. Life History Invariants: Some Explorations of Symmetry in Evolutionary Ecology. Oxford University Press, Oxford.

Chown, S. L. 1997. Speciation and rarity: separating cause from consequence. Pp. 91–109 in: The Biology of Rarity: Patterns, Causes, and Consequences of Rare-Common Differences (W. E. Kunin and K. J. Gaston, eds.). Chapman & Hall, London.

Coleman B. D., M. A. Mares, M. R. Willig, and Y. H. Hsieh. 1982. Randomness, area, and species richness. Ecology, 63:1121–1133.

Colwell, R. K., and G. C. Hurtt. 1994. Nonbiological gradients in species richness and a spurious Rapoport effect. American Naturalist, 144:570–595.

Colwell, R. K., and D. C. Lees. 2000. The mid-domain effect: geographic constraints and the geography of species richness. Trends in Ecology and Evolution, 15:70–76.

Corbet, G. B. 1978. Mammals of the Palearctic Region: A Taxonomic Review. Cornell University Press, Ithaca, N.Y.

Corbet, G B., and J. E. Hill. 1992. The Mammals of the Indomalayan Region: A Systematic Review. Natural History Museum Publications. Oxford University Press, Oxford.

Cristoffer, C. 1990. Nonrandom extinction and the evolution and conservation of continental mammal faunas. Ph.D. Dissertation. University of Florida, Gainesville.

Damuth, J. 1981. Population density and body size in mammals. Nature, 290:699–700.

Eisenberg, J. F. 1980. The density and biomass of tropical mammals. Pp. 35–55 in: Conservation Biology: An Evolutionary-Ecological Perspective (M. E. Soule and B. A. Wilcox, eds.). Sinauer Associates, Sunderland, Mass.

Findley, J. S. 1993. Bats: A Community Perspective. Cambridge University Press, Cambridge.

Fleming, T. H. 1988. The Short-Tailed Fruit Bat. A Study in Plant-Animal Interactions. University of Chicago Press, Chicago.

Gaston, K. J. 1988. The intrinsic rates of increase of insects of different body sizes. Ecological Entomology, 14:399–409.

Gaston, K. J. 1990. Patterns in the geographical ranges of species. Biological Review, 65:105–129.

Gaston, K. J. 1994. Rarity. Population and Community Biology Series 13. Chapman & Hall, London.

Gaston, K. J. 1999. Why Rapoport's rule does not generalize. Oikos, 84:309–312.

Gaston, K. J., T. M. Blackburn, and J. I. Spicer. 1998. Rapoport's rule: time for an epitaph? Trends in Ecology and Evolution, 13:70–74.

Gaston, K. J., and W. E. Kunin. 1997. Rare-common differences: an overview. Pp. 12–29 *in:* The Biology of Rarity: Patterns, Causes and Consequences of Rare-Common Differences (W. E. Kunin and K. J. Gaston, eds.). Chapman & Hall, London.

Gaston, K. J., and J. H. Lawton. 1988a. Patterns in body size, population dynamics and regional distribution of bracken herbivores. American Naturalist, 132:662–680.

Gaston, K. J., and J. H. Lawton. 1988b. Patterns in the distribution and abundance of insect populations. Nature, 331:709–712.

Gentry, A. 1982. Patterns of Neotropical plant species diversity. Evolutionary Biology, 15:1–85.

Gentry, A. 1988. Changes in plant community diversity and floristic composition on environmental and geographic gradients. Annals of the Missouri Botanical Garden, 75:1–34.

Graham, G. L. 1983. Changes in bat species diversity along an elevational gradient up the Peruvian Andes. Journal of Mammalogy, 645:559–571.

Graham, G. L. 1990. Bats versus birds: comparisons among Peruvian volant vertebrate faunas along an elevational gradient. Journal of Biogeography, 17:657–668.

Hagmeier, E. M. 1966. A numerical analysis of the distributional patterns of North American mammals. II. Re-evaluation of the provinces. Systematic Zoology, 15:279–299.

Hagmeier, E. M., and C. D. Stults. 1964. A numerical analysis of the distributional patterns of North American mammals. Systematic Zoology, 13:125–155.

Harvey, P. H., and M. D. Pagel. 1991. The Comparative Method in Evolutionary Biology. Oxford University Press, Oxford.

Heaney, L. R. 1986. Biogeography of mammals in SE Asia: estimates of rates of colonization, extinction, and speciation. Biological Journal of the Linnean Society, 28:127–165.

Heaney, L. R., P. D. Heideman, E. A. Rickart, R. B. Utzurrum, and J. S. H. Klompen. 1989. Elevational zonation of mammals in the central Philippines. Journal of Tropical Ecology, 5:259–280.

Heaney, L. R., M. Louella Dolar, A. C. Alcala, A. T. L. Dans, P. C. Gonzales, N. R. Ingle, M. V. Lepiten, et al. 1998. A synopsis of the mammalian fauna of the Philippine Islands. Fieldiana Zoology, n.s., 88:1–61.

Heithaus, E. R. 1982. Coevolution between bats and plants. Pp. 327–368 *in:* Ecology of Bats (T. H. Kunz, ed.). Plenum Press, New York.

Hofer, U., L. Bersier, and D. Borcard. 1999. Spatial organization of a herpetofauna on an elevational gradient revealed by null model tests. Ecology, 80:976–988.

Holling, C. S. 1992. Cross-scale morphology, geometry, and dynamics of ecosystems. Ecological Monographs, 62:447–502.

Hutchinson, G. E., and R. H. MacArthur. 1959. A theoretical ecological model of size distributions among species of animals. American Naturalist, 93:117–125.

Iniguez Davalos, L. I. 1993. Patrones ecologicos en la comunidad de murcielagos de La Sierra de Manantlan. Pp. 355–370 *in:* Avances en el Estudio de los Mamiferos de Mexico (R. A. Medellin and G. Ceballos, eds.). Publicaciones Especiales, vol. 1. Asociacion Mexicana de Mastozoologia, A.C., Distrito Federal, Mexico.

Jones, K. E., and A. Purvis. 1997. An optimum body size for mammals? Comparative evidence from bats. Functional Ecology, 11:751–756.

Kaufman, D. M. 1995. Diversity of New World mammals: universality of the latitudinal gradients of species and bauplans. Journal of Mammalogy, 76:322–334.

Kaufman, D. M., and M. R. Willig. 1998. Latitudinal patterns of mammalian species richness in the New World: the effects of sampling method and faunal group. Journal of Biogeography, 25:795–805.

Kingdon, J. 1997. The Kingdon Field Guide to African Mammals. Academic Press, San Diego, Calif.

Koopman, K. F. 1978. Zoogeography of Peruvian bats with special emphasis on the role of the Andes. American Museum Novitates, 2651:1–33.

Koopman, K. F. 1984. Bats. Pp. 145–186 in: Orders and Families of Recent Mammals of the World (S. Anderson and J. K. Jones, Jr., eds.). John Wiley & Sons, New York.

Kozlowski, J., and J. Weiner. 1997. Interspecific allometries are by-products of body size optimization. American Naturalist, 149:352–380.

Kunin, W. E., and K. J. Gaston. 1993. The biology of rarity: patterns, causes and consequences. Trends in Ecology and Evolution, 8:298–301.

LaBarbera, M. 1989. Analyzing body size as a factor in ecology and evolution. Annual Review of Ecology and Systematics, 20:97–117.

Lawton, J. H. 1993. Range, population abundance, and conservation. Trends in Ecology and Evolution, 8:409–413.

Lawton, J. H., S. Nee, A. J. Letcher, and P. H. Harvey. 1994. Animal distributions: patterns and processes. Pp. 41–58 in: Large-Scale Ecology and Conservation Biology (P. J. Edwards, R. M. May, and N. R. Webb, eds.). Blackwell Scientific Publications, Oxford.

Lecher, A. J., and P. H. Harvey. 1994. Variation in geographical range size among mammals of the Palearctic. American Naturalist, 144:30–42.

Lyons, S. K., and M. R. Willig. 1997. Latitudinal patterns of range size: methodological concerns and empirical patterns for New World bats and marsupials. Oikos, 80:292–304.

Lyons, S. K., and M. R. Willig. 1999. A hemispheric assessment of scale-dependence in latitudinal gradients of species richness. Ecology, 80:2483–2491.

Lyons, S. K., and M. R. Willig. 2002. Species richness, latitude, and scale sensitivity. Ecology, 83:47–58.

MacArthur, R. H. 1972. Geographical Ecology: Patterns in the Distribution of Species. Harper & Row, New York.

Mares, M. A., and R. A. Ojeda. 1982. Patterns of diversity and adaptation in South American hystricognath rodents. Pp. 393–432 in: Mammalian Biology in South America (M. A. Mares and H. H. Genoways, eds.). Special Publication Series vol. 6. Pymatuning Laboratory of Ecology, University of Pittsburgh, Pittsburgh.

Maurer, B. A. 1999. Untangling Ecological Complexity: The Macroscopic Perspective. University of Chicago Press, Chicago.

Maurer, B. A., J. H. Brown, and R. D. Rusler. 1992. The micro and macro in body size evolution. Evolution, 46:939–953.

McCoy, E. D., and E. F. Connor. 1980. Latitudinal gradients in the species diversity of North American mammals. Evolution, 34:193–203.

Merriam, C. H. 1894. Laws of temperature control the geographic distribution of terrestrial animals and plants. National Geographic, 6:229–238.

Mohr, C. O. 1940. Comparative populations of game, fur, and other mammals. American Midland Naturalist, 24:581–584.

Myers, A. A., and P. S. Giller, eds. 1988. Analytical Biogeography: An Integrated Approach to the Study of Animal and Plant Distributions. Chapman & Hall, London.

Pagel, M. D., R. M. May, and A. R. Collie. 1991. Ecological aspects of the geographical distribution and diversity of mammalian species. American Naturalist, 137: 791–815.

Pastor, J., A. Downing, and H. E. Erickson. 1996. Species-area curves and diversity-productivity relationships in beaver meadows of Voyageurs National Park, Minnesota, USA. Oikos, 77:399–406.

Patterson, B. D., V. Pacheco, and S. Solari. 1996. Distribution of bats along an elevational gradient in the Andes of south-eastern Peru. Journal of the Zoology (London), 240:637–658.

Patterson, B. D., D. G. Stotz, S. Solari, J. W. Fitzpatrick, and V. Pacheco. 1998. Contrasting patterns of elevational zonation for birds and mammals in the Andes of southeastern Peru. Journal of Biogeography, 25:593–607.

Peters, R. H. 1983. The ecological implications of body size. Cambridge University Press, Cambridge.

Peterson, A. T., O. A. Flores, S. Leon-Paniagua, J. Llorente, M. Martinez, A. Navarro-Singuenza, M. Torres-Chavez, and I. Vargas-Fernandez. 1993. Conservation priorities in Mexico: moving up in the world. Biodiversity Letters, 1:33–38.

Rahbek, C. 1995. The elevational gradient of species richness: a uniform pattern? Ecography, 18:200–205.

Rahbek, C. 1997. The relationship among area, elevation, and regional richness in Neotropical birds. American Naturalist, 149:875–902.

Rapoport, E. H. 1975. Areografia: Estrategias Geográficas de las Especies. Fondo de Cultura Económica, Mexico City.

Rapoport, E. H. 1982. Areography: Geographical Strategies of Species. Pergamon Press, Oxford,.

Rapoport, E. H. 1994. Remarks on marine and continental biogeography: an areographical viewpoint. Philosophical Transactions of the Royal Society of London B, 343:71–78.

Robinson, J. G., and K. H. Redford. 1986. Body size, diet, and population density of Neotropical forest mammals. American Naturalist, 128:665–680.

Rohde, K. 1992. Latitudinal gradients in species diversity: the search for the primary cause. Oikos, 65:514–527.

Rohde, K., M. Heap, and D. Heap. 1993. Rapoport's rule does not apply to marine teleosts and cannot explain latitudinal gradients in species richness. American Naturalist, 142:1–16.

Rosenzweig, M. L. 1995. Species Diversity in Space and Time. Cambridge University Press, Cambridge.

Roy, K., D. Jablonski, and J. W. Valentine. 1994. Eastern Pacific molluscan provinces and latitudinal diversity: no evidence for "Rapoport's rule." Proceedings of the National Academy of Sciences of the USA, 91:8871–8874.

Ruggiero, A. 1994. Latitudinal correlates of the sizes of mammalian geographical ranges in South America. Journal of Biogeography, 21:545–559.

Sanchez-Cordero, V. 1993. Biological surveys and conservation in Mexico. Association of Systematics Collections Newsletter, 21:54–58.

Sanchez-Cordero, V. 2001. Elevational gradients of diversity for rodents and bats in Oaxaca, Mexico. Global Ecology and Biogeography, 10:63–76.

Schmidt-Nielsen, K. 1971. How birds breathe. Scientific American, 225:72–88.

Schmidt-Nielsen, K. 1984. Scaling: Why Is Animal Size So Important? Cambridge University Press, Cambridge.

Silva, M., and J. A. Downing. 1995. CRC Handbook of Mammalian Body Masses. CRC Press, Boca Raton, Fla.

Simmons, N. B. 1996. A new species of *Micronycteris* (Chiroptera: Phyllostomidae) from northeastern Brazil, with comments on phylogenetic relationships. American Museum Novitates, 3158:1–34.

Simmons, N. B. 2003. Order Chiroptera. *In:* Mammal Species of the World: A Taxonomic and Geographic Reference. 2d ed. (D. E. Wilson and D. M. Reeder, eds.). Smithsonian Institution Press, Washington, D.C. (in press).

Simmons, N. B., and C. O. Handley, Jr. 1998. A revision of *Centronycteris* Gray (Chiroptera: Emballonuridae) with notes on natural history. American Museum Novitates, 3239:1–28.

Simpson, G. G. 1964. Species density of North American recent mammals. Systematic Zoology, 13:57–73.

Smith, F. D. M., R. M. May, and P. H. Harvey. 1994. Geographical ranges of Australian mammals. Journal of Animal Ecology, 63:441–450.

Sokal, R. R., and F. J. Rohlf. 1995. Biometry: The Principles and Practice of Statistics in Biological Research. 3d ed. W. H. Freeman & Co., New York.

Stevens, G. C. 1989. The latitudinal gradient in geographical range size: how so many species coexist in the tropics. American Naturalist, 133:240–256.

Stevens, G. C. 1992. The elevational gradient in altitudinal range: an extension of Rapoport's latitudinal rule to altitude. American Naturalist, 140:893–911.

Stevens, G. C., and J. F. Fox. 1991. The causes of treeline. Annual Review of Ecology and Systematics, 22:177–191.

Stevens, R. D., and M. R. Willig. 1999. Size assortment in New World bat communities. Journal of Mammalogy, 80:644–658.

Stevens, R. D., and M. R. Willig. 2000. Density compensation in New World bat communities. Oikos, 89:1–11.

Stevens, R. D., and M. R. Willig. 2002. Geographical ecology at the community level: perspectives in the diversity of New World bats. Ecology 83:545–560.

Strahan, R., ed. 1983. The American Museum Complete Book of Australian Mammals. Angus & Robertson, Melbourne.

Strahan, R., ed. 1995. Mammals of Australia. 2d ed. Smithsonian Institution Press, Washington D.C.

Taylor, C. M., and N. J. Gotelli. 1994. The macroecology of *Cyprinella:* correlates of phylogeny, body size and geographical range. American Naturalist, 144:549–569.

Terborgh, J. 1977. Bird species diversity on an Andean elevational gradient. Ecology, 58:1007–1019.

Tokeshi, M. 1993. Species abundance patterns and community structure. Advances in Ecological Research, 24:111–186.

Tuttle, M. D. 1970. Distribution and zoogeography of Peruvian bats, with comments on natural history. University of Kansas Science Bulletin, 49:45–86.

Wetterer, A., M. Rockman, and N. B. Simmons. 2000. Phylogeny of phyllostomid bats (Mammalia: Chiroptera): data from diverse morphological systems, sex chromosomes, and restriction sites. Bulletin of the American Museum of Natural History, 248:1–200.

Whittaker, R. H. 1972. Evolution and measurement of species diversity. Taxon, 21: 213–251.

Willig, M. R. 2000. Latitude, common trends within. Pp. 701–714 *in:* Encyclopedia of Biodiversity. Vol. 3 (S. Levin, ed.). Academic Press, San Diego, Calif.

Willig, M. R., and M. R. Gannon. 1997. Gradients of species density and turnover in marsupials: a hemispheric perspective. Journal of Mammalogy, 78:756–765.

Willig, M. R., and S. K. Lyons. 1998. An analytical model of latitudinal gradients of species richness with an empirical test for marsupials and bats in the New World. Oikos, 81:93–98.

Willig, M. R., and M. A. Mares. 1989. A comparison of bat assemblages from phytogeographic zones of Venezuela. Special Publications, the Museum, Texas Tech University, 28:59–67.

Willig, M. R., and E. A. Sandlin. 1991. Gradients of species density and turnover in New World bats: a comparison of quadrat and band methodologies. Pp. 81–96 *in:* Latin American Mammals: Their Conservation, Ecology, and Evolution (M. A. Mares and D. J. Schmidly, eds.). University of Oklahoma Press, Norman.

Willig, M. R., and K. W. Selcer. 1989. Bat species density gradients in the New World: a statistical assessment. Journal of Biogeography, 16:189–195.

Wilson, D. E., and D. M. Reeder, eds. 2003. Mammal Species of the World: A Taxonomic and Geographic Reference. 2d ed. Smithsonian Institution Press, Washington, D.C. (in press).

Wilson, J. W. 1974. Analytical zoogeography of North American mammals. Evolution, 28:124–140.

CHAPTER 14

Bats, Emerging Virus Infections, and the Rabies Paradigm

Sharon L. Messenger, Charles E. Rupprecht, and Jean S. Smith

Introduction

More than a half century ago, it was stated that the role of diseases in wild-life conservation had been vastly underestimated (Leopold [1933] 1986). This statement is as valid today as it was in 1933, particularly as it relates to bat ecology, which has received scant review beyond Constantine (1970, 1988a) and Sulkin and Allen (1974). Why renew a discussion of bats and disease? The answers are obvious, whether your perspective is one of public health, conservation biology, or bat biology in general. First, there is a heightened anxiety over emerging (i.e., newly evolved or newly recognized), and old but reemerging, infectious diseases (see below, "Bats and the Recent Emergence of New Viral Diseases"), and bats have been implicated in several recent highly publicized disease outbreaks (e.g., Ebola virus [Monath 1999] and Hendra virus [Halpin et al. 1996]). While bat-transmitted human diseases are a public health concern and an occupational hazard to bat researchers, the public health measures for preventing human disease may have the potential to affect bat populations adversely.

Second, conservation biologists must consider disease as a possible reason for any decline in a bat population. Interestingly, those factors often cited as contributing to a species decline and extinction (e.g., environmental degradation, deforestation, and introduction of exotic animal and plant species) are also cited widely as factors playing a role in the emergence of new diseases. Thus, species already threatened by the erosion of their habitats may be particularly sensitive to disease outbreaks. For example, die-offs in animal populations have been documented that appear to have been caused by environmental toxins (e.g., American white pelicans [*Pelicanus erythrorhynchos*], eared grebes [*Podiceps nigricollis*], and double-crested cormorants [*Phalacrocorax auritis*] at the notoriously polluted Salton Sea [Kaiser 1999]), yet in each case an infectious agent (which possibly flourished because of the degraded environment) was identified as the cause. Thus, species other than humans and their domesticated animals can be affected by emerging infections.

A final and perhaps less obvious reason for reviewing disease and bat ecology is to pose the question, How can one ignore the subject of disease ecology as part of any inquiry into the interrelationships between an organism and its environment? A thorough treatise on the ecology of bats should deal with the obvious question, What are the major mortality factors at work in bat populations? Several authors (Keen 1988; Tuttle and Stevenson 1982) have pointed out the inherent dilemmas of accurate measurement and reliable prediction of bat survivorship. Gillette and Kimbrough (1970) indicated that mortality is one of the least understood aspects of the natural history of bats. In addition to infectious diseases, mortality factors that could jeopardize bats include direct or indirect destruction by humans (e.g., loss of critical habitat, pesticides), environmental stressors (e.g., unavailability of water or food, inclement weather), natural catastrophic events (e.g., fires, floods, tornadoes, hurricanes, and other unpredictable phenomena), and predation. The significance of these factors on the overall mortality has rarely been documented for even a single population of any bat species, and infectious disease is almost never mentioned as a major issue (but see Burns et al. 1956b; Constantine 1967a; Constantine et al. 1968b).

Mass die-offs of bats are uncommon but potentially devastating, yet there is almost no thorough investigation and detailed documentation of such events (but see Pybus et al. 1986). Mass bird mortality, in contrast, is documented in quarterly reports of wildlife disease journals. (See the "Quarterly Wildlife Mortality Reports" in supplements to the *Journal of Wildlife Diseases* and U.S. National Wildlife Health Center at http://www.nwhc.usgs.gov.) A 3-mo survey in North America (October–December 1998) documented more than 80,000 avian deaths as a result of diseases such as avian cholera, botulism, enteritis, and toxicosis (Journal of Wildlife Diseases 1999). A review of similar records from 1969 to 1998 found a total of only 50 dead bats submitted for routine diagnosis to the same federal agency (R. G. McLean, personal communication). Not surprisingly, large, visible bird species (e.g., *Pelicanus erythrorhynchos*), particularly seabirds and waterfowl used for game and managed by multiple government and private organizations, receive notable attention. Thus, it is not clear whether this discrepancy in mortality rates between birds and bats is real or results from biased surveillance.

Some mass mortality events of bats have been noted anecdotally. Pierson and Rainey (1992) summarized data on episodes of apparent epidemic disease in Pacific flying fox populations. During survey expeditions in the 1930s to Micronesia, researchers learned from island residents that nearly all of the local *Pteropus mariannus* had died at the same time that an outbreak of measles afflicted the human population. Similarly, an "epidemic" event of unknown origin was suspected in the depletion of *P. tonganus* populations in Fiji and was reported during the 1940s. Flannery (1989) described more recent episodes in the Admiralty Islands, where many *P. neohibernicus* were found dead in 1985. A similar incident in the northern Solomon Islands involving *P. rayneri* suggested a recent disease introduction by domestic animals. In southern Africa in 1980, King et al. (1994) reported notable mortality in a population of epauleted fruit bats (*Epomops dobsoni*) in which 10%–15% of several hundred dead or dying bats were confirmed to be infected with Lagos bat virus. While describing the preservation of bat carcasses and the likelihood of fossil formation under cave conditions, Paradiso (1975, 268) made the following somewhat startling observation regarding *Mormoops:* "Thousands of dead bats have been found on the floors of caves in Mexico. They may have died from rabies." Yet no further information was provided about how this conclusion of disease was reached, what other disease(s) could have been present, or even a sense that this was an unusual event! The only study (Constantine 1967a, 1967c; Constantine et al. 1968b) thoroughly investigating mass mortality in bats is discussed in more detail below (See "Disease Ecology in Bats: The Rabies Paradigm.").

Although a wealth of parasites has been associated with bats, these animals have often been neglected as subjects in the ecology of parasitism (Coggins 1988). In short, as this book and others clearly document, almost all anatomical and physiological systems of bats have been dissected to a detailed level (e.g., skeleton and musculature of the wing for understanding locomotion, or the dentition and digestive tract for inferring diet); however, singular knowledge of the finer structure and function of the immune system of Chiroptera probably lags behind all other courses of related biological study.

The objective of this chapter is to provide a preliminary overview of infectious disease and bats, with a focus on the role of bats in the transmission and maintenance of infectious diseases and the potential impact that these diseases have on bats and other animals. Moreover, we seek to renew a dialogue with professional investigators to consider the unique basic and applied research opportunities available in the field today. Several related questions are posed for discussion. What important microbiological agents have been associated with bats? Are there generalized features or life-history strategies of bats that protect individuals from or predispose them to infection? Do these infectious agents have an impact on bat mortality (or other species) to any significant degree? Are there genuine or potential risks to those working with bats? Regrettably, there are no comprehensive answers to most of these queries.

A Primer of Terms

There is a conventional vocabulary in the field of epizootiology, or the study of occurrence and distribution of infectious disease in animal populations. The typical infectious disease triad includes the host, agent, and environment, where the environment encompasses all of the relevant abiotic (e.g., altitude, climate, season) and biotic components (e.g., nutritional status, stress) affecting the interaction of a host and agent. A host can refer to a species or a single individual that maintains the disease. The natural residence or habitat of a disease-causing agent is the "reservoir," which can be living (e.g., mosquito, bird, mammal) or nonliving (e.g., forest soil, guano deposits). Transmission of infectious agents to hosts can be differentiated as "vertical" (from parent to offspring such as in transplacental or lactogenic transmission) or "horizontal" (from individual to individual in a population). Transmission can occur directly between conspecifics or different species or indirectly via fomites (i.e., infected inanimate vehicles such as cave air or pond water) or living vectors (e.g., frequently arthropods).

Some diseases are worldwide in distribution, irrespective of the actual zoogeographic distribution of any single putative host or reservoir, while others are hidden in secluded niches, restricted by the intimacy of the relationship among the host, agent, and environment. Narrow host ranges (e.g., restriction to a single major host species with fairly discrete geographic boundaries) can lead to compartmentalized distributions of the agent. Other non-reservoir animal species may be incidentally infected through contact with primary reservoir species (i.e., "spillover"). Such cases, however, are relatively rare. Inherent in the definition of a spillover is that such events are evolutionary "dead ends" for the agent (i.e., a long-term cycle of transmission in the spillover host cannot be established). An "enzootic" disease displays only minor shifts in prevalence, or frequency of occurrence over time. In contrast, "epizootics," also known as die-offs or outbreaks, are operationally defined as unusually high disease occurrences that can sporadically affect a large number of animals of one or several species. One proposed operational definition of an epizootic is two or more standard deviations beyond average expected mortality, highlighting the need for baseline data to define what "expected" mortality might be for a population of bats. Would double or triple the normal range of mortality be noticed by current surveillance methods? If not, what is the utility of such a bellwether for environmental sentinels? Such questions pointedly illustrate that the study of bats and infection currently is a qualitative, descriptive endeavor. Providing the necessary quantitative measures of disease rates in bat populations will be a formidable challenge.

Infection (i.e., entry and reproduction of the agent or pathogen into or on the host) that results in any homeostatic disruption or abnormal condition is an "infectious disease." Physical contact between an individual host (e.g., bat) and a particular agent (e.g., virus), however, does not always lead to a successful infection. Mere contact of a pathogen with a susceptible individual may be fruitless if barriers such as fur and intact skin cannot be breached. Blood-feeding ectoparasites may be effective "hypodermic syringes," enabling a pathogen to bypass these barriers, but if the agent is not adapted to the host, a productive infection (i.e., an infection that will ultimately lead to the transmission of viable progeny from one host to another) may not ensue. Despite this, entrance may still generate a specific immune response. For example,

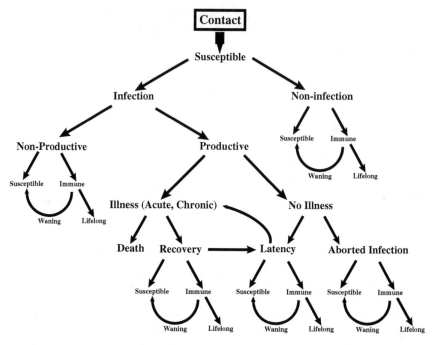

Figure 14.1. Hypothetical outcome of contact between bats and a potential pathogen.

consumption of a fruit laden with plant virus by a *Pteropus* bat does not result in viral replication, but exposure of the host to even a plant virus may initiate a specific immune response. If a pathogen gains access to an appropriate receptor in a permissive host, an infection is possible but not necessarily assured. The agent must be intact, replication competent, and the dose sufficient. A productive infection occurs only after appropriate cellular entry, progeny production, and successful transmission, all the while evading host defenses. The outcome of any contact between host and agent is based on a complex algorithm of probabilities (fig. 14.1), and, in populations, the classes of susceptible, infected, and immune individuals can be dynamic. All of these factors contribute to the difficulty of studying infectious diseases in the field.

Biological Attributes of Bats and Disease Facilitation

Are there common features of Chiroptera as a group that may facilitate their role in pathogen evolution, propagation, or spread? All bats fly, some are abundant and widely distributed, and many are highly gregarious. In theory, these key attributes may contribute to a greater occurrence of disease compared to other mammalian groups (Sulkin and Allen 1974). When contemplating mammalian infectious disease reservoirs throughout past and recent history, rodents are viewed as the most formidable from a public health perspective, as they host many vector-borne and zoonotic diseases. Yet some of the same traits that give rodents this dubious recognition (e.g., large aggregations of high density, species that are commensal with humans) may also apply to bats. High population densities of susceptible individuals present opportunities for disease-causing pathogens to invade and spread rapidly. Some highly gregarious species (e.g., *Tadarida brasiliensis*) comprise the largest aggregations of mammals in terms of absolute numbers of individuals, and they are the predominant biomass in many communities. Also, because of similar life histories, many bat species may share the same habitats. Several species of bats may roost in a single cave or roosting site or may forage in the same immediate area. Use of a site by several species raises the possibility of interspecific transfer of disease-causing agents. For example, there is an anecdotal account of a rabid hoary bat (*Lasiurus cinereus*) attacking other bats (*Lasionycteris noctivagans, Eptesicus fuscus,* and *T. brasiliensis*) in a foraging area (Bell 1980). Unfortunately, the hoary bat cited as rabid by Bell (1980) was only diagnosed by Negri bodies and could not later be confirmed in a Canadian laboratory (D. G. Constantine, personal communication).

Flight, the predominant characteristic distinguishing bats from all other mammals, carries obvious implications for infectious disease transmission. Bats have a greater capacity than nonvolant mammals, humans excepted, to disperse quickly to new areas, such as islands. The migratory habits of some bats provide an opportunity for pathogens to cover long distances and bridge gaps between species assemblages that otherwise might not be in contact. Considering the inherent importance of this overriding characteristic, one can draw parallels to birds, in which notable maladies, such as influenza A, can be tracked along migration routes (Ito et al. 1995).

Despite salient features of bat biology that could enhance their capacity to be hosts of infectious disease, some life-history characteristics of bats would make them particularly vulnerable to the burden of infectious diseases. Bats are relatively long-lived and have late onset of sexual maturity and low fecundity for mammals of their size (see Barclay and Herder, this volume). Recovery from the impact of a highly transmissible infectious pathogen with a high case fatality rate could be much more difficult for bats than for other mammal species with higher fecundity. Thus, the value of studying the ecology of infectious disease in bats is not only to understand the details of how infectious pathogens are maintained in and have an impact on bat populations but also to ask whether, because of their unique suite of life-history traits, bats have evolved novel strategies to contend with infectious diseases.

In fact, few data indicate that bats play a prominent role as reservoirs of infectious diseases. Is this dearth of bona fide disease cases real, or one of perception? The challenges of studying these flying, nocturnal, and often secretive animals limited studies of bats until well into the 1940s and may help to explain why bats are thought to play a minor role in disease maintenance and transmission.

Bats and the Recent Emergence of New Viral Diseases

Currently, some 25 families of viruses (as defined by the International Committee on Taxonomy of Viruses or ICTV [Murphy et al. 1995]) are known to infect vertebrates. Ten of these families have been associated with bats (table 14.1), and most are RNA viruses. Only one of the nine families of DNA viruses associated with vertebrates has a member associated with bats (the Herpesviridae), and this virus is currently restricted to the New World. Conversely, at least nine of the 16 virus families of RNA viruses contain members that infect bats. This bias toward infection with RNA viruses could be an artifact of sampling. Given that RNA viruses cause many serious human and veterinary diseases, there has been quite extensive surveillance for the reservoirs of some of these deadly viruses.

The Mononegavirales (negative-stranded, unsegmented RNA viruses) contains three families: the Filoviridae, Paramyxoviridae, and Rhabdoviridae, all of which contain members with possible, likely, or well-documented associations with bats. In addition, at least one virus within each family fits the criterion for an emerging infectious disease agent that is lethal for humans and other mammals.

Marburg and Ebola Viruses

Besides the human immunodeficiency virus (HIV), the Filoviridae (e.g., Marburg and Ebola) are perhaps the most publicized of modern emerging viruses (Feldmann et al. 1996). Marburg and Ebola have produced sporadic but dramatic outbreaks, characterized by hemorrhagic fevers with high mortality in humans. Marburg virus was first isolated in 1967 in Europe where laboratory workers who processed tissues of green monkeys obtained from Uganda were infected (Martini and Siegert 1971). In 1987, Marburg virus was also identified in one specific episode in which a child, who was known to frequent a cave (on Mount Elgon on the Kenya-Uganda border) that was inhabited by numerous bats, was infected (Peters et al. 1996). During 1976, Ebola virus was discovered in Zaire and Sudan. In the Sudan outbreak, and in another outbreak in 1979, the index human cases were workers in an old cotton factory, where the roof above a storeroom contained large numbers of bats. Over the course of these outbreaks, the source of these viruses remained a mystery. There was only a circumstantial association between these outbreaks and the presence of non-human primates or bats.

More than 30 yr since the initial recognition of Marburg and Ebola, the natural history of these filoviruses still remains unexplained. Despite intense scrutiny, the natural reservoir(s) for these viruses has not been discovered. During 1979–1980, in the Democratic Republic of the Congo and Cameroon (Breman et al. 1999) and again in 1995 in the Democratic Republic of the Congo (Leirs et al. 1999), field investigations were conducted near the sites of previous Ebola outbreaks. Thousands of samples from more than 100 species (including plants, invertebrates, birds, rodents, and bats) were tested for the presence of (e.g., virus isolation) or exposure to (i.e., presence of virus-specific antibody) Ebola virus. All serological tests were negative, and attempts to isolate virus have been unsuccessful.

Investigations have been hampered by unpredictable occurrences of Ebola and the combination of long time lags between disease initiation and clinical recognition, remote locations, imprecise data on primary human exposures, and sample-size limitations for specimen collections, particularly for rarer species. The elusive identity of the reservoir(s) provides a continual source of speculation. Some researchers have suggested that bats are leading contenders as reservoirs (Monath 1999), a belief partially supported by experimental data. Of more than 40 species of plants, vertebrates, and invertebrates inoculated with Ebola virus, only frugivorous (*Epomophorus wahlbergi*) and insectivorous (*Mops condylura* and *Chaerephon pumilus*) bats supported both replication and circulation of virus (Swanepoel et al. 1996). From 41 collective pools of bat viscera tested, 22 different virus isolations were recovered from the bats 1 mo after inoculation. Ebola virus was also isolated from the feces of one *E. wahlbergi,* and virus antigen was observed in the endothelial cells of the lung of a single *C. pumilus*. However, no bats died from Ebola, nor did they demonstrate clinical signs of illness. Further laboratory and field research may elucidate whether any bat species play a role in the epidemiology of filoviruses, such as Ebola.

Hendra Virus

Recently, bats have been associated with the maintenance of three new members of the virus family Paramyxoviridae. The first virus associated with human and domestic animal disease was Hendra virus (Allworth et al. 1995; Murray et al. 1995; Paterson et al. 1998; Selvey and Sheridan 1994; Selvey et al. 1996). It was first identified in Mackay, a coastal town in southeast Queensland, Australia, during August 1994, when two horses and a horse breeder developed respiratory and neurological illness and later died (Hooper et al. 1996; O'Sullivan et al. 1997; Rogers et al. 1996). In September 1994, a second episode of a similar acute respiratory infection affected humans and horses at stables

Table 14.1. Viruses associated with bats

Virus Family/Virus Group (Name)	Bat Taxa	Location	References
DNA viruses:			
Herpesviridae:			
UA (AP)	*Carollia* sp.	Brazil	Karabatsos 1985
UA (CM)	*Myotis lucifugus*	USA	Tandler 1996
RNA viruses:			
Reoviridae:			
Orbivirus (IFE)	*Eidolon helvum*	Nigeria, CAR, Cameroon	Kemp et al. 1988
Orbivirus (IAP)	*Sycomycteris* sp.	Papua New Guinea	Karabatsos 1985
Togaviridae:			
Alphavirus (SIN)	*Hipposideros, Rhinolophus* sp.	Zimbabwe	Blackburn et al. 1982
Alphavirus (VEE)	*Carollia perspicillata*	Colombia	Constantine 1970, table 1
	Artibeus lituratus	Panama	Constantine 1970, table 1
	Desmodus rotundus	Mexico	Correa-Giron et al. 1972
	Uroderma bilobatum, Desmodus rotundus	Guatemala	Seymour et al. 1978
	Desmodus rotundus	Ecuador	Constantine 1988a
Alphavirus (WEE)	*Eptesicus* and other spp.	USA, Brazil	Constantine 1970, table 1
	Artibeus jamaicensis	Haiti	McLean et al. 1979
Alphavirus (VEE, WEE, EEE)	Multiple species	Guatemala	Ubico and McLean 1995
Alphavirus (RR)	*Pteropus* sp.	Australia	Doherty et al. 1966
Alphavirus (CHIK)	*Scotophilus* sp., *Hipposideros* sp.	Senegal	Constantine 1970, table 1; Diallo et al. 1999
Alphavirus (SF)	Multiple species	Uganda	Constantine 1970, table 1
Alphavirus (MAY)	*Mops condylurus?*	Uganda	Constantine 1970, table 1
Alphavirus (MUC)	*Carollia* sp.	Colombia	Sulkin and Allen 1974
Alphavirus (PIX)	*Carollia* sp., *Artibeus* sp.	Brazil	Constantine 1970, table 1
Alphavirus (UNA)	?	Brazil	Constantine 1970, table 1
UA (CAB)	?	Brazil	Constantine 1970, table 1
	?	French Guiana	Digoutte and Girault 1976
Coronaviridae:			
UA (BOC)	*Myotis lucifugus*	USA	Karabatsos 1985

Virus	Host species	Location	Reference
Flaviviridae:			
Flavivirus (YF)	*Glossophaga soricina*	Brazil	Constantine 1970, table 1
	Multiple species	East Africa	Constantine 1970, table 1
	Epomophorus sp.	Ethiopia	Constantine 1970, table 1
Flavivirus (WN)	?	Egypt	Constantine 1970, table 1
	Pteropus sp.	Australia	Constantine 1970, table 1
	Multiple species	East Africa	Constantine 1970, table 1
	Rousettus sp.	Israel	Constantine 1970, table 1
	?	Ethiopia	Constantine 1970, table 1
	Rhinolophus sp.	Italy	Sulkin and Allen 1974
	Rousettus leschenaulti	India	Paul et al. 1970
Flavivirus (MVE)	*Pteropus* sp., *Eptesicus* sp., *Pipistrellus pumilus?*	Australia	Constantine 1970, table 1
Flavivirus (SLE)	Multiple species	USA	Constantine 1970, table 1; Herbold et al. 1983; Karabatsos 1985
Flavivirus (RB)	*Rhinopylla pumilio*	Brazil	Constantine 1970, table 1
	Multiple species	Guatemala	Ubico and McLean 1995
	Tadarida brasiliensis, *Eptesicus fuscus*	USA	Constantine 1970, table 1; Karabatsos 1985
Flavivirus (PPB)	Multiple species	Guatemala	Ubico and McLean 1995
Flavivirus (BUB)	*Cynopterus brachyotis*	Cambodia	Salaun et al. 1974
	Cynopterus brachyotis, *Eonycteris spelaea*	Malaysia	Karabatsos 1985
Flavivirus (DB)	*Chaerephon pumila?*	Uganda	Constantine 1970, table 1
	Chaerephon pumila	Tanzania	Constantine 1970, table 1
	Multiple species	East Africa, Nigeria, Senegal	Constantine 1970, table 1; Karabatsos 1985
Flavivirus (ENT)	*Chaerephon pumila*	Madagascar	Cassel-Beraud et al. 1989
	Multiple species	East Africa	Constantine 1970, table 1; Karabatsos 1985
Flavivirus (SAB)	?	Senegal	Karabatsos 1985; Fontenille et al. 1994
	?	Guinea	Butenko 1996

(continued on next page)

Table 14.1. *(continued)*

Virus Family/ Virus Group (Name)	Bat Taxa	Location	References
Flavivirus (JE)	*Pteropus* sp.	Australia	Constantine 1970, table 1
	Hipposideros armiger terasensis, Miniopterus fuliginosus	Taiwan	Constantine 1970, table 1
	Cynopterus brachyotis	Thailand	Constantine 1970, table 1
	Pteropus vampyrus	Malaysia	Constantine 1970, table 1
	Multiple species	India	Constantine 1970, table 1; Banerjee et al. 1988
Flavivirus (TAB)*	Multiple species	Japan	Constantine 1970, table 1
Flavivirus (KYA)	Multiple species	Trinidad	Price 1978
	Multiple species	India	Constantine 1970, table 1; Bhat et al. 1978
Flavivirus (CI)	*Cynopterus brachyotis, Macroglossus lagochilus*	Malaysia	Karabatsos 1985
Flavivirus (JUG)	*Cynopterus brachyotis*	Malaysia	Karabatsos 1985
Flavivirus (MML)	*Myotis lucifugus*	USA	Constantine 1970, table 1
Flavivirus (BSQ)	?	Brazil	Constantine 1970, table 1
Flavivirus (ILH)	?	Brazil	Constantine 1970, table 1
Flavivirus (SOK)	*Vespertilio?*	USSR	Lvov et al. 1973
UA (BB)	?	Central Africa	Constantine 1970, table 1
UA (TOK)	*Vespertilio superans*	Japan	Miura and Kitaoka 1977
UA (MSB)	*Otomops martiensseni*	East Africa	Constantine 1970, table 1
Negative-stranded RNA viruses:			
Paramyxoviridae:			
UA (BPI)	*Rousettus leschenaulti*	India	Pavri et al. 1971
UA (MAP)	*Sturnira lilium*	Brazil	Zeller et al. 1989
UA (HE)	*Pteropus poliocephalus, Pteropus alecto, Pteropus scapulatus*	Australia	Halpin et al. 1996
UA (ME)	*Pteropus poliocephalus, Pteropus alecto, Pteropus conspicillatus*	Australia	Philbey et al. 1998
Rhabdoviridae:			
Vesiculovirus (VS)	Multiple species	Guatemala	Ubico and McLean 1995
Vesiculovirus (MEB)	*Rhinolophus hildebrandti*	Uganda	Constantine 1970, table 1
Lyssavirus	Multiple species	Worldwide	this chapter
UA (KC)	*Myotis yumanensis*	USA	Karabatsos 1985
UA (GOS)	*Tadarida* sp.	Senegal	Karabatsos 1985

Orthomyxoviridae:			
Influenzavirus (FLU)	*Nyctalus noctula*	Kazakhstan	Lvov et al. 1979
	Multiple species	India	Kelkar et al. 1981
Bunyaviridae:			
Phlebovirus (RVF)	*Micropteropus, Hipposideros* sp.	Guinea	Boiro et al. 1987
Phlebovirus (TOS)	*Pipistrellus kuhlii*	Italy	Verani et al. 1988
UA (YOG)	*Rousettus aegyptiacus*	Senegal	Karabatsos 1985
UA (KASO)	*Rousettus aegyptiacus*	Uganda	Kalunda et al. 1986
UA (KTR)	*Chaerephon plicata, Cynopterus brachyotis, Scotophilus temmencki*	Malaysia	Karabatsos 1985; Varma and Converse 1976
UA (IK)	Multiple species	USSR	Lvov et al. 1973
UA (CUMC)	*Eptesicus serotinus, Rhinolophus ferrumequinum*	Korea	Kim et al. 1994
UA (NEP)	*Artibeus lituratus, Artibeus jamaicensis*	Honduras	Calisher et al. 1971
UA (CATU)	*Molossus* sp.	Brazil	Karabatsos 1985
UA (GMA)	?	Brazil	Karabatsos 1985
UA (KK)	*Chaerephon plicata, Taphozous theobaldi*	Thailand	Williams et al. 1976; Karabatsos 1985
Arenaviridae:			
Arenavirus (TAC)	*Artibeus lituratus, Artibeus jamaicensis*	Trinidad	Constantine 1970, table 1
	Multiple species	Trinidad	Price 1978
	Artibeus lituratus	Guatemala	Ubico and McLean 1995
UC:			
? (MDC)	?	Brazil	Karabatsos 1985
? (FOM)	*Nycteris nana*	Guinea	Boiro et al. 1986

Note. Abbreviations are as follows: AP, Agua Preta; BB, Bangui bat; BOC, Bocas; BPI, bat parainfluenzavirus; BSQ, Bussuquara; BUB, Bukalasa; CAB, Cabassou; CATU, Catu; CHIK, Chikungunya; CI, Carey Island; CM, cytomegalovirus; CUMC, putative hantaan-like isolates; DB, Dakar bat; EEE, eastern equine encephalitis; ENT, Entebbe bat; FLU, influenza H3N2; FOM, Fomede; GMA, Guama; GOS, Gossas; HE, Hendra (formerly equine morbillivirus); IFE, Ife; IK, Issyk-kul; ILH, Ilheus; JAP, Japanaut; JE, Japanese encephalitis; JUG, Jugra; KASO, Kasokero; KK, Kaeng Khoi; KTR, Keterah; KYA, Kyansur forest disease; MAP, Mapuera; MAY, Mayaro; MDC, Mojui dos Campos; ME, Menangle; MEB, Mount Elgon bat; MML, Montana Myotis leukoencephalitis; MSB, Mount Suswa bat; MUC, Mucambo; MVE, Murray Valley encephalitis; NEP, Nepuyo; PIX, Pixuna; PPB, Phnom Pen bat; RBV, Rio Bravo virus; RR, Ross River; RVF, Rift Valley Fever; SAB, Saboya; SF, Semliki Forest; SIN, Sinbis; SLE, Saint Louis encephalitis; SOK, Sokuluk; TAB, Tamana bat; TAC, Tacaribe; TOK, unnamed suspect Japanese flavivirus; TOS, Toscana; UA, unclassified to specific group within the family; UC, unclassified to a specific family; UNA, Una; VEE, Venezuelan equine encephalitis; VS, vesicular stomatitis; WEE, western equine encephalitis; WN, West Nile; YF, Yellow Fever; YOG, Yogue.

*Tentative taxonomic assignment.

in Hendra, a residential suburb of Brisbane, about 1,000 km south of Mackay. A virus isolated from the kidney of one individual who died, a horse trainer, was identical to an isolate from affected horses. Viral morphology suggested Paramyxoviridae (Hyatt and Selleck 1996). It was originally but incorrectly termed equine morbillivirus, based on the presumed host and genetic similarity to other morbilliviruses (e.g., measles, rinderpest, and canine distemper viruses). Interestingly, morbilliviruses typically have narrow host ranges, and no other paramyxovirus is known to cause serious illness in both horses and people.

Because of the distance between Mackay and Brisbane and the lack of known contact between the two affected groups, extensive testing of animals common to the two areas was conducted in an attempt to delineate the source of the virus (Young et al. 1996). In thousands of samples collected throughout Queensland, there was widespread evidence of infection in four flying fox species: *Pteropus alecto, P. poliocephalus, P. scapulatus,* and *P. conspicillatus*) originating from Darwin to Melbourne. Viruses obtained from bats were antigenically and genetically indistinguishable from earlier equine and human isolates (Halpin et al. 1996). Based on these data from serology (12%–53% of bats examined were antibody positive) and virus isolation, flying foxes were believed to be the natural hosts of this agent. Serologic evidence of infection also was reported in other fruit bats (*Dobsonia moluccense, D. andersoni, Pteropus neohibernicus, P. capistratus, P. hypomelanus,* and *P. admiralitatum*) in Papua New Guinea (Mackenzie 1999).

Epidemiological evidence suggests that Hendra virus is of low infectivity for both humans and domestic animals. There was no serological evidence of infection in hundreds of other humans who had contact with stables, sick horses, or ill human patients from the Mackay and Brisbane outbreaks (McCormack et al. 1999). Moreover, there was no evidence of infection in local rehabilitators who annually cared for often hundreds of sick, injured, or orphaned bats (Selvey et al. 1996).

Experimental data support the idea that Hendra virus is not highly contagious (Williamson et al. 1998). *Pteropus poliocephalus* inoculated with Hendra virus did not become ill. Six of eight inoculated bats developed antibody to the virus, and two of them developed vascular lesions containing viral antigen. Neither illness nor antibody was detected in three bats that were not inoculated or two horses housed with infected bats. In a second experiment, four horses inoculated with Hendra virus were placed in contact with three other horses and six cats. Three of the four inoculated horses developed disease. Hendra virus was detected in saliva, urine, and kidneys of affected horses, but neither illness nor antibody developed in noninoculated horses and cats. With a third contact experiment, Hendra virus transmission was shown from 12 infected cats to one of three uninfected horses. In other work, cats were susceptible to infection and infected other naive (i.e., previously unexposed to a pathogen) cats placed in contact with them (Westbury et al. 1996). To date, all human infections with Hendra virus seem to be transmitted directly by horses.

The virus may be associated with mammalian reproductive functions. The index case (i.e., first diagnosed case) in Brisbane was a pregnant mare, and a pregnant mare had been involved in Mackay. Moreover, both incidents occurred during the season of flying fox parturition, and virus was isolated from the uterine fluid of a pregnant animal. In experimentally infected pregnant guinea pigs and fruit bats (*P. poliocephalus*), Williams et al. (2000) reported that although no overt illness was observed with Hendra virus, subclinical disease occurred (i.e., the animals were positive for viral isolation, seroconversion, vascular lesions, and immunostaining for viral antigen).

A third outbreak was identified in January 1999 with the death of another Hendra virus–infected horse near Cairns. Clearly, we need further qualitative and quantitative details of the ecology of Hendra virus infection among bats and among other species (e.g., cats).

Menangle Virus

In 1997, a second paramyxovirus of biomedical importance was identified in Australian bats, when rates of farrowing, piglet survivorship, and deformities unexpectedly changed at a large commercial swine operation in New South Wales (Philbey et al. 1998). The virus isolated from tissues of affected piglets was similar in morphology to viruses in the family Paramyxoviridae, but the new agent, termed Menangle virus, appeared unrelated to any known paramyxovirus. Most swine had titers of serum neutralizing antibody to Menangle virus, as did several farm workers (Chant et al. 1998). These individuals described a flu-like illness at the time of the disease outbreak in swine. The presence of large breeding colonies of *Pteropus poliocephalus* and *P. scapulatus* within 200 km of the affected piggery provided circumstantial evidence to suggest Megachiroptera as a primary source of infection. Serological samples from many species (including rodents, birds, cattle, sheep, pigs, cats, dogs, fruit bats, and humans) were collected throughout Queensland and New South Wales (Mackenzie 1999; Philbey et al. 1998). Serum samples positive for antibodies to Menangle virus were found only in pigs, humans who worked at the piggeries, and *P. alecto* and *P. conspicillatus*. Serum samples collected from other animals were negative.

Nipah Virus

In the Indo-Pacific region, circumstantial evidence links bats to another emerging disease that has had devastating public health and agricultural consequences. From September 1998 through April 1999, more than 250 human cases of encephalitis (including more than 100 fatalities) were reported from Malaysia (CDC 1999a). Cases occurred primarily among male farm workers who had had direct contact with ill swine. Initially, Japanese encephalitis was suspected, but epidemiologic features suggested a different agent. Additionally, electron microscopy of clinical samples showed virus-like structures consistent with a paramyxovirus, and gene sequencing suggested a virus related to, but not identical with, Hendra virus. The virus was termed Nipah after one of the affected villages and is suggested to belong in a new genus, the "Megamyxovirus," together with Hendra virus. Affected swine appeared to be the direct source of infection to people, and nearly a million pigs were culled to stem the outbreak. How swine originally became infected remains a mystery. Studies are ongoing to determine if other domestic species (e.g., horses, dogs, cats, goats) or local wildlife (e.g., rodents, birds, bats) are implicated as reservoirs for this unique virus. Preliminary data suggest that as many as 25% of the more than 300 blood samples obtained from bats throughout Malaysia have antibody to Nipah virus, but virus isolation from bat tissue has not been accomplished. If a link can be established with this previously undescribed agent, it will be intriguing to unravel the thread of host, agent, and environmental factors that led to its initiation. It will be challenging to elucidate realistic prevention and control tactics that strike a reasonable balance between efficacy and conservation.

Regardless of the potential involvement of bats in these and future outbreaks, the remainder of this chapter centers on the third family of Mononegavirales, the Rhabdoviridae. We focus on rabies, an ancient disease, and its causative agents, the lyssaviruses. The rationale for this focus is that these viruses were the first viruses discovered to be associated with bats. Bat are considered to be the primary reservoir for lyssaviruses, with the possible exception of rabies virus and Mokola (reservoir unknown). Rabies viruses in bats, although only relatively recently discovered, are arguably the most thoroughly studied disease of bats, as documented in comprehensive reviews (e.g., Brass 1994). As such, rabies is the only disease for which the more challenging questions of disease ecology can be addressed, if only in a cursory manner.

The Discovery of Rabies-Related Viruses

Historically, rabies virus was considered the etiologic (i.e., causative) agent in all cases of viral encephalitis in which (1) inoculation of laboratory animals with brain material from a case produced a rabies encephalitis, (2) Negri bodies (i.e., eosinophilic intracytoplasmic inclusions) were found in histologic smears of brain material, or (3) fluorescein-labeled antibody to the rabies viral nucleoprotein (N protein) detected intracytoplasmic inclusions in infected neurons (Smith 1999). By the late 1970s, however, viruses related to but distinct from rabies virus had been identified. The genus *Lyssavirus* (*lyssa*, Greek for "rage, rabies" [Shope et al. 1970]) was created to encompass viruses that produce a rabies encephalitis in inoculated mice and react with antibodies to the N protein of rabies virus but that can be differentiated from rabies virus by a markedly reduced cross-reactivity in virus neutralization tests with antibody to the glycoprotein (G protein) of rabies virus and, in some cases, by a failure of rabies virus vaccines to confer protection from infection (table 14.2).

Table 14.2. Currently recognized species in the genus *Lyssavirus*

	Rabies Virus	Australian Bat Virus	Duvenhage Virus	European Bat Lyssavirus 1	European Bat Lyssavirus 2	Mokola Virus	Lagos Bat Virus
Distribution	Worldwide with the exception of Antarctica and some insular regions	Australia	South Africa, Zimbabwe, Guinea	Germany, Denmark, Holland, Poland, Russia, Ukraine, France, Spain	Holland, United Kingdom, Finland, Switzerland	Nigeria, South Africa, Cameroon, Zimbabwe, Central African Republic, Ethiopia	Nigeria, South Africa, Zimbabwe, Central African Republic, Senegal, Ethiopia
Reservoir	Carnivora (Canidae, wild and domestic; Mustelidae, Viverridae) Microchiroptera (*Desmodus rotundus* and almost all species of insectivorous bats that have been adequately sampled)	Megachiroptera (*Pteropus* sp.) and Microchiroptera (*Saccolaimus flaviventris*)	Probably Microchiroptera. Single cases attributed to *Miniopterus schreibersii*, *Nycteris gambiensis*, and *N. thebaica*	Microchiroptera (almost all cases in *Eptesicus serotinus*)	Microchiroptera (almost all cases in *Myotis dasycneme* and *M. daubentonii*)	Probably Insectivora (*Crocidura* sp.) or Rodentia (*Lophyromys sikapusi*), but most reported cases are from domestic cats	Probably Megachiroptera. Cases reported in *Eidolon helvum*, *Micropterus pusillus*, and *Epomophorus wahlbergi*
Reported cases:							
Animals	1000s per year	<100	<10	approx. 10 per year	<10	<25	<25
Humans	1000s per year	2	1	1 confirmed, 1 suspected	1	2	0
Amino acid homology to rabies virus N protein	...	92	86	87	85	76	79
Neutralizing titer of two international units of rabies immune globulin against approx. 50 infectious units of virus	1:940	1:625	1:125	1:125	1:125	1:5	1:12

Note. Data are summarized from recent review articles (Bourhy et al. 1993; Mackenzie 1999; Swanepoel et al. 1993) and unpublished data (Centers for Disease Control and Prevention).

Definitive identification of a viral species within the genus *Lyssavirus* is based on nucleotide sequence analysis (fig. 14.2).

Although originally identified as rabies virus, the first rabies cases involving a lyssavirus species other than rabies virus were probably the few isolated bat rabies cases reported by public health laboratories in Europe and Asia in the early 1950s. The 1953 report of insectivorous bat rabies in the United States generated interest in bats and rabies. In 1954, single cases of rabies were reported in a bat in Germany (unidentified species) and in Yugoslavia (*Nyctalus noctula*), and a rabies death was also reported in a man bitten by a bat (unidentified species) near the village of Kaviti, India (Nikolic and Jelesic 1956, Veeraraghavan 1954; WHO 1982). Over the next decade, only isolated rabies cases were reported in Old World bats, almost all from Germany. Given the nonspecificity of rabies diagnostic tests of the time, little importance was attached to these early reports, and no further work was done on these cases. Boulger and Porterfield (1958) looked for rabies in fruit bats (*Eidolon helvum*) collected in Lagos Island, Nigeria. The single virus isolate they found was registered as a possible arbovirus because the characteristics of the virus did not match those of rabies virus. The virus was distributed widely to viral taxonomists in hopes of identification, but for almost a decade, little work was done on any virus isolate from Old World bats.

By the early 1970s, Shope et al. (1970) had recognized Mokola and Lagos bat virus as related to but distinct from rabies virus. Based on these findings, Meredith et al. (1971) pursued a diagnosis of rabies in a man who died of an encephalitis in Pretoria, South Africa, 5 wk after a bat had bitten him (described by family members as *Miniopterus schreibersi*). The virus, named Duvenhage virus after the victim, had initially been negative in diagnostic tests for rabies but reacted with high-titered antibody to rabies virus (Tignor et al. 1977).

In August 1982, rabies was reported in an unidentified bat caught in the U.S. Army Hospital in Bremerhaven, Germany (WHO 1982). Monoclonal antibody tests on this virus and two earlier isolates from unidentified bats in Germany (Hamburg in 1968 and Stade in 1970) showed that the viruses were closely related antigenically to Duvenhage. An accidental introduction from Africa was suspected because of the small number of cases and the fact that all cases originated from coastal areas. When an identical virus was found in hundreds of rabid bats (primarily *Eptesicus serotinus*) in Denmark in 1985–1986, and rabid bats were identified throughout Europe, it became clear that the virus was well established in bat populations in Europe. The death from rabies in 1985 of a Finnish bat biologist and the retrospective identification of an earlier bat-associated human death in Russia confirmed the virulence of the European bat viruses for humans (Amengual et al. 1997; Lumio et al. 1986). By reaction with panels of monoclonal antibodies and nucleotide sequence analysis, the European bat lyssaviruses were identified as two distinct species (Bourhy et al. 1993), European bat lyssavirus I and II.

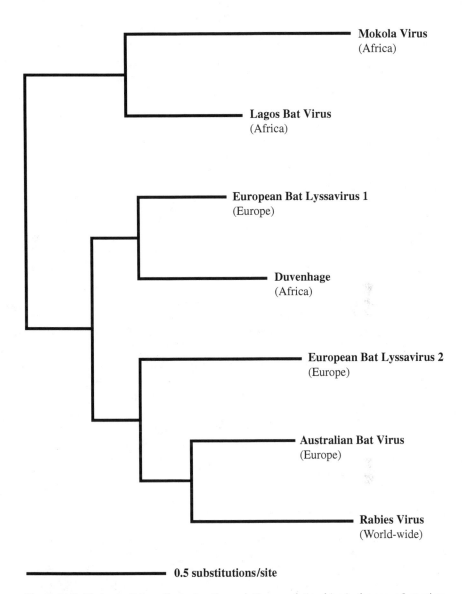

0.5 substitutions/site

Figure 14.2. Phylogenetic tree illustrating the evolutionary relationships in the genus *Lyssavirus*. The tree, recovered from phylogenetic analysis utilizing PAUP*, version 4.0b2 (Swofford 1999), represents the single most parsimonious tree based on a parsimony branch-and-bound search of nucleotide sequences of the nucleoprotein gene (N). Sequence data used in this analysis were taken from Genbank (Genbank accession numbers are as follows: Mokola virus = s59448, Lagos bat virus = u22842, Duvenhage virus = u22848, European bat lyssavirus 1 = u22845, European bat lyssavirus 2 = u22847, Australian bat lyssavirus = af006497, Rabies virus = x03673).

The seventh species (Gould et al. 1998) was not discovered until June 1996, when brain material from a fruit bat (*Pteropus alecto*) submitted as part of a survey for Hendra virus reacted with a rabies virus monoclonal antibody. Cases were subsequently identified throughout eastern Australia in all four indigenous pteropodid species (*P. alecto, P. poliocephalus, P. scapulatus,* and *P. conspicillatus*) and from one insectivorous bat (*Saccolaimus flaviventris*). Two human rabies deaths were associated with Australian bat lyssaviruses. The first case (1996) was identified in a caregiver for insectivorous and fruit bats. A lyssavirus identical to that causing her infection was identified in the insectivorous bat species, *S. flaviventris*. The second case (1998) was in a woman who reportedly was bitten in 1996 by a large flying fox (species unknown) and was infected with the Australian bat lyssavirus found in fruit bats (Mackenzie 1999).

The number of rabies cases attributed to lyssaviruses other than rabies virus has increased in recent years as more laboratories begin to use genetic typing methods to identify lyssavirus species. Rabies virus remains the most "successful" species in the genus, however, with tens of thousands of cases reported each year. Rabies virus is the only lyssavirus species known to use species of both Carnivora and Chiroptera as reservoirs (table 14.2 and fig. 14.3). It can be argued that surveillance bias contributes to the disproportionate involvement of rabies virus in rabies encephalitis. Other lyssavirus species are newly recognized, and specific reagents are not commonly used in rabies diagnostic testing. This is particularly true in Africa, which has the greatest diversity of lyssaviruses but limited diagnostic facilities for rabies. Countering this argument, however, is the lack of terrestrial animal infection with European bat lyssaviruses despite extensive virus typing during the fox rabies vaccination campaigns in Europe. Only a single spillover case is known (a domestic sheep in Jutland, Denmark, August 1998 [see http://www.promedmail.org, archive number 19981012.2010]).

Disease Ecology in Bats: The Rabies Paradigm

Studies in disease ecology are multidisciplinary in nature, incorporating explicitly or implicitly geographic or spatial relationships, genetics, population biology, and the behavior of both pathogen and host (and vectors or intermediate hosts, if appropriate) within the context of a given environment. Such multidisciplinary studies are inherently complex and cumbersome. Bats are difficult study subjects for field ecological studies under the best of circumstances. Thus, it is not surprising that there is a dearth of information on disease ecology in bats. Despite the relatively recent discovery of rabies in bats, rabies is still the most thoroughly characterized infectious disease and will serve as our model for discussion of disease ecology in bats.

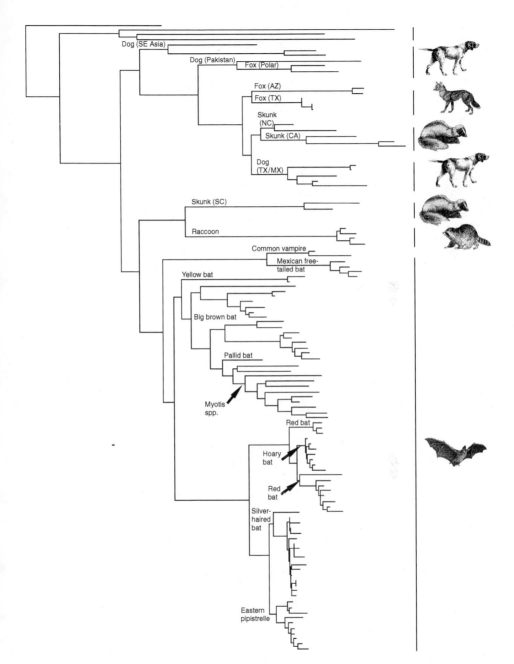

Figure 14.3. Phylogenetic tree representing the relationships among the rabies viruses based on nucleotide sequencing of the nucleoprotein gene (N) from 101 rabies variants sampled worldwide and four non-rabies lyssavirus species (included as outgroups). The tree was recovered from phylogenetic analysis employing the neighbor-joining search algorithm (minimum evolution) using maximum likelihood to estimate the transition:transversion ratio and nucleotide base frequencies.

Historical Perspective: Constantine's Bat Studies

In August 1955, several thousand *Tadarida brasiliensis* died over an 8-d period at Carlsbad Cavern, New Mexico. Limited sampling of the population ($n = 20$) revealed that 11 individuals were rabies positive, leading to speculation that rabies caused the die-off (Burns et al. 1956b). An intensive 2-yr study of the colony of *T. brasiliensis* in Carlsbad Cavern was initiated in August 1956 by the Public Health Service's Communicable Disease Center (now Centers for Disease Control and Prevention or CDC) to assess the potential public health impact. The goal was to understand the potential for disease transmission to other species and with that knowledge to delineate necessary control measures. The work remains the most in-depth study of disease ecology in a population of bats. Data collected at the cavern and in nearby bat colonies included morbidity and mortality summaries (not restricted to rabies), rates of rabies infection and estimates of incidence, measures of serum rabies-neutralizing antibodies (possibly reflecting immunity to rabies), and fundamental ecological data of the bat colonies, such as sex and age structure within the cavern, patterns of migration, and movement of bats among neighboring roosting sites (Constantine 1967a, 1967c; Constantine et al. 1968b).

During the Carlsbad Cavern study, two notable human rabies deaths occurred. (See below, "Transmission of Insectivorous Bat Rabies to Humans.") In the months prior to their deaths, both victims had visited caves with large colonies of bats (Frio Cave, Uvalde County, Texas, and others), but neither reported any animal bites. These deaths raised the specter that rabies might be transmitted among bats and between bats and other mammals through an aerosol route (Constantine 1962; Irons et al. 1957). To gain insights into the potential for aerosol transmission of rabies, Constantine (1962, 1967c) housed numerous species of mammals (e.g., foxes, coyotes, opossums, raccoons, ringtails, skunks, dogs, and cats) within cages and placed them in caves (e.g., Frio and Lava Caves) for varying lengths of time. The transmission potential of bat-associated rabies viruses to terrestrial mammals also was assessed by experimentally infecting mammals by injection and bite routes (Constantine 1962, 1966a, 1966b, 1966c, 1967c; Constantine and Woodall 1966; Constantine et al. 1968a).

Ecology of Bats at Carlsbad Cavern

To put the disease data into perspective, activity patterns of *Tadarida brasiliensis* at Carlsbad Cavern and the neighboring area were characterized (Constantine 1967a). Data on size fluctuations of the Carlsbad Cavern population were critical when interpreting the overall impact of rabies-related mortality.

Constantine and others (Constantine 1967a; Constantine et al. 1968b) monitored changes in the demographic makeup of the cavern colony throughout the season using a variety of sampling methods (e.g., banding and recapture, sampling trapped bats, density estimates in roosting sites) and described associations between bat movement among roosting sites and age, sex, and time of season. They also characterized activity patterns of bats in relation to climatological factors.

Colony size of breeding females at the cavern, not unexpectedly, remained stable throughout the season. Aged females and males moved frequently between the cavern and nearby roosting sites, apparently in response to local temperature changes, fronts, and rain. The numbers of immature individuals remained relatively stable until the fall when dispersal began. In early spring and late fall, colony size in the cavern increased because of temporary influxes of migrating *T. brasiliensis*. Some migrating *T. brasiliensis* cover distances of up to 1,300 km between seasonally occupied roosts (Constantine 1967a; Glass 1958, 1959; Sanborn 1954; Villa-R and Cockrum 1962). Within a season, bats may move relatively short distances to a more or less protected roost or along an elevational gradient to escape inclement weather patterns. Activity patterns at the cavern also varied from year to year. For example, females used Carlsbad Cavern as a maternity roost in warmer years, but in particularly cool years the caves could not provide sufficiently warm rooms for raising young. Also, immature females returned to the cavern at a much greater frequency than immature males. Because *T. brasiliensis* moves regularly to remain active year-round and exhibits highly gregarious behavior, the population structure of this species differs from bat species that form smaller aggregations, remain solitary, or are more sedentary. Such differences in population structure can affect disease transmission.

Rabies-Related Mortality in Bats

Die-offs occurred at Carlsbad Cavern in 1955 and 1956 but not in 1957. Additional short-term increases in mortality were noted after 1957, but they followed the period encompassed by the study and few details were documented. Die-offs were also noted at other localities in Texas (Frio Cave in Uvalde County and a railroad tunnel in Kendall County in 1962 and Frio Cave in 1956), but, again, few data were documented from those cases.

Rabies or pesticide poisoning were hypothesized as the two most likely causes of the die-offs. Analysis of six bats in 1955 did not support the involvement of pesticides; however, 11 of 20 moribund or dead bats collected within 8 d of the August 1955 die-off were rabies positive. Rabies was retrospectively diagnosed as the cause of the die-off. In August 1956, another die-off was documented. A total of 2,228 specimens were retrieved before, during, and after the die-off. Rabies was isolated from 3.9% of the symptomatic bats tested, yet researchers concluded that inclement weather during fall migration rather than rabies likely contributed to the deaths. Constantine (1967a) noted that in 1956 at Carlsbad Cavern, and at Texas localities where die-offs were recorded, ambient temperatures were relatively cool, migration of bats was accompanied by inclement weather, and few bats used the rooms as a roosting site because temperatures within the cavern remained cool. In 1957, no significant increase in mortality was documented.

Prevalence of Rabies in Bat Populations

In addition to sampling bats during die-offs, clinically normal, moribund, and dead bats were collected on a monthly basis throughout the 2-yr study to estimate the number of rabid bats within the population at given time points (i.e., prevalence). The percentage of rabies-positive bats in the clinically normal cavern population was low ($n = 2,192$; mean = 0.23%; range = 0%–1%) and remained stable throughout the study (Constantine 1967a; Constantine et al. 1968b), while asymptomatic bats in a nearby roost exhibited a higher rate of infection ($n = 652$; mean = 0.72%; range = 0%–1.6%). In comparison, 55% of the moribund or dead bat population collected just after the 1955 die-off and 3.9% during the August 1956 die-off tested positive for rabies. Based on their estimates of the population size in Carlsbad Cavern across different months, they estimated a significant increase in the morbidity rate during the August 1956 die-off (0.089% compared to 0.008% from October 1956 to December 1957). Still, in 1956 only 4% of the bats died from rabies, while 96% of the bats died from unknown causes (but suspected to be related to adverse weather).

Antibodies to Rabies Virus

Burns and Farinacci (1955) first reported evidence of rabies-neutralizing antibodies in the serum of healthy-looking *Tadarida brasiliensis* (antibodies were detected in 65% of 35 pooled serum samples, each pool containing approximately six bats), suggesting evidence of recovery from infection or healthy carriers of the disease (Burns et al. 1956b). Constantine et al. (1968b) also detected serum rabies-neutralizing antibody titers in some bats tested for rabies. Prevalence of antibody-positive bats was generally highest in the spring and autumn of 1956, but, in females, peaked in July 1956. Similar percentages of antibodies were found in rabies-free (i.e., brain negative) symptomatic (six out of 23 or 26%) and asymptomatic (44 out of 174 or 25%) bats. Among rabies-infected bats, antibodies were present in one of two symptomatic and none of four asymptomatic bats. Measurements of pregnant females and their fetuses revealed a significant association between maternal and fetal antibodies (χ^2, $P < 0.0005$, $n = 88$ maternal/fetal pairs, df not reported), suggesting transplacental passage of maternal antibody. Among immature bats, prevalence of antibody-positive individuals dropped in September, coincident with an increase of clinical rabies cases in immature bats. These data are consistent with two hypotheses: (1) young bats are protected by passive immunity, which generally only lasts for a short duration, or (2) bats are infected with rabies virus shortly after birth (i.e., no passive immunity) with an average incubation period of 3 mo.

Constantine (1988a) hypothesized that acquired immunity to rabies did exist and suggested that immunodeficient individuals maintain rabies viruses in colonial bat populations. In exploring the role of aerosol transmission of rabies virus in maintenance of rabies, Constantine (1967c) suggested that aerosol transmission of virus may play a greater role immunizing bats than infecting them.

Latency, Stress, and Rabies Infection

Constantine and others (Constantine 1967a; Constantine et al. 1968b) suggested that stress of migration contributed not only to the die-off of 1955 but also to increased rabies-related mortality in 1956. During August 1956, migrating bats arrived at the cavern, but many of these bats did not depart at night to forage, suggesting fatigue and stress. Stress induced by adverse weather conditions might have precipitated the onset of clinical rabies over a relatively compressed time period (Constantine 1967a; Constantine et al. 1968b). Previous work by Soave and others (Soave 1962, 1964; Soave et al. 1961) showed that even after long periods with asymptomatic infections, guinea pigs developed clinical rabies when subjected to stress.

Recovery from Rabies

Because many bats tested positive for rabies-neutralizing antibodies while exhibiting no signs of virus in the tissues, they were believed to have survived rabies virus infection. Constantine et al. (1968b) suggested, alternatively, that bats had latent infections, were incubating virus, or experienced abortive infections. Data were insufficient to test among the alternatives.

Alternative Routes of Rabies Virus Transmission

Aerosol. To assess the potential for aerosol transmission of rabies viruses, Constantine (1967c) placed animals (*Urocyon cinereoargenteus, Canis latrans, Didelphis virginianus, Procyon lotor, Bassaricus astutus, Mephitis mephitis, Canis familiaris, Felis sylvestris, Neotoma* sp., *Peromyscus* sp., and hamsters) in cages within Frio Cave for 1 wk in 1960 (fig. 14.4). Four (one *U. cinereoargenteus*, two *C. latrans*, and one *B. astutus*) of the 13 animals died of rabies (see table 14.3, experiment 1). The experimental design, however, could not rule out physical contact between the experimental subjects and bats or other animals (designated as exposure level I). In a second experiment, a group of animals was divided into two treatments. Both treatment groups excluded contact with other vertebrates, and one treatment also excluded arthropods with mosquito netting (level III), while the other did not (level II). However, the bats vacated that room as the animals were moved in. Of 28 animals (four each of *C. latrans*,

Figure 14.4. Researchers carry animal enclosures into one of the rooms in Frio Cave as part of study to assess the potential for aerosol transmission of rabies virus to terrestrial mammals (from Constantine 1967c).

U. cinereoargenteus, M. mephitis, B. astutus, P. lotor, C. familiaris, and *F. sylvestris)* placed in the cave for 1 wk, none died from rabies.

Subsequent experiments at Frio and Lava Caves in 1961 provided tighter control over the route of exposure to rabies. Animals were allowed one (i.e., aerosol only—level IV) to four levels of exposure (i.e., contact with vertebrates, arthropods, fluids [e.g., guano, urine, and other fluids], and aerosol— level I). Now, 23 of 55 animals at Frio Cave, including 10 *U. cinereoargenteus,* two *Vulpes vulpes,* 10 *C. latrans,* and the only *D. virginianus,* developed rabies. The four *C. latrans* and four *U. cinereoargenteus* that were exposed to rabies virus only through the aerosol route became rabid. At Lava Cave, three of 20 animals (i.e., two *C. latrans* and *one U. cinereoargenteus*) became rabid. In July 1962, four *U. cinereoargenteus,* four *V. vulpes,* 16 native rodents (*Neotoma* sp. and *Peromyscus maniculatus*) and 52 *Tadarida brasiliensis* were held in Frio Cave for 18 d. None of the rodents or bats died of rabies, but seven foxes (all but one *V. vulpes*) died. In a final experiment in July 1963, 62 animals from 12 vertebrate species (two *C. latrans,* four *Lynx rufus,* five *Mustela furo,* eight *D. virginianus,* six *Saimiri sciureus,* four *Aotus trivirgatus,* eight *Spermophilus variegatus,* four *Cathartes aura,* four *Buteo jamaicensis,* four *Falco sparverius,* three *Tyto alba,* and 10 *Elaphe guttata*) were held in Frio Cave for 16 d, but only three of four *D. virginianus* died of rabies. Constantine (1967c) concluded that,

Table 14.3. Summary of rabies aerosol transmission experiments, Frio and Lava Caves, Texas

		Number Positive/Total Number Tested by Exposure Level								
		Experiment 1, July 1960,	Experiment 2, September 1960		Experiment 3, July 1961				Experiment 4, July 1962,	Experiment 5, July 1963,
Species	Common Name	I	II	III	I	II	III	IV	II	II
Canis latrans	Coyote	2/2	0/2	0/2	2/2	2/12 (2/4)	2/2	4/4	...	0/2
Urocyon cinereoargenteus	Gray fox	1/2	0/2	0/2	2/2	2/6 (1/2)	2/2	4/4	4/4	...
Vulpes vulpes	Red or silver fox	...	...	...	...	2/4	...	...	3/4	...
Vulpes macrotis	Kit fox	...	...	...	...	(0/2)	...	...	...	...
Canis familiaris	Domestic dog	0/2	0/2	0/2	...	0/2 (0/2)	...	...	...	...
Mephitis mephitis	Striped skunk	0/1	0/2	0/2	...	0/2 (0/2)	...	...	...	...
Spilogale putorius	Spotted skunk	...	...	...	...	0/2 (0/2)	...	...	...	0/5
Mustela furo	Ferret	...	...	...	...	...	...	...	...	...
Procyon lotor	Raccoon	0/2	0/2	0/2	...	0/2 (0/2)	...	...	...	...
Bassariscus astutus	Ringtail cat	1/2	0/2	0/2	...	0/8 (0/2)	...	...	...	...
Felis sylvestris	Domestic cat	0/2	0/2	0/2	...	0/2 (0/2)	...	...	...	...
Lynx rufus	Bobcat	...	...	...	...	...	...	...	...	0/4
Cricetid rodent	Golden hamster	...	...	...	...	1/20	...	...	...	...
Mus sp.	Albino mouse	...	...	...	...	0/33	...	...	0/16	0/?
Peromyscus sp.	Deer mouse	...	...	...	...	...	...	...	0/12	0/?
Neotoma sp.	Wood rat	...	...	...	...	...	...	...	0/4	...
Spermophilus variegatus	Rock squirrel	...	...	...	...	...	...	...	...	0/8
Didelphis virginianus	Opossum	...	...	...	...	1/1	...	...	...	3/8
Tadarida brasiliensis	Mexican free-tailed bat	...	...	...	...	0/14	...	...	0/52	...
Saimiri sciureus	Squirrel monkey	...	...	...	...	...	...	...	...	0/6
Aotus trivirgatus	Owl monkey	...	...	...	...	...	...	...	...	0/4
Cathartes aura	Turkey vulture	...	...	...	...	...	...	...	...	0/4
Buteo jamaicensis	Red-tailed hawk	...	...	...	...	...	...	...	...	0/4
Falco sparverius	Sparrow hawk	...	...	...	...	...	...	...	...	0/4
Tyto alba	Barn owl	...	...	...	...	...	...	...	...	0/3
Elaphe guttata	Rat snake	...	...	...	...	...	...	...	...	0/10
Total		4/13	0/14	0/14	4/4	11/126	4/4	8/8	7/92	3/62

Note. Exposure levels are: I = bats and other vertebrates, arthropods, bat excreta, and cave atmosphere; II = arthropods, bat excreta, and cave atmosphere; III = smaller arthropods and larvae, bat excreta, cave atmosphere; IV = cave atmosphere. For experiment 3, entries in parentheses indicate animals tested at Lava Cave. All other experiments were performed at Frio Cave. (The total for experiment 3, exposure level II, combines number of animals tested at the two caves.) Ellipses dots indicate that the experiment was not conducted for that species.

although aerosol transmission to carnivores did occur, it was only under very specific conditions (e.g., high temperature and humidity, extremely poor air ventilation) known from only a few caves in the United States (e.g., Frio Cave). While these data suggest canids are particularly susceptible to aerosol transmission, results are difficult to interpret because coyotes and foxes were present in all treatment groups, but most other species were present in only some treatment groups.

Urine. Constantine et al. (1968b) detected rabies virus in kidneys of seven of 23 (30%) rabies-infected bats collected in five Texas caves, suggesting that rabies transmission through urine might be possible. No evidence of virus in urine, however, was found.

Transplacental. Constantine et al. (1968b) did not detect rabies viruses in fetuses, but found evidence of transplacental passage of maternal antibodies to rabies viruses.

General Contributions

The contributions made by Constantine and others to our understanding of disease ecology in bats produced a formidable foundation on which to build. Despite intensive efforts, however, rabies could not be ascribed definitively as the cause of the Carlsbad Cavern die-offs, underscoring the difficulty of such studies (Constantine 1967a, 1967c; Constantine et al. 1968b). Constantine's (1967a) documentation of significant inter- and intraseasonal variation in colony sizes and demographic makeup of *Tadarida brasiliensis* populations, demonstrating their transient nature, contributed to our understanding of disease ecology in bats. Natural history provides context within which data on disease parameters may be interpreted. However, Constantine (1967a) cautioned against liberal extrapolation of data from month to month and year to year, because of the difficulty of defining and working with colonies of *T. brasiliensis*. We assume that individuals within *T. brasiliensis* colonies share more reproductive and ecological interactions with each other than with members of other colonies, but Constantine (1967a) illustrated how misguided this view may be. Variation in colony size and structure is likely true of bat populations in general. Without consideration of these basic ecological, natural history, and population dynamic parameters, studies of disease ecology will likely provide few insights and may even mislead.

Beyond the Carlsbad Cavern Studies

In the years following the Carlsbad Cavern field studies, little emphasis has been placed on ecology of disease in bats. Rabies prevalence data are compiled nationally from passive reporting of the number of rabies-positive animals submitted to state health departments by the general public rather than from active field studies. By the 1970s, some states no longer identified species of bats that were reported as rabid.

Recognition that insectivorous bats contribute to most human rabies cases acquired in the United States has placed renewed emphasis on identification of species of rabies-positive bats. Additionally, new genetic and molecular tools for typing viral variants provide a clearer picture of the patterns of transmission within and between rabies reservoir species, the incidence of spillover of rabies viral variants from reservoir to non-reservoir species (including humans), and the risks associated with transmission of bat rabies to humans and other animals. In light of these technological advances, it is worthwhile to review what has been learned in the 40 yr since Constantine began his studies on the Carlsbad Cavern *Tadarida* colony.

Rabies-Related Mortality in Bats

There still are few relevant data to assess the impact of rabies on bat populations. Following the initial die-offs in Carlsbad Cavern in 1955 and 1956, other large-scale die-offs have not been documented within those cave systems, perhaps, because of lack of surveillance. Die-offs in *Tadarida brasiliensis* colonies were reported in Texas (Frio Cave, Uvalde County, and a railroad tunnel in Kendall County) in July 1962 but were not investigated in the detail of the Carlsbad Cavern study.

Smaller die-offs have been noted, and data, albeit sparse, suggest rabies may have been the cause. For example, between August and September of 1964, Mississippi (a state that had never reported a case of bat rabies) reported that 55 of 419 (13%) bats from 28 counties were rabid (CDC 1964). As many as 10 bats were reported from a single county over a 2-mo period, and all but two were red bats (*Lasiurus borealis*). Increased awareness of bat rabies in the state following the discovery of the first rabid bat may have contributed to increased submissions but is unlikely to explain completely the high rabies-related mortality reported given that no rabid bats were submitted after September 29 despite sustained surveillance efforts. Submissions of rabies-positive bats peak in late summer and early fall (Constantine 1967b). This time period also corresponds with increased movements of bats, such as *L. borealis.* Whether this incident in Mississippi represented an epizootic of rabies in *L. borealis,* or just increased awareness of bat rabies coinciding with the normal peak of rabies-positive bats, is still unknown. These observations were based on passive surveillance that has the unfortunate potential to be biased and insensitive.

Clark et al. (1996) described a die-off in populations of *T. brasiliensis* in four adjacent buildings in Mineral Wells, Texas, in 1993. Of 23 bats collected from June to October, 17 (74%) were diagnosed with rabies. This prevalence of rabies is one of the highest reported. It is not clear, however, how the bats were collected or whether this incident met the definition of a true die-off because the normal level of mortality experienced in these roosts was not documented.

The lack of data on die-offs can be attributed in part to the inherent difficulty of recognizing epizootics in sylvatic animal populations. Detecting increases over the expected mortality rate requires that "normal" mortality rates
be defined for that population. Currently, there is no contextual basis for determining whether a given pattern of rabies-related mortality constitutes an
epizootic or how "finely grained" the data must be to detect an epizootic. Data
from other species suggest that passive surveillance is too insensitive to detect
an epizootic. Greenwood et al. (1997) radio-collared more than 100 striped
skunks (*Mephitis mephitis*) in Stutsman County, North Dakota, just as an epizootic of rabies swept through the population. Overall rabies-related mortality in the radio-collared population was 67%, yet these researchers stated that
this epizootic would have gone unnoticed but for their active surveillance.
Most carcasses were concealed and one-third were recovered from underground. Without samples submitted from the study, they predicted an underestimate of cases by at least 31% in the state and 90% in the county. These results underscore the gaps in our knowledge of the population biology of rabies
and the need for more active surveillance and detailed population-level studies of rabies in sylvatic mammals.

Prevalence of Rabies in Bats

The prevalence of a disease is a fundamental parameter in disease ecology or
population biology studies, yet estimating the prevalence of rabies has been
plagued with difficulties. Most estimates of prevalence for insectivorous bats
in the United States are based on submissions by the public to state and local
health departments for rabies testing. Such estimates are inherently biased in
favor of finding rabies-positive individuals, since animals are submitted because they are behaving abnormally (e.g., nocturnal animals active during
daytime hours) or are sick or dead. These estimates for insectivorous bats in
the United States range from 3% to 25% (Brass 1994; Childs et al. 1994; Constantine 1967b; Schneider et al. 1957; Trimarchi 1998; Trimarchi and Debbie
1977). Variation in the estimates reflects differences in prevalence among species rather than variation within a species over time (Brass 1994; Constantine
1988a; Schneider et al. 1957). Estimates based on collections of moribund or
dead bats from roosting sites also are biased toward overestimating the true
prevalence of rabies and can exceed 50% (Clark et al. 1996; Constantine et al.
1968b). At the same time, sampling asymptomatic bats within a roost may either overestimate actual prevalence (e.g., if rabid bats are easier to capture) or
underestimate it (e.g., if rabid bats are not capable of returning to the roost or
if they roost separately from the uninfected population [Constantine 1988a;
Sullivan et al. 1954]).

Bats captured in flight may provide the most useful survey sample for assessing prevalence of rabies (Constantine 1988a; Yancey et al. 1997). This approach produces estimates from < 1% to 3% for *Tadarida brasiliensis* (Constan-

tine 1967b; Constantine et al. 1968b) and from 1.6% to 4.1% for *Eptesicus fuscus* (Constantine 1967b; Dean et al. 1960; Girard et al. 1965). These estimates may be biased against detecting rabies if rabid bats are less likely to fly, still, they may be more accurate than those predicted by sampling downed, sick, or dead individuals. To date, data are insufficient to address whether changes in prevalence of rabies-positive bats occur in response to stress of migration, as suggested by Constantine (1967a).

Although prevalence rates have remained elusive, an important point to consider is that the prevalence that is of interest for a given question may not be the true prevalence. For example, estimates of prevalence from submissions to state and local health departments are clearly overestimates of the true prevalence of rabies in bat populations, but they are likely to be a more accurate estimate (given sufficient sample sizes) of the prevalence of rabies in the subset of bats encountered by the public—the prevalence that is of interest when considering public health policy and evaluating individual medical decisions (Trimarchi 1999). At issue from the perspective of disease ecology, however, is understanding the relevance of a given estimate of prevalence in terms of how rabies is maintained within a population and what impact rabies has on that population. Attaching some significance to these estimates should be a goal for future studies, yet such studies must address the issue of bias in estimates of prevalence so as not to make spurious associations. For example, estimates of prevalence from submission data suggest that highly gregarious bats have a lower prevalence of rabies than do more solitary species. An explanation for this, and for the suggestion that solitary bats are likely to be more aggressive when rabid, was that higher titers of virus, which translates into higher infection rates per contact, are necessary in solitary bats to ensure transmission (Constantine 1967b). If, however, rabid bats are more likely to be captured and submitted than healthy ones, this bias may be more accentuated in submissions of solitary bats than of colonial bats (Baer 1975, Brass 1994; Childs et al. 1994, Trimarchi 1978).

Antibodies to Rabies Virus

Since initial reports that naturally occurring rabies-neutralizing antibodies were present in insectivorous bats (Burns and Farinacci 1955), numerous studies have confirmed their presence in frugivorous, other insectivorous, and sanguinivorous bats (Burns et al. 1956a, 1956b; Constantine et al. 1968b; Delpietro and Konolsaisen 1991; Lord et al. 1975; Ortega et al. 1985; Price and Everard 1977; Trimarchi and Debbie 1977; Villa-R and Alvarez 1963). Rabies-neutralizing antibodies also have been reported in terrestrial mammals experimentally exposed to bites by rabies-positive bats or inoculations with bat-derived rabies virus, in wild-caught carnivores around bat caves, and in carnivores fed carcasses of rabid animals (Burns et al. 1958, Carey and McLean 1983; Constantine et al. 1968a; Constantine and Woodall 1966; Delpietro et al.

1990; Lawson et al. 1987; Quist et al. 1957; Rosatte and Gunson 1984). Naturally occurring rabies-neutralizing antibodies were detected in terrestrial mammals following an outbreak of rabies in vampire bats (*Desmodus rotundus*) and cattle in Argentina in 1972 (Lord et al. 1975). During this outbreak, high levels of antibodies were found in *D. rotundus* (58%), cattle (33%), foxes (50%), and skunks (100%). No serum-neutralizing antibodies were found in rodents and opossums.

Trimarchi and Debbie (1977) measured rabies-neutralizing antibodies in 314 *Eptesicus fuscus* and *Myotis lucifugus* from several discrete colonies. They found that 9.6% (range = 0%–40% per colony, $n = 7$ colonies) of *E. fuscus* and 2.4% (range = 0%–7.4% per colony, $n = 5$ colonies) of *M. lucifugus* tested positive for the presence of rabies-neutralizing antibodies, despite no evidence of infection in the brain. Among rabies-infected individuals ($n = 6$), none had rabies-neutralizing antibody.

Steece and Altenbach (1989) examined levels of rabies-neutralizing antibodies in a single population of *Tadarida brasiliensis* at Lava Cave, New Mexico, over a 5-mo period. Serum samples of adult female and juvenile bats were divided and tested separately for the presence of rabies-neutralizing immunoglobulins M (IgM) and G (IgG). Brains were also examined for presence of rabies virus. Among adult females, the percentage exhibiting rabies-neutralizing IgG was high (ca. 70%), while IgM (ca. 2%) and rabies virus in the brain ($< 1\%$) were low. These measures did not differ significantly throughout the 5-mo period. Among juveniles, the percentage with rabies-neutralizing IgG fluctuated between a high of 60%–70% (mid-June to early July) and a low of $< 40\%$ (mid-July). The percentage of juveniles with rabies-neutralizing IgM also fluctuated, with increases observed at the end of the 5-mo period. No rabies-positive juveniles were detected until mid-July, at which time the percentage of rabies-positive juveniles (2%) was higher than that observed in adult females ($< 1\%$).

Steece and Altenbach (1989) interpreted the high prevalence of IgG coincident with a low prevalence of IgM as evidence of survival from exposure to rabies virus. It was not possible, however, to determine whether the bats survived clinical disease, experienced abortive infections, or were incubating virus at the time of capture. From an evolutionary or epidemiological perspective, such a distinction is important, especially if certain classes of infection (e.g., clinical disease) permit transmission while others (e.g., abortive or incubating infections) do not. Survival from both clinical and subclinical infection has been documented in controlled laboratory settings (Baer and Bales 1967; Constantine 1967b; Pawan 1948), but no data exist that document recovery from a clinical rabies infection in a natural setting. Data are also needed to verify that antibodies measured in these studies are indeed specific to rabies virus rather than nonspecific, cross-reactive antibodies. An enzyme-linked immunosorbant assay (ELISA) to detect rabies-neutralizing antibodies in bats (Kwiecinski et al. 1993) confirmed both the presence of IgM preceding IgG in a primary immune response to rabies vaccination and a rapid, longer-lived, high-titered IgG response on secondary antigenic challenge in experimentally infected animals (*Eptesicus fuscus, Artibeus jamaicensis,* and *Mus* sp.). Measurement of this immune response in bats on experimental challenge with rabies virus suggests that they were detecting rabies-neutralizing antibodies, but such studies have not been carried out with wild-caught bats. Detection by ELISA, however, is not currently the standard method used, so these results cannot be extrapolated to other studies.

If there were acquired immunity to rabies, it might suggest an alternative mechanism for maintenance of rabies within a population without cyclic population crashes (although this has been dismissed as unlikely in discussions about the maintenance of rabies in wildlife populations—see Anderson et al. [1981]; Bacon [1985]; Ball [1985]; Smith [1985]; but see Coyne et al. [1989] for a model that considers immunity). It is striking that Steece and Altenbach (1989) reported such a high prevalence of rabies-neutralizing antibodies in *Tadarida brasiliensis*. If the high prevalence is confirmed, it could provide an explanation for why large rabies-related die-offs are not observed more often in such a visible and relatively well-studied bat species.

Latency, Stress, and Rabies Infection

The term "latency" has been used to refer to the extended incubation periods seen in some rabies infections. Here, we use the term "extended incubation periods," rather than "latency" (i.e., a state in which the viral nucleic acids exist within the cell but transcription and translation are repressed or at low levels), because there are no data indicating a latent phase of rabies virus. Extended incubation periods should not be confused with the concept of a carrier state of rabies, which implies prolonged, active infection of the central nervous system (CNS), allowing for shedding of virus from the salivary glands without obvious clinical signs of rabies (see below, "Myths, Mysteries, and Misconceptions"). The average incubation period for rabies ranges from 2 wk to 3 mo; however, an *Eptesicus fuscus* held in captivity developed rabies 209 d after capture (Moore and Raymond 1970). This bat apparently had been infected in the wild before capture and showed no clinical signs of rabies before 209 d. Variation in the length of the incubation period of rabies in other mammals has been reported (Baer and Cleary 1972; Bell 1975; Kaplan 1969; Moore and Raymond 1970; Trimarchi 1978). Among humans, there have been three well-documented cases with incubation periods of at least 11 mo, 4 yr, and 6 yr (Smith et al. 1991), and an epidemiological study of human rabies cases determined that 1.2% of 1555 cases had incubation periods of more than 1 yr (Fishbein 1991).

Two issues must be raised when addressing extended incubation periods: (1) where are rabies viruses and what are they doing during the dormant period (i.e., the period between exposure and onset of clinical signs of rabies), and (2) what factors are involved in reactivation of rabies infection?

Sulkin and others (1957, 1959, 1960) hypothesized that rabies virus could remain dormant in brown adipose tissue and that decreased temperatures during hibernation could depress the activity of the virus. When rabies-infected bats (*Tadarida brasiliensis* and *Myotis lucifugus*) were subjected to either normal (29°C) or decreased (5°C or 10°C) environmental temperatures and then assayed for signs of clinical rabies infection, bats held at low temperatures showed a decreased incidence of rabies compared to bats held at higher temperatures. Bats held at 5°C and then transferred to 29°C exhibited a similar incidence of rabies to those bats held at warm temperatures. Sadler and Enright (1959) experimentally infected *Antrozous pallidus* and maintained the bats at 4°C for varying lengths of time before transferring them to 22°C. Incubation lengths following transfer to a warmer temperature were relatively constant among all treatment groups, despite variation in the amount of time that the bats spent at 4°C. Furthermore, rabies virus has been isolated from the brown adipose tissue of naturally infected bats (Bell and Moore 1960).

These data suggest that storage of virus in brown fat, combined with decreased body temperatures and metabolic rates during both seasonal hibernation and daily torpor, contributed to long incubation periods in bats. Yet, these data cannot account for the extended incubation periods seen in non-hibernating mammals, including humans. Additionally, the presence of rabies virus in brown adipose tissue may be an artifact of centrifugal spread of the virus from the CNS late in the infection rather than storage of virus in adipose tissue prior to infection of the CNS (Fischman and Schaeffer 1971). The question remains as to whether rabies viruses reside in nonneural tissue (e.g., muscle cells) or whether the viruses enter the nervous system but remain dormant during protracted incubation periods.

Stress has been suggested as one possible factor in reactivation of rabies virus. Physiological responses to stress were long thought to enhance susceptibility to infection with rabies virus, possibly because of increased cortisone production (Sims et al. 1963; Soave et al. 1961; Sulkin et al. 1957). Soave et al. (1961) reported reactivation of rabies by injecting previously infected guinea pigs with adrenocorticotrophic hormone (ACTH) or corticosteroids. Sulkin et al. (1957) and Sims et al. (1963) also suggested that the stress of pregnancy could trigger reactivation of rabies infection in both naturally and experimentally infected bats. As previously discussed, Constantine (1967a) suggested that stress of migration in *Tadarida brasiliensis* might have precipitated clinical rabies in previously infected but asymptomatic animals. More data, particularly from studies of natural populations, are needed to address this topic.

Alternative Routes of Rabies Virus Transmission

Aerosol. Human rabies deaths associated with Frio Cave prompted numerous studies to determine the potential for aerosol transmission of rabies virus. (On aerosol transmission, see also, below, "Myths, Mysteries, and Misconceptions.") In only the most extreme cave environments with high-density bat populations (e.g., Frio Cave) has rabies virus been detected in the air (Winkler 1968, 1975). At least in these extreme environments, rabies viruses might exist at high enough titers in the air to infect the CNS via the nasal mucosa. Constantine et al. (1972) detected rabies virus in the nasal mucosa of five of 15 naturally infected *Tadarida brasiliensis* (the brains of the five bats also tested positive). Rabies virus also has been reported in the nasal mucosa of laboratory animals (e.g., rodents) infected experimentally (Fischman and Schaeffer 1971; Hronovsky and Benda 1969a, 1969b) and in a cow that had been bitten by a bat (Schaaf and Schaal 1968). While such observations suggest the nasal mucosa may present a route either of entry or of transmission for rabies virus, presence of virus in nasal tissue may be incidental to the natural progression of rabies virus infections. Hronovsky and Benda (1969a, 1969b) indicated that rabies virus replicated in the nasal epithelium before entering the CNS, but these findings have not been confirmed. Most examples of rabies-positive nasal mucosa represent late-stage infections in which rabies virus replicated in the CNS and subsequently traveled to diverse, innervated tissues throughout the body. Additional experimental data will be needed to address the role for the nasal mucosa in rabies transmission or infection.

Urine. Villa-R and Alvarez (1963) detected rabies virus in the kidneys of nine out of 40 *Desmodus rotundus* from western Mexico. Girard et al. (1965) also found rabies virus in the kidneys of two of four rabies-positive bats (*Myotis lucifugus* and *Eptesicus fuscus*). These researchers also detected rabies virus in both kidney and urine samples of a rabid *E. fuscus*. Constantine et al. (1968b) observed that rabies virus was present in the kidneys, but not the urine, of *Tadarida brasiliensis*. No data have confirmed the presence of rabies virus in the urine of bats since the Girard et al. (1965) study, and no recent studies suggest that bat urine is a route of transmission of rabies virus.

Transplacental. Schneider et al. (1957) reported that suckling bats (< 5 d old) were infected with rabies virus, and Sims et al. (1963) showed transplacental transfer of rabies virus from experimentally inoculated pregnant *Tadarida brasiliensis* females to one fetus and three newborn bats. Curiously, the mothers of the three rabies-positive newborns, however, did not test positive for rabies virus. In a study of downed *T. brasiliensis* collected from Frio Cave, Uvalde County, Texas, Constantine (1986) tested 28 fetuses, 284 suckling bats, and 395 newborns (> 5 d old) for rabies infection. None of the fetuses or suckling bats was positive by fluorescent antibody test (FAT) or mouse inoculation while 76 of 395 newborns were positive, suggesting that newborns might become infected (through biting or licking) shortly after birth but not before. Steece and Altenbach (1989) documented rabies-neutralizing IgM levels in very young *T. brasiliensis* (< 1 wk). Because IgM antibodies are not known to pass the placenta, this early development of antibody was presented as evidence of prenatal infection. Steece and Calisher (1989) harvested four *T. brasiliensis* fetuses to establish fetal bat cell lines and found that cells not inoculated with rabies virus exhibited cytopathic effects similar to cells inoculated with rabies virus. Although immunofluorescent antibody assays of these cells did not indicate rabies viral infection, mouse inoculation studies confirmed the presence of rabies virus in the fetal cells. Transplacental passage of rabies virus has been documented from naturally infected striped skunks (Howard 1981) and cows (Martell et al. 1973).

While these studies suggest that prenatal transmission occurs, even in natural settings, the frequency with which this happens is unknown. It appears to be rare (Constantine 1986). From the perspective of disease ecology, however, vertical transmission of rabies virus is not likely to be an important mechanism in maintenance of rabies enzootics. Rabies viruses are adapted for horizontal transmission, via infected saliva delivered through a bite, and have evolved to exploit mammals. Vertical transmission would appear to be a dead-end strategy for the virus unless there are long periods of viral shedding before onset of severe symptoms to allow the mother sufficient time to raise the young before she succumbs to the disease. There are no data to support long-term shedding of rabies virus.

Impact of Bat Rabies on Other Mammal Populations
Vampire Bat Rabies

The first scientific link between bat bites and deaths from rabies was Carini's report (1911) of Negri bodies in the brains of cattle dying of a paralytic disease in southern Brazil. Noting the scarcity of dog rabies cases in the area, the clustering of cases in more forested regions, and the fact that people had seen bats attacking and biting cattle in broad daylight, Carini concluded that bats were the likely source of the cattle disease (cited by Pawan 1936b). Carini's suspicions were confirmed by the isolation of rabies virus from a bat in the same area of Brazil (Haupt and Rehaag 1921).

Carini's report on vampire bat rabies was ignored by those responsible for rabies control efforts. At the turn of the century, rabies was known only as a disease of terrestrial carnivores, and especially in South America and the Caribbean, only as a disease of domestic dogs. Despite the scarcity of dog rabies in Santa Catharina, Brazil, at the time of the outbreak in cattle, more than 6,000 domestic dogs were destroyed as part of the cattle rabies control campaign in the year following Carini's report (cited by Pawan 1936b).

In defense of the rabies control program, Carini's conclusions about vampire bats and rabies were obscured by the fact that the first bats diagnosed with rabies in Brazil were identified as fruit-eating phyllostomids (*Artibeus lituratus*, formerly known as *Phyllostoma superciliatum* [Haupt and Rehaag 1921]). A laboratory diagnosis of rabies was eventually made in *Desmodus rotundus*, but the skeptics had reason to question the test itself. The only laboratory test for rabies available at the time was nonspecific, based on the observation of Negri bodies (i.e., eosinophilic intracytoplasmic inclusions) in histological smears of brain material. According to Kaplan (1985), Negri bodies are seen in only about 85% of rabid animals, and some researchers suggest that bats may exhibit reduced negrigenesis (Burns et al. 1958; Tjalma and Wentworth 1957). Moreover, inclusions similar to Negri bodies are often produced by other diseases or conditions. When confirmatory animal inoculation or cross-immunity tests were performed, the results could be difficult to interpret because of other neurologic disease agents in field samples (e.g., Rio Bravo virus—see Constantine [1988b] and, below, "Bats as Asymptomatic Carriers of Rabies Virus").

Given all of the confounding circumstances, it is indeed remarkable that so many of the assumptions drawn from these early observations have been verified by modern laboratory methods. The initial report in 1908 of rabies in 4,000 cattle and 1,000 horses in Santa Catharina, Brazil, seems accurate, given current estimates of 100,000–500,000 cattle rabies deaths each year in Latin America (Acha 1968). Approximately 4.0% of cattle submitted for slaughter during a 1-yr period in Mexico City showed evidence of rabies infection (Baer 1991). Pawan's (1936b) startling report of 53 human deaths from rabies in Trinidad during 1929–1936 is echoed in the Peruvian Ministry of Health report of 24 human rabies deaths during a 6-mo period in 1989 (Lopez 1991). The deaths in a remote gold-mining community in Madre de Dios, Peru, represented almost 5% of the approximately 500 inhabitants. All had been bitten by *D. rotundus* while sleeping in open, temporary housing. A variant of rabies virus associated with *D. rotundus* was identified in the seven autopsy samples taken during the outbreak (J. Smith, unpublished data).

Insectivorous Bat Rabies

Reported cases of rabies in non-hematophagous bats were so few in the early outbreaks in Brazil and Trinidad that infection of insectivorous and frugivorous bats was generally considered to result from spillover infection from vampire bats. Outside of the range of vampire bats, bites or other contact with bats were not considered potential exposures to rabies, and the bats themselves were not submitted for rabies testing. This scenario changed dramatically in June 1953, when a 7-yr-old Florida boy was attacked and bitten by an insectivorous bat (Scatterday and Galton 1954). The bat, a lactating female *Lasiurus intermedius*, was submitted for rabies testing because the father of the boy was a cattle rancher familiar with vampire bat rabies in Mexico. Finding rabies in the bat prompted submission of bats for testing throughout the United States. Within 5 yr, rabies had been found in 17 states and in 18 of 39 known U.S. bat species (Tierkel 1958). By 1979, the number reached 30 bat species (Constantine 1979), and it seemed likely that all U.S. bat species eventually would be found to be infected with rabies. At the time, bats were being viewed as potential reservoirs for both bat and terrestrial carnivore rabies because transmission of rabies variants from broadly distributed bat populations could explain the pattern of geographically disparate terrestrial mammal enzootics observed in the United States (Burns and Farinacci 1955; Irons et al. 1957; Schneider et al. 1957; Venters et al. 1954). Public concern about bats prompted studies of bat rabies and bat control programs. Research to determine if a relationship existed between insectivorous bat and terrestrial mammal rabies was directed in two ways: (1) experimental lab and field investigations to explore the efficacy of infecting terrestrial mammals with bat rabies variants, and (2) epidemiological studies to explore the relationship between terrestrial mammal and bat rabies in nature.

Experimental inoculation, ingestion, bite, and aerosol studies using viral isolates from experimentally or naturally infected bats demonstrated that terrestrial mammals are susceptible to bat rabies variants in varying degrees (Baer and Bales 1967; Bell et al. 1962; Burns et al. 1958; Constantine 1966a; Constantine et al. 1968a; Constantine and Woodall 1966; Correa-Giron et al. 1970; Delpietro et al. 1990; Fekadu et al. 1988; Fischman and Ward 1968; Soave 1966; Trimarchi et al. 1986). Despite these observations, additional ecological or epidemiological data are necessary to demonstrate that such transmission occurs in natural settings at a frequency that would have an impact on populations of terrestrial mammals.

Epidemiological studies focused on environments perceived as high risk for bat rabies transmission to nonbat species, notably caves. Large numbers of bats, congregating at high densities in and around caves (e.g., maternity roosts, summer roosts, hibernacula) could attract terrestrial predatory mammals. Rabies-neutralizing antibodies have been detected in wild-caught predators (e.g., *Procyon lotor*) around bat caves (Carey and McLean 1983; Quist et al. 1957), consistent with the hypothesis that cave sites were potential foci of bat rabies transmission (Constantine 1962; Eads et al. 1955; Schneider et al. 1957).

Although many terrestrial mammals frequent caves (Harris 1971; Winkler and Adams 1972), foxes became the focus of epidemiological studies because they were thought to be more susceptible to rabies (Constantine 1966c, 1967c) and could act as indicator species for evidence of transmission of bat rabies. Foxes forage in and around caves and eat bats (Constantine et al. 1968a), providing opportunity for infection by bat rabies viruses. One epidemiological study revealed a correlation between the number of caves per area and both the length of fox rabies epizootics and the number of rabies-positive foxes (Frederickson and Thomas 1965), but the associations were criticized as an artifact of a flawed sampling design (Hayne and Neeley 1967). Later studies (Carey 1974; Fischman and Young 1976; Mahan 1973) found a suggestive, but not significant, relationship between caves and fox rabies. Fischman (1976) suggested that the observed associations between bat and terrestrial mammal rabies may be the result of either a causal, indirect, or spurious relationship and that additional data need to be collected to distinguish among these alternatives. Unfortunately, most of the focus has been on the relationship between fox rabies and cave density without establishing other necessary associations, such as relationships between cave density and density of rabid bats. Mahan (1973) gathered data on both the bats and the terrestrial mammals associated with caves in Tennessee. He did not find a significant difference in predator use of caves with or without bats or a significant correlation between the incidence of fox rabies and the percentage of rabies-positive bats by cave. He did find, however, a slightly higher predator use of caves associated with lower rates of fox rabies. Other potentially confounding associations, such as the strong interrelationship between fox rabies and rabies in other terrestrial mammals or the relationship between total fox density and frequency or length of rabies epizootics, were not explored (Hayne and Neeley 1967). Moreover, these studies did not distinguish among the species of foxes studied or account for differences in the behavior and ecology of different fox species when interpreting the observed patterns.

By the late 1970s, viral typing techniques using monoclonal antibodies offered a new source of data to resolve unanswered questions about the relationship between bat and terrestrial mammal rabies (Charlton et al. 1982; Dietzschold et al. 1988; Rupprecht et al. 1987; Schneider et al. 1985; Smith et al. 1986; Webster et al. 1985; Whetstone et al. 1984; Wiktor et al. 1980; Wiktor and Koprowski 1980). Now, it was clear that terrestrial mammal rabies variants were distinct and cycled independently from bat rabies viruses, thus rejecting the hypothesis that bat rabies virus variants are the progenitors of terrestrial mammal rabies epizootics (Smith 1988; Smith et al. 1990). The pattern of evolutionary relationships among rabies viral variants (fig. 14.3) provide additional data. If transmission from bats were responsible for initiating rabies epizootics in terrestrial carnivores, viruses from both the bat species initiating the epizootic and recipient terrestrial carnivore species should be most closely related. The phylogeny shows that all rabies viral variants associated with bats form a monophyletic clade to the exclusion of all terrestrial carnivore variants, verifying that bat rabies enzootics cycle independently of terrestrial carnivore enzootics. Spillover of bat variants into terrestrial mammals, however, does occur. How often spillover occurs and what the impact is on populations of terrestrial mammals are issues of ongoing study.

Spillover of Terrestrial Mammals. The advent of monoclonal antibody typing enabled cases of bat-to-terrestrial transmission to be distinguished from translocations of rabid animals and cases of vaccine-induced rabies (Rupprecht et al. 1987; Smith and Baer 1988; Whetstone et al. 1984). Still, monoclonal antibody typing was limited by the fact that the antibody panels could distinguish among some, but not all, bat variants. When combined with nucleotide sequencing of rabies variants, monoclonal antibody typing is a powerful tool for understanding the epizootiology and ecology of rabies (Rupprecht and Smith 1994; Smith 1996; Smith et al. 1995; Smith and Seidel 1993).

Charlton et al. (1982) reported that a cow bitten by a bat (species unknown) was infected with a rabies variant that shared a similar antigenic profile with local bats. Other studies found bat rabies viruses in a cow in Pennsylvania that 2 mo earlier had been found with a bat (species unknown) clinging to its chest (CDC 1983) and in a horse in New Hampshire that shared a stable with bats (Fuller 1984). The virus variant implicated in both American cases matched the profile of *Eptesicus fuscus* from the eastern United States (Smith et al. 1986). Six additional terrestrial mammals (out of 239 tested) were documented to have been infected with rabies viruses characteristic of bat variants (i.e., *E. fuscus* variant in three gray foxes [*Urocyon cinereoargenteus*] and a red fox [*Vulpes vulpes*], *Lasiurus borealis* variant in a fox, and *Tadarida brasiliensis* variant in a rat). Of 85 terrestrial mammal viral isolates from Canada, Webster et al. (1986) reported a horse from British Columbia infected with a rabies variant characteristic of bats. Rupprecht et al. (1987) tested 25 terrestrial mammals, including humans, and detected non-terrestrial rabies viruses in two cows from Pennsylvania and a cat from Massachusetts. Smith and Baer (1988, table 7) and Brass (1994, tables 16-1, 16-2) summarize 60 U.S. cases and an additional eight Canadian cases between 1976 and 1992 in which domestic or wild terrestrial mammals were infected with bat variants. In Latin America (e.g., Chile), spillover from *T. brasiliensis* to a cat (Smith 1989), a cow and a dog (de Mattos et al. 1997), and four dogs, two cats, and a pig (Favi Cortes et al. 1997) have been reported.

Nonetheless, transmission of rabies viruses from other terrestrial mammals is a more important source of rabies infections for terrestrial mammals than bat rabies spillover. Bat variants contributed to only three of 136 cases of rabies transmission to foxes (one of 70) and cats (two of 66) collected from areas where terrestrial mammal rabies is enzootic (Smith 1988, 1989). Cases of bat-to-terrestrial mammal transmission, however, were discovered through passive surveillance, and most states with enzootic terrestrial mammal rabies do not type all rabies-positive animals submitted (an expensive and laborious task). In states with no terrestrial mammal reservoirs, most, if not all, rabies-confirmed terrestrial mammals are typed, yet these cases are restricted geographically, and lack of awareness of sylvatic rabies in these regions may contribute to fewer submissions to state health departments. Therefore, potential spillover events likely have been overlooked. Until systematic studies are carried out, the frequency of bat to terrestrial mammal spillover events cannot be accurately assessed.

Have Bat Rabies Variants Given Rise to Terrestrial Epizootics? Although spillover of insectivorous bat variants to terrestrial mammals is rare, documented clusters of terrestrial mammal rabies cases of bat origin perpetuate the notion that some terrestrial mammal outbreaks come from bats. A cluster is defined here as two or more cases occurring sufficiently close in space (i.e., within the distances reasonably traveled by the mammals involved) and time (i.e., within a normal incubation period) that a causal link between these cases is possible. A cluster was first noted in New York in 1984, when three gray foxes (*Urocyon cinereoargenteus*) were diagnosed with rabies in two neighboring counties outside the known terrestrial rabies-enzootic region. Monoclonal antibodies indicated that these cases were of bat origin (a variant similar to rabies viruses infecting *Eptesicus fuscus* [Smith et al. 1986]). Three other *U. cinereoargenteus* infected with a bat variant indistinguishable from *Lasiurus borealis* were discovered in Putnam County, New York, in 1987 (J. Smith, unpublished data; Trimarchi 1987). In 1993 on Prince Edward Island, three red foxes (*Vulpes vulpes*), all within 1 km and 1 mo of each other, were infected with a rabies variant also seen in *Myotis lucifugus* in eastern Canada (Daoust et al. 1996; Webster et al. 1989). There are at least three additional cases (J. Smith, unpublished data): four foxes found between November and December of 1995, all within the Mount Charleston area, west of Las Vegas, Nevada; four foxes found in 1993 in nearby counties (Curry and Jackson) in Oregon; and three foxes found within 3 wk of each other in 1994 (one in Coos County and two in Jackson County) in the same area of Oregon. The two foxes from Jackson County were suspected to have been littermates.

Two hypotheses could explain these patterns of transmission: each fox involved in each cluster was infected directly from a bat in independent events, or bat-to-fox transmission led to subsequent fox-to-fox transmission. Only the latter hypothesis implies that bats could contribute to terrestrial mammal outbreaks. Rabies viral antigen was recovered from the salivary glands of the *U. cinereoargenteus* in Putnam County, New York (Trimarchi 1987) and a *V. vulpes* from the Prince Edward Island cluster (Daoust et al. 1996), suggesting that these foxes might have been capable of viral transmission. Additionally, complete virions were recovered from the salivary glands of the Prince Edward Island *V. vulpes* (Daoust et al. 1996). Presence of virus in the salivary glands is not sufficient to prove fox-to-fox transmission occurred, thus there currently are no means by which to distinguish between the two proposed hypotheses.

Transmission of Insectivorous Bat Rabies to Humans. The first known human case of rabies attributed to insectivorous bats was diagnosed retrospectively. Sulkin and Greve (1954) reported that in October 1951 a 43-yr-old woman from west Texas had been bitten on the left forearm by a sick bat she had picked up. Sixteen days after the bite, she developed a generalized illness and pain in her left arm, dying in a coma 8 d later. The initial diagnosis was bulbar poliomyelitis. With the report in 1953 of rabies in insectivorous bats, tissue from the patient was reexamined and Negri bodies found in brain material, but because the tissue had been fixed in formalin, no further tests were performed.

The first well-documented human death associated with insectivorous bats occurred in California in 1958 (Humphrey et al. 1960) when a 53-yr-old woman was bitten by a downed bat that she was attempting to place in a tree. Antirabies treatment was initiated 3 d later and failed to prevent rabies. She died 66 d after the bite. Virus samples from the silver-haired bat (*Lasionycteris noctivagans*) and the human case were preserved and typed by nucleotide sequence analysis. A virus typically found in *L. noctivagans* was identified in both samples.

In recent years, almost all of the one to two human rabies infections acquired each year in the United States (26 of 28 cases since 1981) have been associated with variants of rabies virus transmitted by insectivorous bats (Noah et al. 1998; J. Smith, unpublished data). Since the introduction of potent antirabies biologics in the 1970s, no deaths from rabies have occurred in persons who recognized a risk from rabies in an encounter with a bat and sought treatment. Bat exposure histories from two recent human rabies deaths, however, emphasize that encounters with bats must be evaluated more carefully for risks of infection than encounters with terrestrial animals (CDC 1998). In these two cases, direct physical contact with bats was noted (in one case a bat landed on the patient's shoulder during the night; in the second case, the patient had removed bats from his house). No bite or bite wound was noted in either case, but in both cases the account of bat contact was obtained from family members. By the time rabies was suspected in the human cases, the patient was too ill to be interviewed directly about animal exposures. Although bats were captured in both cases, none was submitted for rabies testing, and no antirabies treatment was sought by the patients. Given the cryptic nature of cases such as these, it is reasonable to assume that insectivorous bat-associated human rabies cases occurred before 1951.

Failure to recognize risk is a reasonable explanation for the disproportionate number of bat-associated human cases, but one additional epidemiologic finding was unexpected. In a CDC summary of human rabies cases from 1980 to 1996, Noah et al. (1998) reported that the majority of bat-associated cases (16 of 22 cases) were infected with a variant of rabies associated with *Lasionycteris noctivagans* and *Pipistrellus subflavus*. Morimoto et al. (1996) found in laboratory experiments that this variant replicates more efficiently in epithelial cells and at lower temperatures than a coyote rabies variant. It is thought that superficial bite wounds are a contributing factor in the failure to recognize a risk of exposure. Superficial contact with a bat or other animal infected with this variant of rabies virus may carry a greater risk of infection than similar contact with a different variant.

By definition, lyssaviruses cause the clinical disease rabies. There is a risk of rabies encephalitis for humans and domestic animals that have had contact with animals infected or potentially infected with these viruses. Despite the occasional official report to the contrary (WHO 1997), if a bat population in a country is a reservoir for a lyssavirus, that country is not "rabies free." Those persons whose vocation or avocation includes physical contact with bats any-where in the world should receive preexposure immunization with rabies vac-cines and postexposure booster immunization as needed (CDC 1999a).

Myths, Mysteries, and Misconceptions

The discovery that many species of bats are reservoirs for rabies brought a re-newed interest in both the disease and its hosts. Biologists and virologists quickly seized on the more unusual characteristics of the natural history of bats to explain aspects of the epidemiology of rabies that had long seemed puzzling. Long distance and seasonal migrations by bats could explain the oc-currence of geographically dispersed outbreaks of rabies in terrestrial animals. Huge maternity colonies of bats in caves could provide an ideal environment for aerosolization of rabies and nonbite transmission of rabies to animals and humans. Hibernation and torpor could provide a mechanism for extremely long incubation periods and high survival rates after infection with rabies virus.

Virologists now agree that migrating bats and bat caves are not the source for persistent disease or recurring outbreaks of rabies in terrestrial animals. Bats respond to rabies virus no differently than other terrestrial mammal reservoirs. In hindsight and with modern methodology, we can attribute many of the faulty conclusions about bats and rabies to insensitive and nonspecific laboratory techniques. Still, many of the early misconceptions about bats and rabies persist, despite efforts to discount them. The two most frequently men-tioned are that bats are asymptomatic carriers of rabies and that aerosol trans-mission of rabies from bats is a source of cryptic rabies in humans and animals.

Bats as Asymptomatic Carriers of Rabies Virus

The concept of asymptomatic rabies in bats arose from experiments done in Brazil and Trinidad when vampire bat rabies was first recognized. In a typical experiment (Pawan 1936a), wild-caught *Desmodus rotundus* were sustained in the laboratory by feeding them blood. After 1 wk in captivity, bats were inoc-ulated subcutaneously with a rabbit brain–passaged rabies virus isolate. One to 3 mo later, bats were permitted to feed on rabbits. After several weeks, some of these rabbits became paralyzed and died, and Negri bodies were observed in their brains. The bats remained well for an additional 1–3 mo, when some became weak and unable to fly. Sick animals died within a few days, and Negri bodies also were found in their brains. Because rabies virus is found in salivary glands and saliva just in the terminal stages of infection, and after replication in the brain, the conclusion was that these *D. rotundus* survived for long periods unaffected clinically by the rabies infection in their CNS.

Sulkin et al. (1959) hypothesized that tissues that provide energy reserves for thermogenesis during hibernation (brown fat) could serve as a storage location for rabies virus during a long preclinical period (see "Latency, Stress, and Rabies Infection" in the section on studies beyond Carlsbad Cavern). Brown fat, as well as brain and salivary gland material, collected from *Tadarida brasiliensis* inoculated with rabies virus, was inoculated intracerebrally into weanling mice. These mice were then observed for signs of rabies. Virus isolated from the brown fat of nine bats with negative brain and salivary gland tissues produced encephalitis in mice. The virus was discovered in the brown fat and salivary glands of a tenth bat, whose brain was negative. No laboratory tests were performed to confirm the presence of rabies in the inoculated mice, since the virus strain used did not reliably produce Negri bodies and immunofluorescent detection of viral antigen was not a commonly performed method of rabies diagnosis at the time. In similar experiments, Tierkel (1958) reported brain-negative/salivary gland–positive *T. brasiliensis* and confirmed the presence of rabies virus in their samples by neutralization with antirabies antisera. The conclusion drawn from these experiments was that rabies virus could be excreted in the saliva of bats in the absence of a CNS infection.

Replication of these early experiments using modern laboratory methods has either failed to duplicate the initial observations or has offered alternative explanations for them. Moreno and Baer (1980) inoculated 42 wild-caught *D. rotundus* by the intracerebral, intramuscular, or subcutaneous route with dilutions of a vampire bat salivary gland virus pool. At intervals after infection, saliva swabs were taken and inoculated into mice and cell cultures. The bats were observed for 1 yr. The brain and other organs of any bat dying during the observation period were examined for rabies by immunofluorescence and animal inoculation. No evidence was found for an asymptomatic carrier status in the inoculated bats. Virus was found in the saliva of 12 out of 16 bats that sickened from the inoculation. The preclinical period of virus excretion was short (i.e., virus was found in the saliva of an inoculated bat 1–8 d before signs of an encephalitis were evident), and no sick animal recovered from the infection. The brains of all sick animals were positive for rabies virus at death. No healthy animal excreted virus in its saliva.

Although these experiments do not explain the observations of earlier work, most virologists now believe that a second contaminating virus from a naturally acquired infection was present in the early experiments, either in the inoculum or in the saliva of the wild-caught bats. Without immuno-diagnostic reagents, the encephalitis produced in mice inoculated with Rio Bravo virus is indistinguishable from rabies (Constantine 1975; Constantine and Woodall 1964). The virus is not lethal for bats, giving the false impression of asymptomatic rabies virus infection. Tierkel et al. (1958) used pooled salivary glands to inoculate laboratory animals for immune serum, which then contained antibody to both Rio Bravo and rabies viruses. Virus neutralization tests identifying the isolate as rabies virus could not be reproduced in a second laboratory using a different antirabies serum (Constantine 1975).

The protracted survival of clinically rabid bats in a laboratory setting (Bell et al. 1969) has been used as evidence for a carrier state. Bats, however, are no different from any other rabies-infected animal in this regard. Protracted survival and shedding of virus over a long period of an obviously ill bat (weak, unable to fly) in a laboratory setting is not evidence of a carrier state, and animals in such a state would not survive over a long period in the wild.

Aerosol Transmission and Cryptic Rabies Infections

Two human rabies deaths during the 1950s (Humphrey et al. 1960; Irons et al. 1957) are notable for the interest they generated in aerosols as a mechanism of rabies transmission by bats. In January 1956, an entomologist with the Texas Department of Health died of rabies. He had been a member of a team studying the ecology of *Tadarida brasiliensis* bats in Frio Cave and other central Texas caves. In the months prior to his death, he had tagged and bled some 10,000 bats in a study of migration and had bled and inoculated bats with rabies virus for laboratory studies. He had worn rubber gloves during all animal work, and when interviewed during his illness, he recalled no bat bite exposure. Because he had chronic skin eruptions on his neck, the hypothesized route of exposure was inadvertent contact of the lesions with the contaminated gloves. In June 1959, a second death from rabies occurred in a consultant mining engineer, who had explored caves in Mexico and Texas, including Frio Cave. He did not recall any bat bite exposure to rabies, but had been observed with blood on his face after one cave trip. Although both infections can be explained by means other than aerosol transmission, they are often mentioned as evidence of aerosol transmission of rabies.

No additional cases of human rabies have been attributed to bat caves, and the risks posed to humans by large natural aggregations of bats is considered low. This is principally because the environmental conditions in caves where aerosolization of virus could be possible (excessive humidity, high levels of ammonia, elevated temperatures) are so unpleasant and unhealthy that humans are unlikely visitors there. Additionally, in almost all recent bat-associated human cases without a clear history of a bite exposure to an animal suspected of rabies, subsequent epidemiologic investigation revealed a history of direct contact with a bat (e.g., capturing bats, handling a downed bat, or a bat landing on a person). The most parsimonious explanation for the rabies infection is an unnoticed bite inflicted by a single bat (See above, "Transmission of Insectivorous Bat Rabies to Humans").

Conclusions

The goal of this chapter was to highlight those studies providing the most thorough information about disease ecology in bats. These studies, however, have generally raised more questions than answers. Much more effort is required if we are to assemble a comprehensive picture of the impact of infectious diseases on populations of bats. Beyond the obvious implications for public health, conservation of bat populations, severely hampered by human encroachment, necessitates a better understanding of the impact of infectious disease in already threatened species.

Acknowledgments

We sincerely thank the many people at U.S. and foreign health departments for their generous contributions to the understanding of disease ecology in bats. Their efforts to collect specimens and monitor local populations of bats have provided an invaluable foundation of knowledge on which much of the research reviewed in this chapter was based. We also thank the staff in the Viral and Rickettsial Zoonoses Branch at the Centers for Disease Control and Prevention (CDC) for their valuable comments and criticism of earlier drafts of this manuscript. Danny Brass, Denny Constantine, Brock Fenton, Tom Kunz, Charles Trimarchi, and three anonymous reviewers also provided insightful criticism that improved the manuscript substantially. Financial support was provided to Messenger by the American Society of Microbiology (ASM) and CDC through the ASM/National Center for Infectious Diseases Postdoctoral Research Associates Program.

Literature Cited

Acha, P. N. 1968. Epidemiology of bat-borne paralytic rabies in cattle. Boletin de la Oficina Sanitaria Panamericana, 64:411–430.

Allworth, T., J. O'Sullivan, L. Selvey, and J. Sheridan. 1995. Equine morbillivirus in Queensland. Communicable Disease Intelligence, 19:575.

Amengual, B., J. E. Whitby, A. King, J. Serra Cobo, and H. Bourhy. 1997. Evolution of European bat lyssaviruses. Journal of General Virology, 78:2319–2328.

Anderson, R. M., H. C. Jackson, R. M. May, and A. M. Smith. 1981. Population dynamics of fox rabies in Europe. Nature, 289:765–771.

Bacon, P. J. 1985. A systems analysis of wildlife rabies epizootics. Pp. 109–130 in: Population Dynamics of Rabies in Wildlife (P. J. Bacon, ed.). Academic Press, London.

Baer, G. M. 1975. Rabies in nonhematophagous bats. Pp. 79–97 *in:* The Natural History of Rabies (G.M. Baer, ed.). Vol. 2. Academic Press, New York.

Baer, G. M. 1991. Vampire bat and bovine paralytic rabies. Pp. 389–404 *in:* The Natural History of Rabies (G. M. Baer, ed.). 2d ed. CRC Press, Boca Raton, Fla.

Baer, G. M., and G. L. Bales. 1967. Experimental rabies infection in the Mexican free-tailed bat. Journal of Infectious Diseases, 117:82–90.

Baer, G. M., and W. F. Cleary. 1972. A model in mice for the pathogenesis and treatment of rabies. Journal of Infectious Diseases, 125:520–527.

Ball, F. G. 1985. Spatial models for the spread and control of rabies incorporating group size. Pp. 197–222 *in:* Population Dynamics of Rabies in Wildlife (P. J. Bacon, ed.). Academic Press, London.

Banerjee, K., H. R. Bhat, G. Geevarghese, P. G. Jacob, and A. S. Malunjkar. 1988. Antibodies against Japanese encephalitis virus in insectivorous bats from Karnataka. Indian Journal of Medical Research, 87:527–530.

Bell, G. P. 1980. A possible case of interspecific transmission of rabies in insectivorous bats. Journal of Mammalogy, 61:528.

Bell, J. F. 1975. Latency and abortive infections. Pp. 331–354 *in:* The Natural History of Rabies (G. M. Baer, ed.). Vol. 1. Academic Press, New York.

Bell, J. F., and G. J. Moore. 1960. Rabies virus isolated from brown fat of naturally infected bats. Proceedings of the Society for Experimental Biology and Medicine, 103:140–142.

Bell, J. F., G. J. Moore, and G. H. Raymond. 1969. Protracted survival of a rabies-infected insectivorous bat after infective bite. American Journal of Tropical Medicine and Hygiene, 18:61–66.

Bell, J. F., G. J. Moore, G. H. Raymond, and C. E. Tibbs. 1962. Characteristics of rabies in bats in Montana. American Journal of Public Health, 52:1293–1301.

Bhat, H. R., M. A. Sreenivasan, M. K. Goverdhan, S. V. Naik, and K. Banerjee. 1978. Antibodies to Kyasanur forest disease virus in bats in the epizootic-epidemic area and neighbourhood. Indian Journal of Medical Research, 68:387–392.

Blackburn, N. K., C. M. Foggin, L. Searle, and P. N. Smith. 1982. Isolation of Sindbis virus from bat organs. Central African Journal of Medicine, 28:201.

Boiro, I., F. M. Fidarov, N. N. Lomonossov, M. B. Linev, V. N. Bachkirsov, and A. Inapogui. 1986. Isolation of the Fomede virus from Chiroptera, Nycteris nana, in the Republic of Guinea. Bulletin de la Societe de Pathologie Exotique et de Ses Filiales, 79:180–182.

Boiro, I., O. K. Konstaninov, and A. D. Numerov. 1987. Isolation of Rift Valley fever virus from bats in the Republic of Guinea. Bulletin de la Societe de Pathologie Exotique et de Ses Filiales, 80:62–67.

Boulger, L. R., and J. S. Porterfield. 1958. Isolation of a virus from Nigerian fruit bats. Transactions of the Royal Society of Tropical Medicine and Hygiene, 52:421–424.

Bourhy, H., B. Kissi, and N. Tordo. 1993. Molecular diversity of the lyssavirus genus. Virology, 194:70–81.

Brass, D. A. 1994. Rabies in Bats: Natural History and Public Health Implications. Livia Press, Ridgefield, Conn.

Breman J. G., K. M. Johnson, G. van der Groen, C. B. Robbins, M. V. Szczeniowski, K. Ruti, P. A. Webb, F. Meier, D. L. Heymann, and the Ebola Virus Study Teams. 1999.

A search for Ebola virus in animals in the Democratic Republic of the Congo and Cameroon: ecologic, virologic, and serologic surveys, 1979–1980. Journal of Infectious Diseases, 179 (suppl. 1):S139–S147.

Burns, K. F., and C. J. Farinacci. 1955. Rabies in nonsanguivorous bats of Texas. Journal of Infectious Diseases, 97:211–218.

Burns, K. F., C. J. Farinacci, and T. G. Murnane. 1956a. Rabies in insectivorous bats of Texas. Journal of the American Veterinary Medical Association, 128:27–31.

Burns, K. F., C. J. Farinacci, T. G. Murnane, and D. F. Shelton. 1956b. Insectivorous bats naturally infected with rabies in southwestern United States. American Journal of Public Health, 46:1089–1097.

Burns, K. F., D. F. Shelton, and E. W. Grogan. 1958. Bat rabies: experimental host transmission studies. Annals of the New York Academy of Science, 70:452–466.

Butenko, A. M. 1996. Arbovirus circulation in the Republic of Guinea. Meditsinskaia Parazitologiia i Parazitarnye Bolezni, 2:40–45.

Calisher C. H., W. A. Chappell, K. S. Maness, R. D. Lord, and W. D. Sudia. 1971. Isolations of Nepuyo virus strains from Honduras. 1967. American Journal of Tropical Medicine and Hygiene, 20:331–337.

Carey, A. B. 1974. An analysis of an apparent rabies epizootic in Rockbridge county, Virginia. M.S. Thesis, Virginia Polytechnic Institute and State University, Blacksburg.

Carey, A. B., and R. G. McLean. 1983. The ecology of rabies: evidence of co-adaptation. Journal of Applied Ecology, 20:777–800.

Carini, A. 1911. Sur une grande épizootie de rage. Annals of the Institute Pasteur (Paris), 25:843–846.

Cassel-Beraud A. M., D. Fontenille, and L. Rabetafika. 1989. Bacterial, viral and parasitological study of a population of *Chaerephon pumila* bats in Anjiro, Madagascar. Archives de L' Institut Pasteur de Madagascar, 56:233–239.

Centers for Disease Control and Prevention (CDC). 1983. Exposure to a rabid cow— Pennsylvania. Morbidity and Mortality Weekly Report, 32:128–129.

Centers for Disease Control and Prevention (CDC). 1964. Rabies in bats—Mississippi. Veterinary Public Health Notes (November), 1–3.

Centers for Disease Control and Prevention (CDC). 1998. Human Rabies—Texas and New Jersey, 1997. Morbidity and Mortality Weekly Report, 47:1–5.

Centers for Disease Control and Prevention (CDC). 1999a. Human rabies prevention— United States, 1999: Recommendations of the Advisory Committee on Immunization Practices (ACIP). Morbidity and Mortality Weekly Report, 48(RR-1):1–21.

Centers for Disease Control and Prevention (CDC). 1999b. Outbreak of Hendra-like virus: Malaysia and Singapore, 1998–1999. Morbidity and Mortality Weekly Report, 48:265–272.

Chant, K., R. Chan, M. Smith, D. E. Dwyer, P. Kirkland, and the NSW Expert Group. 1998. Probable human infection with a newly described virus in the family Paramyxoviridae. Emerging Infectious Diseases, 4:273–275.

Charlton, K. M., G. A. Casey, D. W. Boucher, and T. J. Wiktor. 1982. Antigenic variants in rabies virus. Comparative Immunology, Microbiology, and Infectious Diseases, 5:113–115.

Childs, J., C. Trimarchi, and J. Krebs. 1994. The epidemiology of bat rabies in New York State, 1988–1992. Epidemiology and Infection, 113:501–511.

Clark, D. R., A. Lollar, and D. F. Cowman. 1996. Dead and dying Brazilian free-tailed bats (*Tadarida brasiliensis*) from Texas: rabies and pesticide exposure. Southwestern Naturalist, 41:275–278.

Coggins, R. 1988. Methods for the ecological study of bat endoparasites. Pp. 475–489 *in:* Ecological and Behavioral Methods for the Study of Bats (T. H. Kunz, ed.). Smithsonian Institution Press, Washington D.C.

Constantine, D. G. 1962. Rabies transmission by nonbite route. Public Health Reports, 77:287–289.

Constantine, D. G. 1966a. Transmission experiments with bat rabies isolates: Responses of certain Carnivora to rabies virus isolated from animals infected by nonbite route. American Journal of Veterinary Research, 27:13–15.

Constantine, D. G. 1966b. Transmission experiments with bat rabies isolates: reaction of certain Carnivora, opossum, and bats to intramuscular inoculations of rabies virus isolated from free-tailed bats. American Journal of Veterinary Research, 27:16–19.

Constantine, D. G. 1966c. Transmission experiments with bat rabies isolates: bite transmission of rabies to foxes and coyote by free-tailed bats. American Journal of Veterinary Research, 27:20–23.

Constantine, D. G. 1967a. Activity patterns of the Mexican free-tailed bat. University of New Mexico Publications in Biology, 7:1–79.

Constantine, D. G. 1967b. Rabies transmission by air in bat caves. U.S. Public Health Service Publication 1617. National Communicable Disease Center, Atlanta, 51 pp.

Constantine, D. G. 1967c. Bat rabies in the southwestern United States. Public Health Reports, 82:867–888.

Constantine, D. G. 1970. Bats in relation to the health, welfare, and economy of man. Pp. 319–449 *in:* Biology of Bats (W. A. Wimsatt, ed.). Vol. 2. Academic Press, New York.

Constantine, D. G. 1975. Letter: Rabies virus in bats' brain and salivary glands. New England Journal of Medicine, 292:51.

Constantine, D. G. 1979. An updated list of rabies-infected bats in North America. Journal of Wildlife Diseases, 15:347–349.

Constantine, D. G. 1986. Absence of prenatal infection of bats with rabies virus. Journal of Wildlife Diseases, 22:249–250.

Constantine, D. G. 1988a. Health precautions for bat researchers. Pp. 491–528 *in:* Ecological and Behavioral Methods for the Study of Bats (T. H. Kunz, ed.). Smithsonian Institution Press, Washington, D.C.

Constantine, D. G. 1988b. Transmission of pathogenic microorganisms by vampire bats. Pp. 167–189 *in:* Natural History of Vampire Bats (A. M. Greenhall and U. Schmidt, eds.). CRC Press, Boca Raton, Fla.

Constantine, D. G., R. W. Emmons, and J. D. Woodie. 1972. Rabies virus in nasal mucosa of naturally infected bats. Science, 175:1255–1256.

Constantine, D. G., G.C. Solomon, and D. F. Woodall. 1968a. Transmission experiments with bat rabies isolates: responses of certain carnivores and rodents to rabies viruses from four species of bats. American Journal of Veterinary Research, 29:181–190.

Constantine, D. G., E. S. Tierkel, M. D. Kleckner, and D. M. Hawkins. 1968b. Rabies in New Mexico cavern bats. Public Health Reports, 83:303–316.

Constantine, D. G., and D. F. Woodall. 1964. Latent infection of Rio Bravo virus in salivary glands of bats. Public Health Reports, 79:1033–1039.

Constantine, D. G., and D. F. Woodall. 1966. Transmission experiments with bat rabies isolates: reactions of certain Carnivora, opossums, rodents, and bats to rabies virus of red bat origin when exposed by bat bite or intramuscular inoculation. American Journal of Veterinary Research, 27:24–32.

Correa-Giron, E. P., R. Allen, and S. E. Sulkin. 1970. Infectivity and pathogenesis of rabies virus administered orally. American Journal of Epidemiology, 91:203–215.

Correa-Giron, P., C. H. Calisher, and G. M. Baer. 1972. Epidemic strain of Venezuelan equine encephalomyelitis virus from a vampire bat captured in Oaxaca, Mexico, 1970. Science, 175:546–547.

Coyne, M. J., G. Smith, and F. E. McAllister. 1989. Mathematical model for the population biology of rabies in raccoons in the mid-Atlantic states. American Journal of Veterinary Research, 50:2148–2154.

Daoust, P., A. I. Wandeler, and G. A. Casey. 1996. Cluster of rabies cases of probable bat origin among red foxes in Prince Edward Island, Canada. Journal of Wildlife Diseases, 32:403–406.

Dean, W. D., K. T. Maddy, E. L. Cockrum, and H. G. Crecelius. 1960. Rabies in insectivorous bats of Arizona. Arizona Medicine, 17:69–77.

Delpietro, H. A., and J. F. Konolsaison. 1991. Dinamica da raiva em uma populacao de morecegos hematofagos (Desmodus rotundas) no nordeste Argentino, e sua relacao com a raiva paralitica dos herbivoros. Arquivos de Biologia e Tecnologia 34 (3/4): 381–391.

Delpietro, H. A., L. Segre, N. Marchevsky, and M. Berisso. 1990. Rabies transmission to rodents after ingestion of naturally infected tissues. Medicina (Buenos Aires), 50:356–360.

de Mattos, C. C., C. A. de Mattos, M. Favi Cortes, V. P. Yung, E. V. Ramirez , L. A. Orciari, and J. S. Smith. 1997. Genetic characterization of rabies isolates in Chile. Paper presented at the Rabies in the Americas Conference: Current Issues in Rabies and Rabies Control in the Americas. 8th Annual Rabies in the Americas Conference, November 2–6, 1997, Kingston, Ontario, Canada. http://www.gis.queensu.ca/RReporter/rabies_conference/AuthorIndex.html. (Abstract.)

Diallo, M., J. Thonnon, M. Traore-Lamizana, and D. Fontenille. 1999. Vectors of Chikungunya virus in Senegal: current data and transmission cycles. American Journal of Tropical Medicine and Hygiene, 60:281–286.

Dietzschold, B., C. E. Rupprecht, M. Tollis, M. Lafon, J. Mattei, T. J. Wiktor, and H. Koprowski. 1988. Antigenic diversity of the glycoprotein and nucleocapsid proteins of rabies and rabies-related viruses: implications for epidemiology and control of rabies. Reviews of Infectious Diseases, 10 (suppl. 4):S785–S798.

Digoutte, J. P., and G. Girault. 1976. The protective properties in mice of tonate virus and two strains of cabassou virus against neurovirulent everglades Venezuelan encephalitis virus. Annales de Microbiologie, 127B:429–437.

Doherty, R. L., B. M. Gorman, R. H. Whitehead, and J. G. Carley. 1966. Studies of arthropod-borne virus infections in Queensland. V. Survey of antibodies to group A arboviruses in man and other animals. Australian Journal of Experimental Biology and Medical Science, 44:365–377.

Eads, R. B., J. S. Wiseman, J. E. Grimes, and G. C. Menzies. 1955. Wildlife rabies in Texas. Public Health Reports, 70:995–1000.

Favi Cortes, M., P. Yung, C. B. Pavletic, and E. Ramirez V. 1997. Role of insectivorous bats in the transmission of rabies in Chile: antigenic characterization of field isolates.

Paper presented at the Rabies in the Americas Conference: Current Issues in Rabies and Rabies Control in the Americas. 8th Annual Rabies in the Americas Conference, November 2–6, 1997, Kingston, Ontario, Canada. http://www.gis.queensu.ca/ RReporter/rabies_conference/AuthorIndex.html. (Abstract.)

Fekadu, M., J. H. Shaddock, D. W. Sanderlin, and J. S. Smith. 1988. Efficacy of rabies vaccines against Duvenhage virus isolated from European house bats (*Eptesicus serotinus*), classic rabies and rabies-related viruses. Vaccine, 6:533–539.

Feldmann, H., W. Slenczka, and H. D. Klenk. 1996. Emerging and reemerging of filoviruses. Archives of Virology (suppl.), 11:77–100.

Fischman, H. R. 1976. Consideration of an association between geographic distribution of caves and occurrence of rabies in foxes. Journal of the American Veterinary Medical Association, 169:1207–1213.

Fischman, H. R., and M. Schaeffer. 1971. Pathogenesis of experimental rabies as revealed by immunofluorescence. Annals of the New York Academy of Sciences, 177:78–97.

Fischman, H. R., and F. E. Ward III. 1968. Oral transmission of rabies virus in experimental animals. American Journal of Epidemiology, 88:132–138.

Fischman, H. R., and G. S. Young. 1976. An association between the occurrence of fox rabies and the presence of caves—geographic relationships. American Journal of Epidemiology, 104:593–601.

Fishbein, D. B. 1991. Rabies in humans. Pp. 519–549 in: The Natural History of Rabies (G. M. Baer, ed.). 2d ed. CRC Press, Boca Raton, Fla.

Flannery, T. F. 1989. Flying foxes in Melanesia: populations at risk. Bats, 7:5–7.

Fontenille, D., M. Traore-Lamizana, J. Trouillet, A. Leclerc, M. Mondo, Y. Ba, J.P. Digoutte, and H. G. Zeller. 1994. First isolations of arboviruses from phlebotomine sand flies in West Africa. American Journal of Tropical Medicine and Hygiene, 50:570–574.

Frederickson, L. E., and L. Thomas. 1965. Relationship of fox rabies to caves. Public Health Reports, 80:495–500.

Fuller, Jr., J. E. 1984. A case of equine rabies. New England Journal of Medicine, 310:525–526.

Garrett, L. 1994. The Coming Plague. Farrar, Straus & Giroux, New York.

Gillette, D. D., and J. D. Kimbrough. 1970. Chiropteran mortality. Pp. 262–283 in: About Bats (B. H. Slaughter and D. W. Walton, eds.). Southern Methodist University Press, Dallas.

Girard, K. F., H. B. Hitchcock, G. Edsall, and R. A. MacCready. 1965. Rabies in bats in southern New England. New England Journal of Medicine, 272:75–80.

Glass, B. P. 1958. Returns of the Mexican freetail bats banded in Oklahoma. Journal of Mammalogy, 39:435–437.

Glass, B. P. 1959. Additional returns from free-tailed bats banded in Oklahoma. Journal of Mammalogy, 40:542–545.

Gould, A. R., A. D. Hyatt, R. Lunt, J. A. Kattenbelt, S. Hengstberger, and S. D. Blacksell. 1998. Characterisation of a novel lyssavirus isolated from pteropid bats in Australia. Virus Research, 54:165–187.

Greenwood, R. J., W. E. Newton, G. L. Pearson, and G. J. Schamber. 1997. Population and movement characteristics of radio-collared striped skunks in North Dakota during an epizootic of rabies. Journal of Wildlife Diseases, 33:226–241.

Halpin, K., P. Young, and H. Field. 1996. Identification of likely natural hosts for equine morbillivirus. Communicable Disease Intelligence, 20:476.

Harris, A. H. 1971. Use of Texas caves by terrestrial mammals. Pp. 76–94 *in:* Natural History of Texas Caves (E. L. Lundelius, Jr., and B. H. Slaughter, eds.). Gulf Natural History, Dallas.

Haupt, H., and H. Rehaag. 1921. Durch Fledermause verbreitete seuchenhafte Tollwut unter Viehbestanden in Santa Catharina (Sud-Brasilien). Zeitschrift für Infektionskrankheiten Haustiere, 22:104–127.

Hayne, D. W., and D. L. R. Neeley. 1967. Re-examination of reported relationships between caves and fox rabies in Tennessee. Paper presented at the 47th annual meeting of the American Society of Mammalogists, Nags Head, N.C., June 19.

Herbold, J. R., W. P. Heuschele, R. Berry, and M. A. Parsons. 1983. Reservoir of St. Louis encephalitis virus in Ohio bats. American Journal of Veterinary Research, 44:1889–1893.

Hooper, P. T., A. R. Gould, G. M. Russell, J. A. Kattenbelt, and G. Mitchell. 1996. The retrospective diagnosis of a second outbreak of equine morbillivirus infection. Australian Veterinary Journal, 74:244–245.

Howard, D. R. 1981. Transplacental transmission of rabies virus from a naturally infected skunk. American Journal of Veterinary Research, 42:691–692.

Hronovsky, V., and R. Benda. 1969a. Experimental inhalation infection of laboratory rodents with rabies virus. Acta Virologica (Prague), 13:193–197.

Hronovsky, V., and R. Benda. 1969b. Development of inhalation rabies infection in suckling guinea pigs. Acta Virologica (Prague), 13:198–202.

Humphrey, G. L., G. E. Kemp, and E. G. Wood. 1960. A fatal case of rabies in a woman bitten by an insectivorous bat. Public Health Reports, 75:317–326.

Hyatt, A. D., and P. W. Selleck. 1996. Ultrastructure of equine morbillivirus. Virus Research, 43:1–15.

Irons, J. V., R. B. Eads, J. E. Grimes, and A. Conklin. 1957. The public health importance of bats. Texas Reports on Biology and Medicine, 15:292–298.

Ito, T., K. Okazaki, Y. Kawaoka, A. Takada, R. G. Webster, and H. Kida. 1995. Perpetuation of influenza A viruses in Alaskan waterfowl reservoirs. Archives of Virology, 140:1163–1172.

Journal of Wildlife Diseases. 1999. Supplement to the Journal of Wildlife Diseases. Vol. 35, no. 4.

Kaiser, J. 1999. Battle over a dying sea. Science, 284:28–30.

Kalunda, M., L. G. Mukwaya, A. Mukuye, M. Lule, E. Sekyalo, J. Wright, and J. Casals. 1986. Kasokero virus: a new human pathogen from bats (*Rousettus aegyptiacus*) in Uganda. American Journal of Tropical Medicine and Hygiene, 35:387–392.

Kaplan, C. 1985. Rabies: A worldwide disease. Pp. 1–21 *in:* Population Dynamics of Rabies in Wildlife (P. J. Bacon, ed.). Academic Press, London.

Kaplan, M. M. 1969. Epidemiology of bat rabies. Nature, 221:421–425.

Karabatsos, N. 1985. International catalogue of arboviruses. 3d ed. American Society of Tropical Medicine and Hygiene, San Antonio.

Keen, R. 1988. Mark-recapture estimates of bat survival. Pp. 157–170 *in:* Ecological and Behavioral Methods for the Study of Bats (T. H. Kunz, ed.). Smithsonian Institution Press, Washington D.C.

Kelkar, S. D., S. S. Kadam, and K. Banerjee. 1981. Haemagglutination inhibition antibodies against influenza virus in bats. Indian Journal of Medical Research, 74: 147–152.

Kemp, G. E., G. Le Gonidec, N. Karabatsos, A. Rickenbach, and C. B. Cropp. 1988. IFE: a new African orbivirus isolated from *Eidolon helvum* bats captured in Nigeria, Cameroon and the Central African Republic. Bulletin de la Societe de Pathologie Exotique et de Ses Filiales, 81:40–48.

Kim, G. R., Y. T. Lee, and C. H. Park. 1994. A new natural reservoir of hantavirus: isolation of hantaviruses from lung tissues of bats. Archives of Virology, 134:85–95.

King, A. A., C. D. Meredith, and G. R. Thomson. 1994. The biology of Southern African lyssavirus variants. Current Topics in Microbiology and Immunology, 187:267–295.

Kwiecinski, G. G., D. J. Murphy, and C. E. Rupprecht. 1993. An enzyme linked immunosorbant assay for detection of rabies specific IgG in bats. Bat Research News, 34:116.

Lawson, K. F., J. G. Black, K. M. Charlton, D. H. Johnston, and A. J. Rhodes. 1987. Safety and immunogenicity of a vaccine bait containing the ERA strain of attenuated rabies vaccine. Canadian Journal of Veterinary Research, 51:460–464.

Leirs H., J. N. Mills, J. W. Krebs, J. E. Childs, D. Akaibe, N. Woollen, G. Ludwig, C. J. Peters, and T. G. Ksiazek. 1999. Search for the Ebola virus reservoir in Kikwit, Democratic Republic of the Congo: reflections on a vertebrate collection. Journal of Infectious Diseases, 179 (suppl. 1):S155–S163.

Leopold, A. [1933] 1986. Game Management. University of Wisconsin Press, Madison.

Lopez, A. 1991. Description of the outbreak in Choque, Madre de Dios (Peru). Pp. 1–8 *in:* Expert Consultation on Attention of Persons Exposed to Rabies Transmitted by Vampire Bats: Proceedings of Joint Meeting of Pan American Health Organization and World Health Organization, Washington, D.C., April 1–5.

Lord, R. D., H. Delpietro, E. Fuenzalida, A. M. Oviedo de Diaz, and L. Lazaro. 1975. Presence of rabies neutralizing antibodies in wild carnivores following an outbreak of bovine rabies. Journal of Wildlife Diseases, 11:210–213.

Lumio, J., M. Hillbom, R. Roine, L. Ketonen, M. Haltia, M. Valle, E. Neuvonen, and J. Lahdevirta. 1986. Human rabies of bat origin in Europe. Lancet, 1(8477):378.

Lvov, D. K., B. Easterday, W. Hinshow, I. V. Dandurov, and P. N. Arkhipov. 1979. Isolation of strains of the Hong Kong complex (H3N2) influenza virus from *Nyctalus noctula* bats in Kazakhstan. Voprosy Virusologii, 4:338–341.

Lvov, D. K., F. R. Karas, E. M. Timofeev, Y. M. Tsyrkin, S. G. Vargina, O. V. Veselovskaya, N. Z. Osipova, et al. 1973. "Issyk-Kul" virus, a new arbovirus isolated from bats and *Argas* (Carios) *vespertilionis* (Latr., 1802) in the Ki20rghiz S.S.R. Archiv für die Gesamte Virusforschung, 42:207–209.

Mackenzie, J.S. 1999. Emerging viral diseases: an Australian perspective. Emerging Infectious Diseases, 5:1–8.

Mahan, W. E. 1973. An evaluation of the relationship between cave-dwelling bats and fox rabies in Appalachia. Master's Thesis, College of Veterinary Medicine, University of Georgia, Athens.

Martell, D., M. A. F. Ceron Montes, and R. B. Alcocer. 1973. Transplacental transmission of bovine rabies after natural infection. Journal of Infectious Diseases, 127:291–293.

Martini, G. A., and R. Siegert, eds. 1971. Marburg Virus Disease. Springer-Verlag, Berlin.

McCormack, J. G., A. M. Allworth, L. A. Selvey, and P. W. Selleck. 1999. Transmissibility from horses to humans of a novel paramyxovirus, equine morbillivirus. Journal of Infection, 38:22–23.

McLean, R. G., H. A. Trevino, and G. E. Sather. 1979. Prevalence of selected zoonotic diseases in vertebrates from Haiti, 1972. Journal of Wildlife Diseases, 15:327–330.

Meredith, C. D., A. P. Prossouw, and H. P. Koch. 1971. An unusual case of human rabies thought to be of chiropteran origin. South African Medical Journal, 45:767–769.

Miura, T., and M. Kitaoka. 1977. Viruses isolated from bats in Japan. Archives of Virology, 53:281–286.

Monath, T. P. 1999. Ecology of Marburg and Ebola viruses: speculations and directions for future research. Journal of Infectious Diseases, 179 (suppl. 1):S127–S138.

Moore, G. J., and G. H. Raymond. 1970. Prolonged incubation period of rabies in a naturally infected insectivorous bat, *Eptesicus fuscus.* Journal of Wildlife Diseases, 6:167–168.

Moreno, J. A., and G. M. Baer. 1980. Experimental rabies in the vampire bat. American Journal of Tropical Medicine and Hygiene, 29:254–259.

Morimoto, K., M. Patel, S. Corisdeo, D. C. Hooper, Z. F. Fu, C. E. Rupprecht, H. Koprowski, and B. Dietzschold. 1996. Characterization of a unique variant of bat rabies virus responsible for newly emerging human cases in North America. Proceedings of the National Academy of Sciences of the USA, 93:5653–5658.

Murphy, F. A., C. M. Fauquet, D. H. L. Bishop, S. A. Ghabrial, A. W. Jarvis, G. P. Martelli, M. A. Mayo, and M. D. Summers. 1995. Virus Taxonomy: The Classifications and Nomenclature of Viruses: The Sixth Report of the International Committee on Taxonomy of Viruses. Springer-Verlag, Vienna.

Murray, K., P. Selleck, P. Hooper, A. Hyatt, A. Gould, L. Gleeson, H. Westbury, L. Hiley, L. Selvey, and B. Rodwell. 1995. A morbillivirus that caused fatal disease in horses and humans. Science, 268:94–97.

Nikolic, M., and Z. Jelesic. 1956. Isolation of rabies virus from insectivorous bats in Yugoslavia. Bulletin of the World Health Organization, 14:801–804.

Noah, D. L., C. L. Drenzek, J. S. Smith, J. W. Krebs, L. A. Orciari, J. Shaddock, D. Sanderlin, et al. 1998. Epidemiology of human rabies in the United States, 1980 to 1996. Annals of Internal Medicine, 128:922–930.

Ortega, J. R., M. A. M. Delgado, D. B. Campero, and D. O. Cordova. 1985. Presencia de anticuerpos y virus rabico en *Desmodus rotundus* y otros murcielagos en una region de la zona humeda del Istmo de Tehuantepec. Tecnica Pecuaria en Mexico, 49:9.

O'Sullivan, J. D., A. M. Allworth, D. L. Paterson, T. M. Snow, R. Boots, and L. J. Gleeson. 1997. Fatal encephalitis due to a novel paramyxovirus transmitted from horses. Lancet, 349:93–95.

Paradiso, J. L., ed. 1975. Walker's Mammals of the World. Johns Hopkins University Press, Baltimore.

Paterson, D. L., P. K. Murray, and J. G. McCormack. 1998. Zoonotic disease in Australia caused by a novel member of the paramyxoviridae. Clinical Infectious Diseases, 27:112–118.

Paul, S. D., P. K. Rajagopalan, and M. A. Sreenivasan. 1970. Isolation of the West Nile virus from the frugivorous bat, *Rousettus leschenaulti.* Indian Journal of Medical Research, 58:1169–1171.

Pavri, K. M., K. R. Singh, and F. B. Hollinger. 1971. Isolation of a new parainfluenza virus from a frugivorous bat, *Rousettus leschenaulti,* collected at Poona, India. American Journal of Tropical Medicine and Hygiene, 20:125–130.

Pawan, J. L. 1936a. The transmission of paralytic rabies in Trinidad by the vampire bat (*Desmodus rotundus murinus*, Wagner 1804). Annals of Tropical Medicine and Parasitology, 30:101–129.

Pawan, J. L. 1936b. Rabies in the vampire bat of Trinidad, with special reference to the clinical course and the latency of infection. Annals of Tropical Medicine and Parasitology, 30:401–422.

Pawan, J. L. 1948. Fruit-eating bats and paralytic rabies in Trinidad. Annals of Tropical Medicine and Parasitology, 42:173–177.

Peters, C. J., A. Sanchez, P. E. Rollin, T. G. Ksaizek, and F. A. Murphy. 1996. Filoviridae: Marburg and Ebola viruses. Pp. 1161–1176 *in:* Virology (B. N. Fields, D. M. Knipe, P. M. Howley, R. M. Chanock, J. L. Melnick, T. P. Monath, B. Roizman, and S. E. Straus, eds.). Lippincott-Raven Publishers, Philadelphia.

Philbey, A. W., P. D. Kirkland, A. D. Ross, R. J. Davis, A. B. Gleeson, R. J. Love, P. W. Daniels, A. R. Gould, and A. D. Hyatt. 1998. An apparently new virus (family Paramyxoviridae) infectious for pigs, humans, and fruit bats. Emerging Infectious Diseases, 4:269–271.

Pierson, E. D., and W. E. Rainey. 1992. The biology of flying foxes of the genus *Pteropus:* a review. Pp. 1–17 *in:* Proceedings of the Pacific Island Flying Fox Conservation Conference (D. E. Wilson and G. L. Graham, eds.). U.S. Fish and Wildlife Service Biological Report 90 (23). U.S. Department of the Interior, U.S. Fish and Wildlife Service, Washington, D.C.

Price J. L. 1978. Isolation of Rio Bravo and a hitherto undescribed agent, Tamana bat virus, from insectivorous bats in Trinidad, with serological evidence of infection in bats and man. American Journal of Tropical Medicine and Hygiene, 27: 153–161.

Price, J. L., and C. O. Everard. 1977. Rabies virus and antibody in bats in Grenada and Trinidad. Journal of Wildlife Diseases, 13:131–134.

Pybus, M. J., D. P. Hobson, and D. K. Onderka. 1986. Mass mortality of bats due to probable blue-green algal toxicity. Journal of Wildlife Diseases, 22:449–450.

Quist, K. D., R. B. Eads, and A. Conklin. 1957. Studies on bat rabies in Texas. Journal of the American Veterinary Medical Association, 130:66–68.

Rogers, R. J., I. C. Douglas, and F. C. Baldock. 1996. Investigation of a second focus of equine morbillivirus infection in coastal Queensland. Australian Veterinary Journal, 74:243–244.

Rosatte, R. C., and J. R. Gunson. 1984. Presence of neutralizing antibodies to rabies virus in striped skunks from areas free of skunk rabies in Alberta. Journal of Wildlife Diseases, 20:171–176.

Rupprecht, C. E., L. T. Glickman, P. A. Spencer, and T. J. Wiktor. 1987. Epidemiology of rabies virus variants: differentiation using monoclonal antibodies and discriminant analysis. American Journal of Epidemiology, 126:298–309.

Rupprecht, C. E., and J. S. Smith. 1994. Raccoon rabies: the re-emergence of an epizootic in a densely populated area. Seminars in Virology, 5:155–164.

Sadler, W. W., and J. B. Enright. 1959. Effect of metabolic level of the host upon the pathogenesis of rabies in the bat. Journal of Infectious Diseases, 105:267–273.

Salaun, J. J., J. M. Klein, and G. Hebrard. 1974. A new virus, Phnom-Penh bat virus, isolated in Cambodia from a short-nosed fruit bat, "*Cynopterus brachyotis angulatus*" Miller, 1898. Annales de Microbiologie, 125:485–495.

Sanborn, C. C. 1954. Bats of the United States. Public Health Reports, 69:17–28.

Scatterday, J. E., and M. M. Galton. 1954. Bat rabies in Florida. Veterinary Medicine, 49:133–135.

Schaaf, J., and E. Schaal. 1968. Deutsche Tierarztliche Wochenschrift, 75:315.

Schneider, L. G., B. J. H. Barnard, and H. P. Schneider. 1985. Application of monoclonal antibodies for epidemiological investigations and oral vaccination studies. Pp. 47–59 in: Rabies in the Tropics (E. Kuwert, C. Merieux, H. Koprowski, and K. Bogel, eds.). Springer-Verlag, Berlin.

Schneider, N. J., J. E. Scatterday, A. L. Lewis, W. L. Jennings, H. D. Venters, and A. V. Hardy. 1957. Rabies in bats in Florida. American Journal of Public Health, 47: 983–989.

Selvey, L., and J. Sheridan. 1994. Outbreak of a severe respiratory disease in humans and horses due to a previously unrecognized paramyxovirus. Communicable Disease Intelligence, 18:499.

Selvey, L., R. Taylor, A. Arklay, and J. Gerrard. 1996. Screening of bat carers for antibodies to equine morbillivirus. Communicable Disease Intelligence, 20:477–478.

Seymour, C., R. W. Dickerman, and M. S. Martin. 1978. Venezuelan encephalitis virus infection in Neotropical bats. I. Natural infection in a Guatemalan enzootic focus. American Journal of Tropical Medicine and Hygiene, 27:290–296.

Shope, R. E., F. A. Murphy, A. K. Harrison, O. R. Causey, G. E. Kemp, D. I. Simpson, and D. L. Moore. 1970. Two African viruses serologically and morphologically related to rabies virus. Journal of Virology, 6:690–692.

Sims, R. A., R. Allen, and S. E. Sulkin. 1963. Studies on the pathogenesis of rabies in insectivorous bats. III. Influence of the gravid state. Journal of Infectious Diseases, 112:17–27.

Smith, A. D. M. 1985. A continuous time deterministic model of temporal rabies. Pp. 131–146 in: Population Dynamics of Rabies in Wildlife (P. J. Bacon, ed.). Academic Press, London.

Smith, J. S. 1988. Monoclonal antibody studies of rabies in insectivorous bats of the United States. Reviews of Infectious Diseases, 10 (suppl. 4):S637–S643.

Smith, J. S. 1989. Rabies virus epitopic variation: use in ecologic studies. Advances in Virus Research, 36:215–253.

Smith, J. S. 1996. New aspects of rabies with emphasis on epidemiology, diagnosis, and prevention of the disease in the United States. Clinical Microbiology Reviews, 9(2):166–176.

Smith, J. S. 1999. Rabies virus. Pp. 1099–1106 in: Manual of Clinical Microbiology (P. R. Murray, E. J. Baron, M. A. Pfaller, F. C. Tenover, and R. H. Yolken, eds.). 7th ed. American Society of Microbiology Press, Washington, D.C.

Smith, J. S., and G. M. Baer. 1988. Epizootiology of rabies: the Americas. Pp. 267–299 in: Rabies (J. B. Campbell and K. M. Charlton, eds.). Kluwer Academic, Boston.

Smith, J. S., D. B. Fishbein, C. E. Rupprecht, and K. Clark. 1991. Unexplained rabies in three immigrants in the United States. New England Journal of Medicine, 324: 205–211.

Smith, J. S., L. A. Orciari, and P. A. Yager. 1995. Molecular epidemiology of rabies in the United States. Seminars in Virology, 6:387–400.

Smith, J. S., F. L. Reid-Sanden, L. F. Roumillat, C. Trimarchi, K. Clark, G. M. Baer, and W. G. Winkler. 1986. Demonstration of antigenic variation among rabies virus isolates by using monoclonal antibodies to nucleocapsid proteins. Journal of Clinical Microbiology, 24:573–580.

Smith, J. S., and H. D. Seidel. 1993. Rabies: a new look at an old disease. Pp. 82–106 *in:* Progress in Medical Virology (J. L. Melnick, ed.). Vol. 40, Karger, Basel.

Smith, J. S., P. A. Yager, W. J. Bigler, and E. C. Hartwig, Jr. 1990. Surveillance and epidemiological mapping of monoclonal antibody-defined rabies variants in Florida. Journal of Wildlife Diseases, 26:473–485.

Soave, O. A. 1962. Reactivation of rabies virus infection in a guinea pig with adrenocorticotropic hormone. Journal of Infectious Diseases, 110:129–131.

Soave, O. A. 1964. Reactivation of rabies virus in a guinea pig due to the stress of crowding. American Journal of Veterinary Research, 25:268–269.

Soave, O. A. 1966. Transmission of rabies virus to mice by ingestion of infected tissue. American Journal of Veterinary Research, 27:44–46.

Soave, O. A., H. N. Johnson, and K. Nakamura. 1961. Reactivation of rabies virus infection in a guinea pig with adrenocorticotropic hormone. Science, 133:1360–1361.

Steece, R., and J. S. Altenbach. 1989. Prevalence of rabies specific antibodies in the Mexican free-tailed bat (*Tadarida brasiliensis mexicana*) at Lava Cave, New Mexico. Journal of Wildlife Diseases, 25:490–496.

Steece, R. S., and C. H. Calisher. 1989. Evidence for prenatal transfer of rabies virus in the Mexican free-tailed bat (*Tadarida brasiliensis mexicana*). Journal of Wildlife Diseases, 25:329–334.

Sulkin, S. E., and R. Allen. 1974. Virus infections in bats. Monographs in Virology, 8:1–103.

Sulkin, S. E., R. Allen, R. Sims, P. H. Krutzsch, and C. Kim. 1960. Studies of the pathogenesis of rabies in insectivorous bats. II. Influence of environmental temperature. Journal of Experimental Medicine, 112:595–617.

Sulkin, S. E., and M. J. Greve. 1954. Human rabies caused by bat bite. Texas State Journal of Medicine, 50:620–621.

Sulkin, S. E., P. H. Krutzsch, R. Allen, and C. Wallis. 1959. Studies on the pathogenesis of rabies in insectivorous bats. I. Role of brown adipose tissue. Journal of Experimental Medicine, 110:369–388.

Sulkin, S. E., P. M. Krutzsch, C. Wallis, and R. Allen. 1957. Role of brown fat in pathogenesis of rabies in insectivorous bats (*Tadarida b. mexicana*). Proceedings of the Society for Experimental Biology and Medicine, 96:461–464.

Sullivan, T. D., J. E. Grimes, R. B. Eads, C. C. Menzies, and J. V. Irons. 1954. Recovery of rabies virus from colonial bats in Texas. Public Health Reports, 69:766.

Swanepoel, R., B. J. Barnard, C. D. Meredith, G. C. Bishop, G. K. Bruckner, C. M. Foggin, and O. J. Hubschle. 1993. Rabies in southern Africa. Onderstepoort Journal of Veterinary Research, 60:325–346.

Swanepoel, R., P. A. Leman, F. J. Burt, N. A. Zachariades, L. E. Braack, T. G. Ksiazek, P. E. Rollin, S. R. Zaki, and C. J. Peters. 1996. Experimental inoculation of plants and animals with Ebola virus. Emerging Infectious Diseases, 2:321–325.

Swofford, D. L. 1999. PAUP*: Phylogenetic Analysis Using Parsimony (*and Other Methods). Version 4. Sinauer Associates, Sunderland, Mass.

Tandler, B. 1996. Cytomegalovirus in the principal submandibular gland of the little brown bat, *Myotis lucifugus*. Journal of Comparative Pathology, 114:1–9.

Tierkel, E. S. 1958. Recent developments in the epidemiology of rabies. Annals of the New York Academy of Sciences, 70:445–451.

Tierkel, E. S., V. D. Chadwick, M. J. Cerosaletti, H. R. Cox, E. M. Dwyer, T. J. Grennan, and J. W. Mann. 1958. Report of Committee on Rabies. Pp. 253–259 *in:* Sixty-second

Annual Proceedings of the U.S. Livestock Sanitary Association. U.S. Department of Health, Education and Welfare, Washington, D.C.

Tignor, G. H., F. A. Murphy, H. F. Clark, R. E. Shope, P. Madore, S. P. Bauer, S. M. Buckely, and C. D. Meredith. 1977. Duvenhage virus: morphological, biochemical, histopathological, and antigenic relationships to the rabies serogroup. Journal of General Virology, 37:595–611.

Tjalma, R. A., and B. B. Wentworth. 1957. Bat rabies—report of an isolation of rabies virus from native Ohio bats. Journal of the American Veterinary Medical Association, 30:68.

Trimarchi, C. V. 1978. Rabies in insectivorous temperate-zone bats. Bat Research News, 19:7.

Trimarchi, C. V. 1987. New York State Rabies Annual Summary, 1987. Rabies Laboratory, Wadsworth Center for Laboratories and Research, New York State Department of Health. Albany.

Trimarchi, C. V. 1998. New York State Rabies annual summary, 1998. Rabies Laboratory, Wadsworth Center for Laboratories and Research, New York State Department of Health. Albany.

Trimarchi, C. V. 1999. Bat rabies from a state and national perspective. Paper presented at the 1999 American Veterinary Medical Association Annual Meeting, American Veterinary Medical Association Press, New Orleans, La. (Abstract.)

Trimarchi, C. V., and J. G. Debbie. 1977. Naturally occurring rabies virus and neutralizing antibody in two species of insectivorous bats of New York state. Journal of Wildlife Diseases, 13:366–369.

Trimarchi, C. V., R. J. Rudd, and M. K. Abelseth. 1986. Experimentally induced rabies in four cats inoculated with a rabies virus isolated from a bat. American Journal of Veterinary Research, 47:777–780.

Tuttle, M. D., and D. Stevenson. 1982. Growth and survival of bats. Pp. 105–150 *in:* Ecology of Bats (T. H. Kunz, ed.). Plenum Press, New York.

Ubico, S. R., and R. G. McLean. 1995. Serologic survey of Neotropical bats in Guatemala for virus antibodies. Journal of Wildlife Diseases, 31:1–9.

Varma, M. G., and J. D. Converse. 1976. Keterah virus infections in four species of Argas ticks (Ixodoidea: Argasidae). Journal of Medical Entomology, 13:65–70.

Veeraraghavan, N. 1954. A case of hydrophobia following bat bite. Item D. Pasteur Institute of South India, Conoor: Annual Report of the Director 1953 and Scientific Report 1954.

Venters, H. D., W. R. Hoffert, W. R. Scatterday, and A. V. Hardy. 1954. Rabies in bats in Florida. American Journal of Public Health, 44:182–185.

Verani, P., M. G. Ciufolini, S. Caciolli, A. Renzi, L. Nicoletti, G. Sabatinelli, D. Bartolozzi, G. Volpi, L. Amaducci, and M. Coluzzi. 1988. Ecology of viruses isolated from sand flies in Italy and characterized of a new Phlebovirus (*Arabia virus*). American Journal of Tropical Medicine and Hygiene, 38:433–439.

Villa-R, B., and B. Alvarez. 1963. Rabies virus in the kidney and other tissues of vampire bats in western Mexico. Zoonoses Research, 2:77–82.

Villa-R, B., and E. L. Cockrum. 1962. Migration in the guano bat, *Tadarida brasiliensis mexicana* (Sassure). Journal of Mammalogy, 43:43–64.

Webster, W. A., G. A. Casey, and K. M. Charlton. 1986. Major antigenic groups of rabies virus in Canada determined by anti-nucleocapsid monoclonal antibodies. Comparative Immunology, Microbiology, and Infectious Disease, 9:59–69.

Webster, W. A., G. A. Casey, and K. M. Charlton. 1989. Bat-induced rabies in terrestrial mammals in Nova Scotia and Newfoundland. Canadian Veterinary Journal, 30:679.

Webster, W. A., G. A. Casey, K. M. Charlton, and T. J. Wiktor. 1985. Antigenic variants of rabies virus in isolates from eastern, central, and northern Canada. Canadian Journal of Comparative Medicine, 49:186–188.

Westbury, H. A., P. T. Hooper, S. L. Brouwer, and P. W. Selleck. 1996. Susceptibility of cats to equine morbillivirus. Australian Veterinary Journal, 74:132–134.

Whetstone, C. A., T. O. Bunn, R. W. Emmons, and T. J. Wiktor. 1984. Use of monoclonal antibodies to confirm vaccine-induced rabies in ten dogs, two cats, and one fox. Journal of the American Veterinary Medical Association, 185:285–288.

Wiktor, T. J., A. Flamand, and H. Koprowski. 1980. Use of monoclonal antibodies in diagnosis of rabies virus infection and differentiation of rabies and rabies-related viruses. Journal of Virologic Methods, 1:33–46.

Wiktor, T. J., and H. Koprowski. 1980. Antigenic variants of rabies virus. Journal of Experimental Medicine, 152:99–112.

Williams, J. E., S. Imlarp, F. H. Top, Jr., D. C. Cavanaugh, and P. K. Russell. 1976. Kaeng Khoi virus from naturally infected bedbugs (Cimicidae) and immature free-tailed bats. Bulletin of the World Health Organization, 53:365–369.

Williamson, M. M., P. T. Hooper, P. W. Selleck, L. J. Gleeson, P. W. Daniels, H. A. Westbury, and P. K. Murray. 1998. Transmission studies of Hendra virus (equine morbillivirus) in fruit bats, horses and cats. Australian Veterinary Journal, 76:813–818.

Williamson, M. M., P. T. Hooper, P. W. Selleck, L. J. Gleeson, H. A. Westbury, and R. F. Slocombe. 2000. Experimental Hendra virus infection in pregnant guinea pigs and fruit bats (Pteropus poliocephalus). Journal of Comparative Pathology, 122:201–207.

Winkler, W. G. 1968. Airborne rabies virus isolation. Bulletin of the Wildlife Disease Association, 4:37–40.

Winkler, W. G. 1975. Airborne rabies. Pp. 115–121 in: The Natural History of Rabies (G. M. Baer, ed.). Vol. 2. Academic Press, New York.

Winkler, W. G., and D. B. Adams. 1972. Utilization of southwestern bat caves by terrestrial carnivores. American Midland Naturalist, 87:191–200.

World Health Organization Collaborating Centre for Rabies Surveillance and Research (WHO). 1982. Rabies Bulletin Europe. Rabies Surveillance Report, 6:1–38.

World Health Organization Collaborating Centre for Rabies Surveillance and Research (WHO). 1997. Rabies Bulletin Europe. Rabies Surveillance Report, 21:1–28.

Yancey, F. D., II, P. Raj, S. U. Neill, and C. Jones. 1997. Survey of rabies among free-flying bats from the Big Bend region of Texas. Occasional Papers, the Museum, Texas Tech University, 165:1–5.

Young, P. L., K. Halpin, P. W. Selleck, H. Field, J. Gravel, M. A. Kelly, and J. S. Mackenzie. 1996. Serologic evidence for the presence in Pteropus bats of a paramyxovirus related to equine morbillivirus. Emerging Infectious Diseases, 2:239–240.

Zeller, H. G., N. Karabatsos, C. H. Calisher, J. P. Digoutte, F. A. Murphy, and R. E. Shope. 1989. Electron microscopy and antigenic studies of uncharacterized viruses. I. Evidence suggesting the placement of viruses in families Arenaviridae, Paramyxoviridae, or Poxviridae. Archives of Virology, 108:191–209.

CHAPTER 15

Conservation Ecology of Bats

Paul A. Racey and Abigail C. Entwistle

Introduction

The roots of conservation biology lie in the recognition, a century ago, that populations of game animals were declining in Africa, and there was a need to provide areas where they were protected from hunting (Fitter and Scott 1978). Since then the discipline of conservation biology has grown substantially. It seeks to understand the mechanism of extinctions and the risks faced by populations and to inform policy and practice designed to protect them (Caughley and Gunn 1996). The two paradigms of conservation biology identified by Caughley (1994)—the declining population paradigm and the small population paradigm—are both well illustrated by bats. He pointed out that the former is urgently in need of more theory and the latter needs more practice and that cautious intermixing of the two might lead to a reduction in the rate at which species become extinct.

At present, however, the results of research on bats have had little impact on theories underpinning conservation biology. Lack of knowledge and difficulties in studying bats have often precluded their use as models for extinction processes. Nevertheless, some bat species demonstrate characteristics relevant to a theoretical approach to conservation biology, such as natural rarity and genetic isolation, and all species exhibit slow population growth. There are also many practical examples of bat conservation, demonstrating good practice in the realization of effective solutions, based on extensive autecological research. Such examples are now being used effectively in textbooks of conservation biology (e.g., Primack 1998) showing that bats have much to offer as models of conservation practice, with examples that can be applied to other taxa.

The issue of bat conservation was first brought to light in the early 1950s by researchers in the United States (Mohr 1953) and 20 yr later in Europe (Racey and Stebbings 1972; Stebbings 1971) and Australia (Hamilton-Smith 1974). Since then the bat conservation movement has grown dramatically, with strong commitment from both the professional and voluntary sectors. A number of organizations actively promote bat conservation (including Bat Conservation International and the Bat Conservation Trust, with memberships of 15,000 and 4,000, respectively), and the Chiroptera Specialist Group was established in 1980 as part of International Union for Conservation of Nature and Natural Resources' (IUCN's) Species Survival Commission to provide a focus for information on the conservation status of bats worldwide. Information from the Specialist Group has been compiled in two action plans—for Megachiroptera (Mickleburgh et al. 1992) and Microchiroptera (Hutson et al. 2001). These publications have provided the first comprehensive reviews of the current conservation status of bats and provide important guidance for implementing activities designed for their conservation. It is clear, however, that general information on the conservation requirements of bats is still lacking, and even basic distributional and abundance data are absent for many species. It is in this context that we consider the conservation ecology of bats, recognizing the need to gain an understanding of the ecological requirements of the group in order to develop appropriate conservation policies and management plans. In particular, information is needed on:

• Current status—the distribution and abundance of species, including trends in population size or structure.
• Identification of potential threats or causes of decline—as a result of natural or human-mediated change.
• Ecological requirements—ecological factors essential to the continued persistence of species, including interrelationships with other organisms.
• Conservation applications—research on approaches to avoid or mitigate predicted or actual threats.

This chapter reviews ecological research within the framework of the conservation biology of bats, and demonstrates how such information can be applied successfully to conservation planning and implementation.

The unusual combination of traits manifested by bats (small body size, long life span, nocturnality, and volancy) means that it is difficult to draw parallels with conservation of other taxa. Also, attempts to develop a single approach to bat conservation are hampered by the extent of morphological, ecological, and behavioral variation within this taxon (Altringham 1996; Fenton 1992; Kunz and Racey 1998). In addition, the lack of information available on most species of bats means that conservation planning is difficult to undertake and that decision makers are difficult to convince regarding the most appropriate mechanisms to ensure the persistence of diversity within the order Chiroptera. Effective bat conservation depends on research, particularly in ecology, to elucidate fully the relationships among individual bat species, their environment, and humans.

Much of conservation practice has developed through the analysis of case studies and the application of situation-specific solutions. There have been attempts to identify broad indicators and generalities and to develop a theoretical basis to applied aspects of conservation biology (Caughley and Gunn 1996). While sufficient knowledge has now accrued about some taxa, such as birds and larger carnivores, to enable broader conservation approaches to be implemented across species, the paucity of research that tests conservation theory relating to bats, together with the extent of variation within the group, makes it difficult to draw conclusions across the group as a whole. The only rules for bat conservation are based on general principles, and most have adopted a precautionary approach.

Conservation Status of Bats

The conservation status of bats has recently been assessed using the revised IUCN criteria for assessing threat (IUCN 1994; Mace and Lande 1991). This provides a basic structure for quantifying the likely threats to species, based on natural rarity and evidence of decline. Use of the revised criteria, coupled with a comprehensive assessment of this order, has resulted in many more species of bats being listed in threatened categories, and threat categories have now been attributed to many species that were previously considered data deficient or had not previously been evaluated under the old qualitative crite-ria (IUCN 1994). It was estimated in 1996 that 240 species, or around 24% of the bat fauna, are listed as threatened (critically endangered, endangered, or vulnerable [IUCN 1996]). Subsequent IUCN lists of threatened species are now available only on the World Wide Web (www.iucn.org).

The numbers of threatened bat species is comparable to that in the class Mammalia as a whole (25% [IUCN 1996]) and is higher than that documented for birds (11% [Collar et al. 1994; IUCN 1996])—the only other class that has been studied to the same extent. Species classified as under threat in 1996 for both Megachiroptera and Microchiroptera are listed in table 15.1.

Over the past 400 yr, at least 11 species of bats have become extinct (IUCN 1996). Eight of these were Old World fruit bats (Megachiroptera), representing nearly 5% of this suborder. All but one of the now-extinct fruit bats species were found on islands (e.g., *Pteropus pilosus* in the Palau Islands [Wiles et al. 1997]), which is hardly surprising since the genus *Pteropus,* which accounts for 57 of 161 species of Megachiroptera, is primarily an island taxon (Rainey and Pierson 1992). At least some of these species appear to have been hunted to ex-tinction (Rainey 1998). In addition, almost 8% of megachiropterans are listed as critically endangered (at greatest immediate risk of extinction), and all but two of these species have island distributions. In contrast, fewer than 2% of Microchiroptera are listed as critically endangered. However, if the declines in population size of many cave-dwelling microchiropterans were adequately documented, it is likely that many more would be included in IUCN risk cat-egories (M. D. Tuttle, personal communication).

Indeed, it is not clear whether threat categories (based on population size, distribution, and decline [IUCN 1994]) developed to apply to a wide range of taxa, are equally effective at describing relative extinction risk for specific

Table 15.1. Numbers of Mega- and Microchiroptera listed as threatened under revised IUCN criteria, data revised July 2000

Category	Megachiroptera	Microchiroptera	Total
Extinct	8	3	11
Critically endangered	13	16	29
Endangered	6	30	36
Vulnerable	42	135	177
Lower risk	20	190	210
Data deficient	2	57	59
Least concern	75	404	479
Total	166	835	1001
Total threatened and percent of extant species (not including lower risk species)	61 (39)	181 (22)	242 (24)

Source. Mickleburgh et al. (2002).

Note. The IUCN threat categories are based on quantifiable criteria outlined initially by Mace and Lande (1991) and subsequently modified (IUCN 1996). The categories critically endangered (CR), endangered (EN), and vulnerable (VU) all indicate that there is a clear risk of extinction for that species, with species classified as critically endangered considered to be at the greatest immediate risk of extinction in the wild. Numbers in parentheses are percentages.

groups with unusual life-history traits and distributional profiles. For example, it is not clear how a definition of species range is best applied to highly vagile groups of animals, which may still have highly disjunct distributions. In addition, status assessments of bats are hindered by lack of realistic population estimates for most species and lack of knowledge of the interplay between minimal viable population size and coloniality (since there may be a loss of communal benefits below a threshold colony size). In view of the relative lack of knowledge about bats and their conservation biology, a number of field researchers have suggested that the precautionary principle be applied to assessments of the status of bats. Such an approach may imply that an even greater number of bat species should be considered, at best, data deficient or, at worst, at risk than currently meet strict IUCN Red List criteria.

A number of key declining bat species have been profiled (Hutson et al. 2001) and have clearly demonstrated the variation in threats and responses to environmental change across the order. The production of conservation action plans for apparently widespread and common species (such as *Miniopterus schreibersii* and *Tadarida brasiliensis* [Hutson et al. 2001]) reflects the extent of declines in species that were once found in great abundance and demonstrates that it is not just the rare species that require conservation attention.

Conservation assessments are only as reliable as the data on which they are based with regard to abundance, distribution, and population trends. For numerous species, particularly those with cryptic habits or with limited distributions, providing the necessary basic information on populations is difficult.

For many of these species, the most basic surveys have not yet been conducted. However, it has become clear that even for common and well-studied species, there are problems associated with estimating changes in populations. Although, in some cases, integrated monitoring programs may address this issue, for many species indications of population change are likely to remain based on inexact or anecdotal information.

Baseline Surveys of Distribution and Abundance

Knowledge of bat distribution based on the nineteenth- and early-twentieth-century expeditions and resulting museum collections (where the shotgun was a common collection method) has been enhanced in recent decades by surveys using mist nets and, more recently, harp traps, together with hand-capture in caves—techniques that are fully reviewed by Barlow (2000), Kunz and Kurta (1988), and Kunz et al. (1996). Surveys such as those of Ethiopia (Hill and Morris 1971), Vanuatu (Medway and Marshall 1975), the Solomon Islands (Bowen-Jones et al. 1997), Madagascar (Bayliss and Hayes 1999; Goodman 1996), Thailand (Robinson et al. 1995), Vietnam (Bates et al. 1999; Hendrichsen et al. 2001; Walston 2000), India (Bates and Harrison 1997), Laos (Francis et al. 1999), and French Guiana (Simmons and Voss 1998) have added substantially to our knowledge of bat distribution and have resulted in the discovery of many new species (e.g., Flannery 1991; Flannery and Colgan 1993), along with the rediscovery of species that were thought to have become extinct, such as *Latidens salimalii* in India (Bates et al. 1994), *Pteralopex anceps* on Choiseul (Bowen-Jones et al. 1997), and *Aproteles bulmerae* in Papua New Guinea (Flannery and Seri 1993). However, it is more difficult to draw conclusions about rarity from such surveys since some species are adept at avoiding mist nets, as is clear from the fact that the number of species and individuals captured in some areas increases when harp traps are deployed (Schulz 1999). Further increases occur if the number of frames of the harp trap is doubled from two to four (Francis 1989; Kingston et al. 2000). Examples of the limitations of mist netting are provided in Madagascar, where despite many hundreds of mist-net hours in several separate surveys, only three individuals of the endemic *Myzopoda aurita* were captured (Göpfert and Wasserthal 1995; Pont and Armstrong 1990; Russ and Bennett 1999). Bat detector records, however, suggest that this species is not rare (Russ and Bennett 1999).

Despite the increased availability of bat detectors and their value for identifying some bat species, those involved in inventory work (as opposed to those conducting ecological research) were initially slow to make use of them, a notable exception being Fenton et al. (1983). This is probably because an extra stage of ground truthing is required—a species must be captured, identified, and its calls recorded before it can be subsequently recognized from such calls. However, when this is done, bat detectors add greatly to the distributional data obtained for tropical and temperate zone bats (McCracken et al. 1997; O'Donnell 2000; O'Donnell and Sedgeley 1994; O'Farrell and Gannon 1999; O'Farrell and Miller 1999; Pierson and Rainey 1998; Russ et al. 2001). Barclay (1999) has, however, urged caution in the use of bat detectors in this respect.

Estimating Population Size

Estimating population and colony size for bats presents particular challenges but is an essential element in effective assessments of conservation status and in providing a baseline from which to identify any future population decline. In those families that hang free from their roosting substrate, such as rhinolophids and pteropodids, direct counts are possible and are relatively easy for small colonies. For example, the number of *Rhinolophus ferrumequinum* in the United Kingdom has been determined, by counting roosting bats, to be about 4,000 (Harris et al. 1995). Endemic island populations of *Pteropus* have been counted directly in their tree roosts (Entwistle and Corp 1997), while dispersing from ridge-top roosts to forage at dusk, and from sequences of projected transparencies of photographs traversing through a colony that had taken flight after a prearranged disturbance (Racey 1979; Wiles et al. 1989). Many vespertilionid species form relatively small colonies that cluster inside roof spaces (attics), where they may be captured and counted and where colony size can thus be estimated. Such studies appear to provide good indications of relative colony sizes (e.g., Entwistle et al. 2000), but it is not clear how effective within-roost counts can be for monitoring, unless the numbers of individuals in inaccessible, as well as accessible, parts of the roof spaces can be assessed. Mark-recapture methods have shown that within-roost counts can result in underestimates in the case of *Plecotus auritus,* and, as a result, colony size has been revised upward (Entwistle et al. 2000). Intensive fieldwork has also yielded estimates of the numbers of individuals in metapopulations of *P. auritus* (Entwistle et al. 2000). In situations where disturbance of roosting bats is contraindicated, remote systems are available (Rainey 1995). Use of exit counts and cameras mounted within roosts have demonstrated that counts of *Chalinolobus tuberculatus* using communal tree roosts were poor predictors of total population size as a result of the turnover of individuals between roost sites (O'Donnell and Sedgeley 1999).

A number of studies have estimated population densities of bats on a local scale (Jones et al. 1996; Speakman et al. 1991). Extrapolations from local to national scales based on mean colony size and the density of colonies per unit area have produced estimates of the numbers of bats occurring in the United Kingdom (Speakman 1991). However, the high variance in colony size and assumptions about the uniform distribution of roosts (which are often concentrated in river valleys [Racey 1998b]) means that such extrapolations are likely to be spurious and provide conservation managers with misleading information (Harris et al. 1995).

Determining Changes in Populations

Determining changes in bat populations has become a priority for conservation biologists as an indication of incipient population declines and was the subject of a recent international workshop (O'Shea and Bogan, in press). Data on population changes allows the status of specific bat populations to be assessed—as stable, declining, or increasing.

In temperate regions, bats are commonly counted as they emerge from their maternity roosts at dusk (Kunz and Anthony 1996; Swift 1980), and in the United Kingdom, such counts have been organized on a national scale. The National Bat Monitoring Programme (NBMP), funded by the government of the United Kingdom in response to the requirements of the European Community Habitats and Species Directive and the European Bats Agreement, has developed protocols for monitoring seven European bat species. For example, *Pipistrellus pipistrellus* and *P. pygmaeus*, emerging at dusk from a randomly stratified sample of roosts distributed throughout the United Kingdom, were counted twice each summer over a 5-yr period in an attempt to establish population trends. Similarly, *Myotis daubentonii*, which forages almost exclusively in riparian and lacustrine situations, was counted over a randomly stratified sample of rivers (Walsh et al. 2001).

Despite the difficulty in determining population size in bats and documenting changes, it has nevertheless been done in a number of instances. The decline in bats of the genus *Rhinolophus* is widespread throughout Europe (Ohlendorf 1997), and *R. ferrumequinum* is now the subject of a European Action Plan (Ransome and Hutson 1999). Long-term changes in bat populations in the Netherlands were reviewed by Daan (1980), who concluded that several species were declining. Range contractions and population declines of vespertilionids in the United States, such as *Corynorhinus townsendii* (Harvey 1980; Hensley and Scott 1995; O'Shea and Vaughan 1999) and *Myotis sodalis* (Clawson 1987), have been documented, and dramatic population declines of *Myotis griscescens* and *M. austroriparius* in maternity caves have been reported (Rabinowitz and Tuttle 1980; Tuttle 1979).

Although evidence of a decline in bat numbers is widespread, there are indications that some bat species may be increasing in numbers, such as *Myotis daubentonii* on the Western European mainland. Kokurewicz (1995) suggested this may be due to eutrophication of freshwaters that increases prey availability, and some evidence in support of this hypothesis was obtained by Racey et al. (1998).

In the 1950s and early 1960s, midsummer colonies of *Tadarida brasiliensis* in 20 caves in the southwestern United States were estimated to total 150 million bats, and McCracken (in press) has recently reviewed the population declines in this species. Allison (1937) based his estimate of 8.7 million bats emerging from Carlsbad Cavern in New Mexico on the duration of emergence, visual estimates of flight speed, the cross section of the emerging column, and the density of bats in the column. Twenty years later, Constantine (1967) measured the surface area of the Carlsbad Cavern occupied by bats and multiplied this by the average roosting density to give an estimate of 1–4 million bats. Altenbach et al. (1979) combined high-speed motion pictures and still photography to estimate colony size at Carlsbad at 218,000, or about 5% of Allison's (1937) estimate. The decline has been attributed, at least in part, to the use of pesticides on the farmland over which *T. brasiliensis* foraged (Geluso et al. 1976). The decline of the bat population at Eagle Creek Cave, Arizona, was even more precipitous. Based on the number of bats roosting per unit area at a number of sample sites and the estimated total cave surface area occupied by bats, Cockrum (1969) estimated a peak population size during midsummer 1963 of more than 25 million individuals. Six years later the population had decreased to 30,000 bats.

Despite uncertainties in data linked to the mobility of bats and the lack of understanding of their long-term behavioral and distributional patterns, there is a growing need to monitor bat populations. Increasingly, methods are being devised that account for natural variations in populations and distributions (e.g., the United Kingdom's NBMP [Walsh et al. 2001]). Detecting changes in populations not only allows those species threatened with local or global extinction to be identified but also helps to pinpoint likely causes of declines through correlation and direct experimentation. Monitoring of bats could in turn become a useful tool for detecting far-reaching environmental changes in habitat quality, connectivity, and insecticide use.

Determining Population Structure

There have been a number of demographic studies of banded bats, reviewed by Tuttle and Stevenson (1982). These studies have led to generalizations about sexual segregation during the breeding season, low reproductive rate, long gestation period, and long life span—life-history traits that are relevant to conservation biology (Barclay and Harder, this volume; Racey and Entwistle 2000). However, recent advances in molecular genetics (Avise 1994), combined with developments in metapopulation biology (Hanski 1998), have provided new insights into the population structure of bats.

For example, during summer, *P. auritus* forms stable colonies composed of females and young-of-the-year, and unusually among vespertilionids, adult males. A long-term banding study in northeast Scotland established that there is little movement among colonies and that both sexes are recruited into their natal colony (Entwistle et al. 2000). However, microsatellite DNA markers indicate high gene flow between colonies, although some coancestry among colony members is evident in both sexes (Burland et al. 1999). The banding and genetic data together lead to the conclusion that gene flow occurs through extracolony copulations rather than natal dispersal and that each colony behaves as a distinct subpopulation. These studies provide the first demonstration of microgeographical genetic isolation by distance in a bat species.

Extreme population structuring also has been revealed in Bechstein's bat *Myotis bechsteinii* using mitochondrial DNA (mtDNA [Kerth et al. 2000]). Comparison of markers within and between colonies living a mean distance of 8 km apart revealed little variability within each colony, whereas most colonies were clearly distinguished by colony-specific mitochondrial haplotypes. The data suggested that colonies of *M. bechsteinii* are populated by 20–40 females belonging to one or maximally two matrilines and are socially closed units even in the apparent absence of microgeographical dispersal barriers. These units frequently split into subgroups that occupy different roosts, and there was marked mixing between subgroups. However, females associated according to reproductive status and not according to relatedness, as determined by five nuclear and one mitochondrial microsatellite (Kerth and Koenig 1999).

The use of polymorphic nuclear DNA and mtDNA d-loop sequences from *Myotis myotis* has also revealed that females return to their natal sites, while males disperse (Petri et al. 1997). The higher genetic distance between rather than within colonies suggests appreciable genetic substructuring. Males present in the maternity colonies are not related to the females and fathered few young, and as with *P. auritus*, females appear to seek mating opportunities outside their colony.

These three studies were conducted over a range of tens of kilometers. The most extreme population structuring so far reported to date is seen in the ghost bat *Macroderma gigas*, an Australian endemic studied by Worthington Wilmer et al. (1994, 1999) over thousands of kilometers. Formerly widely distributed in Australia, this species has undergone major range contraction and is now found in only a few highly disjunct maternity sites in the northern part of the country. The mean sequence diversity of mtDNA was 4.5% between populations and was six times higher than within populations for the monomorphic alleles that were evaluated. Such extreme genetic subdivision is thought to be a consequence of long-term female philopatry.

The results of these molecular studies have profound implications for conservation because they indicate that management efforts should be focused on individual bat colonies. In the past, managers have been insufficiently concerned about the loss of specific maternity roosts and have assumed intuitively that the bats excluded from a roost would join conspecifics in another. There is little evidence for this from molecular or banding studies. For example, 99% of 547 little brown bats *Myotis lucifugus* banded in colonies that were excluded from their house roosts in Chautauqua, New York, did not join other colonies or occupy the artificial roosts that were provided (Neilson and Fenton 1994).

More recently, however, a high occupancy rate of several types of artificial roosts (or bat houses) has now been achieved in the United States and Canada (Kiser and Kiser 2000), and excluded bats have taken up residence in these structures (Kiser and Kiser 1999).

Not all molecular evidence points to population structuring. Using data from 22 allozyme loci, McCracken and Gassel (1997) found no significant differences in genetic composition of a migratory population of *Tadarida brasiliensis* from Texas and two populations from California, one of which was migratory. Among Australian pteropodids, protein electrophoresis was used to determine genetic variation within and between populations of *Pteropus alecto*, *P. poliocephalus*, and *P. scapulatus*, together with analysis of randomly amplified polymorphic DNA for the latter species. This revealed that Wright's F_{ST}, an index of between population genetic variation, was uniformly low, reflecting the homogenizing effect of individual movements across the species range, and that these species are effectively panmictic (Sinclair et al. 1996; Webb and Tidemann 1996).

In contrast, protein electrophoresis of the small pteropodid *Cynopterus* in the Indonesian Nusa Tenggara archipelago (which extends from Lombok to Timor) reveals that the genetic distance between populations of the recently described *C. nusatenggara* (Kitchener and Maharadatunkamsi 1991) is strongly correlated with sea-crossing distance. This observation suggests that the sea is a primary and formidable barrier to gene flow in this small pteropodid (Schmitt et al. 1995). Petit and Mayer (1999) also showed that although the overall population structure of *Nyctalus noctula* in Europe was very weak, indicating a high male dispersal rate, there was some evidence that the Alps acted as a geographical barrier to gene flow.

Migration

Evidence of migration has accrued from studies of seasonal changes in distribution (Findley et al. 1964) and from banding (Fleming and Eby, this volume; Griffin 1970; Strelkov 1969). Mass banding is now discouraged because of the incidence of damage to bats by bands and is reserved for intensive local studies involving repeated recaptures, which also allows bands to be removed if necessary. However, contemporary evidence of migration is now emerging from genetic studies. Wilkinson and Fleming (1996) used sequence variation within the control region of mtDNA of *Leptonycteris curasoae* captured in the southern United States, Mexico, and Venezuela. The most compelling evidence of migration was the presence of shared unique haplotypes between distant sites. The results suggest that *L. curasoae* moves along (rather than between) two corridors, one along the Pacific coast and one inland, following the foothills of the Sierra Madre. These results are consistent with the hypothesis that bats use topographic relief to migrate at night (Wilkinson and Fleming

1996). Information from plant phenology is consistent with coastal movements of this nectarivore following a cactus corridor northward in spring, whereas those moving northward inland may follow a corridor of *Agave* flowers.

Petit and Mayer (2000) compared haplotype frequencies derived from the control region of mtDNA of *Nyctalus noctula* in maternity colonies and hibernating aggregations. This revealed that European populations hibernating in Kiel on the Baltic coast of Germany are differentiated from Central and East European populations. The direction of migration was inferred by comparing hibernacula with maternity colonies, and it was clear from the higher haplotype diversity in hibernacula that they consist of individuals from different maternity colonies. Petit and Mayer (2000) also suggested that the Kiel population should receive specific protection, as it is the only large hibernating aggregation known for *N. noctula* in Northern Europe.

These studies contribute to the scientific rationale for initiatives such as the Program for the Conservation of Migratory Bats of Mexico and the United States and the European Bats Agreement, which aims to conserve bats throughout their range. They also illustrate the value of molecular genetic studies in the conservation and management of bats.

Threats to Bat Populations

The IUCN threat categories clearly indicate that extinction risk can be linked to a number of factors: recorded or predicted population decline, small populations, and restricted distributions (IUCN 1994). Each of these factors appears to affect the extinction proneness of different groups, based on broad analyses. While direct population declines reflect the impact of extrinsic factors (generally caused by man's effect on the environment), intrinsic factors affect the likely response of species to such decline and the ability to recover and/or colonize new areas. Intrinsic factors include life-history strategy (particularly important in determining the recovery potential of the population), the implications of colonial living (a possible expression of the Allee effect—see Stephens and Sutherland 1999), and genetic factors such as heterozygosity and gene flow.

Intrinsic Factors

Much research has been conducted over the past 3 decades into what makes some groups more extinction prone than others (Newmark 1995; Pimm et al. 1988; Terborgh and Winter 1980). The key factors include reproductive rate and, for some species, rarity. Such intrinsic factors are not in themselves a threat to bat populations but may exacerbate the consequences of the extrinsic factors that affect bats.

Relative to their size, bats show surprisingly low reproductive rates, resulting in relatively slow population growth and thus limited ability to recover from population crashes—and, hence, increased risks of local extinctions (Barclay and Harder, this volume; Racey and Entwistle 2000). Life-history traits in bats generally are more similar to mammals with larger body sizes. Thus, their low reproductive rate is compensated by low mortality, and they have extended life spans, in some cases more than 2 decades (Racey and Entwistle 2000).

The low reproductive output in bats generally is reflected in a number of life-history traits: litter size, number of litters per year, age at first breeding, and frequency of breeding. The majority of bat species produce only one young per year, and litter size appears to reflect the limitations of wing loading (Barclay and Herder, this volume; Hayssen and Kunz 1996) and the seasonal availability of food to support lactation (Racey 1982; Racey and Entwistle 2000). In tropical latitudes, larger litter sizes and polyestry are more common. However, the record for reproductive output in bats is still only six young per annum in *Pipistrellus mimus* (Isaac et al. 1994).

Lifetime reproductive output (and overall population growth rate) is also affected by other intrinsic factors. The age at first breeding is important in determining the number of years in which bats may produce young. In some species, the age of first breeding is delayed until the third or fourth year of life, although some species become pregnant in their first year (Racey and Entwistle 2000). Moreover, in some species nonbreeding of individuals is common, with up to 50% of some populations failing to breed in any one year (Entwistle 1994; Grindal et al. 1992; Lewis 1993). Failure to breed appears to vary between years (Lewis 1993) and appears to be linked to climatic factors (Grindal et al. 1992; Lewis 1993; Ransome 1995). It can be expected, therefore, that the reproductive success of bats at the northern extreme of their distributional range would be lower than that of conspecifics farther south. That is indeed the case for *Plecotus auritus* in Britain (Entwistle 1994; Stebbings 1976) compared with Spain (Benzal 1991). In contrast, however, postnatal growth rates of young are greater in more northern latitudes (Kunz and Hood 2000; Kunz and Stern 1995).

Thus, a range of factors relating to their life-history traits (and ultimately to their morphology and the constraints associated with carrying a growing fetus and obtaining sufficient nutrients during lactation) may affect the ability of bats to respond to major population changes. Bats lack the "buffer" of high reproductive rate, and the local extinction of bat species, such as *Pteropus niger* on Réunion (Cheke and Dahl 1981) and *Myotis myotis* in the United Kingdom (Stebbings 1992), underlines the difficulty for populations to recover their numbers fast enough to counter mortality.

Rarity generally refers to low abundance and/or small range size (Gaston 1994). Some species are naturally rare and are characterized by a very small but apparently stable distribution. The study of small populations has demonstrated the risks of factors such as demographic stochasticity, and environmental stochasticity or catastrophes, for such species (Caughley 1994). For small populations, there is a danger that such chance events can combine with loss of genetic variation to reduce an ever-dwindling population to the point of extinction (the so-called extinction vortex [Gilpin and Soulé 1986]).

Rarity, however, can be considered from a number of viewpoints, such as total distribution and habitat specificity as well as overall population size (Rabinowitz 1981). Some species of bats appear to have naturally low population density (e.g., *Kerivoula* spp. [Schulz 1999]) and can be considered both locally and globally rare. Often global rarity in bats is related to limited overall distribution (such as island species, e.g., *Coleura seychellensis* [Nicoll and Suttie 1982]), although they may be locally common (such as *Craseonycteris thonglangyai*, which is restricted to a small part of western Thailand [Duanghae 1991]). Many species of bats show habitat specificity, and this may limit their distribution and population (e.g., *Rhinolophus ferrumequinum* [Duvergé and Jones 2003] and *Euderma maculatum* [Pierson and Rainey 1998]). The overall extent of rarity in a species is likely to reflect an interplay among these factors—along a continuum from species with large populations, occurring over wide ranges in many habitats, to the rarest species, with small population sizes, narrow overall ranges, and highly specific habitat requirements, coupled with poor abilities to move to new habitat patches (Laurance 1994). Analysis of these factors for different bat species may give some indication of their likely position in relation to conservation priorities.

Extrinsic Factors

A wide range of factors has been identified that may disturb or kill individual bats or colonies; however, it is not clear how far these can be translated into threats to the species as a whole. Diamond (1984) provides a useful categorization of the threats to species from human impacts. His "evil quartet" refers to (1) habitat loss, including habitat degradation and pollution, (2) overkill, including persecution as well as direct overexploitation, (3) the effects of introduced species, including domestic species, and (4) chains of extinction, where knock-on effects between species can occur. All these factors apply to bat species. In reality, endangerment and extinction are brought about not only by one of these factors acting in isolation but by combinations of threats that reinforce each other to reduce population size.

Habitat Loss and Degradation

Loss of Foraging Habitat. The requirements of increased food production in the last half of the twentieth century have led to major worldwide changes in land use, frequently involving fragmentation and loss of forest habitats (Viña and Cavelier 1999; World Wide Fund for Nature 1996). The loss or reduction in quality of foraging habitat and its fragmentation has become a major threat to bat populations. The United Kingdom's National Bat Habitat Survey has indicated that vespertilionid species strongly avoided foraging over land under cultivation, and some species avoid improved grassland, so that agricultural intensification has reduced available foraging habitat (Walsh and Harris 1996a, 1996b).

Gerrell and Lundberg (1993) studied two populations of *Pipistrellus pipistrellus* that roosted in bat boxes in southern Sweden—one adjacent to intensively farmed land and the other in a pine plantation. The population in farmland declined in numbers during the 8-yr period of study and had higher body burdens of organochloride and cadmium and a lower body mass in September when prehibernal fat accumulation was in progress.

The negative effects of gradual loss of forests on bat populations are poorly documented and can only be inferred. However, historical descriptions of *Pteropus* camps occupying large areas of forests in subtropical and tropical Australia (Ratcliffe 1931) and comparison with the much reduced size of colonies today leaves little doubt that as forested areas have been reduced so have bat numbers. Ratcliffe (1932) reported *Pteropus* camps up to 10 km long and 1.3 km wide, with estimated numbers of up to 30 million individuals. Today, many colonies have disappeared entirely, and only a few contain more than 100,000 individuals (Pierson 1984). Similarly, the disappearance of *Chalinolobus tuberculatus* from coastal and lowland regions in New Zealand coincided with the loss of forest cover (O'Donnell 2000).

Man is not the only cause of forest loss, and catastrophic events such as cyclones have had serious effects on island populations of fruit bats (Rainey 1998). In 1979, Mungroo (cited in Carroll 1988) estimated a 50% reduction in the population of *Pteropus rodricensis* on Rodrigues Island following a cyclone. In the Samoan Islands, *Pteropus samoensis* and *P. tonganus* declined by 80%–90% in the early 1990s due to cyclones and associated hunting by villagers (Craig et al. 1994), while populations of the same species on Tutuila were severely reduced in 1993. Despite a 3-yr moratorium on hunting, modeling projections indicated population recovery would extend for an order of magnitude longer. Research has shown differential impacts of cyclones on the two species, reflecting differences in their habitat use and consequently in their susceptibility to hunting (Pierson et al. 1996). Although hurricanes have a sudden and highly destructive effect on forests and other bat habitats, studies of such events may be a useful tool in predicting, and mitigating, the likely impacts of extensive anthropogenic deforestation on bat populations.

In parts of the Neotropics, some bat species are sensitive to changes in habitat quality following logging and other forestry practices, although it is clear that other species are more tolerant to habitat change. Typically, the species richness of rainforest or old-growth stands of temperate forest declines as these are replaced by secondary growth or plantations (e.g., Brosset et al. 1996;

Heaney and Heideman 1987; Heideman and Heaney 1989; Kalko 1998; Vaughan and Hill 1996; Zubaid 1993), although abundance of individuals is often higher outside primary forest (Brosset et al. 1996; Ochoa 2000; Vaughan and Hill 1996). In temperate-zone forests, the impact of logging has been studied, allowing predictions to be made about the effect of different forestry practices on bat assemblages and populations (Barclay and Brigham 1996; Sedgeley and O'Donnell 1999a, 1999b).

Some studies report increased species richness following disturbance (Estrada et al. 1993), presumably as gap specialists move into cleared areas and edge habitats increase. In southern Arizona, Hovorka (1996) found that although grazing by cattle had a significant impact on the scrub vegetation of the Sonoran Desert, it did not adversely affect the abundance of flying insects and insectivorous bats. Similarly, in the Miombo woodland of the African savanna, the common vespertilionid and molossid bats, which feed on airborne insects, are consistently present even in sites where the tree canopy has been virtually eliminated by elephants. However, the preponderance of bats in intact sites close to degraded sites points to a loss of roosts as a major consequence of elephant damage (Fenton et al. 1997).

Urbanization is one form of habitat degradation, in which almost all natural vegetation is removed, although many roost sites become available to house-dwelling bats. Studies of bats in urban areas indicate that although more adaptable generalist species persist in city centers (Mickleburgh 1987), bat activity is greatest in suburbs and in riparian situations (Gaisler et al. 1998; Kurta and Teramino 1992). Although insect density may be generally lower in urban areas (Geggie and Fenton 1985), white streetlights concentrate the available insects and attract foraging bats (Furlonger et al. 1987; Rydell and Racey 1995). Rarer and more specialized bat species do not persist within urban environments (Gaisler et al. 1998; Mickleburgh 1987), although generalists such as *Artibeus lituratus* roost in palms in gardens and along roadsides in cities (Sazima et al. 1994). Similarly, two Old World pteropodids, *Cynopterus sphinx* in India (Balasingh et al. 1995; Bhat and Kunz 1995; Storz et al. 2000) and *C. brachyotis* in Malaysia (Tan et al. 1998) regularly roost in parks and gardens, where they construct tents. Halstead (1977) reported a large roost of *Eidolon helvum* on the campus of the University of Ife, Nigeria, but the colony of about a million of this species in Kampala, Uganda, has declined with increasing building development (A. M. Hutson, personal communication).

Loss of Roosts. Bats are thought to have evolved in association with trees, and microchiropterans and some megachiropterans found in forested landscapes may still depend on tree cavities for their roosts (Brigham and Barclay 1996; Kunz and Lumsden, this volume; Pierson 1998; Rainey 1998). Some foliage-roosting lasiurine bats in North America together with *Lasionycteris noctiva-*

gans are obligate tree roosters. Although many temperate-zone Microchiroptera utilize anthropogenic structures, where they continue to roost in association with wood, presumably because of its low conductivity, they still roost in trees when available (Brigham and Barclay 1996; Kunz 1982; Kunz and Lumsden, this volume). Current silviculture, which favors even-age monospecific stands, short rotation times, and removal of dead and dying trees, leaves little roosting habitat for most tree-dwelling species (Pierson 1998). There is also convincing evidence that bat densities are greater in old-growth forests of temperate regions where structural diversity provides more roosting options (Crampton and Barclay 1996; Hayes and Adams 1996; Kalcounis et al. 1999; Parker et al. 1996; Sedgeley and O'Donnell 1999a, 1999b). A corresponding concern is that such forests are under threat from timber harvest.

During winter, bats of temperate latitudes require a cool environment in which to hibernate, such as that provided by caves or mines, and some species such as *Myotis grisescens* make use of such environments throughout the year. Most known roosts for *Leptonycteris curasoae* in the United States are in mines. The loss of roosts in buildings has been documented by the use of questionnaires (Kunz and Reynolds, in press; Racey and Stebbings 1972), and there has also been widespread loss of underground sites, particularly mines, as their entrances are sealed to prevent human access (Pierson 1998; Stebbings 1988; Tuttle and Taylor 1994). Other roosts may become effectively lost to bats as a result of excessive human disturbance. Models have predicted that climate change may alter species richness—through its affect on the availability of roosts with appropriate conditions—relative to shifting vegetational and temperature clines (Scheel et al. 1996).

Megachiropterans generally prefer to roost in vegetation where there is less human disturbance. In one study, an inverse relationship has been found between colony size in forests and that in sites closer to villages (Entwistle and Corp 1997). On small oceanic islands, the only trees free from disturbance may be high on steep ridges (Racey and Nicoll 1984), and some species may exclusively forage in such areas (e.g., *Pteropus tonganus* [Brooke et al. 2000]). As horticulture extends higher onto mountainsides on overpopulated islands, suitable roosts become scarcer. However, when protected from disturbance, *Pteropus* will also roost close to man, as in Buddhist monasteries in Thailand, in the Ku-rin-gai roost in the suburbs of Sydney, Australia, and in the Berenty private reserve in southeast Madagascar (P. A. Racey, personal observations).

Pesticides and Pollution. Although the primary purpose of agricultural pesticides is the reduction of insect numbers, direct evidence that such reduction in insect food supply limits the populations of insectivorous bats is not available. There is, however, direct evidence that agricultural pesticides ingested with insects that carry sublethal levels have been responsible for heavy mortality of

Tadarida brasiliensis in New Mexico (Geluso et al. 1976, 1979). Pesticides are mobilized during lactation and transferred to the young in the milk, where they cause death (Geluso et al. 1981).

Jefferies (1972) showed that bats taken from East Anglia, one of the most intensively farmed areas of the United Kingdom, were more heavily contaminated with residues of DDT than either insectivorous or carnivorous birds. Laboratory experiments showed that bats were more sensitive to DDT than other mammals and that DDT was metabolized more slowly than in passerine birds. Bats carried one-third of the lethal level of organochlorine insecticides, but this rose to close to lethal levels following hibernation. These results suggested that organochlorine residues could have caused population declines in bats (Jefferies 1972). Endangered *Myotis grisescens* were found dead with lethal brain levels of dieldrin in two colonies where levels of this pesticide in guano were high relative to other colonies (Clark et al. 1978b). In Australia, high tissue levels of DDT were detected in *Miniopterus schreibersii* even in young bats that had not left the maternity roosts and was the suspected cause of several mass die-offs in this species (Dunsmore et al. 1974). However, in Spain, Hernández et al. (1993) found chlorinated hydrocarbon residues in bats at much lower concentrations than the estimated lower lethal levels.

Historically, chlorinated hydrocarbons were used to exclude bats from buildings in the United States, but this also increased mortality of adults and volant young as fat was mobilized (see Clark et al. 1978a; Kunz et al. 1977). Similarly, these pesticides have been used as remedial timber treatments within roof spaces and have led to the deaths of bats roosting there (Racey and Swift 1986; Voûte 1981). In the United States and the United Kingdom, owing largely to pressure from conservationists and environmentalists, most of these highly toxic chemicals have been replaced by synthetic pyrethroids and other alternatives with much lower mammalian toxicity (Racey 2000). Although Clark (1981, 1988) suggested that pesticides might have subtle but important effects on bat physiology, the sublethal effects of chlorinated hydrocarbons have only recently been documented (Swanepoel et al. 1999). The residues of such pesticides are also commonly found in the tissues of bats in the developing world (McWilliam 1994).

Insofar as pollution is concerned, there is little evidence that eutrophication of freshwaters is harmful to bats (Racey 1998b), and it may be responsible for the apparent increases in *Myotis daubentonii* in Europe (Kokurewicz 1995). In a study designed to test this hypothesis, Racey et al. (1998) found little difference between the number of bats and insects over a eutrophic and an oligotrophic river.

Overkill

Hunting. Although hunting of megachiropterans for food is widespread, there have been few attempts to harvest bats sustainably (Halstead 1977).

Overhunting has resulted in the extinction of *Pteropus subniger* in Mauritius and Réunion (Cheke and Dahl 1981) and contributed to the local extinction of a dozen other congeners from Indo-Pacific islands (Rainey 1998). Hunting continues to be a key threat to pteropodids, including species on islands in the Indian Ocean such as *Pteropus voeltzkowi* (Entwistle and Corp 1997).

Human consumption remains a major factor affecting bat populations on Indo-Pacific Islands and in adjacent areas of Asia, a practice that has led to international trade in fruit bats into Guam and the adjacent Commonwealth of the Northern Marianas (CNMI [Rainey 1998; Wiles 1992]). Hunting by Guamanians reduced the population of *Pteropus mariannus* on the island to a few hundred individuals within the confines of the U.S. military base and, perhaps, contributed to the extinction of *P. tokudae*. The relatively high per capita incomes on Guam also sustain the demand for imported bats. The mean number recorded as imported annually to Guam from 1981 to 1989 was approximately 13,000 (Wiles 1992) and into CNMI from 1986 to 1989 was 3,300 (Stinson et al. 1992). These bats came from other Pacific Islands but also from Papua New Guinea and the Philippines. This led, in 1989, to the listing of seven Central and West Pacific *Pteropus* species in appendix 1 of *The Convention on Trade in Endangered Species of Wild Fauna and Flora* (CITES [http://www.cites.org]), which prohibits signatories from trade in specified species. The remaining species of *Pteropus* are listed in appendix 2, which requires signatories to monitor trade, as are all species of *Acerodon* apart from *A. jubatus* and *A. lucifer* (which were added to app. 1 in 1995).

On the Palau Islands, fruit bats (*Pteropus mariannus*) were intensively harvested until 1994, with between 10,000 and 16,000 bats exported annually to Guam and the CNMI (Wiles et al. 1997). This trade was not prohibited by CITES legislation because it was not regarded as international. Exports were banned due to CITES restrictions in 1994 when Palau became independent, and fortunately, there has been almost no smuggling of animals to Guam or the CNMI since then as a result of effective CITES enforcement by the Palau and U.S. governments. In addition, local laws have been effective in controlling trade in bats between Guam and the CNMI, although limited illegal hunting of *P. mariannus* is thought to continue in the CNMI for local consumption.

The killing of the molossid *Cheiromeles torquatus* for food in Borneo is also causing serious concern (Hutson et al. 2001), and the unsustainable exploitation of microchiropteran bats for food in Laos has been described by Francis et al. (1999), with thousands of *Tadarida* being harvested during exit from a cave and smoked for sale. In one instance, 3,000 individuals were sold to occupants of a single passing truck.

Persecution. As creatures of the night, often associated with witchcraft and superstition, microchiropteran bats do not occupy a favorable place in Western cultures and are often persecuted (McCracken 1992; Tupinier 1989).

However, there are few documented instances of the impact of such persecution on bat populations. Because they eat commercially grown fruit, fruit bats are often killed, particularly in Australia, where licenses are issued to allow such control measures without regard for life-history characteristics and population demography. Wahl (1994) reported that at least 240,000 bats may have been killed in coastal areas between Sydney and the Queensland border from 1986 to 1992. Pierson and Rainey (1992) suggested that annual culling of even 10% of a fruit bat population would reduce the population by one-half in 5 yr and by four-fifths in 20 yr. Using this model, Richards and Hall (1998) estimated that if culling of Australian fruit bats continues at the rate reported by Wahl (1994), a population of 2 million individuals will be reduced to 100,000 in 30 yr. Although a management plan has been developed to address these issues (Eby 1995), it has not been adopted by the range states.

A dietary study of *Rousettus aegyptiacus* in Israel, where this species is considered to be a major pest in commercial orchards, demonstrated that cultivated fruit represented less than 15% of the diet of this species (Korine et al. 1999), suggesting that its pest status clearly needs to be reappraised. Scientific assessments can thus reveal that the condemnation of some bat species is inappropriate.

In Central and South America, *Desmodus rotundus* is extensively persecuted as a vector of rabies, which is transmitted to cattle and other ungulates on which it feeds. The main method of control is the use of anticoagulants applied to individual bats captured in mist nets, which are then dispersed to other individuals in the roost by allogrooming (Brass 1994). Roosts have also been burned, gassed, and dynamited. One such program in Venezuela has resulted in the destruction of 40,000 caves, with the loss of large populations of harmless or beneficial bats as well as other cave fauna (Hutson et al. 2001).

Although rabies occurs in North American bat populations, its incidence is low (< 1%), and there have been few human deaths. Nevertheless, fear of rabies has resulted in much persecution, roost exclusion, and colony extermination (Brass 1994; Messenger et al., this volume). Rabies-related viruses were identified in a European bat in the 1980s, and subsequent investigations revealed a low incidence of infection in bat species in mainland Europe (Amengual et al. 1997; Whitby et al. 1996), with the exception of *Eptesicus serotinus* (Nieuwenhuijs et al. 1992). Despite adverse press coverage, this has not resulted in extensive persecution (Judes 1987).

Introduced Species

Although alien species are one of the greatest threats to the world's biodiversity, there are few well-documented examples of such introduced species having a negative effect on bat populations. Examples of such documented cases, however, include the nonvolant juveniles of *Pteropus mariannus* on Guam, which are taken by the introduced tree snake *Boiga irregularis*, and this has a serious effect on recruitment into the local relict population (Wiles 1987).

In New Zealand, bats represent the only native mammals and appear to be at risk from introduced mammalian species. The disappearance of *Mystacina robusta* from mainland New Zealand coincided with the spread of *Rattus exulans* (Worthy 1997), and *M. robusta* subsequently became extinct during the 1960s when *R. rattus* reached its last known island refuge (Daniel 1990). This implies that alien species may have posed a direct threat to the survival of this bat species.

Feral and domestic cats appear to be common predators of bats and accounted for 28% of reported deaths of *Chalinolobus tuberculatus* in New Zealand (Daniel and Williams 1984). Cats were also responsible for 45% of injured *Chalinolobus gouldi* brought in for care in Victoria, Australia (Dowling et al. 1994). There are also reports of brush-tailed possums (*Trichosurus vulpecula*) attempting to enter roosts of *C. tuberculatus* (C. O'Donnell, personal communication).

Alien species compete with bats for roost sites. In the Fiordland of New Zealand, introduced starlings (*Sturnus vulgaris*) and ship rats (*Rattus rattus*) made nests in *C. tuberculatus* roosts (Sedgeley and O'Donnell 1999a, 1999b), although it is not clear whether the bats were displaced from or had vacated these sites.

Cascades of Extinction

Cascades of extinction are difficult to document but are likely to affect the most highly specialized bat species. Although most bats have a broad dietary niche, there are exceptions. For example, more than 99% by volume of the diet of one of the rarest European species, *Barbastella barbastellus*, consists of moths, both in the steppes of Central Asia and in the wooded areas of the Swiss Alps (Sierro and Arlettaz 1997). Factors that reduce the abundance of moths could thus result in local extinctions. Similarly, a reduction in the abundance of spiders, which constitute over 90% of the diet of *Kerivoula papuensis*, could have a serious impact on populations of this uncommon species (Schulz and Wainer 1997).

Other threats facing bats indicate not only potential losses in the diversity of this order but also increased risk of the loss of community integrity in bats and ultimately key ecological functions in terms of pollination and seed dispersal (Fujita and Tuttle 1991; Medellin and Gaona 2000). This in itself may have consequences for forest regeneration and ecosystem maintenance, which could undermine recovery of declining bat populations.

Ecological Requirements of Bats

Conservation planning relies on the ability to predict the likely response of species to change in the environment and relies on data drawn from autecological studies of species of key interest. From these it may be possible to identify not only species-specific conservation responses but also general trends and likely conservation issues across taxa. However, it is clear that the extent of ecological variation within the Chiroptera makes it difficult to extrapolate conservation advice even between species of the same genus. An attempt at identifying conservation recommendations with some relevance across related species has been made by Hutson et al. (2001).

For most species, even the most basic biological data are often not available, particularly for the rare species, which may require urgent conservation attention. For example, until recently little was known of the 21 recorded species of the genus *Kerivoula* (Hutson et al. 2001). These bats appear to occur at low densities in forest and have proved elusive to capture for many years. Recent studies have now provided data on diet, habitat associations, and echolocation call design (Kingston et al. 1999; Schulz and Wainer 1997). Thus, further research remains a priority to underpin conservation planning for bats, and the following section explores how knowledge of ecological requirements has aided our understanding of bats, both at species-specific and higher taxonomic levels. Much of the information from ecological research has revealed how bats use their environment, allowing predictions to be made (and in a few cases tested) about how environmental change is likely to affect their behavior and survival and generating recommendations for conservation action.

The impact of environmental change often depends on whether there are narrow or broad associations with specific resource types, or with particular species, where changes in one could result in adverse effects on the other. In general, species with more opportunistic and varied use of habitats or food are likely to tolerate, and survive, habitat change better than species that have highly specific requirements. Thus, an investigation of the ecological requirements of a species—and in particular the specificity of these requirements— can help inform conservation policy by predicting the severity of impacts should key resources decline.

Dietary Requirements

Dietary requirements are often expressed through selection of foraging areas, and analyses of dietary components do not generally have a direct bearing on the conservation of bats. An exception is the recent suggestion (Arlettaz et al. 2000) that competition for food by expanding pipistrelle populations (*Pipistrellus pipistrellus*) might contribute to the decline of lesser horseshoe bats (*Rhinolophus hipposideros*). Reviews of the diets of insectivorous bats (Vaughan 1997), New World phyllostomids (Gardner 1977), and megachiropterans (Bhat 1994; Courts 1998; Kunz 1996; Marshall 1985; Mickleburgh et al. 1992; Tan et al. 1998; Utzurrum 1995) in general reveal wide dietary niche breadth for all species that have been studied. However, in some cases knowledge of extreme dietary specialization may need to be incorporated in developing conservation plans. For example, some studies have demonstrated extreme specialization, such as *Corynorhinus townsendii* and *Otomops martiensseni* feeding almost entirely on moths (respectively: Sample and Whitmore 1993; Rydell and Yalden 1997) and *Kerivoula papuensis* with a diet consisting of 92% spiders (Schulz and Wainer 1997).

In the United Kingdom, *Eptesicus serotinus* and *Rhinolophus ferrumequinum* feed mainly on coleopterans, particularly dung beetles of the genus *Aphodius*, which are prevalent over pastures in late summer (Catto et al. 1994; Duvergé and Jones, in press). Juvenile *R. ferrumequinum* forage over pasture close to their roost, and this has been incorporated into management recommendations for this species. Farmers are provided with incentives to maintain their pastures and to avoid practices that may contribute to declines in key prey species such as *Aphodius* (Entwistle et al. 2001; Mitchell-Jones 1995; Ransome 1996).

Although plant-visiting bats appear linked to forest habitats for both roosting and feeding (Kunz and Lumsden, this volume), it is not clear for many species what degree of dietary adaptability they exhibit in response to habitat change. It is possible that specialization of diet and foraging niche contribute to the loss of particular species following deforestation (e.g., Brosset et al. 1996). Studies of the diet of fruit bats have shown that when fruiting trees in primary forest are lost, through selective logging or decline in habitat quality, at least some species (e.g., *Pteropus voeltzkowi*) are flexible and will exploit other sources of food, including agricultural crops and cultivated fruit trees (Entwistle and Corp 1998; Korine et al. 1999; Mickleburgh et al. 1992). Although such flexibility may shield bats from the loss of specific fruit plants, the bat-plant dynamics of natural forests (including pollination and seed dispersal) may become skewed with declining availability of key fruit trees. This may negatively reinforce changes in forest quality—for example, nonforest plants may be introduced to remaining forest fragments as bats bring seeds from other foraging grounds (Entwistle and Corp 1998; Galetti and Morellato 1994).

In general, examples of extreme dietary specialization are rarely documented among bats, and further research is needed to explore the degree of flexibility with regard to diet. However, as with other threatened species, maintaining the food base should be part of conservation plans—through habitat maintenance, reduction in insecticide application, and, where necessary, habitat restoration.

Habitat Requirements

Variation in the diversity of bat assemblages and the occurrence of rare species among different areas or sites may be affected by a range of factors, such as latitude (Rautenbach et al. 1996; Willig et al., this volume) and elevation (Patterson et al. 1996, Patterson et al., this volume; Utzurrum 1998), but often reflects the habitat types available in a given area (Findley 1993; Utzurrum 1998). In addition, studies of habitat requirements are complicated by difficulties in detecting habitat associations when bats may use a mosaic of habitats or may change habitat with season or where the scale of habitat selection (from gross habitat category to microhabitat preference) may not be clear (Fenton 1997; Fleming and Eby, this volume).

Several studies have investigated broad habitat components important for bats (e.g., in the United Kingdom, Walsh and Harris [1996a, 1996b]; in Canada, Grindal et al. [1999]; in Zimbabwe, Fenton and Thomas [1980]; in French Guiana, Simmons and Voss [1998]; and in Panama, Kalko [1998]). In temperate regions, habitats associated with woodland, water, and grassland/woodland edge are important for the broadest range of bat species. Even within forest areas, the diversity and assemblages of bats differ markedly (e.g., Cockle et al. 1998; Patterson et al., this volume). Recent studies using mist nets (Francis 1994; Kalko et al. 1996; Simmons and Voss 1998) and bat detectors (Kalcounis et al. 1999) have also revealed the importance of the forest canopy as an edge habitat. This indicates further structuring of forest habitats and shows that data indicating little bat activity on the floor, particularly of old-growth forest, can be seriously misinterpreted. Studies of disturbed forests demonstrate clearly how bat activity and diversity may be affected by underlying forest structure, the degree of fragmentation, and the impacts of deforestation. Brosset et al. (1996) found that deforestation significantly changed community composition for bats in French Guiana, with the loss of the rarer, forest-dependent species (mainly phyllostomids), although the more common species associated with edge habitats thrived. In such studies it is important to discriminate between bat abundance that may be higher in secondary forest (Vaughan and Hill 1996) or around water (Rautenbach et al. 1996), from patterns of species richness or diversity, and the presence of rarer or threatened species, which may be more meaningful indicators of conservation priorities (Findley 1993).

In temperate regions, old-growth forest appears to be of greater importance for bats than are logged or younger stands, both in terms of overall levels of bat activity (Humes et al. 1999; Jung et al. 1999) and in terms of associations of individual species. However, bat activity also may be greater in thinned areas as accessibility to open-air or gap-using species is increased (Humes et al. 1999). Short-term changes in accessibility of forest blocks (associated with newly cut areas and access roads) may increase bat activity (Grindal and Brigham 1998). This has been explained by increases in accessibility to edge feeders and other bats that commute along edge habitat, as insect availability was unaffected. Studies of bat activity within woodlands have demonstrated an association with bodies of water and forest trails, which may be used for commuting (Krusic et al. 1996; Mayle 1990).

Bat assemblages are often structured according to foraging style (Findley 1993; Schnitzler and Kalko 1998). Individual species appear to specialize in particular types of habitats—for example, forest edge and clearings with vegetation from which insects can be gleaned. The use of habitat can be clearly related to the interplay among diet, wing morphology (and consequently flight abilities), and echolocation calls (and hence ability to navigate in different degrees of clutter). The subdivision of airspace, habitats, and prey species between a range of bat species with some degree of specialization and the structure of bat assemblages with respect to morphological and dietary factors have been considered by a number of investigators (Fenton 1971; Findley 1993; Kalko 1998; McDonald et al. 1990). Factors such as wing shape, maneuverability, and echolocation call structure may affect the ability of bats to exploit cluttered and noncluttered habitats (Brigham et al. 1997a; Kingston et al. 2000), and such factors may underlie contrasting patterns of habitat use found in sympatric species (Fenton 1982). Even seemingly minor factors such as interindividual variation in body mass and wing shape may also affect the use of cluttered habitat by different individuals (Adams 1996; Kalcounis and Brigham 1995; Kingston et al. 2000). Differences in ecomorphology have been cited as an explanation for the differing habitat associations and microhabitat preferences across species, even between different heights and components of forest structure (Jung et al. 1999).

Morphology and call structure, as well as prey and habitat associations, therefore clearly define the niche and foraging behavior of different species (Arita and Fenton 1997; Fenton 1994, 1995; Kalko et al. 1999; Neuweiler 1984; Norberg and Rayner 1987). Investigations of diet and habitat use also reveal differences in the degree of ecological flexibility among species (Arlettaz 1999; Arlettaz et al. 1997). Such information is helpful in distinguishing specialists from generalists and aids in identification of species likely to be most affected by habitat change. Those species with the most specific diets and habitat associations may be least able to adapt to habitat modification. A number of authors have divided bats into closed canopy, edge, and gap/open area specialists (Crome and Richards 1988; Fullard et al. 1991; Schnitzler and Kalko 1998), and as increasing forest loss opens habitats, it seems likely that threats to closed-canopy specialists may increase.

Indeed, the replacement of species after deforestation or fragmentation often reflects a decline in those groups generally associated with closed canopy forest. Evidence points toward particularly strong effects of fragmentation on gleaning insectivores—such as declines in rhinolophids and hipposiderids in Tanzania (Cockle et al. 1998) and gleaning insectivorous phyllostomids in the Neotropics (Brosset et al. 1996; Fenton et al. 1992; Kalko 1998; Ochoa 2000; Simmons and Voss 1998). Similarly, Fenton et al. (1992) also found that phyllostomids, particularly phyllostomines, are sensitive indicators of habitat disruption in Mexico. Gleaning insectivores/carnivores have been proposed as potential habitat indicators (Fenton et al. 1992; Kalko et al. 1999), although the relative proportion of large- to small-bodied frugivores has also been proposed as an alternative and more practical indicator of disturbed habitats (Schulze et al. 2000).

The restriction of the more rarely captured species to primary forest (Brosset et al. 1996; Simmons and Voss 1998) indicates their likely reliance on such undisturbed areas. Since rarer forest-dwelling species are the first to disappear after deforestation (including a range of mature—forests [Brosset et al. 1996]), these species should be of high conservation priority.

The response of bat assemblages to habitat fragmentation—with a decline in diversity, loss of rarer species, and increase in abundance of edge species— mirrors patterns found in other taxa following habitat degradation and fragmentation (Nupp and Swihart 2000). From these studies it becomes apparent that persistence following fragmentation appears to depend on the ecological traits of individual species—including the degree of habitat and dietary specialization, range requirements, and ability to tolerate and cross surrounding modified habitats (Laurance 1991, 1994; Mills 1995; Nupp and Swihart 2000).

Investigations into habitat use by bats have now moved from simply describing where an animal feeds to determining whether animals actively choose particular habitats, especially where habitat types might be limited or threatened. Requirements can be considered in terms of habitat type and specificity, spatial use of the habitat, and its connectivity.

Habitat Specificity

Habitat selection can be determined at a number of levels, from gross associations with a major habitat category (e.g., woodland specialists or riparian feeders) to more specific preferences for particular habitat components. Comparison of habitat use in relation to availability allows some interpretation of degree of preference or selectivity (Neu et al. 1974; Thomas and Taylor 1990). Habitat utilization surveys can be carried out at the level of the assemblage (general bat activity in different habitats) or at an autecological level, using standard bat survey techniques (Kunz et al. 1996). Assemblage survey approaches include assessments of levels of bat activity across different habitats with measured parameters, using bat detectors (transects or point counts) or capture devices (point counts). Autecological studies may rely on data from captures, bat detector assessments, or either intermittent or continuous monitoring of individuals (using light tags or radio transmitters). Each of these approaches has its own intrinsic biases. Data on habitat use can then be compared to habitat availability within the foraging range. In general, the associations considered here are between individual species and habitat types, but these associations are not always clear (Fenton 1997). Some mist netting and bat detector data reveal an association between bats and coastal scrub in Mexico (Fenton et al. 1992), whereas similar studies in riparian situations in Africa failed to establish the importance of the associated vegetation for bats (Rautenbach et al. 1996). In French Guiana, however, Simmons and Voss (1998) found distinct sets of species in different local habitats within a lowland rainforest.

Radio tracking has helped resolve habitat associations of some bat species. For example, during spring in England and Wales, *Rhinolophus ferrumequinum* forages in ancient seminatural woodlands, where the ambient temperature is higher than in the surrounding open countryside and is less likely to fall below the threshold for insect flight. In summer, after cattle dung has accumulated on pasture, this species feeds mainly on dung beetles *Aphodius* (Duvergé and Jones 1994, in press; Jones et al. 1995). This information has been used by conservation managers who employ a mixture of legislation and incentive schemes to protect foraging habitat (Mitchell-Jones 1995; Racey 2000; Ransome 1996).

Radio tracking of *Plecotus auritus* confirmed the association of this species with deciduous woodland, and insect sampling confirmed greater abundance of preferred prey moths within this habitat (Entwistle et al. 1996). Smith (2000) compared habitat use of Natterer's bat *Myotis nattereri* as determined by radio tracking with the availability of different vegetation types and established preferences for broad-leaved woodland and tree-lined river corridors in this foliage-gleaner.

Specific microhabitat associations have been investigated in relation to the distribution of *Barbastella barbastellus* foraging within woodland areas, with activity being related to specific variables such as leaf litter thickness, shrub layer, pine tree cover, and size of oak trees (Sierro 1999). In contrast, some species show greater adaptability and use a range of different habitats for foraging. For example, *Nyctalus leisleri* forages mainly in woodland and shrub-lined roads in southeast England, and over pasture farther west, while arable land and urban areas are consistently avoided (Waters et al. 1999). A separate study in Ireland described use of pasture, drainage canals, streetlamps, streams, beach, and dune habitats (Shiel et al. 1999) and changing patterns of foraging behavior with time of year. This reinforces the influence of seasonality on patterns of habitat selection in temperate regions (de Jong 1994; Jones et al. 1995) and the need to include all necessary habitat components within a conservation plan rather than simply utilizing data collected within a single season.

In some cases, similar approaches have been used to investigate bat activity at the level of the assemblage, building on methodologies developed for autecological studies. The first attempt to establish habitat preferences on a national scale was the National Bat Habitat Survey in the United Kingdom, which compared the incidence of bat passes in random stratified samples of Britain's 32 land classes (Walsh et al. 1995, Walsh and Harris 1996a, 1996b). Bonferroni comparisons between availability and use revealed a far stronger preference of bats for all bodies of water and broad-leaved woodland edge than for any other habitat. Linear vegetation corridors, particularly tree-lined hedgerows and ditches, were selected, indicating the importance of landscape connectivity. Habitats that were more exposed and intensively managed, including arable land, improved grassland, upland and moorland, were strongly and consistently avoided. Thus, bats foraged preferentially in habitats that were comparatively rare in the landscape.

The primary aim of the survey was to provide a means of assessing the significance of habitat change for bat populations. By expanding the scale of previous studies of habitat preferences reviewed by Walsh et al. (1995) to a national scale and using a land classification widely accepted in the United Kingdom, a method was developed for detecting change, establishing its direction, and measuring its magnitude. Analysis of habitat factors associated with high levels of bat activity resulted in equations with high predictive power and of particular value for forecasting the effects of change of land use on bats. The principal caveat in interpreting the results of the survey is that because they are based on recorded bat passes (without species identification) they may reflect the habitat preferences of *Pipistrellus pipistrellus*, the most abundant bat in the United Kingdom (Harris et al. 1995). The goal of the current National Bat Monitoring Programme is for participants to identify bats to species or at least to species groups (Walsh et al. 2001).

Area/Spatial Requirements

Few attempts have been made to investigate the minimum foraging areas required to sustain an individual or bat species. Because they are highly mobile and show relatively patchy use of the environment, it is difficult to categorize the space requirements of bats. The concept of home range is widely used for terrestrial mammals (Harris et al. 1990), and home range size has been usefully correlated with ecological niche across a wide range of body sizes. Computations of home range size become more ecologically meaningful with an analysis of the use that animals make of their range, particularly when visualized in three dimensions (Gorman and Zubaid 1993; Wilkinson and Bradbury 1988). The concept of home range has not found the same currency in studies of avian ecology, and it is interesting in this context that there is an increasing trend for bat ecologists to estimate home range. The results should be used with caution in conservation planning because bats frequently commute between roost and foraging areas along linear landscape features (Verboom and Huitema 1997) and forage by trap lining (Fleming 1988; Racey and Swift 1985). Some species also travel relatively large distances between roosts and foraging areas (Horner et al. 1998; Pierson 1998).

Studies of home range in bats often produce estimates that vary among roost sites and individuals and with season. Home ranges are not fixed but show considerable variability in response to energetic requirements and distribution of prey (de Jong 1994). For example, in one study in Sweden, home range estimates for *Eptesicus nilssoni* increased from 11.6 ha to 757.6 ha as the summer progressed (de Jong 1994). It is clear that some species require a selection of feeding sites to provide sources for food at different times of the night and throughout the season—and the use of space frequently changes markedly over a year. Additionally, it is difficult to assess at what level home ranges should be considered—at the level of the individual or the colony. For example, investigations of *Eptesicus serotinus* in the United Kingdom have estimated home ranges of 24–77 km^2 for maternity colonies (core areas 13–33 km^2), whereas the home ranges of individual bats varied from 0.16 to 47.58 km^2 (Robinson and Stebbings 1997). The nectar-feeding phyllostomid *Leptonycteris curasoae* showed similar interindividual variation, with individual home ranges of 0.29–63 km^2, although in this study the bats also commuted 30 km between roost sites and foraging areas (Horner et al. 1998).

In some species, differences in home range between males and females have been recorded (McCracken and Bradbury 1981). Given this variation, even within species, the interpretation of data from home range analyses is difficult, although an estimate of core area and maximal home range per roost may be useful. For example, group home ranges have been estimated for *Mystacina tuberculata* and *Chalinolobus tuberculatus* (135 km^2 and 117 km^2, respectively [O'Donnell 2001; O'Donnell et al. 1999]).

Some of the smallest home ranges have been recorded from tropical forest-dwelling bats, including the nectar feeder *Syconycteris australis*, with recorded individual ranges of 2.7–13.6 ha, mean 5.5 ha; (Winkelmann et al. 2000) and 3–12 ha for the gleaners *Tonatia silvicola* and *Trachops cirrhosus* (Kalko et al. 1999).

A number of authors have documented the relationship between roost site and foraging areas as a basis for conservation planning, and such approaches have demonstrated clear differences in behavior among species. A number of species concentrate their feeding close to the roost site (e.g., *Plecotus auritus*, < 500 m [Entwistle et al. 1996]; *Eptesicus nilssoni*, < 600 m [de Jong 1994]). In contrast, other species show much greater dispersion in their use of foraging sites (e.g., *Nyctalus noctula* [Kronwitter 1988] and *Nyctalus leisleri* [Shiel et al. 1999; Waters et al. 1999]). There is some indication that the distance traveled between roosts and foraging areas may change with season and reproductive status (e.g., Clark et al. 1993; Shiel et al. 1999) and that intersexual differences may occur—for example, males may travel farther than females (Bradbury and Vehrencamp 1976, Entwistle et al. 1996). However, the discovery that some bat species forage close to the roost site at some times of year (especially during lactation) indicates the conservation importance of foraging opportunities within close range of roost sites (and vice versa). Recommendations based on radio tracking of *Rhinolophus ferrumequinum* suggest the need for protection of suitable habitat up to 3 or 4 km from the roost site (Jones et al. 1995). Parallel results come from Neotropical gleaners, such as *Tonatia silvicola* and *Trachops cirrhosus*, which both use relatively small foraging areas, relatively close to the roost (in the case of *T. silvicola* generally within 500 m [Kalko et al. 1999]). Such species are vulnerable to habitat fragmentation as a result of their foraging behavior.

It has been suggested that variations in the distance that bats travel from the roost to foraging sites (and foraging range) may reflect flight abilities and wing shape (specifically, aspect ratio) of the species concerned (Jones et al. 1995; Kalko 1998; Kalko et al. 1999). A potential relationship between wing shape and foraging range could be an important predictive factor for conservation planning. However, limited studies of such relationships have been conducted, and conflicting evidence has also been reported, especially where distances traveled may increase at particular times of year (e.g., *Corynorhinus townsendii* [Clark et al. 1993]), and some studies have suggested that other factors might affect foraging range, such as body mass (Fenton 1997).

Plecotus auritus with short, broad wings (low aspect ratio) spends the majority of its time within 1 km of the roost site during the summer (Entwistle et al. 1996). Thus, from a conservation perspective, it would appear that such species may rely on spatial proximity between roost and foraging range more than do long-ranging, fast-flying species with higher aspect ratios. The effective foraging range of individuals may in turn affect the grouping of individuals in relation to habitat. Theory suggests that competition close to roost sites may encourage wider-ranging foraging behavior (Hamilton and Watt 1970). However, if morphology constrains the gains from traveling farther (in terms of time and energetic limitations), this may effectively limit the size of the population associated with particular habitat patches. For example, colony size in *P. auritus* is strongly associated with the size of habitat patches (Entwistle 1994), and the relatively small colony size in this species may reflect wing morphology and restricted ability to exploit more distant foraging sites (Entwistle et al. 2000). Species such as *P. auritus* with limited dispersal abilities are also more likely to be isolated to some extent, either physically or genetically. Where dispersal occurs over a small scale, and large-scale migration is rarely recorded, some degree of genetic structuring becomes apparent (Burland et al. 1999).

The extent of seasonal changes in foraging patterns and distances traveled (Clark et al. 1993) raises the question of whether the daily foraging range or complete seasonal/annual foraging range needs to be identified for conservation purposes (see Fleming and Eby, this volume).

Connectivity Requirements

Another factor that appears to affect the use of space and the distribution of feeding patches in relation to the roost is the degree of connectivity between different landscape elements and the relative ability of different bat species to cross open matrix habitats. The use of linear landscape elements by commuting and foraging bats was first described by Limpens et al. (1989) and by Limpens and Kapteyn (1991). These elements include fences, banks, hedges, tree lines, and roadsides. In the semi-industrialized/rural landscape of Western Europe, habitat types are often subdivided by such linear anthropogenic features. Since the use of these features by bats was first described, a more quantitative approach has been used to show interspecific differences: *Pipistrellus pipistrellus* (but not *Eptesicus serotinus*) was recorded more often along linear landscape elements than in open areas, although both were also observed in fragments of woodland isolated by up to 150 m (Downs 2001; Verboom and Huitema 1997). Downs (2001) and Verboom and Spoelstra (1999) found that the activity of *P. pipistrellus* decreased with increasing distance from a tree line. The brown long-eared bat *Plecotus auritus* traveled substantially farther between foraging sites by following landscape element such as fences, tree lines, and ditches than by direct flight (Entwistle et al. 1996). There are three possible functions of such linear elements for bats—they may aid orientation, attract insects, and provide shelter from wind and/or predators—and these functions are not easily separated. Nevertheless, Verboom et al. (1999) provided convincing evidence that commuting *Myotis dasycneme* use canal banks as acoustic landmarks.

At light intensities greater than 1 lux, *Rhinolophus hipposideros,* crossing gaps in hedgerows, dropped to within 1 m of the ground but did not reduce the flight altitude at lower light intensities, suggesting that avoidance of predation was at least one factor in the use of vegetation corridors (Schofield 1996). When using a corridor consisting of a double row of trees, pipistrelles flew between the tree row, regardless of insect abundance or wind speed—behavior interpreted as predator avoidance (Verboom and Spoelstra 1999).

Ekman and de Jong (1996) investigated the use of habitat patches, use of matrix habitat (crop fields and water), and effects of isolation on islands among four bat species: *Myotis brandtii, Eptesicus nilssonii, Plecotus auritus* and *Pipistrellus pipistrellus.* This study showed that bat activity was lower in habitat patches than in continuous tracts of habitat for three of the species. The effects of isolation appeared greatest for *P. auritus* and *M. brandtii,* which were absent from the smaller and more isolated islands. It was suggested that these species would be most at risk from increased patchiness of habitat and isolation of fragments. The effect of patchiness on different species may again relate to habitat associations but may also relate to morphological factors with respect to the use of open areas. Species with relatively short, broad wings and shorter-distance echolocation (such as *P. auritus* and *M. brandtii*) may be less likely to cross open areas and may show greater reliance on linear landscape elements.

In general, it seems that many bat species are reluctant to cross open areas. This has implications beyond the agricultural landscapes of North America and Europe and suggests that fragmentation of key habitats, including forest tracts, is a cause for concern in relation to the conservation of bats and highlights the need to maintain corridors between such habitats.

Discontinuity of habitats or roosts through loss of connecting routes or corridors at a greater geographic scale may also have significant impacts on bat species. For example, loss of key specialized roosting sites within a population's distributional range may lead to the genetic isolation of subpopulations—as observed in *Macroderma gigas* (Worthington Wilmer et al. 1994, 1999). The slow flight and physiological requirements of this species make it unlikely that such population discontinuities can be bridged.

Implications of Habitat Requirements

It is clear from the studies on habitat use, specificity, range, and the use of connective elements in the landscape that the habitat requirements of bats are complex and vary greatly between species. Although information on the foraging behavior and habitat use of different species grows each year, only a tiny percentage of the approximately 1,100 extant bat species have been studied. Nevertheless, patterns are beginning to emerge with respect to the broad habitat requirements across bat assemblages and habitat correlates with species diversity. Furthermore, patterns are emerging that help to identify which species appear least able to adapt to habitat modification and are most in need of conservation interventions.

Given apparent relationships between morphology and both foraging and dispersal patterns, further research is needed to examine what relationship might be found between morphology and the impacts of habitat fragmentation. Such studies would be fundamental to developing and improving models that can predict relative proneness to extinction across bat species.

Roost Requirements

Most bat species show some degree of specificity with regard to their roosting habitats, including the use of particular structures—such as caves, leaves, tree cavities, buildings, or in the case of Old World fruit bats, branches of certain tree species (Barclay and Brigham 1996; Kunz 1982, 1996; Kunz and Lumsden, this volume). Some species of bats have highly specialized roosting requirements—some tent-making bats use certain types of leaf architecture (Kunz et al. 1994) and *Tylonycteris* spp. roost in cavities in bamboo culm in which an opening was formed by beetle larvae (Medway and Marshall 1970). Numerous studies have documented the roosting habits of different bat species and have shown how roost preferences may vary seasonally (Kunz and Lumsden, this volume). As well as showing where bats roost, other studies have demonstrated that bats often select their roosts on the basis of key features, such as height above the ground (Vonhof and Barclay 1996) or amount of adjacent foliage clutter (Kunz et al. 1994), each of which differ from those in other potentially available roosts. Information about such specificity is of particular relevance to conservation, as it indicates the likelihood of a species being adversely affected should the natural habitat associated with these roosts be lost, especially if the bats have a narrow margin of choice and if suitable roosts are limited.

Selection of Roosts

From descriptions of bat roosts in the literature (Barclay and Brigham 1996; Kunz 1982, 1996; Kunz and Lumsden, this volume), it is possible to generalize about the requirements of some species. For example, small vespertilionids, such as *Pipistrellus pipistrellus,* seek contact with the roosting substrate by squeezing into small spaces, whereas others, such as *Plecotus auritus,* roost in open roof spaces. Ambient temperature is often an important factor in roost choice (Humphrey 1975; Kunz 1982), and maternity colonies of some tropical and subtropical species, such as *Macrotus californicus* (Bell et al. 1986) and *Macroderma gigas* (Churchill and Helman 1990; Toop 1985), are found in geothermally heated mines and caves. High humidity is also a requirement of many species, especially during hibernation (Speakman and Thomas, this volume).

Buildings. Few studies on roost preferences of bats in buildings have been conducted. In an attempt to establish such preferences, which could be validated statistically, Entwistle et al. (1997) compared the characteristics of a sample of summer roosts of *Plecotus auritus* with those of random samples of buildings in the study area. Buildings that provided suitable roosts were older and often had roof spaces divided into several compartments that were most likely to be fully lined with rough planking. The temperature inside these roosts (mean 17.9°C) was significantly greater than in random or adjacent houses (mean 16.7°C). Buildings that contained roosts were situated closer to woodland and to bodies of water than randomly selected houses and had greater areas of woodland within a radius of 0.5 km. This suggests that feeding habitat close to the roost is important for roost selection in this foliage-gleaning species. These results also suggest that *P. auritus* selects its roosts relative to other sites that are available. Thus, conservation action should seek to avert significant changes to or destruction of roost sites or woodland in the vicinity of occupied roosts.

Similar methodology was used in a study of the 55 kHz phonic type of *Pipistrellus pipistrellus* (Jenkins et al. 1998). In contrast to *P. auritus, P. pipistrellus* does not select roosts with specific structural attributes. Compared to random buildings, however, roosts selected by *P. pipistrellus* had a great percentage of trees within a radius of 50 m and were more likely to have linear vegetation elements leading away from them. They were closer to and surrounded by a greater area of woodland within a radius of 0.5 km and were more likely to be found within 0.5 km of a major river.

Schofield (1996) surveyed 156 maternity roosts of the lesser horseshoe bat *Rhinolophus hipposideros* in the United Kingdom and found that this species selected predominantly nineteenth-century buildings (77%; $n = 61$) with stone walls (81%; $n = 82$) and slate roofs (88%; $n = 77$) and roosted in the roof spaces or attic rooms (95%; $n = 76$). Access to the roosts was generally through large openings such as open doors or windows that often characterize derelict or semiderelict buildings. The volume of these roosts was frequently greater than 250 m^2. Similar findings are reported for this species from other European countries (Gaisler 1963; McAney and Fairley 1989; Stutz and Haffner 1984). Studies of building use by *Eptesicus fuscus* demonstrated selection for older, taller, and more accessible structures, often having tin roofs, when compared to a random sample (Williams and Brittingham 1997).

Trees. It is now possible to generalize that tree-roosting bats typically select tall trees that are in the early stages of decay and are less cluttered than random wildlife trees (Barclay and Brigham 1996; Kunz and Lumsden, this volume). Although the species of roost tree may differ from forest to forest, general attributes are similar, resulting in higher densities of roosting bats in older forest stands.

In a survey of the roosting requirements of bats in Canadian forests, Vonhof and Barclay (1996) measured 22 tree and site characteristics and found that bats preferred to roost in tall trees close to those used by other wildlife in areas with open canopies. The bats also preferred western white pine (*Pinus monticola*) and to a lesser extent ponderosa pine (*P. ponderosa*) in the intermediate stages of decay. Brigham et al. (1997b) developed the general hypothesis that forest-roosting bats require a number of large dead trees of particular species, in specific stages of decay, that project above the canopy in relatively open areas. Similar conclusions were reached by Betts (1998). In New Zealand, Sedgeley and O'Donnell (1999a, 1999b) found that *Chalinolobus tuberculatus* had specific roosting requirements and roosted only in high tree cavities in *Nothofagus* rainforest, to which they gained access through knotholes where there was little surrounding vegetation.

A number of studies have been conducted to investigate the properties of roost trees selected by bats, focusing on the characteristics both of individual trees and of the surrounding habitat, in relation to random trees (e.g., Rabe et al. 1998; Vonhof and Barclay 1996). For example, *Myotis volans* select taller snags of Douglas fir (*Pseudotsuga menziesii*) as roosts in central Oregon (Ormsbee and McComb 1998). *Eptesicus fuscus* exclusively roosted in south-facing tree cavities in trembling aspen (*Populus tremuloides*), which had previously been excavated by yellow-bellied sapsuckers (*Sphyrapicus varius* [Kalcounis and Brigham 1998]). All suitable cavities that were investigated showed signs of use, suggesting that roosting opportunities were limited for this population. Menzel et al. (1998) demonstrated significant differences in roosting preferences between two tree-roosting species (*Lasiurus borealis* and *L. seminolus*) in southeastern United States. These studies provide clear evidence that management priorities are needed for forest bats at the local level.

Caves and Mines. Natural caves provide some of the most important maternity and hibernating sites for bats. In North America, these sites are used regularly by 21 of the 45 bat species and occasionally by most of the others (Pierson 1998). Abandoned mines are equally important and are used by 28 species (Pierson 1998; Tuttle and Taylor 1994). Shortage of roosts may be limiting for many cave-dwelling bats (Brosset 1966; Humphrey 1975; Kunz 1982; McCracken 1988; Pierson 1998). During hibernation, bats roost on the walls and ceilings of mines and caves or deeper in crevices or may seek refuge beneath rock rubble on the floor (Roer and Egsbaek 1966). They may change position during their occupancy, depending on the microclimate, particularly temperature and airflow. In large underground systems, bats often prefer to roost close to entrances. Apart from the requirements for a constantly cool environment with a high humidity, there have been few studies of the roosting preferences of hibernating bats that will help inform the design of artificial roosts. Nevertheless, two studies suggest potentially adverse effects of disturbance on hibernating bats. In a laboratory-based study, Speakman et al. (1991) showed that auditory or photic stimuli of the type likely to be experienced by bats whose hibernacula are visited by humans failed to cause arousal, whereas even the mildest tactile stimuli did so. In the field, however, Thomas (1995) demonstrated that even nontactile stimuli caused widespread arousal within 24 h of the bats being disturbed. The installation of gates to exclude humans at the entrances to underground sites to prevent such disturbances has allowed populations to recover, both in the United States (Dalton and Dalton 1995; Decher and Choate 1995; Tuttle 1977) and in Europe (Voûte and Lina 1986).

Artificial Roosts. Artificially constructed roosts have been used as conservation tools for bats for many years (Racey 1998a). One of the few studies of the contribution of such boxes to the population ecology of bats is that of Boyd and Stebbings (1989), who analyzed the occupancy by *Plecotus auritus* of 480 wooden bat boxes that were attached to 100 trees in a plantation of mature Corsican (*Pinus nigra*) and Scots pine (*P. sylvestris*). Following the establishment of a population of bats in the boxes, immigration probably accounted for a small proportion of the total recruitment, the remainder coming from reproduction within the population. The total population increased during the study from 73 to 140 bats, giving a doubling time of 10 yr. Benzal (1991) studied the same species, occupying 520 bird boxes in 130 ha *P. sylvestris* forest at the comparatively high altitude of 1,400 m in Central Spain. The boxes were widely used in the first summer after installation, and a total of 197 individuals were found. Although bats arrived in the study area in May and stayed until early November, they left the boxes at the end of May to form maternity colonies in small caves and attics and returned to them in mid July with flying young. These studies provide ecological data that support the promotion of bat boxes as alternative roosts for bats, particularly in areas devoid of such roosts, although the Spanish study suggests that the boxes may not always be appropriate for breeding.

In the United States, considerable success has been achieved with free-standing bat houses (Kiser and Kiser 2000), as well as with structural modification to larger already-existing structures, such as the attics of barns, the undersides of bridges, or the interior of dams, particularly for crevice-dwelling species (Keeley 1999; Tuttle and Hensley 1993).

Effects Linked to Conspecifics

A complicating factor when considering roost selection in bat species is the extent of colony integrity and the effects of conspecifics. In cooler climates, the presence of conspecifics has an important thermoregulatory role since clustering reduces heat loss (Herreid 1967; Kurta 1985; Trune and Slobodchikoff 1976). The high level of roost selection and the high degree of philopatry reported for several bat species (Lewis 1995) suggest that social factors, in addition to physical ones, may affect roost choice (McCracken and Wilkinson 2000; Kunz and Lumsden, this volume). Potentially, there could be a minimum viable colony size below which an individual's reproductive success may be threatened (Tuttle 1979). The extent to which social dynamics of colonies are affected by population declines—and whether local populations decline at a colony or individual level—is not clear (i.e., whether population loss reflects colony extinction or gradual population decline across colonies). The interface between social behavior and population change is an area that invites further study.

Of particular importance in this context are the facts that a colony of bats may use a number of different roosts and that roost switching may be very common in these species (Lewis 1995), such as *Desmodus rotundus* (Wilkinson 1985), *Nyctalus noctula* (Kronwitter 1988) *Antrozous pallidus* (Lewis 1996), and *Myotis nattereri* (Smith 2001). This complicates effective conservation, particularly if it is colony focused.

Landscape Approaches

Effective conservation of bats requires a landscape-based or systems approach, which incorporates the elements discussed above into a holistic management plan. Effective conservation cannot target the requirements of one species (such as roost sites), while ignoring others (such as adjacent foraging areas). Effective plans for threatened and endangered species must draw together available information on roosting and foraging requirements, the spatial relationship between roosts and feeding sites, connectivity requirements, seasonal movement patterns, and, where appropriate, the location of hibernation sites. Only a thorough understanding of these components will ensure that effective and appropriate conservation decisions are made. Thus, we suggest that an effective management unit for many bat species should be the colony and its main roost and peripheral and seasonal roosting sites, along with surrounding foraging areas and connective elements.

Development of Conservation Approaches

Basic ecological research should be the critical first step in all conservation efforts. In the case of bats, this is central to the development of sound management guidelines and policies. In many cases, action plans and species recovery plans have been successful in effectively defining management regimens for declining populations, either at the country level—for example, Australia (Duncan et al. 1999), France (Ministère de l'Environment 1996), New Zealand (Molloy 1995), Switzerland (Moeschler 1991) and the United Kingdom (Hutson 1993)—or at a species level (e.g., *Rhinolophus ferrumequinum* [Ransome and Hutson 1999] or *Myotis dasycneme* [Limpens et al. 1999]).

Research on bats has been incorporated into several programs that have contributed to their conservation. Some of these programs and activities have focused on the assemblage level and some on individual species of concern.

However, additional research is still needed to support the development and refinement of conservation management for bats. This includes:

- further autecological research into key threatened and data deficient species;
- research focusing on commoner species as models to develop generic management guidelines;
- development and testing of predictive models examining relative threat to species in relation to ecological parameters;
- experimental investigations of impacts of different threats on particular species and feedback on relative mitigation resulting from different management regimens;
- further development of robust monitoring techniques for bat populations;
- increased datasets to inform conservation planning tools, such as population and habitat viability analysis (PHVA);
- research to define the influence of colonial living on minimal viable population size in different bat species;
- increased study of the relationships among bat species and between bats and ecosystem components, such as trees;
- research to assist understanding of appropriate management units and spatial extent of protection required (including appropriate sizes for protected areas);
- further research into the true value and costs of bats in a socioeconomic context, particularly at local levels; and
- further research into perceptions of bats and means of improving appeal and use as "flagship species," particularly at local levels.

International Conventions and Legislation
Convention on International Trade in Endangered Species of Wild Fauna and Flora 1973 (CITES)

Currently all species of *Pteropus* and *Acerodon* (more than 60 total) are listed on appendixes 1 or 2 of CITES (http:\\ www.cites.org). However, as illustrated by the situation in the Commonwealth of the Northern Mariannas (CNMI), CITES is an international convention that cannot regulate internal, within-nation trade. Furthermore, in many places, customs officials are faced with a problem of identification. Recently, an illustrated key of island species of *Pteropus* was produced for the U.S. Fish and Wildlife Service (E. D. Pierson and W. E. Rainey, personal communication) to assist customs officials in Guam.

European Bats Agreement

The *Agreement on the Conservation of Bats in Europe* (Foreign and Common-wealth Office 1994) is part of the *Convention on the Conservation of Migratory Species of Wild Animals* (1980). This agreement came into force in January 1994, and by July 2000 had 20 signatories, who have agreed to an action plan with priorities for implementation and with strategies for monitoring populations of selected bat species in Europe. The secretariat in Bonn is currently developing international transboundary programs for the study of migratory species in order to develop appropriate conservation measures. It is also identifying best practice for the management of woodland and underground habitats.

Convention on Biological Diversity

The *Convention on Biological Diversity* was signed by many nations at the United Nations Conference on Environment and Development in Rio de Janeiro in 1992 (United Nations Environment Programme 1992). Many countries that subsequently ratified the *Convention* have developed national action plans. In the United Kingdom, for example, approximately 1,500 species of conservation concern have been listed and action plans prepared for 400, including several bat species (Racey 2000; Wynne et al. 1995). This process has stimulated the production of local biodiversity action plans, some of which also include bat species.

Program for the Conservation of Migratory Bats of Mexico and the United States (PCMM)

Many of North America's most ecologically and economically important bat species migrate seasonally across the U.S.-Mexican border, including *Tadarida brasiliensis, Choeronycteris mexicana, Leptonycteris curasoae,* and *L. nivalis.* As a consequence, PCMM focuses on research, environmental education, and cave conservation in an attempt to reverse the declines in migratory bats and establish a new public conservation ethic in Mexico and the United States. Research projects involve monitoring population trends, studying bat-plant relationships, determining migratory routes, and documenting the economic values of these bats. Population research also has been initiated in populations of cave bats in several Mexican states.

National Legislation

There is legislation to protect bats in many countries of the world, but too often the laws are inadequate because the animals alone are protected and not their roosts or feeding habitats. Although legislation has been in force in some European countries for 50 yr or more, for most of that time there has been little enforcement.

In 1981, the United Kingdom Wildlife and Countryside Act gave bats greater protection than any other group of mammals by making it illegal to disturb them or their roosts. The act requires householders or others who have a problem with a bat roost to notify the statutory nature conservation agency, and the agency, in turn, must visit the roost within a reasonable time interval. This places strain on agency resources, with the result that such visits are often delegated to appropriately trained members of amateur bat groups, thereby providing an opportunity for public education. Subsequent advances in bat conservation in the United Kingdom have resulted from the close working relationships between bat ecologists and those with statutory responsibility for their protection (Racey 2000).

In the United States, the 1973 Endangered Species Act has provisions for protecting species and also for the designation and protection of critical habitat. In addition, it places restrictions on projects in which the United States is involved on foreign soil. The inclusion of bat species on the Endangered Species List (*Myotis sodalis, M. grisescens, Corynorhinus townsendii* (two eastern subspecies), *Leptonycteris curasoae,* and *L. nivalis*) has resulted in funding for research and enhanced roost protection for these species, which has stemmed population decline for *M. grisescens* and the protected eastern subspecies of *C. townsendii.* The key factor for this success appears to be linked to the protection of important cave roosts.

Elsewhere in the world, local legislation to ban hunting has been recommended (e.g., on Palau [Wiles et al. 1997]). In Sarawak, Malaysia, the government has included all bats on the protected species list and has enacted laws reducing the sale of shotgun cartridges and banning the sale of wild meat, both of which are aimed at reducing the hunting pressure on fruit bats (Gumal 2001). On the island of Pemba, community-led bylaws are being developed to ban hunting of bats with shotguns (Entwistle and Corp 1997).

Education and Public Awareness

The establishment and growth of nongovernment organizations (NGOs) dedicated to bat conservation has resulted in major improvements in public attitudes toward bats. In the United States, Bat Conservation International, founded in 1981, has about 15,000 members and has established programs to protect caves and mines used as bat roosts, to encourage bats to roost in bridges and bat houses, and to improve the management of vampire bat populations in Central and South America. It was also instrumental in establishing the PCMM and encouraged the formation of the North American Bat Conservation Partnership, which attempts to organize bat conservation on a federal basis so that the different conservation requirements of species living in widely different environments can be addressed. In the United Kingdom, the Bat Conservation Trust, founded in 1990, has about 4,000 members and supports 90 amateur bat groups with resource materials for roost owners, as well as educational material for school children. It organizes training workshops in the use of bat detectors and has recruited and trained volunteers for the National Bat Habitat Survey and the National Bat Monitoring Program. The success of these NGOs has led to the formation of similar organizations in other European countries, Australia, and South America (Hutson et al. 2001) and in South Africa (Taylor 1999).

An example of a focused education and public awareness campaign is Action Comores, which has concentrated on *Pteropus livingstonii* in the Comoros Archipelago in the western Indian Ocean. Successive expeditions from the United Kingdom have prepared and distributed education materials for use in local schools and communities, including videos, and have also trained local personnel to maintain awareness of the ecological importance of bats and the forest where they roost. These personnel also monitor the bat population, which is distributed among several small roosts (Trewhella et al. 1998). In another example from the western Indian Ocean, the effectiveness of school-based education on bats was assessed in Pemba. The evaluation showed that information had been disseminated effectively to a broader community (A. C. Entwistle and N. Corp, unpublished data), and, as with other smaller mammals, that bats could be used effectively as flagship species (see Entwistle and Stephenson 2000).

Walk-through zoo exhibits, such as the Montreal Biodome and at Chester, United Kingdom, and museum displays have done much to improve the public appeal of bats. In addition, the proliferation of non-technical books has provided much-needed information for the general public (Fenton 1992; Richardson 1985).

Captive Breeding, Restocking, and Translocation

In 1974, *Pteropus rodricensis* was described as perhaps the rarest bat in the world, with an estimated wild population of 75–80 individuals. In 1976, the Jersey Wildlife Preservation Trust (now the Durrell Wildlife Conservation Trust), together with the Mauritian government initiated a captive-breeding program in Jersey and Mauritius. Bats of the genus *Pteropus* generally breed well in captivity, and by 1991 the captive population had increased to about

250 individuals, which had been distributed to nine zoos (Caroll and Mace 1988). By early 1999, the captive population had risen to 700 individuals at 25 locations (D. Wormell, personal communication). In some respects, the captive-breeding program for this species has become a victim of its own success, and contraceptive implants are now used in some zoos to prevent further births (Hayes et al. 1996).

In 1990, the wild population of *P. rodricensis* was estimated as 1,000 individuals, but following a cyclone in 1991 numbers of this species declined to around 350. This pattern is likely to continue and may be exacerbated as climate changes. For this reason, there are no plans to restock Rodrigues Island, although there is mounting pressure for a population and habitat viability analysis. In 1991, surveys revealed that the numbers of *P. livingstonii* on the Comoros had declined to levels that caused concern for the future of the species (Carroll and Thorpe 1991; Reason and Trewhella 1994). Again, the Jersey Wildlife Preservation Trust mounted a series of capture expeditions and established a breeding colony in Jersey, where numbers are now increasing (Clark et al. 1997). However, since *P. livingstonii* is one of the largest fruit bats, it is difficult for zoos to provide an area large enough to allow free flight, which is essential for bats to retain the capacity for flight—a fundamental necessity for restocking should the opportunity arise.

Specialized facilities for the captive breeding and management of Old World fruit bats also exist at the Lubee Foundation, Inc. (Gainesville, Fla.). This is one of the world's premier breeding centers for Megachiropterans, including *P. rodricensis,* and it also maintains breeding colonies of several Asian and African pteropodids as models for threatened and endangered species. Institutions such as this provide important support in developing husbandry techniques for captive bat populations and in supporting captive-based research that contributes to conservation management plans.

Despite the outstanding success of the captive-breeding program for *P. rodricensis* and the involvement of many of the world's finest zoos, which espouse *in situ* and well as *ex situ* conservation, circumstances unfortunately appear to prohibit attempts to establish populations of *P. rodricensis* in the wild. However, this should provide an important lesson for future captive-breeding initiatives involving bats. Translocations and reintroductions should, of course, consider the genetic origin of stocks to be used and should take account of the IUCN position statement on the translocation of living organisms (introduction, reintroduction, and restocking).

Specific Management Guidance

A number of specific guidelines have been produced on the management of habitats for bats, including those aimed at forest management (Mayle 1990). For example, where old-growth forest is under management it is recommended that older, mature trees should be left when selective logging occurs

(Jung et al. 1999). Guidelines have also been produced that distill the findings of research relevant to habitat management in predominantly agricultural areas (e.g., Entwistle et al. 2001) and for individual species (e.g., *Rhinolophus ferrumequinum* [Mitchell-Jones 1995; Ransome 1996; Ransome and Hutson 1999]).

In the United States, recovery plans for other endangered species provide recommendations for management suitable for bats. The U.S. Forest Service has management plans for each of their forests, and there are also regional plans, such as that for forests within the range of the northern spotted owl (*Strix occidentalis caurina*), which has several mandates for bats.

The problem of managing bat-human interactions when bats share a residence with humans is also an area where research has helped to identify the possible courses of action available for resolution (e.g., exclusion), and the potential impacts such actions will have on bat species with different degrees of roost attachment (e.g., Brigham and Fenton 1986; Entwistle et al. 1997; Neilson and Fenton 1994). Where research has indicated that potential roost sites may be limited, or that roosts are about to be lost, the use of artificial roost sites has sometimes proved successful.

Site Protection and Protected Areas

While habitat protection has been identified as one of the key conservation priorities for bats, it is clear that development of site-based protection and, in some cases, the designation of protected areas may be an important step for the protection of key threatened bat species. A national park with a total area of 3,500 ha was designated on American Samoa in 1988 and provides a refuge for the endemic *Pteropus samoensis* and also *P. tonganus*, which were under pressure from hunters (Bats 1988). In Western Samoa, the largest remaining block of continuous lowland rainforest, the Tafua Peninsula, was established as a preserve under indigenous control also as a refuge for bats (Cox and Elmqvist 1991, 1993). Refuges for *P. tonganus* also exist on Tonga (Grant 1996). In the United States, the National Parks Service maintains Carlsbad Caverns as a roost and educational facility for bats and, together with the Forest Service, protects many other roosts on public land.

Of the 6,000 Sites of Special Scientific Interest designated in the United Kingdom by 1994, which are protected from damage by wildlife legislation, 80 in England and 24 in Wales relate specifically to bats or include bats in the wildlife interest of the site. Thus, all maternity roosts of *Rhinolophus ferrumequinum* and all such roosts of *R. hipposideros* that contain more than 100 adults, together with winter roosts of both species with more than 50 individuals, are designated as Sites of Special Scientific Interest (Racey 2000). The guidelines for the designation of such sites include all traditional breeding roosts of *Barbastella barbastellus*, *Myotis bechsteinii*, and *Plecotus austriacus*, together with mixed species hibernacula (Mitchell-Jones et al. 1993).

On the island of Pemba, traditional graveyards represent important roost areas and potential foraging habitat for the endangered *Pteropus voeltzkowi*. As a result of a program of research and education, local villagers have developed their own local bylaws to protect these sites, allowing bat populations to thrive (Entwistle and Corp 1997). Elsewhere in Africa, other traditional sites and sacred groves appear to support rarer species not found in the wider environment (Decher 1997a). Such areas are effectively protected by tradition and taboo rather than by secular law and may have high ecological value for small mammals (Decher 1997b).

In general, however, concern for bats has not resulted in the establishment of focused protected areas. Increasingly, surveys of the bat fauna of established protected areas support continued protection and improved management plans (see above, " Baseline Surveys of Distribution and Abundance" and see also Smith and Kerry [1996]). In the past, few protected areas incorporated conservation plans for bats within their management plans. In such cases, the management of parks for other species (such as elephants) may be detrimental to roosts and foraging requirements of bats (Fenton et al. 1997; Fenton and Rautenbach 1998). There is now a growing movement to consider bats as key components of the mammalian fauna of protected areas, and research is focused toward inventory, monitoring, and development of appropriate site safeguards.

The location of long-established protected areas may not take into consideration the distribution of chiropteran diversity, along with that of other smaller mammals (Williams et al. 2000). Gap analysis for the representation of threatened and rare bats within the existing protected areas is recommended, such as that conducted in South Africa (Gelderblom et al. 1995). Such analyses may help to ensure that good representation of threatened bat species is achieved within existing, and planned, protected areas. In addition, further research is needed to establish appropriate sizes of protected areas for bats, with respect to the distribution of key resources and seasonal movements of individuals.

Epilogue

Conservation biology is a new discipline and continues to emerge as an applied science with a developing theoretical framework. However, despite an active Society for Conservation Biology in the United States, the allocation of the National Science Foundation to conservation biology has not extended beyond a relatively small grants program. In the United Kingdom, the major source of funding for ecological research, the Natural Environmental Research Council, has yet to use the word "conservation" in a funding program.

Notwithstanding, during the past 30 yr, there has been a steady increase in the number of research papers on the conservation biology of bats appearing on the programs of scientific meetings, and the first international symposium on this subject was held in 1995 (Pierson and Racey 1998). The majority of references cited in the present contribution are from the past decade—in contrast to the situation in a recent review of life history of bats (Racey and Entwistle 2000)—demonstrating the growth of interest in the present topic.

In some areas that we have considered, such as the application of molecular techniques to the study of population structure, bat research has the potential to have an important impact on general ecological theory. Most areas we have considered have implications for policy and practice. We hope that this review will stimulate and direct further research that will ultimately lead to effective policy and practice to safeguard remaining bat populations.

Acknowledgments

We are very grateful to M. B. Fenton, A. M. Hutson, T. H. Kunz, S. P. Mickleburgh, C. F. J. O'Donnell, E. D. Pierson, S. B. Piertney, W. E. Rainey, W. Trewhella, M. D. Tuttle, D. E. Wilson, and G. J. Wiles for comments on earlier drafts of this chapter and for suggesting improvements. S. P. Mickleburgh kindly compiled table 15.1.

Literature Cited

Adams, R. A. 1996. Size-specific resource use in juvenile little brown bats, *Myotis lucifugus* (Chiroptera: Vespertilionidae): is there an ontogenetic shift? Canadian Journal of Zoology, 74:1204–1210.

Allison, V. C. 1937. Evening bat flight from Carlsbad Caverns. Journal of Mammalogy, 18:80–82.

Altenbach, J. S., K. N. Geluso, and D. E. Wilson. 1979. Population size of *Tadarida brasiliensis* at Carlsbad Caverns in 1973. Pp. 341–348 *in:* Biological Investigations in the Guadalupe Mountains National Park, Texas (H. H. Genoways and R. J. Baker, eds.). National Parks Service Proceedings and Transactions Series, no. 4. National Park Service, Washington, D.C.

Altringham, J. D. 1996. Bats, Biology and Behavior. Oxford University Press, Oxford.

Amengual, B., J. E. Whitby, A. A. King, J. Serra Cobo, and J. Bourhy. 1997. Evolution of European bat lyssaviruses. Journal of General Virology, 78:2319–2328.

Arita, H., and M. B. Fenton. 1997. Flight and echolocation in the ecology and evolution of bats. Trends in Ecology and Evolution, 12:53–58.

Arlettaz, R. 1999. Habitat selection as a major resource partitioning mechanism between the two sympatric sibling bat species *Myotis myotis* and *Myotis blythii*. Journal of Animal Ecology, 68:460–471.

Arlettaz, R., S. Godat, and H. Meyer. 2000. Competition for food by expanding pipistrelle bat populations (*Pipistrellus pipistrellus*) might contribute to the decline of lesser horseshoe bats (*Rhinolophus hipposideros*). Biological Conservation, 93:55–60.

Arlettaz, R., N. Perring, and J. Hausser. 1997. Trophic resource partitioning and competition between the two sibling bat species *Myotis myotis* and *Myotis blythii*. Journal of Animal Ecology, 66:897–911.

Avise, J. C. 1994. Molecular Markers, Natural History and Evolution. Chapman & Hall, London.

Balasingh, J., J. Koilraj, and T. H. Kunz. 1995. Tent construction by the short-nosed fruit bat *Cynopterus sphinx* (Chiroptera: Pteropodidae) in Southern India. Ethology, 100: 210–229.

Barclay, R. M. R. 1999. Bats are not birds—a cautionary note on using echolocation calls to identify bats: a comment. Journal of Mammalogy, 80:290–296.

Barclay, R. M. R., and R. M. Brigham, eds. 1996. Bats and Forests Symposium: October 19–21, 1995, Victoria, British Columbia, Canada. Working Paper 23/1996. British Columbia Ministry of Forests Research Program, Victoria.

Barlow, K. 2000. Expedition Field Techniques. Expedition Advisory Centre, London, 69 pp.

Bates, P. J. J., and D. L. Harrison. 1997. Bats of the Indian Subcontinent. Harrison Zoological Museum, Sevenoaks, Kent.

Bates, P. J. J., D. L. Harrison, N. M. Thomas, and M. Muni. 1994. The Indian fruit bat *Latidens salimalii* Thonglongya, 1972 (Chiroptera: Pteropodidae) rediscovered in southern India. Bonner Zoologisches Beitrage, 45:89–98.

Bates, P. J. J., D. K. Hendrichsen, J. L. Walston, and B. Hayes. 1999. A review of the mouse-eared bats (Chiroptera: Vespertilionidae: *Myotis*) from Vietnam with significant new records. Acta Chiropterologica, 1:47–74.

Bats. 1988. BCI helps Samoans gain National Park. Bats, 6:3–14.

Bayliss, J., and B. Hayes. 1999. The status and distribution of bats, primates and butterflies from the Makira Plateau, Madagascar. Unpublished report to Fauna and Flora International, 67 pp.

Bell, G. P., G. A. Bartholomew, and K. A. Nagy. 1986. The roles of energetics, water economy, foraging behavior and geothermal refugia in the distribution of the bat *Macrotus californicus*. Journal of Comparative Physiology B, 156:441–450.

Benzal, J. 1991. Population dynamics of the brown long-eared bat (*Plecotus auritus*) occupying bird boxes in a pine forest in central Spain. Netherlands Journal of Zoology, 41:241–249.

Betts, B. J. 1998. Roosts used by maternity colonies of silver-haired bats in north eastern Oregon. Journal of Mammalogy, 79:643–650.

Bhat, H. R. 1994. Observations on the food regimen and feeding behavior of *Cynopterus sphinx* Vahl (Chiroptera, Pteropodidae) at Pune, India. Mammalia, 58:363–370.

Bhat, H. R., and T. H. Kunz. 1995. Altered flower/fruit clusters of kitul palm used as roosts by the short-nosed fruit bat, *Cynopterous sphinx* (Chiroptera: Pteropodidae). Journal of Zoology (London), 235:596–604.

Bowen-Jones, E., D. Abrutat, B. Markham, and S. Bowe. 1997. Flying foxes in Choiseul (Solomon Islands)—the need for conservation action. Oryx, 31:209–217.

Boyd, I. L., and R. E. Stebbings. 1989. Population changes of brown long-eared bats (*Plecotus auritus*) in bat boxes in Thetford Forest. Journal of Applied Ecology, 26: 101–112.

Bradbury, J. W., and S. L. Vehrencamp. 1976. Social organization and foraging in emballonurid bats. I. Field Studies. Behavioral Ecology and Sociobiology, 1:337–381

Brass, D. A. 1994. Rabies in Bats: Natural History and Public Health Implications. Livia Press, Ridgefield, Conn.

Brigham, R. M., and R. M. R. Barclay. 1996. Bats and forests. Pp. xi–xiv *in*: Bats and

Forests Symposium: October 19–21, 1995, Victoria, British Columbia, Canada. Working Paper 23/1996. British Columbia Ministry of Forests Research Program, Victoria.

Brigham, R. M., and M. B. Fenton. 1986. The influence of roost closure on the roosting and foraging behavior of *Eptesicus fuscus* (Chiroptera: Vespertilionidae). Canadian Journal of Zoology, 64:1128–1133.

Brigham, R. M., S. D. Grindal, M. C. Firman, and J. L. Morissette. 1997a. The influence of structural clutter on activity patterns of insectivorous bats. Canadian Journal of Zoology, 75:131–136.

Brigham, R. M., M. J. Vonhof, R. M.R. Barclay, and J. C. Gwilliam 1997b. Roosting behavior and roost-site preference of forest-dwelling California bats (*Myotis californicus*). Journal of Mammalogy, 78:1231–1239.

Brooke, A. P., C. Solek, and A. Tualaulelei. 2000. Roosting behavior of colonial and solitary flying foxes in American Samoa (Chiroptera: Pteropodidae). Biotropica, 32:338–350.

Brosset, A. 1996. La Biologie des Chiroptères. Masson, Paris.

Brosset, A., P. Charles-Dominique, A. Cockle, J.-F. Cosson, and D. Masson. 1996. Bat communities and deforestation in French Guiana. Canadian Journal of Zoology, 74:1974–1982

Burland, T. M., E. M. Barratt, M. A. Beaumont, and P. A. Racey. 1999. Population genetic structure and gene flow in a gleaning bat, *Plecotus auritus*. Proceedings of the Royal Society of London B, 266:975–980.

Carroll, J. B. 1988. The conservation programme for the Rodrigues fruit bat *Pteropus rodricensis*. Pp. 457–475 *in:* Proceedings of the Fifth World Conference on Breeding Endangered Species in Captivity (B. L. Dresser, R. W. Reese, and E. J. Maruska, eds.). Cincinnati Zoo and Botanical Gardens, Cincinnati.

Carroll, J. B., and G. M. Mace. 1988. Population management of the Rodrigues fruit bat *Pteropus rodricensis* in captivity. International Zoo Yearbook, 27:70–8.

Carroll, J. B., and I. C. Thorpe. 1991. The conservation of Livingstone's fruit bat *Pteropus livingstonii*: a report on an expedition to the Comores in 1990. Dodo (Journal of the Jersey Wildlife Preservation Trust), 27:26–40.

Catto, C. M. C., A. M. Hutson, and P. A. Racey. 1994. The diet of *Eptesicus serotinus* in southern England. Folia Zoologica, 43:307–314.

Caughley, G. 1994. Directions in conservation biology. Journal of Animal Ecology, 63:215–244.

Caughley, G., and A. Gunn. 1996. Conservation Biology in Theory and Practice. Blackwell Science, Oxford.

Cheke, A. S., and J. F. Dahl. 1981. The status of bats on western Indian Ocean Islands with special reference to *Pteropus*. Mammalia, 45:205–238.

Churchill, S. K., and P. M. Helman. 1990. Distribution of the ghost bat, *Macroderma gigas* (Chiroptera: Megadermatidae) in Central and South Australia. Australian Mammalogy, 13:149–156.

Clark, B. S., D. M. Leslie, and T. S. Carter. 1993. Foraging activity of adult female Ozark big-eared bats (*Plecotus townsendii ingens*) in summer. Journal of Mammalogy, 74:422–427

Clark, D. R., Jr. 1981. Bats and Environmental Contaminants: A Review. Special Scientific Report—Wildlife, no. 235. U.S. Department of the Interior, Fish and Wildlife Service, Washington D.C., 27 pp.

Clark, D. R., Jr. 1988. Environmental contamination and management of bat populations in the United States. Pp. 409–413 *in:* Management of Amphibians, Reptiles and Small Mammals in North America: Proceedings of the Symposium, July 19–21, 1988, Flagstaff, Arizona (R. C. Szaro; K. E. Severson; D. R. Patton, eds.). General Technical Report RM-166. U.S. Forest Service, Fort Collins, Colo.

Clark, D. R., Jr., T. H. Kunz, and T. E. Kaiser. 1978b. Insecticides applied to a nursery colony of little brown bats (*Myotis lucifugus*): lethal concentrations in brain tissue. Journal of Mammalogy, 59:84–91.

Clark, D. R., Jr., R. K. La Val, and D. M. Swineford. 1978a. Dieldrin—induced mortality in an endangered species, the gray bat (*Myotis grisescens*). Science, 199:1357–1359.

Clark, K. M., J. B. Carroll, M. Clark, S. R. T. Garrett, S. Pinkus, and R. Saw. 1997. Capture and survey of Livingstone's fruit bats *Pteropus livingstonii* in the Comoros Islands: the 1995 expedition. Dodo (Journal of the Wildlife Preservation Trust), 33:20–35.

Clawson, R. L. 1987. Indiana bats: down for the count. Endangered Species Technical Bulletin, 9:9–11.

Cockle, A., D. Kock, L. Stublefield, K. Howell, and N. D. Burgess. 1998. Bat assemblages in Tanzanian coastal forests. Mammalia, 62:53–68

Cockrum, E. L. 1969. Migration in the guano bat *Tadarida brasiliensis.* Miscellaneous Publications, Museum of Natural History, University of Kansas, 51:303–306.

Collar, N. J., M. J. Crosby, and A. J. Stattersfield. 1994. Birds to Watch 2: The World List of Threatened Birds. Birdlife Conservation Series no. 4. Birdlife International, Cambridge.

Constantine, D. G. 1967. Activity patterns of the Mexican free-tailed bat. University of New Mexico, Publications in Biology, 7:1–79.

Convention on the Conservation of Migratory Species of Wild Animals, Bonn, 23 June 1979–22 June 1980. 1980. Her Majesty's Stationery Office, London, 20 pp.

Courts, S. E. 1998. Dietary studies of Old World Fruit Bats (Megachiroptera: Pteropodidae): how do they obtain sufficient protein? Mammal Review, 28:185–194.

Cox, P. A., and T. Elmqvist. 1991. Indigenous control of rainforest preserves. Ambio, 20:317–321.

Cox, P. A., and T. Elmqvist. 1993. Ecocolonialism and indigenous knowledge systems: village controlled rainforest preserves in Samoa. Pacific Conservation Biology, 1:6–13.

Craig, P., P. Trail, and T. E. Morrell. 1994. The decline of fruit bats in American Samoa due to hurricanes and overhunting. Biological Conservation, 69:261–266.

Crampton, L. H., and R. M. R. Barclay. 1996. Habitat selection by bats in fragmented and unfragmented Aspen mixedwood stands of different ages. Pp. 238–259 *in:* Bats and Forests Symposium, October 19–21, 1995, Victoria, British Columbia, Canada (R. M. R. Barclay and R. M. Brigham, eds.). Working Paper 23/1996. British Columbia Ministry of Forests Research Program, Victoria.

Crome, F. H. J., and G. C. Richards. 1988. Bats and gaps: microchiropteran community structure in a Queensland rain forest. Ecology, 69:1960–1969.

Daan, S. 1980. Long term changes in bat populations in the Netherlands: a summary. Lutra, 22:95–105.

Dalton, D. C., and V. M. Dalton. 1995. Mine closure methods including a recommended gate design. Pp. 130–135 *in:* Inactive Mines as Bat Habitat: Guidelines for Research, Survey and Monitoring, and Mine Management in Nevada (R. B. Riddle, ed.). Biological Resources Research Center, University of Nevada, Reno.

Daniel, M. J. 1990. Bats: Order Chiroptera. Pp. 114–137 *in:* The Handbook of New Zealand Mammals (C. M. King, ed.). Oxford University Press, Auckland.

Daniel, M. J., and G. R. Williams. 1984. A survey of the distribution, seasonal activity and roost sites of New Zealand bats. New Zealand Journal of Ecology, 7:9–25.

Decher, J. 1997a. Bat community patterns on the Accra Plains of Ghana, West Africa. Zeitschrift für Säugetierkunde, 62:129–142.

Decher, J. 1997b. Conservation, small mammals, and the future of sacred groves in West Africa. Biodiversity and Conservation, 6:1007–1026.

Decher, J., and J. R. Choate. 1995. Myotis grisescens. Mammalian Species, 510:1–7.

de Jong, J. 1994. Habitat use, home-range and activity pattern of the northern bat, *Eptesicus nilssoni,* in a hemiboreal coniferous forest. Mammalia, 58:535–548

Diamond, J. M. 1984. "Normal" extinctions of isolated populations. Pp. 191–245 *in:* Extinctions (M. H. Nitecki, ed.). University of Chicago Press, Chicago.

Dowling, B., J. H. Seeback, and K. W. Lowe. 1994. Cats and wildlife: results of a survey of wildlife admitted for care to shelters and animal welfare agencies in Victoria. Technical Report Series no. 134. Arthur Rylah Institute for Environmental Research, Department of Conservation and Natural Resources, Melbourne.

Downs, N. C. 2001. The use of landscape elements by bats in N.E. Scotland. Ph.D. Thesis, University of Aberdeen, Aberdeen

Duanghae, S. 1991. Search for Kitti's hog-nosed bat *Craseonycteris thonglongyai* in Western Thailand. Natural History Bulletin of the Siam Society, 39:1–17.

Duncan, A., G. B. Baker, and N. Montgomery. 1999. The Action Plan for Australian Bats. National Parks and Wildlife, Canberra.

Dunsmore, J. D., L. S. Hall, and K. H. Kottek. 1974. DDT in the bent-winged bat in Australia. Search (Carlton), 5:110–111.

Duvergé, P. L., and G. Jones. 1994. Greater horseshoe bats: activity, foraging behaviour and habitat use. British Wildlife 6:69–77.

Duvergé, P. L., and G. Jones. 2003. Use of farm land habitats by greater horseshoe bats (*Rhinolophus ferrum equinum*). *In:* Farming and Mammals (F. Tattersall and W. Manley, eds.). Occasional Publications of the Linnean Society (in press).

Eby, P. 1995. The biology and management of flying foxes in New South Wales. Species Management Report no. 18. National Parks and Wildlife Service of New South Wales, Sydney, 68 pp.

Ekman, M., and J. de Jong. 1996. Local patterns of distribution and resource utilization of four bat species (*Myotis brandti, Eptesicus nilssoni, Plecotus auritus* and *Pipistrellus pipistrellus*) in patchy and continuous environments. Journal of Zoology (London), 238:571–580.

Entwistle, A. C. 1994. Roost ecology of the brown long-eared bat (*Plecotus auritus,* Linnaeus 1758) in North-East Scotland. Ph.D. Thesis, University of Aberdeen, Aberdeen.

Entwistle, A. C., and N. Corp. 1997. Status and distribution of the Pemba flying fox *Pteropus voeltzkowi.* Oryx, 31:135–142

Entwistle, A. C., and N. Corp. 1998. The diet of *Pteropus voeltzkowi*, an endangered fruit bat endemic to Pemba Island, Tanzania. African Journal of Ecology, 35:351–360.

Entwistle, A. C., S. Harris, A. M. Hutson, P. A. Racey, A. L. Walsh, S. Gibson, I. Hepburn, J. Johnson, B. Larkin, and N. Bennett. 2001. Habitat Management for Bats. Joint Nature Conservation Committee, Peterborough.

Entwistle, A. C., P. A. Racey, and J. R. Speakman. 1996. Habitat exploitation by a gleaning bat, *Plecotus auritus*. Philosophical Transactions of the Royal Society of London B, 351:921–931.

Entwistle, A. C., P. A. Racey, and J. R. Speakman. 1997. Roost selection by the brown long-eared bat (*Plecotus auritus*). Journal of Applied Ecology, 34:399–408.

Entwistle, A. C., P. A. Racey, and J. R. Speakman. 2000. Social and population structure of a gleaning bat, *Plecotus auritus*. Journal of Zoology (London), 252:11–17.

Entwistle, A. C., and P. J. Stephenson. 2000. Small mammals and the conservation agenda. Pp. 119–140 *in*: Priorities for the Conservation of Mammalian Diversity: Has the Panda Had Its Day? (A. C. Entwistle and N. Dunstone, eds.). Cambridge University Press, Cambridge.

Estrada, A., R. Coates-Estrada, and D. Meritt. 1993. Bat species richness and abundance in tropical rainforest fragments and agricultural habitats at Los Tuxtlas, Mexico. Ecography, 16:309–318

Fenton, M. B. 1971. The structure of aerial-feeding bat faunas as indicated by ears and wing elements. Canadian Journal of Zoology, 50:287–296.

Fenton, M. B. 1982. Echolocation calls and patterns of hunting and habitat use of bats (Microchiroptera) from Chillagoe, North Queensland. Australian Journal of Zoology, 30:417–425.

Fenton, M. B. 1992. Bats. Facts on File, New York.

Fenton, M. B. 1994. Echolocation: its impact on the behaviour and ecology of bats. Ecoscience, 1:21–30.

Fenton, M. B. 1995. Constraint and flexibility; bats as predators, bats as prey. Symposia of the Zoological Society of London, no. 67:277–290.

Fenton, M. B. 1997. Science and the conservation of bats. Journal of Mammalogy, 78:1–14.

Fenton, M. B., L. Acharya, D. Audet, M. B. C. Hickey, C. B. Merriman, M. K. Obrist, D. M. Syme, and B. Adkins. 1992. Phyllostomid bats (Chiroptera: Phyllostomidae) as indicators of habitat disruption in the Neotropics. Biotropica, 24:440–446.

Fenton, M. B., H. M. Cumming, I. L. Rautenbach, G. S. Cumming, M. S. Cumming, G. Ford, R. D. Raylor, et al. 1997. Bats and the loss of tree canopy in African Woodlands. Conservation Biology, 12:399–407.

Fenton, M. B., H. G. Merriam, and G. L. Holroyd. 1983. Bats of Kootenay, Glacier and Mount Revelstroke National Parks in Canada: identification by echolocation calls, distribution and biology. Canadian Journal of Zoology, 61:2503–2508.

Fenton, M. B., and I. L. Rautenbach. 1998. Impacts of ignorance and human and elephant populations on the conservation of bats in African woodlands. Pp. 261–270 *in*: Bat Biology and Conservation (T. H. Kunz and P. A. Racey eds.). Smithsonian Institution Press, Washington, D.C.

Fenton, M. B., and D. W. Thomas. 1980. Dry-season overlap in activity patterns, habitat use and prey selection by sympatric African insectivorous bats. Biotropica, 12:81–90.

Findley, J. S. 1993. Bats—a Community Perspective. Cambridge University Press, Cambridge

Findley, E. G., J. S. Jones, and C. Jones. 1964. Seasonal distribution of the hoary bat. Journal of Mammalogy, 45:461–470.

Fitter, R., and P. Scott. 1978. The Penitent Butchers. Fauna Preservation Society, London, 48 pp.

Flannery, T. F. 1991. A new species of *Pteralopex* (Chiroptera: Pteropodidae) from montane Guadal canal, Solomon Islands. Records of the Australian Museum, 43: 123–130.

Flannery, T. F., and D. J. Colgan. 1993. A new species and two new subspecies of *Hipposideros* (Chiroptera) from Western Papua New Guinea. Records of the Australian Museum, 45:43–57.

Flannery, T. F., and L. Seri. 1993. Rediscovery of *Aproteles bulmerae* (Chiroptera: Pteropodidae): morphology, ecology and conservation. Mammalia, 57:19–25.

Fleming, T. H. 1988. The Short-Tailed Fruit Bat: A Study of Plant-Animal Interactions. University of Chicago Press, Chicago.

Foreign and Commonwealth Office. 1994. Agreement on the Conservation of Bats in Europe, London, 4 December 1991. Her Majesty's Stationery Office, London, 16 pp.

Francis, C. 1994. Vertical stratification of fruit bats (Pteropodidae) on lowland dipterocarp rainforests in Malaysia. Journal of Tropical Ecology, 6:421–431.

Francis, C. M. 1989. A comparison of mist nets and two types of harp traps for capturing bats. Journal of Mammalogy, 70:865–870.

Francis, C. M., A. Guillén, and M. F. Robinson. 1999. Order Chiroptera: Bats. Pp. 225–235 *in:* Wildlife in Lao PDR: 1999, Status Report (W. J. Duckworth, R. E. Salter and K. Khounboline, eds.). IUCN/WCS/Centre for Protected Areas and Watershed Management (CPAWM), Vientiane, Laos.

Fujita, M. S., and M. D. Tuttle. 1991. Flying foxes (Chiroptera: Pteropidiae): threatened animals of key ecological and economic importance. Conservation Biology, 5: 455–463.

Fullard, J. H., C. Koehler, A. Surlykke, and N. L. McKenzie. 1991. Echolocation ecology and flight morphology of insectivorous bats (Chiroptera) in south-western Australia. Australian Journal of Zoology, 39:427–438

Furlonger, C. L., H. J. Dewar, and M. B. Fenton. 1987. Habitat use by foraging insectivorous bats. Canadian Journal of Zoology, 65:284–288.

Gaisler, J. 1963. The ecology of lesser horseshoe bat (*Rhinolophus hipposideros hipposideros* Bechstein 1800) in Czechoslovakia. Pt. 2. Ecological demands, problems of synanthropy. Véstník Czechoslovenske Spolecnosti Zoologické, 22:322–327.

Gaisler, J., J. Zukal, Z. Rehak, and M. Homolka. 1998. Habitat preference and flight activity of bats in a city. Journal of Zoology (London), 244:439–445.

Galetti, M., and L. P. C. Morellato. 1994. Diet of the large fruit-eating bat *Artibeus literatus* in a forest fragment in Brazil. Mammalia, 58:661–665.

Gardner, A. L. 1977. Feeding habits. Pp. 293–350 *in:* Biology of Bats of the New World Family Phyllostomatidae (R. J. Baker, J. K. Jones, Jr., and D. C. Carter, eds.). Vol. 2. Special Publications, the Museum, Texas Tech University, 13. Texas Tech Press, Lubbock.

Gaston, K. J. 1994. Rarity. Chapman & Hall, London.

Geggie, J. F., and M. B. Fenton. 1985. A comparison of foraging by *Eptesicus fuscus*

(Chiroptera: Vespertilionidae) in urban and rural environments. Canadian Journal of Zoology, 63:263–266.

Gelderblom, C. M., G. N. Bronner, A. T. Lombard, and P. J. Taylor. 1995. Patterns of distribution and current protection status of the Carnivora, Chiroptera and Insectivora in South Africa. South African Journal of Zoology, 30:103–114.

Geluso, K. N., J. S. Altenbach, and D. E. Wilson. 1976. Bat Mortality: pesticide poisoning and migratory stress. Science, 194:184–186.

Geluso, K. N, J. S. Altenbach, and D. E. Wilson. 1978. The ontogeny of fat deposition in *Tadarida brasiliensis*. Pp.15–20 *in:* Proceedings of the Fourth International Bat Research Conference (R. J. Olembo, J. B. Castelino, and F. A. Mutere, eds.). Kenya National Academy for Advancement of Arts and Sciences, Nairobi.

Geluso, K. N, J. S. Altenbach, and D. E. Wilson. 1979. Population size of *Tadarida brasiliensis* at Carlsbad Caverns in 1973. Pp. 341–348 *in:* Biological Investigations in the Guadalupe Mountains National Park, Texas (H. H. Genoways and R. J. Baker, eds.). Proceedings and Transactions Series no. 4. U.S. National Park Service, Washington D.C.

Geluso, K. N, J. S. Altenbach, and D. E. Wilson. 1981. Organochlorine residues in young Mexican free-tailed bats from several roosts. American Midland Naturalist, 105:249–257.

Gerrell, R., and K. Lundberg. 1993. Decline of a bat, *Pipistrellus pipistrellus*, population in an industrialised area in southern Sweden. Biological Conservation, 65:153–157.

Gilpin, M. E., and M. E. Soulé. 1986. Minimum viable populations: processes of species extinction. Pp. 19–34 *in:* Conservation Biology: The Science of Scarcity and Diversity (M. E. Soulé, ed.). Sinauer Associates, Sunderland, Mass.

Goodman, S. M. 1996. Results of a bat survey of the eastern slopes of the Reserve Naturelle Integral d'Andringitra, Madagascar. Fieldiana (Zoology) (Special issue: A Flora and Faunal Inventory of the Eastern Slopes of the Reserve Naturelle Integral d'Andringitra, Madagascar with Reference to Elevational Zonation, ed. S. M. Goodman), 85:284–288.

Gopfert, M. C., and L. T. Wasserthal. 1995. Notes on echolocation calls, food and roosting behavior of the Old World sucker-footed bat *Myzopoda aurita* (Chiroptera, Myzopodidae). Zeitschrift für Säugetierkunde, 60:1–8.

Gorman, M. L., and A. Zubaid. 1993. A comparative study of the ecology of woodmice *Apodemus sylvaticus* in two contrasting habitats—deciduous woodland and maritime sand-dunes. Journal of Zoology (London), 229:385–396.

Grant, G. 1996. Kingdom of Tonga: safe haven for flying foxes. Bats, 14:16–17.

Griffin, D. R. 1970. Migration and homing of bats. Pp. 233–246 *in:* Biology of Bats (W. A. Wimsatt, ed.). Vol. 1. Academic Press, New York.

Grindal, S. D., and R. M. Brigham. 1998. Short-term effects of small-scale habitat disturbance on activity by insectivorous bats. Journal of Wildlife Management, 62: 996–1003.

Grindal, S. D., T. S. Collard, R. M. Brigham, and R. M. R. Barclay. 1992. The influence of precipitation on reproduction in *Myotis* bats in British Columbia. American Midland Naturalist, 128:339–344.

Grindal, S. D., J. L. Morisette, and R. M. Brigham. 1999. Concentration of bat activity in riparian habitats over an elevational gradient. Canadian Journal of Zoology, 77: 972–977.

Gumal, M. 2001. Tree walking to the flying fox in Sarawak. Wildlife Conservation, 104:46–51.

Halstead, L. B. 1977. Fruit bats—an example of wildlife management. Nigerian Field, 42:50–56.

Hamilton, W. J., III., and K. E. F. Watt. 1970. Refuging. Annual Review of Ecology and Systematics, 1:263–286.

Hamilton-Smith, E. 1974. The present knowledge of Australian Chiroptera. Australian Mammalogy, 1:95–108.

Hanski, I. 1998. Metapopulation dynamics. Nature, 395:41–49.

Harris, S., W. J. Cresswell, P. G. Forde, W. J. Trewhella, T. Woollard, and S. Wray. 1990. Home-range analysis using radio-tracking data—a review of problems and techniques particularly as applied to the study of mammals. Mammal Review, 20: 97–123.

Harris, S., P. Morris, S. Wray, and D. Yalden. 1995. A Review of British Mammals: Population Estimates and Conservation Status of British Mammals Other Than Cetaceans. Joint Nature Conservation Committee, Peterborough.

Harvey, M. J. 1980. Status of the endangered bats, *Myotis sodalis*, *M. grisescens* and *Plecotus townsendii ingens*, in the southern Ozarks. Pp. 221–223 *in:* Proceedings of the Fifth International Bat Research Conference (D. E. Wilson and A. L. Gardner, eds.). Texas Tech University Press, Lubbock.

Hayes, J. P., and M. D. Adam. 1996. The influence of logging riparian areas on habitat utilization by bats in Western Oregon. Pp. 228–237 *in:* Bats and Forests Symposium, October 19–21, 1995, Victoria, British Columbia, Canada (R. M. R. Barclay and R. M. Brigham, eds.). Working Paper 23/1996. British Columbia Ministry of Forests Research Program, Victoria.

Hayes, K. T., A. T. C. Feistner, and E. C. Halliwell. 1996. The effect of contraceptive implants on the behaviour of female Rodrigues fruit bats, *Pteropus rodricensis.* Zoo Biology, 15:21–36.

Hayssen, V., and T. H. Kunz. 1996. Allometry of litter mass in bats: comparisons with respect to maternal size, wing morphology, and phylogeny. Journal of Mammalogy, 77:476–490.

Heaney, L. R., and P. D. Heideman. 1987. Philippine fruit bats: endangered and extinct. Bats, 5:3–5.

Heideman, P. D., and L. R. Heaney. 1989. Population biology and estimates of abundance of fruit bats (Pteropodidae) in Philippine submontane rainforest. Journal of Zoology (London), 218:565–586.

Hendrichsen, D. K., P. J. J. Bates, B. D. Hayes, and J. L. Walston. 2001. Recent records of bat (Mammalia: Chiroptera) from Vietnam with six species new to the country. Myotis, 39:35–122.

Hensley, S., and C. Scott. 1995. Ozark Big-Eared Bat, *Plecotus, townsendii ingens* (Handley): Revised Recovery Plan. U.S. Fish and Wildlife Service, Region 2, Albuquerque, N.Mex., 79 pp.

Hernández, L. M., C. Ibáñez, M. A. Fernández, A. Guillén, Ma. J. González, and J. L. Pérez. 1993. Organochlorine insecticide and PCB residues in two bat species from four localities in Spain. Bulletin of Environmental Contamination and Toxicology, 50:871–877.

Herreid, C. F., II. 1967. Temperature regulation, temperature preferences and tolerance, and metabolism of young and adult free-tailed bats. Physiological Zoology, 40:1–22.

Hill, J. E., and P. Morris. 1971. Bats from Ethiopia collected by the Great Abbai Expedition, 1968. Bulletin of the British Museum (Natural History) Zoology, 1:25–49.

Horner, M. A., T. H. Fleming, and C. T. Sahley. 1998. Foraging behaviour and energetics of a nectar feeding bat, *Leponycteris curasoae* (Chiroptera: Phyllostomidae). Journal of Zoology (London), 244:575–586.

Hovorka, M. D. 1996. The impacts of livestock grazing on Sonora desert sand vegetation, insect and insectivorous bat communities. M.S. Thesis, York University, North York, Ontario, 84 pp.

Humes, M. L., J. P. Hayes, and M. W. Collopy. 1999. Bat activity in thinned, unthinned and old-growth forests in western Oregon. Journal of Wildlife Management, 63: 553–561.

Humphrey, S. R. 1975. Nursery roosts and community diversity of Nearctic bats. Journal of Mammalogy, 56:321–346.

Hutson, A. M. 1993. Action Plan for the Conservation of Bats in the United Kingdom. Bat Conservation Trust, London, 49 pp.

Hutson, A. M., S. P. Mickleburgh, and P. A. Racey, eds. 2001. Global Action Plan for Microchiropteran Bats. IUCN, Gland, Switzerland.

Isaac, S., G. Marimuthu, and M. K. Chandrashekaran. 1994. Fecundity in the Indian pygmy bat (*Pipistrellus mimus*). Journal of Zoology (London), 234:665–668.

IUCN. 1994. IUCN Red List Categories. Prepared by IUCN Species Survival Commission. IUCN, Gland, Switzerland, 21 pp.

IUCN. 1996. 1996 IUCN Red List of Threatened Animals. IUCN, Gland, Switzerland.

Jefferies, D. J. 1972. Organochlorine insecticide residues in British bats and their significance. Journal of Zoology (London), 166:245–263.

Jenkins, E. V., T. Laine, S. E. Morgan, K. R. Cole, and J. R. Speakman. 1998. Roost selection in the pipistrelle bat *Pipistrellus pipistrellus* (Chiroptera: Vespertilionidae) in north-east Scotland. Animal Behaviour, 56:909–917.

Jones, G., P. D. Duvergé, and R. D. Ransome. 1995. Conservation biology of an endangered species: field studies of greater horseshoe bats. Symposia of the Zoological Society of London, no. 67:309–324.

Jones, K. E., J. D. Altringham, and R. Deaton. 1996. Distribution and population densities of seven species of bat in northern England. Journal of Zoology (London), 240:777–780.

Judes, U. 1987. Zum problem der "Tollwut" bei Fledermäusen. Myotis, 251:41–64.

Jung, T. S., I. D. Thompson, R. D. Titman, and A. P. Applejohn. 1999. Habitat selection by forest bats in relation to mixed-wood stand types and structure in central Ontario. Journal of Wildlife Management, 63:1306–1319.

Kalcounis, M. C., and R. M. Brigham. 1995. Intraspecific variation in wing loading affects habitat use by little brown bats (*Myotis lucifugus*). Canadian Journal of Zoology, 73:89–95.

Kalcounis, M. C., and R. M. Brigham. 1998. Secondary use of aspen cavities by tree-roosting big brown bats. Journal of Wildlife Management, 62:603–611.

Kalcounis, M. C., K. A. Hobson, R. M. Brigham, and K. R. Hecker. 1999. Bat activity in the boreal forest: importance of stand type and vertical strata. Journal of Mammalogy, 80:673–682.

Kalko, E. K. V. 1998. Organization and diversity of tropical bat communities through space and time. Zoology, 101:281–297.

Kalko, E. K. V., D. Friemel, C. O. Handley, Jr., and H.-U. Schnitzler. 1999. Roosting and foraging behavior of two Neotropical gleaning bats, *Tonatia silvicola* and *Trachops cirrhosus* (Phyllostomidae). Biotropica, 31:44–353.

Kalko, E. K. V., C. O. Handley, Jr., and D. Handley. 1996. Organization, diversity and long-term dynamics of a Neotropical bat community. Pp. 503–553 *in:* Long-Term Studies in Vertebrate Communities (M. Cody and J. Smallwood, eds.). Academic Press, Los Angeles.

Keeley, B. 1999. Bat houses in bridges. Bat House Researcher, 7:4–6.

Kerth, K., and B. Koenig. 1999. Fission, fusion and non random associations in female Bechstein's bat (*Myotis bechsteini*). Behaviour, 136:1187–1202.

Kerth, G., F. Mayer, and B. Koenig. 2000. MtDNA reveals that female Bechstein's bats live in closed societies. Molecular Ecology, 9:793–800.

Kingston, T., G. Jones, A. Zubaid, and T. H. Kunz. 1999. Echolocation signal design in Kerivoulinae and Murininae (Chiroptera: Vespertilionidae) from Malaysia. Journal of Zoology (London), 249:359–374.

Kingston, T., G. Jones, A. Zubaid, and T. H. Kunz. 2000. Resource partitioning in rhinolophoid bats revisited. Oecologia, 124:332–242.

Kiser, M., and S. Kiser. 1999. Bat houses and exclusion in British Columbia. Bat House Researcher, 7:3–4.

Kiser, M., and S. Kiser. 2000. 1999 season results. Bat House Researcher, 8:1–3.

Kitchener, D. J., and Maharadatunkamsi. 1991. Description of a new species of *Cynopterus* (Chiroptera: Pteropodidae) from Nusa Tenggara, Indonesia. Records of the Western Australian Museum, 15:307–363.

Kokurewicz, T. 1995. Increased population of Daubenton's bat (*Myotis daubentonii* [Kuhl, 1819]) (Chiroptera: Vespertilionidae) in Poland. Myotis, 32–33:155–161.

Korine, C., I. Izhaki, and Z. Arad. 1999. Is the Egyptian fruit bat *Rousettus aegyptiacus* a pest in Israel? an analysis of the bat's diet and implications for its conservation. Biological Conservation, 88:301–306.

Kronwitter, F. 1988. Population structure, habitat use and activity patterns of the noctule bat, *Nyctalus noctula* Schreb., 1774 (Chiroptera: Vespertilionidae) revealed by radio tracking. Myotis, 26:23–85

Krusic, R. A., M. Yamasaki, C. D. Neefus, and P. J. Pekins. 1996. Bat habitat use in White Mountain National Forest. Journal of Wildlife Management, 60:625–631.

Kunz, T. H. 1982. Roosting ecology of bats. Pp. 1–56 *in:* Ecology of Bats (T. H. Kunz, ed.). Plenum Press, New York.

Kunz, T. H. 1996. Obligate and opportunistic interactions of Old World tropical fruit bats and plants. Pp. 38–65 *in:* Conservation and Faunal Biodiversity in Malaysia (Z. A. A. Hasan and Z. Akbar, eds.). Penerbit Universiti Kebangsaan Malaysia, Bangi.

Kunz, T. H., and E. L. P. Anthony. 1996. Variation in nightly emergence behavior in the little brown bat, *Myotis lucifugus* (Chiroptera: vespertilionidae). Pp. 225–236 *in:* Contributions in Mammalogy: A Memorial Volume Honoring J. Knox Jones, Jr. (H. H. Genoways and R. J. Baker, eds.). Texas Tech University Press, Lubbock.

Kunz, T. H., E. L. P. Anthony, and W. T . Rumage. 1977. Mortality of little brown bats following multiple pesticide applications. Journal of Wildlife Management, 41: 476–483.

Kunz, T. H., M. S. Fujita, A. P. Brooke, and G. F. McCracken. 1994. Convergence in tent architecture and tent-making behavior among Neotropical and Palaeotropical bats. Journal of Mammalian Evolution, 2:57–78.

Kunz, T. H., and W. R. Hood. 2000. Parental care and postnatal growth in the Chiroptera. Pp. 415–468 *in:* Reproductive Biology of Bats (E. G. Crichton and P. H. Krutzsch, eds.). Academic Press, San Diego, Calif.

Kunz, T. H., and A. Kurta. 1988. Capture methods and holding devices. Pp. 1–29 *in:* Ecological and Behavioral Methods for the Study of Bats (T. H. Kunz, ed.). Smithsonian Institution Press, Washington, D.C.

Kunz, T. H., and P. A. Racey, eds. 1998. Bat Biology and Conservation. Smithsonian Institution Press, Washington, D.C.

Kunz, T. H., and D. S. Reynolds. In press. Bat colonies in buildings. *In:* Monitoring Trends in Bat Populations of the U.S. and Territories: Problems and Prospects. (T. J. O'Shea and M. A. Bogan, eds.). Information and Technology Report. U.S. Geological Survey, Biological Resources Division, Washington, D.C.

Kunz, T. H., and A. A. Stern. 1995. Maternal investment and postnatal growth in bats. Symposia of the Zoological Society of London, no. 67:63–77.

Kunz, T. H., D. W. Thomas, G. C. Richards, C. D. Tidemann, E. D. Pierson, and P. A. Racey. 1996. Observational techniques for bats. Pp. 105–114 *in:* Measuring and Monitoring Biological Diversity: Standard Methods for Mammals (D. E. Wilson, F. R. Cole, J. D. Nichols, R. Rudran, and M. S. Foster, eds.). Smithsonian Institution Press, Washington, D.C.

Kurta, A. 1985. External insulation available to a non-nesting mammal, the little brown bat (*Myotis lucifugus*). Comparative Biochemistry and Physiology A, 82:413–420.

Kurta, A., and J. A. Teramino. 1992. Bat community structure in an urban park. Ecography, 15:257–261.

Laurance, W. F. 1991. Ecological correlates of extinction proneness in Australian tropical rain forest mammals. Conservation Biology, 5:79–89.

Laurance, W. F. 1994. Rainforest fragmentation and the structure of small mammal communities in tropical Queensland. Biological Conservation, 69:23–32.

Lewis, S. E. 1993. Effect of climatic variation on reproduction by pallid bats (*Antrozous pallidus*). Canadian Journal of Zoology, 71:1429–1433.

Lewis, S. E. 1995. Roost fidelity of bats: a review. Journal of Mammalogy, 76:481–496.

Lewis, S. E. 1996. Low roost-site fidelity in pallid bats: associated factors and effect on group stability. Behavioral Ecology and Sociobiology, 39:335–344.

Limpens, H. J. G. A., W. Helmer, A. van Winden, and K. Mostert. 1989. Vleermuizen (Chiroptera) en lint-ormige landschapselementen. Lutra, 32:2–20.

Limpens, H. J. G. A., and K. Kapteyn. 1991. Bats, their behavior and linear landscape elements. Myotis, 29:39–48.

Limpens, H. J. G. A., P. H. C. Lina, and A. M. Hutson. 1999. Draft Action Plan for Conservation of the Pond Bat (*Myotis dascycneme*) in Europe. T-PVS (99) 12. Council of Europe, Strasbourg, 57 pp.

Mace, G. M., and R. Lande. 1991. Assessing extinction threats: toward a re-evaluation of IUCN threatened species categories. Conservation Biology, 5:148–157.

Marshall, A. G. 1985. Old World phytophagous bats (Megachiroptera) and their food plants: a survey. Zoological Journal of the Linnean Society, 83:351–369.

Mayle, B. A. 1990. A biological basis for bat conservation in British woodlands—a review. Mammal Review, 20:159–195

McAney, C. M., and J. S. Fairley. 1989. Observations at summer roosts of the lesser horseshoe bat in County Clare. Irish Naturalists Journal, 23:1–6.

McCracken, G. F. 1988. Who's endangered and what can we do? Bats, 6:5–9.

McCracken, G. F. 1992. Bats in magic, potions and medicinal preparations. Bats, 10: 14–15.

McCracken, G. F. In press. Estimates of population sizes in summer colonies of Brazilian free-tailed bats (*Tadarida brasiliensis*). *In:* Monitoring Trends in Bat Populations of the U.S. and Territories: Problems and Prospects. (T. J. O'Shea and M. A. Bogan, eds.). Information and Technology Report. U.S. Geological Survey, Biological Resources Division, Washington, D.C.

McCracken, G. F., and J. W. Bradbury. 1981. Social organization and kinship in the polygynous bat, *Phyllostomus hastatus*. Behavioral Ecology and Sociobiology, 8: 11–34.

McCracken, G. F., and M. F. Gassel. 1997. Genetic structure in migratory and nonmigratory populations of Brazilian free-tailed bats. Journal of Mammalogy, 78: 348–357.

McCracken, G. F., J. P. Hayes, J. Cevallos, S. Z. Guffey, and F. Carlos Romero. 1997. Observations on the distribution, ecology and behaviour of bats in the Galalpagos Islands. Journal of Zoology (London), 243:757–770.

McCracken, G. F., and G. S. Wilkinson. 2000. Bat mating systems. Pp. 321–362 *in:* Reproductive Biology of Bats (E. G. Crichton and P. H. Krutzsch, eds.). Academic Press, San Diego, Calif.

McDonald, J. T., I. L. Rautenbach, and J. A. J. Nel. 1990. Foraging ecology of bats observed at De Hoop Provincial Nature Reserve, southern Cape Province. South African Journal of Wildlife Research, 20:133–145

McWilliam, A. N. 1994. Nocturnal animals. Pp. 103–133 *in:* DDT in the Tropics: The Impact on Wildlife in Zimbabwe of Ground-Spraying for Tsetse-Fly Control (J. R. Douthwaite and C. C. D. Tingle, eds.). Natural Resources Institute, Chatham.

Medellin, R. A., and O. Gaona. 2000. Seed dispersal by bats and birds in forest and disturbed habitats of Chiapas, Mexico. Biotropica, 31:478–485.

Medway, Lord, and A. G. Marshall. 1970. Roost site selection among flat-headed bats (*Tylonycteris* spp.). Journal of Zoology (London), 161:237–245.

Medway, Lord, and A. G. Marshall. 1975. Terrestrial vertebrates of the New Hebrides: origin and distribution. Philosophical Transactions of the Royal Society of London B, 272:423–456.

Menzel, M. A., T. C. Carter, B. R. Chapman, and J. Larem. 1998. Quantitative comparison of tree roosts used by red bats (*Lasiurus borealis*) and Seminole bats (*Lasiurus seminolus*). Canadian Journal of Zoology, 76:630–634.

Mickleburgh, S. P. 1987. Distribution and status of bats in the London area. London Naturalist, 66:41–91.

Mickleburgh, S. P., A. M. Hutson, and P. A. Racey, eds. 1992. Old World Fruit Bats—an Action Plan for Their Conservation. IUCN, Gland, Switzerland.

Mickleburgh, S. P., A. M. Hutson, and P. A. Racey. 2002. A review of the global conservation status of bats. Oryx, 36:18–34.

Mills, L. S. 1995. Edge effects and isolation: red-backed voles on forest remnants. Conservation Biology, 9:395–403.

Ministère de l'Environment. 1996. Plan d'action pour la conservation de la biodiversité cas de Chiroptères. Ministère de l'Environment, Sous-Direction de la chasse, de la faune et de la flore sauvage, 30 pp.

Mitchell-Jones, A. J. 1995. The status and conservation of horseshoe bats in Britain. Myotis, 32–33:271–284.

Mitchell-Jones, A. J., A. M. Hutson, and P. A. Racey. 1993. The growth and development of bat conservation in Britain. Mammal Review, 23:139–148.

Moeschler, P. 1991. Concept national pour la protection et l'étude des chauves-souris. Publication Speciale du Rhinolophie no. 1. [Geneve : Muséum d'histoire naturelle], 102 pp.

Mohr, C. E. 1953. Possible causes of an apparent decline in wintering populations of cave bats. National Speleological Society News (November), 4–5.

Molloy, J. 1995. Bat (peka peka) Recovery Plan (Mystacina, Chalinolobus). Threatened Species Recovery Plan Series no. 15. Department of Conservation, Wellington.

Neilson, A. L., and M. B. Fenton. 1994. Responses of little brown Myotis to exclusion and to bat houses. Wildlife Society Bulletin, 22:8–14.

Neu, C. W., C. R. Byers, and J. M. Peek. 1974. A technique for analysis of utilization-availability data. Journal of Wildlife Management, 38:541–545.

Neuweiler, G. 1984. Foraging, echolocation and audition in bats. Naturwissenschaften, 71:446–455.

Newmark, W. D. 1995. Extinction of mammal populations in western North American national parks. Conservation Biology, 9:512–526

Nicoll, M. E., and J. M. Suttie. 1982. The sheath-tailed bat, Coleura seychellensis (Chiroptera: Emballonuridae) in the Seychelles Islands. Journal of Zoology (London), 197:421–426.

Nieuwenhuijs, J., J. Haagsma, and P. Lina. 1992. Epidemiology and control of rabies in bats in the Netherlands. Revue Scientifique et Technique de l'Office of International des Epizootics, 11:1155–1166.

Norberg, U. M., and J. M. V. Rayner. 1987. Ecological morphology and flight in bats (Mammalia: Chiroptera): wing adaptations, flight performance, foraging strategy and echolocation. Philosophical Transactions of the Royal Society B, 316:335–427.

Nupp, T. E., and R. K. Swihart. 2000. Landscape-level correlates of small-mammal assemblages in forest fragments of farmland. Journal of Mammalogy, 81:512–526.

Ochoa, J. 2000. Effectos de la extraccíon de maderas sobre la diversidad de mammiferos pequeños en bosques de tierras Bajas de la Gguayana Venezolana. Biotropica, 32:146–164.

O'Donnell, C. F. J. 2000. Conservation status and causes of decline of the threatened New Zealand long-tailed bat Chalinolobus tuberculatus (Chiroptera: Vespertilionidae). Mammal Review, 30:89–106.

O'Donnell, C. F. J. 2001. Home range and use of space by Chalinolobus tuberculatus, a temperate rainforest bat from New Zealand. Journal of Zoology (London), 253:253–264.

O'Donnell, C. F. J, J. Christie, C. Corben, J. A. Sedgeley, and W. Simpson. 1999. Rediscovery of short-tailed bats (Mystacina sp.) in Fiordland, New Zealand: preliminary observations of taxonomy, echolocation calls, population size, home range and habitat use. New Zealand Journal of Ecology, 23:21–30.

O'Donnell, C. F. J., and J. A. Sedgeley. 1994. An automatic monitoring system for recording bat activity. Department of Conservation Technical Series no. 5. Department of Conservation, Wellington, 16 pp.

O'Donnell, C. F. J., and J. A. Sedgeley. 1999. Use of roosts by the long-tailed bat, Chalinolobus tuberculatus, in temperate rainforest in New Zealand. Jounral of Mammalogy, 80:913–923.

O'Farrell, M. J., and M. L. Gannon. 1999. A comparison of acoustic versus capture techniques for the inventory of bats. Journal of Mammalogy, 80:24–30.

O'Farrell, M. J., and B. W. Miller. 1999. Use of vocal signatures for the inventory of free-flying Neotropical bats. Biotropica, 31:507–516.

Ohlendorf, B., ed. 1997. Zur Situation der Hufeisennasen in Europa. Arbeitskreis Fledermause Sachsen-Anhalt e.V. IFA Verlag, Berlin.

Ormsbee, P. C., and W. C. McComb. 1998. Selection of day roosts by female long-legged myotis in the central Oregon Cascade Range. Journal of Wildlife Management, 62:596–603.

O'Shea, T. J., and M. A. Bogan, eds. In press. Monitoring Trends in Bat Populations of the U.S. and Territories: Problems and Prospects. Information and Technology Report. U.S. Geological Survey, Biological Resources Division, Washington, D.C.

O'Shea, T. J., and T. A. Vaughan. 1999. Population changes in bats from Central Arizona: 1971 and 1997. Southwestern Naturalist, 44:495–500.

Parker, D. I., J. A. Cook, and S. W. Lewis. 1996. Effects of timber harvest on bat activity in Southeastern Alaska's temperate rainforests. Pp. 277–292 in: Bats and Forests Symposium, October 19–21, 1995, Victoria, British Columbia, Canada (R. M. R. Barclay and R. M. Brigham, eds.). Working Paper 23/1996. British Columbia Ministry of Forests Research Program, Victoria.

Patterson, B. D., V. Pacheco, and S. Solari. 1996. Distributions of bats along an elevational gradient in the Andes of South-eastern Peru. Journal of Zoology (London), 240:637–658.

Petit, E., and F. Mayer. 1999. Male dispersal in the noctule bat (Nyctalus noctula): where are the limits? Proceedings of the Royal Society of London B, 266:1717–1722.

Petit, E., and F. Mayer. 2000. A population genetic analysis of migration: the case of the noctule bat (Nyctalus noctula). Molecular Ecology, 9:683–690.

Petri, B., S. Pääbo, A. von Haeseler, and D. Tautz. 1997. Paternity assessment and population subdivision in a natural population of the larger mouse-eared bat Myotis myotis. Molecular Ecology, 6:235–242.

Pierson, E. D. 1984. Can Australia's flying foxes survive? Bats, 1:1–4.

Pierson, E. D. 1998. Tall trees, deep holes and scarred landscapes: conservation biology of North American bats. Pp. 309–325 in: Bat Biology and Conservation (T. H. Kunz and P. A. Racey, eds.). Smithsonian Institution Press, Washington, D.C.

Pierson, E. D., and P. A. Racey. 1998. Conservation biology. Pp. 247–248 in: Bat Biology and Conservation (T. H. Kunz and P. A. Racey, eds.). Smithsonian Institution Press, Washington, D.C.

Pierson, E. D., and W. E. Rainey. 1992. The biology of flying foxes of the genus Pteropus: a review. Pp. 1–17 in: Pacific Island Flying Foxes: Proceedings of an International Conservation Conference (D. E. Wilson and G. L. Graham, eds.). Biological Report 90(23). U.S. Fish and Wildlife Services, Washington, D.C.

Pierson, E. D., and W. E. Rainey. 1998. Distribution of the spotted bat, Euderma maculatum, in California. Journal of Mammalogy, 79:1296–1305.

Pierson, E. D., T. Elmqvist, W. E. Rainey, and P. A. Cox. 1996. Effects of tropical cyclonic storms on flying fox populations on the South Pacific Islands of Samoa. Conservation Biology, 10:438–451.

Pimm, S. L., H. L. Jones and J. Diamond. 1988. On the risk of extinction. American Naturalist, 132:757–785

Pont, S. M., and J. D. Armstrong. 1990. A study of the bat fauna of the reserve Naturelle Integral de Marojejy in north-east Madagascar. Report of the Aberdeen University

Expedition to Madagascar 1989. Department of Geography, University of Aberdeen, Aberdeen, 57 pp.

Primack, R. B. 1998. Essentials of Conservation Biology. Sinauer Associates, Sunderland, Mass.

Rabe, M. J., T. E. Morrell, H. Green, J. C. deVos, and C. R. Miller. 1998. Characteristics of ponderosa pine snag roosts used by reproductive bats in northern Arizona. Journal of Wildlife Management, 62:612–621

Rabinowitz, D. 1981. Seven forms of rarity. Pp. 205–215 in: The Biological Aspects of Rare Plant Conservation (H. Synge, ed.). Wiley, New York.

Rabinowitz, A., and M. D. Tuttle. 1980. Status of summer colonies of the endangered gray bat in Kentucky. Journal of Wildlife Management, 44:955–960.

Racey, P. A. 1979. Two bats in the Seychelles. Oryx, 15:148–152.

Racey, P. A. 1982. Ecology of bat reproduction. Pp. 57–104 in: Ecology of Bats (T. H. Kunz, ed.). Plenum Press, New York.

Racey, P. A. 1998a. Ecology of European bats in relation to their conservation. Pp. 249–259 in: Bat Biology and Conservation (T. H. Kunz and P. A. Racey, eds.). Smithsonian Institution Press, Washington, D.C.

Racey, P. A. 1998b. The importance of the riparian environment as a habitat for European bats. Symposia of the Zoological Society of London, no. 71:69–91.

Racey, P. A. 2000. Does legislation conserve and does research drive policy—the case of bats in the UK? Pp. 159–175 in: Priorities for the Conservation of Mammalian Diversity: Has the Panda Had Its Day? (A. C. Entwistle and N. Dunstone, eds.). Cambridge University Press, Cambridge.

Racey, P. A., and A. E. Entwistle. 2000. Life history and reproductive strategies of bats. Pp. 363–414 in: Reproductive Biology of Bats (E. G. Crighton and P. H. Krutzsch, eds.). Academic Press, New York.

Racey, P. A., and M. E. Nicoll. 1984. Mammals. Pp. 607–626 in: Biogeography and Ecology of the Seychelles Islands (D. R. Stoddart, ed.). Dr. W. Junk, The Hague.

Racey, P. A., and R. E. Stebbings. 1972. The status of bats in Britain: a report commissioned by the Fauna Preservation Society from the Mammal Society. Oryx, 11:319–327.

Racey, P. A., and S. M. Swift. 1985. Feeding ecology of Pipistrellus pipistrellus (Chiroptera: vespertilionidae) during pregnancy and lactation. I. Foraging behaviour. Journal of Animal Ecology, 54:205–215.

Racey, P. A., and S. M. Swift. 1986. Residual effects of remedial timber treatments on bats. Biological Conservation, 35:205–14.

Racey, P. A., S. M. Swift, J. Rydell, and L. Brodie. 1998. Bats and insects over two Scottish rivers with contrasting nitrate status. Animal Conservation, 1:195–202.

Rainey, W. E. 1995. Tools for low-disturbance monitoring of bat-activity. Pp. 62–71 in: Inactive Mines as Bat Habitat: Guidelines for Research, Survey Monitoring and Mine Management in Nevada (B. R. Riddle, ed.). Biological Research Center, University of Nevada, Reno.

Rainey, W. E. 1998. Conservation of bats on remote Indo-Pacific Islands. Pp. 326–341 in: Bat Biology and Conservation (T. H. Kunz and P. A. Racey, eds.). Smithsonian Institution Press, Washington, D.C.

Rainey, W. E., and E. D. Pierson. 1992. Distribution of Pacific Island flying foxes: implications for conservation. Pp. 111–122 in: Pacific Island Flying Foxes: Proceedings of an International Conservation Conference (D. E. Wilson and G. L. Graham, eds.). Biological Report 90 (23). U.S. Fish and Wildlife Service, Washington, D.C.

Ransome, R. D. 1995. Earlier breeding shortens life in female greater horseshoe bats. Philosophical Transactions of the Royal Society B, 350:153–161.

Ransome, R. 1996. The management of feeding areas for greater horseshoe bats. English Nature Research Reports no. 174. English Nature, Peterborough, 74 pp.

Ransome, R., and A. M. Hutson. 1999. Revised action plan for conservation of the greater horseshoe bat (Rhinolophus ferrumequinum) in Europe. T-PVS (99) 11 (rev.). Council of Europe, Strasbourg, 48 pp.

Ratcliffe, F. N. 1931. The flying fox (Pteropus) in Australia. Council of Scientific and Industrial Research Bulletin, 53:1–81.

Ratcliffe, F. N. 1932. Notes on the fruit bats (Pteropus spp.) of Australia. Journal of Animal Ecology, 1:32–57.

Rautenbach, I. L., M. B. Fenton, and M. J. Whiting. 1996. Bats in riverine forests and woodlands: a latitudinal transect in southern Africa. Canadian Journal of Zoology, 74:312–322.

Reason, P. F., and W. J. Trewhella. 1994. The status of Pteropus livingstonii (Gray 1866) in the Comores. Oryx, 28:107–114.

Richards, G. C., and L. S. Hall. 1998. Conservation biology of Australian bats: are recent advances solving our problems? Pp. 271–281 in: Bat Biology and Conservation (T. H. Kunz and P. A. Racey, eds.). Smithsonian Institution Press, Washington, D.C.

Richardson, P. 1985. Bats. Whittet, London.

Robinson, M. F., A. L. Smith, and S. Bumrungsri. 1995. Small mammals of Thung Yai Naresuan and Huai Kha Khaeng Wildlife Sanctuaries in Western Thailand. Natural History Bulletin of the Siam Society, 43:27–54.

Robinson, M. F., and R. E. Stebbings. 1997. Home range and habitat use by the serotine bat, Eptesicus serotinus, in England. Journal of Zoology (London), 243:117–136.

Roer, H., and H. Egsbaek. 1966. Zur biologie einer Skandinavischen population der wasserfledermause (Myotis daubentoni). Zeitschrift für Säugetierkunde, 31:440–453.

Russ, J., and D. Bennett. 1999. The bats of the Masoala Peninsular, K. Madagascar and the use of time expansion ultrasound detectors in surveying microchiropteran communities. Final Report of Queens University Belfast Masoala Bat Project, 1999. Viper Press, Glossop, England.

Russ, J. M., A. M. Hutson, W. I. Montgomery, P. A. Racey, and J. R. Speakman. 2001. The status of Nathusius' pipistrelle, Pipistrellus nathusii (Keyserling and Blasius 1839) in the British Isles. Journal of Zoology (London), 254:91–100.

Rydell, J., and P. A. Racey. 1995. Street lamps and the feeding ecology of insectivorous bats. Symposia of the Zoological Society of London, no. 67:291–307.

Rydell, J., and Yalden, D. 1997. The diets of two high-flying bats from Africa. Journal of Zoology (London), 242:69–76.

Sample, B. E., and R. C. Whitmore. 1993. Food habits of the endangered Virginia big-eared bat in West Virginia. Journal of Mammalogy, 74:428–435.

Sazima, I., W. A. Fischer, M. Sazima, and E. A. Fischer. 1994. The fruit bat Artibeus lituratus as a forest and city dweller. Ciência e Cultura, 46:164–168.

Scheel, D., T. L. S. Vincent, and G. N. Cameron. 1996. Global warming and the species richness of bats in Texas. Conservation Biology, 10:452–464.

Schmitt, I. H., D. J. Kitchener, and R. A. How. 1995. A genetic perspective of mammalian variation and evolution in the Indonesian archipelago: biogeographic correlates in the fruit bat genus *Cynopterus*. Evolution, 49:399–412.

Schnitzler, H.-U., and E. K. V. Kalko. 1998. How echolocating bats search and find food. Pp. 183–196 *in:* Bat Biology and Conservation (T. H. Kunz and P. A. Racey, eds.). Smithsonian Institution Press, Washington, D.C.

Schofield, H. W. 1996. The ecology and conservation biology of *Rhinolophus hipposideros*, the lesser horseshoe bat. Ph.D. Thesis, University of Aberdeen, Aberdeen.

Schulz, M. 1999. The conservation ecology of the rare golden-tipped bat *Kerivoula papuensis* and flute-nosed bat *Murina florium* (Chiroptera: vespertilionidae) in Australia. Ph.D. Thesis, Southern Cross University, Lismore.

Schulz, M., and J. Wainer. 1997. Diet of the golden-tipped bat *Kerivoula papuensis* (Microchiroptera) from north-eastern New South Wales, Australia. Journal of Zoology (London), 243:653–658.

Schulze, M. D., N. Seavy, and D. F. Whitacre. 2000. A comparison of the phyllostomid bat assemblages in undisturbed Neotropical forest and in forest fragments of a slash-and-burn farming mosaic in Petén, Guatamala. Biotropica, 32:174–184.

Sedgeley, J. A., and C. F. J. O'Donnell. 1999a. Factors influencing the selection of roost cavities by a temperate rainforest bat (Vespertilionidae: *Chalinolobus tuberculatus*) in New Zealand. Journal of Zoology (London), 249:437–446.

Sedgeley, J. A., and C. F. J. O'Donnell. 1999b. Roost selection by the long-tailed bat, *Chalinolobus tuberculatus*, in temperate New Zealand rainforest and its implications for the conservation of bats in managed forests. Biological Conservation, 88: 261–276.

Shiel, C. B., R. E. Shiel, and J. S. Fairley. 1999. Seasonal changes in the foraging behaviour of Leisler's bats (*Nyctalus leisleri*) in Ireland as revealed by radiotelemetry. Journal of Zoology (London), 249:347–358.

Sierro, A. 1999. Habitat selection by barbastelle bats (*Barbastella barbastellus*) in the Swiss Alps (Valais). Journal of Zoology (London), 248:429–432.

Sierro, A., and R. Arlettaz. 1997. Barbastelle bats (*Barbastella* spp.) specialize in the predation of moths: implications for foraging tactics and conservation. Acta Oecologica, 18:19–106.

Simmons, N. B., and R. S. Voss. 1998. The mammals of Paracou, French Guiana: a Neotropical lowland rainforest fauna. Pt. 1. Bats. Bulletin of the American Museum of Natural History, 237:1–219.

Sinclair, E. A., N. J. Webb, A. D. Marchant, and C. R. Tidemann. 1996. Genetic variation in the little red flying-fox *Pteropus scapulatus* (Chiroptera: Pteropodidae): implications for management. Biological Conservation, 7:45–50.

Smith, P. G. 2001. Habitat preference, range use and roosting ecology of Natterer's bats *Myotis nattereri* in a grassland-woodland landscape. Ph.D. Thesis, University of Aberdeen, Aberdeen,.

Smith, P. G., and S. M. Kerry. 1996. The Iwokrama Rain Forest programme for sustainable development: how much of Guyana's bat (Chiroptera) diversity does it encompass? Biodiversity and Conservation, 5:921–942.

Speakman, J. R. 1991. The impact of predation by birds on bat populations in the British Isles. Mammal Review, 2:123–142.

Speakman, J. R., P. A. Racey, C. M. C. Catto, P. I. Webb, S. M. Swift, and M. Burnett. 1991. Minimum summer populations and densities of bats in N.E. Scotland, near the northern borders of their distributions. Journal of Zoology (London), 225:327–345.

Speakman, J. R., P. I. Webb, and P. A. Racey. 1991. Effects of disturbance on the energy expenditure of hibernating bats. Journal of Applied Ecology, 28:1087–1104.

Stebbings, R. E. 1971. Bats in danger. Oryx, 10:311–312.

Stebbings, R. E. 1976. Studies on the population ecology of British bats. Ph.D. Thesis, University of East Anglia, Norwich.

Stebbings, R. E. 1988. The Conservation of European Bats. Christopher Helm, London.

Stebbings, R. E. 1992. Mouse-eared bat—extinct in Britain? Bat News, 26:2–3.

Stephens, P. A., and W. J. Sutherland. 1999. Consequences of the Allee effect for behaviour, ecology and conservation. Trends in Ecology and Evolution, 14:401–405.

Stinson, D. W., P. O. Glass, and E. M. Tasaican. 1992. Declines and trade in fruit bats on Saipan, Tiniam, Aguijan and Rota. Pp. 61–67 in: Pacific Island Flying Foxes: Proceedings of an International Conservation Conference (D. E. Wilson and G. L. Graham, eds.). Biological Report 90 (23). U.S. Fish and Wildlife Service, Washington, D.C.

Storz, J. F., H. R. Bhat, and T. H. Kunz. 2000. Social structure of a polygynous tent-making bat, Cynopterus sphinx (Megachiroptera). Journal of Zoology (London), 252: 151–165.

Strelkov, P. P. 1969. Migratory and stationary bats (Chiroptera) of the European part of the Soviet Union. Acta Zoologica Cracoviensia, 24:394–440.

Stutz, H. P., and M. Haffner. 1984. Arealverlust und Bestandesrückgang der kleinen Hufeisennase Rhinolophus hipposideros (Bechstein, 1800) (Mammalia: Chiroptera) in der Schweiz. Jahresbericht der Naturforschenden Gesellschaft Graubünden, 101: 169–178.

Swanepoel, R. E., P. A. Racey, R. F. Shore, and J. R. Speakman. 1999. Energetic effects of sub-lethal exposure to lindane on pipistrelle bats (Pipistrellus pipistrellus). Environmental Pollution, 104:169–177.

Swift, S. M. 1980. Activity patterns of pipistrelle bats (Pipistrellus pipistrellus) in northeast Scotland. Journal of Zoology (London), 190:285–295.

Tan, K. H., A. Zubaid, and T. H. Kunz. 1998. Food habits of Cynopterus brachyotis (Muller) (Chiroptera: Pteropodidae) in peninsular Malaysia. Journal of Tropical Ecology, 14:299–307.

Taylor, P. J. 1999. The role of amateurs in the growth of bat conservation and research in South Africa. South African Journal of Zoology, 34:19–26.

Terborgh, J., and B. Winter. 1980. Some causes of extinction. Pp. 119–134 in: Conservation Biology: An Evolutionary and Ecological Perspective (M. E. Soulé and B. A. Wilcox, eds.). Sinauer Associates, Sunderland, Mass.

Thomas, D. L., and E. J. Taylor. 1990. Study designs and tests for comparing resource use and availability. Journal of Wildlife Management, 54:322–330.

Thomas, D. W. 1995. Hibernating bats are sensitive to non-tactile disturbance. Journal of Mammalogy, 76:940–446.

Toop, J. 1985. Habitat requirements, survival strategies of the ghost bat Macroderma gigas Dobson (Microchiroptera, Megadermatidae) in central coastal Queensland. Macroderma, 1:37–41.

Trewhella, W. J., P. F. Reason, K. M. Clark, and S. R. T. Garrett. 1998. The current status of Livingstone's flying fox (*Pteropus livingstonii*) in the Federal Islamic Republic (RFI) of the Comores. Phelsuma, 6:32–40.

Trune, D. R., and C. N. Slobodchikoff. 1976. Social effects of roosting on the metabolism of the pallid bat (*Antrozous pallidus*). Journal of Mammalogy, 57:656–663.

Tupinier, D. 1989. La Chauve-Souris et L'Homme. L'Harmatan, Paris.

Tuttle, M. D. 1977. Gating as a means of protecting cave dwelling bats. Pp. 77–82 *in:* National Cave Management Symposium Proceedings 1976 (T. Ailey and D. Rhodes eds.). Speleobooks, Albuquerque, N.Mex.

Tuttle, M. D. 1979. Status, cause of decline, and management of endangered gray bats. Journal of Wildlife Management, 43:1–17.

Tuttle, M. D., and D. Hensley. 1993. Bat houses: the secrets of success. Bats, 11:3–14.

Tuttle, M. D., and D. Stevenson. 1982. Growth and survival of bats. Pp. 105–150 *in:* Ecology of Bats (T. H. Kunz, ed.). Plenum Press, New York.

Tuttle, M. D., and D. Taylor. 1994. Bats and Mines. Resource Publication no. 3. Bat Conservation International, Austin, Tex., 42 pp.

United Nations Environment Programme. 1992. Convention on Biological Diversity, 5 June 1992. United Nations Conference on Environment and Development, Rio de Janeiro, 24 pp.

Utzurrum, R. C. B. 1995. Feeding ecology of Philippine fruit bats: patterns of resource use and seed dispersal. Symposia of the Zoological Society of London, no. 67:63–77.

Utzurrum, R. C. B. 1998. Geographic patterns, ecological gradients, and the maintenance of tropical fruit bat diversity—the Philippine model. Pp. 342–353 *in:* Bat Biology and Conservation (T. H. Kunz and P. A. Racey, eds.). Smithsonian Institution Press, Washington, D.C.

Vaughan, N. 1997. The diet of British bats (Chiroptera). Mammal Review, 27:77–94.

Vaughan, N., and J. E. Hill. 1996. Bat (Chiroptera) diversity and abundance in banana plantations and rainforest, and three new records for St. Vincent, Lesser Antilles. Mammalia, 60:441–447.

Verboom, B., A. M. Boonman, and H. J. G. A. Limpens. 1999. Acoustic perception of landscape elements by the pond bat (*Myotis dasycneme*). Journal of Zoology (London), 248:59–66.

Verboom, B., and H. Huitema. 1997. The importance of linear landscape elements for the pipistrelle *Pipistrellus pipistrellus* and the serotine bat *Eptesicus serotinus.* Landscape Ecology, 12:117–125.

Verboom, B., and K. Spoelstra. 1999. Effects of food abundance and wind on the use of tree lines by an insectivorous bat, *Pipistrellus pipistrellus.* Canadian Journal of Zoology, 77:1391–1401.

Viña, A., and J. Cavelier. 1999. Deforestation rates (1938–1988) of tropical lowland forests on the Andean foothills of Colombia. Biotropica, 31:31–36.

Vonhof, M. J., and R. M. R. Barclay. 1996. Roost-site selection and roosting ecology of forest-dwelling bats in southern British Columbia. Canadian Journal of Zoology, 74:1797–1805.

Voûte, A. M. 1981. The conflict between bats and wood preservatives. Myotis, 19:41–44.

Voûte, A. M., and P. H. C. Lina. 1986. Management effects on bat hibernation in the Netherlands. Biological Conservation, 38:167–177.

Wahl, D. E. 1994. The management of flying foxes (*Pteropus spp.*) in New South Wales. Master's Thesis, University of Canberra, Canberra.

Walsh, A., C. M. C. Catto, T. M. Hutson, S. Langton, and P. A. Racey. In press. The United Kingdom Bat Monitoring Programme: turning conservation goals into tangible results. *In:* Monitoring Trends in Bat Populations of the U.S. and Territories: Problems and Prospects (T. J. O'Shea and M. A. Bogan, eds.). Information and Technology Report. U.S. Geological Survey, Biological Resources Division, Washington, D.C.

Walsh, A., C. M. C. Catto, T. M. Hutson, P. A. Racey, P. Richardson, and S. Langton. 2001 The UK's National Bat Monitoring Programme—Final Report 2001. Department for Environment, Food and Rural Affairs, Bristol.

Walsh, A. L., and S. Harris. 1996a. Foraging habitat preferences of vespertilionid bats in Britain. Journal of Applied Ecology, 33:508–518.

Walsh, A. L., and S. Harris. 1996b. Factors determining the abundance of vespertilionid bats in Britain: geographic, land class, and local habitat relationships. Journal of Applied Ecology, 33:519–529.

Walsh, A. L., S. Harris, and A. M. Hutson. 1995. Abundance and habitat selection of foraging vespertilionid bats in Britain: a landscape-scale approach. Symposia of the Zoological Society of London, no. 67:325–344.

Walston, J. L. 2000. The Bats of Vietnam. M.Sc. Thesis, University of Aberdeen, Aberdeen.

Waters, D., G. Jones, M. Furlong. 1999. Foraging ecology of Leisler's bat (*Nyctalus leisleri*) at two sites in southern Britain. Journal of Zoology (London), 249:173–180

Webb, N. J., and C. R. Tidemann. 1996. Mobility of Australian flying-foxes, *Pteropus* spp. (Megachiroptera): evidence from genetic variation. Proceedings of the Royal Society of London B, 263:497–502.

Whitby, J. E., P. Johnstone, G. Parsons, A. A. King, and A. M. Hutson. 1996. Ten year survey of British bats for the existence of rabies. Veterinary Record, 139:491–493.

Wiles, G. J. 1987. Current research and future management of Marianas fruit bats (Chiroptera: Pteropodidae) on Guam. Australian Mammalogy, 10:93–95.

Wiles, G. J. 1992. Recent trends in the fruit bat trade in Guam. Pp. 53–60 *in:* Pacific Island Flying Foxes: Proceedings of an International Conservation Conference (D. E. Wilson, and G. L. Graham, eds.). Biological Report 90 (23). U.S. Fish and Wildlife Service, Washington, D.C.

Wiles, G. J., J. Engbring, and D. Otobed. 1997. Abundance, biology and human exploitation of bats in the Palau Islands. Journal of Zoology (London), 241:203–227.

Wiles, G. J., T. O. Lemke, and N. H. Payne. 1989. Population estimates of fruit bats (*Pteropus mariannus*) in the Mariana Islands. Conservation Biology, 3:66–76.

Wilkinson, G. S. 1985. The social organization of the common vampire bat. 1. Pattern and cause of association. Behavioral Ecology and Sociobiology, 17:111–121.

Wilkinson, G. S., and J. W. Bradbury. 1988. Radiotelemetry: techniques and analysis. Pp. 105–124 *in:* Ecological and Behavioural Methods for the Study of Bats (T. H. Kunz, ed.). Smithsonian Institution Press, Washington, D.C.

Wilkinson, G. S., and T. H. Fleming. 1996. Migration and evolution of lesser long-nosed bats *Leptonycteris curasoae*, inferred from mitochondrial DNA. Molecular Ecology, 5:329–339.

Williams, L. M., and M. C. Brittingham. 1997. Selection of maternity roosts by big brown bats. Journal of Wildlife Management, 61:359–368.

Williams, P., N. Burgess, and C. Rahbek. 2000. Assessing large "flagship species" for representing the diversity of sub-Saharan mammals. Pp. 85–99 *in:* Priorities for the Conservation of Mammalian Diversity: Has the Panda Had Its Day? (A. C. Entwistle and N. Dunstone, eds.). Cambridge University Press, Cambridge.

Winkelmann, J. R., F. J. Bonaccorso, and T. L. Strickler. 2000. Home range of the southern blossom bat *Syconycteris australis,* in Papua New Guinea. Journal of Mammalogy, 81:408–414.

Worthington Wilmer, J., L. Hall, E. Barratt, and C. Moritz. 1999. Genetic structure and male mediated gene flow in the ghost bat (*Macroderma gigas*). Evolution, 55: 1582–1591.

Worthington Wilmer, J., C. Moritz, L. Hall, and J. Toop. 1994. Extreme population structuring in the threatened ghost-bat, *Macroderma gigas:* evidence from mitochondrial DNA. Proceedings of the Royal Society of London B, 257:193–198.

Worthy, T. H. 1997. Quaternary fossil fauna of South Canterbury, South Island, New Zealand. Journal of the Royal Society of New Zealand, 27:67–162.

World Wide Fund for Nature. 1996. Forests for Life—the WWF/IUCN Forest Policy Book. WWF-UK, Godalming, England, 62 pp.

Wynne, G., M. Avery, L. Campbell, S. Gubbay, S Hawkswell, T. Juniper, M. King, et al. 1995. Biodiversity Challenge. 2d ed. Royal Society for the Protection of Birds, Sandy.

Zubaid, A. 1993. A comparison of the bat fauna between a primary and fragmented secondary forest in Peninsular Malaysia. Mammalia, 57:201–206.

CONTRIBUTORS

John D. Altringham
School of Biology
Louis Compton Miall Building
University of Leeds, Leeds
West Yorkshire, England LS2 9JT, UK

Robert M. R. Barclay
Department of Biological Sciences
University of Calgary
Calgary, Alberta, T2N 1N4, Canada

Tenley M. Conway
Department of Mammalogy
American Museum of Natural History
Central Park West at 79th Street
New York, New York 10024, USA

Elizabeth R. Dumont
Department of Biology
221 Morrill Science Center
University of Massachusetts
Amherst, Massachusetts 01003, USA

Peggy Eby
Department of Ecosystem Management
University of New England
Armidale, New South Wales 2351, Australia

Abigail C. Entwistle
Fauna and Flora International
Great Eastern House
Tenison Road
Cambridge, England CB1 2DT, UK

Theodore H. Fleming
Department of Biology
University of Miami
Coral Gables, Florida 33124, USA

M. Brock Fenton
Department of Biology
York University
Toronto, Ontario M3J 1P3, Canada

Patricia W. Freeman
University of Nebraska State Museum
W-436 Nebraska Hall
University of Nebraska—Lincoln
Lincoln, Nebraska 68588-0514, USA

Lawrence D. Harder
Department of Biological Sciences
University of Calgary
Calgary, Alberta T2N 1N4, Canada

Otto von Helversen
Institute of Biology II
University of Erlangen
Saudstrasse 5
Erlangen D-91058, Germany

Gareth Jones
School of Biological Sciences
Bristol University
Woodland Road
Bristol, England BS8 1UG, UK

Thomas H. Kunz
Department of Biology
Boston University
Boston, Massachusetts 02215, USA

Linda F. Lumsden
Arthur Rylah Institute
Department of Natural Resources and
   Environment,
Heidelberg, Victoria 3084, Australia

Gary F. McCracken
Department of Ecology and Evolutionary
   Biology
University of Tennessee
Knoxville, Tennessee 37996, USA

Sharon L. Messenger
CDC Bioterrorism Lab, MS-C18
Centers for Disease Control and Prevention
1600 Clifton Road
Atlanta, Georgia 30333, USA

Bruce D. Patterson
Department of Zoology
Field Museum of Natural History
1400 South Lake Shore Drive
Chicago, Illinois 60605-2496, USA

Paul A. Racey
Department of Zoology
University of Aberdeen
Tillydrone Avenue
Aberdeen, Scotland AB24 2TZ, UK

Jens Rydell
Department of Zoology
Göteborg University
Box 463
SE-405 30 Göteborg, Sweden

Charles E. Rupprecht
Division of Viral and Rickettsial Diseases,
    MS-G33
Centers for Disease Control and Prevention
1600 Clifton Road
Atlanta, Georgia 30333, USA

Nancy B. Simmons
Department of Mammalogy
American Museum of Natural History
Central Park West at 79th Street
New York, New York 10024, USA

Jean S. Smith
Division of Viral and Rickettsial Diseases,
    MS-G33
Centers for Disease Control and Prevention
1600 Clifton Road
Atlanta, Georgia 30333, USA

John R. Speakman
Aberdeen Centre for Energy Regulation
    and Obesity
Department of Zoology
University of Aberdeen
Aberdeen, Scotland AB24 2TZ, UK

Richard D. Stevens
Program in Ecology
Department of Biological Sciences
Texas Tech University
Lubbock, Texas 79409-3131, USA

Elizabeth F. Stockwell
Department of Biology
Dalhousie University
Halifax, Nova Scotia B3H 4J1, Canada

Sharon M. Swartz
Department of Ecology and Evolutionary
    Biology
Brown University, Box G-B206
Providence, Rhode Island 02912, USA

Donald W. Thomas
Groupe de Recherche en Ecologie,
    Nutrition et Energétique
Département de Biologie
Université de Sherbrooke
Sherbrooke, Québec J1K2R1, Canada

Gerald S. Wilkinson
Department of Biology
University of Maryland
College Park, Maryland 20742, USA

Michael R. Willig
Program in Ecology
Department of Biological Sciences
Texas Tech University
Lubbock, Texas 79409-3131, USA

York Winter
Zoological Institute
University of Munich
Luisenstraße 14
80333 Munich, Germany